STATISTICS IN CONTEXT

BARBARA BLATCHLEY
Agnes Scott College

NEW YORK   OXFORD
OXFORD UNIVERSITY PRESS

Oxford University Press is a department of the University of Oxford. It furthers
the University's objective of excellence in research, scholarship, and education by
publishing worldwide. Oxford is a registered trade mark of Oxford University Press in
the UK and certain other countries.

Published in the United States of America by Oxford University Press
198 Madison Avenue, New York, NY 10016, United States of America.

© 2019 by Oxford University Press

For titles covered by Section 112 of the US Higher Education Opportunity Act, please
visit www.oup.com/us/he for the latest information about pricing and alternate formats.

All rights reserved. No part of this publication may be reproduced, stored in a retrieval
system, or transmitted, in any form or by any means, without the prior permission
in writing of Oxford University Press, or as expressly permitted by law, by license, or
under terms agreed with the appropriate reproduction rights organization. Inquiries
concerning reproduction outside the scope of the above should be sent to the Rights
Department, Oxford University Press, at the address above.

You must not circulate this work in any other form and you must impose this same
condition on any acquirer.

Library of Congress Cataloging-in-Publication Data

Names: Blatchley, Barbara, author.
Title: Statistics in context / Barbara Blatchley.
Description: First Edition. | New York : Oxford University Press, 2018.
Identifiers: LCCN 2017036139 (print) | LCCN 2017052156 (ebook) | ISBN
  9780190278991 (ebook) | ISBN 9780190278953 (hardback) | ISBN 9780190864682
  (looseleaf)
Subjects: LCSH: Statistics—Textbooks. | BISAC: PSYCHOLOGY / Assessment,
  Testing & Measurement. | PSYCHOLOGY / Statistics.
Classification: LCC QA276 (ebook) | LCC QA276 .B614 2018 (print) | DDC
  519.5—dc23
LC record available at https://lccn.loc.gov/2017036139

9 8 7 6 5 4 3 2 1

Printed by Sheridan Books, Inc., United States of America
Printed in the United Stated of America

CONTENTS IN BRIEF

CHAPTER 1	INTRODUCTION: STATISTICS—WHO NEEDS THEM?	2
CHAPTER 2	TYPES OF DATA	32
CHAPTER 3	A PICTURE IS WORTH A THOUSAND WORDS: CREATING AND INTERPRETING GRAPHICS	100
CHAPTER 4	MEASURES OF CENTRAL TENDENCY: WHAT'S SO AVERAGE ABOUT THE MEAN?	150
CHAPTER 5	VARIABILITY: THE "LAW OF LIFE"	190
CHAPTER 6	WHERE AM I? NORMAL DISTRIBUTIONS AND STANDARD SCORES	252
CHAPTER 7	BASIC PROBABILITY THEORY	294
CHAPTER 8	THE CENTRAL LIMIT THEOREM AND HYPOTHESIS TESTING	322
CHAPTER 9	THE z-TEST	354
CHAPTER 10	t-TESTS	388
CHAPTER 11	ANALYSIS OF VARIANCE	442
CHAPTER 12	CONFIDENCE INTERVALS AND EFFECT SIZE: BUILDING A BETTER MOUSETRAP	512
CHAPTER 13	CORRELATION AND REGRESSION: ARE WE RELATED?	550
CHAPTER 14	THE CHI-SQUARE TEST	598
CHAPTER 15	NONPARAMETRIC TESTS	638
CHAPTER 16	WHICH TEST SHOULD I USE, AND WHY?	686
APPENDIX A	THE PROPORTIONS UNDER THE STANDARD NORMAL CURVE	697
APPENDIX B	THE STUDENT'S TABLE OF CRITICAL t-VALUES	702
APPENDIX C	CRITICAL F-VALUES	705
APPENDIX D	CRITICAL TUKEY HSD VALUES	707
APPENDIX E	CRITICAL VALUES OF CHI SQUARE	709
APPENDIX F	THE PEARSON CORRELATION COEFFICIENT: CRITICAL r-VALUES	712
APPENDIX G	CRITICAL r_s VALUES FOR THE SPEARMAN CORRELATION COEFFICIENT	713
APPENDIX H	MANN–WHITNEY CRITICAL U-VALUES	716
APPENDIX I	CRITICAL VALUES FOR THE WILCOXON SIGNED-RANK, MATCHED-PAIRS t-TEST	719

CONTENTS

Figures, Tables, and Boxes xi
Preface xix
Introducing . . . *Statistics in Context* xx
Acknowledgments xxii
Contents Overview xxiii

CHAPTER 1 INTRODUCTION: STATISTICS—WHO NEEDS THEM? 2

Overview 3
Learning Objectives 3
Everyday Statistics 3
What Are Statistics? 4
Types of Statistics 4

THE HISTORICAL CONTEXT Roll Them Bones 4

Descriptive Statistics 5
Inferential Statistics 6
Variables 7
Independent and Dependent Variables 8

THINK ABOUT IT . . . How a Tan Affects Attractiveness 10

Chance Error 11
Using Statistics 12
Some Cautionary Notes About Statistics 13
Statistics in Context 15
Summary 16
Terms You Should Know 17
Writing Assignment 17
Practice Problems 17
Think About It . . . 26
References 27

INTRODUCING SPSS: THE STATISTICAL PACKAGE FOR THE SOCIAL SCIENCES 28

CHAPTER 2 TYPES OF DATA 32

Overview 33
Learning Objectives 33
Everyday Statistics 33
Data, Data Everywhere 34
Scales of Measurement 34
Qualitative Data 34

THE HISTORICAL CONTEXT S. S. Stevens and His Power Law 35

Quantitative Data 39

THINK ABOUT IT . . . Scales of Measurement 40

Organizing Data 42
Frequency Distributions 42
Ungrouped Frequency Distributions 45

THINK ABOUT IT . . . Frequency Distributions (Part 1) 50

Grouped Frequency Distributions 51

THINK ABOUT IT . . . Frequency Distributions (Part 2) 58

Scales of Measurement in Context 59
. . . And Frequency Distributions 62
Summary 62
Terms You Should Know 62
Glossary of Equations 63
Writing Assignment 63
Practice Problems 63
Think About It . . . 88
References 91

GETTING DATA INTO YOUR STATS PROGRAM 92

Reading in Data with SPSS 92
Reading in Data with R 97

iv

CHAPTER 3   A PICTURE IS WORTH A THOUSAND WORDS: CREATING AND INTERPRETING GRAPHICS   100

Overview   101
Learning Objectives   101
Everyday Statistics   101
Visualizing Patterns in Data   102

THE HISTORICAL CONTEXT William Playfair and the Use of Graphics in Publishing   102

Bar Charts and Histograms   103
  Discrete Data   104
  Continuous Data   105
Stem-and-Leaf Graphs   106
Frequency Polygons   107
Pie Charts   108

THINK ABOUT IT . . . Interpreting Graphics   110

Other Graphics   111
  Graphing Means   111
  Graphing Relationships   112
  Graphing Time   113
Graphics in Context: Rules for Creating Good Graphs   114
Rules for Creating a Good Graph   114
Summary   115
Terms You Should Know   115
Writing Assignment   115
A Note on Graphing with Statistical Software   116
Practice Problems   116
Think About It . . .   129
References   129

GRAPHING WITH SPSS AND R   130

Graphing with SPSS   130
Graphing with R   143

CHAPTER 4   MEASURES OF CENTRAL TENDENCY: WHAT'S SO AVERAGE ABOUT THE MEAN?   150

Overview   151
Learning Objectives   151
Everyday Statistics   151
Measures of Center   152
Measures of Center: What Is Typical?   152

THE HISTORICAL CONTEXT Adolphe Quetelet and the "Average Man"   153

The Mode and the Median   154
  Finding the Position of the Mode   154
  Finding the Position of the Median   155
The Mean   156
Mode, Median, and Mean: Which Is the "Best" Measure of Center?   158

THINK ABOUT IT . . . Estimating Measures of Center   160

Shapes of Distributions   161
  Normal Distributions   162
  Finding Center with Grouped Data   163

THINK ABOUT IT . . . Shapes of Distributions   165

Measures of Center in Context   166
Summary   168
Terms You Should Know   168
Glossary of Equations   168
Writing Assignment   168
Practice Problems   169
Think About It . . .   188
References   189

CHAPTER 5   VARIABILITY: THE "LAW OF LIFE"   190

Overview   191
Learning Objectives   191
Everyday Statistics   191
Measuring Variability   192
Consistency and Inconsistency in Data   192

THE HISTORICAL CONTEXT What Is the Shape of the Earth?   193

Measures of Variability   194

CONTENTS   v

The Range   194
  The Interquartile Range   195
  Graphing the IQR   197
  The Variance   198
  Average Deviation from the Mean   198
The Standard Deviation   200
  Finding the Variance in a Population   200
  Finding the Standard Deviation in a Population   202
  Finding Standard Deviation in a Population versus a Sample   204
  Finding Variance and Standard Deviation: An Example   204

THINK ABOUT IT . . . The Range Rule   206

Standard Deviation in Context   207

Descriptive Statistics in Context   210
Summary   212
Terms You Should Know   212
Glossary of Equations   213
Writing Assignment   213
Practice Problems   215
Think About It . . .   231
References   231

DESCRIPTIVE STATISTICS WITH SPSS AND R   232

Descriptive Statistics with SPSS   232
Descriptive Statistics with R   243

CHAPTER 6 WHERE AM I? NORMAL DISTRIBUTIONS AND STANDARD SCORES   252

  Overview   253
  Learning Objectives   253
  Everyday Statistics   253
Statistics So Far   254
Standard Scores   254

THE HISTORICAL CONTEXT Alfred Binet and Intelligence Testing   255

  The z-Score   256
  The "3-Sigma Rule"   259
  Proportions in the Standard Normal Curve   262
The Benefits of Standard Scores   270
  Comparing Scores from Different Distributions   270
  Converting a z-Score into a Raw Score   272

  Converting a Percentile Rank into a Raw Score   273

THINK ABOUT IT . . . The Range Rule Revisited   276

Standard Scores in Context   277
Summary   277
Terms You Should Know   278
Glossary of Equations   278
Writing Assignment   278
Practice Problems   279
Think About It . . .   292
References   293

CHAPTER 7 BASIC PROBABILITY THEORY   294

  Overview   295
  Learning Objectives   295
  Everyday Statistics   295
Probability   296
Probability and Frequency   296
  Basic Set Theory   296

THE HISTORICAL CONTEXT The Gambler's Fallacy   297

  Conditional Probability   300
  Combining Probabilities   301
Using Probability   306

THINK ABOUT IT . . . Probability Theory and Card Games   309

Probability in Context   310
Summary   312
Terms You Should Know   313
Glossary of Equations   313
Practice Problems   313
Think About It . . .   321
References   321

CHAPTER 8 — THE CENTRAL LIMIT THEOREM AND HYPOTHESIS TESTING 322

Overview 323
Learning Objectives 323
Everyday Statistics 323
Introduction: Error in Statistics 324
   Inferential Statistics 324
   The Scientific Method 324
THE HISTORICAL CONTEXT Kinnebrook's Error and Statistics 325

The Central Limit Theorem 325
   Measuring the Distribution of Large Sets of Events 325
   The Law of Large Numbers and the Central Limit Theorem 327
THINK ABOUT IT . . . The Law of Large Numbers and Dice Games 328

Drawing Samples from Populations 329
   The Sampling Distribution of the Means 329
   The Three Statements That Make Up the Central Limit Theorem 330
   Random Sampling 333
Using the Central Limit Theorem 337
Estimating Parameters and Hypothesis Testing 338
   The Null and Alternative Hypotheses 338
   Directional Hypotheses 339
Hypothesis Testing in Context 342
Summary 344
Terms You Should Know 345
Glossary of Equations 345
Practice Problems 345
Think About It . . . 352
References 352

CHAPTER 9 — THE z-TEST 354

Overview 355
Learning Objectives 355
Everyday Statistics 355
Error Revisited 356
THE HISTORICAL CONTEXT The Trial of the Pyx 358

How Different Is Different Enough? Critical Values and *p* 359
   Assumptions in Hypothesis Testing 359
   The Outer 5%: The Rejection Region 360
THINK ABOUT IT . . . *p*-Values and Alpha Levels 362

   Finding the z-Value in a Nondirectional Hypothesis 363
The *z*-Test 364
   Another Example 369
   Statistics as Estimates 370
On Being Right: Type I and Type II Errors 371
   An Example 371
   p-Values 373
Inferential Statistics in Context: Galton and the Quincunx 373
Summary 375
Terms You Should Know 376
Glossary of Equations 376
Writing Assignment 376
Practice Problems 376
Think About It . . . 387
References 387

CHAPTER 10 — t-TESTS 388

Overview 389
Learning Objectives 389
Everyday Statistics 389
Inferential Testing So Far 390
William Gosset and the Development of the *t*-Test 390
"Student's" Famous Test 392
THE HISTORICAL CONTEXT Statistics and Beer 393

   The Single-Sample t-Test 394
   Degrees of Freedom 396
When Both σ and μ Are Unknown 400
Independent-Samples *t*-Test 402
   The Standard Error of the Difference 403
   Finding the Difference Between Two Means: An Example 404

CONTENTS vii

Finding the Difference Between Means with Unequal Samples 406
Assumptions 411

THINK ABOUT IT . . . *t*-Tests and Sample Size 411

Dependent-Samples *t*-Tests 413
   Introduction 413
   Using a Dependent-Samples t-Test: An Example 413
   Calculations and Results 414

THINK ABOUT IT . . . *t*-Tests and Variability 417

t-Tests in Context: "Garbage In, Garbage Out" 419
Summary 421
Terms You Should Know 422
Glossary of Equations 422
Writing Assignment 423
Practice Problems 423
Using the Formula for Unequal *n*'s with Equal *n*'s 433
Think About It . . . 435
References 436

CONDUCTING *t*-TESTS WITH SPSS AND R 437

t-Tests with SPSS 437
t-Tests with R 440

CHAPTER 11   ANALYSIS OF VARIANCE 442

   Overview 443
   Learning Objectives 443
   Everyday Statistics 443
Comparing More Than Two Groups 444
Analysis of Variance: What Does It Mean? 444

THE HISTORICAL CONTEXT Fertilizer, Potatoes, and Fisher's Analysis of Variance 445

   A Hypothetical Study of Blood Doping 446
   Assessing Between-Group and Within-Group Variability in Our Hypothetical Results 447
ANOVA Terminology 448
   The One-Way ANOVA Procedure 449
   Sums of Squares 449

THINK ABOUT IT . . . *t* for Two and *F* for Many 456

THINK ABOUT IT . . . The *F*-Statistic 460

Post-hoc Testing 460
   The Tukey HSD Test with Equal n's 462
   The Tukey HSD Test with Unequal n's 466
Models of *F* 470
One-Way ANOVA Assumptions 471

Factorial Designs or Two-Way ANOVAs 471
   An Example: Albert Bandura's Study of Imitating Violence 472
   Graphing the Main Effects 474
   The Logic of the Two-Way ANOVA 476
   Using the Two-Way ANOVA Source Table 477
   Interpreting the Results 479
ANOVA in Context: Interpretation and Misinterpretation 480
Summary 482
Terms You Should Know 482
Glossary of Equations 483
Writing Assignment 484
Practice Problems 485
Think About It . . . 495
References 496

USING SPSS AND R FOR ANOVA 496

One-Way ANOVA in SPSS 497
Two-Way ANOVA in SPSS 501
One-Way ANOVA in R 504
Two-Way ANOVA in R 508

CHAPTER 12   CONFIDENCE INTERVALS AND EFFECT SIZE: BUILDING A BETTER MOUSETRAP 512

   Overview 513
   Learning Objectives 513
   Everyday Statistics 513
Using Estimations 514
Estimates and Confidence Intervals 515

THE HISTORICAL CONTEXT Jerzy Neyman: What's a Lemma? 516

   CIs and the *z*-Test 518

THINK ABOUT IT . . . What Does a Confidence Interval Really Mean? (Part 1) 520

   CIs and the Single-Sample t-Test 522
   CIs and Independent- and Dependent-Samples t-Tests 524

THINK ABOUT IT... What Does a Confidence Interval Really Mean? (Part 2)   528

Effect Size: How Different Are These Means, Really?   529
Effect Size and ANOVA   533
Statistics in Context: The CI versus the Inferential Test   535
Summary   537

Terms You Should Know   537
Glossary of Equations   537
Writing Assignment   538
Practice Problems   538
Think About It . . .   547
References   548

CHAPTER 13 CORRELATION AND REGRESSION: ARE WE RELATED?   550

Overview   551
Learning Objectives   551
Everyday Statistics   551
The Correlation Coefficient   552

THE HISTORICAL CONTEXT Statistics and Sweet Peas   552

Positive and Negative Correlations   554
Weak and Strong Correlations   555
Calculating r   557
Testing Hypotheses About *r*   560
The Least Squares Regression Line, a.k.a. the Line of Best Fit   561
Connecting the Dots   561
Finding the Regression Line   562

THINK ABOUT IT . . . What Does Perfect Mean?   565

Some Cautionary Notes   566
Linear versus Curvilinear Relationships   566
Truncated Range   567

Coefficient of Determination   569

THINK ABOUT IT . . . Shared Variability and Restricted Range   570

Statistics in Context: Correlations and Causation   571
Summary   573
Terms You Should Know   573
Glossary of Equations   573
Writing Assignment   574
Practice Problems   575
Think About It . . .   587
References   588

USING SPSS AND R FOR CORRELATION AND REGRESSION   589

Pearson Correlation Coefficient in SPSS   589
Conducting a Regression in SPSS   591
Pearson Correlation Coefficient in R   595
Conducting a Regression in R   596

CHAPTER 14 THE CHI-SQUARE TEST   598

Overview   599
Learning Objectives   599
Everyday Statistics   599
The Chi-Square Test and Why We Need It   600
Parametric versus Nonparametric Testing   600

THE HISTORICAL CONTEXT The Questionnaire   601

Making Assumptions   602
The One-Way Chi-Square Test for Goodness of Fit   604
The Two-Way Chi-Square Test of Independence   609
An Example of the Two-Way Test: Americans' Belief in Ghosts by Region   609
A Shortcut   611
A Special Case for Chi Square: The "2 by 2" Design   615

THINK ABOUT IT . . . The Risk Ratio   617

Nonparametric Tests in Context: Types of Data Revisited   619
Summary   620
Terms You Should Know   621
Glossary of Equations   621
Practice Problems   621
Think About It . . .   629
References   629

CONDUCTING CHI-SQUARE TESTS WITH SPSS AND R   629

Chi-Square Test for Goodness-of-Fit in SPSS   629

CONTENTS   ix

Two-Way Chi-Square Test for Independence in SPSS   631

Chi-Square Test for Goodness-of-Fit in R   635
Two-Way Chi-Square Test for Independence in R   636

CHAPTER 15 NONPARAMETRIC TESTS   638

Overview   639
Learning Objectives   639
Everyday Statistics   639
Nonparametric Statistics   640
   Review: How Parametric and Nonparametric Tests Differ   640

THE HISTORICAL CONTEXT The Power of Statistics in Society   641

   Performing Nonparametric Tests   642
Ranking the Data   643
The Spearman Correlation (r_s)   644
   The Spearman Correlation: An Example   644
   The Results   647
The Mann–Whitney U-test   648
The Wilcoxon Signed-Rank, Matched-Pairs t-Test   652
   The Wilcoxon Test: An Example   653
   Our Results   655
Non-normal Distributions   656

THINK ABOUT IT . . . Thinking Ahead   659

Nonparametric Tests in Context   661
Summary   662
Terms You Should Know   663
Glossary of Equations   663
Writing Assignment   663
Practice Problems   664
Think About It . . .   675
References   676

CONDUCTING NONPARAMETRIC TESTS WITH SPSS AND R   677

   Spearman's Correlation in SPSS   677
   Mann–Whitney U-Test in SPSS   679
   Spearman's Correlation in R   683
   Mann–Whitney U-Test in R   684

CHAPTER 16 WHICH TEST SHOULD I USE, AND WHY?   686

Overview   687
Statistics in Context   687
Examples   690

One Last Word   696
References   696

APPENDIX A	THE AREA UNDER THE NORMAL CURVE: CRITICAL z-VALUES 697
APPENDIX B	THE STUDENT'S TABLE OF CRITICAL t-VALUES 702
APPENDIX C	CRITICAL F-VALUES 705
APPENDIX D	CRITICAL TUKEY HSD VALUES 707
APPENDIX E	CRITICAL VALUES OF CHI SQUARE 709
APPENDIX F	THE PEARSON CORRELATION COEFFICIENT: r-VALUES 712
APPENDIX G	CRITICAL r_s VALUES FOR THE SPEARMAN CORRELATION COEFFICIENT 713
APPENDIX H	MANN–WHITNEY CRITICAL U-VALUES 716
APPENDIX I	CRITICAL VALUES FOR THE WILCOXON SIGNED-RANK, MATCHED-PAIRS t-TEST 719

Answers to Odds End-of-Chapter Practice Problems   720
Credit   766
Index   770

FIGURES, TABLES, AND BOXES

FIGURES

Figure 1.1. An astragalus, or "knucklebone," from the ankle of a hoofed animal.

Figure 1.2. The cover page from the *Bills of Mortality* for 1664.

Figure 1.3. Pierre Beauvallet's portrait of Galileo Galilei (1564–1642). Galileo was found guilty of heresy for insisting that the earth revolved around the sun and spent the final 12 years of his life living under house arrest.

Figure 1.4. A model of the apparatus used by Galileo for his inclined plane experiment..

Figure 1.5. Magazine advertisement for Cavalier Cigarettes, dated 1943.

Figure 2.1. Qualitative and quantitative data viewed in terms of nominal, ordinal, interval, and ratio scales.

Figure 2.2. All intervals between adjacent categories are equal.

Figure 2.3. Options for presenting data in frequency distributions.

Figure 3.1. William Playfair's first bar chart: Exports and imports of Scotland to and from different parts for 1 year, from Christmas 1780 to Christmas 1781.

Figure 3.2. Bar chart showing responses to the question "Do you smoke?"

Figure 3.3. Frequency histogram of sleep duration data (in minutes).

Figure 3.4. Stem-and-leaf graph showing walking times (in minutes) of 39 cardiac patients.

Figure 3.5. Stem-and-leaf graph shown as a frequency histogram.

Figure 3.6. Frequency polygon showing sleep duration data (in minutes).

Figure 3.7. Pie chart showing the relative frequency of answers to the question "Do you smoke?"

Figure 3.8. "Clock face" method for creating a pie chart.

Figure 3.9. How to show a missing block of data on the x-axis.

Figure 3.10. Exaggeration of a category.

Figure 3.11. Epidemiological curve showing the number of patients admitted to two referral hospitals with unexplained acute neurological illness, by date of admission, in Muzaffarpur, India, from May 26 to July 17, 2014.

Figure 3.12. Spot map of deaths from cholera in the Golden Square area of London, 1854. Redrawn from the original.

Figure 3.13. Deaths from cholera during 1854 outbreak in London.

Figure 4.1. Adolphe Quetelet.

Figure 4.2. The mode.

Figure 4.3. Negative skew caused by very low outliers.

Figure 4.4. Positive skew caused by very high outliers.

Figure 4.5. Normal distribution showing a balance of high and low outliers and all three measures of center holding the same value.

Figure 4.6. Shapes of distributions: positively skewed, normal, and negatively skewed.

Figure 4.7. Histograms to accompany question 4.

Figure 4.8. Stem-and-leaf plots of average driving distance on two professional golf tours.

Figure 4.9. Box-and-whisker plot of height (in inches) by sex of high school basketball players.

Figure 5.1. Two theories of the shape of the earth: (a) Newton's theory that the earth is flattened at the poles; (b) the Cassinis' theory that the earth is flattened at the equator.

Figure 5.2. Problem with the range as a measure of variability.

Figure 5.3. Box-and-whisker plot of the data in Box 5.3.

Figure 5.4. Deviations from the mean.

Figure 5.5. Box-and-whisker plot of IQ scores for five groups of adolescents.

Figure 6.1. Position of an observation within a normal distribution.

Figure 6.2. The 3-sigma rule (above the mean).
Figure 6.3. The 3-sigma rule (above and below the mean).
Figure 6.4. Calculation of proportion above a *z*-score.
Figure 6.5. Types of standard scores.
Figure 7.1. The French roulette wheel.
Figure 7.2. Probability distribution of Mah Jong cards.
Figure 7.3. Distributions of (a) PSQI scores and (b) sleep durations in the population.
Figure 8.1. Representation of the results of De Moivre's experiment throwing sets of coins.
Figure 8.2. The average value of the roll of one die against the number of rolls made.
Figure 8.3. Comparison of the elements in a population and in a sampling distribution of the means.
Figure 8.4. The five steps involved in creating a sampling distribution of the means.
Figure 8.5. The central limit theorem (CLT) illustrated.
Figure 8.6. The logic of hypothesis testing.
Figure 8.7. Comparing null and alternative hypotheses.
Figure 9.1. Commemorative coins undergo conformity testing during the Trial of the Pyx at Goldsmith's Hall in London, England, in January 2017.
Figure 9.2. Probability in a normal distribution (the SDM).
Figure 9.3. The outer 16% (on the left) and the outer 5% (on the right) of the sampling distribution of the means.
Figure 9.4. Relationship between null and alternative populations.
Figure 9.5. Results of *z*-test for effects of stress on sleep duration.
Figure 9.6. Examples of Galton's Quincunx.
Figure 10.1. William Sealy Gosset (1908).
Figure 10.2. Relationship between sample size and shape of the *t*-distribution.
Figure 10.3. APA Style formatting for reporting statistical test results.
Figure 11.1. Ronald Aylmer Fisher.
Figure 11.2. The *t*-distribution and the *F*-distribution.
Figure 11.3. Bandura's famous Bobo doll experiment. The female model is shown in the top row; children from the experimental group are shown in the middle and bottom rows.
Figure 11.4. Graphic representation of data from Bandura's Bobo doll study.
Figure 11.5. Graphs for question 9.
Figure 12.1. Jerzy Neyman.
Figure 12.2. The null population.
Figure 12.3. Distributions for the null and alternative populations in our study of stressed-out sleeping students.
Figure 12.4. Confidence interval versus the *z*-test.
Figure 13.1. Sir Francis Galton.
Figure 13.2 Galton's original scatterplot (to the nearest 0.01 inch)
Figure 13.3. Regression to the mean.
Figure 13.4. Two directions for correlations.
Figure 13.5. The strength of the correlation.
Figure 13.6. Scatterplot showing the relationship between height and cancer risk in women.
Figure 13.7. Possible "lines of best fit" for the data from Figure 13.6.
Figure 13.8. Residuals, or deviations, from the line of best fit.
Figure 13.9. Drawing the line of best fit on the graph.
Figure 13.10. Curvilinear relationship between anxiety and test performance.
Figure 13.11. Scatterplots for question 1.
Figure 14.1. Chi-square distributions.
Figure 14.2. Regions of the United States.
Figure 14.3. Typical Likert scales.
Figure 15.1. Charles Edward Spearman.
Figure 15.2. Held and Hein's (1963) experimental setup: The "kitten carousel."
Figure 15.3. Histograms of girls' and boys' height data.
Figure 15.4. Histograms of ranked height data for boys and girls
Figure 15.5. Cartoon pain rating scale for children.
Figure 16.1. Decision tree: Which test should I use?

TABLES

Table 1.1 Sleep duration and quality for 10 fictitious college undergraduates
Table 2.1 Smoking behavior of six students
Table 2.2 Ratings of female beauty by 10 participants
Table 2.3 Standard women's clothing sizes (misses)
Table 2.4 Student sleep study: Night 1 sleep duration for 34 students (in minutes)
Table 2.5 Student sleep study: Night 1 sleep duration (in order)
Table 2.6 Student sleep study: Night 1 results, displayed as an ungrouped simple frequency distribution
Table 2.7 Student sleep study: Night 1 results, displayed as an ungrouped relative frequency distribution

Table 2.8 Student sleep study: Night 1 results, displayed as an ungrouped cumulative frequency distribution

Table 2.9 Student sleep study: Night 1 results, displayed as an ungrouped cumulative relative frequency distribution

Table 2.10 Sleep duration for 20 college undergraduates (in minutes)

Table 2.11 Student sleep study: Seven-night results, displayed as a simple ungrouped frequency distribution

Table 2.12 Student sleep study: Seven-night results, displayed as a grouped frequency distribution

Table 2.13 Student sleep study: Seven-night results, displayed as a grouped relative, cumulative, and cumulative relative frequency distribution

Table 2.14 Time-estimation errors (in milliseconds) by age

Table 2.15 Calorie count by type of food at popular fast-food restaurants

Table 2.16 Memory span (number of stimuli recalled) for control and experimental groups

Table 2.17 Cholesterol level (in mg/dl) in adolescents who attempted suicide (experimental group) and depressed adolescents who have not attempted suicide (control group)

Table 2.18 Survival time in motion sickness

Table 2.19 Scores of 14 participants on perceived change in creativity (PCC) scale

Table 2.20 Cigarette smoking in 20 schizophrenic and 20 nonschizophrenic men

Table 2.21 Impulsivity scores for 25 rats

Table 2.22 Grip strength (in pounds) in two different lighting conditions

Table 2.23 Selection ratios as a function of age of person portrayed

Table 2.24 Attractiveness ratings for line drawings of women with four WHRs

Table 2.25 Difference scores between personality ratings for self and ideal romantic partner

Table 2.26 Data sets A through D

Table 2.27 Deaths in London for the year 1662, from the *Bills of Mortality* by John Graunt

Table 2.28 Tracking the 2014 Ebola outbreak

Table 2.29 MARS scores for 30 students enrolled in statistics

Table 2.30 Number of "knocks" for each of 11 questions ("T" indicates target question)

Table 2.31 Need to knock on wood on a four-point scale

Table 3.1 Participant responses to the question "Do you smoke?"

Table 3.2 Simple frequency distribution of smoking data

Table 3.3 Grouped frequency distribution of sleep time data

Table 3.4 Number of minutes on treadmill for 39 cardiac patients

Table 3.5 Hair cortisol level (in ng/g) and sleep quality ratings (1–5)

Table 3.6 Sleep duration

Table 3.7 Heart rate (in beats per minute)

Table 3.8 Weight (in grams) for 20 mice

Table 3.9 Surface area of the corpus callosum (in cm^2) in 10 twin pairs

Table 3.10 Time needed (in seconds) to complete visual search task while smelling unpleasant or pleasant odor

Table 3.11 Recidivism rates for 500 male convicts, by offense

Table 3.12 Total number of lugs per harvest, 1983–1991

Table 3.13 Average driving distance (in yards) of the top 25 drivers on the PGA tour

Table 3.14 Frequency distribution of responses to the question "Do you smoke?" for schizophrenic and nonschizophrenic men

Table 3.15 Response to the statement "I am a lucky person," by birth month

Table 3.16 Superstitious beliefs reported by 100 individuals in Britain

Table 3.17 Perception of the "illusory dot" as a function of color

Table 3.18 Mean CU scores and mean time-estimation accuracy scores for 40 participants

Table 3.19 Age and time-estimation accuracy for 15 participants

Table 3.20 Number of male inmates with and without a history of alcohol abuse problems, by eye color

Table 3.21 Average alcohol consumption of light-eyed and dark-eyed females

Table 4.1 The mode in a frequency distribution

Table 4.2 Effect of changing a data point on measures of center

Table 4.3 Changing a data point to the data set

Table 4.4 Midpoints of intervals in a grouped frequency distribution

Table 4.5 Measures of center and scales of measurement

Table 4.6 Symbols for descriptive statistics in APA Style

Table 4.7 Sleep times

Table 4.8 Error (in milliseconds) on time estimation task

Table 4.9 Reaction time (in milliseconds) with and without TMS

Table 4.10 Weights (in grams) for 20 mice
Table 4.11
Table 4.12 Salary at "Chocolate Daydreams" Bakery for all 10 employees
Table 4.13 Grouped frequency distribution
Table 4.14 Hypothetical data set for calculating the 10% trimmed mean
Table 4.15 Number of faces recalled by depressed, manic, and healthy individuals
Table 4.16 Diurnal type score (DTS) and maternal response to depression question for 12 children
Table 4.17 Life expectancy (in years) by race, gender, and level of education
Table 4.18 Data from the fastest shot competition
Table 4.19 Data from the fastest shot competition: Data missing
Table 4.20 Ages (in years) of 30 fitness program participants
Table 4.21 Reaction time (RT; in seconds) of participants with and without caffeine
Table 4.22 Nutritional information from fast-food restaurants
Table 5.1 Sleep data
Table 5.2 APA Style symbols for descriptive statistics
Table 5.3 Time-estimation data
Table 5.4 Estimating variability in three data sets
Table 5.5 Body weight of the domestic cat (in pounds)
Table 5.6 Time needed (in seconds) to complete visual search task while smelling an unpleasant or a pleasant odor
Table 5.7 Scores on statistics final exam with and without a lucky charm
Table 5.8 Body weight of Sprague-Dawley rats (in grams)
Table 5.9 Volume of the amygdala and size of social network
Table 5.10 Attractiveness ratings
Table 5.10a A small data set
Table 5.10b Results
Table 5.11 Another small data set
Table 5.12 Error (in millimeters) during a motor-control test performed under unperturbed and perturbed conditions
Table 5.13 Earned run averages (ERAs) of three pitchers against 10 different opponents
Table 5.14 Weights of 30 cereal boxes (in kilograms)
Table 5.15 Starting reaction times (RT; in milliseconds) of finalists in the 100-meter sprint
Table 5.16 T-scores and education status of 30 patients at a fracture clinic
Table 6.1 Percentile rank
Table 6.2 Converting from percentile rank to raw score

Table 6.3 APA Style symbols for descriptive statistics
Table 6.4 Data for question 1
Table 6.5 Scores on statistics final exam, 1992 and 2017
Table 6.6 Weights of 20 domestic cats (in pounds)
Table 6.7 Average BMD measures for women (in mg/cm^2) by age and ethnic group
Table 6.8 National averages on the SAT in 2013 (with *SD*)
Table 6.9 Average levels of DBH in two groups of schizophrenic males
Table 6.10 Number of hours slept by infants in neonatal nursery
Table 6.11 Class grades for five students
Table 6.12 Mean spelling bee scores at three district elementary schools
Table 6.13 Normal distributions for question 25
Table 6.14 General science exam scores at three colleges
Table 6.15 Depression scores for five young adults
Table 6.16 Mean family incomes in five counties
Table 7.1 Distribution of cards in Mah Jong deck
Table 7.2 Results of 20,000 rolls of a die
Table 7.3 Snoring and heart disease in a sample of 2,484 adult males
Table 7.4 Age, smoking, and heart disease in a sample of male physicians
Table 7.5 PSQI scores and sleep durations for 20 medical students
Table 7.6 Emotional disturbance in the population of US adults, 18 years or older
Table 7.7 Favorite team sport named by 81 adult Americans
Table 8.1 Results of 20,000 rolls of a die
Table 8.2 Selection process for four samples
Table 8.3 Results of 25 rolls of a die
Table 9.1 Critical values for *z*-tests
Table 9.2 Type I and Type II errors
Table 9.3 Total cholesterol in 25 patients after taking new statin drug
Table 9.4 IQ score for 15 patients with hypocholesterolemia
Table 9.5 List of population parameters
Table 9.6 GPAs for RTC students
Table 9.7 Score on the Stanford–Binet IQ test for 10 students
Table 9.8 Body temperature for a sample of alcoholics during withdrawal
Table 9.9 Birth weight of babies born to mothers who smoked during pregnancy

Table 9.10 Tutees' overall academic averages
Table 10.1 Abbreviated table of critical *t*-values
Table 10.2 Matched-pairs techniques
Table 10.3 Which test should I use? Four situations
Table 10.4 Number of words recalled from middle 20 words on the list
Table 10.5 Number of minutes of REM sleep during an 8-hour sleep period
Table 10.6 Number of minutes of REM sleep in patients matched for insomnia
Table 10.7 Scores on standardized test of anxiety
Table 10.8 Number of unfavorable traits attributed to minorities
Table 10.9 Amount of alcohol-spiked milk consumed (in milliliters)
Table 10.10 Compliance scores (0–25)
Table 10.11 Number of errors made on a driving simulator
Table 10.12 Number of minutes needed to solve a mechanical puzzle
Table 10.13 Supine diastolic pressure (in mm Hg)
Table 10.14 Perceived fairness scores
Table 10.15 Daily donation totals (in dollars)
Table 11.1 Table of significant *F*-values, α = .05
Table 11.2 Partial table of critical Tukey HSD values
Table 11.3 ANOVA terminology
Table 11.4 Number of aggressive acts toward Bobo as a function of AUDIT score
Table 11.5 Data for question 7
Table 11.6 Scores on manual dexterity test
Table 11.7 Number of colds experienced in four schools
Table 11.8 Time needed (in seconds) to learn the list of criteria for four levels of association
Table 11.9 Number of errors made on three different keyboards
Table 11.10 Depression scores for three trial groups
Table 11.11 Number of trials required to learn a maze for three groups of rats
Table 11.12 Memory scores for three groups
Table 11.13 Mean heart rates recorded during a stressful task for four groups
Table 11.14 Final grades of 30 students
Table 12.1 Interpretation of Cohen's *d*
Table 12.2 Cortisol levels (in mg/dl) in children and adults after a traumatic event
Table 12.3 Levels of OPG (in ng/ml) in healthy controls and individuals with schizophrenia
Table 12.4 Cholesterol levels (in ng/ml) in suicide attempters and healthy controls
Table 12.5 Male versus female chimpanzee CPM scores
Table 12.6 PC scores for 12 students in academic difficulty
Table 12.7 Time (in seconds) needed to find food
Table 12.8 Walking times (in seconds) for primed and unprimed participants
Table 13.1 High school and college GPAs of four students
Table 13.2 Data for the Chapter 13 writing assignment
Table 13.3 Belief in luck and religiosity scores
Table 13.4 BMI measures and number of siblings for 12 kindergarten-age children
Table 13.5 Head circumference, Bayley Infant IQ Score, and duration of smoking during pregnancy
Table 13.6 Maternal age at birth and birth weight (in grams) of the newborn
Table 13.7 Hours spent studying and hours spent playing video games for five volunteers
Table 13.8 Exercise and milk consumption
Table 13.9 ACHA and mania scores for 20 individuals
Table 13.10 Overall quality score and price of nine cordless phones
Table 13.11 Height, weight, and head circumference for 20 female college students
Table 13.12 Hours worked per week and GPA for 15 college students
Table 13.13 Depression and rumination data
Table 14.1 Sample questions from Galton's questionnaire
Table 14.2 Thirty consecutive roles of a single die
Table 14.3 Political party affiliation by region of the country
Table 14.4 Attitude toward abortion
Table 14.5 Attitude toward abortion: Revised questionnaire
Table 14.6 Worldwide distribution of blood type
Table 14.7 Student performance on the state achievement test using the new and the old textbook
Table 14.8 Distribution of flavors in a five-flavor roll of Life Savers
Table 14.9 Grade distribution for professor in his first year of teaching
Table 14.10 Student performance in statistics as a function of algebra prerequisite
Table 14.11 Methods of combating fatigue, by sex
Table 14.12 Number of years of college completed and residence
Table 14.13 Importance of statistics to chosen specialization in psychology
Table 14.14 Maternal encouragement and change in IQ scores
Table 14.15 Memory for father's profanity as a function of censoring the father

Table 14.16 Lottery prize frequencies
Table 14.17 Average number of hours per week spent watching television, by season
Table 15.1 Parametric versus nonparametric tests
Table 15.2 Attitude toward online dating for men and women
Table 15.3 Proficiency of visually guided behavior of kittens
Table 15.4 Height (in inches) of 18 boys and 18 girls
Table 15.5 Data for the Chapter 15 writing assignment
Table 15.6 Estimated weekly TV viewing times
Table 15.7 Height (in inches) of 18 boys and 18 girls
Table 15.8 Teacher's ranking of student intellectual ability and intelligence test scores
Table 15.9 Scores on the DPS before and after treatment for chronic pain
Table 15.10 Rating of pain by men and women
Table 15.11 Rating of pain by first-time and third-time patients of pediatrician
Table 15.12 Length of postoperative hospital stay (in days)
Table 15.13 Number of bar presses by eight rats
Table 15.14 Number of trials needed to learn criterion by two groups of rats
Table 15.15 Age and score on the HS for 10 individuals
Table 15.16 Weights (in pounds) of cats owned by families with children and by single people
Table 15.17 Average rank order for 10 events for novice and experienced observers
Table 15.18 Number of cigarettes smoked in a day by women and men
Table 15.19 IQ score and number of hours spent listening to Mozart
Table 15.20 Level of sexual arousal
Table 15.21 Movie enjoyment
Table 16.1 Sleep quality of 200 college students, by year in school

BOXES

Box 2.1 The mean and nominal data
Box 2.2 Calculating percentages
Box 2.3 Calculating percentages: Example
Box 2.4 Finding interval width
Box 2.5 Even multiples of the width
Box 2.6 Finding the upper limit
Box 2.7 Finding the midpoint
Box 2.8 Measuring beauty

Box 3.1 Place values in Arabic numbers
Box 3.2 Converting to relative frequency
Box 3.3 Converting relative frequency to minutes
Box 3.4 Graphing the mean
Box 3.5 Scatterplots
Box 3.6 Time-series graphs
Box 4.1 Finding the median
Box 4.2 Calculating the mean
Box 4.3 Calculating the mean for the sleep data in Table 4.1
Box 4.4 Finding the midpoint of a 60-minute interval
Box 4.5 Calculating the mean from grouped data
Box 5.1 Calculating the range
Box 5.2 Finding the quartiles in a set of data
Box 5.3 Using quartiles to find the IQR
Box 5.4 Averaging deviation from the mean
Box 5.5 Calculating average deviation using a small data set
Box 5.6 Definition and calculation formulas for finding variance
Box 5.7 Written instructions for using the calculation formula to find variance
Box 5.8 Finding the variance of a small data set using definition and calculation formulas
Box 5.9 Definition and calculation formulas for finding standard deviation
Box 5.10 Finding the standard deviation of a small data set
Box 5.11 Finding the variance and standard deviation of sleeping college students
Box 5.12 Calculating standard deviation (SD) and mean deviation (MD)
Box 6.1 Calculating IQ
Box 6.2 Calculating the z-score
Box 6.3 Calculating the z-score: Example
Box 6.4 Calculating the z-score: Example using a different SD
Box 6.5 Calculating the z-scores of three different results
Box 6.6 Using the z-table: Example
Box 6.7 Sketch of the normal curve showing students who slept 480 minutes or longer
Box 6.8 Sketch of the normal curve showing students who slept between 240 and 480 minutes
Box. 6.9 Sketch of the normal curve showing students who slept 360 minutes or less
Box 6.10 Sketch of the normal curve showing students who slept between 120 and 180 minutes
Box 6.11 Calculating the percentage of students who slept between 120 and 180 minutes: First method

Box 6.12 Calculating the percentage of students who slept between 120 and 180 minutes: Second method

Box 6.13 Using z-scores to compare results from different distributions

Box 6.14 Solving for x

Box 6.15 Sketch of the normal curve showing a score in the 10th percentile

Box 6.16 Calculating a percentile rank into a raw score: Example

Box 7.1 The probability equation

Box 7.2 The probability equation: Example

Box 7.3 Finding the probability of two events separately

Box 7.4 Using the AND rule to find the probability of dependent events

Box 7.5 Using the OR rule to find the probability of independent events

Box 7.6 Combining the AND & OR rules

Box 7.7 Working out the solution to our AND & OR rule equation

Box 7.8 Calculating the probability of being dealt four aces

Box 7.9 The distribution of PSQI scores in the population

Box 8.1 A tiny population

Box 8.2 Calculating the mean and standard deviation of our tiny population

Box 8.3 10 samples of $n = 4$ each

Box 8.4 Calculating the mean and standard deviation of the sample means

Box 8.5 Samples and their means

Box 8.6 Null, directional alternative, and nondirectional alternative hypotheses

Box 9.1 The inferential z-formula

Box 9.2 The z-formula

Box 9.3 Research question: What is the effect of stress on sleep duration in college students?

Box 9.4 Calculation of z

Box 9.5 Research question: Does sleep deprivation affect digit span?

Box 9.6 Solving for z

Box 10.1 Calculating the sample standard deviation

Box 10.2 The z-test and t-test compared

Box 10.3 The z-score, z-test, and t-test compared

Box 10.4 Research question: Does regular daily exercise change sleep time?

Box 10.5 Does daily exercise improve sleep time? Calculating t

Box 10.6 Degrees of freedom: An example

Box 10.7 Checking our observed t-value against the critical t-value

Box 10.8 Formula for calculating t

Box 10.9 Error terms in single-sample and independent-samples t-tests

Box 10.10 Research question: Does warm milk decrease the time needed to fall asleep?

Box 10.11 Calculating t for our soporific milk study

Box 10.12 Comparing the observed and critical t-values for our sleep study

Box 10.13 Adjusting the t-formula for unequal n's

Box 10.14 Our research question remains the same

Box 10.15 The test

Box 10.16 Formula for a dependent-samples t-test

Box 10.17 Are college students more distracted in the spring than in the fall?

Box 10.18 Calculating t

Box 10.19 Data when $n_1 = n_2 = 20$

Box 11.1 Hypothetical effect of different methods of blood doping

Box 11.2 Averaging out the effects of random chance between and within groups

Box 11.3 Definition formula for variance: The sum of squares

Box 11.4 General procedure for calculating F

Box 11.5 If the IV has no effect at all

Box 11.6 Finding the sum of squares between and within groups

Box 11.7 Finding the mean sum of squares between and within groups

Box 11.8 Calculating the total variability between groups (SS_b) in our doping study

Box 11.9 Calculating the overall variability within groups (SS_w) in our doping study

Box 11.10 Finding the total sum of squares

Box 11.11 Source table

Box 11.12 Calculating MS and F

Box 11.13 Completed source table

Box 11.14 Possible pairwise comparisons

Box 11.15 Calculating Tukey HSD when n's are equal for each sample

FIGURES, TABLES, AND BOXES

Box 11.16 Possible pairwise comparisons
Box 11.17 Tukey HSD statistics
Box 11.18 Performing a Tukey HSD test when n's are unequal
Box 11.19 Changes to the experiment when one subject drops out
Box 11.20 Tukey HSD comparison with unequal sample size
Box 11.21 Results of a Bobo doll experiment: Number of aggressive imitations
Box 11.22 Graphs showing (a) significant and (b) nonsignificant main effects of the IV
Box 11.23 Graphs showing (a) significant and (b) nonsignificant interactions of two IVs
Box 11.24 The logic of two-way ANOVA
Box 11.25 SPSS source table for a two-way ANOVA
Box 11.26 Questions to ask before selecting an inferential test
Box 12.1 Point versus interval estimates and confidence versus precision
Box 12.2 Results of our stressed-out students study
Box 12.3 Calculating the confidence interval around our sample mean
Box 12.4 Results of our exercise and sleep study
Box 12.5 Calculating a confidence interval around our sample mean
Box 12.6 Estimates of variability in an independent-samples t-test
Box 12.7 Calculating CIs for the z-test and single-sample t-test
Box 12.8 Calculating a CI for an independent-samples t-test
Box 12.9 Calculating a confidence interval for our sleepy milk study
Box 12.10 Results of our spring–fall distractedness study
Box 12.11 Calculating a confidence interval for a dependent-samples t-test
Box 12.12 Basic format of Cohen's d
Box 12.13 Formulas for calculating effect size
Box 12.14 Calculating the size of the effect of stress on sleep time
Box 12.15 Eta squared: The effect size statistic
Box 12.16 Blood doping study: ANOVA and effect size source table
Box 12.17 Partial eta squared
Box 12.18 Calculating the effect size for the two IVs in the Bobo doll study
Box 13.1 Two formulas for calculating r
Box 13.2 Height and risk of cancer in 15 women (hypothetical data)
Box 13.3 Null and alternative hypotheses concerning height and cancer risk
Box 13.4 Calculating r using the definition formula and the raw data formula
Box 13.5 The formula for a straight line (the regression line)
Box 13.6 Calculating the slope and y-intercept
Box 13.7 Calculating two Y-values
Box 13.8 Definition formula for finding r
Box 13.9 Calculating r^2
Box 14.1 Research question: Are Americans sleep-deprived?
Box 14.2 Responses to Galton's question, "What is your father's occupation?"
Box 14.3 Expected versus observed frequency
Box 14.4 The formula for chi square (χ^2)
Box 14.5 Calculating $\chi^2 = \Sigma \frac{(f_o - f_e)^2}{f_e}$
Box 14.6 Determining the significance of our observed chi square
Box 14.7 Survey responses: Do you believe in ghosts?
Box 14.8 Determining the expected frequencies for our regional belief in ghosts study
Box 14.9 A shortcut for finding expected frequencies in a two-way chi-square test
Box 14.10 Observed and expected frequencies
Box 14.11 Calculating the chi square
Box 14.12 Data and chi-square calculation for a 2 by 2 design
Box 15.1 Ranking three tennis players by age
Box 15.2 Ranking tennis players with tied scores
Box 15.3 Calculating the Spearman correlation coefficient
Box 15.4 Data for our age and competitiveness study
Box 15.5 Calculating the difference in rank between age and competitiveness scores
Box 15.6 Ranking the data in our survey on online dating
Box 15.7 Using the MWU formula to calculate U-values
Box 15.8 Steps in the Wilcoxon t-test
Box 15.9 Calculating the Wilcoxon t-value

PREFACE

It is an unfortunate fact that students often come to their first statistics class with a built-in aversion to the subject matter. The research process—asking a question and finding an answer—can be incredibly rewarding and a lot of fun. However, it often does not look that way when you run into your first statistical conundrum. When students look at research for the first time, they see a confusing array of jargon, mathematics, and numbers. As a consequence, they regard doing research, reading research reports, and, most of all, studying statistics with a great deal of trepidation.

To make matters worse, many statistics textbooks use an academic tone that can be both uninviting and intimidating. I've seen many a student with that "deer in the headlights" look on the first day of classes—and heard many more bemoan the necessity of taking research statistics at all. Truth be told, I was one of those students, back in the day. So, when I sat down to write a textbook of my own, I decided to write about statistics using everyday language and (I hope) a good dose of humor. The language may be colorful, but the goal is always to get the basic idea across, clearly and without ambiguity.

In addition, when I looked back on my own experiences as a student, I remembered that discovering an idea's history and purpose made it much less scary (even when it involved taking the square root). Many textbooks ignore the context of the idea, focusing instead just on the mechanics of the process. To me, these books miss out on the benefits of knowing *why*. So, my emphasis in this book is on where statistics came from and why we use them. The theme running through the entire book is the presentation of statistics in their context—the *who*, *what*, *when*, *where*, and *why*, as well as the *how*, of statistics.

To students, statistics may never be your favorite class—I was astonished to discover it was one of mine. But I hope you always remember that statistics are an incredibly useful tool. Learn how to use all of them, and have fun asking questions. Don't be scared—it's just statistics.

THANK YOU

I first started thinking about this book almost 10 years ago. Along the way, I had the immense pleasure of working with many smart, talented, and inspired people. They've made this process fascinating. I need to say thank you to so many.

I want to thank my agent, Barbara Rosenberg, for not letting me give up when the going got tough and I started to wonder if this was ever going to happen. I also want to thank both Jane Potter and Eric Sinkins, my editors at Oxford University Press. (This manuscript took a sort of circuitous route to publication. Fortunately, all of the people I've had the pleasure of working with at OUP have been, like Jane, patient, perceptive and receptive.) Eric's vision for what this book could be, and his excellent advice and good humor, were exactly what this "newbie" author needed. I also want to thank my colleagues and the amazing array of students I have encountered over my almost 30 years of teaching. I am deeply grateful for everything I've learned from each of you and only wish I had enough space to list you all here.

Finally, this book is dedicated to my husband, Christopher. You've been here from the beginning, offering love, support, advice, patience, emergency cookies and tea (and the occasional Scotch), as well as your invaluable services as my own personal in-house, holler-down-the-stairs with questions about semicolon placement and spelling editor. You even forgave me for forgetting where the comma goes. I would not have been able to do this without you. "*Mo Anam Cara,*" Dude—"*Vous et nul autre.*"

Barbara Blatchley
December 2017

INTRODUCING... *Statistics in Context*

Oxford University Press is delighted to introduce *Statistics in Context*, a fresh approach to teaching statistics. Designed to reduce student fear of numbers, the book aims to put the stats-wary reader at ease with uncomplicated explanations and practical examples drawn from real research and everyday life. Written in lively, accessible prose, the narrative tells the *who*, *what*, *when*, and *where*, as well as the *how*, of statistics.

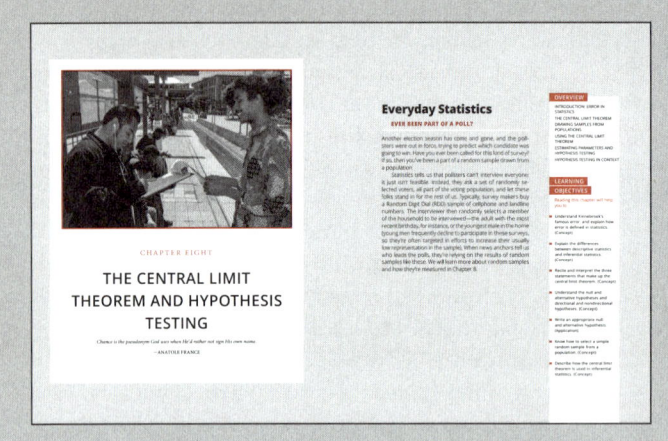

OUTSTANDING FEATURES

Everyday Statistics boxes examine some practical applications of the topics to be discussed—from predicting earthquakes to winning at poker—capturing the attention of readers and drawing them into the themes of the chapter.

The Historical Context tells the story of how different statistical procedures developed from such varied activities as assaying gold, brewing beer, and growing potatoes.

CheckPoint exercises and answers give students the chance to make sure they've thoroughly understood the material before moving on to the next section.

XX   INTRODUCING... *STATISTICS IN CONTEXT*

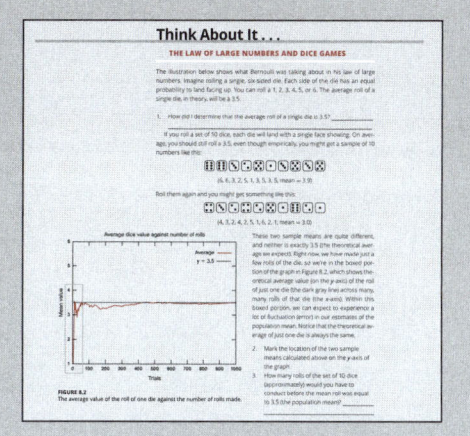

Think About It... boxes consolidate student understanding by challenging readers to apply what they have learned to a more difficult problem.

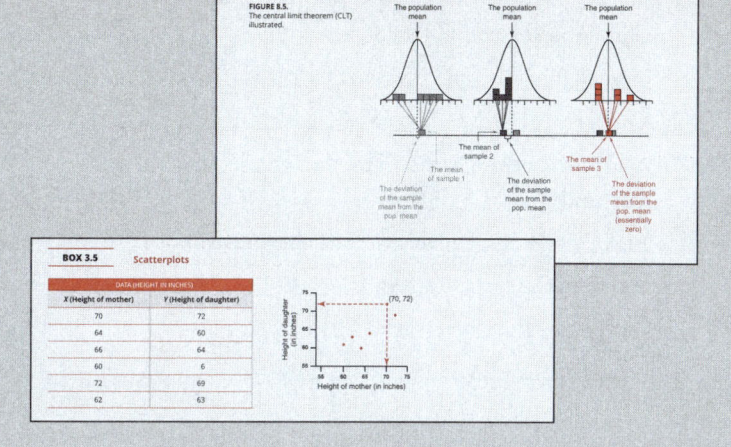

Figures, tables, and worked-examples boxes guide readers step by step through the calculations described in the text.

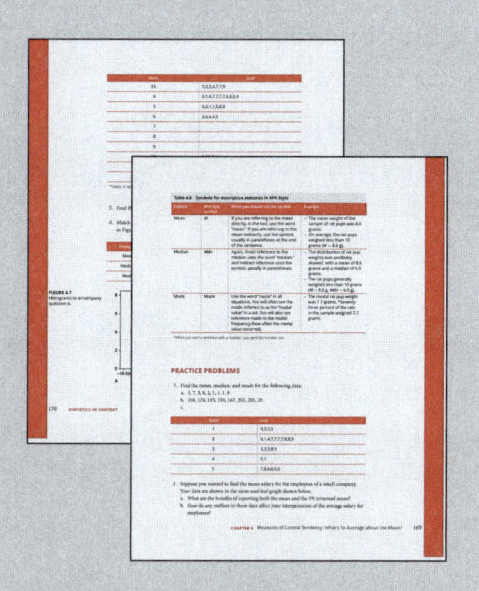

Abundant end-of-chapter practice problems, supplied with detailed answers at the back of the book, give students many opportunities to test their mastery of the procedures described in the chapter.

Using Statistical Software, a supplement following select chapters, instructs readers on how to perform statistical analysis using either SPSS or R.

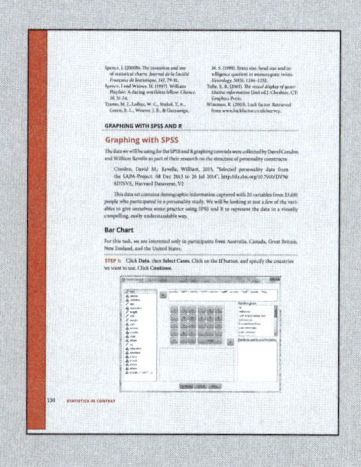

ADDITIONAL RESOURCES FOR INSTRUCTORS AND STUDENTS

The online **Instructor Resource Center** contains a test generator, detailed solutions for all end-of-chapter exercises, links to relevant streaming videos, and additional tools for enhancing lectures and engaging students.

The **Student Resource Center** features chapter summaries, self-testing quizzes, and other material to solidify your understanding of statistics.

INTRODUCING... STATISTICS IN CONTEXT   xxi

ACKNOWLEDGMENTS

Oxford University Press would like to thank the following reviewers, as well as several anonymous reviewers, whose thoughtful suggestions and feedback have helped to shape the first edition of *Statistics in Context*:

Chris Aberson, Humboldt State

Jorge Ballinas, Temple University

Kimberly L. Barrett, Eastern Michigan University

Agatha E. Carroo, North Carolina Central University

Arlo Clark-Foos, University of Michigan Dearborn

Meredith C. Frey, Otterbein University

Gang Guo, The University of Mississippi

Brian J. Hock, Austin Peay State University

Dan Ispas, Illinois State University

Duane M. Jackson, Morehouse College

Sean Laraway, San Jose State University

Julia Lechuga, The University of Texas at El Paso

Melanie Leussis, Emmanuel College

Stuart Marcovitch, University of North Carolina at Greensboro

Tina Miyake, Idaho State University

Beverly J. Moore, SUNY Sullivan

Osvaldo F. Morera, University of Texas at El Paso

Angela Pirlott, University of Wisconsin – Eau Claire

Altovise Rogers, San Jose State University

Lisa Rosen, Texas Woman's University

N. Clayton Silver, University of Nevada, Las Vegas

Steven Specht, Utica College

Cheryl Terrance, University of North Dakota

Justin N. Thorpe, Idaho State University

Marc Zucker, Nassau Community College

CONTENTS OVERVIEW

CHAPTER 1 — INTRODUCTION: STATISTICS—WHO NEEDS THEM?

This chapter begins with a definition of statistics and a discussion of the two types of statistics (descriptive and inferential) that will be covered in the book. It introduces the idea of statistics in context and provides several examples of what that means.

CHAPTER 2 — TYPES OF DATA

This chapter discusses S. S. Stevens' four levels of measurement (nominal, ordinal, interval, and ratio) and why understanding the type of data you are working with is important to the statistical technique you use. The second half of the chapter discusses how data can be organized (using frequency distributions) to emphasize important patterns in the data.

CHAPTER 3 — A PICTURE IS WORTH A THOUSAND WORDS: CREATING AND INTERPRETING GRAPHICS

This chapter introduces the most frequently used methods for presenting data in graphic form—from bar charts and histograms to stem-and-leaf graphs and pie charts—along with less frequently used but still useful graphic techniques. There are also some cautionary notes about how to interpret the kinds of graphics encountered in academic journals and mainstream news.

CHAPTER 4 — MEASURES OF CENTRAL TENDENCY: WHAT'S SO AVERAGE ABOUT THE MEAN?

This chapter discusses the three measures of center—the mean, the median, and the mode—along with what the relationship between these three measures tells us about the shape of a distribution. The chapter ends with a discussion of how to determine which measure of center is appropriate for your data.

CHAPTER 5 — VARIABILITY: THE "LAW OF LIFE"

This chapter describes how the range, variance, and standard deviation are measured. The relationship between the measures of center and the measures of variability is also described. Look for a helpful review of how to write about descriptive statistics near the end of the chapter.

CHAPTER 6 — WHERE AM I? NORMAL DISTRIBUTIONS AND STANDARD SCORES

This chapter describes measures of location within a normal distribution (z-scores) and how z-scores are related to event probability. *The Historical Context* feature discusses intelligence tests and measurement in psychology, focusing on Alfred Binet and the origin of standardized tests of human abilities.

CHAPTER 7 — BASIC PROBABILITY THEORY

The basics of measuring probability, including set theory, conditional probability, independent and dependent events, and the rules for combining probability, are described. These ideas are then linked to a discussion of theoretical and empirical distributions and how probability is used in statistics to help scientists draw conclusions from their data. The chapter ends with a discussion of probability as the heart of inferential statistics.

CHAPTER 8 — THE CENTRAL LIMIT THEOREM AND HYPOTHESIS TESTING

This chapter focuses on the central limit theorem and how it is used to allow scientists to draw conclusions about the population from a sample. Topics include the sampling distribution of the means, the standard error, making hypotheses, the null and alternative hypotheses, and the alpha level.

CHAPTER 9 — THE z-TEST

This chapter discusses comparisons of single-sample means with population means. Topics include using the z-test to locate the position of a sample mean in the sampling distribution of the means, critical values, rejection regions, p-values and how they are used to make decisions, assumptions inherent in hypothesis testing, and Type I and II errors. *The Historical Context* feature describes an ancient use of inferential testing called the "Trial of the Pyx."

CHAPTER 10 — t-TESTS

We begin with a review of inferential testing to this point before exploring three different types of t-test: single-sample t-tests as well as independent-samples and dependent-samples t-tests. Topics include the relationship between z- and t-tests, degrees of freedom, what makes samples independent and dependent, and the standard error of the difference. *The Historical Context* feature introduces William Gosset as the creator of the famous "Student's t-test."

CHAPTER 11 — ANALYSIS OF VARIANCE

This chapter focuses on analysis of variance, or ANOVA, the most commonly used method for comparing more than two sample means. Topics include sources of variability between and within groups, calculation of the sum of squares, mean sums of squares, degrees of freedom between and within groups, and calculation of the F-statistic. The chapter closes with a discussion of what we mean when we use the word "significant."

CHAPTER 12 — CONFIDENCE INTERVALS AND EFFECT SIZE: BUILDING A BETTER MOUSETRAP

This chapter describes the calculation and use of confidence intervals (CIs) and Cohen's d-statistic, a measure of effect size. Topics include the difference between precision and confidence, point versus interval estimates, and the relationship between inferential testing and CIs.

CHAPTER 13 — CORRELATION AND REGRESSION: ARE WE RELATED?

This chapter describes correlation and regression. Topics include calculation of the Pearson product–moment correlation coefficient, least squares regression, finding the slope and the y-intercept of the regression line, and interpretation of linear regression. There is also a warning about mistaking correlation for causation.

CHAPTER 14 — THE CHI-SQUARE TEST

This chapter describes how parametric and nonparametric statistics differ and how the chi-square test (both goodness of fit and the test of independence) is calculated. A section near the end reviews different kinds of qualitative and quantitative data.

CHAPTER 15 — NONPARAMETRIC TESTS

This chapter describes several more nonparametric tests used frequently in the sciences. Topics include the Spearman correlation coefficient, the Mann–Whitney U-test, and the Wilcoxon t-test.

CHAPTER 16 — WHICH TEST SHOULD I USE, AND WHY?

This chapter focuses on how to determine which test is appropriate for a given purpose. Topics discussed include the assumptions behind the tests discussed in earlier chapters; practice in matching assumptions, tests, and research designs is also provided.

APPENDIX A — THE AREA UNDER THE NORMAL CURVE: CRITICAL z-VALUES

APPENDIX B — THE STUDENT'S TABLE OF CRITICAL t-VALUES

APPENDIX C — CRITICAL F-VALUES

APPENDIX D — CRITICAL TUKEY HSD VALUES

APPENDIX E — CRITICAL VALUES OF CHI SQUARE

APPENDIX F — THE PEARSON CORRELATION COEFFICIENT: r-VALUES

APPENDIX G — CRITICAL r_s VALUES FOR THE SPEARMAN CORRELATION COEFFICIENT

APPENDIX H — MANN–WHITNEY CRITICAL U-VALUES

APPENDIX I — CRITICAL VALUES FOR THE WILCOXON SIGNED-RANK, MATCHED-PAIRS TEST

STATISTICS
IN CONTEXT

CHAPTER ONE

INTRODUCTION: STATISTICS—WHO NEEDS THEM?

Statistics: The only science that enables different experts using the same figures to draw different conclusions.

—EVAN ESAR

Everyday Statistics

HOW TO WIN AT POKER

Want to win at the tables in Vegas? Statistics can tell us the odds of walking away richer. The odds of being dealt a royal flush—an unbeatable hand in poker—are 2,598,960 (the number of five-card hands you could be dealt) divided by four (the number of royal flush hands possible in a deck with four suits). That means we have one chance in 649,739 of keeping all the money on the table.

Should we switch to the slot machine? The odds are not much better here. Your chances of hitting the megabucks jackpot on a slot machine have been calculated at 1 in 49,836,032. That's a lot of coins funneled into the slot—we might be here a while.

Human beings have been interested in knowing the odds since the beginning of our species. The field of statistics was invented, in part, as an attempt to understand probability, especially when gambling. In Chapter 1, we will talk about the history of statistics and how we humans have used them.

OVERVIEW

WHAT ARE STATISTICS?
TYPES OF STATISTICS
VARIABLES
USING STATISTICS
SOME CAUTIONARY NOTES ABOUT STATISTICS
STATISTICS IN CONTEXT

LEARNING OBJECTIVES

Reading this chapter will help you to . . .

- Distinguish between the two main types of statistics: descriptive and inferential. (Concept)

- Recognize descriptive and inferential statistics in examples of research. (Application)

- Understand the difference between the two main types of variables used in statistics: independent and dependent. (Concept)

- Identify independent and dependent variables in an example of research that uses statistics. (Application)

- Recognize the context in an example of research that uses statistics. (Application)

- Understand what we mean by *context* and why context is so important in the study of statistics. (Concept)

WHAT ARE STATISTICS?

statistics A branch of mathematics concerned with the collection, analysis, interpretation, and presentation of masses of numeric data.

data Measurements or observations; an individual measurement is a **data point**, or **datum**; a collection of measurements is a **data set**.

So, what are statistics anyway? **Statistics** are a set of procedures that allow the user to collect, analyze, and interpret the meaning of usually large sets of numeric data. **Data** are measurements or observations. A **data set** is a set, or collection, of measurements. A single observation in that set is called a **data point**, or **datum**. Statistical procedures come from mathematics, and at least part of statistics involves applying mathematical techniques to the data in order to determine what the data have to tell us.

There are a number of other definitions of statistics, including one of the best humorous definitions I've ever heard: that statistics are a bunch of numbers looking for an argument. This is actually fairly accurate. Statistics are, in fact, just numbers. We who use statistics hope that they are numbers that can tell us about patterns or points of interest in the data set—that they can help us make our argument. After all, we went to all the trouble to collect the observations, and it would be nice to know that the effort was well spent. Our objective is to apply statistical techniques to the data in order to formulate and then present an argument or interpretation of what the data mean.

TYPES OF STATISTICS

The word *statistics* is derived from the Latin word *status*, meaning "state." Statistics were originally "state numbers"—numbers that told us something about the state of the state—although some historians argue statistics may have begun even earlier, with gamblers trying to gain a competitive advantage in games of chance (see *The Historical Context*, page 4). Whenever their actual origin, state numbers like the census, the unemployment rate, the mortality rate, the national debt, and the number of cars on the road (or carts, if you happen to be in ancient Rome) are good examples of the first of two types of statistics we will be discussing in this book: descriptive statistics.

The Historical Context

ROLL THEM BONES

In her book about the history of statistics, F. N. David (1962) says that modern statistics developed out of two desires. The first was fairly prosaic. The study of probability, an essential part of statistics, developed because people wanted to win at games of chance, specifically (at least in the beginning) dice. The ancient Greeks laid claim to inventing games that featured random chance or luck as an essential element (David, 1962). And the game of dice is probably as old as humanity. Before the cubic dice we're familiar with came along, players threw bones (hence at least one modern nickname for dice) called *astragali*.

The astragalus bone comes from the ankle of a hoofed animal and is roughly cubic in shape (see Figure 1.1). Players would toss the bone and bet on which side would land face up. The shape of the bone introduced random chance into the results of the toss, and betting kept things interesting. The ancient Egyptians played a version of dice called "Hounds and Jackals," and the Roman emperors were prodigious

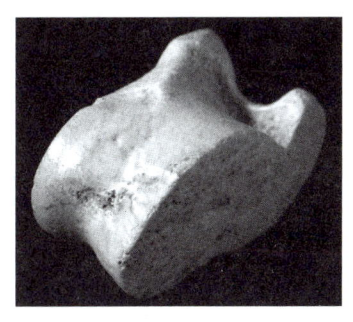

FIGURE 1.1
An astragalus, or "knucklebone," from the ankle of a hoofed animal.

players, rolling the bones any time they got the chance. The study of probability developed (at least in part) out of a desire to win more money more often when playing dice.

According to David (1962), the second reason for the development of statistics was the need to collect and record information about the health of the state. Politicians and others interested in the state of the state have used numbers (a.k.a. descriptive statistics) to keep track of the populace since the dawn of politics. For example, in 1532, King Henry VIII of England needed to know how his people were faring in the midst of yet another outbreak of bubonic plague. He had his grand chancellor, Thomas Cromwell, order the creation of a "bill of mortality," an account of christenings, weddings, burials, and deaths obtained from all the churches in London.

The *Bills of Mortality* became a regular weekly publication available to the public in 1592 (see Figure 1.2). In 1662, John Graunt, a London haberdasher (a seller of women's toiletries and cloth) who had pulled together several years of these bills, began the first attempt to see patterns in the lives and deaths of his fellow citizens. He published his results in *Natural and Political Observations on the Bills of Mortality*, a book that was influential in getting other countries to follow England's example and track births, deaths, diseases, and accidents in their own cities. The modern version of a bill of mortality is published weekly by the Centers for Disease Control and Prevention (CDC), headquartered in Atlanta, Georgia: It is called the *Morbidity and Mortality Report*, and it details illnesses and deaths state by state. You can see these modern bills of mortality by visiting the CDC website at www.cdc.gov.

FIGURE 1.2
The cover page from the *Bills of Mortality* for 1664.

DESCRIPTIVE STATISTICS

Descriptive statistics are so named for a fairly obvious reason: They are numbers that describe some aspect of a larger data set. In the case of state numbers, the unemployment rate describes the employment (or lack thereof) of the larger data set—namely, the current population of the United States.

descriptive statistics Techniques used to describe, in numbers, some aspect of a data set.

[Sticky note: Examples of descriptive stats - Unemployment rate.]

Statistics become very important when we want to measure something, like the health of the economy. If you have taken a course in political science or economics, you will remember that the unemployment rate is a descriptive statistic that is very closely watched because it can tell us a great deal about our economic health. The Bureau of Labor Statistics (BLS) defines the unemployment rate as the percentage of the civilian workforce who could be employed, have actively sought work in the past 4 weeks, but are unemployed; the BLS tracks this measure for each state across the year. Monitoring the unemployment rate to gauge economic health is analogous to monitoring your body temperature to gauge your own physical health: A rising body temperature indicates poor health, and a rising unemployment rate indicates an unhealthy economy.

Descriptive statistics are used to describe trends and patterns in a large data set. They make describing or summarizing any data set much easier because they can summarize the entire data set with just one number. This is why descriptive statistics are so often the first type of statistic used by anyone trying to talk about data. Imagine having to describe the unemployment rate of even a relatively small data set (like the residents of your neighborhood) without descriptive statistics. You would have to go door to door, asking all of your neighbors if they had a job; then you would have to write down every one of the responses. Without descriptive statistics, in order to talk about the data you have collected, you would have to read off your list—who said "yes" and who said "no." You could not report that two of your fifteen neighbors were unemployed, because that **ratio** (a figure that represents the relative size of two values) is a descriptive statistic. You could not report how often your neighbors threatened to call the police to stop you from ringing their doorbells and bugging them with questions that are none of your business, because the **frequency** of an event (a number that represents how often an event or observation occurs in a given unit of time) is a descriptive statistic. You couldn't even report that the typical answer to your question "Are you employed?" was "yes," because a number that represents the **typical observation** (a figure that describes or represents all of the observations in a data set) is a descriptive statistic.

Now imagine having to describe the unemployment rate of a much larger data set—say, the population of the state where you live. It would terribly difficult, not to mention inefficient, to knock on each and every door in your state in the same way that you polled your neighborhood. This is where the second type of statistics—inferential statistics—comes in.

INFERENTIAL STATISTICS

Of more recent origin, **inferential statistics** developed as science (one of the principal users of inferential statistics) evolved. These statistics were developed to help scientists make good conclusions about the **population** (a data set defined as *all* of the individual people, objects, or events that are of interest to the person carrying out the study) based on the study of a **sample** (a subset of the whole population) taken from that population.

Think about the name *inferential statistics*. When you make an inference, you come to a conclusion based on past experience, on logic, or on facts (data). Modern science involves making inferences based on systematic and careful observation of nature so

---

ratio A figure that represents the relative size of two values, such as two out of fifteen (written as 2/15, or 13.33%).

frequency A figure that represents the number of times an event or observation occurs within a given unit of time.

typical observation A single number that describes or represents all of the observations in a data set.

inferential statistics Techniques that allow researchers to draw conclusions or make generalizations about the population based on a sample drawn from that population.

population All members of a specified group that are being studied or observed. The U.S. population refers to all people living within the borders of the United States of America. In statistics, the population refers to all members of the group you want to draw an inference about.

sample A subset of the population studied or measured in order to make inferences about the population as a whole.

that the laws governing nature can be described. When science makes a proposition relating to, say, the lifespan of African pygmy hedgehogs, it is not based on a study of every African pygmy hedgehog that ever lived; instead, it is an inference based on a sample of African pygmy hedgehogs. A reliable proposition cannot be based on any old sample of hedgehogs, though. It has to be a **representative sample**. By representative, we mean that our sample must resemble the makeup of the entire population being studied. We cannot draw reliable conclusions about the lifespan of all hedgehogs by studying only female hedgehogs, or only hedgehogs living in zoos or kept as pets: None of these subsets of the population would constitute a representative sample.

FIGURE 1.3
Pierre Beauvallet's portrait of Galileo Galilei (1564–1642). Galileo was found guilty of heresy for insisting that the earth revolved around the sun and spent the final 12 years of his life living under house arrest.

representative sample A sample with a composition similar to that of the entire population being studied.

While we are considering populations and samples, let's take a moment to note the difference between a *statistic* and a *parameter*. With apologies in advance for the redundancy, a **statistic** is a descriptive statistic that describes a characteristic of a sample, and a **parameter** is a descriptive statistic that describes a characteristic of a population. If we said the lifespan of African pygmy hedgehogs in your local pet store was 2 to 3 years, this would be a statistic, since it is clearly based on a sample and not the entire population of African pygmy hedgehogs living in the wild. If we said the average birth weight of all African pygmy hedgehogs currently living in zoos in the United States and Canada was 2 to 3 ounces, that would be a parameter because it is based on all members of a specified population.

statistic A description, in numbers, of some characteristic of a sample.

parameter A description, in numbers, of some characteristic of a population.

The scientific methods we use today (note the plural—there are many methods that can be used to systematically observe nature) were first proposed by the Italian astronomer and physicist Galileo Galilei (see Figure 1.3). Galileo proposed that the mathematical truth of the universe and its laws could be discovered via **experimentation**. He advocated that we carefully observe and study samples rather than whole populations, and then use our findings to draw conclusions about the population the sample came from—in other words, make inferences.

experimentation A test of a principle or hypothesis for the purpose of discovering something unknown.

VARIABLES

Galileo proposed that scientists carry out their experimentation by controlling variables. A **variable** is anything that can take on more than one value, or vary. If you were interested in studying the movement of a ball on an inclined plane (as Galileo was), you would want to devise experiments that let you control, and therefore study, the influence of variables like the angle of the inclined plane, the weight of the ball, and the surface texture of the plane on the movement of the ball.

variable Anything that can take on more than one value, or vary.

target variable The variable the researcher is interested in investigating; the target of the study.

extraneous (or confounding) variable A variable that influences the outcome of an experiment even though it is not the variable of interest to the experimenter. Extraneous variables add error to an experiment.

error Uncertainty in the data stemming from limitations of the measuring equipment or technique, mistakes made in sampling or in the act of observation itself, or the influence of extraneous variables.

hypothesis A tentative explanation that a scientist sets out to test through research.

levels The varying amounts of the independent variable presented in a study.

independent variable (IV) The variable that is manipulated, changed, delivered to the experimental participants in varying amounts, or used to define groups.

dependent variable (DV) The variable that is measured in order to see the effect of the independent variable.

FIGURE 1.4
A model of the apparatus used by Galileo for his inclined plane experiment.

Galileo also proposed that we exert as much control on these variables as we could, studying the effect of one or two variables at a time.

The variable that we are interested in studying is called the **target variable**. Other variables might be influencing our target variable, but these variables (called **extraneous variables**, or **confounding variables**) are of secondary importance to the situation we happen to be studying. By attempting to *control for* these extraneous variables—in other words, by attempting to hold them constant—and then carefully manipulating the one target variable at hand, we can see the effect of the target variable and minimize the risk of **error**.

In an experiment, the researcher is testing an explanation (a **hypothesis**) about why or how something happens. Let's say this something is the effect of a drug on the intensity of the common cold. The researcher might frame her hypothesis about the drug by saying something like "Zinc, dissolved under the tongue, will reduce the duration of the cold." She would then administer zinc to one group of cold sufferers, give another group a placebo, and carefully measure the intensity and duration of the illnesses in each group. She would also control for potential confounding variables that might also influence the length of the illness in each group by making sure that all the participants started out at similar levels of physical fitness, that they all ate the same healthy diet during testing, and so on. If the researcher can control for extraneous variables, she has a much better chance of seeing the effect of her target variable (the zinc).

INDEPENDENT AND DEPENDENT VARIABLES

By tradition, the target variable that is manipulated, changed, delivered to the experimental participants in varying amounts or **levels**, or used to define the experimental groups is called the **independent variable** (**IV**). The variable that we measure in order to see what effect the manipulation of the independent variable has had is called the **dependent variable** (**DV**). Think of it this way: Change in the dependent variable *depends on* what we did with the independent variable.

The IV is the variable that is the focus of the study. In the case of our medical researcher, the independent variable is zinc. We start the study off with the assumption that the IV has an effect on the world (if we thought the IV had no effect at all, why would we bother to study it?) The DV is a variable we think is affected or changed by the IV. The value of the dependent variable is a measurement we take—the weight of something in ounces, the number of times something happened, the time it takes something to happen in microseconds, and so on. In our study of zinc, it is the duration of the cold, measured in days. Any change in the measurement of the DV reflects the influence of the IV.

To illustrate independent and dependent variables, let's consider Galileo's experiment with the inclined plane (described beautifully by Johnson, 2008). Galileo was testing several of Aristotle's ideas about motion, and he wanted to control as many extraneous variables as he possibly could. He built a ramp that could be adjusted to a variety of incline angles and rolled a brass ball down this ramp, timing its descent (see Figure 1.4).

He considered possible extraneous variables, including the weight of the ball, the effect of room temperature on the ramp or the ball, and the resistance of the material the ramp was made of to the rolling of the ball, and he controlled for these by using the same ramp and ball in all of his tests and by running all of his tests at the same time of the year. Timing the descent with a water clock, he weighed the amount of water that dripped into a collection container after each run—more water equaled more weight equaled more time—so that he could minimize timing errors as well. He could then change the angle of the inclined ramp and see what altering the angle did to the time it took the ball to run down the ramp.

Let's suppose Galileo tested the time it took the ball to roll down the ramp at five different incline angles. His independent variable would be the ramp angle, and his experiment would be described as having five levels. His dependent variable would be the time it took the ball to roll down the ramp, measured in terms of the weight of the water collected.

Notice that Galileo is actively changing one variable and trying to measure the effect of that change by examining another variable. Notice, too, that he is not studying *all* balls and *all* inclined planes or even *all* angles. He is studying a specific inclined plane and a specific ball, and then making an inference about how factors like gravity, weight, resistance, and so on affect motion on inclined planes *in general*. In modern terms, we'd say that Galileo is making an inference about a population based on careful study, observation, and measurement of a sample (a subset of the population). Inferential statistics include procedures for predicting error, which make inferences about the whole, huge population made from one tiny sample drawn from it possible.

Let's look at another example of independent and dependent variables, this one from research performed more recently than the sixteenth century. This study measures the effectiveness of "pet therapy." This unique form of treatment is based on research showing that pet ownership is a better predictor of long-term survival for patients recovering from cardiovascular disease than either marital status or family contacts (Friedmann, Katcher, Lynch, & Thomas, 1980).

Several researchers from the University of Northern Iowa wanted to know if pet therapy would be effective in children who suffered from attention-deficit/hyperactivity disorder, or ADHD (Somervill, Swanson, Robertson, Arnett, & MacLin, 2009). Specifically, they wanted to know if the opportunity to interact with a friendly, furry, and well-behaved Shih Tzu would calm the autonomic nervous systems of children with ADHD. They measured blood pressure and heart rate in 22 children with ADHD after 5 minutes of interaction with an adult (the experimenter) and again after 5 minutes of interaction with the lap dog. Each child was tested several times, with and without the dog, across several days (a technique called a **repeated-measures design**, because each participant in the study is tested "repeatedly"—for more about this design, see Chapter 10). Researchers controlled for the possible extraneous variable of the type of dog by using the same dog with all of the children and by making sure that the dog was accustomed to interacting with a variety of children and adults.

The researchers' independent variable was the presence or absence of the dog (so there were two levels of their IV), and their DV was the reaction of the child's autonomic nervous system. They found that the presence of the dog actually

repeated-measures design A study that tests or measures the same participants repeatedly in every condition.

stimulated the children—the opposite of the effect seen in adults—resulting in an increase in blood pressure rather than the decrease usually reported in adults.

Notice also that Somervill et al. did not examine all children with ADHD, just a subset of them—a sample rather than the entire population. Studying a sample rather than the whole population is an important innovation. Think back to our example of studying unemployment in your neighborhood, or the lifespan of African pygmy hedgehogs. As long as the population that we want to study is small, we can measure each and every element of that population. When populations get very large, however, observing each element of the population becomes unwieldy, costly, and sometimes downright impossible. Studying a sample drawn from the population is more affordable, less time-consuming, and generally more doable. However, studying a sample rather than the whole population also introduces a potential problem: chance error.

Think About It . . .

HOW A TAN AFFECTS ATTRACTIVENESS

Periodically, I'm going to ask you to think about the issues, vocabulary, statistical tests, and so on that we will discuss in the text. I encourage you to use these *Think About It* questions to see how well you understand the topics being covered. Some—like the one here—won't require any mathematical operations to answer. Others will. But all of them should help you acquire a better understanding of the material. So, here we go: your first *Think About It* exercise.

The risk of skin cancer notwithstanding, many Caucasians are convinced that having a tan will make them more attractive to others. Banerjee, Campo, and Greene (2008) asked 135 men and 226 women to rate the attractiveness of a female Caucasian model whose image had been digitally altered to have a light, a medium, or a dark tan. All participants were of college age (19–25 years) and from a variety of racial and ethnic backgrounds (64% *Caucasian*, 16% *Asian/Pacific Islander*, 5% *Hispanic/Latino*, 3% *Bi- or Multiracial*, 2% *African American*, and 2% *Other*). All participants rated attractiveness on a scale from 0 (*very unattractive*) to 100 (*very attractive*).

The findings: Male participants in the study rated the model with the darkest tan as more attractive than the same model shown with a light or a medium tan. The attractiveness ratings of female participants were not affected by the depth/darkness of the tan shown in the photograph.

Think about this study, and answer the following questions:

1. There are two independent variables in this study. One is the sex of the person judging the photograph. What is the other IV?
2. How many levels did the second IV have?
3. What is the dependent variable used in this study?
4. Several social variables are often controlled for in a study like this one. I've listed a number of them below. Do you think these variables might be potential

extraneous (or confounding) variables in this study? How might you control for them? Which of these variables did Banerjee et al. control for in their study, and how did they do it?
a. Age
b. Social class
c. Region of origin
d. Region of residence
e. Employment status
f. Socioeconomic status
5. Can you think of any other potentially confounding variables not listed in question 4?

Answers to this exercise can be found on page 26.

CHANCE ERROR

Once you decide to focus on just a sample rather than the entire set of observations, you introduce the possibility that any conclusion you reach based on your study of that sample is contaminated by chance errors.

Chance errors are just what they sound like: They are errors that result from chance, or bad luck. If you use a sample to draw a conclusion about a population, you instantly risk unwittingly drawing a sample that is not representative of the entire population. Your chance error may be unavoidable; you may not even be aware of it. It occurs because you have not included each element of the population in your sample—you have not checked the employment status of every single resident of your state; you have not measured the life of every African pygmy hedgehog. The problem is that by leaving some things out of your sample, you have introduced the possibility that you've left out something important. This is one reason why the general rule in statistics is that *large samples are better than small ones*. The very best samples are, from a statistical standpoint, large and representative of the parent population they were taken from. The need to represent the population is why you must pay careful attention to how a sample is selected. We will discuss sampling in more detail in Chapter 8.

Random error in measurement, or inconsistency in the universe, has been a favorite subject of mathematicians, philosophers, and scientists for generations. The history of statistics is full of mathematicians trying to figure out how to measure, control, reduce, predict, or in a pinch, simply use error. Mathematicians like Karl Pearson (for whom the Pearson correlation statistic is named), R. A. Fisher (for whom the F-statistic is named), William Gossett (who doesn't have a statistic named after him directly, but who is famous in statistics nonetheless), and many others all created *inferential statistics*, which allow us to measure the effects of chance in an experiment and, thereby, allow scientists to come to better conclusions about populations.

chance error An error or difference in measurement that is the result of chance.

CheckPoint

Indicate whether the following statements are *true* or *false*.

1. An experimenter's target variable is typically the independent variable. _____
2. A variable used to group participants before testing is called the dependent variable. _____
3. A confounding variable is an independent variable that is not the variable of interest to the researcher and that does not affect the outcome of the experiment. _____
4. An independent variable is manipulated by the researcher. _____
5. A dependent variable is measured by the researcher. _____
6. Descriptive statistics are used to generalize the results from a sample to a population. _____

Answers to these questions are found on page 14.

USING STATISTICS

Statistics, both descriptive and inferential, are tools, used by everyone, every day. Even if you don't want to do research and don't see yourself drawing samples and making predictions about populations, you still need to understand statistics. You can't open a newspaper, turn on the television, or read a magazine without encountering statistics. I've found the following statistics without even trying in my local newspaper.

In an article bemoaning some trades made by the Atlanta Braves baseball team, I found that "[the Cardinals would] still be in first place but not 25 games over .500 without the contributions of the pitchers they got from Atlanta, starter Jason Marquis (4.03 ERA) . . ." (O'Brien, 2004). The statistic .500 is a proportion representing the number of games won or lost. If your team is at .500, they are winning as often as they are losing. A team that is 25 games over .500 is winning more games than they're losing. A pitcher like Jason Marquis with an earned run average (ERA) of about 4 is doing well—his ERA represents the average number of earned runs given up by the pitcher per nine innings. Since pitchers don't want the opposing team to hit their pitches and get runs, the closer the ERA is to zero, the happier the pitcher will be. An ERA of 4.03 is respectable.*

In an article about changes in religious affiliation in the U.S. population, I found that "[b]etween 1993 and 2002, the share of Americans who said they were Protestants dropped from 63 percent to 52 percent, after years of remaining stable, according to . . . the National Opinion Research Center at the University of Chicago" (Quinn, 2008). These statistics (63% and 52%) are obvious—they represent the section of the population who describe themselves as Protestant on a

---

* To calculate the ERA, divide earned runs by innings pitched and multiply by 9. For a pitcher in Major League Baseball, an ERA less than 4.00 is very good, and an ERA between 4 and 5 is average. An ERA of 6.00 or more means the pitcher is in danger of losing his job.

survey. Interestingly, the number of people who describe themselves as "Christian" has gone up—perhaps fewer people are making distinctions between denominations of Protestantism?

Statistics are also used in every academic discipline out there, not just the sciences (social and otherwise). To cite a few examples from fields not typically associated with statistics: A friend working on a dissertation about 2 years in the life of the French poet Charles Baudelaire found himself writing about statistics. He needed to describe what was happening in publishing during the mid-nineteenth century and did so by using statistics to describe publishing trends (number of books published in science, social science, fiction, etc.) in the early years of the reign of Napoleon III.

Another example is what is known as *stylometrics*—literally, "measuring style." Using inferential statistics, researchers interested in stylometrics have attempted to determine the probable author of a number of disputed works. For example, they have counted and compared the number of words used, the number of times a particular word occurs, and the mean length of the words, and then used those statistics to conclude who might have written that text (did Shakespeare actually write all those plays, or was Sir Francis Bacon the author of some of them?).

SOME CAUTIONARY NOTES ABOUT STATISTICS

In just about every case where statistics have been used to bolster an argument (regardless of the area of investigation), there is disagreement about what the statistics are telling us. Students are often surprised to find out that statistics don't necessarily tell you what is *true*.

I'm sure that at some point you've read about the results of an experiment and then, maybe only a few days later, read about another experiment on the same topic that came to the opposite conclusion. It isn't surprising that people start to wonder what the heck statistics really tell us if a variable is important one day and insignificant the next. In my opinion, this is likely because human beings tend to think that if math or numbers are involved, the resulting numeric answer will be true, constant, and unchanging.

So, here is the first cautionary note about statistics: Statistics tell you only what is *probably* true. Any statistical result you calculate can be *accurate*, if you used the proper formula in calculating it and didn't make any mistakes. But it can also be *false*, if it indicates a conclusion about your sample and the population that is untrue, or at least untrue outside of the particular situation you examined. Remember, we said that inferential statistics allow us to measure the influence of chance in our study. What this means in practice is that we can calculate the chances that *our conclusion is a mistake*. We will never be able to completely eliminate the possibility that we are drawing an incorrect conclusion about our data—we can only reduce the risk of being wrong. We will consider how any conclusion can be wrong, statistically speaking at least, in more detail in Chapter 9.

> **CheckPoint**
> *Answers to questions on page 12*
>
> 1. TRUE ☐
> 2. FALSE ☐ The *independent variable* would be used to assign subjects to groups.
> 3. FALSE ☐ Confounding variables *do* affect the outcome of the experiment.
> 4. TRUE ☐
> 5. TRUE ☐
> 6. FALSE ☐ *Inferential* statistics allow generalization of results from sample to population.
>
> **SCORE:** /6

It is the uncertainty about the "truth" of a statistic that creates the second cautionary note about statistics: Statistics can be, and often are, used to mislead you, and mislead you very effectively. There are a number of interesting, and often extremely funny, books written about what Benjamin Disraeli famously referred to as the "three kinds of lies—lies, damned lies, and statistics." These books detail the long history of the misuse and abuse of statistics by politicians; advertising agencies; state, local, federal, and foreign governments; and the average Josephine or Joe with a point to prove.

It is easy to use statistics to mislead people, both because statistics look impressive and because people often do not question numbers. For example, you can mislead people by simply concentrating on the positive aspects of your data and not reporting any negatives. Cigarette manufacturers funded research, replete with statistics, to show that the addition of menthol to cigarettes created the perception of a smoother, less irritating smoke and seemed to reduce the scratchy throat that people often said they got from smoking cigarettes. However, they omitted mention of the fact that menthol itself caused serious health risks when smoked (see Foley, Payne, & Raskino, 1971, for an example of this research).

You can also mislead people by playing around with definitions. Ivory Soap used to advertise that their soap was "99 and 44/100ths percent pure," obviously hoping that the consumer would look at this statistic and be impressed by the sincerity the 44/100ths implied. How many consumers asked how it was possible to calculate a percentage of an absolute, like purity? Something is either pure or impure—it can't be "sort-of" pure. How often have you heard someone described as 99 and 44/100ths percent dead? There's an old saying in philosophy: "he [or she] who defines the terms wins the argument." In this case, the marketing people were using their own definition of purity to win consumers over to their side of the soap aisle.

CheckPoint

Indicate whether the following statements are *true* or *false*.

1. Inferential statistics always have error. _____
2. When two or more distinct groups are measured on the same dependent variable, this is a repeated-measures design. _____
3. Statistics always represent the truth. _____
4. When measuring weight for men and women, height could be a confounding variable. _____
5. Sets of measurements must have at least 100 units (individuals) to be considered a population. _____

Answers to these questions are found on page 16.

STATISTICS IN CONTEXT

Most people don't know very much about statistics. And this lack of knowledge is what makes misleading people with statistics so easy. Please remember, however, that it isn't the *statistics* that are the lie—statistics are relatively harmless little numbers happily doing exactly what we ask of them. The misleading part comes from the people who use the statistics to make their point. Remember also that statistics are not *always* misleading. The difficult part of reading any statistic is deciding if it makes sense to you.

All authors of statistics have a **context** in mind: They are using their statistics to make a point. Some authors intentionally bend the "facts"; others do so without realizing it, often in the course of conveying what they sincerely believe to be an important truth. Your job as the reader of a statistic is to consider the context, evaluate what you are being told, and make a decision about the value of that message. Statistics are used quite often as the definitive endpoint of the discussion. They get slapped down and pointed to with vehemence, authority, or anger. Then, listeners who don't understand statistics often do exactly what the persuasive messenger wanted—they back off and stop asking questions.

context The point that the author of the statistic is trying to make.

After reading just this first chapter, you now know more than the average person does about statistics, and you have also taken the first step in acquiring some very effective decision-making tools. Learning about statistics, and understanding that statistics are *always* a part of an argument for or against something, will allow you to evaluate the research and the conclusions reached by others so that you may decide for yourself what to take from their arguments. The key point is to trust your own intelligence and, above all, keep asking questions.

You should also begin to consider your role as an author of statistics. Here, your job is to make sure you know what your context is and that you are being

> **CheckPoint**
> *Answers to questions on page 15*
>
> 1. TRUE ☐
> 2. FALSE ☐ When the same participants are tested or measured repeatedly in every condition, this is a repeated-measures design.
> 3. FALSE ☐ Statistics can represent the truth, but they don't always represent it.
> 4. TRUE ☐
> 5. FALSE ☐ Populations are user-defined, so there is no required number of elements that a particular population must have in order to be considered a population.
>
> **SCORE:** /5

clear about what you say when you write about statistics. Writing effectively about statistics is a skill that can be learned—and one that I think *must* be learned by students like you, who will be called upon to use statistics. Keep in mind that your job as the interpreter of statistics is to tell your audience what the numbers mean. Keep your writing simple and clear, and you will be much more effective and much less likely to mislead. I'll talk more about how to write with statistics and about context as we go on. For now, though, remember the KISS rule: Keep It Short and Simple.

SUMMARY

Statistics are numbers. *Descriptive statistics* are numbers that *describe* some aspect of a larger data set, and *inferential statistics* allow us to predict the effects of chance in a study and so make good inferences about a population.

Both types of statistics can be applied to a *population* (all members of a specified group) or to a *sample* (a subset of the population). Inferential statistics are used to draw conclusions about a population from a sample.

Science attempts to make these conclusions better by controlling secondary variables and then carefully manipulating one variable (the *independent variable*) and observing the effects of that manipulation on another variable (the *dependent variable*).

Statistics can and should always be interpreted in terms of their *context*, or the point the author of the statistic is trying to make. Being mindful of the context is crucial to deciding for yourself whether the statistics support the argument the author is trying to make.

TERMS YOU SHOULD KNOW

chance error, p. 11
confounding variable, p. 8
context, p. 15
data, p. 4
data point, p. 4
data set, p. 4
datum, p. 4
dependent variable (DV), p. 8
descriptive statistics, p. 5
error, p. 8
experimentation, p. 7
extraneous variable, p. 8
frequency, p. 6
hypothesis, p. 8

independent variable (IV), p. 8
inferential statistics, p. 6
levels, p. 8
parameter, p. 7
population, p. 6
ratio, p. 6
repeated-measures design, p. 9
representative sample, p. 7
sample, p. 6
statistic, p. 7
statistics, p. 4
target variable, p. 8
typical observation, p. 6
variable, p. 7

WRITING ASSIGNMENT

College libraries are practically bursting at the seams with journals in an enormous variety of disciplines. Each of these journals provides example after example of how to write with statistics in the particular format used by that discipline. For example, in psychology, the format we follow in writing research reports is called American Psychological Association (or APA) style. Most sciences have their own "style" used in research writing, and students of a particular discipline should learn how their science reports statistics. Once you start looking at these styles, you'll quickly notice that they are all fairly similar. This is not so surprising, since all of the sciences are trying to do the same thing: They are all trying to report the results of studies and experiments in a way that is effective, efficient, and interesting. So, for your first writing assignment, your instructions are to go to the library and pick out a current issue of a journal in your discipline. Read an article that interests you and then, in a paragraph, answer the following questions about that article:

1. What were the independent and dependent variables in the study you read?
2. What were the authors trying to show in their article? (In other words, what was the *context* of the study they are reporting on?)
3. What was the most important result of the study? (Or, what was the authors' "take home" message?)

PRACTICE PROBLEMS

For each example below, identify the independent variable (IV) and the dependent variable (DV). Remember that the independent variable is the variable researchers are interested in seeing the effect of, and the dependent variable is the one measured to see the effect of the IV.

1. To test the effectiveness of a new supplement designed to strengthen the immune system, 25 elementary school teachers (a group of people who are routinely exposed to an array of cold and flu viruses) were asked to take the new supplement daily for one semester (3 months). The number of colds these teachers got during the subsequent 3 months was counted. At the same time, a second group of 25 elementary school teachers were given a placebo, and the number of colds they experienced over the same period of time was counted.
   a. The IV was _dose of supplement_.
   b. The DV was _# of colds during subsequent in 3 months_.

2. In a study of the effects of stress on test performance, 20 student volunteers enrolled in an undergraduate statistics class are separated into two groups. Group 1 is told that that their performance on an upcoming 100-point test will be monitored via hidden cameras, and that points will be deducted from their test grade if they take more than 20 seconds to answer any one question. Group 2 is not told anything except that the experimenter will have access to their grades on the upcoming test. The number of questions correctly answered on the test is determined.
   a. The IV was _stress_.
   b. The DV was _questions correctly answered (Test performance)_.

3. The effect of different types of exercise on fitness was evaluated. Several new gym members were selected and assigned at random to three groups. Group 1 took a 6-week yoga class, Group 2 took a 6-week water aerobics class, and Group 3 took a 6-week weight training class. Overall physical fitness was evaluated at the end of the training period for each group. (Think about this example. There are a number of ways you might measure physical fitness. What method would you use?)
   a. The IV was _different classes_.
   b. The DV was _physical fitness_.

4. The influence of age on the ability to perceive the passage of time was evaluated. Five different age groups (comprising participants ranging in age from 7 to 75 years) were compared on a time-estimation task. Accuracy in estimating the passage of a 15-second interval of time without the aid of a clock or a counting system was calculated for each of the age groups.
   a. The IV was _____.
   b. The DV was _____.

5. A popcorn manufacturer is interested in determining what kind of movie—comedy, drama, action, or romance—generates the greatest consumption of its product. A local megaplex showing one movie of each type serves as the popcorn company's laboratory. To get a full house for each type of movie, the company places an ad in the local paper offering free tickets until every seat in each theater is taken. Tubs of popcorn—all of the same size—are weighed before being given to each audience member in each

STATISTICS IN CONTEXT

of the megaplex's four theaters. After the movie, the tubs are weighed again, and the amount of popcorn consumed is calculated.
   a. The IV was ___movie seeing___.
   b. The DV was ___amount of popcorn consumed___.

6. The ability of children to accurately perceive the passage of time is examined in two different settings, one very noisy (the playground) and one very quiet (the library), to see if noise influences time perception. The children are tested individually and given the same instructions: Beginning as soon as the experimenter says "GO," they must estimate an 11-second interval and say "STOP" as soon as they think 11 seconds have elapsed. The experimenter times the intervals with a stopwatch.
   a. The IV was ___different settings___.
   b. The DV was ___if noise influences time perception___.

7. Does the environment in which alcohol is consumed affect the way the human body reacts to alcohol? To find out, student volunteers of legal drinking age are given exactly 1 ounce of alcohol to consume first in a social setting (in a local hangout surrounded by people and music) and then 1 week later in their dormitory room with no one else around. In each environment, small blood samples are drawn 30 minutes after the alcohol is consumed in order to measure alcohol metabolism. (There's at least one extraneous variable here—that is, a variable other than alcohol that might very well affect the results of this study. What extraneous variables can you identify, and what might you do to reduce their effects on the data?)
   a. The IV was _____.
   b. The DV was _____.

8. Health and fitness experts often tout the benefits of exercise in helping us cope with stress. To find out if exercise really does reduce stress, college students are recruited to participate in a study of the effects of exercise during a very stressful period of time in college life: the 2 days before final exams begin. Blood samples are taken from all participants at the beginning of the experiment, and levels of cortisol (a hormone secreted by the body in response to stress) are measured. Half the students (selected at random) are asked to participate in an hour-long exercise program with a personal trainer followed by 30 minutes of rest. The other half of the students are told to go to the library and study for 90 minutes. Both groups then donate another blood sample, and cortisol levels are measured again.
   a. The IV was _____.
   b. The DV was _____.

9. J. B. Watson and R. Rayner (1920) conducted one of the most famous experiments in the history of psychology. They wanted to find out if a phobia could be learned. They exposed a 9-month-old child (called "Little Albert") to a white rat, something that initially Little Albert did not fear at all. Subsequently, however, each time Little Albert saw the rat, Watson, standing out of Albert's sight, would bang two steel pipes

CHAPTER 1   Introduction: Statistics—Who Needs Them?

together creating a loud and unexpected noise, startling the infant. The repeated pairing of the loud noise with the sight of the white rat eventually created a fearful response: Albert would start crying and try to crawl away as soon as he saw the rat, even without the loud noise being introduced.
a. The IV was _____.
b. The DV was _____.

10. Can worms learn? To find out, Thompson and McConnell (1955) examined a sample of flatworms, *Dugesia dorotocephala* (a good model for a learning experiment because it is a very simple animal with true synaptic nerve conduction and a proto-brain). The worms were placed individually in petri dishes filled two-thirds full with water. A lightbulb was suspended 6 inches above the dish. A mild shock, sufficient to make the animal change its direction of travel, was delivered as soon as the lightbulb was turned on. After several pairings of light and shock, the light coming on alone, without any shock administered, elicited a turning response from the worm.
a. The IV was _____.
b. The DV was _____.

11. In a similar study of learning in flatworms, Corning (1964) placed a circle of cardboard under the clear glass petri dish. The cardboard was painted black on the left side and white on the right. In the center of the cardboard was a small "start box" circle. The worm was placed in this circle at the start of each trial. Corning first established the worm's "preferred side" of the dish by counting the number of times the animal moved to the black side and to the white side in a 10-minute period. Then, he began training the animal to avoid its preferred side by pairing an electric shock with movement to that side. After 30 minutes of training, the worm learned to avoid (it no longer traveled to) its previously preferred side.
a. The IV was _____.
b. The DV was _____.

12. Does having a stable social group help you (or any primate) survive? Cohen, Kaplan, Cunnick, Manuck, and Rabin (1992) examined 43 healthy adult male cynomolgus monkeys (*Macaca fascicularis*). These monkeys were raised for 14 months in stable, unchanging social groups. Then, 21 of them were assigned to a "socially unstable" living situation: Their social groups—the three to four other monkeys they lived with—were changed every month, introducing three to four new monkeys to the environment. The remaining 22 monkeys continued living in their "socially stable" groups: There were no changes in the three to four other monkeys that each test monkey lived with. The researchers then measured the immune system function of the monkeys in each group. Monkeys subjected to chronic social stress (the "socially unstable" group) showed significantly suppressed immune system function compared to the "socially stable" group.
a. The IV was _____.
b. The DV was _____.

13. James Olds and Peter Milner carried out an experiment in 1954 considered by many to be one of the most important experiments ever conducted in psychology or neuroscience. Working at McGill University in Montreal, Canada, Olds and Milner were testing the "drive reduction" theory of learning, which suggested that food worked as a reinforcer because eating when we're hungry reduces the unpleasant drive to find food (negative reinforcement). Most animals, including humans, will work to repeat this feeling. Olds and Milner were trying to stimulate the brain's reticular activating system (RAS), theorizing that increased activity here should motivate a test animal—a rat, in this case—to work to reduce stimulation to this area. They inserted an electrode into what they thought was the rat's RAS (they actually missed and hit a bundle of nerve fibers called the medial forebrain bundle, which runs through the hypothalamus) and then tested the rat to see what its reaction to electrical stimulation would be. They placed the rat in a large box with the corners designated A, B, C, and D. Whenever the rat wandered into corner A, it was given a brief electrical stimulation of the electrode. They then counted the number of times it subsequently visited corners A, B, C, and D. The rat very quickly learned to go to corner A—exactly the opposite of what the theory predicted. Apparently, the electrical tickle felt good, and the rat would work to repeat it.
    a. The IV was _____.
    b. The DV was _____.

14. In another set of experiments, Olds and Milner placed the rat with the electrode into a T-maze and tried to teach it to turn right in order to obtain electrical stimulation to its brain (Study a) (Olds, 1956). Very quickly, the animal learned to turn to the right. Olds and Milner then switched things around and stimulated the brain if the rat turned left (Study b). Again, the animal very quickly learned to stop turning right and start turning left. Finally, they set up the T-maze so that food was available in both the right and left arms, but the animal would receive electrical stimulation halfway down the alley leading to the arms, before it got to the food (Study c). The rats learned to stop halfway down the alley and never went on to find the food, despite the fact that they had been food-deprived for 24 hours prior to this experiment and were very hungry.

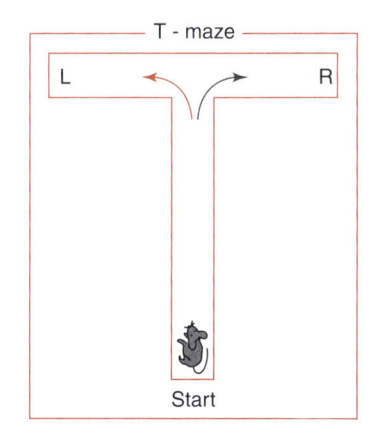

a. The IV (the variable they wanted to see the effect of) was:
   Study a: _____
   Study b: _____
   Study c: _____
b. The DV was:
   Study a: _____
   Study b: _____
   Study c: _____

15. We humans have evolved to use facial cues to determine who in a social situation is dominant and who is submissive. In an effort to find out where in the brain information about social dominance is being processed, Chiao et al. (2008) asked 14 male participants to make an assessment of the sex of the person in a series of pictures. Each picture showed human faces with dominant, angry, submissive, fearful, or neutral expressions. Accuracy in recognizing the sex of the person in the photograph was measured. Males were significantly less accurate at recognizing sex in photos of female angry faces compared to male angry faces.
    a. The IV was _____.
    b. The DV was _____.

16. Chiao et al. (2008) also measured functional magnetic resonance images (fMRIs) in their study. In an fMRI, activity in the brain is color-coded by a computer so that very active areas show up as red and less active areas show up as cool blue. They found that each facial expression—dominance, submission, anger, fear, or neutral—was associated with high levels of activity in distinct regions of the brain. They concluded that different regions of the brain are processing specific information about the emotional expressions of the people that surround us.
    a. The IV was _____.
    b. The DV was _____.

17. Quick: Name a number between zero and nine. Did you pick seven? If you did, you're not alone. Kubovy and Psotka (1976) asked 558 people to do just this, and almost a third of them picked seven. Kubovy and Psotka wondered if the context of the question was driving this tendency to pick the number seven, so they changed the context and asked another set of 237 people to pick the first number that comes to mind between 6 and 15. Sure enough, no one number stuck out as most frequently picked. Instead, people tended to pick single-digit numbers over two-digit ones.
    a. The IV was _context of the question_.
    b. The DV was _tendency to pick a number_.

18. Do you feel lucky? Maltby, Day, Pinto, Hogan, & Wood (2013) tested the Dysexecutive Luck hypothesis, which suggests that believing yourself to be unlucky is associated with deficits in cognition, specifically in what is called "executive functioning." Participants in the study completed two questionnaires. The first was the Darke and Freedman Belief in Good Luck Scale, which measures how lucky or unlucky you

believe yourself to be. High scores on this questionnaire indicate a stronger belief in being unlucky. Participants also completed the Dysexecutive Questionnaire, an instrument usually given to patients suffering from frontal lobe damage. This questionnaire measures executive function (and dysfunction), assessing things like whether you have difficulty in showing emotion, an inability to keep your mind on a task, or difficulty planning for the future. High scores here indicate high levels of dysfunction. Maltby et al. found that people who scored very high on the Darke and Freedman scale also tended to score high on the Dysfunction Questionnaire.
    a. The IV was __Questionnaires__
    b. The DV was __association__ between being unlucky and deficits in cognition.

19. Students in my lab last year wanted to find out if believing in luck improved performance. To find out, we recruited volunteers to play the card game Concentration, where the goal is to flip cards over two at a time in order to make matches as quickly as possible. Participants were randomly assigned to one of three groups: Group 1 was told that the deck of cards they were playing with had been lucky for others in the past; Group 2 was told that the deck of cards they were playing with was unlucky; and Group 3 was simply told to make as many matches as they could in the allotted time. The number of correct matches made and the time needed to make them were measured for each group. Believing that the cards were lucky did not affect performance on the task—participants in all three groups made approximately the same number of matches.
    a. The IV was _____.
    b. The DV was _____.

20. A statistics professor wants to investigate the effects of math anxiety on test performance in her statistics class. Using the students' scores on the Mathematics Anxiety Rating Scale (the MARS: see Richardson & Suinn, 1972) administered during the first day of class, the professor separates students into two groups: those *high* in math anxiety (Group 1) and those *low* in math anxiety (Group 2). Grades on all exams during the semester were averaged for each student in each group, and groups were compared at the end of the semester.
    a. The IV was _____.
    b. The DV was _____.

21. Now go back through each study in questions 1 through 20, and determine the number of levels of the IV in each study.

22. For each of the statements listed below, indicate whether the statement is descriptive (**D**) or inferential (**I**).
    a. The average GPA for American College and University students in 1955 was 2.30 on a 4.00-point scale. _____
    b. The latency to the onset of the first episode of rapid eye movement (REM) in depressed individuals was significantly shorter than latency in non-depressed persons. _____

c. On average, office workers at the BigEye Insurance Company spent 20 minutes per day playing desktop football during the last quarter. _____
d. Females were significantly faster at categorizing emotional facial expressions than were males. _____
e. Cholesterol level did not predict heart disease. _____
f. Spending by politicians in the next election is predicted to surpass the total expenditures seen in the last election. _____
g. The average height of women in the United States in 1940 was 62 inches. In 2010, the average height of U.S. women was 64 inches. _____
h. Women in the United States are getting taller. _____
i. The regular practice of meditation is associated with a decrease in blood pressure. _____
j. In 2012, professional golfer "Bubba" Watson was ranked number one in longest average drive on the PGA tour. _____

23. Cigarette manufacturers spent a great deal of time and money on advertising designed to sell people on the benefits and pleasures of smoking. Not all of their ads featured shining examples of good science. Take a look at the magazine ad in Figure 1.5, and think about the science being discussed here.

FIGURE 1.5
Magazine advertisement for Cavalier Cigarettes, dated 1943.

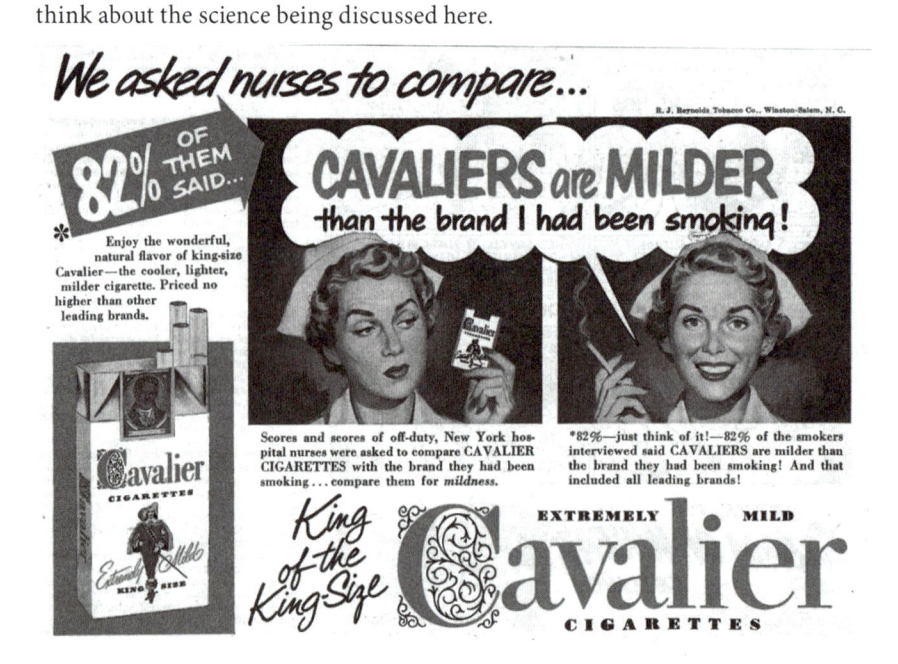

a. What problems do you see with the data that are being presented? Is there any information missing?
b. Has any information been added to the bare statistic of 82% in order to sway your opinion about the mildness of Cavalier cigarettes?

Suppose you were asked to replicate the study described here:

c. What would your independent and dependent variables be?
d. How might you measure the "mildness" of the smoke?

24. Complete the crossword puzzle shown below.

[Crossword grid with filled answers: 6 Across = "Dependent", 8 Across = "Independent"]

ACROSS
- 3. Measurements or observations
- 6. The variable that is measured
- 8. The variable that is manipulated
- 9. Anything that can take on more than one value
- 10. A test of a principle or hypothesis
- 13. All members of a specified group
- 14. The point the author of a statistic is trying to make

DOWN
- 1. Statistics used to describe some aspect of a larger data set
- 2. A subset of a population
- 4. The number of times an observation or event occurs
- 5. The type of statistic that takes into account the effects of random chance
- 7. The varying amounts of an IV presented
- 11. A number representing the relative size of two values
- 12. A branch of mathematics dealing with the collection, analysis, interpretation & presentation of numerical data

CHAPTER 1   Introduction: Statistics—Who Needs Them?   25

Think About It...

HOW A TAN AFFECTS ATTRACTIVENESS

SOLUTIONS

1. There are two independent variables in this study. One is the sex of the person judging the photograph. What is the other IV?   **The darkness of the tan.**
2. How many levels did the second IV have?   **Three: light, medium, and heavy tans.**
3. What is the dependent variable used in this study?   **The attractiveness rating scale.**
4. Several social variables are often controlled for in a study like this one. I've listed a number of them below. Do you think these variables might be potential extraneous variables in this study? How might you control for them? Which of these variables did Banerjee et al. control for in their study, and how did they do it?
    a. Age   **The authors controlled this by limiting the age of the participants to "college age," or 19 to 25 years. Personally, I think age might be an important extraneous variable here, as personal definitions of beauty most likely do change with age.**
    b. Social class   **This variable is likely important (many sociological studies have suggested that definitions of beauty are dependent on social class), but with the information provided here, it's impossible to tell if this variable was controlled for.**
    c. Region of origin   **Both region of origin and the next variable, region of residence, would probably matter, at least in terms of exposure to tans (unlikely in some regions, and unwelcome in others). Definitions of beauty might well be regionally dependent, but again, we can't tell if the authors controlled for this variable.**
    d. Region of residence   **As mentioned above, region of residence might well matter in terms of exposure to tans: For instance, the closer to the equator one lives, the more exposure to the sun one experiences, and so the more likely one is to have had a tan or to have considered the pros and cons of tanning in general. We can't tell if the authors have controlled for this variable.**
    e. Employment status and f. Socioeconomic status   **Both employment status and socioeconomic status are related to social class and so might well influence the perception of beauty, but again, we can't tell from the information given if they were controlled for. All of these variables could be controlled for by surveying the participants ahead of testing and asking them about these variables.**
5. Can you think of any other potentially confounding variables not listed in question 4?   **The authors of the study might want to consider level of education attained and sexual preference, as both variables might be related to the perception of beauty.**

STATISTICS IN CONTEXT

REFERENCES

Banerjee, S. C., Campo, S., & Greene, K. (2008). Fact or wishful thinking? Biased expectations in "I think I look better when I'm tanned." *American Journal of Health Behavior, 32*, 243–252.

Best, J. (2001). *Damned lies and statistics: Untangling numbers from the media, politicians, and activists.* Berkeley & Los Angeles: University of California Press.

Bills of Mortality. Retrieved from faculty.humanities.uci.edu/bjbecker/plaguesandpeople/lecture12.html

Chiao, J. Y., Adams, R. B., Tse, P. U., Lowenthal, W. T., Richeson, J. A., & Ambady, N. (2008). Knowing who's boss: fMRI and ERP investigations of social dominance perceptions. *Group Process and Intergroup Relations, 11*(2), 201–214.

Cohen, S., Kaplan, J. R., Cunnick, J. E., Manuck, S. B., & Rabin, B. S. (1992). Chronic social stress, affiliation, and cellular immune response in nonhuman primates. *Psychological Science, 3*(5), 301–304.

Corning, W. C. (1964). Evidence of right-left discrimination in planarians. *The Journal of Psychology, 58*, 131–139.

David, F. N. (1998). *Games, gods and gambling: A history of probability and statistical ideas.* Mineola, NY: Dover.

Foley, M. G., Payne, G. S., & Raskino, L. M. A. (1971). Micro-encapsulation of menthol & its use as a smoke smoothing additive at "sub-recognition" threshold. Retrieved 8/30/2017 from Tobacco Documents Online. https://www.industrydocumentslibrary.ucsf.edu/tobacco/docs/#id=gyck0141

Friedmann, E., Katcher, A. H., Lynch, J. J., & Thomas, S. A. (1980). Animal companions and one-year survival of patients after discharge from a coronary care unit. *Public Health Reports, 95*(4), 307–312.

Johnson, G. (2008). *The ten most beautiful experiments.* New York, NY: Alfred A. Knopf.

Kubovy, M., & Psotka, J. (1976). The predominance of seven and the apparent spontaneity of numerical choices. *Journal of Experimental Psychology: Human Perception and Performance, 2*(2), 291–294.

Maltby, J., Day, L., Pinto, D. G., Hogan, R. A., & Wood, A. M. (2013). Beliefs in being unlucky and deficits in executive functioning. *Consciousness and Cognition, 22*, 137–147.

O'Brien, D. (2004, July 21). Baseball: Braves report. *The Atlanta Journal–Constitution*, B3.

Olds, J., & Milner, P. (1954). Positive reinforcement produced by electrical stimulation of septal area and other regions of the brain. *Journal of Comparative and Physiological Psychology, 47*(6), 419–427.

Olds, J. (1956). Pleasure centers in the brain. *Scientific American, 195*, 105–116.

Quinn, C. (2008, June 28). Religious affiliation in the U.S. *The Atlanta Journal–Constitution*, E7.

Richardson, F. C., & Suinn, R. M. (1972). The mathematics anxiety rating scale: Psychometric data. *Journal of Counseling Psychology, 19*, 551–554.

Thompson, R., & McConnell, J. V. (1955). Classical conditioning in planarian, *Dugesia dorotocephala. Journal of Comparative and Physiological Psychology, 48*, 65–68.

Somervill, J. W., Swanson, A. M., Robertson, R. L., Arnett, M. A., & MacLin, O. H. (2009). Handling a dog by children with attention-deficit/hyperactivity disorder: Calming or exciting? *North American Journal of Psychology, 11*(1), 111–120.

Watson, J. B., & Rayner, R. (1920). Conditioned emotional reactions. *Journal of Experimental Psychology, 3*(1), 1–14.

Introducing SPSS: The Statistical Package for the Social Sciences

When I was a student, lo these many years ago, the professor teaching research statistics had only recently allowed what was then a brand-new technological innovation into his classroom: the desktop calculator. I vividly remember consigning my slide rule (look it up if you don't know what a slide rule is) to the bottom drawer of my desk and eagerly learning how to make my new calculator add, subtract, multiply, and divide for me. I still had to use my table of square roots, however, because calculating square roots was too much for that simple little box to handle.

My first calculator would be as foreign to students today as isinglass windows on a horse-drawn buggy would have been to me as a college student. Today, most data are analyzed via computer software, and one of the most popular statistical programs out there is SPSS—the Statistical Package for the Social Sciences. Originally developed for social scientists (as the name of the program suggests), SPSS is now used in the social and natural sciences, marketing, education, business, health science, and other fields. The program handles descriptive and inferential statistics via pull-down menus that are easy to use, once you get comfortable with the lingo. The program will also produce beautiful graphics that are (or can be) "camera-ready," or suitable for publication. It is one of the programs you'll likely run into if you continue down the research and data analysis path, so let's start learning how to use it.

The first thing you'll want to learn is how to create a data file that SPSS can use. Let's start with something simple. Table 1.1 presents some data for you to get started with in SPSS. The data are based on a study of the sleeping behavior of college undergraduates. In the table, each row represents a (fictitious) individual participant in my study, and each column represents something I measured about that participant. Here, I measured sleep duration (in minutes) and sleep quality (on a scale from 1 to 5).

SPSS expects data to be entered in a specific format of columns and rows. Each variable in your data file will be entered in a column, and all of the data for a given participant will be entered into a row. So, take a quick look at the data in Table 1.1, and determine how many columns and rows will be required.

You should see that we have 10 participants, so we'll have 10 rows in the data table we create. We also have three variables (ID number, sleep duration, and sleep quality), so we'll need to set up three columns, one for each variable. Notice that two of these variables (sleep duration and quality) are dependent variables: They are measures we've taken from our participants. ID number is a sort of "bookkeeping" variable, useful in keeping track of who is who in our data set.

When you open SPSS, the first screen you see is split down the center. On the left-hand side, you'll see two sets of options for opening up stored files. The boxes on the right are called "What's New," "Modules," and "Tutorials." We'll talk about these later. For now, look at the options on the left side of the screen. At the top of the left-hand side, you'll see **NEW FILE** and **DATABASE QUERY**. The **NEW FILE** option is what we want (we're going to create a new file). **DATABASE QUERY** would be selected if you wanted to open a non-SPSS file—for example, a file you created in another program like Excel that you wanted to move

Table 1.1 Sleep duration and quality for 10 fictitious college undergraduates

ID	Sleep Duration (in min.)	Sleep Quality (1 = *poor*, 5 = *excellent*)
1	480	3
2	400	4
3	120	1
4	400	5
5	356	5
6	200	2
7	490	5
8	485	5
9	600	4
10	477	2

into SPSS. At the bottom left, there's a list of **RECENT FILES**, which would have the names of existing SPSS files you've already created if, in fact, there were existing files to open. For now, click to highlight **NEW FILE**, and then click **OK** (lower right-hand corner).

You should see an empty spreadsheet composed of rows and columns, all blank for the moment. If you look at the lower left-hand corner of this screen, you'll see two "tabs" that look somewhat like the tabs on the edges of manila folders. One tab reads **DATA VIEW**, and the other reads **VARIABLE VIEW**. By default, we open into data view, and because we don't have any data in the empty spreadsheet, we don't see much. The columns are all labeled "var" (for "variable"), and the cells are all blank. To enter data into a file, we first need to go to **VARIABLE VIEW** so that we can define the variables that will be in our file. So, click to highlight the **VARIABLE VIEW** tab, and take a look at what happens to the screen. You should see columns labeled as follows:

NAME: Here's where you enter the name of each variable (column) in your data file. We'll have three columns. Name the first one **ID**, the second **DUR**, and the third **QUAL**. There's a limit of eight (8) characters for any name that you use, which accounts for the short names I've used here, and you shouldn't use any punctuation or any other special characters in the name you select.

TYPE: As soon as you enter a name, you'll notice that several other columns automatically fill with default values. This is to save you time in entering data. There a several types of data you might want to use, and SPSS will default to the most commonly used type, which is numeric data. Click on the word **NUMERIC**, and you should see three small dots at the right of the word. Click on the dots, and you'll see your other data type options. Notice that on the right side of this window, the **WIDTH** and number of **DECIMAL** places are defined. By default, SPSS expects the numbers you enter to be eight digits long, with two digits to the right of the decimal

CHAPTER 1   Introduction: Statistics—Who Needs Them?   29

place. The data in our first column (labeled **ID**) are between 1 and 10, and there are no decimal places needed. So, type the number "2" in the **WIDTH** box and "0" in the **DECIMALS** box, and then click **OK**. You should see that the next two columns (**WIDTH** and **DECIMALS**) have changed and now read "2" and "0," respectively.

WIDTH: The width of the data that will be entered into the columns.

DECIMALS: The number of decimal places for the data entered into the columns.

LABEL: Here you can write almost anything, but users typically type in a label for the variable being entered. For example, the first column of data will contain participant ID numbers, so the label you attach to it could read "ID Number" or "Identification Number" or something else of your choosing. Lots of people use this as a sort of reminder about what the data in a particular column represent. However, you should keep in mind that whatever you enter here will be used later on in output that involves the data in the column, so spelling and punctuation matter. There are no limits on length or type of characters entered here.

VALUES: Click here, and notice that the default value is **NONE**. Once again, you should also see three small dots at the right of the word "None." Click on the dots to see what your options are. Sometimes, data are represented by a numeric code, and this column will let you define what these codes mean. For example, suppose you want to compare sleep time for females and males. You'll need a way to define the sex of each participant. Typically, this information is entered into a column of data (perhaps labeled "Sex" or "Gender") so that a code number is assigned to each participant and each sleep time. Suppose further that we want to assign the number "1" to females and "2" to males. We'll have a column full of 1s and 2s, and we want SPSS to "know" that all of the 1s indicate females and the 2s males. So, we can enter the number "1" into the **VALUE** box and then the word "Female" into the **LABEL** box. Then, we can click **ADD** to add the new code to the list being created for SPSS. We're not using any special codes, so let's cancel out of here and go back to defining our variables.

COLUMNS: The function of the next two columns has to do with the way the data will look in the data file. The **COLUMNS** column will let you define the physical width of the columns in the data file. The default width is eight. You can make the column wider or narrower (it's up to you), but keep in mind that if you use a long name on a column (e.g., if you use the word "Duration" as a column name) and then define the physical width of the column as only two or three digits, you'll cut the name of the column down as well. Go ahead and try changing this, and see what happens. You really can't hurt anything.

ALIGNMENT: This also deals with the physical appearance of the data in the column. You can tell SPSS to display the data aligned with the left side, right side, or center of each column.

MEASURE: Here you can define the type of data (Nominal, Ordinal, Interval, or Ratio) to the data you've entered into a given column. We'll talk more about this in the next chapter, so for now, you can use the default value of **UNKNOWN**.

ROLE: Here, you can define the data in a column as an independent variable or a dependent variable. The default value of **INPUT** indicates that the numbers in this column represent independent variables. If you want to assign the role of dependent variable to the data, click on the three small dots at the right of the word input, and select the role of **TARGET**. All of our variables are IVs, so let's use the default role.

So far, we've defined just the first column of data (named **ID**). You'll need to define each of the three columns, telling SPSS about the variable that will be entered into each column as you go. When you've defined all three, click on the **DATA VIEW** tab to go back to the spreadsheet and make note of what has changed.

You should see that instead of the vague name "var" at the top of each column, you now have the words **ID**, **DUR**, and **QUAL**. And now you can enter the actual values into the spreadsheet. Type in the values for each variable, and when you're done entering the data, *save the data file!* To save the data, click on the word **FILE** at the upper left-hand corner of the screen. Select **SAVE AS** so that you can name the file as you'd like to and so that you can specify where the file should be saved.

At the top of the **SAVE DATA AS** screen, you should see a box called **LOOK IN**. Below this box is a window showing the available folders on the default drive. Saving to the hard drive of the computer you're on is usually the default option, and if that's where you want to save the file, then select the folder you want to save to. If you want to save to a thumb drive or some other removable device, click on the downward-facing arrow at the right of the **LOOK IN** box, and highlight the drive you want to save to.

You can name the file anything you want, but I'd advise using something descriptive and short so that you can find it again easily. Let's try calling this file "Sleep1." To save with this name, type the words "Sleep1" into the box labeled **FILE NAME**. Notice that another box, labeled **SAVE AS TYPE**, is just below the file name box. SPSS expects to save data files using the extension ".sav" and so this should be the default type listed. You don't have to do anything here to tell SPSS that this is an SPSS data file. We'll talk about other options for file types later on in the book.

CHAPTER TWO

TYPES OF DATA

Everybody gets so much information all day long that they lose their common sense.

—GERTRUDE STEIN

Everyday Statistics

"LIVE LONG AND PROSPER"

There's a very old saying that the only certain things in life are death and taxes. How do we know? If you guessed that statistics are involved, you're correct. For instance, how are predictions about life expectancy made? Graphs like the one shown here are constructed by counting the number of people (frequency) in a given country who live to a given age and then determining the average age at death by country.

According to the World Health Organization's 2016 report, Japan has the longest life expectancy, at 83.7 years, while the African nation of Sierra Leone occupies the bottom of the list, with a life expectancy of just over 50 years. Canada comes in at number 12 on the list, and the United States at number 31: Why do you think there is such a big difference between these two close neighbors? Graphs like this one have led to research on the factors that influence life, disease, and life expectancy, such as temperature, the amount of available light and its effect on circadian rhythms, economic prosperity, and so on. In Chapter 2, we will look at frequency distributions as a first step in organizing our data.

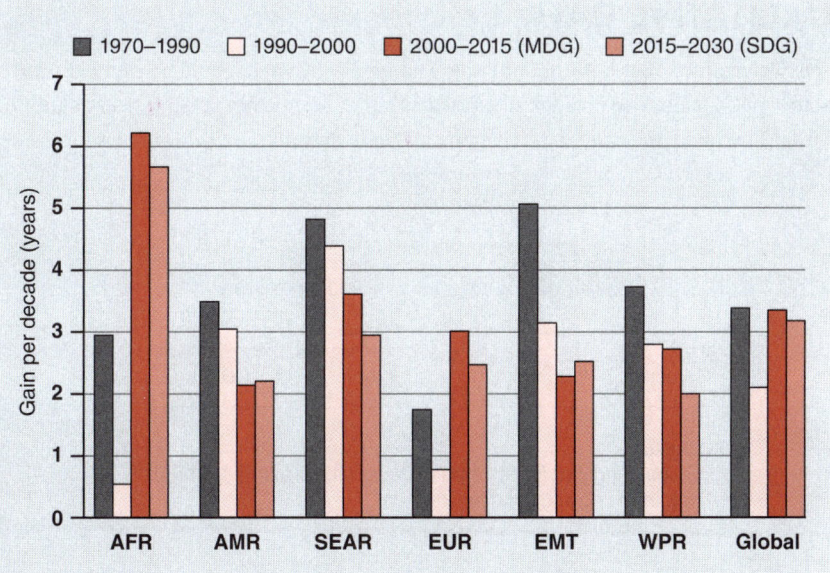

OVERVIEW

DATA, DATA EVERYWHERE
SCALES OF MEASUREMENT
ORGANIZING DATA
SCALES OF MEASUREMENT IN CONTEXT
. . . AND FREQUENCY DISTRIBUTIONS

LEARNING OBJECTIVES

Reading this chapter will help you to . . .

- Explain the differences between quantitative and qualitative measurement scales. (Concept)

- Understand the scales of measurement proposed by S. S. Stevens. (Concept)

- Identify nominal, ordinal, interval, and ratio scales of measurement in research examples. (Application)

- Understand how to organize data using four types of frequency distributions: simple, relative, cumulative, and cumulative relative (or percentile rank). (Concept)

- Organize data using frequency distributions, both grouped and ungrouped. (Application)

- Know which mathematical operations are appropriate for each of the four scales of measurement. (Concept)

DATA, DATA EVERYWHERE

Why does time seem to pass more quickly when we're having fun? How come time slows down during finals week, yet spring break seems to pass by in a heartbeat? To answer these questions, we need to collect some data. For example, we might ask a group of students waiting to see the dentist to estimate the number of minutes that have elapsed while they sat in the waiting room; then, we might ask a second set of students to do the same thing while playing the newest video game. Now we have measurements of time, and we can use descriptive statistics to describe how time passes in these two situations. However, not all descriptive statistics work with all types of data. The very first thing we need to do is determine what kind of data we actually have. The type of data we have collected determines the type of descriptive statistic we can use.

SCALES OF MEASUREMENT

According to Harvard psychophysicist S. S. Stevens (whom we will meet in *The Historical Context* on page 35), there are two basic characteristics of an object or event that we can measure: *quantity* and *quality*. If our measurements tell us *how much* of something there is (amount) or *how many* of something there are (count), then our measurements are measurements of quantity, and we have what is known as **quantitative data**. If we were to ask the people at the dentist's office to estimate how many minutes they'd been waiting, we would be obtaining a quantitative measure of time.

On the other hand, if our measurements tell us what *category* (a general division of a variable) our event or object belongs to, or if the measurements put those categories in some kind of logical order (from least to most, for example), then we have measurements of quality, otherwise known as **qualitative data**. For example, we might measure elapsed time categorically by asking our participants at the dentist's office to declare whether they had been waiting a "long time" or a "short time" and allowing only those answers. Now our measurement of time is categorical, or qualitative.

quantitative data A measure of the amount or count of an event or object.

qualitative data A measure of a quality of an event or object; a measure that places that object or event in a category and/or places those categories in meaningful order.

QUALITATIVE DATA

Stevens determined that both quantitative and qualitative data could be divided into two subtypes. Qualitative data are described as either *nominal* or *ordinal*, and quantitative data are said to be either *interval* or *ratio*. These divisions are shown in Figure 2.1.

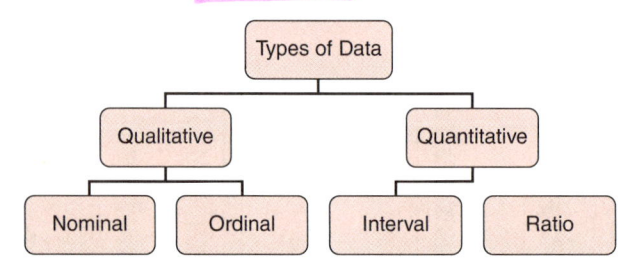

FIGURE 2.1
Qualitative and quantitative data viewed in terms of nominal, ordinal, interval, and ratio scales.

34 STATISTICS IN CONTEXT

The Historical Context

S. S. STEVENS AND HIS POWER LAW

Each of us understands the world around us in our own unique way, despite the fact that we all live in the same physical world. In the behavioral sciences in particular, we find ourselves having to measure variables like feelings, emotions, or memories: stimuli that cannot be picked up, weighed, or measured in the traditional sense. Stanley Smith (S. S.) Stevens is remembered in statistics for his efforts to create a means to measure intangibles like perceptions, thoughts, and feelings.

Born in Ogden, Utah, in 1906, Stevens attended Stanford University and graduated with a bachelor's degree in 1931. He had apparently taken so many courses as an undergraduate that "the identification of a major was somewhat problematic" (Teghtsoonian, 2001, p. 15105). He went on to Harvard, where he earned a Ph.D. in philosophy in only 2 years. In 1936, he was appointed to the position of instructor in Harvard's Department of Psychology, which was established as an independent department the year after Stevens completed his degree. He remained at Harvard until his death in 1973.

Stevens specialized in the study of psychophysics. He was particularly interested in the difference between the physical magnitude of a stimulus and our perception of that magnitude. Stevens showed that this relationship was not linear (we don't see a room lit by two candles as being twice as bright as a room lit by just one); instead, the relationship can be described with what has come to be called Stevens' power law. This law ($S = KI^b$) says that our perception of the magnitude of the stimulus (S) is equal to the physical intensity of the stimulus (I) raised to the power b (K is a constant that is different for each sensory system).

This research led Stevens to propose his new scale of measurement, and if you think about what his power law is really describing, you can see why he needed to propose it. Measuring the intensity of a physical stimulus can be done with instruments. Most people agree about what is being measured, and anyone using the same instrument will get the same measure of intensity. But measuring a *perception* of that intensity is much more subjective and difficult. In fact, it was generally seen as so difficult that it could not be done at all, and the psychologists trying to measure this sort of thing were accused of engaging in pseudoscience. Stevens was very much involved in the debate about what science was and whether psychology was one. He wanted to provide support for his measurements of psychological events like perceptions.

His proposal generated immediate response from statisticians, mathematicians, and scientists of all stripes—not all of it favorable. Stevens' biographers point out that "[t]he work that Stevens began has now evolved into a mathematical specialty in its own right, yielding results that confirm some of what Stevens intuited, and disconfirm other aspects of his work" (Teghtsoonian, 2001, p. 15106). The question of how best to measure a purely psychological event (like how loud a sound is, or how saturated a color is) continues to be debated in research labs across the world.

Nominal Scales

An example or two of each type of data will help you understand how they differ. Let's start with the most basic sort of measurement: one based on a **nominal scale**. A nominal measurement does not actually use numbers at all. Instead, nominal data are represented by *numerals*. The distinction between numbers and numerals is important here. **Numerals** are the figures or symbols we have adopted to represent **number**, which is the amount or count of objects present. If we wanted

nominal scale A qualitative measurement using numerals to represent different categories.

number The amount or count of objects or events present.

numerals The figures or symbols used to represent number.

to communicate the number of chairs in a room, we could choose to do so with Arabic numerals (1, 2, 3, 4, . . .) or with Roman numerals (I, II, III, IV, . . .)—two different systems of symbols (numerals) that both represent the idea of amount or count (number).

Numerals are essentially labels we put on categories to make each category distinct from all others. They tell us about what category a response belongs to. However, numerals do not tell us about amount or count: They are not actually numbers at all, and so they cannot be treated as numbers. *Numbers* can be added, subtracted, multiplied, or divided; *numerals* cannot.

Suppose you asked the students in your statistics class whether or not they smoked cigarettes. Table 2.1 shows the hypothetical responses from participants 1 through 6.

These are qualitative data. Each student's response has been placed into one of two categories so that it can be measured. The numeral 1 under "Student ID Number" does not represent how much of that person was there, nor does it represent how many people were in the room. It is qualitative: it tells you *which* person was there. The response to the question "Do you smoke?" is also measured using qualitative, nominal data. The symbols used in the second column represent, or substitute for, the words "yes" or "no." The data tell you which answer the students gave to your question. Your measurement tool simply puts observations into categories, representing the categories with numerals.

Suppose you were within earshot of the people filling out the questionnaire and you happened to overhear person 1 tell person 3 that she smokes about a pack a day. This information is quantitative and *can* be represented with numbers: the number 1, indicating *how many* packs per day, and the number 20, indicating the *count* of cigarettes per pack.

Ordinal Scales

The second kind of qualitative data is *ordinal*, and as the name suggests, it can be used to put categorical data in order. Here's an example: Imagine a study of the physical characteristic of women considered by others to be "beautiful." A number

TABLE 2.1 Smoking behavior of six students

Student ID Number	Response (1 = *Yes*, 2 = *No*)
1	1
2	2
3	2
4	2
5	1
6	1

of studies have shown that both women and men tend to equate thin with beautiful (see, e.g., Cawley, 2001; Paxton, Norris, Wertheim, Durkin, & Anderson, 2005). So, let's imagine a study designed to compare male and female assessment of overall beauty in women. We'll give two groups of participants a series of pictures of women and ask each group to tell us how beautiful they think each woman pictured is. There are established measurement scales of beauty, but we will use our brand-new "Beauty Measurement Scale" (or BMS), where beauty can range from 7 (the most gorgeous person you ever saw) to 1 (not in the least beautiful). Table 2.2 shows some hypothetical data from this study, representing 10 participants' assessments of one of the pictures.

Obviously, a picture rated a 7 is more beautiful, in some sense of that word, than a picture rated a 5, but how much more beautiful? Is the difference in "amount of beauty" for a picture rated a 7 and a picture rated a 5 the same as that between a picture rated a 3 and a picture rated a 5?

Because we don't have a quantitative measurement of beauty, we cannot answer either of these questions—we have not measured "amount of beauty" at all. Instead, we have an **ordinal scale** of numerals that separates participants into seven categories (just as nominal data would do) and then puts those categories into order from most to least (first, second, third, etc.). So, in an ordinal scale of measurement, the numeral 1 means both first category and the least beautiful, while the numeral 6 means both more beautiful than 1 and less beautiful than 7, and so on.

With ordinal-level data, we don't know how big the interval between any two adjacent categories actually is. And since we don't know how big any one category

ordinal scale A qualitative measurement using numerals to represent both category membership and order.

TABLE 2.2 Ratings of female beauty by 10 participants

Participant ID	Rating of Beauty (7 = Extremely Beautiful, 1 = Not at all Beautiful)
1	1
2	5
3	6
4	2
5	1
6	7
7	6
8	7
9	3
10	2

is, we can't meaningfully add the categories together—we have no idea how much beauty a rating of 7 plus a rating of 3 actually is.

Does it make any sense at all to calculate the average when you have qualitative data? Those of you who have already been introduced to the idea of calculating an average (in this case, the mean) know that to find the mean you add up the observations in your set and then divide by the number of observations you had. Consider the data from our hypothetical study of smoking behavior. We asked people to answer the question "Do you smoke?" and coded the "yes" answers as 1 and the "no" answers as 2. Box 2.1 shows how we would go about calculating the mean response for the smoking data.

BOX 2.1 — **The mean and nominal data**

$$1 + 2 + 2 + 2 + 1 + 1 = 9$$

$$9/6 = 1.5$$

So, we can confidently say that the mean answer to the question "Do you smoke?" is 1.5, and we would be absolutely correct—but we would also be saying something absolutely meaningless. Our measurement scale consisted of the numerals 1 and 2, representing the responses "yes" and "no." What does 1.5 represent on this scale? It is a meaningless statistic. It describes nothing and, when it gets reported as somehow descriptive of qualitative data, happens to be one of the reasons why statistics have the reputation of being murky and misleading.

When we have qualitative data, we simply cannot meaningfully add, subtract, multiply, or divide the measurements. Because we can't use these familiar mathematical operations with qualitative data, many users of statistics think of qualitative data as less valuable than quantitative data (though it isn't), and some researchers will go to extremes to find a quantitative measurement scale rather than use the easier and more appropriate qualitative one.

For example, we could probably come up with a *quantitative* measurement of beauty to use in the next beauty study we conduct (I'll let you decide if we should even be trying to quantify beauty). Kowner (1996) proposed that human beings see beauty in symmetry. Suppose that instead of measuring relative beauty with our ordinal scale, we actually measured, in millimeters, the symmetry of the facial features of the contestants in our beauty contest. Now we have quantitative data that represent an amount or count, and we can calculate means, measures of variability, and anything else that requires we add, subtract, multiply, or divide our data.

QUANTITATIVE DATA

The two types of quantitative data, interval and ratio, share an important characteristic. Both scales have equal intervals between any two adjacent measurements on the scale. That means that the amount of change in the thing being measured is constant as we go from one point on the scale to the adjacent point. Suppose we were measuring facial symmetry with a ruler marked off in inches, as shown in Figure 2.2. The distance between 1 and 2 on the ruler is 1 inch, just as the distance between 5 and 6 is 1 inch, and between 10 and 11 is 1 inch, and so on into infinity or the end of your ruler, whichever comes first.

Because the interval between adjacent categories is a constant, we can now, finally, treat our data as number (quantitative) data. We can add categories together, both because we now know exactly how big each category is and because we know that all the categories are exactly the same size. Putting them together makes sense, mathematically speaking.

Interval and Ratio Scales

The distinction between ratio-level measurement and interval-level measurement has to do primarily with the meaning of zero in each measurement system. Ratio scales of measurement start at a true zero, while interval scales start at an arbitrary zero point.

Zero, or a symbol indicating the absence of the thing being measured, appeared in several cultures around the world somewhere between 400 BCE and 500 CE, depending on which culture you look at and which historian you read. Regardless of exactly when it became common to refer to "nothing" with a particular symbol, having a true zero point means that all ratio scales start at the same point. All measurements taken with a **ratio scale** are then relative to that common zero point.

Examples of ratio-level data include measurements in inches, centimeters, micrograms, kilograms, miles per hour, minutes, seconds, microns—really, any measurement where we have equal intervals between adjacent categories and a real zero point that signifies an absence of the thing we are measuring represents measurement on a ratio scale.

On the other hand, if we used a scale where zero was arbitrarily determined, and where all measurements were relative to that arbitrary zero point, then we would have an **interval scale** of measurement.

It's a bit harder to come up with examples of interval-level measurement, but there are some classics. Temperature measurement on a Fahrenheit scale is probably the first that everyone thinks of. The zero point on the Fahrenheit scale is arbitrary. Anyone who has been outdoors on a deep winter evening will tell you that

ratio scale A quantitative measurement with a real zero point.

interval scale A quantitative measurement with an arbitrary zero point.

FIGURE 2.2
All intervals between adjacent categories are equal.

when the thermometer reads 0 degrees, it is without question cold, but it certainly is not what science calls absolute zero (the complete absence of heat). In fact, Gabriel Fahrenheit arbitrarily decided that 0 degrees indicated the point at which a slurry of water, ice, and salt would remain stable (neither melt nor freeze). Take the salt away, and the temperature at which water froze was 32 degrees above the zero point on Fahrenheit's scale. Anders Celsius, equally arbitrarily, set the zero point on his scale at the temperature at which plain water froze and then measured the temperature of everything else relative to that point. The zero point in both temperature interval scales does not indicate the absence of heat. Instead, both zeros are arbitrary.

Please note that the fact the zero is arbitrary on an interval scale does not prevent you from treating it as if it were a real zero and adding, subtracting, multiplying, and dividing the measurements you collect with it. For instance, if it's 64 degrees in Atlanta and 32 degrees in Chicago, we could say that the temperature in Atlanta is twice as warm.

It is difficult to think of another interval-level measurement scale that is so often used by so many people, although a former student of mine did point out another one that I had not thought of. She says that shoes and clothing both come in size zero. Now if we assume some kind of meaningful relationship between the size of an article of clothing and the size of the person it is intended to fit (an assumption challenged by many), then this zero must be arbitrary (unless the clothing manufacturers of the world intended to fit a nonexistent person). There is probably an interesting debate in there somewhere, but let's move on.

CheckPoint

Indicate whether the following statements are *true* or *false*.

1. Interval-level measurement has a real zero. _____
2. The Fahrenheit scale is a good example of a ratio scale of measurement. _____
3. Ordinal data have equal spacing between values. _____
4. Numerals are symbols used to represent number. _____
5. You cannot perform mathematical operations with nominal data. _____

Answers to these questions are found on page 42.

Think About It . . .

SCALES OF MEASUREMENT

When we measure something, we are assigning a number to that object in the world. The number that we assign should represent the attributes of that object. For example, if we measure how much an object weighs, the number we assign to represent the weight of that thing should maintain all the attributes of the weight of that object that exists in the real world. If it is a heavy object, the number should be large, and it should be larger than the number representing some other, lighter object.

Measurement scales are often described as satisfying one or more of the properties of measurement. These properties are as follows:

- *Identity:* Each value on the scale should have a unique meaning.
- *Magnitude:* When an object has more of the thing being measured than another object, the number assigned to it is bigger.
- *Equal intervals:* All the units on a given scale are equal to one another, so the difference between any two adjacent points on the scale are all equal. Therefore, 1 pound differs from 2 pounds by the same amount that 20 pounds differs from 21 pounds, and that 67 pounds differs from 66 pounds, and so on.
- *Rational zero point:* An object that has none of the attribute being measured would be assigned a value of zero on the scale. An object that has no weight (despite the basic impossibility of a *thing* having no weight) would be assigned the value zero.

Think about these properties and the scales of measurement developed by S. S. Stevens as you answer the following questions:

1. Which properties does a NOMINAL scale of measurement possess?
2. Which properties does an ORDINAL scale of measurement possess?
3. Which properties does an INTERVAL scale of measurement possess?
4. Which properties does a RATIO scale of measurement possess?
5. Suppose I were to measure height using the following scale:
   1 = *shorter than me*
   2 = *the same height as me*
   3 = *taller than me.*
   Which properties does my scale have?
6. Which properties differentiate QUALITATIVE from QUANTITATIVE scales?
7. Which properties does the clothing size scale shown in Table 2.3 possess? (FYI, I could not find any "standard sizes" for men, so I've not included them in the table. There are several different "standard sizing" charts for women, all loosely based on height.)

TABLE 2.3 Standard women's clothing sizes (misses)

| 00 | 0 | 2 | 4 | 6 | 8 | 10 | 12 | 14 | 16 | 18 | 20 |

Source: American Society for Testing and Materials (ASTM), 2011.

Answers to this exercise can be found on page 88.

> **CheckPoint**
> *Answers to questions on page 40*
>
> 1. FALSE ☐ Interval-level measurement has an *arbitrary* zero point.
> 2. FALSE ☐ The Fahrenheit scale is a good example of an *interval* scale of measurement
> 3. FALSE ☐ Ordinal data do *not* necessarily have equal spacing between values
> 4. TRUE ☐
> 5. TRUE ☐
>
> **SCORE:** /5

ORGANIZING DATA

Descriptive statistics are numbers that describe some aspect of the data set. Earlier in this chapter, I referred to the mean response to the question "Do you smoke?" as a *statistic* (a bad statistic, but a statistic nonetheless). The mean is a single number that describes some aspect of the data set—in this case, the average response to the question we asked our subjects about smoking.

Descriptive statistics (numbers used to describe a set of measurements) were designed to make life easier for the people using them. Imagine that you've collected your data and found something amazing, and now you want to tell other people about it. You could go up to people you meet on the street and start listing the variables and observations you found in your experiment, but the probability that you would be allowed to pursue this technique for disseminating your data for very long is awfully low. At some point, you're likely to become a data point in someone else's study of personality disorders. So, if running around listing all 200 observations one by one is out, what should you do? The answer is to *summarize*, which descriptive statistics allow you to do.

Descriptive statistics allow us to look for patterns or consistencies in the measurements, places where the observations seem to clump together and places where they spread out as thinly as butter on toast. Reporting patterns takes less time than reporting each individual data point, and it focuses our audience's attention on what is important in the data.

So, the next questions to ask are: What does a pattern look like, and how do we find one? The first thing we'll want to do in our hunt for a pattern is to get our data organized so that we can see what we've got and what we're looking for.

FREQUENCY DISTRIBUTIONS

I've been teaching for a while now, and I've noticed that it isn't completely out of the realm of possibility for students to fall asleep in class (especially a class like statistics). That revelation got me wondering about the sleeping habits of the typical college sophomore. So, I asked 34 of my students to keep a sleep log for 1 week. Every day, as soon as they woke up, they were to record the time of day, along with the time they had gone to bed the previous night (or, in some cases, earlier that morning). I then calculated the duration of a night's sleep for each student (in minutes). Table 2.4 shows sleep duration for all 34 students for the first night of the study.

Take a look at the table and see if you can determine how long a typical night's sleep lasted. Were the students getting the recommended 8 hours of sleep per night ($8 \times 60 = 480$ minutes)? The data are in no particular order, and that can make it very hard to see what's going on.

There is a simple, and probably very obvious, reorganization of the data that will help immensely, and that is to put the data in order from least to most. There is nothing sacred about the order the data are currently in (subject 1 is first only because subject 1 was the first person to volunteer to participate). It won't disrupt anything to reorder the data according to the length of time each person spent sleeping rather than the completely arbitrary subject identification number.

Once we have the data reorganized from least to most, it will become much easier to see what's going on. Take a look at Table 2.5, and use the data displayed there to answer the following questions:

- What was the shortest night's sleep recorded by these subjects?
- What was the longest night's sleep?
- What, in your opinion, was a "typical" night's sleep for participants in this data set?

TABLE 2.4 Student sleep study: Night 1 sleep duration for 34 students (in minutes)

ID	Sleep Duration	ID	Sleep Duration	ID	Sleep Duration	ID	Sleep Duration
1	455	11	450	21	120	31	470
2	450	12	120	22	360	32	340
3	445	13	495	23	380	33	Missing
4	375	14	390	24	450	34	405
5	480	15	520	25	420		
6	525	16	450	26	420		
7	330	17	385	27	Missing		
8	405	18	620	28	Missing		
9	275	19	430	29	440		
10	405	20	450	30	Missing		

TABLE 2.5 Student sleep study: Night 1 sleep duration (in order)*

Minutes			
120	390	450	520
120	405	450	525
275	405	450	620
330	405	450	Missing
340	420	450	Missing
360	420	455	Missing
375	430	470	Missing
380	440	480	
385	445	495	

I've stripped off the ID numbers here.

I'm betting that your definition of "typical" in this data set was based on how often a particular sleep duration showed up in the data. In other words, typicality can be defined by the *frequency* of an event, represented by the symbol *f*. Typical events are frequent events, and atypical events are infrequent or rare. Describing a set of data in terms of frequency is a very popular way to start talking about patterns in the data. It should be no surprise, then, that the first set of methods for reorganizing the data that we're going to talk about involve **frequency distributions**, or tables that list events and event frequencies. There are eight different ways to display data in frequency distributions (shown in Figure 2.3). They differ in how the data are arranged (grouped or ungrouped) and/or how the frequency is represented (as a count, as a cumulative count, or as a percentage).

frequency distribution A table showing the frequency of occurrence of observations.

Our first decision when we're creating a frequency distribution involves how to list our observations: We can either group our data or leave it ungrouped. Once we have made that decision, we can represent either grouped or ungrouped data in any of four ways:

- as simple count
- as relative frequency (percentage)
- as cumulative frequency
- as cumulative relative frequency (percentile rank).

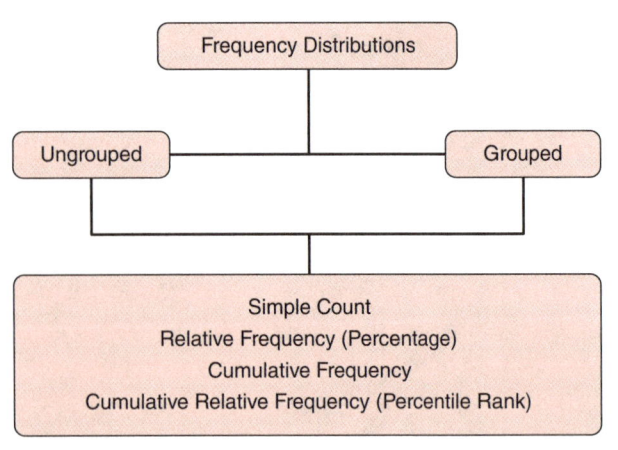

FIGURE 2.3
Options for presenting data in frequency distributions.

Frequency Distributions
- Ungrouped
- Grouped

Simple Count
Relative Frequency (Percentage)
Cumulative Frequency
Cumulative Relative Frequency (Percentile Rank)

UNGROUPED FREQUENCY DISTRIBUTIONS

The simplest way to reorganize a small data set is as an **ungrouped frequency distribution**. To create this table, set up two columns:

- In column 1, list each unique data point.
- In column 2, next to each data point, list the frequency of the data point.

Basically, if the data point "120 minutes" occurs twice in the data set, then instead of listing each observation separately, we list the observation once and, in the second column, note that it happened twice. The ungrouped frequency (*f*) distribution for the data we're working with should look something like what you see in Table 2.6.

Now it's much easier to see what the data are telling us. For example, relatively few students got a full 8 hours of sleep. In fact, only 1 person out of the 30 who remembered to fill out the sleep log for the first night of the experiment got exactly 8 hours of sleep, and only 5 people out of the 30 got 8 hours of sleep or more. The most frequently occurring sleep duration was 450 minutes (about 7.5 hours). Two people got only 2 hours of sleep (120 minutes), and one lucky devil got a bit more than 10 hours of sleep. I can look at these data and start to draw some conclusions about the students in my class. Maybe the reason students fall asleep in class is because they are sleep-deprived.

ungrouped frequency distribution A table showing the frequency of individual events or observations; the raw data are not grouped.

TABLE 2.6 Student sleep study: Night 1 results, displayed as an ungrouped simple frequency distribution

Minutes	f	Minutes	f
120	2	430	1
275	1	440	1
330	1	445	1
340	1	450	5
360	1	455	1
375	1	470	1
380	1	480	1
385	1	495	1
390	1	520	1
405	3	525	1
420	2	620	1

Relative Frequency

With small data sets, ungrouped frequency distributions can be quite helpful. Reorganizing the data this way lets the researcher immediately see several interesting features of the data: minimum, maximum, and most frequently occurring observation, for example. We just created an ungrouped frequency distribution using a simple count of each data point.

relative frequency Frequency expressed as a percentage.

We could also choose to display frequency as a *percentage* (also called **relative frequency**). To convert a simple count to relative frequency, we need to divide the number of times an observation occurred (its frequency) by the total number of observations in the set. This gives us the *proportion* (the part out of the whole) of times the event occurred. Suppose the box of 12 donuts in the break room has 3 of your favorite blueberry cake donuts in it. The proportion of blueberry donuts in the box is 3:12, which can be written as 3/12, or (dividing 3 by 12) as 0.25/1, or as 0.25. To convert from proportion to percentage, we multiply the proportion by 100 (percentage indicates the part out of a whole of 100). The percentage of blueberry cake donuts in the box is 0.25 × 100, or 25%. Box 2.2 shows the formula for calculating percentage.

BOX 2.2 **Calculating percentages**

$$\text{percentage} = \left(\frac{\text{frequency}}{N}\right) \times 100$$

Remember we said earlier that 2 out of the 30 students who filled out the survey slept only 2 hours the previous night. That means that 6.67% of the participants reported sleeping 120 minutes on the previous night: Box 2.3 shows this calculation.

BOX 2.3 **Calculating percentages: Example**

$$\text{percentage} = \left(\frac{2}{30}\right) \times 100$$

$$= (0.0667) \times 100$$

$$= 6.67\%$$

When we convert from frequency (f) to relative frequency (%), the frequency distribution should look something like what is shown in Table 2.7.

TABLE 2.7 Student sleep study: Night 1 results, displayed as an ungrouped relative frequency distribution

Minutes	f	%	Minutes	f	%
120	2	**6.67%**	430	1	**3.33%**
275	1	**3.33%**	440	1	**3.33%**
330	1	**3.33%**	445	1	**3.33%**
340	1	**3.33%**	450	5	**16.67%**
360	1	**3.33%**	455	1	**3.33%**
375	1	**3.33%**	470	1	**3.33%**
380	1	**3.33%**	480	1	**3.33%**
385	1	**3.33%**	495	1	**3.33%**
390	1	**3.33%**	520	1	**3.33%**
405	3	**10.00%**	525	1	**3.33%**
420	2	**6.67%**	620	1	**3.33%**

Cumulative Frequency

Cumulative frequency (CF) is also a count, but in this case, it is a count of the frequency of observations at or below a given point. To create cumulative frequency, we start at the lowest observation in our set, and in a separate column, we count all observations equal to or less than that starting observation.

The lowest observation (the minimum value) in this data set is 120 minutes, so we would count all the observations at 120 minutes or less than 120 minutes. This is a snap: There are no observations less than 120, and there are two observations equal to 120 minutes. That means the cumulative frequency of the observation 120 minutes is two. The next highest observation is 275 minutes. We would then count the number of observations equal to or less than 275. Again, easy to do: There were two observations less than 275 and one observation equal to 275, so the cumulative frequency of 275 is three.

You're adding together frequencies as you move up the table from lowest observation to highest. When you get to the largest observation (the maximum value), you should find that all 30 observations are equal to or less than the largest value in the set (in this case, 620 minutes). If the cumulative frequency of the largest value in the set does not equal N (the number of observations in the set), then you've done something wrong. I've added the cumulative frequencies to our ever-expanding table, producing Table 2.8.

cumulative frequency The frequency of events at or below a particular point in an ordered set.

TABLE 2.8 Student sleep study: Night 1 results, displayed as an ungrouped cumulative frequency distribution

Minutes	f	%	CF	Minutes	f	%	CF
120	2	6.67%	**2**	430	1	3.33%	**16**
275	1	3.33%	**3**	440	1	3.33%	**17**
330	1	3.33%	**4**	445	1	3.33%	**18**
340	1	3.33%	**5**	450	5	16.7%	**23**
360	1	3.33%	**6**	455	1	3.33%	**24**
375	1	3.33%	**7**	470	1	3.33%	**25**
380	1	3.33%	**8**	480	1	3.33%	**26**
385	1	3.33%	**9**	495	1	3.33%	**27**
390	1	3.33%	**10**	520	1	3.33%	**28**
405	3	10.00%	**13**	525	1	3.33%	**29**
420	2	6.67%	**15**	620	1	3.33%	**30**

Using the cumulative frequencies, I can add to my description of the data, saying that 24 of the 30 students who filled out the sleep log slept 8 hours or less on the first night of the study, 15 of the 30 participants slept 7 hours or less, and 6 of the 30 participants slept 6 hours or less on the first night. It should be obvious that we could also represent cumulative frequency as a percentage and report the percentage of participants at or below a particular data point. Table 2.9 shows the **cumulative relative frequency (CRF)**, or the percentage of data points at or below a given point, added to our table.

cumulative relative frequency Cumulative frequency expressed as a percentage. Also known as **percentile rank**.

Percentile Rank

The cumulative relative frequency of an event is a statistic that has become very popular, especially when organizations are reporting the results of standardized tests. Most college students become (sometimes painfully) familiar with cumulative relative frequency when they receive their SAT scores (the SAT being a standardized test), although many probably know cumulative relative frequency by its other name: **percentile rank**.

percentile rank The percentage of observations at or below a particular value in the frequency distribution. Also known as **cumulative relative frequency**.

Percentile rank tells you the percentage of observations in a set at or below a particular point. If you happened to score in the 90th percentile on the SAT, then you know that 90% of the students who took the same test scored as well as or worse than you did on the test. By default, then, 10% of the students taking the SAT scored higher than you did—not bad. In the data set we've been

TABLE 2.9 Student sleep study: Night 1 results, displayed as an ungrouped cumulative relative frequency distribution

Minutes	f	%	CF	CRF	Minutes	f	%	CF	CRF
120	2	6.67%	2	**6.67%**	430	1	3.33%	16	**53.33%**
275	1	3.33%	3	**10%**	440	1	3.33%	17	**56.67%**
330	1	3.33%	4	**13.33%**	445	1	3.33%	18	**60%**
340	1	3.33%	5	**16.67%**	450	5	16.67%	23	**76.67%**
360	1	3.33%	6	**20%**	455	1	3.33%	24	**80%**
375	1	3.33%	7	**23.3%**	470	1	3.33%	25	**83.33%**
380	1	3.33%	8	**26.67%**	480	1	3.33%	26	**86.67%**
385	1	3.33%	9	**30%**	495	1	3.33%	27	**90%**
390	1	3.33%	10	**43.33%**	520	1	3.33%	28	**93.33%**
405	3	10%	13	**43.33%**	525	1	3.33%	29	**96.67%**
420	2	6.7%	15	**50%**	620	1	3.33%	30	**100%**

working with, we can report that a night's sleep of 495 minutes was at the 90th percentile, 420 minutes was the 50th percentile, and 360 minutes was at the 20th percentile.

CheckPoint

Indicate whether the following statements are *true* or *false*.

1. The number of observations at or below a particular point in an ordered set is called relative frequency. _____

2. Relative frequency is the frequency of an observation expressed as a percentage. _____

3. A police officer records the speeds of 50 cars passing through a school zone. The lowest observed speeds are 16 miles per hour (two cars), 15 miles per hour (one car), and 14 miles per hour (three cars). The cumulative frequency of the observation 16 miles per hour is therefore 3. _____

4. Fatima, who is 3 years old, is weighed during a routine checkup. The doctor informs Fatima's father that his daughter is in the 20th percentile for her age group. This means that Fatima's weight is greater than that of 80% of the population of 3-year-olds. _____

5. Another term for *percentile rank* is *cumulative relative frequency*. _____

Answers to this exercise are found on page 51.

Think About It...

FREQUENCY DISTRIBUTIONS (PART 1)

Because sleep deprivation has been linked to an increased risk of accidents on the job, as well as to poorer physical and mental health and decrements in the overall quality of life, many researchers have investigated the relationship between sleep and performance at work or in school. For example, Lund, Reider, Whiting, and Prichard (2010) asked a very large group of college students (a total of 1,125 undergraduates) about their sleep duration and sleep quality. The data set in Table 2.10 is representative of their results (I used an imaginary and much smaller data set to make life easier for you). Using the data I've provided, create an ungrouped frequency distribution of the sleep durations, and answer the questions posed below.

TABLE 2.10 Sleep duration for 20 college undergraduates (in minutes)

x	f	CF	Percentile Rank (CRF)
328	1		
367	1		
368	1		
373	1		
376	1		
377	1		
379	1		
383	2		
385	1		
387	1		
388	1		
390	1		
425	1		
488	3		
523	2		
600	1		
Total n	20		

1. Fill in the blanks in the table by calculating the cumulative frequency and the percentile rank for the sleeping students in this study.

2. What is a typical sleep duration for the 20 (hypothetical) students in this sample? (Think about how to define "typical" here before you answer.) _____
3. What sleep duration defines the 50th percentile? _____
4. In this sample, 30% of the participants slept at least _____ minutes.
5. How many students slept more than 8 hours? _____ How many slept less than 6 hours? _____

Answers to this exercise may be found on page 88.

CheckPoint
Answers to questions on page 49

1. FALSE ☐ The number of observations at or below a particular point in an ordered set is called *cumulative frequency*.
2. TRUE ☐
3. FALSE ☐ The cumulative frequency of the observation 16 miles per hour is therefore 6.
4. FALSE ☐ This means that Fatima's weight is greater than that of *20%* of the population of 3-year-olds.
5. TRUE ☐

SCORE: /5

GROUPED FREQUENCY DISTRIBUTIONS

With small data sets, ungrouped frequency distributions can be very helpful as we look for patterns in the data. So far, we have seen that arranging the data into an ungrouped frequency distribution allows us to discover the minimum, maximum, and typical sleep duration for the students in the data set. We can also describe several interesting "break points" in the data set—the 50th percentile, the 90th percentile, and so on.

There is, however, a drawback to ungrouped frequency distributions, and that is their size. If your original data set is relatively small and compact (meaning there are not too many observations in the set and the difference between the largest and smallest observation is small), then ungrouped frequency distributions are easy to read and use. However, with a very large data set or one with a great deal of variability, the ungrouped frequency distribution quickly becomes unmanageable. There's simply too much to look at. Table 2.11 shows the complete data set for all 7 nights of the study for all 34 students, arranged in an ungrouped frequency distribution.

The minimum and maximum values are easy to pick out, but there is so much going on in between that it becomes difficult to see consistencies and patterns in

TABLE 2.11 Student sleep study: Seven-night results, displayed as a simple ungrouped frequency distribution

Minutes	f	Minutes	f	Minutes	f
120	2	370	1	500	1
150	1	375	5	505	2
190	1	380	4	510	9
215	1	385	1	515	1
225	1	400	9	520	3
240	2	405	2	525	7
260	2	410	1	530	3
270	5	415	2	540	3
275	2	420	11	570	5
280	1	425	2	585	1
285	2	430	3	590	1
300	5	435	7	600	3
315	1	440	4	615	1
320	1	445	1	620	1
330	5	450	12	650	1
335	1	455	1	660	1
340	2	460	5	675	1
345	4	465	7	690	2
350	3	470	2	720	1
355	1	480	13	730	1
360	14	490	1		
365	1	495	6		

grouped frequency distribution A table showing the frequency of events or observations; the raw data have been grouped into specific intervals of equal size.

the bulk of the data. What can we do to make it easier to see what's going on in a large or complex data set? The easiest solution is to use a **grouped frequency distribution**. We'll reorganize the data into groups or intervals and then report the frequency for all of the scores within any interval.

There are some things to consider in creating our intervals. First, when should we use a grouped frequency distribution instead of an ungrouped one? Second, how many intervals do we want? Third, how big (or wide) should each interval be?

The general rule is that grouped frequency distributions are a good idea *when the difference between the largest and smallest observation (the range) is greater than 20*. Grouping small data sets may actually work against us in our quest to make the data easier to read. If we were to group small data sets, we would run the risk of clumping all our observations into two or three very large groups, thereby obscuring any patterns in the data.

The question of how many groups to use is equally easy to answer: We can use as many as we'd like. I'm not trying to be difficult here, I promise. The overriding rule when creating a frequency distribution is that these distributions should make looking at the data easier, not harder. If we use too few groups (say, we create just two groups for the seven-night sleep data), then we risk collapsing the data so much that everything unusual gets squished into one of two boxes and disappears from view. On the other hand, if we were to use too many intervals (say, we break the sleep data into 100 intervals), our groups would be so tiny, and there would be so many of them, that we wouldn't have simplified the ungrouped frequency distribution at all.

The general rule about the number of groups is to start with 10, and if that doesn't help make the data clearer, then try something else. Using 10 groups with the sleep data should work out just fine, however, because it will separate the data into intervals of about an hour each. I hope you're asking how I figured that out, since I'm about to tell you and answer the third question at the same time.

There is a formula for determining how wide each group should be in order to create the number of groups we've selected. We simply divide the *range* (the difference between the largest and smallest observation) by the number of groups we'd like to have in our distribution, as shown in Box 2.4.

BOX 2.4 **Finding interval width**

$$\frac{\text{range}}{\text{groups}} = \text{width}$$

$$\frac{730 - 120}{10} = 61$$

So, if we use 10 groups for our sleep data, each group will then be 61 minutes wide—in other words, the difference between the largest and smallest score in any group will be 61 minutes. I'm going to advocate that we make another executive decision here. Since we're dealing with time data, why not use the convenient units of hours (60 minutes) to break our data into groups? So, instead of making each of our groups 61 minutes wide, we'll make them all 60 minutes wide. After all, it's our data, and we're trying to make it easier to talk about, not harder.

Once we've determined the number of groups and the width of each group, we need to determine what observation should start the first group (in other words, what will be the first group's *lower limit*, or *LL*) and what observation will end the interval (i.e., the *upper limit*, or *UL*). There is (of course) a standard procedure for determining the *LL* and *UL* of the first group. The general rule is to start at a value equal to an even multiple of the width and less than or equal to the lowest observation in the data set, or at zero. Our width is 60 minutes, so the rule says that we should set the *LL* of the first group at an even multiple of 60 (or at zero), remembering to include our lowest value. Even multiples of 60 are easy to figure out—check out Box 2.5.

BOX 2.5 **Even multiples of the width**

$$60 \times 1 = 60$$
$$60 \times 2 = 120$$
$$60 \times 3 = 180$$

We want an even multiple that is equal to or less than our lowest observation (120 minutes). We can start our first interval at 60, but I don't recommend it. If the *LL* of the first interval is 60, we'll have an empty interval at the bottom of our table: No one slept less than 120 minutes, so the frequency would be zero—a waste of space and potentially confusing to the reader. We cannot start at 180 minutes because we would cut off the poor, bleary-eyed people who only slept 120 and 150 minutes. We can, however, set the *LL* of the first group to 120 and then work our way up. Take a look at Box 2.6 to see what I mean.

BOX 2.6 **Finding the upper limit**

$$UL = LL + (\text{width} - 1)$$
$$= 120 + (60 - 1)$$
$$= 120 + 59$$
$$= 179$$

Each group should be 60 minutes wide, so if we start at 120 minutes, what would the *UL* of the first interval be? The rule says that the *UL* of each interval should be equal to the *LL* of that interval plus the width minus one.

Why does our first interval end at 179 minutes instead of 180? If we made the first interval extend from 120 to 180 minutes, it would actually be 61 minutes wide instead of 60. Remember, the observations that define the limits of each interval are included in the interval. The second group in our distribution will start at 180 minutes and end at 180 + 59, or 239 minutes, and so on.

You may be asking yourself why I'm making this complicated. Why not just start at the bottom of the data set and work our way up? Well, mostly for aesthetic reasons. Starting at an even multiple of the width rather than the smallest value in the data set will make the table easier to read. If you don't believe me, try creating a grouped frequency distribution out of the data set we're working with if we add just one sleep time to the set: Suppose the smallest value in the data set happened to be 103 minutes rather than 120 minutes. What would happen to the frequency distribution if we started at 103 minutes rather than an even multiple of the width?

By the way, what even multiple of the width *should* we use if the lowest value in the set were 103? According to our rule, if the lowest value in the set is 103, we need an even multiple of the width (60) that would include 103. So, our first interval would go from 60 to 119, and then up from there.

Try constructing the entire grouped frequency distribution using the rules we've been talking about here. (I recommend you do this before reading on to see my version.) You should wind up with something that looks like Table 2.12.

TABLE 2.12 Student sleep study: Seven-night results, displayed as a grouped frequency distribution

Interval	*f*
120–179	3
180–239	3
240–299	14
300–359	23
360–419	40
420–479	55
480–539	47
540–599	10
600–659	6
660–719	4
720–779	2

TABLE 2.13 Student sleep study: Seven-night results, displayed as a grouped relative, cumulative, and cumulative relative frequency distribution

Interval	f	%	CF	CRF	Interval	f	%	CF	CRF
120–79	3	1.45%	3	1.45%	480–539	47	22.71%	185	89.37%
180–239	3	1.45%	6	2.90%	540–599	10	4.83%	195	94.20%
240–299	14	6.76%	20	9.66%	600–659	6	2.90%	201	97.10%
300–359	23	11.11%	43	20.77%	660–719	4	1.93%	205	99.03%
360–419	40	19.32%	83	40.10%	720–779	2	0.97%	207	100%
420–479	55	26.57%	138	66.67%					

Please keep in mind that you don't have to use 10 groups to create your distribution. You may even want to use 2-hour intervals and break up the data into even larger groups. Whatever your final decision, the frequency distribution you create should make it easier to see what's going on in the data. (If you do decide to use something other than 10 intervals in your distribution, your table will not look like mine.)

Is Table 2.12 easier to read than the ungrouped distribution in Table 2.9? Was 8 hours of sleep typical in this data set? What sleep duration was typical? How many students slept 10 hours or more? Do you have any explanation for why these students are able to sleep so long?

Expanding the grouped frequency distribution to include relative, cumulative, and cumulative relative frequencies is easy. We just use the frequencies of the observations within each group as our starting point. Try adding the relative, cumulative, and cumulative relative frequencies onto the table before you go on to see my version of the complete table (see Table 2.13).

Because the data are represented in groups, it is often easiest to label and talk about each group using its midpoint. The *midpoint* (*MP*) is simply the observation that is in the middle of the group. You find the midpoint by dividing the absolute value of the range of each group by 2 and then adding that value to the *LL* of the interval (see Box 2.7).

So, we can refer to the first group (sleep durations from 120 to 179 minutes) by the *MP* for that group, 149.5 minutes. You will, I hope, notice that all the groups have the same range, so to find the *MP* of each group, we simply add 29.5 to each *LL*.

BOX 2.7 Finding the midpoint

$$MP = \frac{|\text{range}|}{2} + LL$$

$$= \frac{|120 - 179|}{2} + 120$$

$$= \frac{59}{2} + 120$$

$$= 29.5 + 120$$

$$= 149.5$$

CheckPoint

1. Indicate whether the following statements are *true* or *false*.
   a. The general rule is to use grouped data when the range is greater than 10. _____
   b. When creating grouped data, a general rule is to start with 10 groups and adjust to make the data easier to understand. _____
2. Calculate the width of each group based on the following information:

Minimum	Maximum	Number of Groups
30	78	8
480	620	20
7	139	4

3. Determine the lower limit (LL) of the first group based on the following information:

Minimum	Width
21	10
1	3
430	80

4. Calculate the upper limit (UL) for each group based on the following information:

LL	Width
40	16
620	50
5	5

5. Calculate the midpoint (*MP*) of each group based on the given information:

LL	Range
15	5
0	7
310	35

Answers to this exercise are found on page 60.

Think About It . . .

FREQUENCY DISTRIBUTIONS (PART 2)

In Part 1 of this chapter's *Think About It* exercise (p. 50) we created a simple, ungrouped frequency distribution for our hypothetical sleepers. Now let's try grouping the data. We've met the criterion for using a grouped frequency distribution: The difference between the smallest value in our data set and the largest is more than 20 (600 − 328 = 272). So, let's create a grouped frequency distribution with 10 intervals.

x	f	Intervals	f	CF	Percentile Rank (CRF)
328	1				
367	1				
368	1				
373	1				
376	1				
377	1				
379	1				
383	2				
385	1				
387	1				
388	1				
390	1				
425	1				
488	3				

523	2			
600	1			
Total *n*	20			

1. If you use 10 intervals, how wide will each interval be? _____
2. Where should the first interval start? (Use the even multiple of the width rule to determine the starting point for the first interval.) _____
3. Where should the first interval end? _____
4. Create the grouped frequency distributions in the space provided on the table.
5. Reassess the data. Has your definition of a typical night's sleep for the participants changed after creating the grouped frequency distribution? If so, what is the change?

Answers to this exercise may be found on page 89.

SCALES OF MEASUREMENT IN CONTEXT

The debate about how to measure the nonphysical, the qualitative, or the purely psychological, begun by S. S. Stevens in 1946, has not stopped. In fact, it is as strongly contested as ever, and if you want to start an argument among behavioral scientists, bring up Stevens and what is known as the "qualitative versus quantitative debate."

Stevens proposed that because of the nature of qualitative data (nominal and ordinal scales measure category membership rather than amount or count), traditional mathematical operations (like adding, subtracting, multiplying, and dividing) cannot—and should not—be applied to it. Quantitative data, on the other hand, use cardinal numbers (cardinal numbers are numbers that measure amount or count), and so mathematical operations can be appropriately be used. Stevens went on to list the statistical measures that were appropriate for each type of data, and in doing so, he began the argument.

The other side of the debate (see Gaito, 1980, for a readable review of this argument) says that Stevens, and those who agreed with him, had misunderstood both measurement theory and statistical theory. Statistical methods don't require that a particular scale of measurement be used, only that a set of assumptions about the data (that are quite different than what the numbers do or do not "represent") be valid. As Frederick M. Lord (1953, p. 751) famously said, "[T]he numbers don't remember where they came from." The interpretation of what the statistical analysis of the data means is up to the scientists doing the study, not the scale of measurement they used.

CheckPoint
Answers to questions on page 57

1. a. FALSE ☐ The general rule is to use grouped data when the range is greater than *20*.
   b. TRUE ☐

2.

Minimum	Maximum	Number of Groups	Width of Each Group
30	78	8	<u>6</u>
480	620	20	<u>7</u>
7	139	4	<u>33</u>

3.

Min.	Width	LL
21	10	<u>20</u>
1	3	<u>0</u>
430	80	<u>400</u>

4.

LL	Width	UL
32	16	<u>47</u>
600	50	<u>649</u>
5	5	<u>9</u>

5.

LL	Range	MP
15	5	<u>17.5</u>
0	7	<u>3.5</u>
310	35	<u>327.5</u>

SCORE: /15

Let me give you an example of how this debate applies to psychology today. Suppose we were to measure "attractiveness" with an ordinal-level scale of measurement. We might ask a group of male students and a group of female students to assess the attractiveness of a person shown in a photo, using a five-point scale. The scale is shown in Box 2.8, along with a numeric measurement of the average amount of attractiveness seen by the male and female participants.

BOX 2.8   **Measuring beauty**

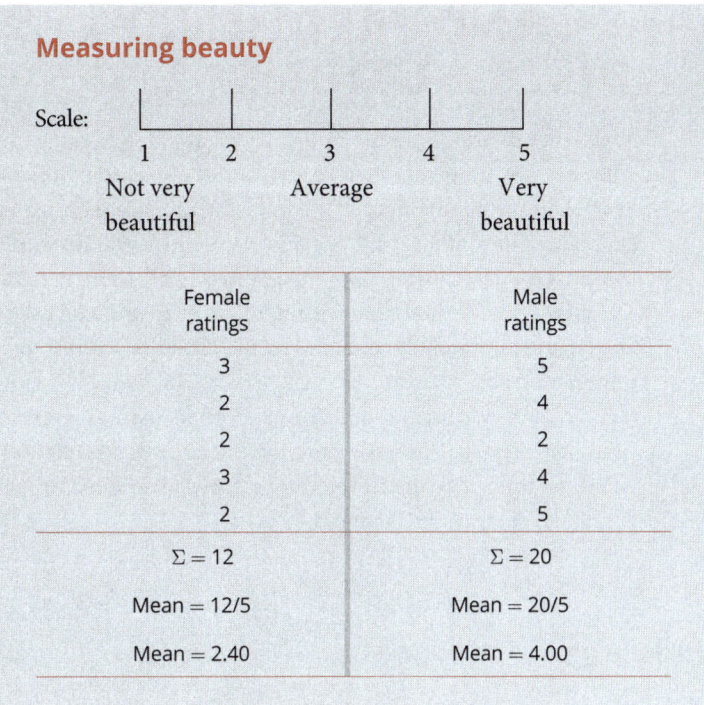

Scale:

1	2	3	4	5
Not very beautiful		Average		Very beautiful

Female ratings	Male ratings
3	5
2	4
2	2
3	4
2	5
$\Sigma = 12$	$\Sigma = 20$
Mean = 12/5	Mean = 20/5
Mean = 2.40	Mean = 4.00

What can we say about how these two groups of people assessed attractiveness? Stevens would argue that because we have an ordinal-level scale, about all we can say is that the women in the study used ratings of 2 and 3 more often than did men, while men used measures of 4 or 5 more often than did women. Gaito and Lord would counter that it is perfectly appropriate to say that on average women in the study rated the attractiveness of the person in the picture as 2.40, which is lower than the average rating among the men of 4.00, and that the difference in average rating is 1.60—even though neither value (2.4 nor 1.6) exists on this scale of measurement.

The beginning statistics student should probably not be too concerned about the finer philosophical points of this argument. But you should be aware that the argument exists, and that it can be used to suggest qualitative measurement is somehow less desirable, or less scientific, than is quantitative measurement. Please don't fall victim to this idea: It is just not true. There are perfectly valid statistical methods that can be applied to qualitative data, just as there are valid and useful statistical methods that can be applied to quantitative measurements. In fact, we will be considering these methods later on in this text, in Chapters 14 and 15.

...AND FREQUENCY DISTRIBUTIONS

Frequency distributions are incredibly useful in our search for patterns in the data. They are not, however, often reported in the scientific literature. If you haven't already had the opportunity to see how statistics are reported in a professional journal, take a look at a report or two in any of the hundreds of professional science journals (social, natural, or extraterrestrial—all sciences report statistics in essentially the same way). I'd be willing to bet that you won't see too many frequency distributions presented as tables in the "Results" section of whatever article you choose to read.

Frequency distributions are most often used as a first step only by the researcher interested in answering the initial research question. However, you will often see frequency distributions presented in graphic form, and that is the topic we'll take up next, in Chapter 3.

SUMMARY

S. S. Stevens proposed four scales of measurement: nominal, ordinal, interval, and ratio. Nominal and ordinal scales measure *quality* by putting observations into categories (nominal) and by putting those categories into some kind of logical order (ordinal). Stevens also proposed that interval and ratio scales were appropriate to measurement of *quantity*—amount or count. Interval scales begin at an arbitrarily determined zero point, while ratio scales begin at a true or real zero.

Frequency distributions are tables listing unique observations and the frequency of each observation. The tables can list the observations individually or in intervals or groups. Frequency can be displayed simply as the frequency of each individual observation or group, but it can also be shown as a percentage (relative frequency), as the frequency at or below a given observation (cumulative frequency), or as the percentage of observations at or below a given observation (cumulative relative frequency, also known as percentile rank).

TERMS YOU SHOULD KNOW

cumulative frequency, p. 47
cumulative relative frequency, p. 48
frequency distribution, p. 44
grouped frequency distribution, p. 52
interval scale, p. 39
nominal scale, p. 35
number, p. 35
numeral, p. 35

ordinal scale, p. 37
percentile rank, p. 48
qualitative data, p. 34
quantitative data, p. 34
ratio scale, p. 39
relative frequency, p. 46
ungrouped frequency distribution, p. 45

GLOSSARY OF EQUATIONS

Formula	Name	Symbol
$\text{percentage} = \dfrac{\text{frequency}}{N} \times 100$	Percentage	%
$\text{width} = \dfrac{\text{range}}{\text{groups}}$	Interval width in grouped frequency distribution	
$UL = LL + (\text{width} - 1)$	Upper limit of interval	UL
$MP = \dfrac{|\text{range}|}{2} + LL$	Midpoint of interval	MP

WRITING ASSIGNMENT

Select one of the data sets listed in the practice problems, and describe what the frequency distributions you've created tell you about the data in writing. Remember that your readers are relying on you to tell them which statistics support any conclusion you make, so don't forget to provide the reader with this important information.

PRACTICE PROBLEMS

1. Define the four scales of measurement proposed by S. S. Stevens.

2. Define percentile rank.

3. What problems, if any, do you see with each of the frequency distributions shown in Figure 2.4?

x	f		Interval	f		Interval	f
0	0		0–5	3		60–69	5
23	1		5–10	8		70–79	7
25	5		11–16	9		80–89	2
27	9		17–22	10		90–129	22
30	14					130–139	3
43	3					140–149	1
50	2						

FIGURE 2.4

4. Using the frequency distributions we created for the sleep duration data (Tables 2.12 and 2.13), answer the following questions:
   a. What is the most frequently occurring sleep duration in minutes?
   b. What is the most frequently occurring sleep duration in hours?
   c. Is 8 hours of sleep typical in this data set?
   d. How many students are getting 10 to 12 hours of sleep a night?
   e. Do you have any explanation for why these students are able to sleep so long? Remember, these data represent sleep durations over seven consecutive days.

5. In a senior research class, several students decided to see if the perception of the passage of time was influenced by age. They asked volunteers to participate in a time-estimation task. Participants were first asked their age. Then, they were told that as soon as a green light flashed on the computer monitor in front of them, they were to begin estimating a 27-second time interval, without counting or using any kind of timing device. As soon as the participant felt that the indicated interval had passed, she or he pressed the space bar on the keyboard. The actual duration of the time interval that had passed was then presented on the monitor, followed by a warning tone and the next trial. Table 2.14 shows the mean errors (in milliseconds) made by the participants for each time interval. Positive errors indicate an overestimation of the time interval, and negative errors indicate an underestimation of the time interval. Participants in age group 1 were between 10 and 21 years of age. Participants in age group 2 were over the age of 50.

Table 2.14 Time-estimation errors (in milliseconds) by age

ID	Age Group	Error	ID	Age Group	Error
1	1	9.00	11	2	−3.00
2	1	13.00	12	2	3.00
3	1	13.00	13	2	7.00
4	1	1.00	14	2	4.00
5	1	5.00	15	2	5.00
6	1	−11.00	16	2	5.00
7	1	11.00	17	2	10.00
8	1	1.00	18	2	1.00
9	1	−1.00	19	2	11.00
10	1	−16.00	20	2	14.00

   a. What kind of data (nominal, ordinal, interval, or ratio) does each of the following variables represent?
      i) Age group
      ii) Error in time estimation (in milliseconds)
      iii) Participant ID number

b. Use the data shown in the table to construct a grouped frequency distribution (you'll need to determine how big your intervals should be and where they will start) as well as relative and cumulative frequency distributions for each age group. Use the data from your frequency distributions to answer the following questions:
   i) What percentage of over-50-year-olds made errors of 5 milliseconds or less?
   ii) How many of the participants in age group 1 made underestimations of the time interval?
   iii) How many of the over-50-year-olds made underestimations?
   iv) How many participants (considering both groups) made overestimations?

c. What kind of data (nominal, ordinal, interval, or ratio) does each of the following represent?
   i) Age group (infant, toddler, school-age)
   ii) Errors (in milliseconds)
   iii) Coffee serving size (short, tall, grande, venti, trenta)
   iv) Score on the Myers-Briggs Personality Inventory (Extroverted/Introverted, Sensing/Intuiting, Thinking/Feeling, Judging/Perceiving)
   v) Quality of sleep (from 1 [*very poor*] to 5 [*excellent*])

6. Construct a grouped and relative frequency distribution for the data shown in Table 2.15. Use the information in the frequency distribution to answer the following questions:
   a. What percentage of fast-food hamburgers have 600 calories or more?
   b. What calorie count for milkshakes is at the 50th percentile?
   c. What calorie count for french fries is at the 90th percentile?
   d. Issued jointly by the Dept. of Health and Human Services and the Dept. of Agriculture, the "Dietary Guidelines for Americans" recommends 2,000 calories per day for women and 2,400 calories per day for men aged 19 to 30 with a sedentary lifestyle. Consider the fast-food restaurants where you might order a burger, fries, and a shake, and answer the following questions:
      i) Given the recommended total calorie intakes listed in Table 2.15, how many calories per meal must women eat to meet the recommended total (assume three meals per day)?
      ii) How many calories per meal should men consume if their total daily calorie intake should be no more than 2,400 (again, assume three meals per day)?
      iii) If you're female, what should you order if you want to consume only 1/3 of the total calories in a given meal?
      iv) What percentage of restaurant meals (burger, fries, and a shake) exceed the total daily recommended calorie intake for women in just one meal?

CHAPTER 2   Types of Data   65

Table 2.15 Calorie count by type of food at popular fast-food restaurants

Restaurant	Hamburger Type	Hamburger Calories	Hamburger Weight (in Grams)	French Fries (Medium) Calories	Chocolate Shake (Large) Calories
A&W	Deluxe Hamburger	461	212	311	1,120
Dairy Queen	Classic Grill Burger	540	213	380	1,140
Hardee's	¼-lb. Double Cheeseburger	510	186	250	N/A
In-N-Out	Cheeseburger w/onions	390	243	400	690
Krystal	BA Burger	470	202	470	N/A
McDonald's	McLean Deluxe	340	214	380	1,160
Burger King	Whopper	670	290	360	950
Wendy's	Deluxe Double Stack	470	211	490	393

7. Every year, students in a sensation and perception class are asked to design and carry out pilot studies on a topic of their choice. Here are some of their experiments. For each, identify the independent variable (IV), the dependent variable (DV), and the scale of measurement used (nominal [N], ordinal [O], interval [I], or ratio [R]).
   a. Are musicians better than nonmusicians at auditory scene analysis? Two sets of volunteers (those who had at least 2 years of formal musical training and those with none) were asked to listen to a series of natural sounds. The sounds consisted of a dog barking, a doorbell ringing, applause, and a car horn honking. The sounds were presented either singly or in "layers" (all four sounds played simultaneously). Both groups of participants were asked to identify the sounds they heard in the stimuli. Accuracy (the number of sounds correctly identified) and elapsed time were measured to assess their ability to identify the sounds in the auditory scene.
   b. Can you taste the difference between a McDonald's Big Mac and one created at home? Blindfolded subjects were asked to eat a small piece of a "real" Big Mac and one created from a recipe that purports to mimic the taste of a Big Mac right in your own kitchen. Participants were asked which of the two they preferred and which one was homemade.
   c. How well can you identify objects by touch alone? Two groups of blindfolded participants were asked to identify a series of objects. One group had been shown a series of slides that identified some, but not all, of the objects they would later be asked to identify by touch. The other group did not have the opportunity to see the objects before handling them. The number of objects that were correctly identified and the time it took to identify them were recorded.

STATISTICS IN CONTEXT

d. Your two-point threshold is a measure of tactile or touch acuity. It is defined as the smallest distance between two points simultaneously touching your skin that you can distinguish as two points rather than one. What is your two-point threshold in the palm of your hand? Is it the same as it is on the bottom of your feet? A pair of small calipers were used to provide a two-point touch (touching the skin very lightly at two points simultaneously). Participants were touched either on their palm or the bottoms of their feet and asked if they felt one touch point or two. Their responses were recorded.

8. Researchers studying memory often use a model of human memory proposed by Atkinson and Shiffrin (1968). The Atkinson and Shiffrin model proposes three stages of human memory: sensory memory, short-term memory (STM), and long-term memory (LTM). Students in a cognitive neuroscience class participate in a test of the storage capacity of STM. The class is divided into two groups. Half of the class is presented with 25 sequences of numbers ranging in length from four numbers in a sequence to nine numbers in a sequence. They're shown the numbers in a given sequence, one number at a time, each presentation lasting 2 seconds. After all the numbers in the sequence are shown, the students are asked to recall the numbers in the sequence and write them down on a data sheet. The second group has their memory tested in exactly the same way, except that before they begin testing, they're taught a mnemonic method designed to help them remember more information. They're advised and allowed to practice "chunking" the numbers: They're told to imagine that the numbers are a phone number and to "chunk" them together in this way. The number of numbers, in order, correctly recalled for each group is shown in Table 2.16.
    a. Create an ungrouped frequency distribution for both data sets.
    b. Use the frequency distributions to answer the following questions:
        i) Was there a difference in the memory span for the two groups? Describe the differences in performance on the task that you see.
        ii) What percentage of students in each group were able to recall all nine numbers in a sequence?

Table 2.16 Memory span (number of stimuli recalled) for control and experimental groups

Control Group	Experimental Group
5	5
6	6
3	9
4	2
4	9
2	6
2	8

c. What kind of measurement scale is being used to measure memory span?
d. What is the dependent variable in this study? What was the independent variable?

9. Plana et al. (2010) examined the levels of cholesterol in children and adolescents who had been hospitalized for depression and who had attempted suicide. They compared these levels to those in other depressed and hospitalized adolescents who had not attempted suicide. The data shown in Table 2.17 reflect the data that Plana et al. found.

Table 2.17 Cholesterol level (in mg/dl) in adolescents who attempted suicide (experimental group) and depressed adolescents who have not attempted suicide (control group)

Experimental Group	Control Group
126	126
149	202
152	157
170	129
161	150
126	188
166	145
116	140
150	156
184	203
165	208
153	174
101	134
166	111
136	123
162	188
164	104

a. Construct a grouped frequency distribution for each group.
b. Use the frequency distribution to answer the following questions:
   i) Was there a difference in the serum cholesterol levels for the two groups? Describe any differences that you see.
   ii) Doctors often advise adults to keep their serum cholesterol levels below 200 mg/dl. What percentage of adolescents in each group had cholesterol levels below 200 mg/dl?
   iii) What was a typical cholesterol level for adolescents in each group?

STATISTICS IN CONTEXT

c. What kind of measurement scale is being used to measure cholesterol level?
d. What is the dependent variable in this study? What was the independent variable?

10. Identify the scale of measurement as quantitative or qualitative in the following:

Measure	Quantitative	Qualitative
Classification of personality as introverted or extroverted		
Time needed to press a key on the computer when signaled to do so		
Rank of the quality of graduate programs in U.S. college and universities		
Latency (in seconds) for a rat to find the goal box in a maze		
Grip strength of participants in a pink room compared to grip strength in a blue room		
Wavelength of light at five different saturation levels		
Memory span measured in terms of number of items on a list recalled 1 day after learning the list.		
The more "beautiful" of two portraits presented simultaneously		
Preferred movie type (comedy, romance, suspense, adventure)		
Jersey number for players on a soccer team		

11. Ever get seasick? If you have experienced this particular form of movement-related disorder, you know that it is often triggered by a "choppy" sea—irregular waves that bounce the boat around in unpredictable ways. The data in Table 2.18 come from a book by Hand, Daly, Lunn, McConway, and Ostrowski (1994) that contains a number of small data sets from classic, but often little-known, experiments. In this study, Burns (1984) examined the causes of motion sickness at sea. Volunteers were placed in a cubical cabin mounted on hydraulic pistons and subjected to wavelike motion for 2 hours. Researchers recorded the time that elapsed until the first subject vomited (called "survival time"). In experiment 1, participants (shown on the left side of Table 2.18) experienced motion at a frequency of 0.167 Hz and 0.111g acceleration (a slow roll or swell). In experiment 2 (shown on the right side of Table 2.18), participants moved at a frequency of 0.333 Hz and an acceleration of 0.222g (a choppier sea). Answer the following questions:
    a. Create grouped, relative, cumulative, and cumulative relative frequency distributions for the two experiments.
    b. What is the independent variable for this experiment?
    c. What is the dependent variable for this experiment?

d. What scale of measurement is being used to measure the dependent variable?
e. What time is at the 90th percentile in both experiments? What time is at the 50th percentile?
f. Does the frequency/acceleration of the up-and-down motion matter? Describe any important differences you see in the data from the two experiments.

Table 2.18 Survival time in motion sickness

| \multicolumn{2}{c|}{Experiment 1} | \multicolumn{2}{c}{Experiment 2} |

ID	Survival Time (in Minutes)	ID	Survival Time (in Minutes)
1	120	1	11
2	30	2	6
3	51	3	69
4	120	4	69
5	120	5	102
6	50	6	82
7	120	7	115
8	50	8	11
9	120	9	120
10	120	10	120
11	120	11	79
12	120	12	5
13	92	13	120
14	120	14	63
15	120	15	65
16	120	16	120
17	66	17	120
18	82	18	120
19	120	19	120
20	120	20	120
21	120	21	120
		22	13
		23	120
		24	120
		25	82

STATISTICS IN CONTEXT

		26	120
		27	120
		28	24

12. Are artists who have suffered great upset and upheaval in their personal lives more creative that those who have not? To find out, Marie Forgeard (2013) asked people if they had experienced any traumatic events (surprisingly, almost all of them said that they had experienced at least one) and then asked them to take a series of questionnaires designed to measure a variety of cognitive and perceptual aspects of their thinking. The data shown in Table 2.19 are similar to what Professor Forgeard found. Perceived creativity increase that results from a traumatic experience is measured on an eight-item, five-point scale (1 = *not at all*, 5 = *extremely*). The higher the score on this scale, the more the participant perceived their creativity to have been changed as a result of their experience.

Table 2.19 Scores of 14 participants on perceived change in creativity (PCC) scale

ID	Perceived Change in Creativity
1	8
2	22
3	21
4	17
5	23
6	8
7	23
8	21
9	26
10	12
11	12
12	8
13	10
14	23

a. Create a grouped frequency distribution where each interval has a width of three, as well as relative and cumulative relative frequency distributions for the data.
b. What would be the maximum score on this scale? What would be the minimum?
c. Forgeard concludes that while there is a tendency for creativity to be changed by the experience of a traumatic event, her results "did not show that adversity is needed

for creativity" (2013, p. 257). What possible confounding variables do you see in this study? What else, in addition to or instead of trauma, might explain her results?

13. Many people suffering from schizophrenia also smoke cigarettes. In fact, the difference in cigarette-smoking behavior between schizophrenics and nonschizophrenics is so pronounced that several studies have attempted to find out why this behavior is so much more common in the population of those with this disorder. One prominent explanation suggests that nicotine not only might help the body clear antipsychotic medication from the body, but also reduce some of the very unpleasant side effects of antipsychotic medication (like involuntary body movement). Researchers think cigarettes manage to do this by increasing the release of the neurotransmitter dopamine in the prefrontal cortex (Jiang, Mei See, Subramaniam, & Lee, 2013). The data shown in Table 2.20 replicate this pattern.

Table 2.20 Cigarette smoking in 20 schizophrenic and 20 nonschizophrenic men

Schizophrenics (ID)	Do You Smoke?	Nonschizophrenics (ID)	Do You Smoke?
1	Y	1	Y
2	Y	2	N
3	Y	3	N
4	N	4	N
5	Y	5	Y
6	Y	6	N
7	N	7	N
8	Y	8	N
9	Y	9	N
10	Y	10	N
11	Y	11	N
12	Y	12	N
13	Y	13	N
14	N	14	Y
15	Y	15	N
16	Y	16	Y
17	Y	17	Y
18	Y	18	N
19	Y	19	N
20	Y	20	N

a. Create simple and relative frequency distributions for the data from the schizophrenic and nonschizophrenic men.
b. Describe cigarette smoking in these two groups of men. What percentage of men suffering from schizophrenia smoke? What percentage of nonschizophrenic men smoke?
c. Which scale of measurement (nominal, ordinal, interval, or ratio) is being used to measure cigarette-smoking behavior?

14. Impulsivity is defined as "the tendency for rapid, unplanned reactions to stimuli with diminished regard for the negative consequences of that behavior" (Cottone et al., 2012, p. 127). This personality construct can be broken down into two parts: impulsive action (the inability to refrain from action) and impulsive choice (the preference for an immediate and small reward over a delayed but large one). Cottone and colleagues studied impulsive choice in rats. First, they trained the rat to press a bar to receive a sugar water reward (rats love sweets and will work hard for a sweet reinforcement). Then, the rats learned that if they pressed one lever in the test box, they got a "small" but immediate reward (a 1.5% glucose solution), but if they pressed the other lever, they got a much sweeter reward (glucose plus 0.4% saccharine so that it tasted much sweeter but had the same calorie content as the glucose-only reward). The "bigger," sweeter reward was delayed. Researchers measured impulsivity on a scale that could range from 0.5 (*the rat always went for the immediate reward and never went for the bigger but delayed one*) to 21.5 (*the rat always went for the big but delayed reward and never went for the small but immediate one*). Next, they administered a drug that blocks NMDA receptors in the brain. NMDA receptors are involved in learning and memory, and recent research has suggested that they are also involved in controlling impulsive behavior. Table 2.21 shows impulsivity scores similar to the results Cottone et al. saw.
a. Predict what Cottone et al. should see if their hypothesis about the role of NMDA receptors is correct.

Table 2.21 Impulsivity scores for 25 rats

ID	Baseline Impulsivity Score	NMDA Antagonist
1	16	7
2	15	6
3	15	5
4	15	10
5	14	7
6	11	4
7	17	9
8	14	8
9	12	7

Table 2.21 Continued

10	17	6
11	15	7
12	18	6
13	14	7
14	15	5
15	12	8
16	14	7
17	15	8
18	15	7
19	15	9
20	14	6
21	14	5
22	12	7
23	12	8
24	15	10
25	14	6

b. What are the independent and dependent variables in this study? Which scale of measurement (nominal, ordinal, interval, or ratio) are the researchers using to measure impulsive choice?
c. Create a grouped frequency distribution and a cumulative relative frequency distribution for the baseline and postdrug conditions.
d. What impulsivity score sits at the 50th percentile in each condition?
e. Describe the results of this study. Does blocking NMDA receptors affect impulsive choice?

15. Does color affect your physical strength? Crane, Hensarling, Jung, Sands, and Petrella (2008) examined the effect of the color of light in a room on hand-grip strength. Eighteen young men volunteered to have their grip strength tested under blue and white (neutral) light using a hand dynamometer that measures grip strength in pounds. Data representative of their results are shown in Table 2.22.

Table 2.22 Grip strength (in pounds) in two different lighting conditions

ID	Grip Strength in Neutral Light	Grip Strength in Blue Light
1	100	120
2	110	120

3	140	110
4	100	130
5	150	110
6	150	120
7	120	110
8	120	110
9	130	120
10	140	120
11	140	130
12	150	150
13	110	110
14	110	120
15	94	65
16	120	120
17	140	120
18	150	110

a. Create grouped frequency distributions for the grip strength data under both conditions.
b. What is the independent variable for this experiment?
c. What is the dependent variable for this experiment?
d. Which scale of measurement is being used to measure the dependent variable?
e. Does the color of the light in the room affect grip strength? Describe any important differences you see in the data from the two conditions.

16. Kowner (1996) actually carried out four studies of the relationship between facial symmetry and perceived beauty. In one study, he created symmetrical portraits by splitting a photograph of a person's face in half (digitally, of course), duplicating the half-face image, and flipping it vertically to form a whole, symmetrical, composite portrait. Kowner then varied the age of the person in the portrait (his three age groups were children, young adults, and elderly). He presented his participants with two portraits, one symmetrical and one asymmetrical (unaltered), and asked them to select the more "beautiful" of each pair. He then counted the number of symmetrical faces selected (a measure he called the "selection ratio," or the number of symmetrical faces selected divided by the number of asymmetrical faces selected). The data shown in Table 2.23 are representative of the results of this portion of Kowner's research. Create distributions for the data in the table, and use the results to answer the following questions:

a. What would a selection ratio of 0.50 indicate about the participants' preference for facial symmetry in assessments of beauty?
b. Does it appear that preference for facial symmetry depends on the age of the person being portrayed? How does age affect the preference for facial symmetry?
c. What percentage of the ratings was above 0.50 for the:
Children _____
Young adults _____
Elderly _____
d. What kind of measurement scale (nominal, ordinal, interval, or ratio) is the measurement used in Kowner's study?

Table 2.23 Selection ratios as a function of age of person portrayed

Child	Young Adult	Elderly
0.25	0.77	0.26
0.23	0.27	0.27
0.53	0.30	0.22
0.64	0.30	0.50
0.26	0.30	0.63
0.88	0.32	0.96
0.59	0.33	0.94
0.48	0.85	0.50
0.13	0.41	0.49
0.20	0.82	0.63
0.75	0.34	0.46
0.22	0.53	0.75
0.64	0.45	0.53
0.87	0.15	0.40
0.53	0.50	0.66
0.37	0.55	0.62

17. Kowner is by no means the only researcher interested in how and why we deem some things (and people) to be beautiful and others not. Hans Deiter Schmalt (2006) wondered if the waist-to-hip ratio, or WHR, of women mattered in assessment of their beauty by others. So, he presented 133 participants (both male and female) with line drawings of women of varying weight (normal weight, underweight, and overweight) and WHRs (ratios of 0.70, 0.80, 0.90, and 1.00). The facial and body features of each image were held constant within each weight category. Only the WHR varied within a given weight category. The participants were asked to rate the attractiveness of each

image on a scale from 1 (*very low attractiveness*) to 8 (*very high attractiveness*). The data shown in Table 2.24 is representative of what Schmalt found.

Create an ungrouped frequency distribution for the data in each WHR category and use the data to answer the following questions:

a. What kind of measurement scale is being used to assess attractiveness?
b. What WHR is perceived as the least attractive?
c. Does it appear that WHR affects perception of attractiveness? If so, in what way?
d. Schmalt discusses in some detail his reasons for investigating the effect of WHR on the perception of attractiveness. Why do you think WHR might be related to perceived attractiveness?

Table 2.24 Attractiveness ratings for line drawings of women with four WHRs

ID	WHR 1.00	WHR 0.90	WHR 0.80	WHR 0.70
1	4	5	5	8
2	2	5	6	8
3	3	4	8	7
4	4	6	6	7
5	4	6	6	6
6	4	5	7	7
7	5	4	6	8
8	3	6	8	7
9	5	5	5	8
10	4	6	7	7

18. Do opposites really attract? Or, when looking for a mate, do we look for someone whose personality is very similar to ours? To find out, Figueredo, Sefcek, and Jones (2006) asked volunteers to first assess their imaginary "ideal romantic partner" using the NEO-FFI (Neuroticism-Extraversion-Openness Five Factor Inventory; Costa & McCrae, 1992). Normally, the NEO-FFI presents a series of 60 statements about personality characteristics, and participants are asked to rate each statement in terms of how well it describes their personality. They rate the accuracy of each statement on a scale from 1 (*very accurate*) to 5 (*very inaccurate*). To get an assessment of an "ideal romantic partner" (someone who doesn't actually exist), Figueredo et al. substituted the phrase "my ideal romantic partner" for the first-person pronoun in each statement. Participants were then asked to assess their own personality on the NEO-FFI (without the substituted phrase). The authors next created a "Difference Score"—the rating for the personality characteristic for the participant minus the rating for the ideal romantic partner (D = Self − Ideal). The data in Table 2.25 are representative of the results of this study. Create an ungrouped frequency distribution of the difference scores and use the data to answer the following questions:

Table 2.25 Difference scores between personality ratings for self and ideal romantic partner

ID	Difference Score
1	1
2	0
3	0
4	1
5	0
6	0
7	−3
8	2
9	3
10	−4
11	−1
12	−1
13	1
14	−4

    a. What kind of difference score (positive, negative, or zero) would indicate that the person is looking for someone very similar?
    b. What kind of difference score (positive, negative, or zero) would indicate that the person is looking for someone who scores higher on the personality characteristics measured than he or she does?
    c. Summarize the results of this study in your own words. Are we looking for a mirror when we look for a romantic partner, or do opposites attract?

19. Table 2.26 shows four data sets. For each set, create a grouped frequency distribution with the number of intervals (groups) as indicated. Once you've created the frequency distribution, find the midpoint of each interval for all four data sets

Table 2.26 Data sets A through D*

Data Set A	Data Set B	Data Set C	Data Set D
2	67	15	100
3	68	15	109
4	70	15	115
5	72	15	117
5	73	15	120

STATISTICS IN CONTEXT

7	74	17	124
9	79	19	132
9	81	25	133
9	90	27	147
14	100	29	150
23	104	34	166
26	104	35	173
26	104	39	182
30	110	50	192

Hint: For all of the data sets, round the width to the nearest whole number.

a. Data Set A:
   Range = _____   # of Groups = 5   Width = _____   Midpoint = _____
b. Data Set B:
   Range = _____   # of Groups = 6   Width = _____   Midpoint = _____
c. Data Set C:
   Range = _____   # of Groups = 7   Width = _____   Midpoint = _____
d. Data Set D:
   Range = _____   # of Groups = 10   Width = _____   Midpoint = _____

20. Indicate the type of measurement used (nominal [N], ordinal [O], interval [I], or ratio [R]) with a checkmark in the appropriate box.

Example	N	O	I	R
Gender (1 = female, 2 = male)				
Height (in centimeters)				
Weight (in pounds)				
Favorite soft drink				
Year of birth				
Pulse rate (in beats per minute)				
Fruit or vegetable?				
Residence (dorm or apartment?)				
Time needed to complete an exam (in minutes)				
Grade on exam (A, B, C, D, or F)				
Grade on exam (in number correct out of 100)				
Reaction time (in milliseconds)				

CHAPTER 2   Types of Data

Continued

Pizza sales (in U.S. dollars)			
Age (in years)			
Political party you voted in the last election			
Race			
Happiness on a scale from 1 (*very unhappy*) to 5 (*very happy*)			
Movie ratings (G, PG, PG-13, R, X)			
Taxonomic rank (Species, Genus, Family, Class, etc.)			

21. Use the data in Table 2.27 (listing causes of death for the citizens of London in 1662) to answer the following questions:
    a. In 1662 London, what would you most likely die from?
    b. How many people died from "Quinsie"?
    c. How many people died from syphilis?
    d. What were the five least common causes of death in 1662 London?
    e. What PERCENTAGE of the total number of citizens who died perished of "The King's Evil"?
    f. What PROPORTION of the total perished from cancer?
    g. What PERCENTAGE of the total died of plague?

Just FYI:

Rising Of The Lights: croup; any obstructive condition of the larynx or trachea (windpipe) characterized by a hoarse, barking cough and difficult breathing, occurring chiefly in infants and children.

Headmouldshot: a condition when the sutures of the skull, generally the coronal suture, "ride" (that is, have their edges shot over one another), which is frequent in infants and results in convulsions and death.

French Pox: venereal disease; a former name of syphilis.

Stangury: restricted urine flow; difficulty in urinating attended with pain, sometimes caused by bladder stones.

Scowery (or Scowring): probably diarrhea.

Impostuhume: a collection of purulent matter in a bag or cyst.

Kings Evil: scrofula, a tubercular infection of the throat lymph glands; the name originated in the time of Edward the Confessor, with the belief that the disease could be cured by the touch of the King of England.

Quinsie: an acute inflammation of the soft palate around the tonsils, often leading to an abscess; also called suppurative tonsillitis.

Spotted feaver: likely typhus or meningitis.

Table 2.27 Deaths in London for the year 1662, from the *Bills of Mortality* by John Graunt

The Diseases and Casualties This Year					
Abortive and Stillborn	617	Executed	21	Murthered and Shot	9
Aged	1,545	Flox and Small-pox	653	Overlaid and Starved	45
Ague and Feaver	5,257	Found dead in streets, fields etc	20	Palsie	30
Apoplexie and Suddenly	116			Plague	68,596
Bedrid	10	French Pox	84	Plannet	6
Blasted	5	Frighted	23	Plurisie	15
Bleeding	16	Gout and Sciatica	27	Poysoned	1
Bloudy Flux, Scowring and Flux	185	Grief	46	Quinsie	35
		Gripping In The Guts	1,288	Rickets	557
Burnt and Scalded	8	Hanged and Made away themselves	7	Rising Of The Lights	397
Calenture	3			Rupture	34
Cancer, Gangrene and Fistula	56	Headmouldshot and Mouldfallen	14	Scurvy	105
				Shingles and Swine Pox	2
Canker and Thrush	111	Jaundices	110	Sores, Ulcers, broken limbs	82
Childbed	625	Impostuhume	227	Spleen	14
Chrisomes and Infants	1,258	Kild by several accidents	41	Spotted Feaver and Purples	1,929
Cold and Cough	68	King's Evil	86	Stopping Of The Stomach	332
Collick and Winde	134	Leprosie	2	Stone and Stangury	98
Consumption and Tissick	4,808	Lethargy	14	Surfet	1,251
Convulsions and Mother	2,036	Livergrowne	29	Teeth and Worms	2,614
Distracted	5	Meagrom and Headach	12	Vomiting	51
Dropsie and Timpany	1,478	Measles	7	Wenn	1
Drowned	50				
				Total	97,306

22. In 2014, there was another terrifying outbreak of Ebola in Africa. The Ebola viruses (there are several, the first of them discovered near the Ebola River in the Democratic Republic of the Congo) have been found in multiple African countries since the first recorded case in 1976. Since then, outbreaks have appeared sporadically in Africa. According to the Centers for Disease Control and Prevention, "The 2014 Ebola

outbreak is the largest Ebola outbreak in history and the first Ebola outbreak in West Africa. This outbreak is the first Ebola epidemic the world has ever known" (http://www.cdc.gov/vhf/ebola/outbreaks/index.html). Table 2.28 illustrates the deadly extent of the disease.

Table 2.28   Tracking the 2014 Ebola outbreak*

Country Affected	Number of Cases	Number of Deaths
Guinea	3,808	2,536
Italy	1	0
Liberia	10,666	4,806
Mali	8	6
Nigeria	20	8
Senegal	1	0
Sierra Leone	14,067	3,955
Spain	1	0
United Kingdom	1	0
United States	4	1

*Data from www.cdc.gov/vhf/ebola/outbreaks/2014-west-africa/case-counts.html

Use the data in the table to answer the following questions:
a. How many people, in total, were infected with the Ebola virus in the 2014 outbreak?
b. How many people died of Ebola in the 2014 outbreak?
c. Which country suffered the highest number of cases of Ebola during the 2014 outbreak?
d. What percentage of the people who contracted Ebola in Mali died of the disease?
e. What percentage of people who contracted Ebola died of the disease?
f. For each country, calculate the percentage of sufferers of Ebola who died (number of deaths divided by number of cases, multiplied by 100). Which country had the highest risk of death by Ebola in this outbreak?
g. Sierra Leone was one of the hottest of the hot spots during the outbreak. What percentage of Ebola sufferers in Sierra Leone died of the disease?

23. Statistics are used in many fields, and you've probably encountered them far more often than you may realize. Take a look at the following claims, each made with statistics. How might you test these claims to see if they're true? Propose an independent variable (IV) and a dependent variable (DV) you might use to test the claim. Identify the type of measurement scale used (nominal [N], ordinal [O], interval [I], or ratio [R]) with your proposed DV.*

STATISTICS IN CONTEXT

Claim	Variables
Pom Wonderful claims that their pomegranate juice can reduce the risk of cancer, heart disease, and impotence.	IV: DV: Scale:
Reebok claims that its Easy Tone and Run Tone shoes are "proven to work your hamstrings and calves up to 11% harder, and tone your butt up to 28% more than regular sneakers . . . just by walking!"	IV: DV: Scale:
Airborne Herbal Supplement claims that their tablets boost the immune system and prevent colds.	IV: DV: Scale:
Dannon Yogurt says that its Activia line of yogurt relieves irregularity and helps the body digest food.	IV: DV: Scale:

Every one of these companies has had to stop advertising their remarkable claims—not one has been backed up by scientific evidence.

24. Frank Richardson and Richard Suinn created a rating scale (the MARS or Mathematics Anxiety Rating Scale) to measure math anxiety in 1972. There are 98 questions on the MARS, assessing the anxiety felt when confronted with a situation that involves numbers. Participants rate each question on a scale from 1 to 5, according to how much they are frightened by math (5 indicates the most fear and anxiety). The higher the score, the more math anxiety the respondent feels. Suppose the MARS data shown in Table 2.29 were collected from the students enrolled in statistics last year.

Table 2.29 MARS scores for 30 students enrolled in statistics

ID	MARS Score	ID	MARS Score
1	120	16	210
2	250	17	270
3	202	18	250
4	210	19	240
5	190	20	192
6	190	21	280
7	210	22	210
8	340	23	256
9	240	24	150
10	345	25	190

CHAPTER 2 Types of Data

Table 2.29 *Continued*

11	270	26	270
12	270	27	211
13	295	28	180
14	280	29	290
15	250	30	210

Put the data in order from least to most. Create an ungrouped frequency distribution along with cumulative and relative frequency distributions. Use the data to answer the following questions:

a. Assume that the average score on the MARS for typical college students is 215. How many students in your sample scored less than average on the MARS?
b. What scores sit at the following percentile ranks?
   _____ 50th percentile
   _____ 80th percentile
   _____ 90th percentile
c. What is the highest score possible on the MARS? What is the lowest score possible?
d. In one or two sentences, describe the results of the MARS for the students in your sample.

25. Do you believe in magic? Several researchers have suggested that reliance on magical thinking (believing that energy might transfer from one object to another because the two objects are similar or happen to occur together in time and space), once thought to happen mostly in primitive cultures, is also prevalent in more "advanced" cultures like our own. Keinan (2002) wondered if stress might make us more likely to engage in magical thinking. To find out, he interviewed 28 students, asking them a series of 11 questions. Three of the questions were "target questions" designed to elicit an expression of magical thinking—specifically, Keinan was interested in the behavior of knocking on wood to ward off potential danger. The other eight questions were diversion questions, designed to conceal the real purpose of the study. The three target questions were:
    1) "Has anyone in your immediate family suffered from lung cancer?"
    2) "Is your health alright on the whole?"
    3) "Have you ever been involved in a fatal road accident?"
    Diversionary questions asked about mundane and ordinary, nonstressful situations like "What is your favorite TV program?" Table 2.30 illustrates the kind of results Keinan saw.
    a. Create a simple frequency distribution for the data that shows frequency of knocking on wood at least once for the two basic types of questions asked (target and non-target questions).
    b. Describe when the superstitious behavior of knocking on wood was elicited in this sample. Under what conditions was superstitious behavior not elicited?

STATISTICS IN CONTEXT

c. Is there evidence that magical thinking might be related to stress?

Table 2.30 Number of "knocks" for each of 11 questions ("T" indicates target question)

ID	Q1	Q2	Q3T	Q4	Q5	Q6	Q7T	Q8	Q9T	Q10	Q11
1			✓						✓		
2			✓								
3							✓				
4							✓		✓		
5			✓								
6			✓								
7											
8											
9			✓								
10											
11			✓				✓				
12											
13											
14											
15											
16											
17			✓				✓		✓		
18											
19											
20			✓				✓				
21											
22											
23											
24											
25			✓								
26											
27											
28											

26. Keinan (2002) went on to ask another group of students to answer the same series of 11 interview questions described above. The interviewer recorded whether the participant knocked on the wooden table during the interview. After the interview was over, each participant was asked how strongly she or he had felt the urge or need to knock on wood during the interview. Strength of the desire to knock was measured on a four-point scale (1 = *no need*, 2 = *slight need*, 3 = *moderate need*, 4 = *great or strong need*). Table 2.31 shows the kind of results Keinan saw in this study. Use the data in the table to answer the following questions:
   a. Create an ungrouped frequency distribution for the answers on the "Need to knock" scale. What was the most common response on this scale?
   b. Convert the frequency data to relative frequency. What percentage of students in the study reported that they had a strong need to knock on wood during the interview?
   c. Do the data support Keinan's hypothesis that the need to engage in superstitious, ritual behavior like knocking on wood is related to stress?

Table 2.31 Need to knock on wood on a four-point scale

ID	Need	ID	Need
1	1	16	1
2	1	17	4
3	4	18	2
4	2	19	1
5	1	20	3
6	1	21	4
7	1	22	2
8	2	23	2
9	1	24	1
10	2	25	1
11	2	26	1
12	3	27	2
13	3	28	3
14	1	29	3
15	1	30	3

27. Complete the crossword puzzle shown below.

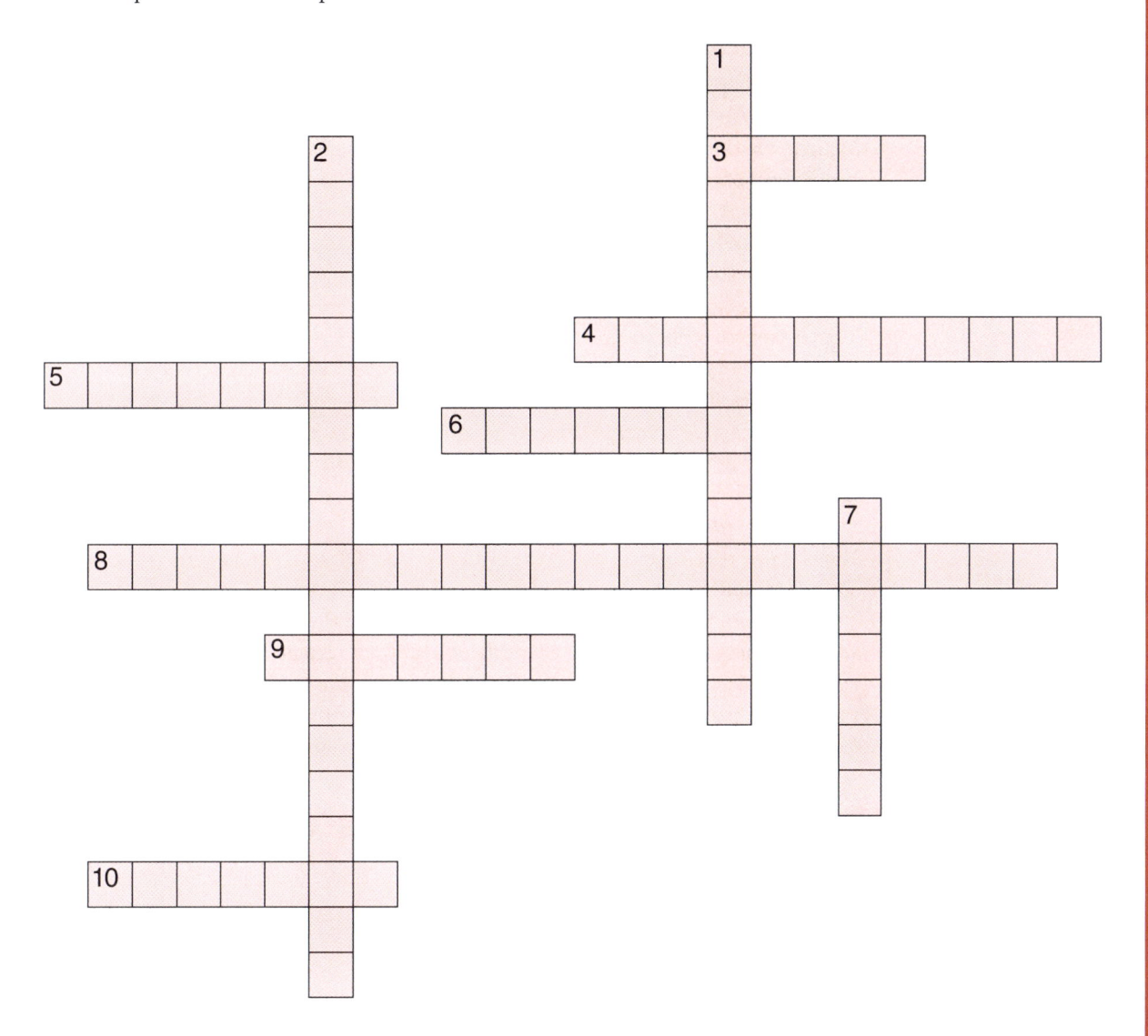

ACROSS
3. Quantitative measure with real zero
4. Measurement of amount or count
5. Quantitative measure with arbitrary zero
6. Qualitative measure of category membership
8. (2 words) Table showing frequency of occurrence of events
9. Qualitative measure with ordered categories
10. Creator of N.O.I.R. measurement scales

DOWN
1. (2 words) Cumulative frequency expressed as %
2. (2 words) Number of events at or below a given point
7. Figure or symbol used to represent number

Think About It . . .

SCALES OF MEASUREMENT

SOLUTIONS

1. Which properties does a NOMINAL scale of measurement possess? **Only the property of Identity.**
2. Which properties does an ORDINAL scale of measurement possess? **The properties of Identity and Magnitude.**
3. Which properties does an INTERVAL scale of measurement possess? **The properties of Identity, Magnitude, and Equal Intervals.**
4. Which properties does a RATIO scale of measurement possess? **All four properties—Identity, Magnitude, Equal Intervals, and a Rational Zero Point.**
5. Suppose I measure height using the following scale: 1 = *shorter than me*, 2 = *the same height as me*, and 3 = *taller than me*. Which properties does my scale have? **Identity (there's no overlap between categories on my scale; in other words, the heights I'm measuring can't be in more than one category at a time) and magnitude—there's meaningful order to the categories on my scale.**
6. Which properties differentiate QUALITATIVE from QUANTITATIVE scales? **Qualitative scales have identity and magnitude but lack equal intervals and a rational zero point. Quantitative scales have at least three of the four properties—they don't necessarily have the property of a rational zero point. This is often because the "objects" measured with qualitative scales don't physically exist—opinion, perspective, perception are all mental "objects" rather than physical ones.**
7. Which properties does the clothing size scale shown in Table 2.3 possess?

Standard Women's Clothing Sizes (Missus)											
00	0	2	4	6	8	10	12	14	16	18	20

Clothing size would appear to be measured on an interval scale. The property of identity is present (no article of clothing could simultaneously be both a size 0 and a size 00) as well as magnitude (a size 2 dress would be smaller than a size 4 dress). However, we don't know if the intervals between adjacent sizes are all equal (is the difference between a size 8 pair of pants and a size 10 pair of pants the same as the difference between a size 18 and a size 20?). And not only is there a zero point, there's also a "double zero" measure (indicating less than no size at all, I assume).

FREQUENCY DISTRIBUTIONS (PART 1)

SOLUTIONS

1. Fill in the blanks on the table by calculating the cumulative frequency and the percentile rank for the sleeping students in this study.

Sleep duration for 20 college undergraduates (in minutes)			
x	f	CF	Percentile Rank (CRF)
328	1	1	5%
367	1	2	10%
368	1	3	15%
373	1	4	20%
376	1	5	25%
377	1	6	30%
379	1	7	35%
383	2	9	45%
385	1	10	50%
387	1	11	55%
388	1	12	60%
390	1	13	65%
425	1	14	70%
488	3	17	85%
523	2	19	95%
600	1	20	100%
Total n	20	20	

2. What is a typical sleep duration for the 20 (hypothetical) students in this sample? **There are several answers you might choose, but I would say that since 488 minutes was the most frequent sleep duration in the set, it works as a typical duration.**
3. What sleep duration defines the 50th percentile? **385 minutes of sleep.**
4. In this sample, 30% of the participants slept at least **377** minutes.
5. How many students slept more than 8 hours? **8 hours = 480 minutes, so six students slept more than 8 hours.** How many slept less than 6 hours? **6 hours = 360 minutes, so only one student.**

FREQUENCY DISTRIBUTIONS (PART 2)

SOLUTIONS

x	f	Intervals	f	CF	Percentile Rank (CRF)
328	1	324–350	1	1	5%
367	1	351–377	5	6	30%

CHAPTER 2 Types of Data   89

Continued					
368	1	378–404	7	13	65%
373	1	405–431	1	14	70%
376	1	432–458	0	14	70%
377	1	459–-485	0	14	70%
379	1	486–512	3	17	85%
383	2	513–539	2	19	95%
385	1	540–566	0	19	95%
387	1	567–593	0	19	95%
388	1	594–620	1	20	100%
390	1	Total *n*	20		
425	1				
488	3				
523	2				
600	1				
Total *n*	20				

1. If you use 10 intervals, how wide will each interval be? **272/10 = 27.2, or 27 minutes wide**
2. Where should the first interval start? (Use the even multiple of the width rule to determine the starting point for the first interval.) **27 × 12 = 324, so start the first interval at 324. It is less than the smallest data point (327) yet still close to that point in value.**
3. Where should the first interval end? **324 + 26 = 350, so the first interval should start at 324 and end at 350 to be 27 minutes wide.**
4. Create the grouped frequency distributions in the space provided on the table.
5. Reassess the data. Has your definition of the typical night's sleep for the participants changed after creating the grouped frequency distribution? If so, what is the change? **Actually, it might change now that the data have been grouped. Most of the students (65%) slept less than 404 minutes (6.73 hours), and only six slept more than 8 hours so I would adjust the typical night's sleep down to about 404 minutes.**

REFERENCES

Atkinson, R. C., & Shiffrin, R. M. (1968). Chapter: Human memory: A proposed system and its control processes. In Spence, K. W., & Spence, J. T. *The psychology of learning and motivation*(Volume 2). New York: Academic Press. pp. 89–195.

Burns, K. C. (1984). Motion sickness incidence: Distribution of time to first emesis and comparison of some complex motion conditions. *Aviation Space and Environmental Medicine, 55*(6), 521–527.

Cawley, J. (2001). Body weight and the data and sexual behaviors of young adolescents. In R.T. Michael (Ed.), *Social awakening: Adolescent behavior as adulthood approaches* (pp. 174–198). New York, NY: Russell Sage Foundation.

Crane, D. K., Hensarling, R. W., Jung, A. P., Sands, C. D., & Petrella, J. K. (2008). The effect of light color on muscular strength and power. *Perceptual and Motor Skills, 106*(3), 958–962.

Costa, P. T., & McCrae, R. R. (1992). An introduction to the five-factor model and its application. *Journal of Personality, 60*(2), 175-215.

Cottone, P., Iemolo, A., Narayan, A. R., Kwak, J., Momeny, D., & Sabino, V. (2012). The uncompetitive NMDA receptor antagonist ketamine and memantine preferentially increase the choice for small, immediate reward in low-impulsive rats. *Psychopharmacology, 226*, 127–138.

Figueredo, A. J., Sefcek, J. A., & Jones, D. N. (2006). The ideal romantic partner personality. *Personality and Individual Differences, 41*, 431–441.

Forgeard, M. J. C. (2013). Perceiving benefits after adversity: The relationship between self-reported posttraumatic growth and creativity. *Psychology of Aesthetics, Creativity and the Arts, 7*(3), 245–264.

Gaito, J. (1980). Measurement scales and statistics: Resurgence of an old misconception. *Psychological Bulletin, 87*(3), 564–567.

Hand, D. J., Daly, F., Lunn, A. D., McConway, K. J., & Ostrowski, E. (Eds.) (1994). *A handbook of small data sets*. London, UK: Chapman & Hall.

Jiang, J., Mei See, Y., Subramaniam, M., & Lee, J. (2013). Investigation of cigarette smoking among male schizophrenia patients. *PLoS ONE, 8*(8), e71343.

Keinan, G. (2002). The effects of stress and desire for control on superstitious behavior. *Personality and Social Psychology Bulletin, 28*, 102–108.

Kowner, R. (1996). Facial asymmetry and attractiveness judgment in developmental perspective. *Journal of Experimental Psychology: Human Perception and Performance, 22*(3), 662–675.

Lord, F. M. (1953). On the statistical treatment of football numbers. *The American Psychologist, 8*, 750–751.

Lund, H. G., Reider, B. A., Whiting, A. B., & Prichard, J. R. (2010). Sleep patterns and predictors of disturbed sleep in a large population of college students. *Journal of Adolescent Health, 46*(2), 124–132.

Paxton, S. J., Norris, M., Wertheim, E. H., Durkin, S. J., & Anderson, J. (2005). Body dissatisfaction, dating and importance of thinness to attractiveness in adolescent girls. *Sex Roles, 53*(9/10), 663–675. DOI: 10.1007/s11199-005-7732-5.

Plana, T., Gracia, R., Mendez, I., Pintor, L., Lazaro, L., & Castro-Fornielas, J. (2010). Total serum cholesterol levels and suicide attempts in child and adolescent psychiatric inpatients. *European Child and Adolescent Psychiatry, 19*, 615-619.

Schmalt, H. D. (2006). Waist-to-hip ratio and female physical attractiveness: The moderating role of power motivation and the mating context. *Personality and Individual Differences, 41*, 455–465.

Stevens, S. S. (1946). On the theory of scales of measurement. *Science, 103*, 677–680.

Teghtsoonian, R. (2001). Stevens, Stanley Smith (1906–73). Retrieved from web.mit.edu/epl/StevensBiography.pdf

U.S. Standard Clothing Size. (n.d.). In *Wikipedia*. Retrieved from en.wikipedia.org/wiki/US_standard_clothing_size

World Health Organization (WHO). (2016). *World health statistics 2016: Monitoring health for the DSGs, sustainable development goals*. Geneva: WHO.

Getting Data into Your Stats Program

Raw data come in many different forms. For the purpose of this book, we will assume that your raw data are stored in a specific type of text file: the comma-separated value file, or CSV file.

This file type is extremely flexible in that every common (and many less common) statistics program as well as every spreadsheet program can both read and write these files. This is important particularly when you collaborate with colleagues who may not have access to the same software you do.

Figure 1 shows the contents of a CSV file. The first row of information contains the column headings or variable labels. Each subsequent row contains the actual data, one case/observation per row.

Figure 1.
A CSV file opened in a text editor.

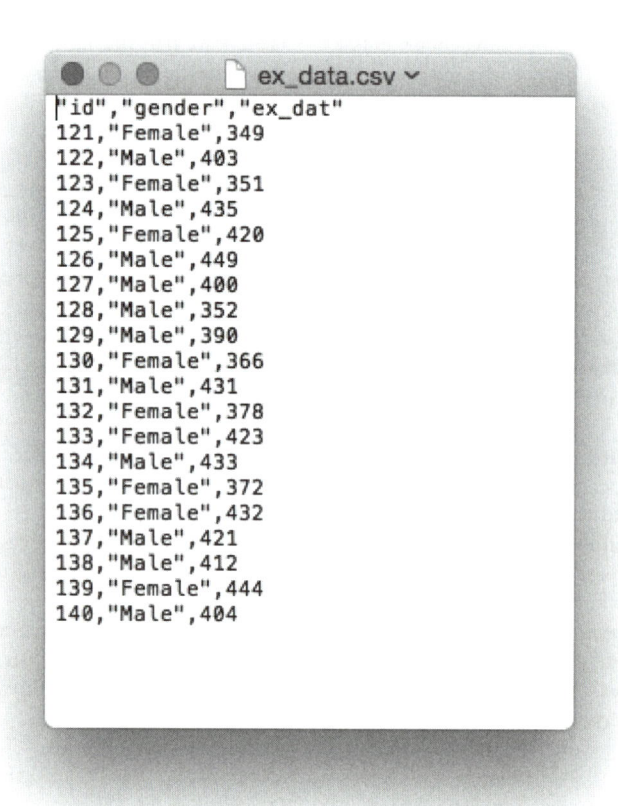

```
ex_data.csv
"id","gender","ex_dat"
121,"Female",349
122,"Male",403
123,"Female",351
124,"Male",435
125,"Female",420
126,"Male",449
127,"Male",400
128,"Male",352
129,"Male",390
130,"Female",366
131,"Male",431
132,"Female",378
133,"Female",423
134,"Male",433
135,"Female",372
136,"Female",432
137,"Male",421
138,"Male",412
139,"Female",444
140,"Male",404
```

Reading in Data with SPSS

Before we can do anything with our data, we have to read it into our statistics program. In this section, we will go through how to import a CSV file containing the sample data shown in Figure 1 into SPSS.

STEP 1: Open the SPSS Statistics application on your computer.

STEP 2: Click **File** in the menu bar and select **Read Text Data...**

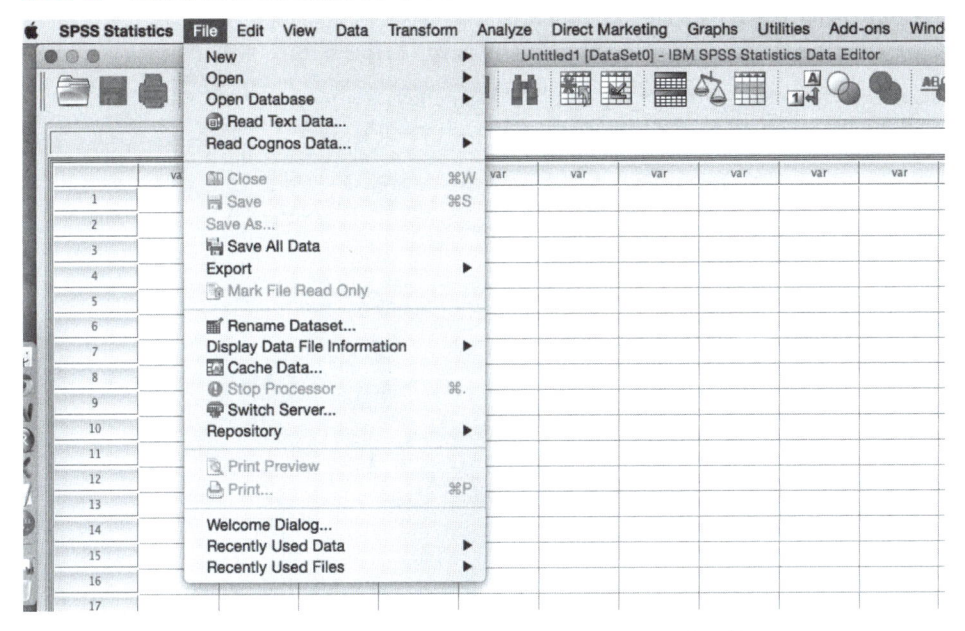

An **Open Data** dialog box will open, asking you to select the file you want to open (**ex_data.csv**). After selecting your CSV file, the **Text Import Wizard** will start.

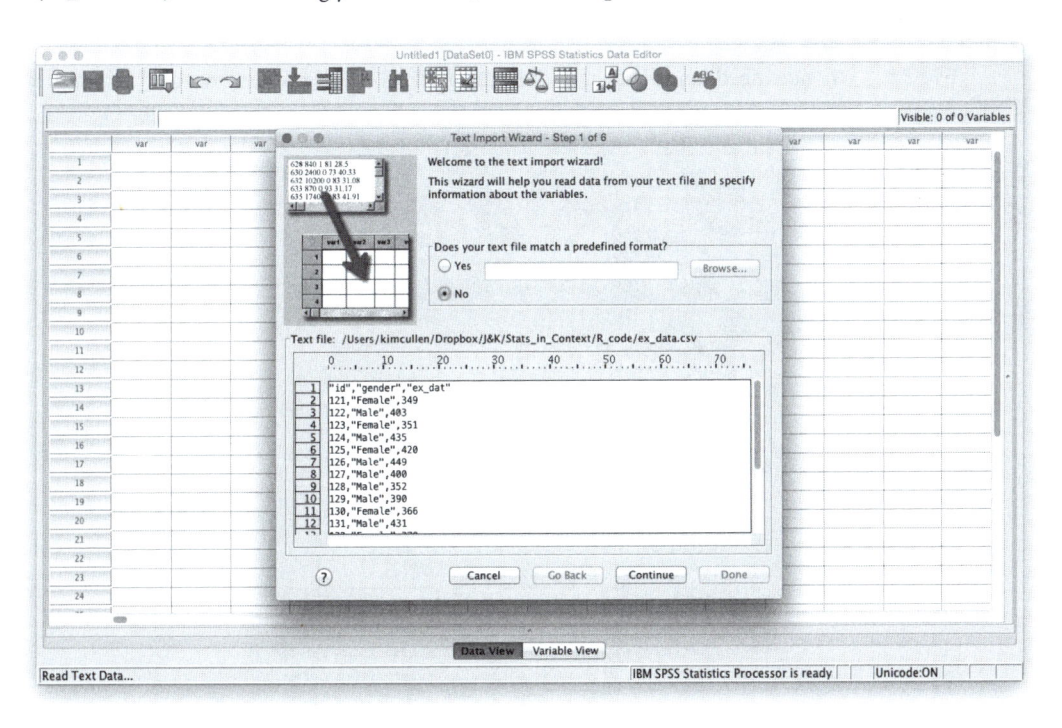

STEP 3: A new dialog box for the **Text Import Wizard** will appear, labeled **Step 1 of 6.** Leave the existing defaults and click **Continue** to advance to **Text Import Wizard - Step 2 of 6**.

CHAPTER 2 Types of Data 93

STEP 4: In **Text Import Wizard Step 2 of 6** select **Delimited** for "How are your variables arranged" and **Yes** to indicate that the variable names are at the top of the CSV file. Click **Continue** to advance to **Text Import Wizard - Step 3 of 6**.

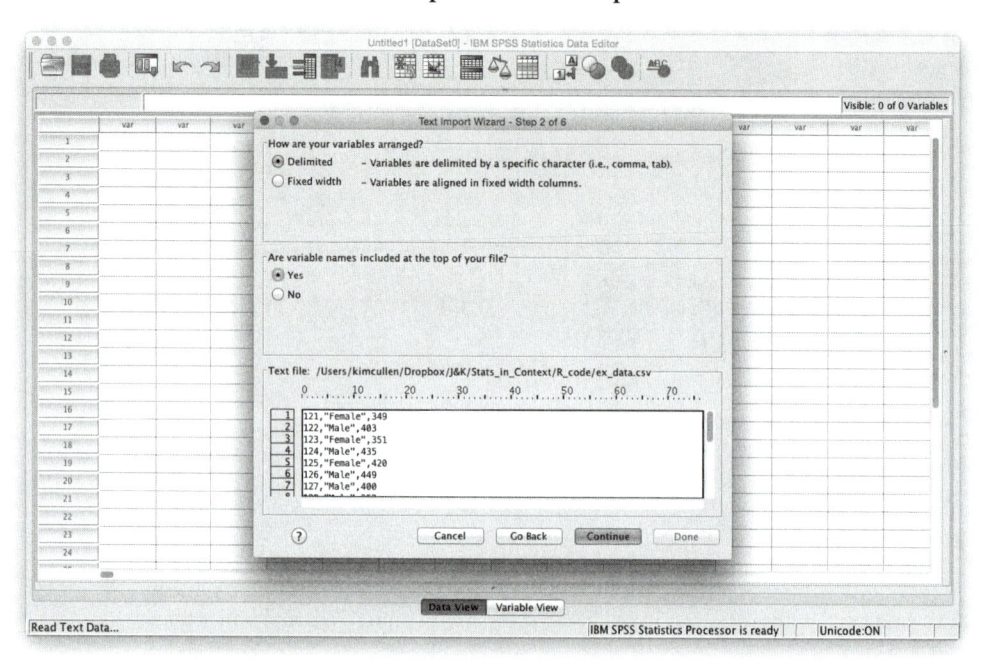

STEP 5: We now need to indicate that our data start on the second line of the CSV file (our variable names are on the first). Select **Each line represents a case** to describe how the cases are represented and that we want to import **All of the cases**. Click **Continue**.

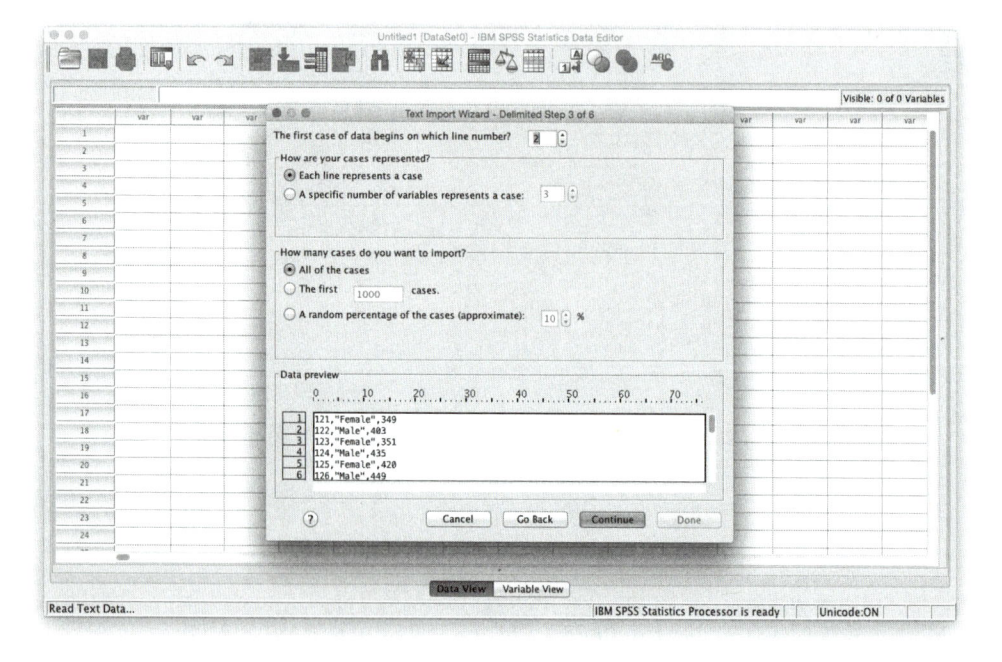

STEP 6: In **Text Import Wizard - Step 4 of 6**, we select **Comma** as our delimiter and choose **Double quote** as the text qualifier. The text qualifier is important if you collect character/string data. By selecting the Double quote, it prevents apostrophes and other punctuation from confusing the computer. Click **Continue** to advance to **Text Import Wizard - Step 5 of 6**.

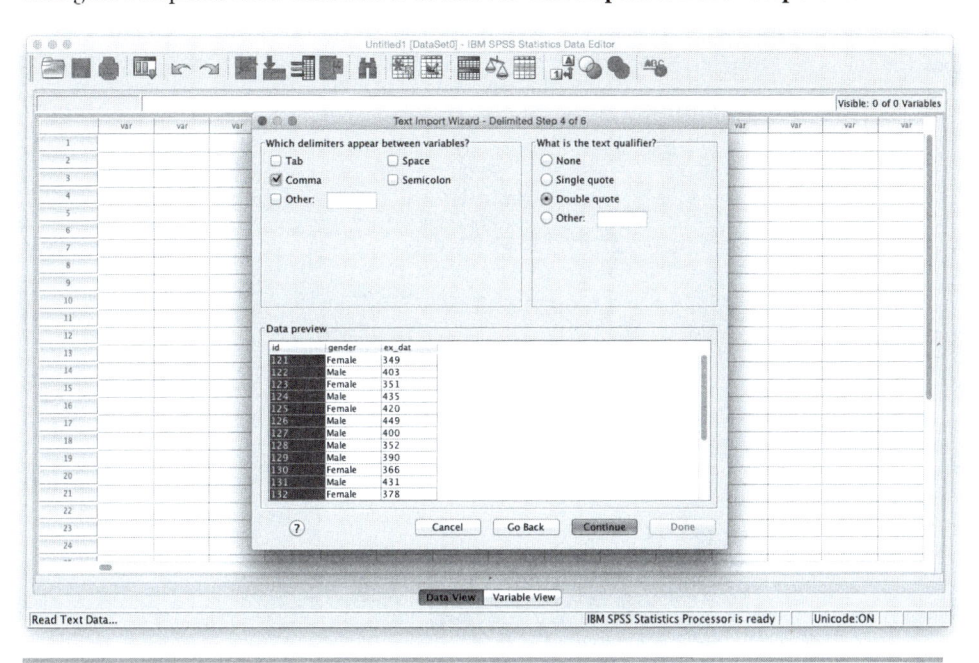

STEP 7: Step 5 of 6 gives us the opportunity to rename our variables if we want. This is a good time to give your variables meaningful names. It's very difficult for colleagues to understand your data if you name variables var01, var02, etc.

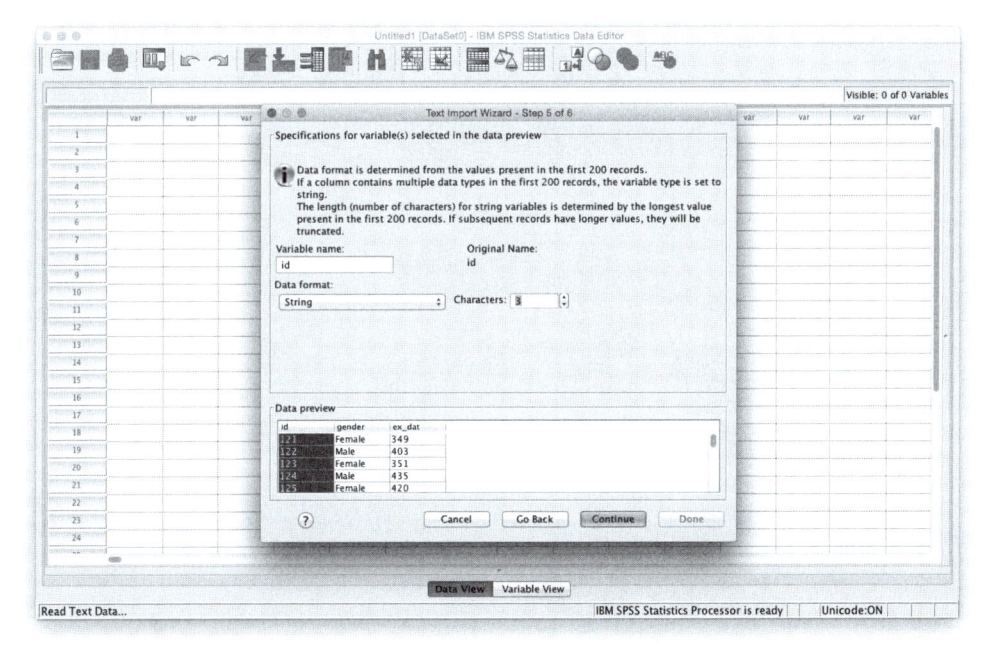

CHAPTER 2 Types of Data   95

We can also check and correct any misclassification of our variable types in our data set. In our case, I changed the id variable from numeric to string. While the id values were numbers, these values were labels for the participants and thus are categorical (nominal) data. Click **Continue** to advance to **Text Import Wizard - Step 6 of 6**.

STEP 8: In the final step, we leave the defaults (**No**, to saving this format, and **No**, to pasting the syntax) and click **Done**.

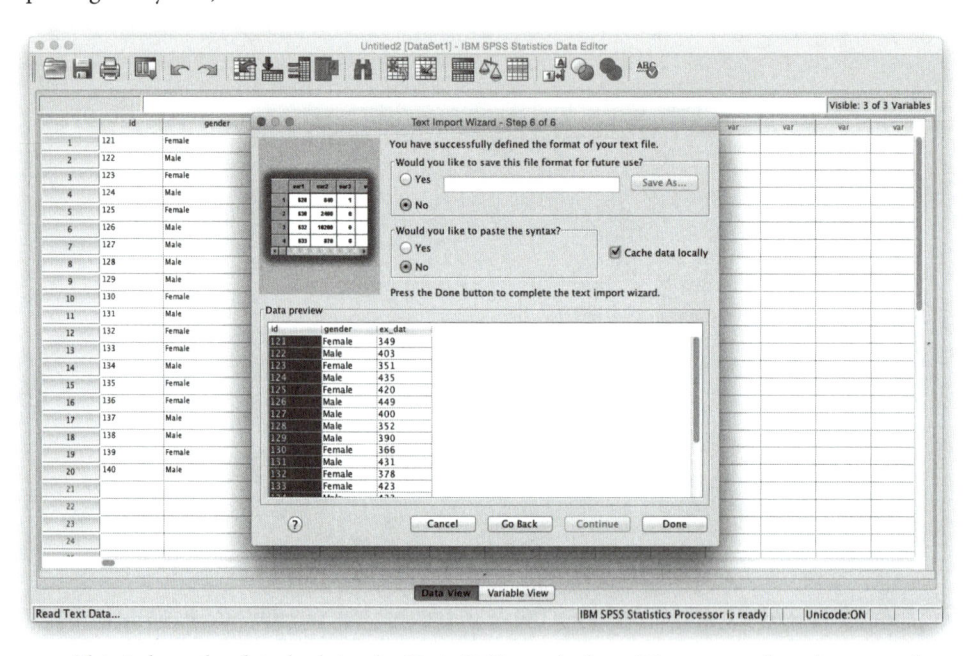

This is how the data look in the **Data Editor** window. We can see that there are three variables and 20 cases. We can now get to work and doing some analyses.

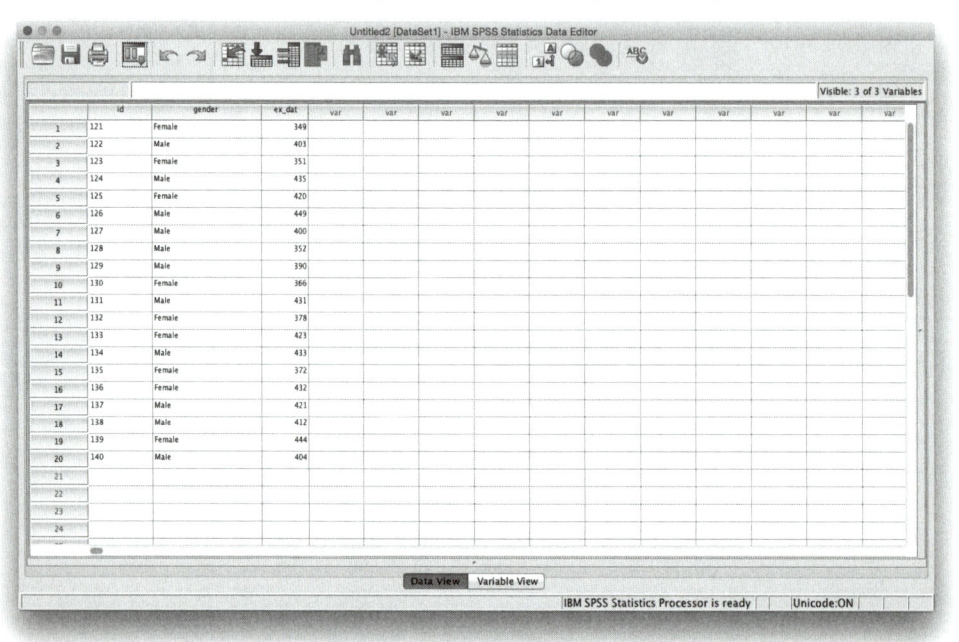

96 **STATISTICS IN CONTEXT**

Reading in Data with R

In this section, we will be reading in our CSV file to the statistical program called R using RStudio IDE. This software is free and open-source (www.r-project.org, www.rstudio.com).

STEP 1: Open the RStudio application on your computer.

STEP 2: Select the **Import Dataset** button on the **Environment** tab in the top-right window.

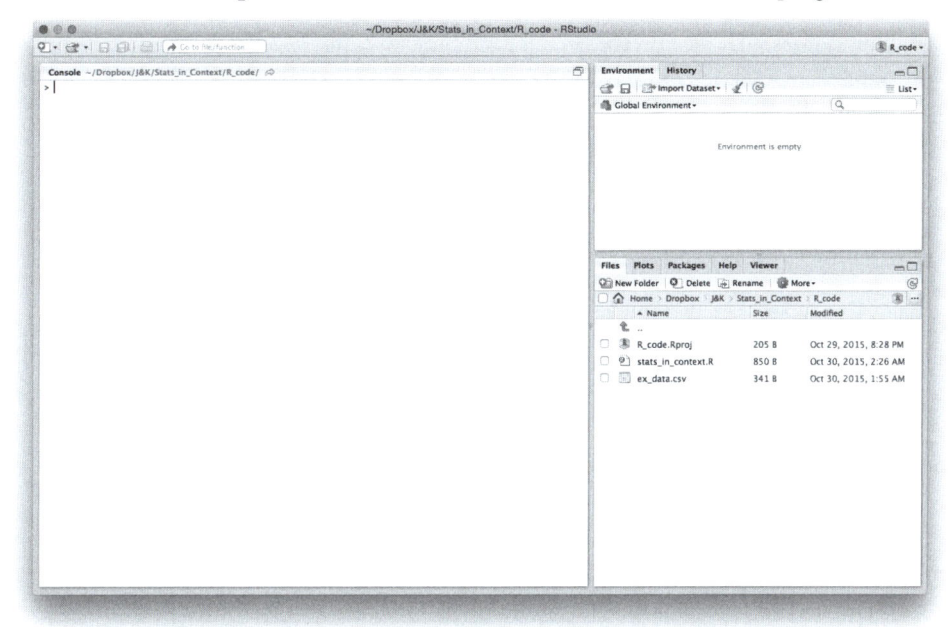

STEP 3: Select **From Text File...** from the drop-down menu. An **Import** dialog box will open, and you can select your CSV file (**ex_data.csv**) that you want to bring into R.

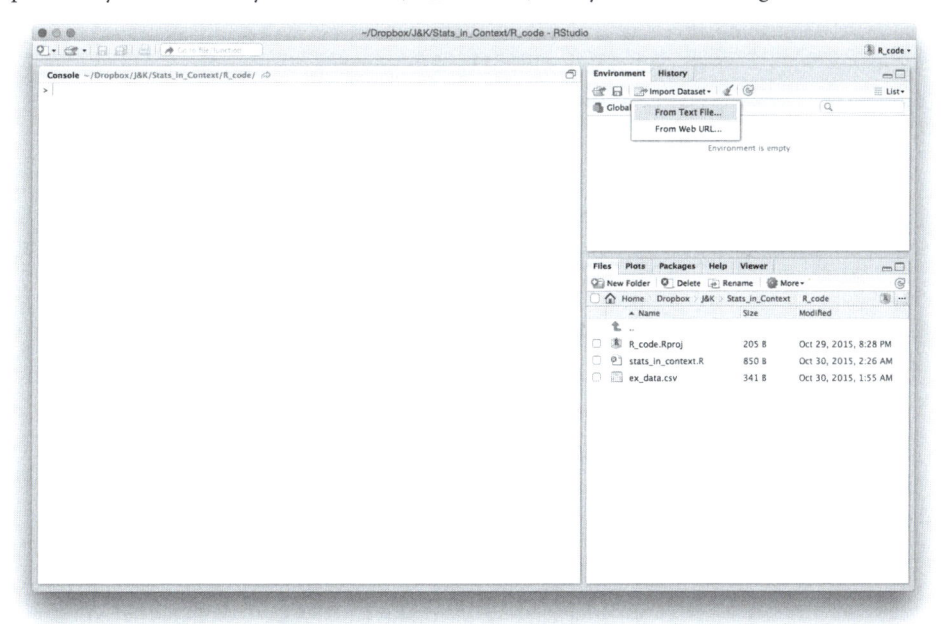

CHAPTER 2 Types of Data 97

STEP 4: Once you select your data file, the **Import Dataset** dialog box opens. All the options needed to successfully import your data are contained here. These are the settings that will bring in this data set:

- **Name** – R works with objects. Your data set will be an object. Name it something that is descriptive but concise. R is primarily a text interface, so you will be typing the name of your data set frequently.
- **Encoding** – Leave it on the default Automatic setting.
- **Row names** – Leave it on the default Automatic setting.
- **Separator** – Select Comma.
- **Decimal** – Select Period.
- **Quote** – Select Double quote (").
- **Comment character** – Select None.
- **na.strings** – Leave it as the default NA.
- **Strings as factors** – Check.

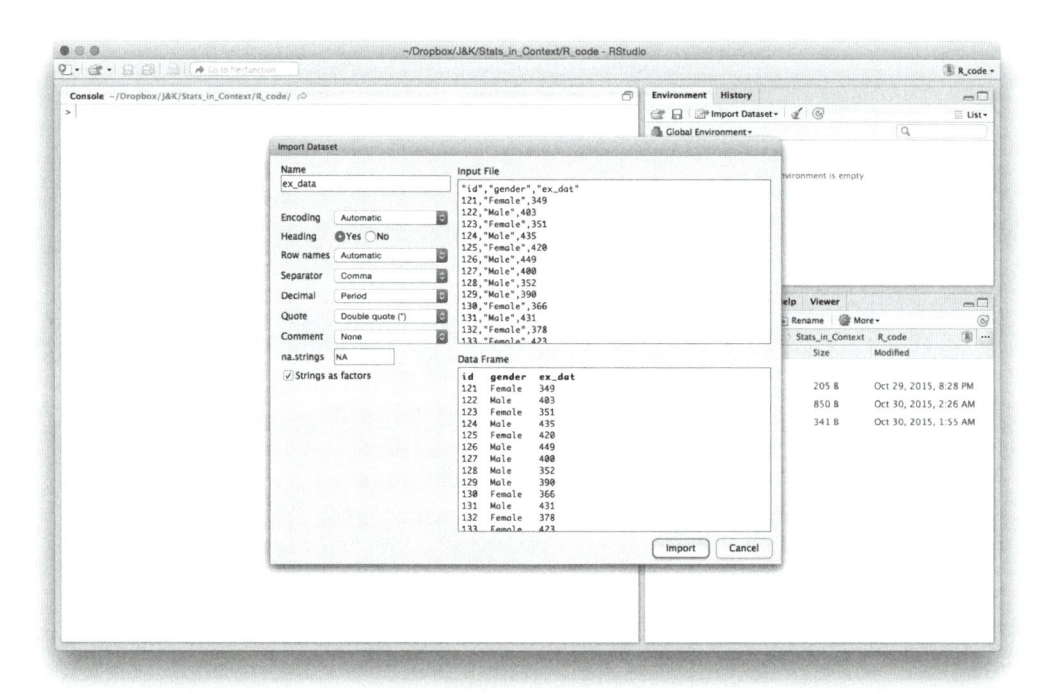

STEP 5: As you make adjustments to the settings, you can see how the input file shown in the top-right side of the **Import Dataset** dialog box will look inside R as a Data Frame object. Click **Import**.

STATISTICS IN CONTEXT

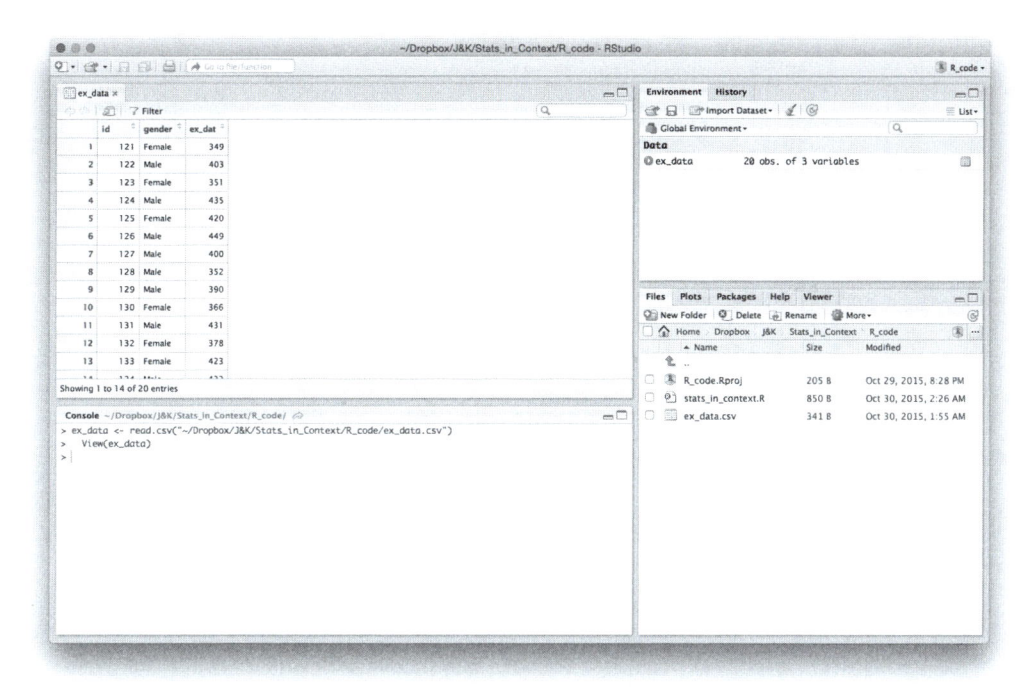

Your data have now been successfully imported into R and are ready to be analyzed.

CHAPTER THREE

A PICTURE IS WORTH A THOUSAND WORDS: CREATING AND INTERPRETING GRAPHICS

—"Dirty Harry" Callahan, *Sudden Impact* (1983)

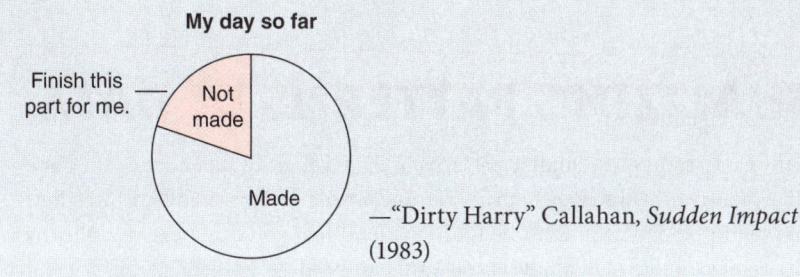

Everyday Statistics

HOW'S THE WEATHER UP THERE?

Did you know that adding 40 to the number of cricket chirps in 14 seconds (allegedly) predicts the temperature?

Modern meteorologists usually rely on statistics instead of bugs, and collect massive amounts of data on wind, humidity, temperature of the air and the ocean, and so on, to generate graphics in their efforts to predict the weather. Here, we have a plot of the tracks of every tropical storm and hurricane between 1851 and 2012. Do you see a pattern? The typical Atlantic hurricane sweeps up the east coast, curving from the southwest to the northeast. Hurricane Sandy—a very unusual and very destructive storm—did not stick to the script. That break in the pattern made Sandy very hard to prepare for and likely contributed to the estimated $50 billion cost of the storm—until recently second only to that of Hurricane Katrina (estimates of the damage caused by hurricanes Harvey and Irma range from $150 to $200 billion, making the two storms together the most expensive to date; "Billion dollar weather," 2017).

In later chapters, we will look at what it means when a finding defies our expectations. In Chapter 3, we will consider graphics and how useful they can be when we're trying to summarize sometimes very complex data.

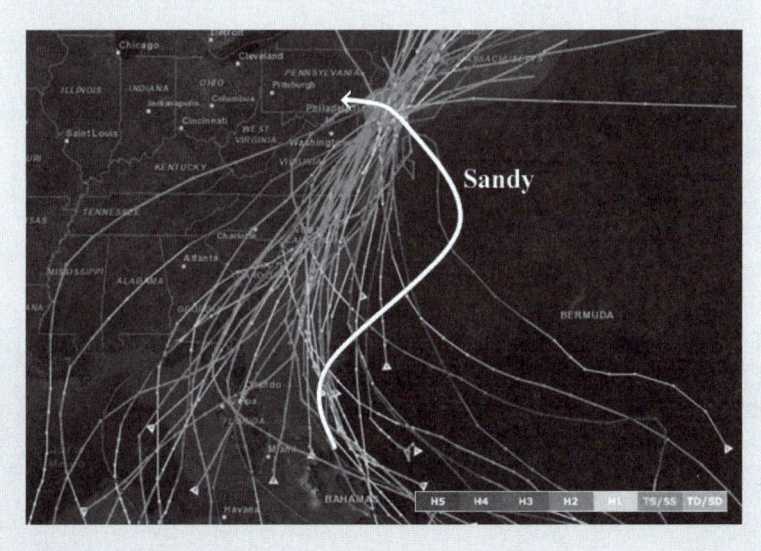

OVERVIEW

VISUALIZING PATTERNS IN DATA
BAR CHARTS AND HISTOGRAMS
STEM-AND-LEAF GRAPHS
FREQUENCY POLYGONS
PIE CHARTS
OTHER GRAPHICS
GRAPHICS IN CONTEXT: RULES FOR CREATING GOOD GRAPHS

LEARNING OBJECTIVES

Reading this chapter will help you to . . .

- Appreciate the contributions made to modern statistics by William Playfair. (Concept)

- Recognize the features, strengths, and limitations of five basic types of graphs: bar chart, frequency histogram, frequency polygon, stem-and-leaf graph, and pie chart. (Concept)

- Understand how these types of graphs are related to the scales of measurement discussed in Chapter 2. (Concept)

- Know the difference between discrete and continuous data. (Concept)

- Select the appropriate graphic representation of data you wish to present. (Application)

- Create graphic representations of data using the guidelines for creating good graphs. (Application)

VISUALIZING PATTERNS IN DATA

One of the easiest ways to visualize patterns in data is to draw a picture of the data; after all, a picture is said to be worth a thousand words. In this chapter, I introduce a few of the more popular graphs, including several that you can create roughly by hand. Nowadays, most data analysis software can produce gorgeous graphics, but it is worth becoming familiar with the principles of good graphing design for times when you don't have a computer handy. When computers are called for, I will be using SPSS (the Statistical Package for the Social Sciences) to produce the graphs. SPSS is one of the most powerful and most popular statistical analysis software packages on the market, but it is by no means the only one available. Open-source software such as R is increasingly popular. Look around and see what programs you have access to and what they can do for you.

The benefits of presenting numbers in graphs may seem so obvious you'll be surprised to learn that using visual representations of data is a relatively recent innovation. In fact, as modern science developed, using graphs was frowned upon. Spence (2000) suggests that early scientists viewed graphs very skeptically, perhaps because the tradition at the time was to present any data in a table and let the reader determine what patterns might be there. Graphs didn't show the raw data, so scientists didn't trust them. But in the late 1700s, a hapless jack-of-all-trades by the name of William Playfair began to use graphics in one of his many entrepreneurial pursuits: writing and publishing. While he was not commercially successful, he did help to establish the use of charts and graphs as an effective way to present complex data (see *The Historical Context*).

The Historical Context

WILLIAM PLAYFAIR AND THE USE OF GRAPHICS IN PUBLISHING

Several of the statistical graphs that we are familiar with today—the bar chart, the pie chart, the line graph, and time-series charts—were invented by William Playfair in the late eighteenth century. A bit of an oddball, Playfair was

> in turn, millwright, engineer, draftsman, accountant, inventor, silversmith, merchant, investment broker, economist, statistician, pamphleteer, translator, publicist, land speculator, convict, banker, ardent royalist, editor, blackmailer, and journalist. Some of his business activities were questionable, if not downright illegal, and it may be fairly said that he was something of a rogue and scoundrel. (Spence & Wainer, 1997, p. 133)

William Playfair was born near Dundee, Scotland, in 1759. At age 15, he was apprenticed to a millwright, and at 18, he was hired by James Watt (scientist and developer of the steam engine) to create and copy drawings of the steam engines that Watt's company produced. In 1781, Playfair left Watt's firm and, with a coworker, set up shop as a silversmith and platemaker. After he was accused by his partner of stealing patent ideas, the silversmithing business failed. In another attempt to earn a living, Playfair published a book on economics, called

Regulation of the Interest of Money (1785), and a year later a study of English trade, entitled *A Commercial and Political Atlas*—the first publication to contain graphs of statistical data. The *Atlas* featured 43 time-series charts and one lone bar chart (shown in Figure 3.1). It did not make Playfair any money.

In 1787, William left for Paris, where he became involved with Joel Barlow, the Parisian representative of the Scioto Land Company. The company was selling American land on the Ohio River to French settlers. Suspected of embezzling the money that prospective settlers had paid for their land, Playfair left France abruptly in 1793. Because the Scioto Company did not actually own the land it was "selling," the settlers found themselves abandoned in America, without money and without a way to return to France. They were eventually settled in the town of Gallipolis, Ohio.

Playfair was a poor businessman. Between 1793 and 1814, he engaged in a number of mostly unsuccessful schemes to make money. He even tried to blackmail Lord Archibald Douglas with a letter allegedly related to the question of Lord Douglas' parentage, which had famously and scandalously been questioned nearly 50 years earlier. (Lord Douglas did not pay.)

Playfair also continued to publish on a range of topics, from mathematics (*Lineal Arithmetic,* 1798) and economics (*An Inquiry into the Permanent Causes of the Decline and Fall of Powerful and Wealthy Nations,* 1805) to history (the nine-volume *British Family Antiquity Illustrative of the Origin and Progress of the Rank, Honours, and Personal Merit, of the Nobility of the United Kingdom,* 1809–1811). All of these works featured charts to illustrate his data. In 1801, he published *A Statistical Breviary*, which also used graphs to illustrate statistics on trade, population, and so on from

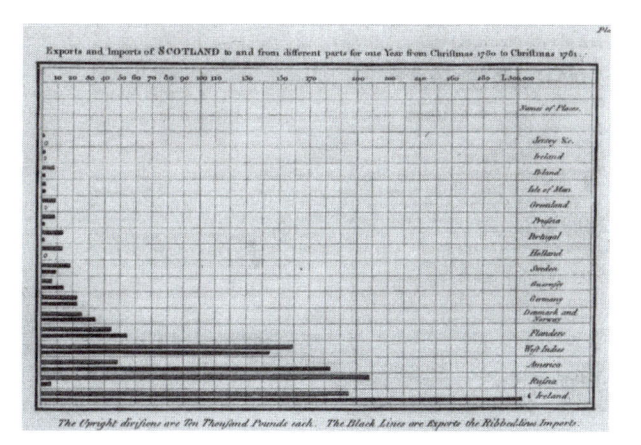

FIGURE 3.1
William Playfair's first bar chart: Exports and imports of Scotland to and from different parts for 1 year, from Christmas 1780 to Christmas 1781.

several European countries. Here, Playfair introduced two new graphs, the pie chart and the circle diagram, both attempting to make comparison of the statistical information across countries easier for the reader.

Playfair believed that graphs would make complex data easier for everyone to understand, remarking that "[n]o study is less alluring or more dry and tedious than statistics, unless the mind and the imagination are set to work" (Playfair, 1801, p. 16). He hoped his methods for the graphic display of information would provide the necessary spark to the mind and imagination of the reader, yet Playfair's bumpy background did not help make his inventions acceptable to established science. Despite the fact nearly everyone agreed that graphs made it easier to understand the data, his methods did not catch on for another 50 years. Playfair died penniless, in Covent Garden, on February 11, 1823.

BAR CHARTS AND HISTOGRAMS

The graphic displays of information that Playfair created and championed have become a standard part of statistics. Bar charts and histograms, like the ones Playfair used in his economic writings, use a bar to illustrate frequency. In both, the observations are displayed along the horizontal *x*-axis (more formally known as the **abscissa**); the frequency of each observation is shown along the vertical *y*-axis (the **ordinate**). The height of the bar above the *x*-axis represents the frequency of

abscissa The horizontal axis (or *x*-axis) on a graph.

ordinate The vertical axis (or *y*-axis) on a graph.

TABLE 3.1 Participant responses to the question "Do you smoke?"

ID	Response (1 = Yes, 2 = No)	ID	Response (1 = Yes, 2 = No)
1	1	6	2
2	2	7	1
3	2	8	2
4	2	9	2
5	2	10	1

TABLE 3.2 Simple frequency distribution of smoking data

Response	f
1 (yes)	3
2 (no)	7

each observation (or groups of observations, if you are illustrating a grouped frequency distribution).

Here are some data to play with: Ten college student volunteers were asked to respond with either "yes" or "no" to the question "Do you smoke?" Table 3.1 shows the responses to this question. The frequency distribution for these data is shown in Table 3.2.

We can represent the ungrouped frequency distribution using a **bar chart**, as shown in Figure 3.2. The characteristics of a bar chart are as follows:

- Frequency is represented by the height of the bar above the *x*-axis.
- The bars do not touch one another so as to indicate that what is being graphed is **discrete data**. This statement should bring up another question: What is discrete data?

bar chart A graph where individual observations in a data set are represented on the *x*-axis and frequency is represented on the *y*-axis. The height of a bar above the *x*-axis represents observation frequency. The bars do not touch one another to indicate the use of discrete data.

discrete data Data sets where there are no possible measurements in between any two adjacent data points on the *x*-axis.

DISCRETE DATA

The term *discrete* describes data sets where there are no possible measurements in between any two adjacent data points. In the case of the data graphed in Figure 3.2, there are two—and only two—possible answers to our question "Do you smoke?" The participant can answer with either a "yes" or a "no." There are no possible answers between those two—a participant cannot respond to the survey question with "I don't know" or "Sometimes" or "None of your business." This is a forced-choice situation: Only two answers are permitted, and each category of this nominal-level variable is discrete.

The space between each bar in the bar chart is meant to illustrate the discrete nature of the data. You will often have discrete data when you use surveys to gather information from your participants. Notice how quickly you can determine which answer was given most frequently from the bar chart.

CONTINUOUS DATA

The sleep duration data from Chapter 2 (see Table 3.3) are a bit different from the smoking data shown in Table 3.2. Here, we have ratio-level, **continuous data**, meaning that it is possible for someone to report that they slept 120 minutes, 120.5 minutes, or 120.6789 minutes, depending on the accuracy of the timing device being used. With continuous data, any number of observations between any two adjacent categories on the *x*-axis are possible. The actual number of observations that you have in your data depends on the accuracy of the measuring device.

As the data stand right now, I have measured sleep duration to the nearest minute. A subject might report 120 or 121 minutes of sleep. If I'd measured sleep duration down to the second instead of to the minute, it would be possible to have data between 120 and 121 minutes. I might have had a subject who slept 120 minutes and 3 seconds, and another who slept 120 minutes and 16 seconds, and so on. We'd like the graphic representation of this kind of data to illustrate the continuous nature of the data, so we'll have the bars on the graph touch one another.

When the bars on the graph touch, it is called a **frequency histogram** (or a histogram, for short). In a histogram, the height of the bars above the *x*-axis represents the frequency of an *x*-axis response, just as it does in a bar graph. The bars touch one another to indicate that the data are continuous. Figure 3.3 shows a frequency

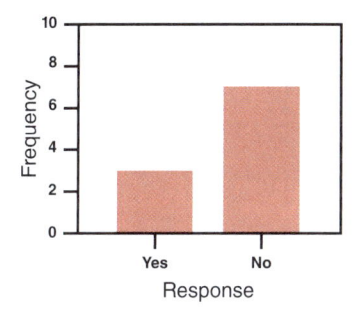

FIGURE 3.2
Bar chart showing responses to the question "Do you smoke?"

continuous data Data sets where any number of observations between any two adjacent categories on the *x*-axis are possible.

frequency histogram A graph where individual observations in a data set are represented on the *x*-axis and frequency is represented displayed on the *y*-axis. The height of the bar above the *x*-axis represents observation frequency. The bars touch one another to indicate the use of continuous data.

TABLE 3.3 Grouped frequency distribution of sleep time data

Interval	*f*
120–179	3
180–239	3
240–299	14
300–359	23
360–419	40
420–479	55
480–539	47
540–599	10
600–659	6
660–719	4
720–779	2

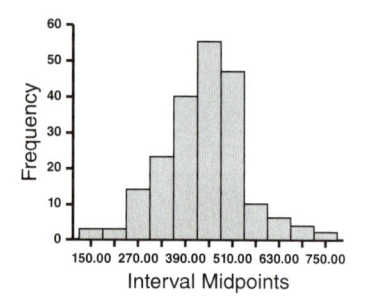

FIGURE 3.3
Frequency histogram of sleep duration data (in minutes).

stem-and-leaf graph A graph showing the frequency of individual observations in a data set. The individual observations are split into "leaves" (typically the portion of the number in the "ones" place) and "stems" (the remaining numbers in the remaining places).

histogram for our sleep data. Notice that we can save space on the *x*-axis by reporting the midpoint of each interval instead of both the lower limit and the upper limit.

STEM-AND-LEAF GRAPHS

We have a number of other options for displaying frequency. My personal favorite, the **stem-and-leaf graph**, was invented by John Wilder Tukey (1915–2000). Tukey, whose name we'll come across several more times throughout this book, has been an important contributor to statistics. He is also credited with coining the terms *bit* (which stands for binary digit) and *software*.

Stem-and-leaf graphs are particularly useful when we want a quick look at a grouped frequency distribution and a histogram all at the same time. First, we need a quick review of the way we write numbers and why we do so. Take a number like 2,475 (shown in Box 3.1): The place that each number occupies in the sequence has meaning. By tradition, the numeral just to the left of the decimal point represents the number of "ones" we have in our number (in the number 2,475 there are five ones). Moving left through the number, the next numerals represent the number of tens, the number of hundreds, the number of thousands, and so on. Therefore, in this number we have two thousands, four hundreds, seven tens, and five ones.

BOX 3.1 **Place values in Arabic numbers**

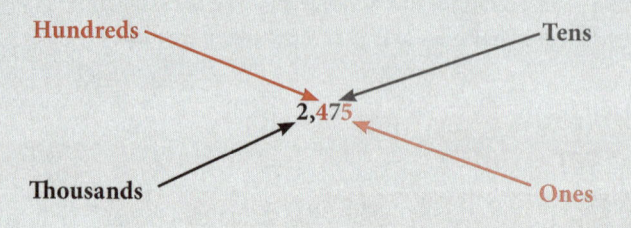

Hundreds · Tens · 2,475 · Thousands · Ones

Now consider the ungrouped frequency distribution shown in Table 3.4, which shows the data for 39 cardiac patients who were asked to walk on a treadmill for as long as they could without stopping. To describe these data with a stem-and-leaf graph, we'll consider the numerals in the tens place to be the *stems*, and we'll use the numerals in the ones place as the *leaves*. The number of leaves for each stem will represent the frequency of the observations: We'll have one leaf for every instance of that observation in the set.

First, we create two columns, labeling the first "Stem" and the second "Leaf." We write each individual stem (12 minutes and 15 minutes both have the same stem: a one) in the left-hand column. There were three people who stayed on the treadmill for 10 minutes. Each of these three observations has a stem of one and a leaf of zero. We'll write down three zeros (separated by commas) in the right-hand "Leaf" column next to the stem of one.

TABLE 3.4 Number of minutes on treadmill for 39 cardiac patients

Minutes on Treadmill	f
10	3
15	4
20	8
25	10
30	6
40	4
45	2
50	1
60	1

Stem	Leaf
1	0,0,0,5,5,5,5
2	0,0,0,0,0,0,0,0,5,5,5,5,5,5,5,5,5,5
3	0,0,0,0,0,0
4	0,0,0,0,5,5
5	0
6	0

FIGURE 3.4
Stem-and-leaf graph showing walking times (in minutes) of 39 cardiac patients.

There are four observations of 15 minutes: Each has a stem of one and a leaf of five. So, we'll write down four fives to the right of the zeros we just entered into the leaf column, and so on. Figure 3.4 illustrates the final stem-and-leaf graph.

Notice the advantages of a stem-and-leaf graph over both the histogram and the bar chart. First, we've grouped the data into sets of 10 (all observations between 0 and 9 will have a stem of zero, those between 10 and 19 will have a stem of one, and so on). Second, though we've grouped the data, we haven't lost any of the original information as we did when we used a traditional grouped frequency distribution. We can still tell that we had three people stay on the treadmill for 10 minutes (stem of one and leaf of zero) and two people who managed 45 minutes (stem of four and leaf of five).

In a traditional grouped frequency distribution, we would know how many observations were in a given interval, but we would not be able to pull a specific, individual observation out of any interval. Finally, the stem-and-leaf graph also serves as a histogram (albeit one lying on its side; see Figure 3.5): The length of each row indicates the frequency of observations within any one stem.

What do you suppose we'd do if our raw scores were larger than 100? When we were working with two-digit numbers, we decided that the numeral in the ones' place would be the stem and the numeral in the tens' place would be the leaf. With a three-, four-, or five-digit number, it isn't immediately apparent. The general rule is to consider the numeral in the ones place as the leaf and any remaining numerals as the stem.

1	0,0,0,5,5,5,5
2	0,0,0,0,0,0,0,0,5,5,5,5,5,5,5,5,5,5
3	0,0,0,0,0,0
4	0,0,0,0,5,5
5	0
6	0

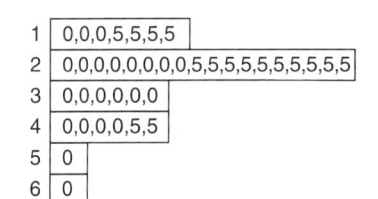

FIGURE 3.5
Stem-and-leaf graph shown as a frequency histogram.

FREQUENCY POLYGONS

In a **frequency polygon**, the bars of a histogram are replaced with dots (or some other symbol) and then joined with a straight line. The height of each dot above the *x*-axis again indicates the frequency of a particular observation. Like histograms,

frequency polygon A graph where individual observations in a data set are represented on the *x*-axis and the frequency of each observation is represented on the *y*-axis. The symbols are joined by a straight line to indicate the use of continuous data.

CHAPTER 3   A Picture Is Worth a Thousand Words: Creating and Interpreting Graphics

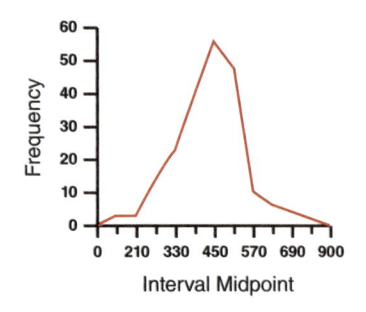

FIGURE 3.6
Frequency polygon showing sleep duration data (in minutes).

pie chart A graph representing relative frequency with a circle (the "pie") divided into sections (the "slices" of the pie). The relative size of each section illustrates the proportion of the quantity being measured.

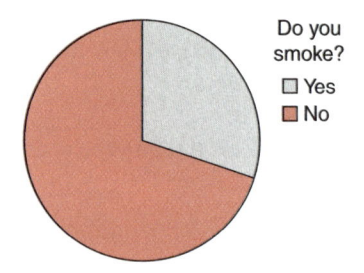

FIGURE 3.7
Pie chart showing the relative frequency of answers to the question "Do you smoke?"

frequency polygons indicate that the data are continuous. Figure 3.6 shows the sleep duration data from Table 3.3 as a frequency polygon. Just as the name "polygon" (literally a multisided figure) implies, the line drops to zero at either end of the distribution, connecting with the horizontal axis to create a many-sided shape.

PIE CHARTS

Pie charts are a favorite of the media for presenting frequency distributions quickly and easily. Pie charts are particularly useful in illustrating percentages (relative frequency) for nominal-level data. In a pie chart, the whole distribution is represented by the whole "pie," or circle. The relative frequency of any category is represented by a slice of the pie. The answer to our question about smoking behavior in students (Table 3.2) is shown as a pie chart in Figure 3.7. The slice of the pie representing the answer "yes" to our question "Do you smoke?" is smaller than the slice representing the answer "no," which was the most frequently occurring answer.

There are two ways to come up with your own pie chart should you find yourself without a computer or suitable graphics software. You can treat the "pie" as a 360-degree circle, then get yourself a compass and protractor to figure out the appropriate angle to correspond to each individual slice. The alternative, which does not require anything but some simple calculations (and maybe a calculator to help get them right), is to consider the "pie" as a clock face and then determine the number of minutes that should be devoted to each slice of the pie. Since this is the easier method, we'll focus on it.

We start by drawing a circle and then marking it off like the face of a clock, with 12 noon at the top, 6 o'clock opposite it at the bottom, 3 and 9 o'clock at the quarters, and the rest of the numbers as close to their appropriate positions as we can get. Remember that we're treating each slice of pie as a representation of the relative size of a category on our survey question. We'll need to convert from frequency (how many people answered our question with a "yes" and how many with a "no") to relative frequency, or percentage, and then figure out how much of the clock face each slice should take up. There are 60 minutes represented on this clock, so our whole distribution (all 10 respondents) should take up all 60 of the available minutes. Three out of the 10 respondents answered "yes," and 7 out of the 10 responded "no." So, we convert 3/10 and 7/10 into percentages, as shown in Box 3.2.

BOX 3.2 **Converting to relative frequency**

$$\left(\frac{3}{10}\right) \times 100 = 0.30 \times 100 = 30\%$$

$$\left(\frac{7}{10}\right) \times 100 = 0.70 \times 100 = 70\%$$

STATISTICS IN CONTEXT

The slice of the pie representing "yes" should take up 30% of the total, and the slice representing "no" should take up the remaining 70%. Now we need to figure out how many minutes 30% of 60 is. All we need to do to find that out is multiply 60 minutes by 0.30 (see Box 3.3).

BOX 3.3   **Converting relative frequency to minutes**

$$60 \times 0.30 = 18 \text{ min.}$$
$$60 \times 0.70 = 42 \text{ min.}$$

Our first slice should take up the first 18 minutes of the clock, and the second slice should take up the remaining 42 minutes. If we start by drawing a straight line from the center point of the clock face to 12 noon, and then draw a second line from the center to 18 minutes after the hour, we'll have our first slice of the pie (see Figure 3.8).

You have probably already figured out that if the first slice takes up 18 minutes, and if there are only two slices of the distribution to represent, then the second slice pretty much by default should take up the rest of the available space. You can check yourself (and figure out what to do with more complex data having more than just two categories) by calculating what should happen to the next slice of pie and making sure you're right. If the first slice ends at 18 minutes after the hour and the second slice should occupy 42 minutes out of the 60, where should the next slice start? Where should it end? Obviously, the next slice does not start back at 12 noon: If it did, our slices would overlap, and that would make this a very bad graph. So, we'll start the next slice where the first one left off, at 18 minutes after the hour.

The next slice ends 42 minutes later, so it should take up the rest of the available space (18 + 42 = 60) and we're back at 12 noon. I hope you see the advantage of using a clock face instead of a 360-degree circle: It makes figuring out the relative size of each slice of the pie very simple.

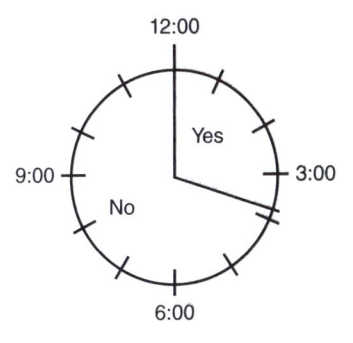

FIGURE 3.8
"Clock face" method for creating a pie chart.

CheckPoint

Indicate whether the following statements are *true* or *false*.

1. Pie charts are particularly useful in representing simple frequency. _____
2. Histograms are used to represent continuous data by having bars that touch. _____
3. William Playfair invented the stem-and-leaf chart. _____
4. The frequency of different varieties of flowers (rose, daisy, tulip, etc.) entered in a local flower show should be graphed using a frequency polygon. _____
5. The number of "yes" and "no" responses to a survey question is an example of discrete data. _____

Answers to these questions are found on page 111.

Think About It...

INTERPRETING GRAPHICS

In the summer of 2015, after years of contentious disagreement and debate, the U.S. Supreme Court announced that same-sex couples had a constitutional right to marry, and that the 13 states which had previously banned same-sex marriage by constitutional amendment would now have to reverse those bans.

The New York Times discussed the changes in the law and in public opinion using an interesting and very effective graphic, shown below.

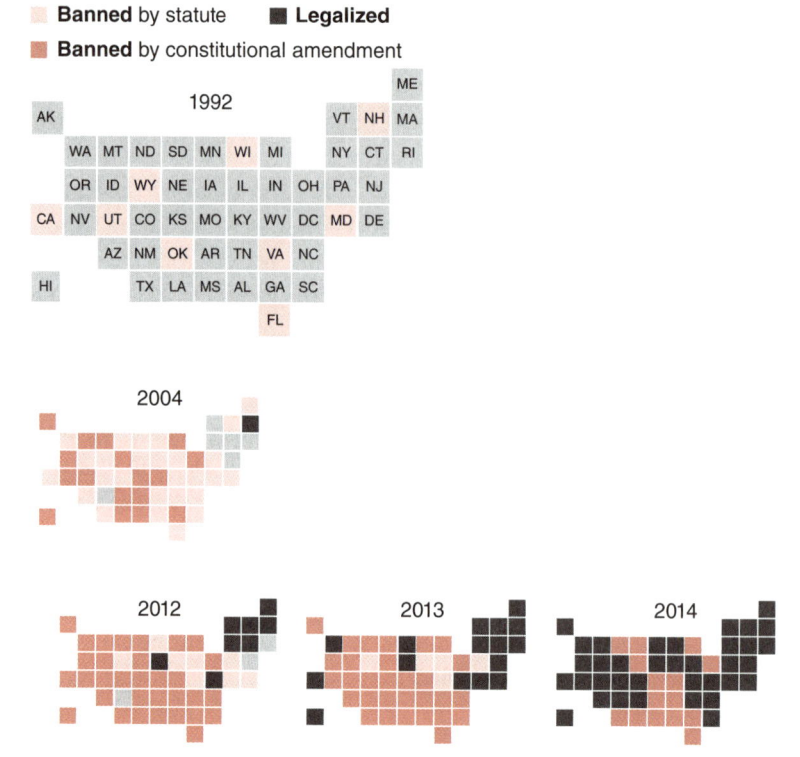

Think about the way this information is conveyed, and answer the following questions:

1. Which state was the first to legalize same-sex marriage?
2. How would you describe the pattern of change in the law/public opinion between 2012 and 2014?
3. The graph resembles a map of the United States, but only just. What is different about the map shown here compared to a conventional map of America?
4. In your opinion, do the differences in this map help or hinder your understanding of how the law and public opinion changed over time?

Answers to this exercise can be found on page 129.

STATISTICS IN CONTEXT

> **CheckPoint**
> Answers to questions on page 109
>
> 1. FALSE ☐ Pie charts are particularly useful in representing *relative frequency*.
> 2. TRUE ☐
> 3. FALSE ☐ *John Wilder Tukey* invented the stem-and-leaf chart.
> 4. FALSE ☐ The frequency of different varieties of flowers (rose, daisy, tulip, etc.) entered in a local flower show are discrete categories and should be graphed using a *bar chart*.
> 5. TRUE ☐
>
> SCORE: /5

OTHER GRAPHICS

All of the graphs that we've discussed in this chapter have been pictures of frequency distributions. On each of the graphs, we had frequency on the *y*-axis (showing how often measurements in the set occurred) and the measurements themselves on the *x*-axis.

Suppose, however, we wanted to do something else with our graph, like compare group performance on a test, or show how something changed over time, or how two sets of measurements were related to one another. There's at least one graph available to show each of these options.

GRAPHING MEANS

Let's consider means first. To show averages graphically, we would first put the independent variable on the *x*-axis; the *y*-axis would then show the mean of the dependent variable. Suppose we wanted to compare the average number of minutes of sleep during the week (Monday through Friday) with average sleep times on the weekend (Saturday and Sunday) for the students in our sleep study. Our independent variable (the variable we're interested in seeing the effect of) would be shown on the *x*-axis, and it would have two levels (weekday and weekend). The mean of our dependent variable (sleep duration) would be shown on the *y*-axis. The height of the bars above the *x*-axis would indicate the value of the mean for each group. Box 3.4 shows the hypothetical data as well as a bar chart illustration of the data.

BOX 3.4 **Graphing the mean**

Data
Mean sleep time on weekdays = 480 minutes
Mean sleep time on weekends = 300 minutes

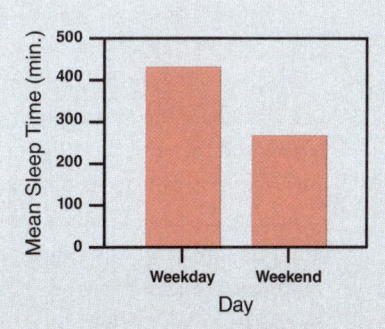

Notice that we're using a bar chart to show the means of these two groups. The bars don't touch because group is discrete data: There are no possible days of the week between "weekdays" and "weekends," and we want the graph to reflect this. We will explore ways to graph other statistics in Chapter 5.

GRAPHING RELATIONSHIPS

scatterplot A graph showing the relationship between two variables.

Another pioneer in the history of statistics developed a method for graphically representing the relationship between two variables. Sir Francis Galton (Charles Darwin's cousin) created a graph called the **scatterplot** to illustrate relationships between variables. Let's consider a classic: the height of parents and the height of their children. A number of studies have shown that children of tall parents tend to be tall themselves, while children of short parents tend to be short. There is a regular and predictable relationship between these two variables.

Box 3.5 shows some sample data and a scatterplot that illustrates the relationship between them. We've measured the heights of six moms and their adult daughters. Each pair of measurements represents a parent and her child. The scatterplot illustrates each pair of values in the set with a single dot for each pair.

BOX 3.5   Scatterplots

Data (Height in Inches)	
x (Height of Mother)	y (Height of Daughter)
70	72
64	60
66	64
60	6
72	69
62	63

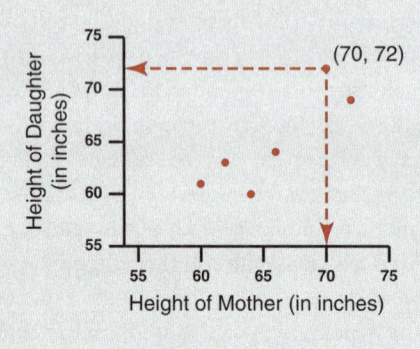

To create the scatterplot, we first choose one variable to be displayed on the x-axis and then put the other on the y-axis. It does not matter which one we choose for each axis as long as both variables are represented. Consider the first pair of values in the data: We have a mom who is 5 feet 10 inches tall (70 inches) and her adult daughter, who is an even 6 feet (72 inches) tall. The red dashed lines on the graph show the dot that represents this pair of values. The set of dots overall tends to rise from left to right, illustrating the pattern in the data: Short parents (to the left on the x-axis) tend to have short daughters (toward the bottom of the y-axis).

As we move to the right on the *x*-axis (taller moms), the dot showing the pair of values tends to move up the *y*-axis (taller daughters). We'll encounter scatterplots again in Chapter 13.

GRAPHING TIME

A **time-series graph**, as the name implies, shows how a variable changes over time. Typically, the variable on the *x*-axis is time (in whatever units we've used to measure it). The dependent variable (the variable we're interested in—the one we predict that time will have an effect on) is shown on the *y*-axis.

For example, suppose we want to track change in body temperature across time. We measure body temperature (in degrees Fahrenheit) at the same time (9:00 a.m.) each day for 39 consecutive days. Time is displayed on the *x*-axis and body temperature on the *y*-axis. Box 3.6 shows the time-series graph for these data. The person whose temperature is depicted here might have had a low-grade fever at the beginning of the measurement period. Temperature drops over time and then fluctuates around the normal body temperature of 98.6 degrees.

time-series graph A graph showing change in a quantitative variable over time

---

BOX 3.6 **Time-series graphs**

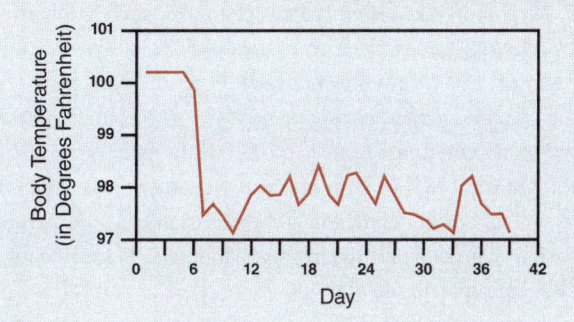

Most software programs will quickly and easily create graphic displays of your data. You can customize those graphs with color, three-dimensional images, background texture, pattern overlays, and so on. Don't go overboard with the bells and whistles, though. As Edward Tufte (2001) eloquently says,

> [S]tatistical graphics, just like statistical calculations, are only as good as what goes in to them. An ill-specified or preposterous model or a puny data set cannot be rescued by a graphic (or by calculation), no matter how clever or fancy. A silly theory means a silly graphic. (p. 114)

CHAPTER 3   A Picture Is Worth a Thousand Words: Creating and Interpreting Graphics   113

GRAPHICS IN CONTEXT: RULES FOR CREATING GOOD GRAPHS

Graphic representations of data provide the researcher with an easy-to-read, eye-catching way to summarize large amounts of information. However, graphs can also be used to misrepresent information. To prepare you to interpret a graph, consider the following general rules for creating a good graph. Violations of these rules are usually a tip-off that the graph may be being used to direct or misdirect your attention.

RULES FOR CREATING A GOOD GRAPH

1. Make sure that the graph is necessary. Is a graph the best way to display the data? Would a table or a description in the text make more sense or be easier to understand?
2. If you decide to use a graph, make sure that it is clear and complete. Your graph should display the data unambiguously: The point you are trying to illustrate should be immediately obvious to the reader. You should also include all of the relevant information so that the graph is self-explanatory. In a complete graph,
    a) the axes are labeled;
    b) the title and caption are clear, informative, and brief; and
    c) the legend is clear and informative.
3. The *x*-axis and *y*-axis should intersect at zero, or if the starting point is something other than zero, the start point should be clearly marked. In addition, sometimes it is a more efficient use of the space available in your graph to omit blocks of missing data. For example, if you give your subjects an intelligence test where scores can range from 1 to 100 and none of your subjects scores below 75, you might want to start your *x*-axis at something other than zero. Typically, two parallel lines are drawn through the axis to indicate that a block of possible observations has been omitted in this fashion. See Figure 3.9 for an example.
4. Use an appropriate scale. If you make the scale on the *x*-axis too broad, you risk "smearing" your graph across too wide an expanse and distorting your message. On the other hand, if you use a scale that isn't broad enough, you risk smashing your data into just a small corner of the graph and, again, distorting your message.
5. Don't add emphasis to your graph over and above what the data support. For example, consider Figure 3.10, showing the frequency of cigarette smoking in a sample of college students. Do you think the person who created this graph might have had an axe to grind? The data show that about the same number of people answered in the negative as in the affirmative (slightly more said "yes"), but the author clearly wanted to focus the reader's attention on the people who said "yes." By itself, this graph is attempting to mislead the reader into thinking that many more people smoke than don't smoke. Once again,

FIGURE 3.9
How to show a missing block of data on the *x*-axis.

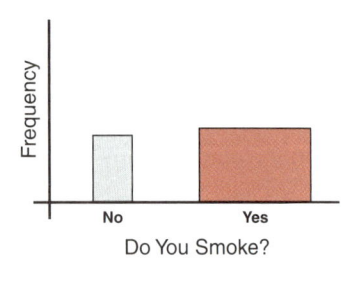

FIGURE 3.10
Exaggeration of a category.

the K.I.S.S. rule (*Keep It Short and Simple*) applies. If you keep your graphs simple, easy to read, and unambiguous, you will avoid being accused of trying to mislead people with your science.

SUMMARY

Graphs make visualizing patterns in your data easy to do. Data can be described as either discrete or continuous, and this distinction will help you determine which type of graph is most appropriate.

Bar charts (where bars are separated by a space) and pie charts work best with discrete data. Pie charts are most often used to represent relative frequency. Frequency histograms and frequency polygons are most appropriate for continuous data.

All four types of graphs show patterns in frequency data. However, other types of graphs are also effective. For example, the stem-and-leaf graph simultaneously provides you with a grouped frequency distribution and a frequency histogram. You can also create graphs to represent group data, time data, and the relationships between variables.

Graphs can be used to clarify your data, but they can also mislead the reader if they are used inappropriately.

TERMS YOU SHOULD KNOW

abscissa, p. 103
bar chart, p. 104
continuous data, p. 105
discrete data, p. 104
frequency histogram, p. 105
frequency polygon, p. 107

ordinate, p. 103
pie chart, p. 108
scatterplot, p. 112
stem-and-leaf graph, p. 106
time-series graph, p. 113

WRITING ASSIGNMENT

Edward Tufte is internationally known for his books on graphic design. In *The Visual Display of Quantitative Information* (2001, p. 40), he nominates one particular graph as possibly "the best statistical graph ever drawn." The tour de force in statistical graphs that he is referring to is a chart drawn by Charles Joseph Minard (1781–1870) showing, in brutal and chilling detail, the fate of Napoleon's army as they marched on Moscow during the campaign of 1812. One of your jobs as the author of a scientific article is to explain to your reader what you did and what you found, including explanations of any graphs you use to represent your data. Keep in mind that you never just throw a graph into your article, letting your reader figure out what it means. You *must* explain, clearly and effectively in the text that you write, what the graph means, why it's there, and what story it is telling the reader. Writing about graphs effectively, like anything else, takes practice. So, find a copy of this famous graph (it's available readily online) and examine it. Then, write a paragraph explaining the story this graph is telling.

A NOTE ON GRAPHING WITH STATISTICAL SOFTWARE

One of the greatest benefits of computers and their software (in my opinion) is that they make drawing a graph by hand obsolete. No more slaving over a hot lightbox with a finicky artist pen that for some reason isn't flowing smoothly, trying to draw a straight line or evenly spaced hash marks. SPSS, Excel, R, and other specialized programs will produce publication-quality graphics at the push of a button (or two). All it takes is for you to tell the computer what to draw.

You will find an introduction to graphing using SPSS and R at the end of this chapter. You may want to jump ahead to that section now so that you can apply some of the techniques presented there to the *Practice Problems*. Of course, if you would rather create your graphs the old-school way, then sharpen your colored pencils and dive right into the problems.

PRACTICE PROBLEMS

1. Students in an introductory psychology course were asked to participate in an experiment examining the relationship between quality of sleep and perceived stress. Each student in the class kept track of the quality of his or her sleep for seven consecutive days using a 5-point scale where 1 indicated very poor quality, 3 indicated average quality, and 5 indicated excellent sleep quality for the previous night. At the end of the week, each student donated a small hair sample consisting of approximately 20 to 30 strands of hair. The level of cortisol in the hair sample was obtained (in nanograms per gram of hair). Higher levels of cortisol indicate higher levels of stress. Table 3.5 shows the raw data from the 50 students in the study.

Table 3.5 Hair cortisol level (in ng/g) and sleep quality ratings (1–5)

ID	Sleep Quality	Cortisol	ID	Sleep Quality	Cortisol	ID	Sleep Quality	Cortisol	ID	Sleep Quality	Cortisol
1	3	136	16	4	103	31	3	79	46	1	35
2	3	74	17	4	31	32	2	105	47	5	166
3	3	143	18	4	81	33	3	145	48	5	83
4	4	137	19	4	97	34	1	48	49	1	165
5	1	107	20	2	151	35	1	194	50	2	51
6	2	136	21	3	140	36	2	174			
7	1	151	22	4	97	37	2	99			
8	1	177	23	4	109	38	2	28			
9	5	140	24	3	199	39	4	130			
10	3	61	25	3	69	40	4	148			
11	2	41	26	4	26	41	5	179			
12	5	189	27	1	112	42	5	189			
13	2	79	28	3	201	43	3	129			
14	2	43	29	2	34	44	2	80			
15	2	170	30	2	110	45	1	79			

STATISTICS IN CONTEXT

a. Create frequency distributions of both the sleep quality and cortisol levels.
b. Draw a graph showing the distribution of sleep quality in the sample.
c. Draw a graph showing the distribution of cortisol in the hair samples.

2. Table 3.6 shows the sleep data we worked with in Chapter 2. Construct a bar chart and a stem-and-leaf graph for the data. How will you handle the large gaps in the data in these two types of graphs?

Table 3.6 Sleep duration

Minutes	f	Minutes	f
120	22	430	1
275	1	440	1
330	1	445	1
340	1	450	5
360	1	455	1
375	1	470	1
380	1	480	1
385	1	495	1
390	1	520	1
405	3	525	1
420	2	620	1

3. Graph the data shown in Table 3.7 using an appropriate graph type. Then, answer the questions about the data that follow.
   a. Describe the main point of the graph. Construct a grouped frequency distribution of the data (use 10 beats per minute as your interval size), and graph the grouped frequency distribution. Which of these two graphs is clearer?
   b. Women typically have a faster heart rate than men do. Does the graph suggest that both men and women are in this data set? What aspect of the graph supports your conclusion?
   c. If you have not yet done so, create a stem-and-leaf graph of the data.

Table 3.7 Heart rate (in beats per minute)

ID	Heart Rate
1	65
2	67
3	85
4	88
5	85
6	56

Table 3.7   Continued

7	96
8	98
9	57
10	92
11	84
12	73
13	80
14	76
15	70
16	56
17	69
18	51
19	75
20	77
21	65
22	74
23	92
24	65
25	71

4. Graph the data shown in Table 3.8 using a frequency polygon. (Note that in order to do this, you'll have to create a frequency distribution of the data.)

Table 3.8   Weight (in grams) for 20 mice

ID	Weight	ID	Weight
1	19	11	21
2	17	12	21
3	20	13	20
4	20	14	22
5	16	15	21
6	19	16	17
7	19	17	18
8	16	18	17
9	19	19	19
10	19	20	18

Describe the typical weight of a mouse.

STATISTICS IN CONTEXT

5. Tramo et al. (1998) examined the surface area of the corpus callosum, a large bundle of nerve fibers that connects the left and right hemispheres of the brain, in newborn twins. The data in Table 3.9 are similar to what they found. Create a graph of the frequency distributions for the males and the females.

Table 3.9 Surface area of the corpus callosum (in cm^2) in 10 twin pairs

Pair	Males	Females
1	7.01	5.88
2	6.61	7.34
3	7.84	6.85
4	6.64	7.10
5	6.68	7.66
6	6.59	6.30
7	7.52	7.39
8	7.67	7.08
9	6.22	7.16
10	7.95	6.55

The mean surface area of the corpus callosum for males is 7.073 cm^2, and the mean surface area for females is 6.931 cm^2. Create a graph of these two means. Do you think these two groups of measurements are meaningfully different?

6. Suppose you participated in a study about the effects of stimulating the olfactory system on your ability to concentrate during a visual search task. Subjects are asked to complete a visual search task while smelling either an unpleasant or a pleasant odor. The time needed to complete the visual search task (in seconds) is recorded. Table 3.10 shows this hypothetical data.

Table 3.10 Time needed (in seconds) to complete visual search task while smelling unpleasant or pleasant odor

Unpleasant Odor	Pleasant Odor
65	80
55	70
82	63
42	77
48	75
55	71
71	58
93	80
83	71

Table 3.10 *Continued*

41	72
88	85
78	155
38	75
48	66
91	71
56	122
43	69
60	84
40	95
84	70
57	86
81	120
50	96
68	72

a. Graph the data using the graph type of your choice.
b. Describe the results of this study. Was there a difference in the time needed to complete the visual search in the two conditions (unpleasant odor vs. pleasant odor)?

7. Create a pie chart for the data shown in Table 3.11. If a convict is going to reoffend, what type of crime is he most likely to commit?

Table 3.11 Recidivism rates for 500 male convicts, by offense

Violent Offenses	Property Offenses	Drug Offenses	Public Disorder Offenses
50	100	335	15

8. Do you know what a "lug" is? Nope, it's not a big guy in a movie from the 1940s. It's a way to measure grape harvests in a vineyard. A "lug" is a large basket (1.5 feet by 3 feet by 6 inches) that is placed at the end of a row of grapevines. Workers move down each row of grapevines and dump the clumps of grapes they pick into the nearest lug. Lug counts are then used to measure grapevine production for a given harvest. Table 3.12 shows the lug count from a nine-year span at the *Château du Plonk* winery. Describe how the harvest changed over these nine years.

Table 3.12 Total number of lugs per harvest, 1983–1991

1983	1984	1985	1986	1987	1988	1989	1990	1991
534	552	401	266	514	377	170	502	940

Source: Data are from Chatterjee, Handcock, and Simonoff (1995).

9. How far can the average golfer on the PGA tour drive a golf ball? Here are the average driving distances for the top 25 drivers on the tour. If you go looking for this statistic, you'll quickly realize that it changes almost weekly as players vie for the top spot on this list in each tournament.
    a. Are there any outliers in this distribution?
    b. Create a graph of the data. (To make it easier, round the distances to the nearest whole number.)
    c. What is the length of a "typical" drive for these 25 players?

Table 3.13 Average driving distance (in yards) of the top 25 drivers on the PGA tour

Rank) Name	Distance	Rank) Name	Distance
1) B. Watson	315.1	14) K. Bradley	298.6
2) J. Lovemark	309.9	15) A. Cabrera	297.7
3) R. Garrigus	306.6	16) B. Gates	297.6
4) D. Johnson	305.9	17) S. Piercy	297.3
5) J. Kokrak	304.4	18) T. Matteson	297.0
6) C. Beljan	303.6	19) J. Teater	296.6
7) J. B. Holmes	302.8	20) S.-Y. Noh	296.5
8) K. Stanley	302.6	21) C. Hoffman	296.3
9) H. English	301.3	22) T. Kelly	296.1
10) J. Vegas	300.9	23) B. Jobe	295.9
11) G. DeLaet	300.8	24) M. Laird	295.4
12) R. Palmer	300.3	25) T. Gainey	295.2
13) G. Woodland	300.3		

10. Graph the driving distances in Table 3.13 using a stem-and-leaf graph.

11. Remember the question about smoking and mental health from Chapter 2? Table 3.14 shows the frequency distribution data. Create a pie chart for each group (schizophrenics and nonschizophrenics). Write a short description of the smoking behavior in these two groups of people.

Table 3.14 Frequency distribution of responses to the question "Do you smoke?" for schizophrenic and nonschizophrenic men

Schizophrenics	f	Nonschizophrenics	f
Yes	17	Yes	5
No	3	No	15

12. When is your birthday? Are you a midwinter baby, or were you born in the heat of summer? Several studies have suggested a relationship between the season of birth and psychiatric or neurological disorders. The relationship might be the result of the influence of the environment and seasonal characteristics like the amount of light available, temperature, and prevalence of infectious agents on brain development. Chotai and Wiseman (2005) surveyed almost 30,000 people from 67 countries (75% from Britain) and asked what month they were born in and how they would respond to a simple statement: "I am a lucky person." Their response options were on a scale from 1 to 5, where 1 = *strong disagreement* and 5 = *strong agreement* with the statement. Higher numbers on this scale indicate a stronger belief in luck. The data shown in Table 3.15 are similar to what Chotai and Wiseman found. Does belief in being lucky depend on the month in which a person was born?

Table 3.15 Response to the statement "I am a lucky person," by birth month

Birth Month	Belief in Luck	Birth Month	Belief in Luck	Birth Month	Belief in Luck
Jan.	3.22	May	3.33	Sep.	3.19
Feb.	3.22	Jun.	3.20	Oct.	3.18
Mar.	3.28	Jul.	3.24	Nov.	3.20
Apr.	3.30	Aug.	3.21	Dec.	3.20

13. Let's divide the data in Table 3.15 into two sections: Fall/Winter (September through February) and Spring/Summer (March through August). We will calculate the average "Belief in Luck" score for each section. To do this, we need to add up the scores for each month in a given section and then divide the total by the number of months in that section.

Fall/Winter		Spring/Summer	
Birth Month	Belief in Luck	Birth Month	Belief in Luck
Sep.	3.19	Mar.	3.28
Oct.	3.18	Apr.	3.30
Nov.	3.20	May	3.33
Dec.	3.20	Jun.	3.20
Jan.	3.22	Jul.	3.24
Feb.	3.22	Aug.	3.21
Total	19.21	Total	19.56
Average	**3.20**	**Average**	**3.26**

Now plot the means using a bar chart. Does this new graph change in your interpretation of the data?

14. Professor Richard Wiseman has been studying belief in luck for quite a while. In 2003, Professor Wiseman surveyed residents of Britain about their superstitions. Table 3.16 shows results similar to what Professor Wiseman found.

a. Notice that the total number of responses is more than twice the total number of people surveyed. Why? What does this tell you about belief in superstition?
b. Describe superstitious behavior in British citizens.

Table 3.16 Superstitious beliefs reported by 100 individuals in Britain

Belief	Number of Respondents Who Believed in a Given Superstitious Behavior
Touching wood	74
Crossing fingers	65
Avoiding ladders	50
Smashing mirrors	39
Carrying a lucky charm	28
Avoiding the number 13	26

15. The Hermann Grid is a visual illusion created by Ludimar Hermann in 1870. Hermann found that when observers looked at a grid made of black squares on a white background, they saw an "illusory dot"—a fuzzy gray dot that actually was not there—at the intersection of the white lines. One of my students had a great idea: She wanted to know if the color of the Hermann grid affected the visual illusion (Blatchley & Moses, 2012). Does the illusory dot disappear if the grid is something other than black and white? She presented 50 participants with Hermann grids in six colors—black, red, yellow, green, blue, and purple, all on white backgrounds—and asked the participants if they saw the dot. The data in Table 3.17 are representative of what we found.

Table 3.17 Perception of the "illusory dot" as a function of color

Color of the Grid	Number of "Yes" Responses
Black	50
Red	43
Yellow	10
Green	46
Blue	48
Purple	47

16. How well do you perceive the passage of time? Students in my senior research seminar wanted to know if the perception of time was affected by experience with all of the labor-saving devices we're so accustomed to using. We asked participants ranging in age from 6 to 71 years to complete a short survey measuring their daily computer use and then to judge the duration of a series of time intervals (3 to 27 seconds in length) without using a counting method. "Computer Use" (CU) was measured on a scale from 9 to 74: The higher the number, the more time the participant spent using the computer on a daily basis. Accuracy in the ability to estimate time intervals was measured by subtracting the estimated duration of the time interval from the actual time interval: the smaller this

error, the more accurate the time estimation. The data in Table 3.18 are representative of what we found (Blatchley et al., 2007). What can you say about computer use in these age groups? What can you say about accuracy in time estimation as a function of age?

Table 3.18 Mean CU scores and mean time-estimation accuracy scores for 40 participants

Age	Mean CU Score	Mean Estimation Score
6–20 years	41.88	6.38
21–49 years	52.42	4.18
≥50 years	41.75	4.84

17. Create a scatterplot for the data shown in Table 3.19. The data come from the time-estimation study referenced in question 16. Each row represents two measurements (age and time-estimation error) taken from 15 of our participants. What can you say about the relationship between age and accuracy in time estimation?

Table 3.19 Age and time-estimation accuracy for 15 participants

Age (in Years)	Error in Time Estimation
13	4.29
20	2.15
35	1.38
17	6.02
20	2.01
16	4.29
22	1.35
12	7.89
11	2.08
31	1.06
19	1.70
6	8.87
11	3.44
6	9.72
7	5.28

18. Create stem-and-leaf graphs for the grip strength data shown in Table 2.22 (in question 15 of the Chapter 2 practice problems). We want to compare grip strength in "neutral" light and in blue light, so you will need a way to distinguish between the two conditions in your graph (or graphs).

19. Researchers have found that eye color and risk of alcoholism are related. Jonathan Bassett and James Dabbs (2001) speculated that because light-eyed people are generally less responsive to drugs in general, they might drink more alcohol before they felt its effects. This increased exposure might make light-eyed people more likely to become

dependent on alcohol. So, they examined a very large "archival" sample—a set of records that been collected in previous research—and looked at this "old" data in a new way. First, they went to the records of the Georgia Board of Pardons and Paroles and looked at the data on eye color and history of problems with alcohol abuse for a set of Caucasian male inmates of the Georgia Prison System. They separated the sample into two groups: "light-eyed" (blue, gray, green, and hazel eyes) and "dark-eyed" (brown and black eyes). They then counted the number of inmates in each group who had been identified in the prison records as having a history of problems with alcohol. The data in Table 3.20 are representative of what they found. Use the data to answer the questions that follow.

Table 3.20 Number of male inmates with and without a history of alcohol abuse problems, by eye color

History of Alcohol Abuse Problems	Light-colored Eyes	Dark-colored Eyes	Row Totals
With	2,633	1,797	
Without	3,637	2,933	
Column totals			

   a. Why did Bassett and Dabbs look at the records of only Caucasian inmates?
   b. How many inmate records, in total, were examined?
   c. How many inmates with light-colored eyes were in the sample?
   d. How many inmates with dark-colored eyes were in the sample?
   e. How many inmates had a history of alcohol abuse?
   f. Construct two pie charts, one for each group (light-eyed and dark-eyed), showing the percentage of inmates in each group with and without a history of alcohol abuse.
   g. Is there evidence of a relationship between eye color and alcohol dependency?

20. Let's stick with Bassett and Dabbs (2001) for a moment. They also examined archival records from the Bureau of Labor Statistics (collected in 1979) and wondered if women with light-colored eyes consumed more alcohol overall than did women with dark-colored eyes. The data in Table 3.21 are representative of the results these authors found.
    a. Use the data in the table to construct a bar graph displaying the mean number of drinks consumed "last week" for light- and dark-eyed women in this sample.
    b. Graph the mean number of drinks consumed in the previous month for both groups.
    c. Graph the mean number of days in the previous month for which the women reported drinking more than six drinks in that 24-hour period.
    d. In your opinion, is it easier to understand the data when presented as numbers in a table or when presented graphically?

Table 3.21 Average alcohol consumption of light-eyed and dark-eyed females

	Light-eyed	Dark-eyed
Mean number of drinks last week	1.39	1.29
Mean number of drinks last month	5.78	4.91
Mean number of days in last month drank more than 6 drinks	1.02	0.75

21. All of the data collected by Bassett and Dabbs (2001) were archival. Consider this method for collecting data as you answer the following questions:
    a. Do you see any problems with using archival data? Would it be better to go directly to the source and collect data via interview or questionnaire from people today rather than to rely on data from the archives?
    b. The data in the second study regarding number of drinks consumed were obtained via "self-report." The women were asked about their drinking habits, and they then wrote their answers down. Do you see any problems with this method of data collection? Should the data, graphic or numeric, be looked at with suspicion or caution because of the way the data were collected?

22. The *Morbidity and Mortality Weekly Report* (*MMWR*) from the Centers for Disease Control and Prevention describes an unexplained acute neurological illness that has been affecting young children in the Muzaffarpur district of Bihar state in India. Reports of this strange, and unfortunately often fatal, illness began in 1995 and continue to this day. In an effort to determine the cause of this disorder, researchers at the CDC began tracking the rates of admission for this disease across the course of a year. Figure 3.11 comes from the *MMWR* for January 30, 2015 (Shrivastava et al., 2015). Use the graph to answer the following questions:
    a. Is there a relationship between number of patients admitted with symptoms of this illness and month? What does the relationship look like?
    b. Is there a relationship between the number of deaths from this illness and time of year?
    c. Speculate about why rates of admission and time of year might be related. Is there something happening in June in India that might be related to outbreaks of this illness?

23. In the mid-1800s, John Snow, now considered the "Father of Epidemiology," studied a deadly outbreak of cholera. At the time, cholera was thought to be caused by "bad air"—apparently, London smelled bad and doctors thought this bad smell caused disease. Snow suspected there might be another cause, and he was determined to find it in order to prevent the disease from occurring in the future. He created a new kind of graph to see if cholera might be caused by something in the public water supply, which was

FIGURE 3.11
Epidemiological curve showing the number of patients admitted to two referral hospitals with unexplained acute neurological illness, by date of admission, in Muzaffarpur, India, from May 26 to July 17, 2014.

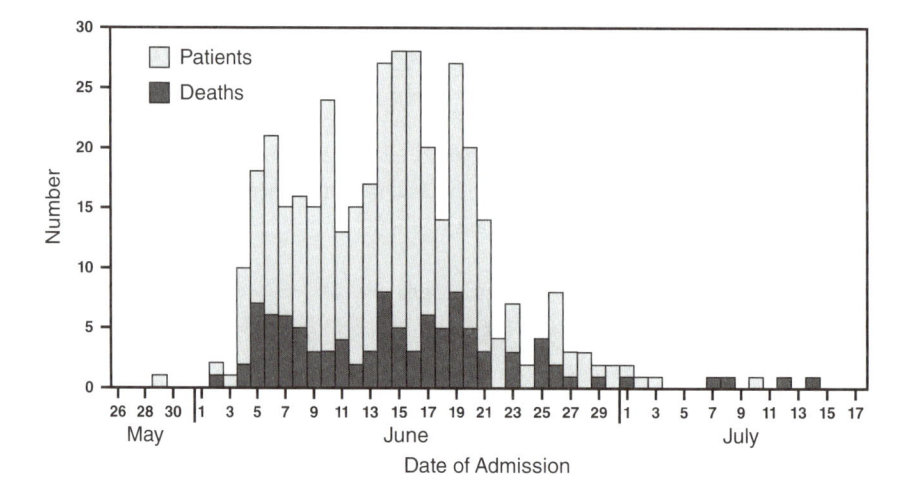

STATISTICS IN CONTEXT

provided to neighborhoods in London by a series of public pumps. Snow focused on the neighborhood where the outbreak started and watched the behavior of the women in the neighborhood, paying attention to the specific pump where most households got their water for the day. Snow then marked each residence on a map of the neighborhood and indicated the number of deaths in that building with a dot—one dot for each death in a given house. Mark Monmonier is one of many cartographers who have redrawn Snow's resulting graph (now called a "spot map") to show how cases of illness are distributed across a given region. His version is presented below in Figure 3.12.
   a. Is there a relationship between the Broad Street Pump and deaths from cholera?
   b. What aspects of Snow's spot map led you to your conclusion?

24. After collecting the data on pump usage and deaths from cholera, Snow decided that the problem was localized to one specific pump: the one on Broad Street, at the center of the outbreak. It took some effort to convince the powers-that-be, but eventually, Snow got permission to shut the pump down. On September 8, 1854, Snow had the handle removed

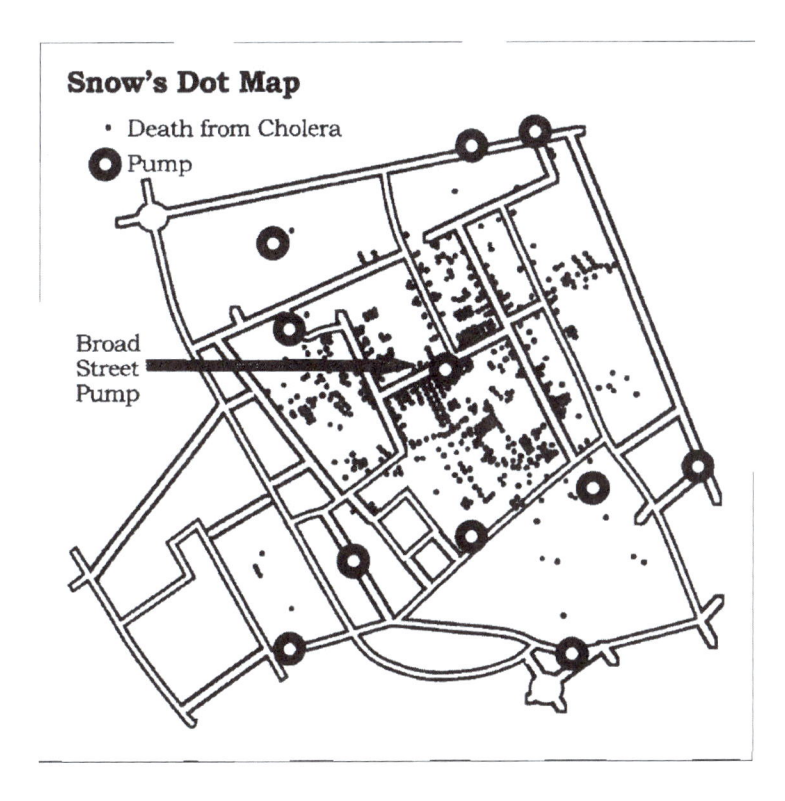

FIGURE 3.12
Spot map of deaths from cholera in the Golden Square area of London, 1854. Redrawn from the original.

from the Broad Street pump. Figure 3.13 is a graph created by Edward Tufte to show the number of deaths from cholera before and after the Broad Street pump was disabled.
   a. Did removal of the pump handle work?
   b. Is the change in number of deaths after removal of the pump handle enough for you (or Mr. Snow) to conclude that cholera is caused by something in the water?
   c. What would you do next in your hunt for the cause of this disease?

CHAPTER 3 A Picture Is Worth a Thousand Words: Creating and Interpreting Graphics

FIGURE 3.13
Deaths from cholera during 1854 outbreak in London.

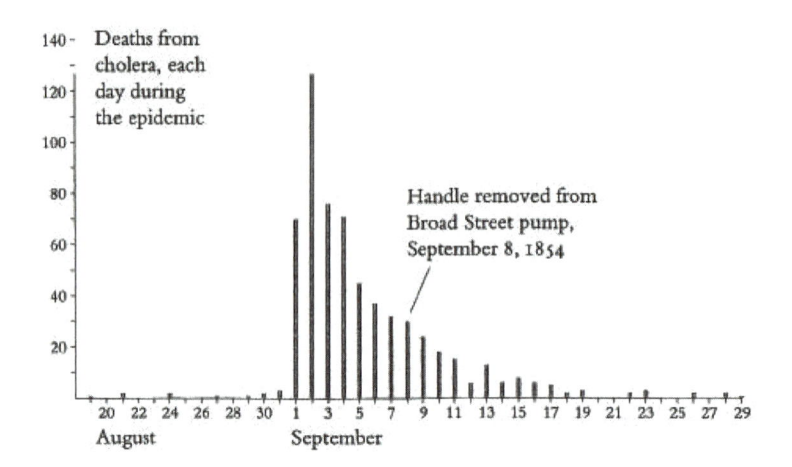

25. Complete the crossword puzzle shown below.

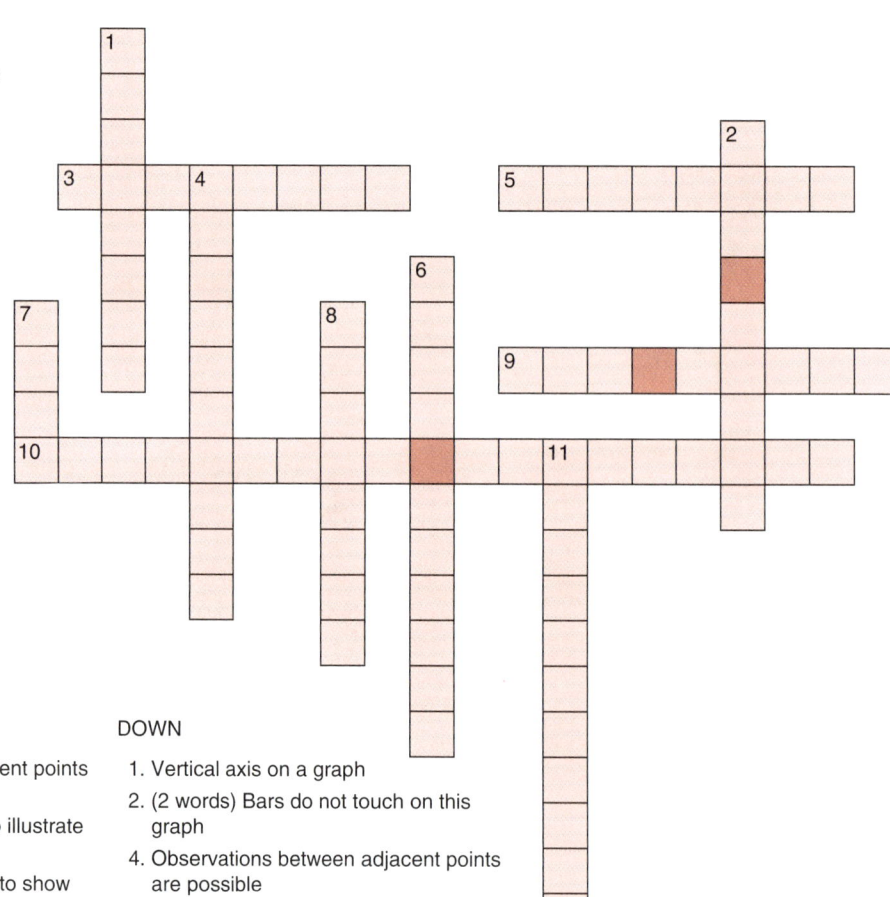

ACROSS

3. Observations between adjacent points not possible
5. First person to use graphs to illustrate data
9. (2 words) Graph best suited to show relative frequency
10. (2 words) Bars touch on this graph

DOWN

1. Vertical axis on a graph
2. (2 words) Bars do not touch on this graph
4. Observations between adjacent points are possible
6. (2 words) Graph showing change in a quantitative variable over time
7. Observation in the "one's" place
8. Horizontal axis on a graph
11. Graph showing relationship between two variables

128   STATISTICS IN CONTEXT

Think About It...

INTERPRETING GRAPHICS

SOLUTIONS

1. Which state was the first to legalize same-sex marriage? **Massachusetts.**
2. Describe the pattern of change in the law/public opinion between 2012 and 2014. **This was a period of rapid change in both public opinion and the law. In 2012, the majority of states either had constitutional amendments banning same-sex marriage or banned it by statute. Only nine states allowed legal same-sex marriage, and these states tended to be concentrated in the northeastern United States. By the following year, same-sex marriage had been legalized in several more states, and by 2014, the majority had shifted. In that year, most states allowed same-sex marriage, and only 13 states still had constitutional bans on it.**
3. The graph resembles a map of the United States, but only just. What is different about the map shown here compared to a conventional map of America? **Each state is represented by a box, and all the boxes are the same size.**
4. In your opinion, do the differences in this map help or hinder you understanding of how the law and public opinion changed over time? **In making all the states the same size, the creators of this graph eliminated a potential distractor variable and gave equal weight to very large state (Texas, for example) and very small ones (say, Rhode Island). Since the size of each individual state is irrelevant to this data set, making all states equal in size helps clarify the patterns of change over time. In my opinion, this helps make the change in public opinion and in state law over time easier to see. This method of graphing also emphasizes the pattern so often seen in public opinion in the United States—change starts (generally) on the coasts and moves inland. It certainly did here.**

REFERENCES

Bassett, J. F., & Dabbs, J. M. (2001). Eye color predicts alcohol use in two archival samples. *Personality and Individual Differences, 31*(4), 535–539.

Blatchley, B., & Moses, H. (2012). Color and saturation effects on perception: The Hermann Grid. *North American Journal of Psychology, 14*(2), 257–267.

Billion dollar weather (Sept. 13, 2017). Retrieved from http://news.nationalgeographic.com/2017/09/hurricane-irma-harvey-damage-graphic.

Blatchley, B., Dixon, R., Purvis, A., Slack, J., Thomas, T., Weber, N., & Wiley, C. (2007). Computer use and the perception of time. *North American Journal of Psychology, 9*(1), 131–142.

Chatterjee, S., Handcock, M. S., & Simonoff, J. S. (1995). *A casebook for a first course in statistics and data analysis.* Hoboken, NJ: John Wiley and Sons.

Chotai, J., & Wiseman, R. (2005) Born lucky? The relationship between feeling lucky and month of birth. *Personality and Individual Differences, 39*, 1451–1460.

Playfair, W. (1801). *The commercial and political atlas and statistical breviary* (Reprint of the third edition, 2005, Edited and Introduced by Howard Wainer and Ian Spence). New York, NY: Cambridge University Press.

Shrivastava, A., et al. (2015, Jan. 30). Outbreaks of unexplained neurologic illness—Muzaffarpur, India, 2013–2014. *Morbidity and Mortality Weekly Report, 64*(03), 49–53.

Spence, I. (2000b). The invention and use of statistical charts. *Journal de la Société Française de Statistique, 141*, 79-81.

Spence, I., & Wainer, H. (1997). William Playfair: A daring worthless fellow. *Chance, 10*, 31-34.

Tramo, M. J., Loftus, W. C., Stukel, T. A., Green, R. L., Weaver, J. B., & Gazzaniga, M. S. (1998). Brain size, head size and intelligence quotient in monozygotic twins. *Neurology, 50*(5), 1246–1252.

Tufte, E. R. (2001). *The visual display of quantitative information* (2nd ed.). Cheshire, CT: Graphics Press.

Wiseman, R. (2003). Luck factor. Retrieved from www.luckfactor.co.uk/survey.

GRAPHING WITH SPSS AND R

Graphing with SPSS

The data we will be using for the SPSS and R graphing tutorials were collected by David Condon and William Revelle as part of their research on the structure of personality constructs:

Condon, David M.; Revelle, William, 2015, "Selected personality data from the SAPA-Project: 08 Dec 2013 to 26 Jul 2014", http://dx.doi.org/10.7910/DVN/SD7SVE, Harvard Dataverse, V2

This data set contains demographic information captured with 20 variables from 23,681 people who participated in a personality study. We will be looking at just a few of the variables to give ourselves some practice using SPSS and R to represent the data in a visually compelling, easily understandable way.

Bar Chart

For this task, we are interested only in participants from Australia, Canada, Great Britain, New Zealand, and the United States.

STEP 1: Click **Data**, then **Select Cases**. Click on the **If** button, and specify the countries we want to use. Click **Continue**.

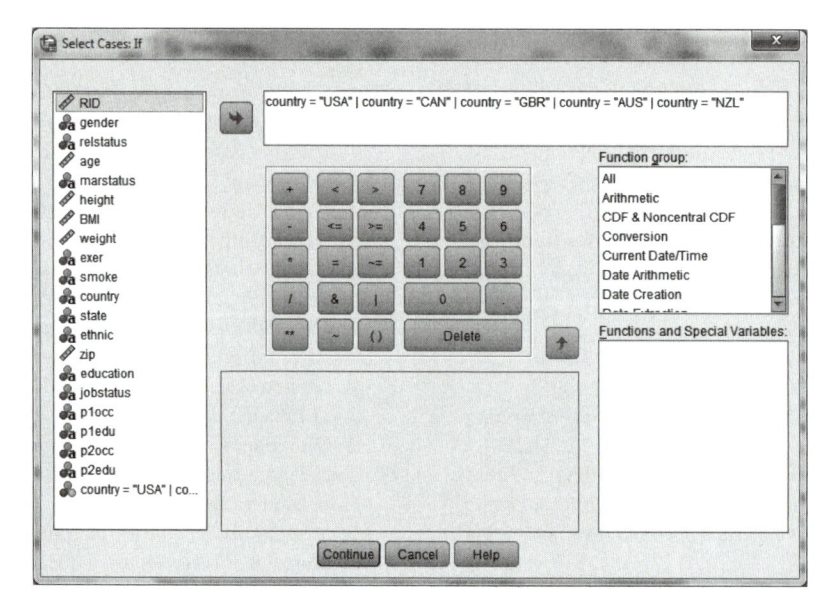

Now that we have the information we want, we are going to see how often participants in the study exercise.

STEP 2: Select **Graphs**, then **Legacy Dialogs**, then **Bar**.

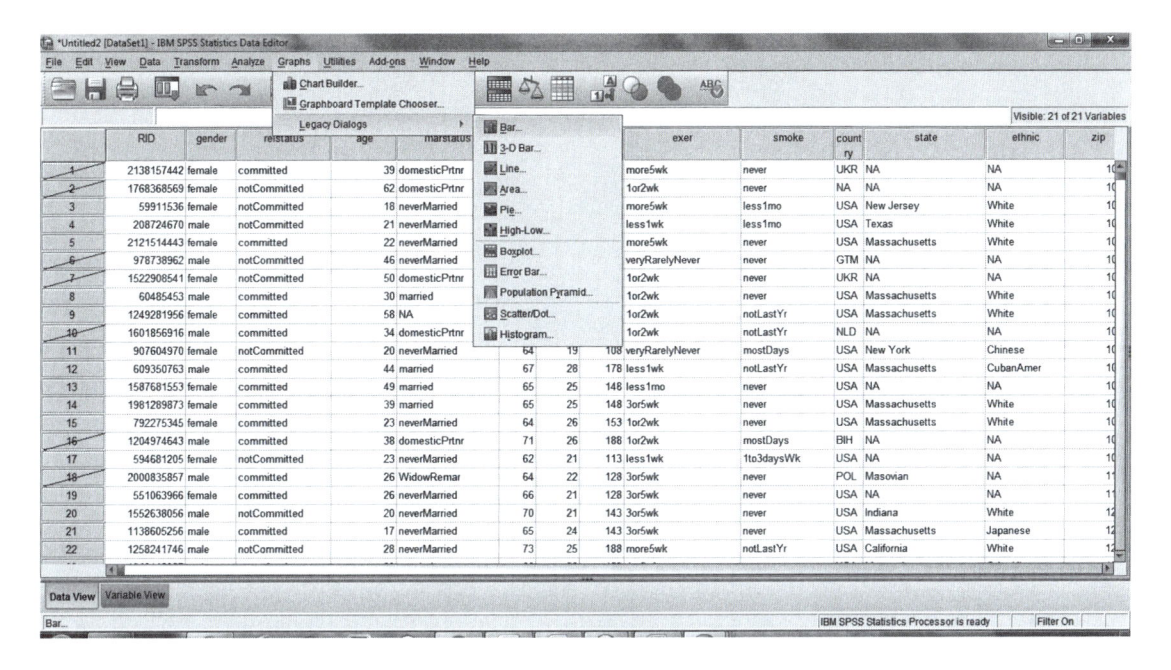

STEP 3: In the **Bar Charts** dialogue box, click **Simple,** and select **Summaries for groups of cases**. Click **Define**.

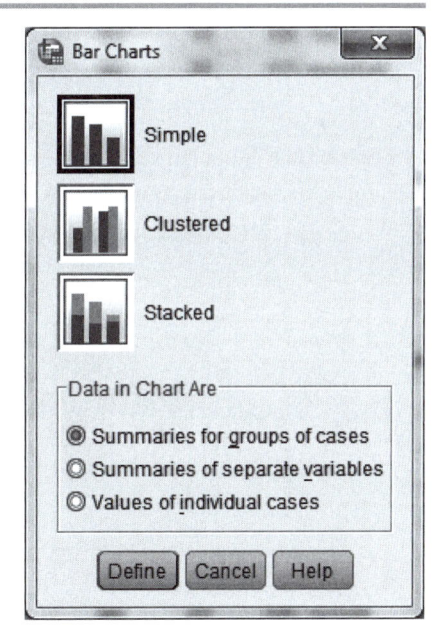

CHAPTER 3   A Picture Is Worth a Thousand Words: Creating and Interpreting Graphics   131

STEP 4: In the **Define Simple Bar: Summaries for Groups of Cases** dialogue box, we will set the bars to represent **% of cases** and then specify the **exer** variable as the **Category Axis**. Click **OK**.

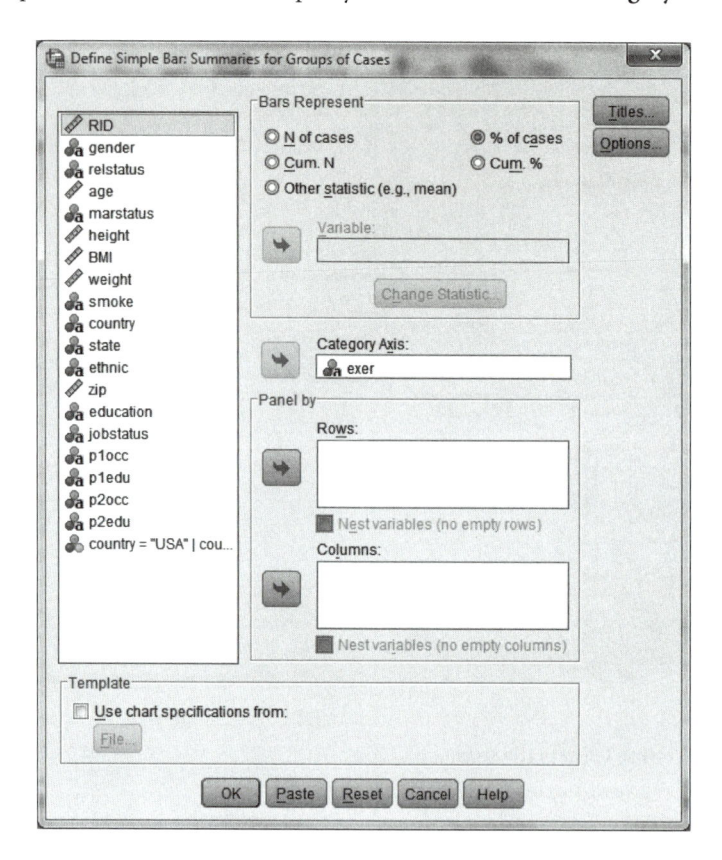

This produces our bar chart in the output window. By double-clicking on the plot, we can customize the various parts using the Chart Editor to change colors, scale the axes, and so on.

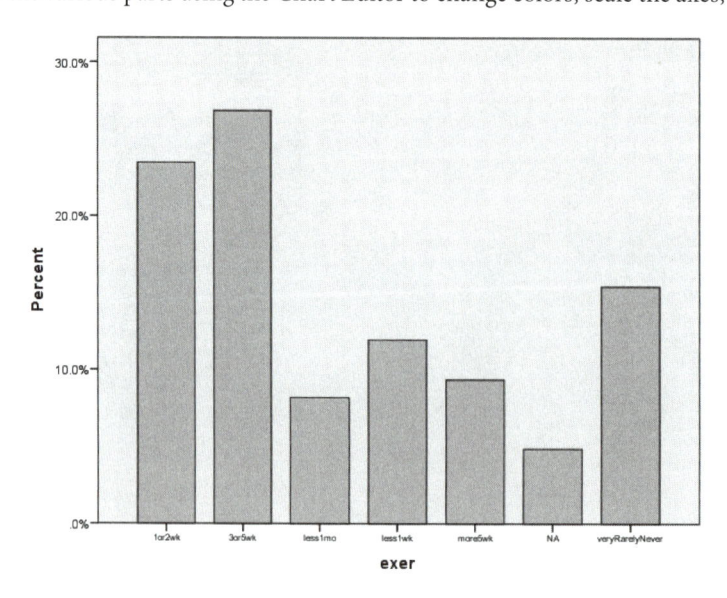

132   STATISTICS IN CONTEXT

Frequency Polygon

Now we will look at the distribution of the age variable using all 23,681 cases.

STEP 1: Return to the **Data, Select Cases** dialogue box. Choose **All Cases**, and click **OK**. On the main menu bar, select **Graphs** and then **Chart Builder**.

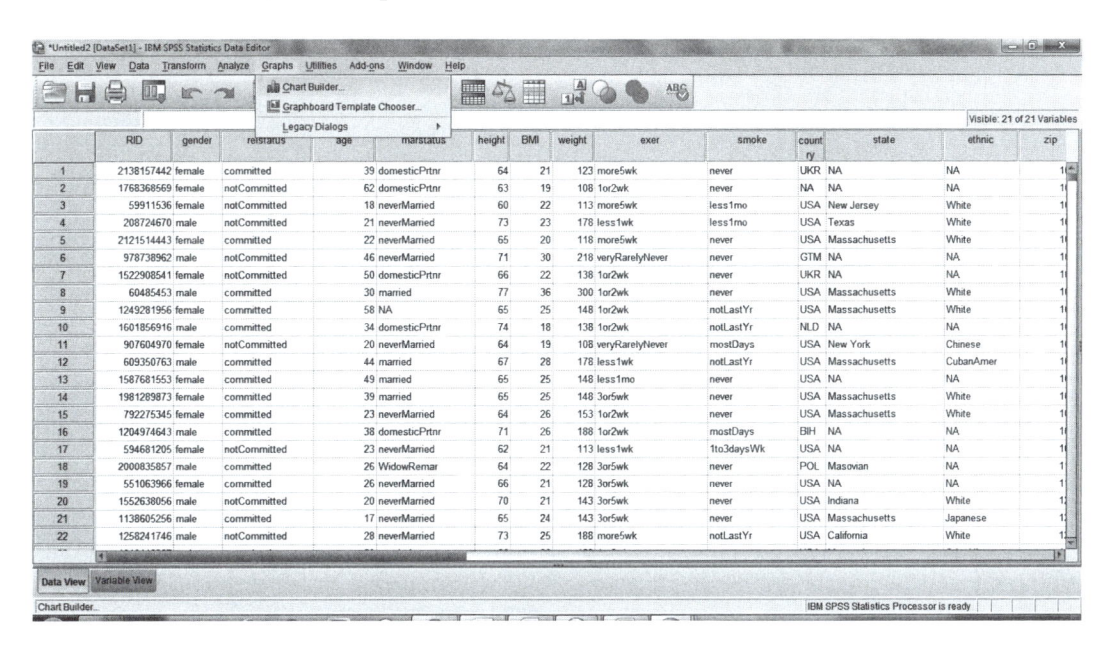

STEP 2: In the **Chart Builder** dialogue box, select the **Gallery** tab, and then select **Histogram** from the list on the left-hand side. Double-click on the **Frequency Polygon** icon to load the preview in the **Chart Preview** area of the dialogue box.

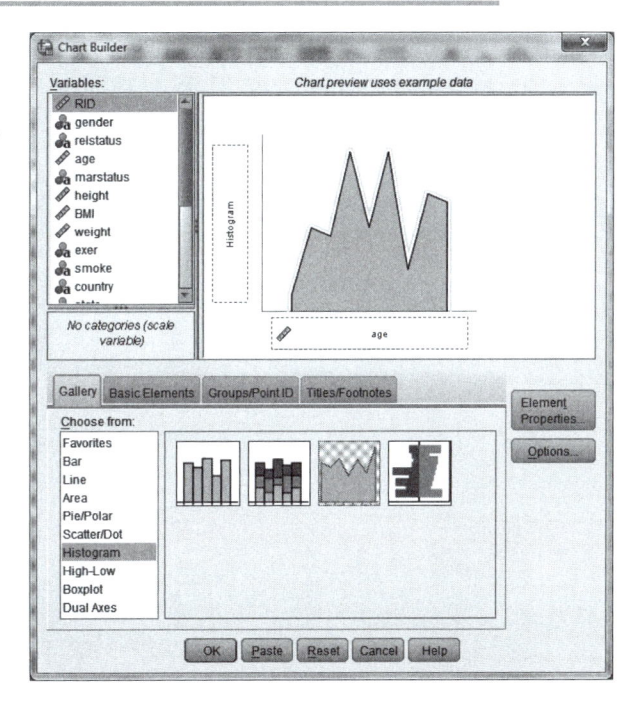

CHAPTER 3   A Picture Is Worth a Thousand Words: Creating and Interpreting Graphics   133

STEP 3: Next, click and drag the **age** variable to the *x*-axis. **Histogram** is the default for the *y*-axis. Click **OK**.

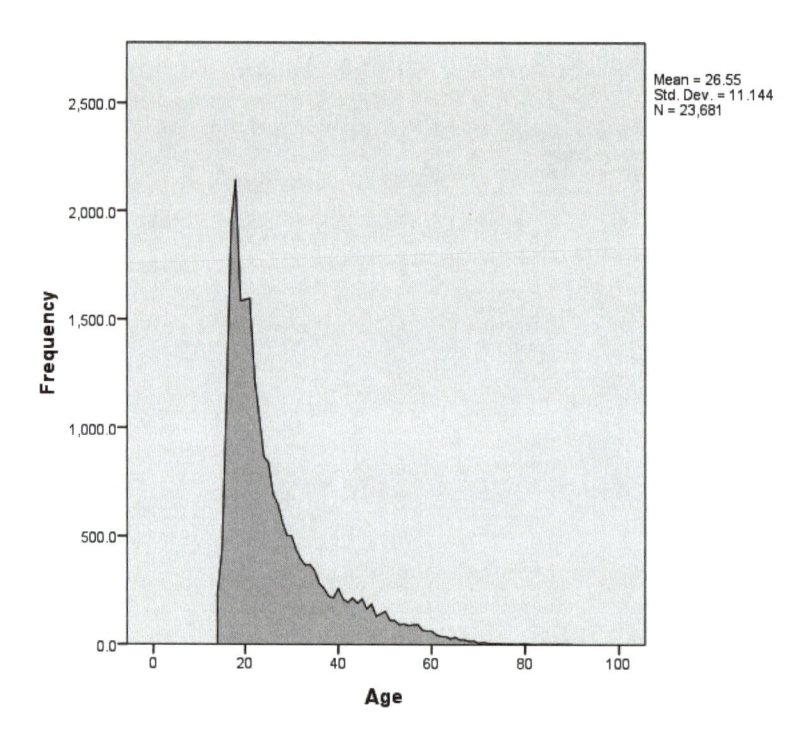

Histogram

A histogram presents similar information as the frequency polygon plot. We'll use the same variable (age) for this visualization.

STEP 1: On the main menu bar, select **Graphs**, then **Legacy Dialogs**, and then **Histogram**.

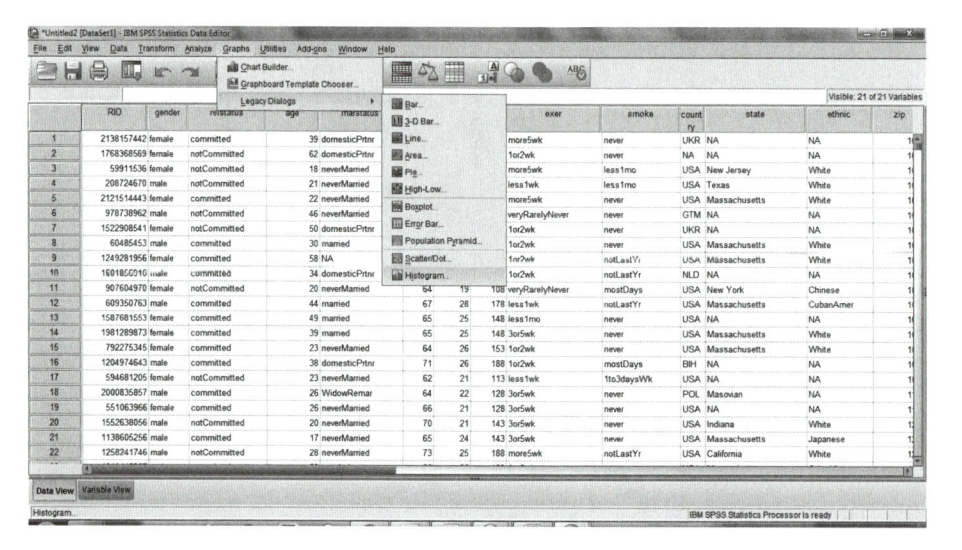

134 STATISTICS IN CONTEXT

STEP 2: In the **Histogram** dialogue box, select **age** and add it to the **Variable** input box. Click **OK**.

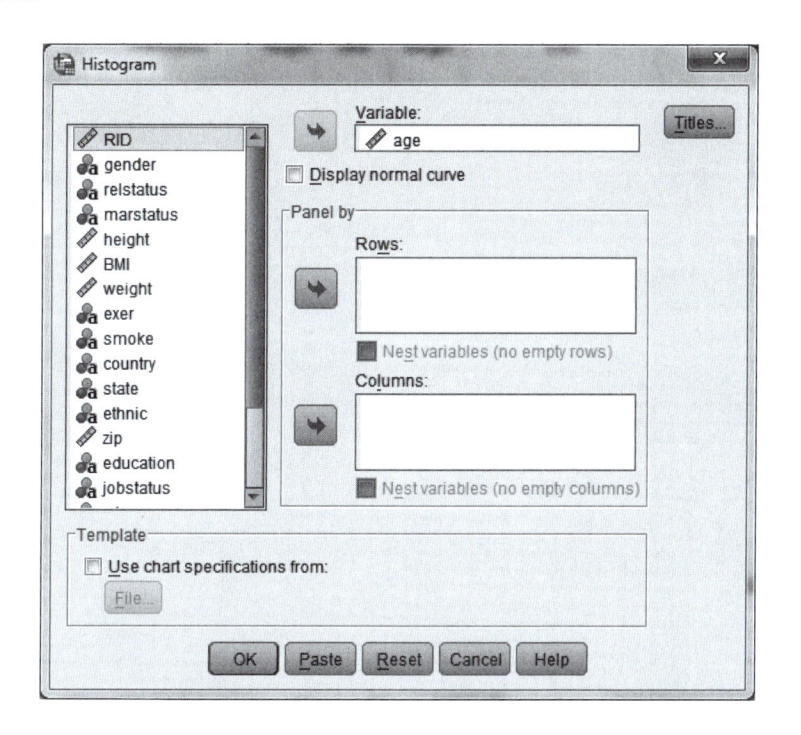

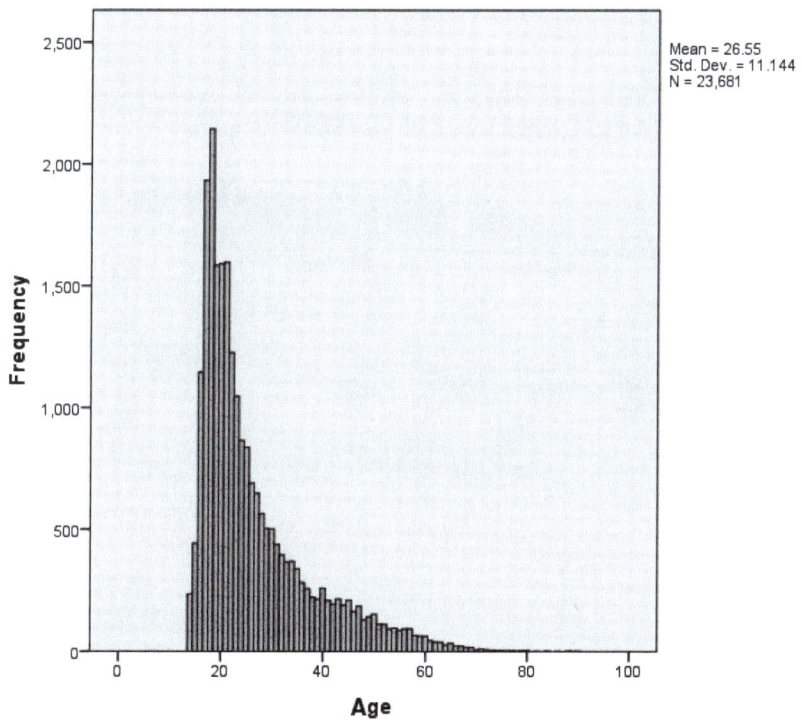

Scatterplot

We are interested in the relationship between height and weight in our sample. For this example, we've selected 200 cases at random to investigate.

STEP 1: Start by selecting **Graphs** from the main menu bar. Click **Legacy Dialogs** and then **Scatter/Dot**.

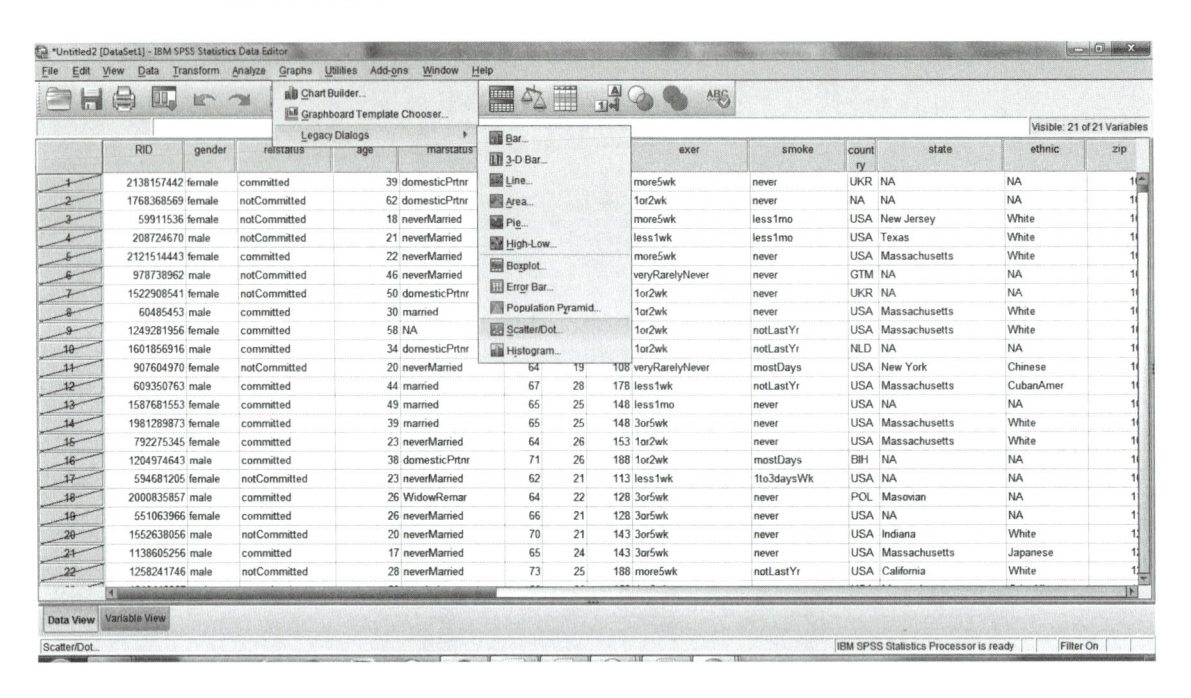

STEP 2: In the **Scatter/Dot** dialogue box, choose **Simple Scatter**, and then click **Define**.

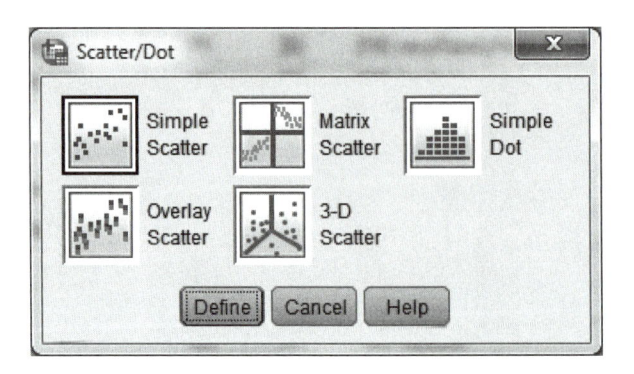

136 **STATISTICS IN CONTEXT**

STEP 3: In the **Simple Scatterplot** dialogue box, specify **height** in the **Y Axis** input box and **weight** in the **X Axis** input box. Then, click **OK**.

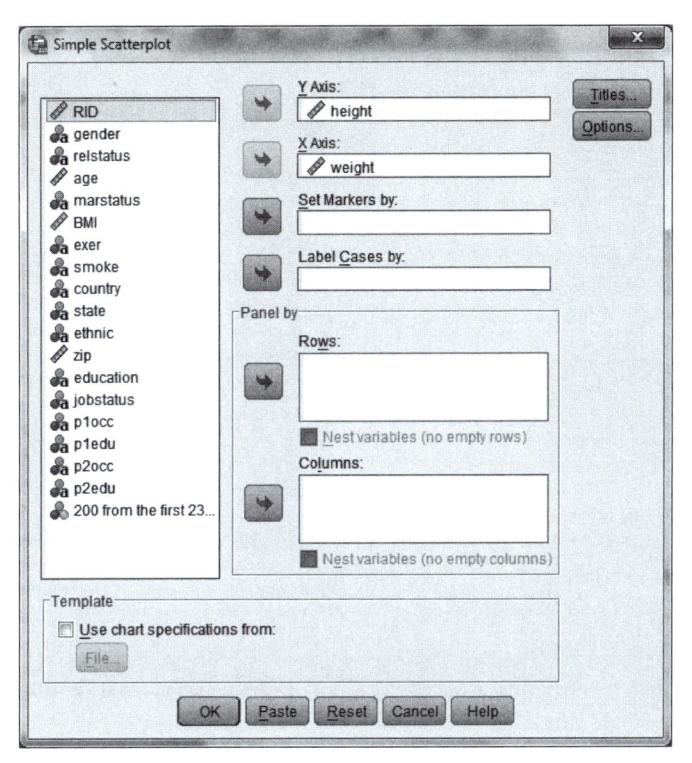

This produces a basic scatterplot.

Graph

[DataSet1]

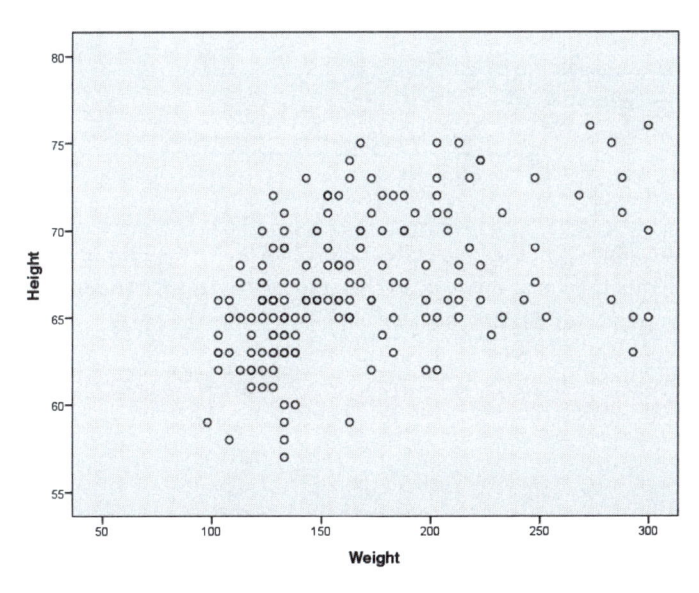

CHAPTER 3   A Picture Is Worth a Thousand Words: Creating and Interpreting Graphics

STEP 4: Let's say we want to add a fitted line to better see the relationship between the two variables. Double-click on the plot in the output window to open up the **Chart Editor**.

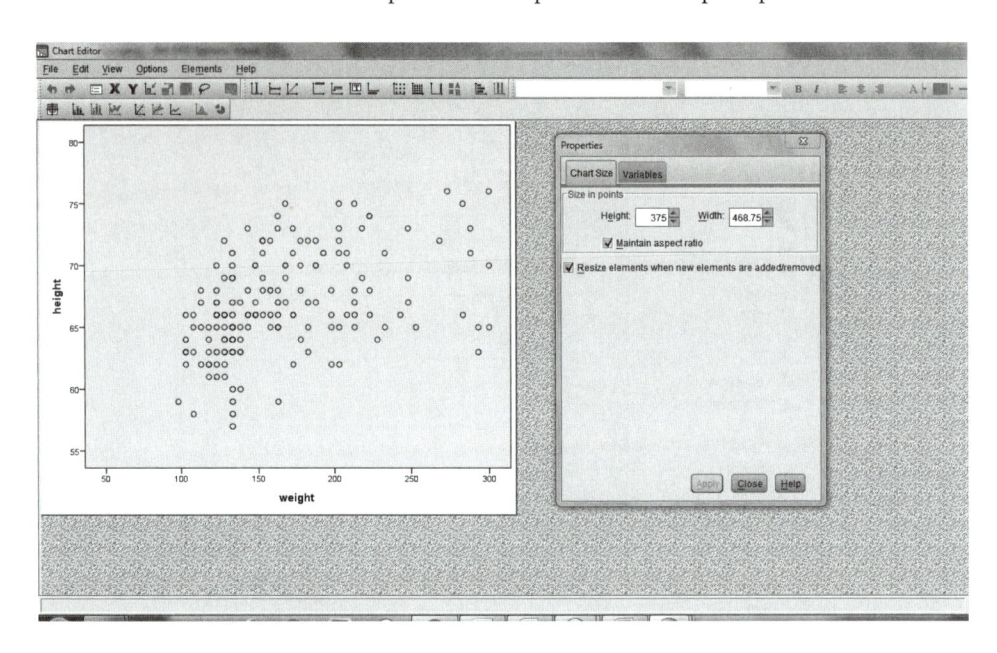

STEP 5: In the top left-hand corner of the **Chart Editor**, click the **Add Fit Line at Total** button in the toolbar. This adds a fitted line to the scatterplot.

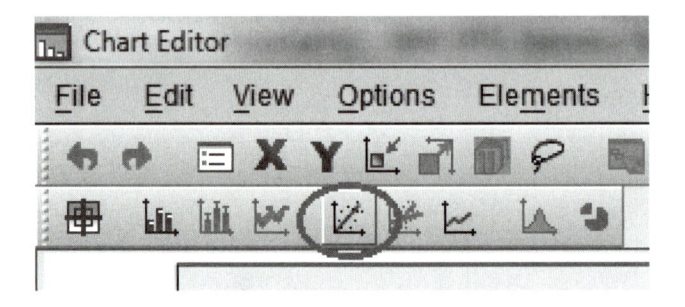

STEP 6: You can adjust it in the **Properties** dialog box under the **Fit Line** tab. For this graph, we want to use the linear method, so click **Linear**. For the **Confidence Intervals** section of the **Properties** dialog box, we will add **95%** around the **Mean**.

STATISTICS IN CONTEXT

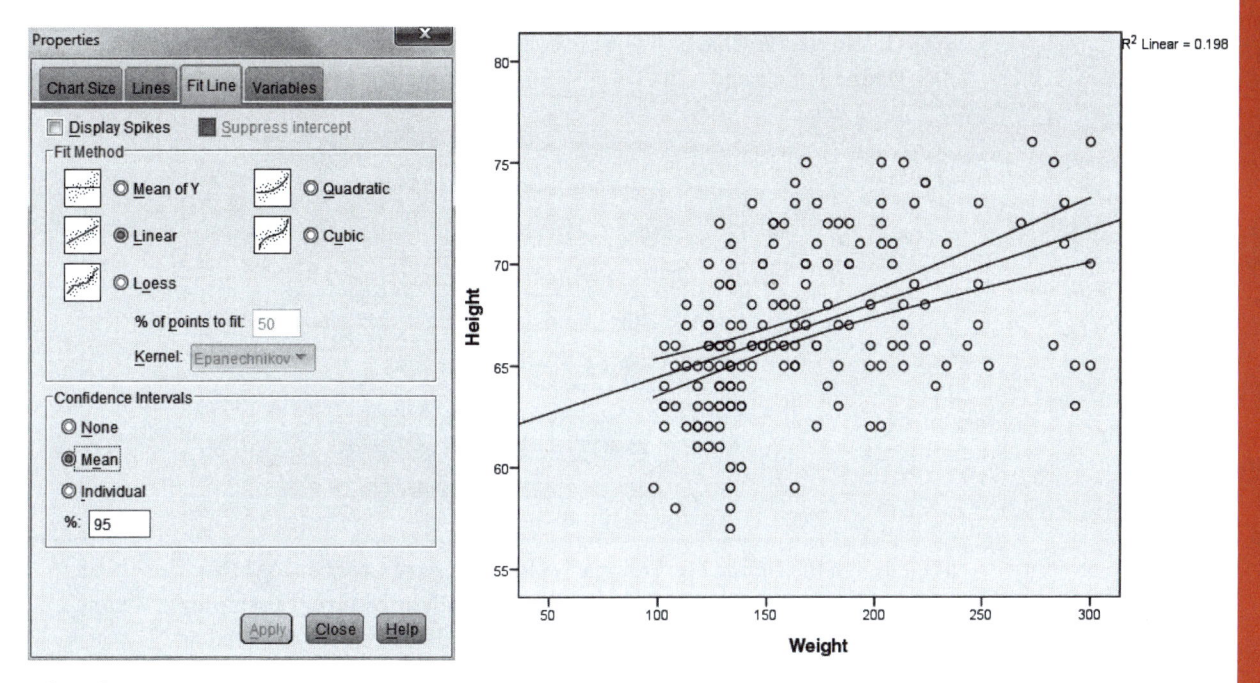

Pie Chart

The next graph we will try our hand at is the pie chart. This plot is usually used to visualize relative proportions. Here, we are going to look at the breakdown of nonsmokers in the survey across Australia, Canada, Great Britain, and New Zealand.

STEP 1: To get started, click **Data** and then **Select Cases**. Click the **If** button, and specify the countries of interest.

STEP 2: Next, select **Graph** from the main menu bar, then **Legacy Dialogs**, and then **Pie**.

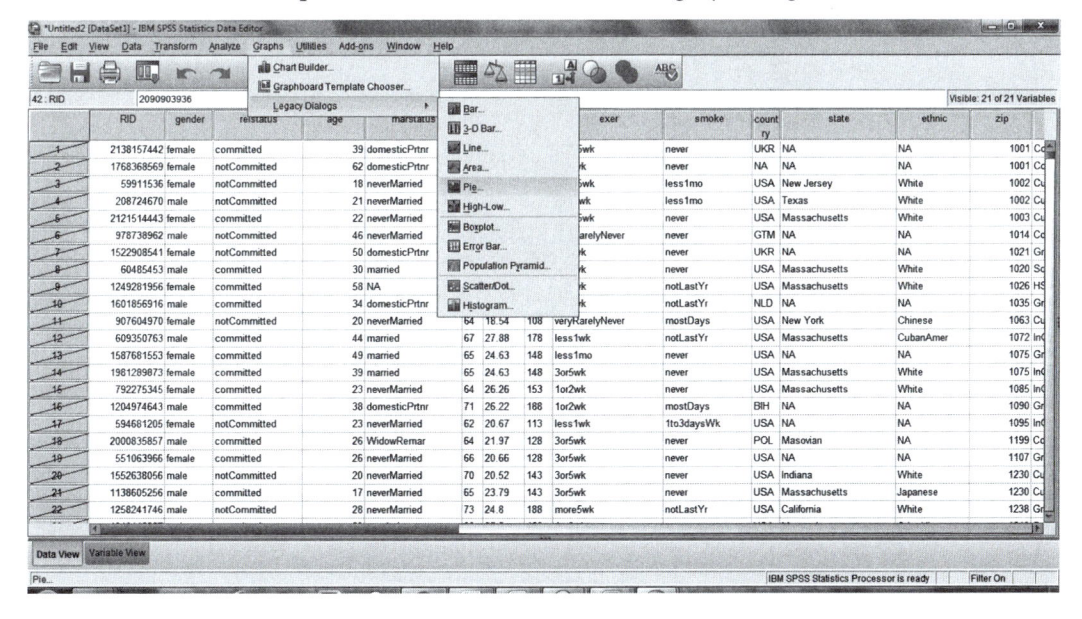

CHAPTER 3   A Picture Is Worth a Thousand Words: Creating and Interpreting Graphics   139

STEP 3: In the **Pie Charts** dialog box, select **Summaries for groups of cases** and then click **Define**.

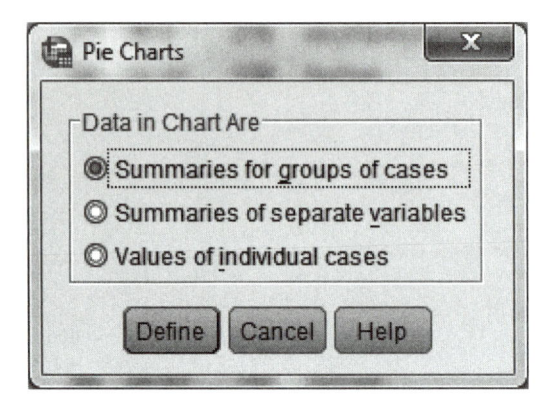

STEP 4: In the **Define Pie: Summaries for Groups of Cases** dialog box, choose **N of cases** under the **Slices Represent** section. Next, add the **country** variable to the **Define Slices by** input box. Click **OK** to produce the pie chart in the output window.

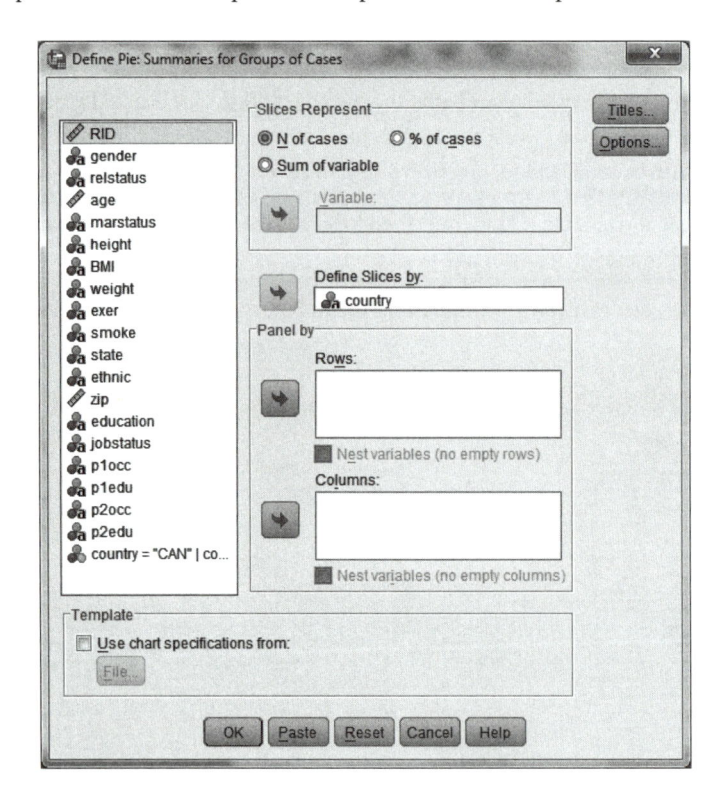

Again, you can customize the look of your plot by double-clicking on it in the output window and editing it in the **Chart Editor**.

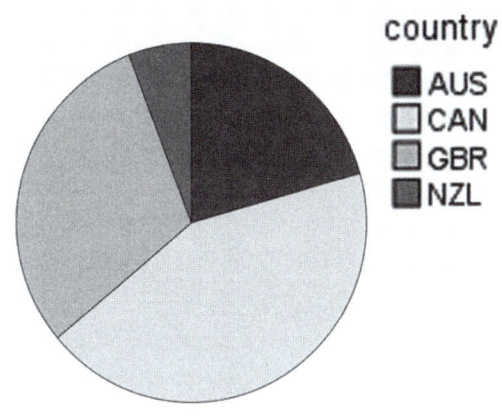

Time Series

As the name implies, time-series plots are helpful in visualizing how a measure changes over time. In this example, we will be using a different data set to demonstrate how to create a time-series graph. The data come from Environment Canada and contain hourly average temperature readings (in Celsius) for January 18, 2016, from a weather station in Yukon. There are two variables (time and temperature) and 24 observations.

STEP 1: Open the data set (**wthr18.csv**) using the text import tool. Define the **Time** variable as **type Date**. On the main menu bar, select **Graph**, then **Legacy Dialogs**, and then **Line**.

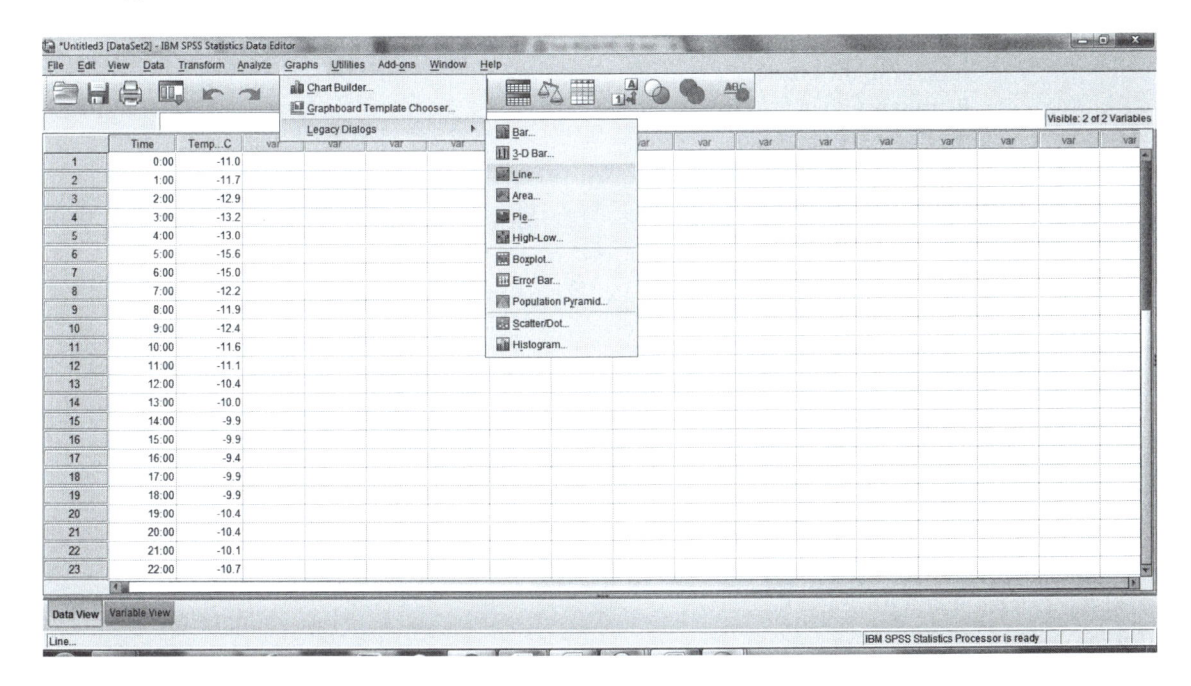

STEP 2: In the **Line Charts** dialog box, select **Simple** and choose **Values of individual cases** under the **Data in Chart Are** section. Click **Define**.

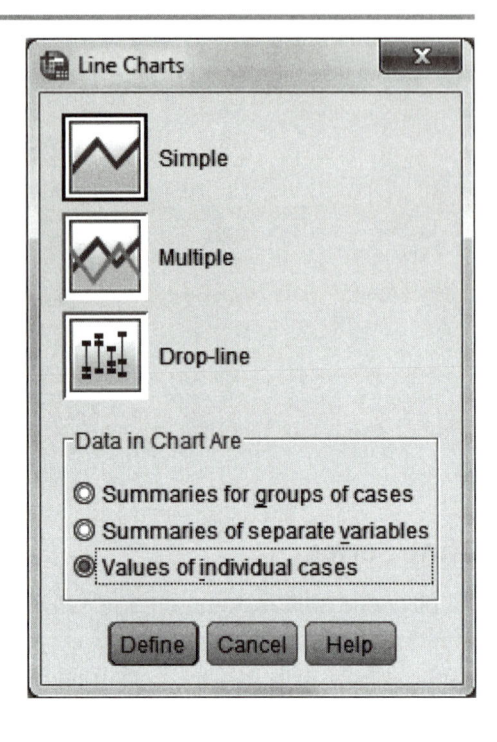

STEP 3: In the **Define Simple Line: Values of Individual Cases** dialog box, add the **Temp...C** variable to the **Line Represents** input box. In the **Category Labels** section, choose **Variable** and add **Time** to that input box. Click **OK**.

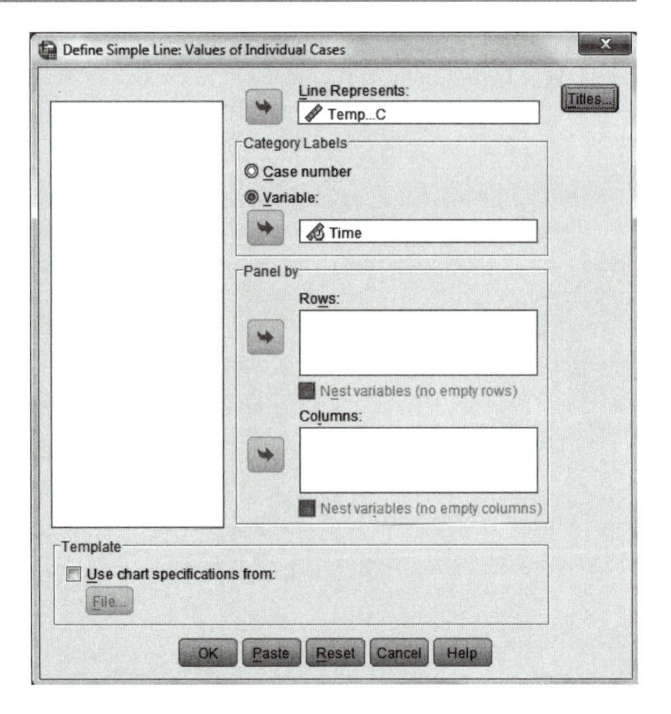

Your time-series plot is produced in the output window. Adjustments and customization can be performed through the **Chart Editor**.

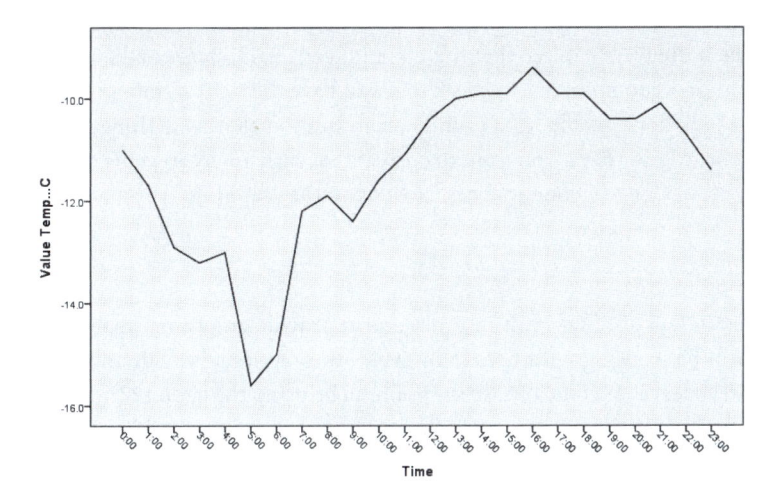

Graphing with R

We will be using the same data sets as in the SPSS graphing tutorial. Let's get started with a bar chart.

We will begin by loading the **dplyr** package to help with data management. This package allows us to easily manipulate the data using easy-to-understand commands that can be linked together in logical steps. After that, we'll read in the data set and take a quick look at what it contains.

```
library(dplyr) # load data manipulation package

dem_samp <- read.csv("dem_samp.csv", stringsAsFactors = FALSE)
# read in data from CSV file
dem_samp <- tbl_df(dem_samp) # make dem_samp into a dplyr
table which makes printing it easier dem_samp # print the data

Source: local data frame [23,681 x 20]
   RID       gender  relstatus    age  marstatus      height  BMI    weight
  (int)      (chr)   (chr)        (int) (chr)         (int)   (dbl)  (int)
1  2138157442 female    committed  39 domesticPrtnr   64     21.11    123
2  1768368569 female notCommitted  62 domesticPrtnr   63     19.13    108
3    59911536 female notCommitted  18  neverMarried   60     22.07    113
4   208724670   male notCommitted  21  neverMarried   73     23.48    178
5  2121514443 female    committed  22  neverMarried   65     19.63    118
6   978738962   male notCommitted  46  neverMarried   71     30.40    218
7  1522908541 female notCommitted  50 domesticPrtnr   66     22.27    138
8    60485453   male    committed  30       married   77     35.57    300
9  1249281956 female    committed  58            NA   65     24.63    148
10 1601856916   male    committed  34 domesticPrtnr   74     17.72    138
..        ...    ...         ...  ...          ...  ...      ...     ...
```

CHAPTER 3 A Picture Is Worth a Thousand Words: Creating and Interpreting Graphics   143

```
Variables not shown: exer (chr), smoke (chr), country (chr), state (chr),
   ethnic (chr), zip (int), education (chr), jobstatus (chr), p1occ (chr),
   p1edu (chr), p2occ (chr), p2edu (chr)
```

From the output, we can see that there are indeed 23,681 cases and 20 variables in the data set. It's always good practice to check that you have the right number of cases and variables. When they don't match, it is usually an indication that something has gone wrong with the import of your data. You want to correct that early on to prevent compounding the error as you move along in your analysis. Now on to the graphing!

ggplot2

The base installation of R has built-in functions to provide graphing capabilities for all the graphs we will be working with here. However, other options provide more functionality and are easier to learn. Specifically, we are going to be using the ggplot2 (http://ggplot2.org) package by Hadley Wickham. As you will see, it's an easy way to build publication-quality plots with precise control over all aspects of the graphic.

Let's start by loading the **ggplot2** package.

```
library(ggplot2)
```

Bar Charts

The first type of graph we will be creating is the bar chart. We want to show the relative frequency of nonsmokers in our data set who are from Australia, Canada, Great Britain, New Zealand, and the United States.

STEP 1: The first step is to calculate the numbers we want to plot. In this case, it is the frequency and relative frequency of nonsmokers for our countries of interest.

```
dem_samp %>%
  filter(country %in% c("AUS","CAN", "USA", "GBR", "NZL")) %>%
# select only these countries listed
  group_by(country, smoke) %>%
  summarise(n = n()) %>% # frequencies for each country
  mutate(rel_freq = n / sum(n)) %>% # relative frequency of
smokers vs. non-smokers for each country
  filter(smoke %in% "never") -> non_smokers.df # keep only non-smokers
```

STEP 2: Next, let's print out a table that summarizes our aggregate data.

```
library(pander)
panderOptions('digits', 2)
panderOptions('round', 3)
panderOptions('keep.trailing.zeros', TRUE)
pander(non_smokers.df)
```

country	Smoke	n	rel_freq
AUS	Never	226	0.48
CAN	Never	573	0.58
GBR	Never	367	0.53
NZL	Never	87	0.66
USA	Never	8071	0.55

STEP 3: Finally, we will use the **ggplot** function to create our graph. We specify our data set **non_smokers.df** and map the variables from the data set to the appropriate axes. Here, we want country across the *x*-axis and the relative frequency of nonsmokers on the *y*-axis.

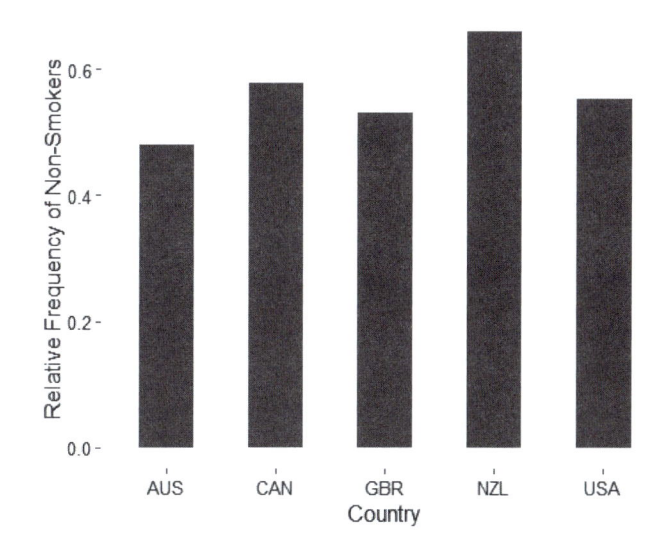

```
p_smoke <- ggplot(data = non_smokers.df, aes(x=country, y=rel_
freq)) + # map variables to axes
        geom_bar(fill="#aa1111", width=.5, stat = "identity")
+ # specify bar graph
        ylab("Relative Frequency of Non-Smokers") + # label y-axis
        xlab("Country") + # label x-axis
        theme_classic() # apply classic theme
p_smoke # print the graph
```

Frequency Polygon

For our second graph, we will create a frequency polygon to look at the distribution of age found in our data set. Once again, we need to specify our data set and map our variable of interest, **age**, to an axis. Once that is done, we specify that we want a frequency polygon and add the options we want. Finally, we apply the classic theme to standardize the look of the graph.

```
p_age_fp <- ggplot(data = dem_samp, aes(x=age)) +
        geom_freqpoly(colour = "green", binwidth = 3) +
        theme_classic()
p_age_fp
```

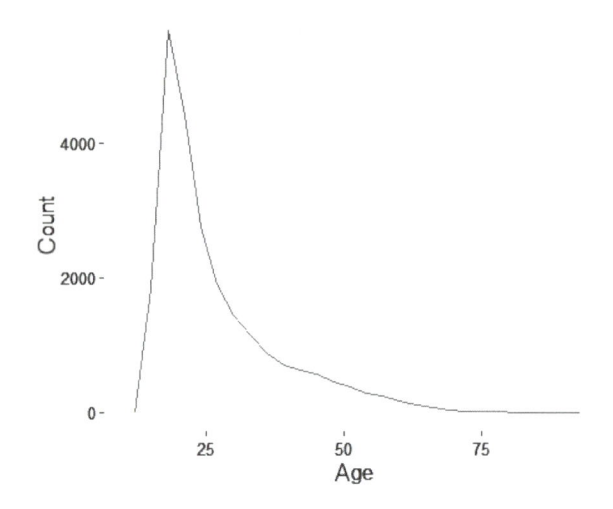

Histogram

We can also look at the distribution of age from our data set using a histogram.

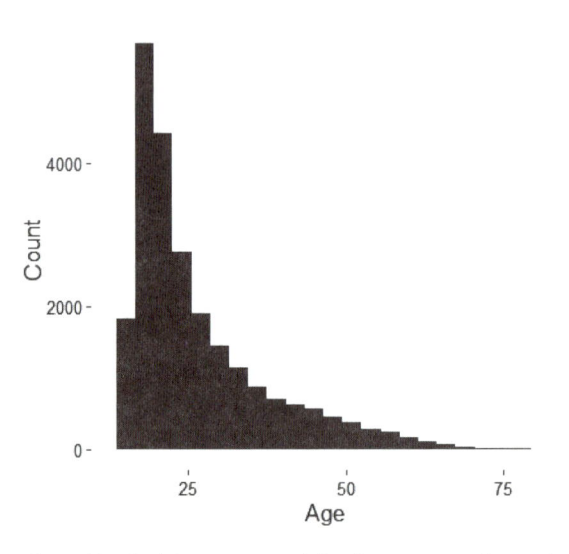

```
p_age_h <- ggplot(data = dem_samp, aes(x=age)) +
           geom_histogram(fill = "#AA1111", binwidth = 3) +
           theme_classic()
p_age_h
```

The information found in the histogram and the frequency polygon is the same. We can very easily plot both together. If we take the frequency polygon plot **p_age_fp** and add the histogram with options, we can see that they present the same data.

```
p_age_fp + geom_histogram(alpha = .2, fill = "#AA1111",
binwidth = 3, colour = "red")
```

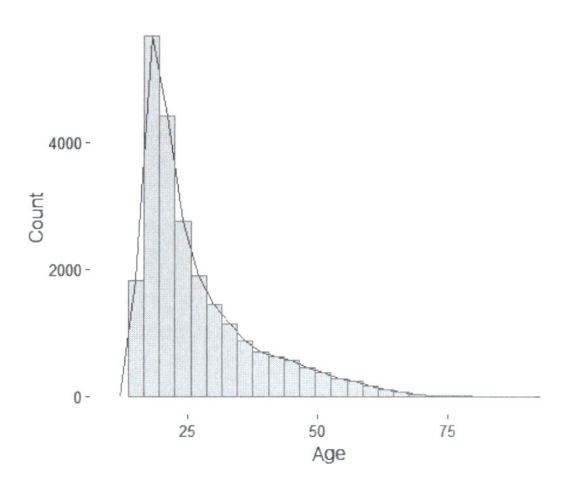

Scatterplot

If we are interested in the relationship between two variables, a scatterplot is a great way to visualize it. Here we investigate the relationship between height and weight.

STEP 1: To start, we'll take a random sample of 200 cases and map height to the *y*-axis and weight to the *x*-axis.

STEP 2: Next, we specify that we want points to represent each case, **geom_point**, and add a fitted line **geom_smooth** to represent the conditional mean. By default, a .95 confidence interval is shown around the line.

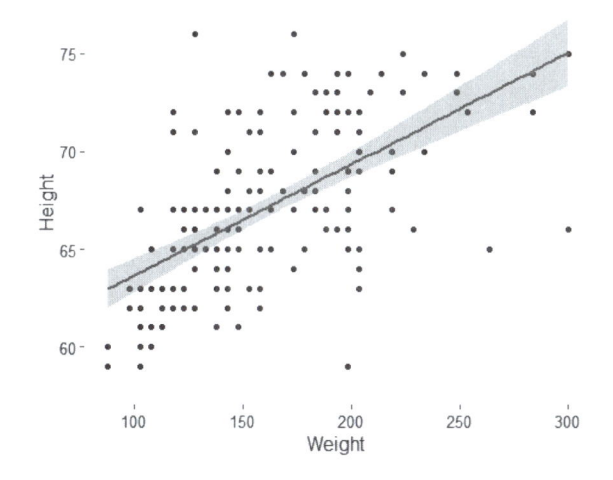

```
set.seed(54321) # used to create a reproducible sample
p_htwt <- ggplot(data = sample_n(dem_samp, 200), aes(x=weight,
y=height)) +
          geom_point() +
          geom_smooth(method = lm, colour = '#AA1111') +
          theme_classic()
p_htwt
```

CHAPTER 3 A Picture Is Worth a Thousand Words: Creating and Interpreting Graphics

When we examine the plot, we see that there definitely does appear to be a relationship between height and weight: As weight increases, so does height.

Pie Chart

We are now going to use a pie chart to look at the relative contribution of each country from our database to the total number of nonsmokers from Australia, Canada, Great Britain, and New Zealand. Since the U.S. contributes many more cases than the other countries, we are removing Americans from this plot so that we can see how the remaining countries compare.

STEP: We use the **pie** function and specify the variable to label the plot, **country**. The **col** argument sets the colors that will be used.

```
no_us_non_smokers.df <- non_smokers.df %>%
  filter(country != "USA") %>% # remove US
  group_by(country) %>%
  summarise(n = n) %>% # frequency count
  mutate(prop = n/sum(n)) # proportion
pie(no_us_non_smokers.df$n, labels = no_us_non_smokers.df$country, col = rainbow(length(no_us_non_smokers.df$country)))
```

We can see that Canada contributes the most nonsmokers to our sample, followed by Great Britain, Australia, and then New Zealand.

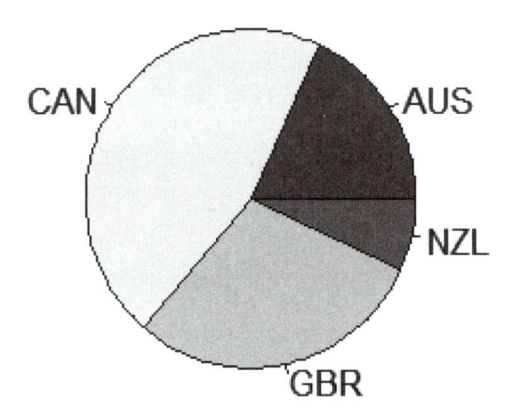

Time Series

We are going to use a different data set to demonstrate how to create a time-series graph. The data come from Environment Canada and contain hourly average temperature readings (in Celsius) for January 18, 2016, from a weather station in Yukon. There are two variables (**Time, Temp...C**) and 24 observations.

As you will have noticed, there is a common process to creating plots with **ggplot2**.

STEP 1: First, we map our variables to the *x*- and *y*-axes.

STEP 2: Then, we specify how we want our data represented (**geom_line**).

STEP 3: Finally, we add some labels (**ylab(expression(paste("Temperature ",degree,"C"))))** and formatting (**theme_classic() + theme(axis.text.x = element_text (angle = 90)))** to make the plot a little more appealing to the eye.

```
wthr_samp <- read.csv("wthr18.csv") # read in data
p.ts <- ggplot(data = wthr_samp, aes(x = Time, y = Temp...C., group = 1)) +
        geom_line() +
        ylab(expression(paste("Temperature ",degree,"C"))) +
        theme_classic()+
        theme(axis.text.x = element_text(angle = 90))
p.ts
```

CHAPTER 3   A Picture Is Worth a Thousand Words: Creating and Interpreting Graphics   149

CHAPTER FOUR

MEASURES OF CENTRAL TENDENCY: WHAT'S SO AVERAGE ABOUT THE MEAN?

Children laugh an average of three hundred or more times a day; adults laugh an average of five times a day. We have a lot of catching up to do.

—HEATHER KING

Everyday Statistics

HOT ENOUGH FOR YOU?

According to scientists, the average temperature of the United States in September is 52.02 degrees Fahrenheit. But, you say, how can that be? It's 95 today, and it was 100 yesterday!

Averages are statistics that summarize large sets of measurements, usually with just one number. In the graph below, each black dot represents the average temperature for a given year (the sum of the daily temperatures for the entire year, divided by 365 days). Our figure of 52.02°F is the overall average from 1895 to 2016, indicated by the gray horizontal line that bisects the graph.

Do you see any patterns here? You should notice a trend toward higher average temperatures that begins in the 1990s and continues today. Scientists are now trying to determine how best to reverse this trend (if possible) and predicting what we can expect in the coming years. Chapter 4 is all about the statistics that are used to summarize patterns in data.

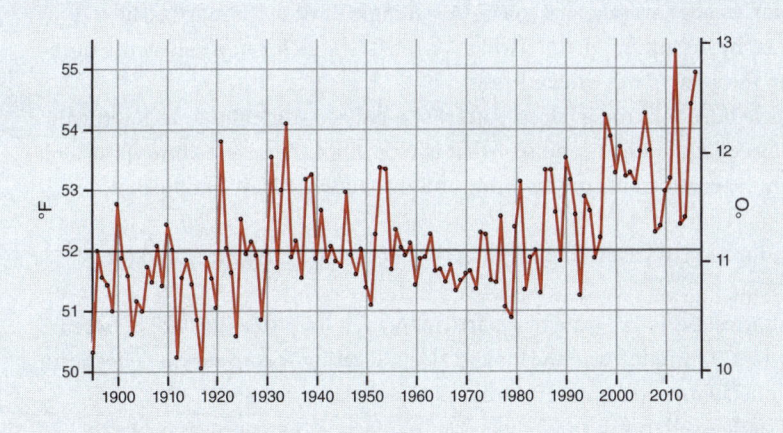

Average annual temperature in the contiguous United States, 1895–2016

OVERVIEW

MEASURES OF CENTER
MEASURE OF CENTER: WHAT IS TYPICAL?
THE MODE AND THE MEDIAN
THE MEAN
MODE, MEDIAN, AND MEAN: WHICH IS THE "BEST" MEASURE OF CENTER?
SHAPES OF DISTRIBUTIONS
MEASURES OF CENTER IN CONTEXT

LEARNING OBJECTIVES

Reading this chapter will help you to . . .

- Appreciate the contributions to modern statistics made by Adolphe Quetelet. (Concept)

- Understand what the mean, median, and mode describe in a distribution and how these measures differ from one another. (Concept)

- Calculate the mean, median, and mode. (Application)

- Know the difference between normal, positively skewed, and negatively skewed distributions. (Concept)

- Understand how the relationship between the mean, median, and mode describes the shape of a distribution. (Concept)

- Explain how the measures of center (mean, median, and mode) are related to data type (nominal, ordinal, interval, and ratio). (Concept)

MEASURES OF CENTER

I'm sure you've heard about "the average person." Politicians and the media especially love to talk about what the average person does, thinks, earns, buys, and so on. But have you ever wondered just who the average American actually is? According to the U.S. Census Bureau (2010) the average American lives in an urban area (81% of us live in a city or suburb) in the western or southwestern United States. The average American is 37.2 years of age, has 1.86 children, makes about $53,482 a year in a white-collar job, and can be classified as working to middle class. The average American does not have a college degree but does own a home and will marry—probably more than once. Have you ever met this person? Does he or she sound like you? If you do not resemble this portrait of the "average American," are you abnormal in some way? Just who is this average American, and what makes him or her so "average"?

MEASURES OF CENTER: WHAT IS TYPICAL?

The concept of an average man, whose characteristics describe the typical resident of a particular place, state, region, or country, originated with Adolphe Quetelet, a Belgian mathematician and statistician, whose work is described in *The Historical Context*, starting on page 153. Quetelet observed that many measurable human characteristics tend to be *normally distributed*: In other words, they tend to be close in value to the arithmetic average, or "mean," with few measurements that are much larger or much smaller than this average. His observation supported a tendency that prevailed in the social sciences then just as it does today: We generally look for what is average or typical in a sample of observations because "typical," almost by definition, defines what is probably true for most elements in the population the sample was drawn from.

We have been looking at sleep patterns in college students, so let's return to the sleeping college students and see what is typical for them. We know quite a bit about both the participants in this study and the study itself. We know that:

- Sleep duration, measured in minutes, is *quantitative data* measured on a *ratio scale*.
- The shortest period of time that participants reported sleeping was between 120 and 179 minutes, and the longest sleep duration was somewhere between 720 and 779 minutes (around 12 hours of sleep in one night).
- Most of the students in the sample (89.37%) slept 8 hours or less per night.

We have some useful statistics here, but if we were to present what we have to an audience interested in the effects of sleep deprivation on performance in the classroom, we might be accused of dancing around the main issue: Just what was a typical night's sleep for the college students in the sample?

The Historical Context

ADOLPHE QUETELET AND THE "AVERAGE MAN"

Lambert Adolphe Jacques Quetelet (see Figure 4.1) was born in 1796 in Ghent, Belgium. He earned his Ph.D. in mathematics at the University of Ghent in 1819 and was a teacher, an astronomer, a census taker, and a prolific writer until his death in 1874.

In the late eighteenth and early nineteenth centuries, statistics and probability were being used primarily in astronomy to try to understand measurement errors in one of the major scientific questions of the day: How can ship captains accurately find their longitude and latitude? Quetelet wanted to apply the same statistical techniques and probability theory to the social sciences in order to understand complex social phenomena. For example, he wanted to understand the social factors (like age, gender, education, income, climate, alcohol consumption, and so on) that were associated with criminal behavior. Quetelet proposed the term *social physics* to describe his work, and he introduced the concept of the "average man" (*l'homme moyen* in French) in his 1835 book *Sur l'homme et le développement des ses facultés, ou essai de physique sociale* (*A Treatise on Man and the Development of his Faculties*). Quetelet intended the "average man" as a summary of some of the characteristics he had found in the national population, but somewhere along the way,

> [the average man] took on a life of his own, and in some of Quetelet's later work he is presented as an ideal type, as if nature were shooting at the average man as a target and deviations from this target were errors (Stigler, 1997, p. 64).

Nine years later, Quetelet published his second book on social physics (*Letters Addressed to H.R.H. the Grand Duke of Saxe Coburg and Gotha, on the Theory of Probabilities as Applied to the Moral and Political Sciences*, 1849), where he noted that many measurements of human characteristics in the population—height, weight, chest circumference, and so

FIGURE 4.1
Adolphe Quetelet.

on—were approximately *normally distributed*. In other words, the measurements tended to cluster around the mean, and measurements that were much larger or much smaller than the mathematical average did not happen very often. This pattern in the way characteristics were distributed in populations suggested that perhaps deviation from the average was perfectly normal and not an "error" at all.

Quetelet went on to write dozens of articles and books on his new field of social physics, along the way creating a statistic that is used quite frequently today: the "Quetelet Index," a.k.a. the Body Mass Index, or BMI, which he used to categorize a person's weight relative to his or her height. Quetelet's ideas were influential in the development of social science (psychology, sociology, and political science) as we know it today.

Statisticians have developed several statistics that describe "typical" in a set of observations. The three measures of the typical observation in a set are called *measures of center* because they describe (each with one number) the center of a distribution of observations. The three measures of center are the following:

- *Mean:* the arithmetic average of the set of measurements in the sample.
- *Median:* the exact center of the set of measurements, or the observation (either real or theoretical) that splits the distribution of measurements in half; 50% of the observations will be higher in value than the median, and 50% will have a value lower than the median.
- *Mode:* the observation in the data set that occurs more often than any other.

THE MODE AND THE MEDIAN
FINDING THE POSITION OF THE MODE

Take a look at Table 4.1, which shows the ungrouped frequency distribution of the sleep durations for 1 day of our week-long study of sleeping college students. We'll use the three measures of center just introduced to describe this data set.

The **mode** is by far the easiest measure of center to find. It is *the value or measurement in the set that happened more often than any other*. In the data set shown in Table 4.1, the mode is 450 minutes. The sleep duration of 450 minutes was reported five times, more often than any other measurement.

mode The most frequently occurring observation in the data set.

TABLE 4.1 The mode in a frequency distribution

Minutes	f	Minutes	f
120	2	430	1
275	1	440	1
330	1	445	1
340	1	**450**	**5**
360	1	455	1
375	1	470	1
380	1	480	1
385	1	495	1
390	1	520	1
405	3	525	1
420	2	620	1

This distribution can be described as *unimodal*—having only one mode—because there is one measurement (450 minutes) that clearly stands out as happening more often than any other. However, you might come across sets of measurements that have more than one mode. For example, in a *bimodal* distribution there are two measurements that happen more often than the others, in a *trimodal* distribution there would be three measurements that stand out, and so on. Distributions with more than three modes are described as *multimodal*, probably to save time and counting.

If you created a frequency histogram of the data in this set (you should recognize that a histogram, rather than a bar chart, is appropriate to create a picture of this data because sleep data are measured on a continuous scale), the single mode should be obvious. The mode corresponds to the observation at the peak of the graph. Figure 4.2 illustrates the shape of the sleep duration distribution and its single mode.

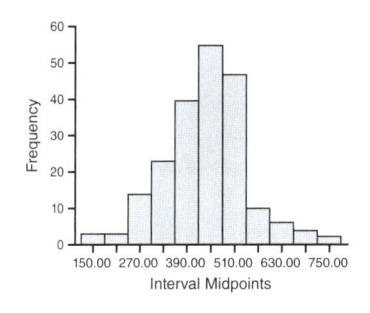

FIGURE 4.2
The mode.

FINDING THE POSITION OF THE MEDIAN

The **median** is the measurement (either real or theoretical) that splits the set of measurements exactly in half, so that 50% of the values in the set are greater than the median value and 50% are less than the median value. To put it another way, the median is the 50th percentile.

To find the median, we need to figure out where the middle of the distribution is. The easy way to do this is to arrange the measurements in our set in order from least to most (as I've already done in Table 4.1). Then, we use the formula in Box 4.1 to find the *position* of the median measurement in the set (remember that the median is the measurement or observation in the data that splits the distribution in half—we're trying to find the position of that measurement in this ordered set).

median The observation at the center of the distribution so that 50% of the observations in the set are below it and 50% are above it. The median can be a theoretical observation in the set.

BOX 4.1 **Finding the median**

$$\text{position of the median} = \frac{n+1}{2}$$

So, we first need to know how many observations we have in our data set (the symbol for the number of observations in a set is the letter *n*). To find the position of the median value in this set, we add 1 to our *n* of 30 (30 + 1 = 31); we then divide the total by 2 in order to find the position of the exact center of the distribution. The median value in this set is in the fifteen-and-one-half-th position, or halfway between the value in the 15th position and the value in the 16th position. So, what observation sits in the 15th position? We start at one end of the ordered set (either at the lowest or the highest value) and then count up (or down) 15 observations to find it. Table 4.2 shows how to find the observation at the median (15½th) position.

TABLE 4.2 Effect of changing a data point on measures of center

Observation	Frequency	Raw Value	Position in the Set
⋮		⋮	⋮
405	3 →	405 405 405	11th position 12th position 13th position
420	2 →	420 420	14th position 15th position
430	1 →	430	16th position
440	1 →	440	17th position
445	1 →	435	18th position
⋮		⋮	⋮

The value in the 15th position in this set is 420 minutes, and the value at the 16th position is 430. The median would be the value, *either real or theoretical*, that sits in the 15½th position, halfway between the values in the 15th and the 16th positions. So, what value would be halfway between 420 and 430? To find out, we just add the two values together (420 + 430 = 850) and then divide that sum by 2 (850 ÷ 2 = 425). The value of the median in this case would be 425: It is the value that is exactly halfway between 420 and 430.

The thing to keep in mind here is that you should use the formula in Box 4.1 to find the *position* of the median, not the value of the median itself. The *value of the data point* that we would report as the median observation in the data set is the value at that position—in this case, 425. Notice also that this value is *theoretical*. No one in our sample reported sleeping 425 minutes. However, 425 is the value that splits the distribution exactly in half, so 425 is the median in this set.

THE MEAN

So, we know that the mode in this data set is 450, and that the median is 425, somewhat less than the mode. The last measure of center, and the one that is reported most often, is the **mean**, or *the arithmetic average of the observations in the set*. Before we talk about what the phrase "arithmetic average" means, take a look at the formula (shown in Box 4.2) for calculating the mean. This process (formulas are sets of instructions about what to do with your data) of calculating the average or mean value in a set is going to come up repeatedly in this text. Take a moment to make sure that you understand how to calculate an average value.

mean The arithmetic average of a set of observations obtained by summing the observations in the set and dividing that sum by the number of observations in the set (*n*).

BOX 4.2 — Calculating the mean

$$\bar{X} = \frac{(\sum x)}{n}$$

The formula tells us to carry out the following procedure to calculate a mean. First, add up all of the values in the set ($\sum x$), and then divide that sum by the number of values in the set (n). The process of summing the elements of a set and dividing that sum by the number of elements in the set to find the mean applies to finding the mean of both a sample and the population. The result is the mean value in the set and is labeled with the symbol $\bar{X}$ (an X with a bar over it, or "X-bar") when you are referring to the mean of a sample; and "μ" (the Greek letter "mu") when you are referring to the mean of the population. Be aware that when you are *writing* about the mean, this symbol is not used. Instead, American Psychological Association (APA) Style requires the use of an italicized capital *M* (probably because "X-bar" isn't a feature of very many keyboards).

The mean, or average, value in a set is special in statistics. It will appear in almost everything you do with statistics from this point onward, and it is often referred to as the "best" measure of center—the one you probably should use, whenever possible (we'll talk more about when you should or should not use the mean later), to describe a set of data. When you think about why it is called the "arithmetic average" of a set of values, you can see why it is so useful in describing sets of data.

The word *arithmetic* comes from the Greek *arithmos*, for "number," and basically refers to anything that involves arithmetic operations. An *operation* is a fancy word for a procedure you can apply to numbers. The basic operations used by everyone just about every day are addition, subtraction, multiplication, and division. When we apply these operations to numbers, we are doing arithmetic. The formula in Box 4.2 tells us that we are applying two of these operations (addition and division) to numbers when we calculate the mean, hence the use *arithmetic*.

The arithmetic average of a set of observations is a number that typifies the center of the set, one we calculated using arithmetic operations. It is a measure of center if ever there was one, but please note that *all three measures of center—the mean, median, and mode—are "averages."*

Let's calculate the mean for the sleep data shown in Table 4.1. Remember that we first have to add up all of the values in the data set. Once we get the sum, we divide it by the number of values in the set (n) to find the mean. If you try this yourself, you should wind up with something that looks like Box 4.3. The average number of minutes of sleep for the 30 people represented in this sample is 408.67 minutes (a little less than 7 hours).

BOX 4.3 Calculating the mean for the sleep data in Table 4.1

$$\bar{X} = \frac{(\sum x)}{n}$$
$$= \frac{12,260}{30}$$
$$= 408.67$$

MODE, MEDIAN, AND MEAN: WHICH IS THE "BEST" MEASURE OF CENTER?

We now have three numbers that describe the center of the distribution of our values for sleep duration: the mode (450 minutes), the median (425 minutes), and the mean (408.67 minutes). Notice that these three measures of the center of this distribution are not the same value. The mode is the largest measure of center, the mean is the smallest, and the median is between the two in value. Why are they not all the same if they're all describing the center of the same distribution?

The answer is that each of these measures describes the center of the distribution a bit differently. The mode is simply the one measurement that occurred more often than any other measurement in the set. The mode then reflects only this one observation. You might say that the mode "ignores" all the other values in the set and only "pays attention to" the one frequently occurring value. The median essentially does the same thing: It simply reflects the exact center of the distribution, ignoring everything else.

The mean is different. The mean reflects each and every value in the data set because each and every value is used in its calculation. To illustrate this, let's change one, and only one, value in the data set (see Table 4.3). Suppose the 30th person in our set slept around the clock (24 hours, or 1,440 minutes) instead of the 620 minutes shown in Table 4.1. What does this one change do to each of the measures of center?

What is the new mode? Well, because our change did not affect the observation with the highest frequency, the mode remains unchanged: It is still 450. What is the new median? Again, our change did not affect the total number of observations in our set, so the position of the median does not change. And since the change in the data did not affect the spread of the data and happened "outside the view" of the median, the value of the median also does not change.

Now consider what happens to the mean when we change just that one data point. Here, we have a big change. The sum of all the values increases from 12,260 to 13,080. Our n didn't change, so we still divide the sum of the x values by 30.

STATISTICS IN CONTEXT

TABLE 4.3 Changing a data point to the data set

Minutes	f	Minutes	f
120	2	430	1
275	1	440	1
330	1	445	1
340	1	450	5
360	1	455	1
375	1	470	1
380	1	480	1
385	1	495	1
390	1	520	1
405	3	525	1
420	2	**1,440**	**1**

Our new mean then becomes $13,080/30 = 436$. So, the mean increases from a little less than 7 hours of sleep (408.67 minutes) up to about 7 hours and 30 minutes of sleep—quite a large change in our description of what is "typical" in this set.

Because the mean takes into account each and every data point when it is calculated, this measure reflects each and every data point in its value. Changing one value in the set will change the mean, but it might not change the median or the mode. The mean is the most *sensitive* measure of center, meaning it is the one that will most likely respond to even a small change in the set. The mode and median are relatively insensitive to change in the data. Because the mean reflects all of the data points in the set, it is considered the "best" measure of center.

CheckPoint

Indicate whether the following statements are *true* or *false*.

1. Any one of the three measures of center can be used to tell what is typical in a set of observations. _____
2. Mode refers to the way an observation is measured. _____
3. In a bimodal distribution, one or more measurements occur twice. _____
4. The median of the data set 1, 2, 3, 6, 8, 11, 15 is 8. _____
5. There is no median in a data set featuring an even number of observations. _____
6. The mean is the arithmetic average of the observations in the set. _____

Answers to these questions are found on page 161.

Think About It...

ESTIMATING MEASURES OF CENTER

Consider distributions A, B, and C shown below.

A		B		C	
x	f	x	f	x	f
1	1	66	5	24	2
2	3	72	3	25	2
4	7	74	2	26	2
5	2	76	1	27	2
9	1	80	1	28	2
				29	2
				30	2

1. First, find the mode for each distribution.

   A _____ B _____ C _____

2. Now, without doing any calculations, determine if the mean and median will be smaller, larger, or the same as the mode (check the box that you think applies for each distribution).

   A
   ☐ Smaller
   ☐ Larger
   ☐ Same

   B
   ☐ Smaller
   ☐ Larger
   ☐ Same

   C
   ☐ Smaller
   ☐ Larger
   ☐ Same

3. Now check yourself. Find the mean and the median for each distribution. Were your estimations of these two measures of center correct?

   A
   Median _____
   Mean _____

   B
   Median _____
   Mean _____

   C
   Median _____
   Mean _____

   Answers to this exercise are found on page 188.

> **CheckPoint**
> *Answers to questions on page 159*
>
> 1. TRUE ☐
> 2. FALSE ☐ Mode is the most frequently occurring observation in a data set.
> 3. FALSE ☐ In a bimodal distribution, two observations occur more often than the others.
> 4. FALSE ☐ The median is 6. (In a set of seven observations, the median will be the observation in the fourth position.)
> 5. FALSE ☐ In a data set with an even number of observations, the median is a theoretical value sitting between the upper 50% of observations and the lower 50%.
> 6. TRUE ☐
>
> **SCORE:** /6

SHAPES OF DISTRIBUTIONS

There is one more thing that the mean, median, and mode can tell us about the distribution of data we're working with. The relationship between these three measures of center can tell us about the *shape* of the distribution. This works because the mean is the most sensitive measure of center, and it will move toward what we call **outliers** in the data set. An outlier is an unusual data point, either very high or very low in value, and "pulls" the mean toward itself like a magnetic field pulls on iron filings.

When there are very low outliers, much smaller in value than all the other values in the set, the mean is pulled down, away from the center and toward the low outliers. In the case of our sleep study, a low outlier would be a person who didn't get very much sleep at all—like the two individuals who slept for only 2 hours. If we were to graph a distribution with very low outliers, it would look something like what you see in Figure 4.3.

Notice the long left-hand tail of this distribution in Figure 4.3, pointing toward low values. Notice, too, that most of the values in this distribution (note the location of the mode) are clustered at the high end of the *x*-axis. And finally, notice that the mean has been pulled "down" (i.e., the mean has a lower value than either the median or the mode) by the outliers. The mean is the most sensitive measure of center, so it is the one that responds to the presence of outliers in the data: It moves toward the outliers.

When there are high outliers, we get the same process, but in the opposite direction. The mean is pulled up, again away from the center, toward these unusual high values. A graph of this distribution would have a long right-hand tail, and most of the values in the set would be at the low end of the *x*-axis, as you can see in Figure 4.4.

Now imagine a distribution with the same number of very low and very high outliers in the set. The mean, responding to the outliers, would wind up in the

outlier An observation in a data set that is much larger or smaller than most other scores in the distribution.

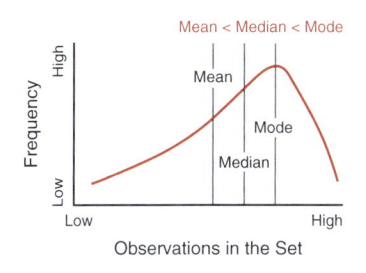

FIGURE 4.3
Negative skew caused by very low outliers.

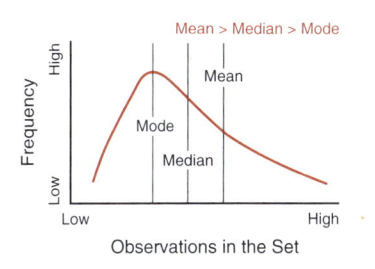

FIGURE 4.4
Positive skew caused by very high outliers.

CHAPTER 4 Measures of Central Tendency: What's So Average About the Mean?

negatively skewed distribution
A non-normal distribution where most observations have high values and the outliers have low values; a distribution with a long left-hand tail, where the mean is less than the median and the mode.

positively skewed distribution
A non-normal distribution where most observations have low values and the outliers have high values; a distribution with a long right-hand tail, where the mean is greater than the median and the mode.

normal distribution A distribution that is bell-shaped and symmetrical about the mean; a distribution where the mean, median, and mode all have the same value and high-value outliers are balanced by proportionally the same number of low-value outliers.

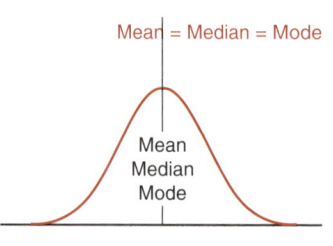

FIGURE 4.5
Normal distribution showing a balance of high and low outliers and all three measures of center holding the same value.

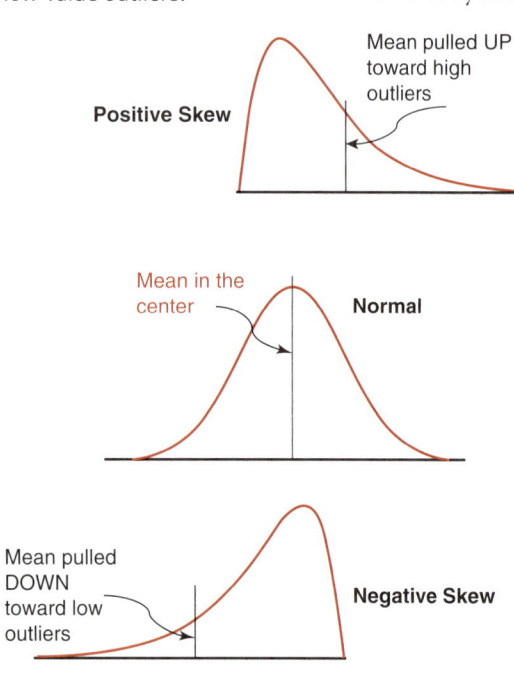

FIGURE 4.6
Shapes of distributions: positively skewed, normal, and negatively skewed.

center of the distribution, with a value either very close or identical to those of the median and the mode, our insensitive measures of center. When the outliers in a distribution are "balanced" high and low, the pull on the mean evens out, and the three measures of center have the same value: They reflect the same spot in the distribution. A graph of this distribution would probably look very familiar. There would be a few very low values, a few very high values, and most of the values in the set would be in the middle, as shown in Figure 4.5.

The three general shapes of distributions happen so often in statistics that each one has its own name. When very low outliers are pulling the mean down, so that the mean is less than the median and/or the mode, we have a **negatively skewed distribution**. When very high outliers are pulling the mean up, so that the value of the mean is greater than the median and/or the mode, we have a **positively skewed distribution**.

The most famous distribution shape of all—one that will show up in our discussions of statistics throughout the remainder of this book—is the so-called **normal distribution**. In a normal distribution, the mean, the median, and the mode are all the same value, and the low outliers are balanced out by an equal number of high outliers. Figure 4.6 reviews all three shapes. (You may recall that Quetelet found that many human characteristics, like height and weight, were normally distributed. Most of us are about average in height, but some of us are very tall and others very short.)

NORMAL DISTRIBUTIONS

Normal distributions are extremely popular in statistics. If you know that a distribution of observations is normal in shape and not skewed, you automatically know several things about it, regardless of the value of the mean or any other descriptive statistic. You know that:

1. The mean, the median, and the mode are all the same value.
2. The distribution is *symmetrical about the mean*, meaning that if you folded any normal distribution at the mean, the right and left halves of the distribution would overlap one another perfectly. The half of the distribution that is above the mean would be a mirror image of the half that is below the mean in value.
3. The outliers in the distribution balance one another out: If you have a few low outliers in a normal distribution, you will have approximately the same number of high outliers.

These characteristics of a normal distribution will come in very handy later on.

162 STATISTICS IN CONTEXT

FINDING CENTER WITH GROUPED DATA

So far, we have been calculating the mean, median, and mode using essentially raw data. However, you will likely need to find these three measures of center when you have no access to the raw data—for instance, when you're working with a large data set and (as a result) with a grouped frequency distribution. No need to panic, though—you don't have to un-group the data set and add up very long lists of numbers in order to find the mean, median, and mode. Instead, we can use the *midpoints* of each of the grouping intervals we've created in our grouped frequency distribution.

The values for the mean, median, and mode that we come up with using the grouped data will be close to the values we calculated using the raw data, but they will not necessarily be exactly the same. Think of these values as estimates of the real mean, median, and mode.

In Table 4.4, I've rearranged the data we've been playing with into a grouped frequency distribution, using 60 minutes as my interval. Then, I've added the midpoints of each interval.

Remember from Chapter 2 that the midpoint of an interval is simply the value at the center of the interval. (You might think of the midpoint as the median of the interval, if that helps.) The formula for finding the midpoint of each of these 60-minute intervals is fairly straightforward: We take the absolute value of the range of each interval (120–179 for the first interval), divide it by 2, and then add that result to the lower limit of the interval. So, for the first interval, the midpoint would be 149.5. (Box 4.4 shows the calculations for this midpoint.) From this point onward, we're going to treat the midpoints of each interval as our data.

Finding the mode in a grouped frequency distribution remains as simple as can be. The mode is the midpoint of the *most frequently occurring interval* in our set. In Table 4.4, the 420- to 479-minute interval has the highest frequency (12), so the mode is the midpoint of that interval (449.5). Compare this "calculation" of the

TABLE 4.4   Midpoints of intervals in a grouped frequency distribution

Interval	f	Midpoint
120–179	2	149.5
180–239	0	209.5
240–299	1	269.5
300–359	2	329.5
360–419	8	389.5
420–479	12	449.5
480–539	4	509.5
540–599	0	569.5
600–659	1	629.5

BOX 4.4 **Finding the midpoint of a 60-minute interval**

$$\frac{|120-179|}{2} + 120 = 29.5 + 120$$
$$= 149.5$$

mode using grouped data to the mode we described in the ungrouped data (450): There is very little difference in these two values.

Finding the median is also easy. We use the same basic procedure (n divided by 2 to find the position of the median), so our median should be in the 15½th position (grouping the data did not change n, so the position of the median does not change). We then report the midpoint of the interval that contains the median (in this case the 15th and 16th positions). Counting up from the lowest interval, we find that the interval containing the median is 420–479. So, the median is the midpoint of this interval, or 449.5. Again, compare this "calculation" of the median to the value we obtained using the ungrouped raw data (425). Now we have a difference. Why did the value of the median change?

Grouping has changed the shape of the distribution a bit. Large gaps between raw scores will tend to be eliminated (or at least reduced) when we group the data, as if the distribution has "pulled in its skirts," so to speak. Grouping the data might also change the location of the middle of the distribution. Differences between the values of the mode and median in the grouped and ungrouped distributions are generally not terribly important because, as we discovered earlier, the mode and the median are insensitive measures of center. Grouping should not, however, create any sizeable difference between the means calculated for the two different distributions. If we do see a large difference, we're in trouble, so let's see what happens when we calculate the mean using the grouped data.

BOX 4.5 **Calculating the mean from grouped data**

$$\bar{X} = \frac{\sum MP}{n}$$
$$= \frac{12,405}{30}$$
$$= 413.5$$

Again, we're going to use the midpoints (MP) of the intervals to calculate the mean; the formula stays the same. When we used the raw data, we calculated a mean of 408.67. Using the grouped data, our mean is 413.5. So, we did see a difference, but only a small one (4.83 minutes), between these two calculations of the mean.

STATISTICS IN CONTEXT

CheckPoint

Consider the following grouped frequency distribution.

Interval	f	Midpoint
20–39	0	29.5
40–59	1	49.5
60–79	1	69.5
80–99	3	89.5
100–119	9	109.5
120–139	8	129.5
140–159	3	149.5
160–179	3	169.5
180–199	2	189.5

1. What is the mode? _____
2. What is the median? _____
3. What is the mean? _____
4. What shape—positively skewed, negatively skewed, or normal—would you expect a graph of the data to have? _____

Answers to these questions are found on page 166.

Think About It . . .

SHAPES OF DISTRIBUTIONS

The shape of a distribution can tell you quite a bit about the data that make up the distribution. Consider the following distributions as you answer the questions below:

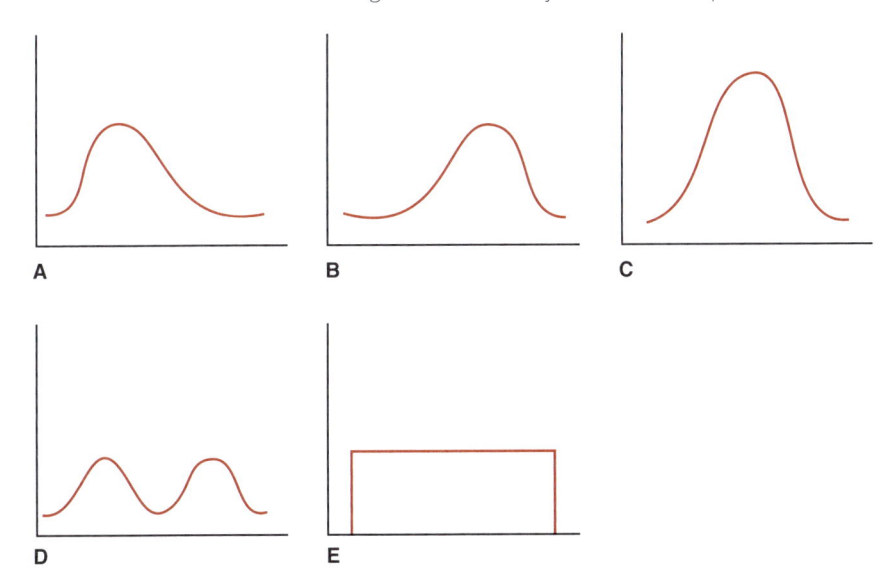

A

B

C

D

E

CHAPTER 4 Measures of Central Tendency: What's So Average About the Mean?

Think About It... continued

Let's suppose that these distributions show grades on the statistics final exam for the past five years. The lowest grade in the class was a 0, and the highest was a 100.

1. Describe the shape of each distribution and estimate the mean, median, and mode for each.

   A) Mean _____ Median _____ Mode _____ Shape _____
   B) Mean _____ Median _____ Mode _____ Shape _____
   C) Mean _____ Median _____ Mode _____ Shape _____
   D) Mean _____ Median _____ Mode _____ Shape _____
   E) Mean _____ Median _____ Mode _____ Shape _____

2. If you had the choice, which distribution would you like your final exam grade to be a part of, and why?

3. Which distribution would you *not* want your final grade to be a part of, and why?

Answers to this exercise are found on page 188.

CheckPoint Answers
Answers to questions on page 165

1. 109.5 ☐
2. 129.5 ☐
3. 125.4 ☐
4. Negatively skewed ☐

SCORE: /4

MEASURES OF CENTER IN CONTEXT

The mean, the median, and the mode are quite useful in describing where the "weight" of the data is, and because they are so useful, they are used quite often. However, these three measures of center are not, as we've just discovered, all created equal, and they should not be used interchangeably. The mean is considered the best and most sensitive measure of center, one that reflects all of the data. But this does not mean that we should *always* use the mean to describe the center of our data. The measure of center we should use depends on what kind of data we're trying to describe.

With *qualitative* data, a measure of center that requires us to add, subtract, multiply, or divide the data in order to derive it is not appropriate. Think back to the example of the survey asking students about their smoking habits. The survey was set up so that a "yes" answer to the question "Do you smoke?" would be coded as a 1 and a "no" answer coded as a 2. These are numerals, not numbers, and they do not indicate amount or count. They only label the category each response belongs to. Because the data here are nominal, it does not make sense to add these numerals together, and because we cannot meaningfully sum the data, calculating a mean is equally meaningless. What does a mean response of 1.67 tell us anyway? It means neither "yes" nor "no."

The median is also less than helpful here. We could find the (imaginary) data point that splits the "yes" and "no" answers in half, but what would it tell us? With *nominal data*, the appropriate and most descriptive measure of center is the mode.

If we had a data set with some order to it, then both the median and the mode would be appropriate. With ordinal-level data, we might want to describe the data point that happened most often and the data point that separates the data set into upper and lower halves. The mean is again inappropriate with ordinal-level data for the same reason that it was inappropriate with nominal-level data: We cannot meaningfully use arithmetic operations with ordinal data, so we cannot calculate a meaningful mean.

If our data are quantitative, either interval or ratio, the mean becomes the best bet for describing the center of our distribution. Now we can add, subtract, multiply, or divide all we want, and the resulting measure of center actually tells us something meaningful about our data. Table 4.5 lists the appropriate measures of center for each type of data we've discussed so far.

Many statistical analysis software programs will calculate a slightly different form of the mean for you that helps in determining the shape of your distribution. In SPSS, for example, this is called the *trimmed mean*. To calculate the trimmed mean, SPSS cuts off the upper and lower 5% of the distribution, then recalculates the *5% trimmed mean* without these extreme values. If there are outliers in the data, a comparison of the mean and the 5% trimmed mean will show that difference. If the trimmed mean is higher than the mean value, it suggests that a low outlier has been pulling down on the mean. Cutting off that outlier lets the mean "rebound" back up. Similarly, a trimmed mean that is lower in value than the mean indicates the presence of a high outlier in the data. Trimming this high value lets the mean rebound back down. When you're describing your statistics, you can use a comparison of the trimmed mean and the mean to support your discussion of outliers in the data.

TABLE 4.5 Measures of center and scales of measurement

Type of Data	Appropriate Measure of Center
Nominal	Mode
Ordinal	Mode and median
Interval	Mean, median, and mode
Ratio	Mean, median, and mode

SUMMARY

There are three measures of the typical value in a set of data: the *mode* (the most frequently occurring measurement in the set), the *median* (the measurement that splits the set of data exactly in half), and the *mean* (the arithmetic average of all measurements in the set). The relationship between these three measures describes the shape of the distribution that the measurements are a part of. In a normal distribution, the mean, median, and mode all have the same value. When the distribution is skewed, the mean is pulled in the direction of outliers. In a positively skewed distribution, high outliers pull the mean up, and the mean has a higher value than the mode and the median. When there are low outliers, the distribution is said to be negatively skewed, and the mean has a lower value than either the median or the mode.

TERMS YOU SHOULD KNOW

mean, p. 156
median, p. 155
mode, p. 154
negatively skewed distribution, p. 162

normal distribution, p. 162
outlier, p. 161
positively skewed distribution, p. 162

GLOSSARY OF EQUATIONS

Formula	Name	Symbols	
$\dfrac{n+1}{2}$	The position of the median in an ordered set		
$\bar{X} = \dfrac{\Sigma x}{n}$	The mean	$\bar{X}$	"X-bar" (in the sample)
		μ	"Mu" (in the population)

WRITING ASSIGNMENT

In most scientific articles, the statistical analysis of the data (both descriptive and inferential) can be found in what is known as the results section. This is where the author explains the patterns seen in the data (the descriptive statistics) as well as the results of any inferential tests that were run on the data. APA style uses a standard format for presenting descriptive statistics. Your job here is to write an APA style paragraph (two paragraphs at most) describing the major patterns in the data from the time perception study (shown in Table 4.8 in the *Practice Problems* section). Feel free to use graphs of the data to help make your point and to back up any conclusions you draw from the data with the appropriate statistics. Table 4.6 below lists the symbols used in an APA style article to designate the mean, median, and mode, and it indicates when it is appropriate to use the symbol instead of the spelled-out word.

Table 4.6 Symbols for descriptive statistics in APA Style

Statistic	APA Style Symbol	When you Should Use the Symbol	Example
Mean	*M*	If you are referring to the mean directly, in the text, use the word "mean." If you are referring to the mean indirectly, use the symbol, usually in parentheses at the end of the sentence.	– The mean weight of the sample of rat pups was 8.6 grams. – On average, the rat pups weighed less than 10 grams ($M = 8.6$ g).
Median	*Mdn*	Again, direct reference to the median uses the word "median," and indirect reference uses the symbol, usually in parentheses.	– The distribution of rat pup weights was positively skewed, with a mean of 8.6 grams and a median of 6.0 grams. – The rat pups generally weighed less than 10 grams ($M = 8.6$ g, $Mdn = 6.0$ g).
Mode	Mode	Use the word "mode" in all situations. You will often see the mode referred to as the "modal value" in a set. You will also see reference made to the modal frequency (how often the modal value occurred).	– The modal rat pup weight was 7.7 grams. *Seventy-three percent of the rats in the sample weighed 7.7 grams.

*When you start a sentence with a number, you spell the number out.

PRACTICE PROBLEMS

1. Find the mean, median, and mode for the following data:
   a. 5, 7, 3, 8, 2, 3, 3, 1, 9
   b. 104, 139, 145, 150, 167, 205, 205, 20
   c.

Stem	Leaf
1	0,3,5,5
2	0,1,4,7,7,7,7,8,8,9
3	3,3,5,8,9
4	0,1
5	7,8,8,8,9,9

2. Suppose you wanted to find the mean salary for the employees of a small company. Your data are shown in the stem-and-leaf graph shown below.
   a. What are the benefits of reporting both the mean and the 5% trimmed mean?
   b. How do any outliers in these data affect your interpretation of the average salary for employees?

Stem	Leaf
33	0,3,3,4,7,7,9
4	0,1,4,7,7,7,8,8,8,9
5	0,0,1,1,5,8,9
6	3,4,4,4,6
7	
8	
9	
10	3,9,9,9
11	4,7
12	9

*Salary is reported in tens of thousands.

3. Find the mean, median, and mode of the salary data in question 2.

4. Match the following statistics (mean, median, and mode) to the distributions pictured in Figure 4.7.

Histogram	?	?	?	?
Mean	5.00	4.06	3.81	4.00
Median	3.00	4.00	4.50	4.00
Mode	1.00	6.00	6.00	4.00

FIGURE 4.7
Histograms to accompany question 4.

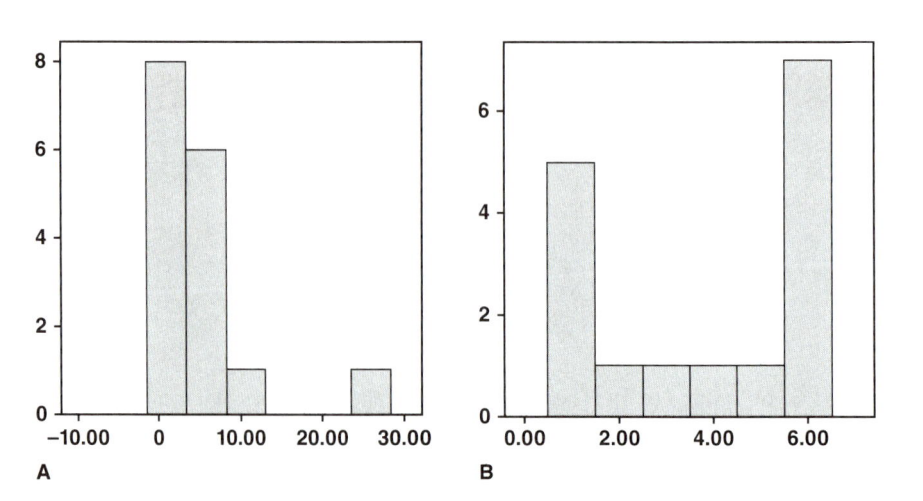

170   STATISTICS IN CONTEXT

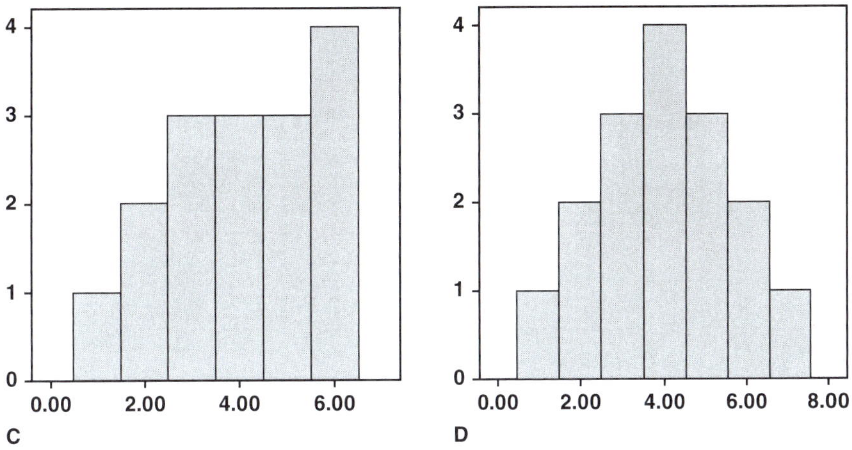

FIGURE 4.7 *Continued*

5. Suppose your data were distributed normally. What difference, if any, would you expect to see between the mean and the trimmed mean?

6. Find the mean, median, and mode of the sleep data shown in Table 4.7, first computing these values by hand (and calculator) and then using SPSS to find the averages. Graph the data using an appropriate graph type, and describe the shape of the distribution: Is it normal or skewed, and if it is skewed, is it positively or negatively skewed?

Table 4.7 Sleep times

Minutes	f	Minutes	f	Minutes	f
120	2	370	1	500	1
150	1	375	5	505	2
190	1	380	4	510	9
215	1	385	1	515	1
225	1	400	9	520	3
240	2	405	2	525	7
260	2	410	1	530	3
270	5	415	2	540	3
275	2	420	11	570	5
280	1	425	2	585	1
285	2	430	3	590	1
300	5	435	7	600	3
315	1	440	4	615	1
320	1	445	1	620	1

CHAPTER 4 Measures of Central Tendency: What's So Average About the Mean?

Table 4.7 Continued

330	5	450	12	650	1
335	1	455	1	660	1
340	2	460	5	675	1
345	4	465	7	690	2
350	3	470	2	720	1
355	1	480	13	730	1
360	14	490	1		
365	1	495	6		

7. Find the mean, median, and mode of the time perception data shown below. Are the errors normally distributed? If not, how is the distribution skewed?

Table 4.8 Error (in milliseconds) on time-estimation task

ID	Age Group	Error	ID	Age Group	Error
1	1	9.00	11	2	−3.00
2	1	13.00	12	2	3.00
3	1	13.00	13	2	7.00
4	1	1.00	14	2	4.00
5	1	5.00	15	2	5.00
6	1	−11.00	16	2	5.00
7	1	11.00	17	2	10.00
8	1	1.00	18	2	1.00
9	1	−1.00	19	2	11.00
10	1	−16.00	20	2	14.00

8. Golf is a lot of fun but has a very steep learning curve. It can take years of practice to be able to drive the ball long and straight and play well enough to earn a place on the professional circuit. Figure 4.8 shows two stem-and-leaf plots, generated by SPSS. On the left are the average driving distances for the top 20 players in the Ladies Professional Golf Association (LPGA). On the right are the driving distances for the top 20 men on the PGA tour. (The precise driving distances will vary depending of who is currently in the top 20 of each tour. These values came from 2014.)

STATISTICS IN CONTEXT

```
LPGA Stem-and-Leaf Plot

Frequency      Stem &  Leaf

    2.00       262  .  00
    3.00       263  .  000
    3.00       264  .  000
    2.00       265  .  00
    4.00       266  .  0000
    2.00       267  .  00
     .00       268  .
    1.00       269  .  0
    3.00   Extremes
 (>=273.0)

Stem width:          1
Each leaf:      1 case(s)
```

```
PGA Stem-and-Leaf Plot

Frequency      Stem &  Leaf

    7.00       29  .  7777889
    9.00       30  .  001113344
    2.00       30  .  67
    1.00       31  .  0
    1.00   Extremes
 (>=315)

Stem width:         10
Each leaf:     1 case(s)
```

FIGURE 4.8
Stem-and-leaf plots of average driving distance on two professional golf tours.

a. Find the mean driving distances for players on each tour.
b. Find the median and modal driving distances for players on both tours.
c. Do these two distributions overlap at all? What is the longest distance driven on the LPGA tour? Compare this distance to the shortest distance driven on the PGA tour.
d. Are these two distributions normal in shape? If not, what kind of skew do you see in them?
e. Notice that these two stem-and-leaf plots are not constructed in exactly the same way. What is the difference between the two plots, and why do you think SPSS built them this way?

9. van Loon, van den Wildenberg, van Stegeren, Ridderinkhof, and Hajcak (2010) wanted to know if it was possible to prime the brain to make it more willing, ready, and able to respond to stimuli in the world around us. They used transcranial magnetic stimulation (TMS) to get the cells of the brain fired up and ready to go. In TMS, a magnetic pulse is delivered to the cortex (they aimed the pulse at the motor cortex, the area of the brain responsible for making us move). Previous research had shown that TMS caused neurons to depolarize, making them ready to fire. Participants in this study were asked to squeeze a force sensor as soon as a picture appeared on a screen in front of them. Reaction time (the time between presentation of the picture and a squeeze to the force sensor) was measured both when TMS was presented and without TMS. The data shown in Table 4.9 are similar to what van Loon et al. found.
   a. Find the mean, median, and mode for both conditions (with TMS and without TMS)
   b. Are these two distributions normal or skewed? If skewed, are they positively or negatively skewed?
   c. Did TMS affect reaction time? If so, in what way?

Table 4.9 Reaction time (in milliseconds) with and without TMS

ID	TMS	No TMS
1	490	520
2	580	550
3	460	540
4	520	500
5	490	520
6	490	520
7	520	480
8	500	530
9	490	500
10	460	500
11	500	520

10. Table 4.10 shows the weights of 20 mice. Find the mean, median, and mode. Is the distribution of weights normal in shape or skewed?

Table 4.10 Weights (in grams) for 20 mice

ID	Weight	ID	Weight
1	19	11	21
2	17	12	21
3	20	13	20
4	20	14	22
5	16	15	21
6	19	16	17
7	19	17	18
8	16	18	17
9	19	19	19
10	19	20	18

11. In a set of students studying for the final exam in this class, the average time spent studying was 18 hours, and the median time spent studying was 5 hours. Indicate whether the following statements are *true* (T) or *false* (F), or if you are

unable to determine (UD) whether the statement is true or false. Circle the appropriate answer.

a. Some students in the group are probably studying slightly less than the average amount of time.   T   F   UD
b. There are some students who study significantly more than average in the set.   T   F   UD
c. This data set has outliers at the low end of the distribution.   T   F   UD
d. This data set is positively skewed.   T   F   UD
e. This data set is normally distributed.   T   F   UD
f. The mean is being pulled up in value by outliers at the high end of the distribution.   T   F   UD
g. Fifty percent of the students in the sample study 5 hours or more.   T   F   UD
h. The high value outlier in this data set happened only once.   T   F   UD

12. Consider the distribution shown in Table 4.11 below.
    a. Calculate the measures of center that we discussed in this chapter:
       Mean   = _____
       Median = _____
       Mode   = _____
    b. Create a grouped frequency distribution of the data using five groups:
       What is the width of each group? _____
       How many groups do you actually have in the grouped frequency distribution?
       _____
    c. Recalculate the measures of center for the grouped frequency distribution:
       Mean   = _____
       Median = _____
       Mode   = _____
    d. How different are these two sets of measures of center?

Raw Data		Grouped Data	
Mean =	_____	Mean =	_____
Median =	_____	Median =	_____
Mode =	_____	Mode =	_____

Table 4.11

x	Interval	f
48		
42		
48		

CHAPTER 4   Measures of Central Tendency: What's So Average About the Mean?

Table 4.11	Continued		
54			
57			
55			
50			
50			
50			
49			
50			
50			
53			
52			
51			
50			
55			
45			
53			
54			
60			

13. Suppose you had the following data given in Table 4.12.

Table 4.12	Salary at "Chocolate Daydreams" Bakery for all 10 employees									
Staff ID	1	2	3	4	5	6	7	8	9	10
Salary	15k	18k	16k	14k	15k	15k	12k	17k	90k	95k

  a. What is the mean salary at the bakery? _____
  b. Is the mean the best measure of central tendency for this data set? Why or why not?
  c. What measure of center might provide a better representation of the center of this data set?
  d. Calculate the other measure of center. _____
  e. How much does this other measure of center differ from the mean?

14. Calculate the mean, median, and mode for the grouped frequency distribution shown in Table 4.13.

Table 4.13

Interval	f
500–600	3
600–700	6
700–800	5
800–900	5
900–1,000	0
1,000–1,100	1

15. The trimmed mean is a descriptive statistic that is easy to calculate and will help you determine the shape of the distribution you're dealing with. The trimmed mean is the mean calculated after you have first removed a certain percentage of the largest and smallest values. Let's calculate the 10% trimmed mean for the data in Table 4.14.
    a. We want to remove the highest and the lowest 10% of the data. So, first find the total number of observations in the data set: _____
    b. We want to remove the upper and lower 10% of the observations, so determine what 10% of the total number of observations in the set is: _____
    c. List the lowest and highest 10% of the observations in the set: _____
    d. Calculate the mean when all of the observations are included: _____
    e. Calculate the mean when the upper and lower 10% have been removed: _____
    f. Did removing the upper and lower 10% of the observations change the mean? Why or why not?

Table 4.14 Hypothetical data set for calculating the 10% trimmed mean

x	f
7	1
8	1
9	1
10	3
11	4
12	6
13	6
14	13
15	15

16. One characteristic of depression is that it tends to make us pay more attention to negative stimuli (sadness, pain, death, etc.) than to happy or positive ones. Sardaripour (2014) examined this characteristic of depression by comparing the responses of depressed, manic, and healthy individuals. The participants in this study were asked to view pictures of smiling and happy individuals and pictures of sad, crying, and angry individuals (a total of nine pictures of each type). The number of pictures each group could recall was measured. Table 4.15 presents the results.

Table 4.15 Number of faces recalled by depressed, manic, and healthy individuals

Depressed Pleasant	Depressed Unpleasant	Manic Pleasant	Manic Unpleasant	Control Pleasant	Control Unpleasant
5	8	7	1	4	5
2	5	4	1	6	5
2	5	5	5	6	5
5	7	6	5	4	4
4	6	4	3	5	4
3	6	4	4	5	4
4	4	6	5	2	3
4	7	4	5	5	5
2	4	5	6	6	5
3	5	4	5	5	6

a. Fill in the table below with the measures of center.

	Depressed Pleasant	Depressed Unpleasant	Manic Pleasant	Manic Unpleasant	Control Pleasant	Control Unpleasant
Mean						
Median						
Mode						
Mean overall total						

b. Did depressed, manic, and control individuals differ in the number of faces they could recall overall (when pleasantness or unpleasantness is disregarded)?
c. Was there a difference across groups in the number of pleasant faces recalled?
d. Was there a difference across groups in the number of unpleasant faces recalled?

17. "Milk," the commercial used to say, "it does a body good," and mothers the world over took that saying to heart. As it turns out, it may do your brain good as well. Milk consumption may aid in the production of serotonin (5-HT) and so perhaps contribute to better mental health. Takeuchi et al. (2014) compared "diurnal type" (i.e., whether you're a morning person whose energy spikes early in the day or an evening person who doesn't really get going until late in the day) for two groups of children. Group 1 consisted of kids who drank cow milk for breakfast. Kids in group 2 did not drink milk at breakfast. Diurnal type was measured on a survey developed by Torsvall and Akerstedt (1980). Scores on this scale could range from 7 to 28, with low scores indicating a preference for the evening and high scores indicating a preference for the morning. The researchers also asked the mothers of these children "How frequently does your child become depressed?" Answers to this question were on a 4-point scale, where 4 = *frequently*, 3 = *sometimes*, 2 = *rarely*, and 1 = *not at all*. The results in Table 4.16 are similar to the results Takeuchi et al. (2014) found.

Table 4.16 Diurnal type score (DTS) and maternal response to depression question for 12 children

ID	DTS Milk	DTS No Milk	Depression Milk	Depression No Milk
1	19	15	4	1
2	22	18	1	1
3	20	14	1	2
4	26	16	1	3
5	23	23	2	3
6	22	21	1	4
7	18	20	1	2
8	20	28	2	2
9	18	15	1	1
10	19	24	2	1
11	24	12	3	1
12	22	22	1	1

a. Fill in the blanks in the table shown below.

	DTS		Depression	
	Milk	No Milk	Milk	No Milk
Mean				
Median				
Mode				

b. In your opinion, did drinking milk affect diurnal preference (whether the child was active in the morning or the evening)?

c. In your opinion, did drinking milk affect frequency of depression in these children?

18. At the University of Illinois at Chicago, Professor S. Jay Olshansky studies our aging society. Professor Olshansky has found that life expectancy and level of education are related to one another. The data shown in Table 4.17 comes from a story in *The Atlanta Journal Constitution* of Saturday, April 19, 2014, describing the results of Professor Olshansky's research.

Table 4.17 Life expectancy (in years) by race, gender, and level of education

Education	White Women		Black Women		Hispanic Women		White Men		Black Men		Hispanic Men	
	1990	2008	1990	2008	1990	2008	1990	2008	1990	2008	1990	2008
Less than HS	78.5	73.5	72.7	73.6	79.1	82.6	70.5	67.5	62	66.2	70.3	77.2
HS diploma	77.6	78.2	70.3	74.0	73.3	82.0	70.0	72.1	61.2	67.3	75.1	76.7
Some college	82.1	83.4	76.7	79.6	84.1	85.5	76.8	79.1	69.7	75.1	82.1	81.2
College degree or more	80.4	83.9	74.6	80.1	82.7	85.5	75.4	80.4	68	75.9	75.6	82.7

a. Compute the mean life expectancy by race and year (disregarding level of education for the moment), and fill in table shown below. Use the data in this table to answer the questions.

	White Women	Black Women	Hispanic Women	White Men	Black Men	Hispanic Men
1990						
2008						
Overall mean	Mean life expectancy for all women in 2008			Mean life expectancy for all men in 2008		

      i. What is the effect of race on life expectancy for women?
      ii. What is the effect of race on life expectancy for men?
      iii. Did life expectancy for women change between 1990 and 2008? If so, how did it change?
      iv. Did life expectancy for men change between 1990 and 2008? If so, how did it change?
      v. Historically, women live longer than men. Was that the case in 2008?

  b. Compute the mean life expectancy by level of education and year (disregarding race). Use these data to answer questions c through e.

	1990	2008	Overall Average
			Women, 1990 and 2008, All Education Levels
Women, some HS			
Women, HS diploma			
Women, some college			
Women, college degree or more			
			Men, 1990 and 2008, All Education Levels
Men, some HS			
Men, HS diploma			
Men, some college			
Men, college degree or more			

  c. What is the effect of level of education on life expectancy for women?
  d. What is the effect of level of education on life expectancy for men?
  e. Generally speaking, do women live longer than men?

19. According to a 2010 survey by the Organisation for Economic Co-operation and Development (OECD), the average number of student contact hours (time spent teaching) for primary school teachers in the United States is 30.47 hours per week. Suppose that at the school where you teach you found the mean number of contact hours per week was 27, the median was 30, and the mode was 32.
    a. Is the distribution of student contact hours skewed, and if so, in which direction?
    b. Which of the following statements is likely true about the teachers in your school?
        i. A few teachers have significantly fewer contact hours than do the majority.
        ii. A few teachers have significantly more contact hours than do the majority.

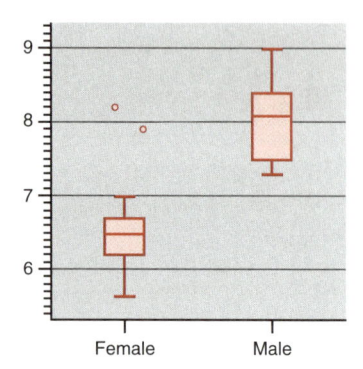

FIGURE 4.9
Box-and-whisker plot of height (in inches) by sex of high school basketball players.

20. Indicate the measure or measures of center that should be used for each of the following sets of data:

Data	Mode	Median	Mean	Mode & Median	All Three
a. Average clothing size worn by a set of 25 women					
b. Average attractiveness rating of 100 men on a 5-point scale (1 = *unattractive*, 3 = *neutral*, and 5 = *very attractive*)					
c. Average body weight of a set of 30 undergraduates					
d. Average shoe size in a set of 40 nine-year-olds					
e. Average political party preference for a set of 100 first-time voters					
f. Average finishing position for a set of 20 hurdlers when running on a wet track					
g. Average lap time (in milliseconds) for the same set of 20 hurdlers running on a wet track					

21. Given the following descriptive statistics (parameter estimates), determine if the value of the missing data point is greater than or less than the median.

    Mode = 6

    Median = 5.5

    Data: 6, 2, 8, ?, 2, 6, 5, 1, 7

182   STATISTICS IN CONTEXT

22. Suppose you were really interested in ice hockey and you attended an All-Star skills competition. During the competition, you record each player's shot speed in the fastest shot competition. Given the following data, calculate the mean and median for everyone in the competition. Then, calculate the mean and the median for each team, for each position, and for each position on each team. With data we can determine the mode, but would you recommend it?

Table 4.18 Data from the fastest shot competition

ID	Team	Position	Shot Speed mph
1	Green	Forward	92
2	Red	Forward	96
3	Green	Defense	93
4	Green	Forward	101
5	Red	Defense	98
6	Red	Forward	89
7	Red	Forward	91
8	Green	Forward	92
9	Red	Defense	88
10	Green	Defense	88
11	Red	Defense	90
12	Red	Defense	88
13	Green	Forward	98
14	Green	Defense	100
15	Red	Forward	96
16	Green	Forward	89
17	Red	Forward	91
18	Green	Defense	89
19	Red	Defense	100
20	Green	Defense	97

23. It's not uncommon to find yourself missing some data in your set. Let's say you stepped away from the All-Star skills competition in question 22 to get some hot dogs and a drink. Unfortunately, that means you missed some of the players in the fastest shot competition. Your data now look like this:

Table 4.19 Data from the fastest shot competition: Data missing

ID	Team	Position	Shot Speed mph
1	Green	Forward	92
2	Red	Forward	96
3	Green	Defense	93
4	Green	Forward	101
5	Red	Defense	98
6	Red	Forward	89
7	Red	Forward	—
8	Green	Forward	—
9	Red	Defense	—
10	Green	Defense	—
11	Red	Defense	—
12	Red	Defense	88
13	Green	Forward	98
14	Green	Defense	100
15	Red	Forward	96
16	Green	Forward	89
17	Red	Forward	91
18	Green	Defense	89
19	Red	Defense	100
20	Green	Defense	97

Recalculate the mean and the median for everyone in the competition, for each team, for each position, and for each position on each team. How have your distributions changed? What measure of center is affected the most?

24. Suppose you really don't like all that missing data in question 23. What value could you put in place of the missing data points that would change your distributions the least?

STATISTICS IN CONTEXT

25. You have a large data set containing the average household income for 76,024 households. Unfortunately, there are a substantial number of missing data points (~30%). You are told that the missing data are coded with a special missing numeric code, but you don't know what that code is. Assuming a normal distribution of the available data, how could you use a measure of center to discover the special missing code? Is there a way to find it graphically?

26. Below are data containing the ages of 30 participants in a community fitness program. The program coordinator has asked you to provide a brief summary of who is taking part in the program and would like the data presented in a grouped format. Group the participants in 10-year increments (e.g., 20–29, 30–39, etc.), and provide the grouped mean, median, and mode for each increment.

Table 4.20 Ages (in years) of 30 fitness program participants

ID	Age	ID	Age	ID	Age
1	36	11	22	21	66
2	31	12	43	22	49
3	55	13	33	23	41
4	71	14	22	24	34
5	64	15	65	25	65
6	29	16	21	26	50
7	61	17	38	27	69
8	29	18	70	28	25
9	27	19	25	29	43
10	33	20	50	30	37

27. A researcher is very interested in how caffeine can affect commuters' reaction time while driving to work early in the morning. Is it better to drink your coffee before you jump in the car, or is it better to wait until you get into the office before having your java fix? She decides to run a quick experiment to see if, indeed, drinking coffee before heading out is the way to go. Recruiting 20 participants from one of her classes, she has each perform a complex video reaction task. After collecting this baseline, she lets half the participants relax for 30 minutes and then runs them through the test again. The other half of the participants are asked to consume the equivalent of three cups of coffee in 5 minutes and then wait for 25 minutes before doing the reaction time test again. Below you will find the results from the experiment. The reaction time values (in seconds) are for the second test. Assume there was no difference between groups in the first test.

Table 4.21 Reaction time (RT; in seconds) of participants with and without caffeine

ID	Group	RT	ID	Group	RT
1	Control	3.14	11	Caffeine	2.17
2	Control	4.57	12	Caffeine	5.29
3	Control	3.66	13	Caffeine	1.83
4	Control	3.77	14	Caffeine	2.74
5	Control	4.99	15	Caffeine	19.34
6	Control	7.71	16	Caffeine	4.86
7	Control	6.84	17	Caffeine	4.05
8	Control	7.1	18	Caffeine	5.17
9	Control	3.56	19	Caffeine	4.59
10	Control	7.72	20	Caffeine	3.3

The researcher concludes that there wasn't much difference between reaction times for the two groups based on their means. Calculate the mean and median for each group. What conclusion would you draw?

28. Find the measures of center for the calorie count data shown below. Are calories distributed normally for hamburgers? What about for milk shakes and French fries?

Table 4.22 Nutritional information from fast-food restaurants

Restaurant	Hamburger Type	Hamburger Calories	Hamburger Weight (in grams)	French Fries (Medium) Calories	Chocolate Shake (Large) Calories
A&W	Deluxe Hamburger	461	212	311	1120
Dairy Queen	Classic Grill Burger	540	213	380	1140
Hardee's	¼-lb. Double Cheeseburger	510	186	250	N/A
In-N-Out	Cheeseburger w/ onions	390	243	400	690
Krystal	BA Burger	470	202	470	N/A
McDonald's	McLean Deluxe	340	214	380	1160
Burger King	Whopper	670	290	360	950
Wendy's	Deluxe Double Stack	470	211	490	393

29. Graph the mean calorie count for the eight restaurants described in Table 4.22. Are the data continuous or discrete?

30. Complete the crossword below.

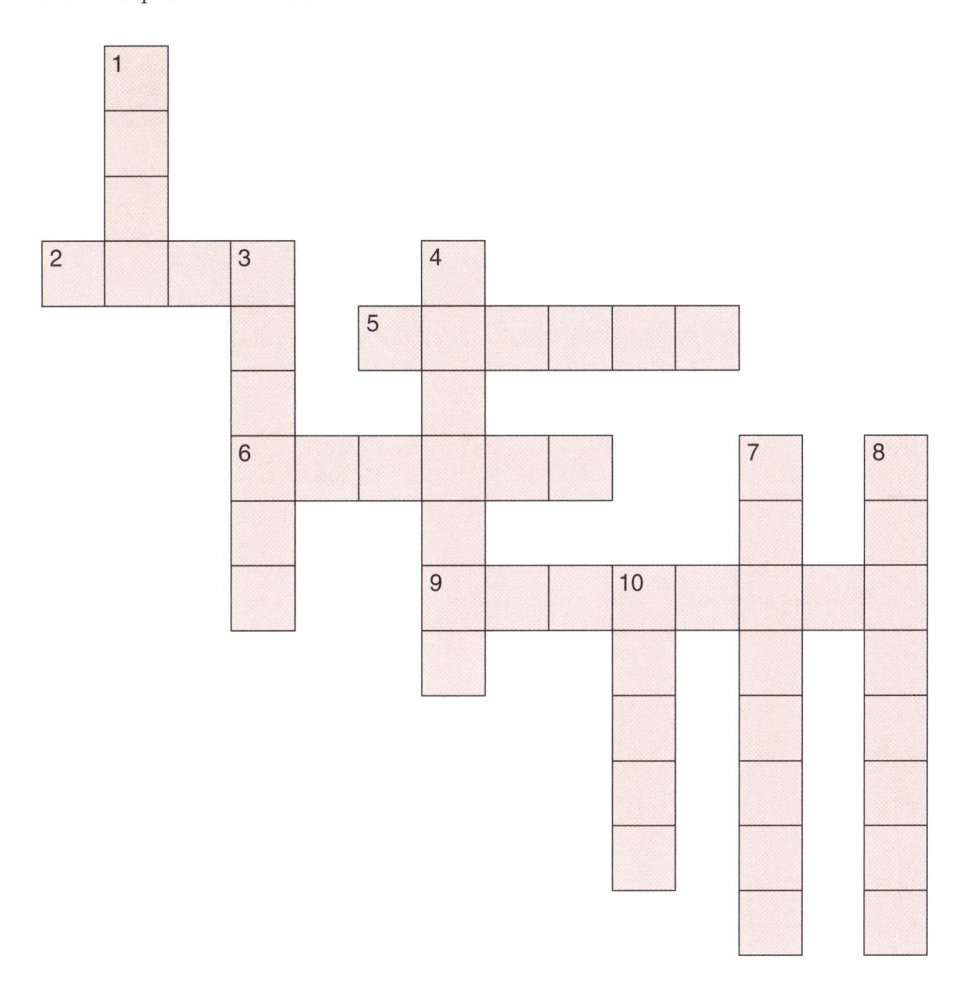

ACROSS

2. Arithmetic average of the set of observations
5. Mean is the most _____ measure of center
6. Observation that splits distribution in half
9. All three measures of center are examples of this

DOWN

1. The most frequently occurring
3. Mean, median, and mode are all the same value in this type of distribution
4. Mode is appropriate measure of center for this kind of data
7. Mean is smaller than median in a distribution with _____ skew
8. Mean is larger than median in a distribution with _____ skew
10. Mean is appropriate measure of center for this kind of data

Think About It . . .

ESTIMATING MEASURES OF CENTER

SOLUTIONS

1. Find the mode for each distribution.

A	B	C
4	66	No Mode

2. Now, without doing any calculations, determine if the mean and median will be smaller, larger, or the same as the mode (check the box that you think applies for each distribution).

A	B	C
☐ Smaller	☐ Smaller	☐ Smaller
☐ Larger	☑ Larger	☐ Larger
☑ Same	☐ Same	☑ Same

3. Now check yourself. Find the mean and the median for each distribution. Were your estimations of these two measures of center correct?

A	B	C
Median 4.0	Median 72.00	Median 27.00
Mean 3.86	Mean 70.83	Mean 27.00

SHAPES OF DISTRIBUTIONS

SOLUTIONS

1. Describe the shape of each distribution, and estimate the mean, median, and mode for each.

   A Mean __40__ Median __55__ Mode __30__ Shape __Positive skew__

   B Mean __60__ Median __45__ Mode __55__ Shape __Negative skew__

   C Mean __50__ Median __50__ Mode __50__ Shape __Normal__

   D Mean __50__ Median __50__ Mode __25 & 75__ Shape __Bimodal__

   E Mean __50__ Median __50__ Mode __No mode__ Shape __Rectangular__

2. If you had the choice, which distribution would you like your final exam grade to be a part of, and why?   **I would pick distribution B: Most of the grades are at the high end of the scale, and only a few are low outliers. Then again, if you are especially confident that you have done well on the exam, you might prefer distribution A, where your score, as one of the outliers, would be well above the mean.**

3. Which distribution would you *not* want your final grade to be a part of, and why?
I would pick distribution A. Most of the grades are fairly low, and the few outliers are high value—there are not many of these outliers, so my chances of having one of those good grades are fairly low.

REFERENCES

Organisation for Economic Cooperation and Development. (2010). Education at a glance 2010: OECD indicators. Retrieved from http://www.oecd-ilibrary.org/docserver/download/9610071e.pdf?expires=1504380811&id=id&accname=guest&checksum=83BCC80412761F469F10AADB319ED509.

Quetelet, L. A. J. (2013). *A treatise on man and the development of his faculties.* T Smibert (Ed.) and R. Knox (Trans.). New York, NY: Cambridge University Press. (Original work published 1835).

Quetelet, L. A. J. (2012). *Letters Addressed To H.R.H. The Grand Duke Of Saxe Coburg and Gotha: On the Theory of Probabilities, as Applied to the Moral and Political Sciences*, Charleston, SC: Nabu Press. (Original work published 1849.)

Sardaripour, M., (2014). Comparing the recall to pleasant and unpleasant face pictures in depressed and manic individuals. *Psychology, 5*(1), 15–19.

Stigler, S. M. (1997). Quetelet, Lambert Adolphe Jacques. In N. L. Johnson and S. Kotz (Eds.) *Leading personalities in statistical sciences* (pp. 64–66). New York, NY: John Wiley and Sons.

Takeuchi, H., Wada, K., Kawasaki, K., Krejci, M., Noji, T., Kawada, T., Nakada, M., & Harada, T. (2014). Effects of cow milk intake at breakfast on the circadian typology and mental health of Japanese infants aged 1-6 years. *Psychology, 5*(2), 172–176.

Torsvall, M. D., & Akerstedt, T. A. (1980). A diurnal type scale: Construction, consistency and validation in shift work. *Scandinavian Journal of Work and Environmental Health, 6*, 283–290.

U.S. Census Bureau. (2010). Retrieved from http://factfinder.census.gov/faces/nav/jsf/pages/index/xhtml.

van Loon, A. M., van den Wildenberg, W. P., van Stegeren, A. H., Hajcak, G., & Ridderinkhof, K. R. (2010). Emotional stimuli modulate readiness for action: A transcranial magnetic stimulation study. *Cognitive, Affective and Behavioral Neuroscience, 10*(2), 174–181.

CHAPTER FIVE

VARIABILITY: THE "LAW OF LIFE"

*Consistency is contrary to nature, contrary to life.
The only completely consistent people are dead.*

—ALDOUS HUXLEY

Everyday Statistics

PREPARING FOR "SNOWMAGEDDON," A.K.A. THE "SNOWPOCALYPSE"

We all know that the time it takes to get to work or to school varies, depending on the weather, the road conditions, and all those other drivers. Change in one of these variables, like an unexpected storm or a minor accident, can introduce tremendous variability in the time it takes to get to where you're going.

According to statisticians, the average commute time in metropolitan Atlanta (population 5.7 million) is 28.7 minutes. Here is what happened in 2014 to Atlanta commuters when one variable, the weather, didn't cooperate. A minor daytime snow storm (producing an accumulation of just 2.5 inches of snow and ice) in an area of the country that almost never sees snow triggered a mass exodus from the city. Thousands of people all tried to head home at the same time, and the result was a traffic jam of epic proportions. Some drivers reportedly spent more than 18 hours stuck on the interstate, and traveling just 4 miles took as much as 6 hours!

Statistics can provide you with a measure of how consistent, or inconsistent, a particular measure can be. How do we quantify variability? We will find out in Chapter 5!

OVERVIEW

MEASURING VARIABILITY
CONSISTENCY AND INCONSISTENCY IN DATA
MEASURES OF VARIABILITY
THE STANDARD DEVIATION
STANDARD DEVIATION IN CONTEXT
DESCRIPTIVE STATISTICS IN CONTEXT

LEARNING OBJECTIVES

Reading this chapter will help you to . . .

- Understand how the term *variability* is used in science and how averaging over repeated measurement is used to minimize variability. (Concept)

- Know what the range, interquartile range, variance, and standard deviation describe in a distribution and how these measures differ from one another. (Concept)

- Calculate the range, interquartile range, variance, and standard deviation. (Application)

- Explain the difference between the standard deviation and the mean deviation. (Concept)

- Describe at least one reason why modern science has adopted standard deviation as its measure of variability. (Concept)

- Use the descriptive statistics discussed in Chapters 1 through 5 to write a description of data in APA Style. (Application)

MEASURING VARIABILITY

Science involves measurement; in fact, to many people, science *is* measurement. Speaking about science, Lord Kelvin, creator of the Kelvin temperature scale, said,

> When you can measure what you are speaking about, and express it in numbers, you know something about it; but when you cannot measure it, when you cannot express it in numbers, your knowledge is of a meager and unsatisfactory kind. (Kelvin, cited in Stigler, 1986, p. 1)

The problem is that even the most carefully made measurements are often inconsistent. A single individual measuring a single event or object more than once will probably come up with two slightly different measurements. When you add in multiple observers making multiple observations, you can wind up with as many measurements as you have observers and uncertainty piled on top of doubt. Measuring uncertainty has been a major problem in science since its inception. Handled the wrong way, it can lead to academic embarrassment, which is the moral of the story of Giovanni and Jacques Cassini, told in *The Historical Context* on page 193.

CONSISTENCY AND INCONSISTENCY IN DATA

The tale of the Cassinis and Maupertuis is a story about measurement error in science. Most of the time, when we think of an "error," we think of a mistake that a person has made—someone wasn't paying attention, or placed the ruler in the wrong spot, or let the tape measure slip while measuring. But in science, we actually mean something just a bit different by the term *error*.

Suppose I asked you to measure a degree of latitude. You would take out your sextant, line it up with the horizon and the sun (at exactly 12-o'clock noon), and measure the angle of elevation of the sun above the horizon in degrees. The difference between the angle as you measure it and the true angle of elevation is an example of measurement error. But wait, you say: If I already know the true angle of elevation, why am I measuring it to begin with? The problem is that we *don't* know the true measure of *anything*, and we never can know it. It is impossible to know what the true measure of the angle really is, because all measurement devices, as well as all of the people who use these devices to measure things, are imperfect. So, all measurements are imperfect, and measurement error, or variability, is an inherent part of measuring anything.

The other problem with measuring just about anything on earth is that things on earth change. Perfect consistency (or *constancy*) in the size of something, or in the duration of an event, or in the appeal of something, just doesn't happen

The Historical Context

WHAT IS THE SHAPE OF THE EARTH?

The ancient Greeks knew that individual observers differed in what they reportedly saw, but the Greeks did not figure out how to fix it. The idea that averaging over a series of measurements to reduce observational or systematic error first shows up in the work of the famous astronomer Tycho Brahe toward the end of the sixteenth century. Brahe, recognizing that even his most careful measurements of the position of a star did not always agree, took many measurements over 6 years. He then found the average position of this star and reported that mean.

Variability shows up in another astronomical problem, this time dealing with the shape of the earth itself. In 1735, the French launched an expedition to Lapland in order to determine which of two theories about the shape of the earth was correct. The first theory, proposed by Sir Isaac Newton, was that the earth was flattened at the poles. The alternative theory, proposed by French–Italian astronomers the Cassinis (Giovanni, the father [1625–1712], and Jacques, the son [1677–1756]), was that the earth was flattened at the equator. Figure 5.1 illustrates the two sides of this argument.

The mission of the expedition, led by French mathematician Pierre Louis Maupertuis (1698–1759), was to measure the size of a single degree of latitude (lines of latitude run east–west on a map) in Lapland, selected because it is situated very close to the North Pole. You can see in Figure 5.1 how a degree of latitude would differ for these two theoretical earths. They researchers planned to then compare their measurement with a degree of latitude already made in Paris by the Cassinis.

To get an accurate measurement of this degree of latitude, several observers took several individual measurements, and then these measurements were averaged. When the Lapland expedition later returned with a measurement different from the one obtained in France, the French Enlightenment thinker Voltaire congratulated Maupertuis for having "flattened both the earth and the Cassinis" (Plackett, 1958, p. 133). The idea that averaging over a series of measurements reduced error in those measurements took hold and became a part of science.

In case you're interested, according to the Royal Astronomical Society of Canada, one degree of latitude measured at the equator is 0.69569 miles (1.1196 km) wider than one degree of latitude measured at the pole (http://calgary.rasc.ca/latlong.htm). As the Society's website explains—confirming Newton's theory—"The distance between latitudes increases toward the poles because the earth is flatter the further you are from the equator."

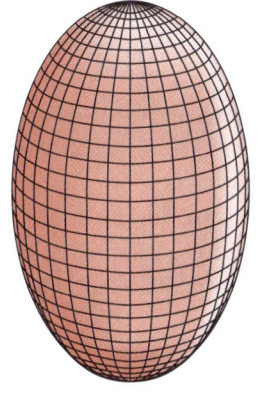

FIGURE 5.1
Two theories of the shape of the earth: (a) Newton's theory that the earth is flattened at the poles; (b) the Cassinis' theory that the earth is flattened at the equator.

very often in the real world. John Fitzgerald Kennedy, thirty-fifth president of the United States, is quoted as saying that "change is the law of life." ("Address in the Assembly Hall at the Paulskirche in Frankfurt," 1963). That law is evident when we go to measure almost anything.

Think about our sleeping college students in the previous chapters for a moment. If I measure the sleep duration for just one student, I will find that she might report sleeping 2 hours the first night, 3 hours the next, and no sleep at all on the third night (it's probably finals time and she's pulling all-nighters). She is fairly consistent across these three nights (she doesn't sleep anywhere near enough), but there is variability in this sample. Another student might sleep 2 hours the first night, 15 hours the next, and 4 hours the night after that. Clearly, this student is much more variable than the first one. Even if we all agree about the number of minutes each student spent sleeping on a given night, the data change from one night to the next. Being able to describe the variability in my data would help me understand the behavior I'm interested in.

Speaking of our sleepy students, let's consider what we know about them and think about error in measurement. We know that:

- Sleep duration, measured in minutes, is quantitative data measured on a ratio scale.
- The shortest period of time participants reported sleeping was somewhere between 120 to 179 minutes, and the longest night's sleep was somewhere between 720 and 779 minutes (around 12 hours of sleep in one night).
- Most of the students (89.37%) in the sample slept 8 hours or less on the night in question.
- The mean time spent sleeping on the first night of the study was 408.67 minutes, the median sleep duration was 425 minutes, and the mode was 450 minutes.
- Looking at the relationship between mean, median, and mode, we can say that we have a negatively skewed distribution (mean < median < mode), telling us that we have some low-value outliers in our sample (in this case, the two students who slept only 2 hours).

Along with knowing what the center of the distribution looks like (what the mean, median, and mode are, and how the set of measurements are distributed), we might also want to know how much *variability* was there is in this set of measurements—in other words, how consistent (or inconsistent) were the measures.

MEASURES OF VARIABILITY

THE RANGE

range The difference between the largest and the smallest values in a set of data. The range provides a relatively insensitive measure of variability.

So, how might you go about measuring variability? One of the easiest ways to do so is to calculate the **range** of the data. Doing so is easy: You simply subtract the minimum value from the maximum. Box 5.1 shows the "formula" for calculating the range.

> **BOX 5.1** **Calculating the range**
>
> $$\text{range} = \text{maximum} - \text{minimum}$$

The range is a quick and easy way to measure variability. There's a problem with the range, however, and you can see it in the two distributions (and their frequency polygons) shown in Figure 5.2.

Suppose these two distributions show the grades on a 10-point quiz from two classes of 17 students each. Notice that Distribution A has a good bit of variability. Students in this class were all over the map in terms of their grades on the quiz. One person got nine points; several only got one, two, or three points; and curiously, no one was in the middle in terms of score on the quiz. Distribution B is quite different in that students in this class were remarkably consistent: All but one student flunked the quiz (16 out of the 17 got only one point on the quiz). Only one student did well, receiving nine points. There is very little variability in these scores (and one unhappy professor in this classroom).

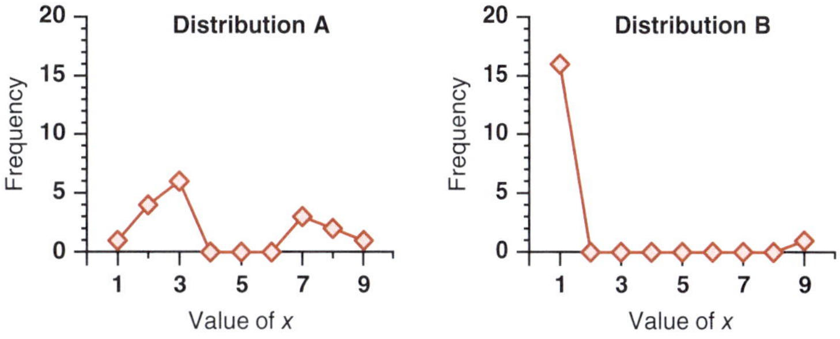

So, the problem with measuring variability using the range becomes obvious quite quickly. Both of these distributions have exactly the same range, but they clearly do not have the same amount of variability. The reason for this discrepancy is probably equally obvious: The range responds only to two extreme values in the set, functionally "ignoring" any value other than the minimum and maximum in the set.

Distribution A

x	f
1	1
2	4
3	6
4	0
5	0
6	0
7	3
8	2
9	1

Range: 9 − 1 = 8

Distribution B

x	f
1	16
2	0
3	0
4	0
5	0
6	0
7	0
8	0
9	1

Range: 9 − 1 = 8

FIGURE 5.2
Problem with the range as a measure of variability.

THE INTERQUARTILE RANGE

There is another kind of range that is also easy to calculate and gets around the problem with values in the middle of the distribution. The **interquartile range (IQR)** is considered a more "robust" statistic than the range because it overcomes the problem the range has of ignoring the values in the middle of the set. We should probably begin with a short discussion of what a *quartile* is. Think back to our discussion of the median as the value that splits the distribution in half—into the

interquartile range (IQR) The range of the central 50% of the data; the difference between the first quartile and the third quartile.

upper and lower 50%. A quartile splits the data into quarters: the *first quartile* marks the lower 25%; the *second quartile* marks the lower 50%; and the *third quartile* marks the lower 75% and, therefore, the upper 25% of the data. The median is also known as the second quartile.

We will need to find three values in order to split the data set into four parts. The easiest way is to use what we already know about finding the position of the median in order to find the position of the three values that split the data into quartiles. Box 5.2 illustrates this process.

BOX 5.2   **Finding the quartiles in a set of data**

$$\frac{n+1}{2} = \text{position of the median}$$

Step 1: Find the position of the median, splitting the data into halves.
Step 2: Find the median of the lower half (the first quartile).
Step 3: Find the median of the upper half (the third quartile).

Let's use a small data set to try out this process (shown in Box 5.3)

BOX 5.3   **Using quartiles to find the IQR**

Data Set (in order)

Position		Value		
Position	1st	5		
Position	2nd	7		
Position	3rd	9	Median of lower 50% = 1st quartile	$\frac{n+1}{2} = \frac{11+1}{2}$
Position	4th	20		= 6th position
Position	5th	21		
Position	6th	35	Median of the whole set = 2nd quartile	The median value is 35
Position	7th	39		
Position	8th	40		
Position	9th	42	Median of the upper 50% = 3rd quartile	$\frac{n+1}{2} = \frac{5+1}{2}$
Position	10th	45		= 3rd position
Position	11th	49		

The **IQR** is the range between the 1st quartile (25th percentile) and 3rd quartile (75th percentile). So, for this data set, the IQR would be the range between 42 and 9. The IQR (the range of the middle 50% of this data) would be 42 − 9 = **33**.

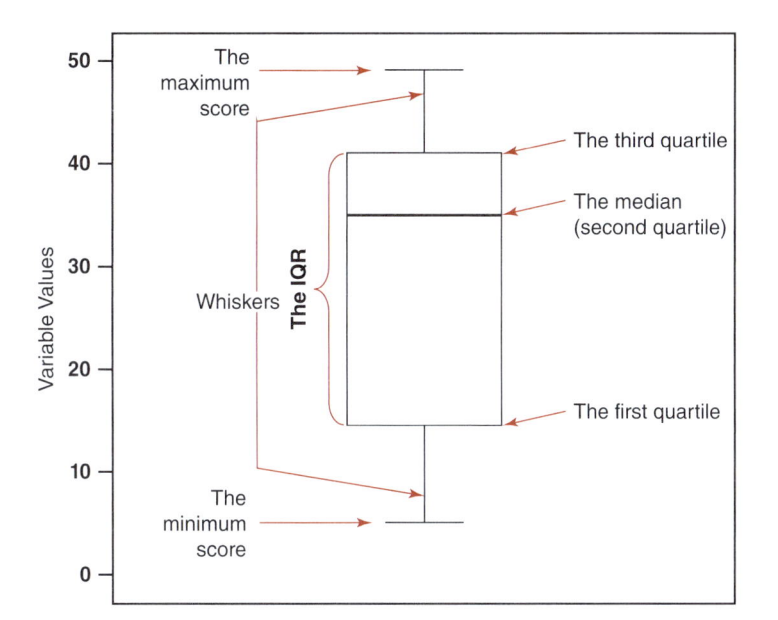

FIGURE 5.3
Box-and-whisker plot of the data in Box 5.3.

GRAPHING THE IQR

A graph called a *box-and-whisker plot* describes the dispersion of the data using quartiles and the IQR. For each group depicted on the *x*-axis, a rectangle is drawn. The height of the rectangle shows the IQR, and a horizontal line drawn across the rectangle shows the median or second quartile. The "whiskers" on a box-and-whisker plot extend from the top and bottom of the rectangle and show the minimum value in the set (below the rectangle) and the maximum value in the set (above the rectangle). Extreme outliers are often shown on the box-and-whisker plot using a symbol (an asterisk for example) either above or below the whiskers. Obviously, an outlier at the extreme high end of the data set would be indicated by a symbol above the rectangle, while a low outlier would be below the rectangle. Figure 5.3 shows a box-and-whisker plot for the data shown in Box 5.3.

CheckPoint

1. Indicate whether the following statements are *true* or *false*.
   a. In statistics, an error is the difference between a value obtained by measurement and the true value of the thing measured. _____
   b. The range equals the maximum value minus the minimum value. _____
   c. The range provides a more robust measure of variability than does the IQR because it ignores values in the middle of a set. _____
2. a. Find the range for the following data set: 4, 5, 8, 13, 14, 15, 18, 19, 21, 22, 27. _____
   b. Find the interquartile range for the same set. _____

Answers to these questions are found on page 198.

THE VARIANCE

If you'll recall, we said that the mean is considered the best measure of center. This is because it is the only measure of center that is influenced by each and every data point in the set. So, it would be nice to have a measure of variability that also took into consideration each and every data point in its calculation.

A number of mathematicians are responsible for the brilliant solution to the problem of measuring variability. Collectively, they came up with a measure of variability that takes into account each and every data point in the set. The solution involves incorporating the mean into the calculation of variability.

The mean defines the center of a distribution. These mathematicians realized that variability could be measured in terms of the distance from the mean of each measurement in the set. Using the distance of each observation in the set from the mean, also known as the **deviation** from the mean or the **deviation score**, it would be possible to calculate a single number that represents the average deviation from the mean.

Imagine a distribution of scores for a set of 50 students on the statistics final exam (see Figure 5.4). The average score on the exam was a 77.5. We'll treat the mean as a sort of flagpole that indicates the center of the distribution and then calculate how far each test score is from that flagged point. Some students scored above average on this test, and others scored below average.

Using the mean score on the test as a sort of baseline, we can determine how far away from the baseline all of the scores in the set are. A score of 95 on this test is 95 − 77.5, or 17.5 points above average; a score of 72 is 72 − 77.5, or 5.5 points below average.

deviation (or deviation score) The distance of an observation in a data set from the mean of the set. Negative deviations indicate that the observation was smaller than the mean; positive deviations indicate that the observation was larger than the mean.

CheckPoint
Answers to questions on page 197

1. a. TRUE ☐
   b. TRUE ☐
   c. FALSE ☐   The range, because it ignores the values in the middle of a set, is a weaker measure of variability than the IQR.
2. a. 23 ☐
   b. 13 ☐

   SCORE: /5

AVERAGE DEVIATION FROM THE MEAN

When we calculated the value of the average observation in a set, we used a fairly simple procedure: We added up each observation, then divided the total by the number of observations we had, and this gave us the mean, or average. We can use the same basic procedure to find the average of the deviations from the mean.

By dividing the sum of our measures of deviation by the number of deviations we have, we should wind up with an average deviation from the mean. Box 5.4 shows the mathematics behind averaging deviations.

BOX 5.4   **Averaging deviation from the mean**

Definition of deviation from the mean: $(x - \bar{X})$ ← Add up all the deviations from the mean.

Average deviation from the mean: $\dfrac{\sum(x - \bar{X})}{n}$ ← Divide the sum of the deviations by n.

Try calculating the average deviation from the mean with the small data set shown in Box 5.5. Subtract each x value from the mean (already calculated for you), and then add up the deviation scores. Values in our data set that are less than the mean turn into negative deviations, while those larger than average turn into positive deviations. Now we need to add up all of the deviation scores. If you do that, you'll notice a peculiar thing—*the sum of the deviation scores equals zero*. And unfortunately for us, this will *always* be the case. Because the mean was used as our center point, and because the mean is the sum of all the individual x values divided by n, we will always get zero when we add up all the deviations from the mean.

So, what do we do? The solution devised by the creators of the method for calculating variance was to *square the deviations around the mean*. This procedure gets rid of any negative deviations (because a negative number squared is a positive number) and allows us to get a nonzero sum when we add up the squared deviation scores. We can then divide the sum of

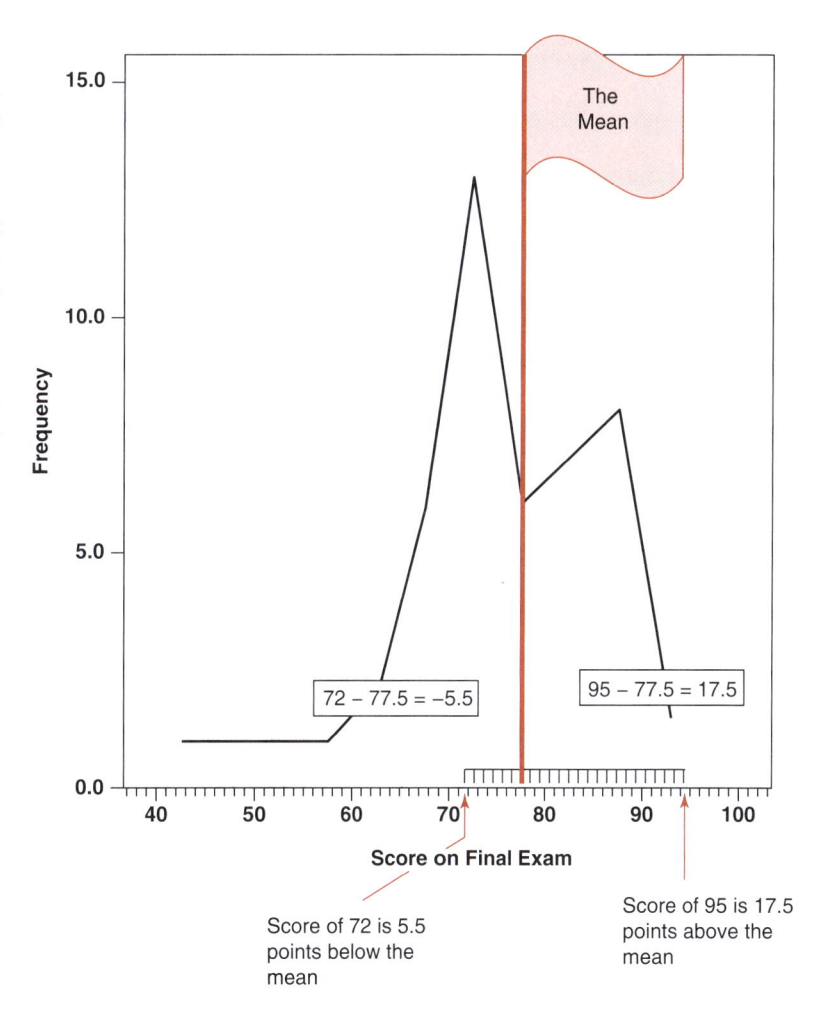

FIGURE 5.4
Deviations from the mean.

CHAPTER 5   Variability: The "Law of Life"

BOX 5.5 Calculating average deviation using a small data set

x	$(x - \bar{X})$	
2	$2 - 6 = -4$	
6	$6 - 6 = 0$	Total of negative deviation score is -4
6	$6 - 6 = 0$	
7	$7 - 6 = 1$	Total of positive deviation scores is $+4$
9	$9 - 6 = 3$	

$$\bar{X} = \frac{\sum x}{n} = \frac{30}{5} = 6 \qquad \sum(x - \bar{X}) = 0$$

variance The average squared deviation of the observations in a set from the mean of that set.

the squared deviations from the mean by n to get an average squared deviation from the mean, also known as the **variance**.

THE STANDARD DEVIATION
FINDING THE VARIANCE IN A POPULATION

The variance is famous in the history of statistics, and you're going to see reference to it from now on. Conventionally, the symbol for the variance in a sample is s^2; for variance in the population, the symbol is the Greek letter σ, or sigma, squared (i.e., $σ^2$). This measure of variability is also known as the *sum of squares* (*SS*), which is shorthand for the *sum* of the *squared* deviations, which becomes the numerator in the equation for variance (see Box 5.6).

Box 5.6 shows the definition formula for calculating the variance and a formula for calculating variance that does not require you to determine the mean first and then subtract the mean from every number (if you remember number properties and simplifying formulas from algebra class, you'll see how we get from the definition formula to the calculation formula). Both formulas will give you the variance. I prefer the calculation formula because I find it easier to use, but you should use the one that works best for you. And, as a reminder, the symbol "n" refers to the number of elements in the sample. If we were to calculate the standard deviation of the population, we would divide the squared deviation from the mean by the total number of elements in the population ("N"). Capital letters, in Greek or English, almost always refer to the population, while the same letters in lower case refer to characteristics of the sample.

BOX 5.6 — Definition and calculation formulas for finding variance

Definition formula	Calculation formula
SS = The sum of the square deviations from the mean $$\sigma^2 = \frac{\overbrace{\sum(x-\bar{X})^2}^{}}{n}$$	$$\sigma^2 = \frac{\sum x^2 - \frac{(\sum x)^2}{n}}{n}$$

Every year, I have students who tell me that they are not comfortable reading mathematical formulas, and since a mathematical formula is just a series of instructions, couldn't we write them out in English just as well? The answer is (of course) yes, and if you prefer your math written out in sentences, try Box 5.7 below.

BOX 5.7 — Written instructions for using the calculation formula to find variance

Step 1: $\sum x^2$ means square each x value and then add them all up.

Step 2: $(\sum x)^2$ means add up all the x values and then square the total. (Note: You will *not* get the same result for Step 1 and Step 2.)

Step 3: Divide the result of Step 2 by n (the number of observations in your set).

Step 4: Subtract the result of Step 3 from the result of Step 1.

Step 5: Divide the result of Step 4 by n, and you have calculated the variance.

Please note: These instructions are for the *calculation formula*, shown in the right-hand portion of Box 5.6; the procedure used in the *definition formula* is simple enough that writing it out in English seems superfluous. Box 5.8 shows the calculation of the variance using both versions of the formula for a small sample set of data.

BOX 5.8 Finding the variance of a small data set using definition and calculation formulas

The Data
2, 6, 6, 7, 9

Definition Formula

x	$(x - \bar{X})^2$
2	$2 - 6 = -4^2 = 16$
6	$6 - 6 = 0^2 = 0$
6	$6 - 6 = 0^2 = 0$
7	$7 - 6 = 1^2 = 1$
9	$9 - 6 = 3^2 = 9$

$$\bar{X} = \frac{\sum x}{N} = \frac{(2+6+6+7+9)}{5} = \frac{30}{5} = 6$$

$$\sigma^2 = \frac{\sum (x - \bar{X})^2}{N} = \frac{(16+0+0+1+9)}{5} = \frac{26}{5} = 5.2$$

Calculation Formula

$$\sigma^2 = \frac{\sum x^2 - \frac{(\sum x)^2}{N}}{N} = \frac{(4+36+36+49+81) - \frac{(30)^2}{5}}{5}$$

$$= \frac{\left(206 - \frac{900}{5}\right)}{5} = \frac{206 - 180}{5} = \frac{26}{5} = 5.2$$

Do you see the mean in this step?

Notice that these ratios are the same.

FINDING THE STANDARD DEVIATION IN A POPULATION

We have a solution to our original problem (the fact that deviations from the mean sum to zero), which is great. However, although we solved one problem by squaring the deviations from the mean, in doing so we created another problem: We squared all the deviations from the mean. We now know the average *squared* deviation, or distance, of any data point from the mean. And our original units of measurement, whatever they were, are now squared. So, for example, if we were measuring height in inches, we would now know how much variability there was in the set of measurements *in inches squared*.

How do we fix this problem? It has probably occurred to you that all we have to do is "undo" the fact that we squared all the deviations by taking the square root of the variance. If we take the square root of the variance, we get *a measure of average deviation from the mean that is in our original units of measure*—and we have what is known as the **standard deviation**.

Box 5.9 shows the definition and calculation formulas for finding the standard deviation, along with the instructions written out in English for the calculation formula. Please note that the standard deviation is a statistic (a number) that reflects the average deviation of any observation from the mean. The symbol for the standard deviation of a sample is simply the letter *s*; the symbol for the standard deviation of a population is the Greek letter sigma (σ). The bigger the value of the standard deviation, the more variable the observations in set were.

If we were to apply the formulas for variance and standard deviation to the mini data set shown in Box 5.8, we would see the calculations shown in Box 5.10.

standard deviation The average deviation of the observations in a set from the mean of that set; the square root of the variance.

BOX 5.9 **Definition and calculation formulas for finding standard deviation**

Definition Formula	Calculation Formula
$\sigma = \sqrt{\dfrac{\sum(x-\bar{X})^2}{N}}$	$\sigma = \sqrt{\dfrac{\sum x^2 - \dfrac{(\sum x)^2}{N}}{N}}$

Step 1: Calculate the variance (see Box 5.6).

Step 2: Take the square root of the variance.

BOX 5.10 **Finding the standard deviation of a small data set**

x	$(x - \bar{X})^2$
2	$2 - 6 = -4^2 = 16$
6	$6 - 6 = 0^2 = 0$
6	$6 - 6 = 0^2 = 0$
7	$7 - 6 = 1^2 = 1$
9	$9 - 6 = 3^2 = 9$

$$\sigma = \sqrt{\dfrac{\sum x^2 - \dfrac{(\sum x)^2}{N}}{N}} = \sqrt{\dfrac{26}{5}} = \sqrt{5.2} = 2.28$$

The variance is 5.2.

The standard deviation is 2.28.

CHAPTER 5 Variability: The "Law of Life" 203

FINDING STANDARD DEVIATION IN A POPULATION VERSUS A SAMPLE

The formula shown in Box 5.9, where the sum of all the squared deviations from the mean is divided by N, is the formula for finding the standard deviation in a *population*. We will see, in Chapter 10, that we can use the standard deviation of a *sample* as an estimate of the variability in the whole population the sample came from. However, if we want to find the standard deviation of a sample, we need to make one very small change to this formula. Why? Because without it, small samples do not provide a great estimate of the population variability. Is your curiosity fired up? Want to know what that very small change is? Here's a preview: Instead of dividing the average squared deviation from the mean by N, we will divide by $n - 1$.

The measures of variability we calculate using these two formulas don't differ by much, but they do differ. If we were to calculate the standard deviation for the example shown in Box 5.10 using $n - 1$ (or 4) instead of N (5), we would get a result that was slightly larger. Try it for yourself and see. You should also be aware that many calculators that come preprogrammed and ready to calculate means and variability offer the user *both* the "divide by n" and the "divide by $n - 1$" options. For the moment, we will be using the "divide by n" formula. In Chapter 10, we will work with the $n - 1$ version, and explain a little more why we do so.

FINDING VARIANCE AND STANDARD DEVIATION: AN EXAMPLE

Just for practice, let's try calculating the variance and standard deviation of our sleeping college students. Here again are the 30 students we've been describing right along. Notice that I'm using the raw data here (x values are the minutes of sleep reported by each participant) rather than a frequency distribution of the data: We will need to calculate the deviation of each score from the mean, and this will be much easier if we can see all the scores at once. You may remember that the mean duration of sleep for this example was 408.67 minutes. Table 5.1 shows

TABLE 5.1   Sleep data

x	x^2	x	x^2	x	x^2	x	x^2
120	14,400	385	148,225	440	193,600	470	220,900
120	14,400	390	152,100	445	198,025	480	230,400
275	75,625	405	164,025	450	202,500	495	245,025
330	108,900	405	164,025	450	202,500	520	270,400
340	115,600	405	164,025	450	202,500	525	275,625
360	129,600	420	176,400	450	202,500	620	384,400
375	140,625	420	176,400	450	202,500		
380	144,400	430	184,900	455	207,025		

the data that we'll be working with, and Box 5.11 shows the calculations for both variance and standard deviation. Notice that I'm using the calculation formula.

BOX 5.11 **Finding the variance and standard deviation of sleeping college students**

$$\sigma^2 = \frac{\sum x^2 - \frac{(\sum x)^2}{N}}{N} = \frac{5,311,550 - \frac{(12,260)^2}{30}}{30}$$

$$= \frac{5,311,550 - \frac{150,307,600}{30}}{30} = \frac{5,311,550 - 5,010,253.3}{30}$$

$$= \frac{301,296.7}{30} = \mathbf{10,043.22} \quad \text{The } variance.$$

$$\sigma = \sqrt{\sigma^2} = \sqrt{10,043.223} = \mathbf{100.22} \quad \text{The } standard\ deviation.$$

The variance is 10,043.22 squared minutes (not all that helpful unless you regularly think about time in terms of squared minutes). The standard deviation is 100.22 minutes, and that tells us the average deviation from the mean in this sample was 100.22 minutes. If we were asked to write about these descriptive statistics, we might say something like this:

> The average sleep duration for the 30 participants in this set was 408.67 minutes ($SD = 100.22$ min), where the symbol SD stands for the standard deviation.

Now we've given the reader of our report some useful information. They now know that a typical night's sleep was 408.67 minutes for these 30 college students, and they know that typically observations deviated or varied from that mean by 100.22 minutes.

Standard deviation gives us a window of typical values around the mean. If we pulled a student out of this set, asked how long he had slept on this particular night and he told us "I slept seven-and-a-half hours," we would know that this was a fairly typical night's sleep in this set of data. After all, the typical night's sleep was 408.67 minutes (or 6.8 hours), and the typical deviation from that point was 100.22 minutes. This means that sleep durations within 100.22 minutes of the mean are basically in the center of the distribution and so can be called typical of the observations in this set. The mean plus 100.22 is 508.89 (8.48 hours of sleep), and the mean minus 100.22 minutes is 308.45 (5.14 hours of sleep). The 7.5 hours of sleep claimed by the student we pulled out of the group is right in this window of typical.

Since we know what *typical* is, we now also know what *atypical* is. Sleeping less than 5.14 hours or more than 8.48 hours is atypical in this set of observations. If we add what we've just learned to our list at the beginning of the chapter, we find that:

- Looking at the relationship between mean, median, and mode, we can say that we have a negatively skewed distribution (mean < median < mode), telling us that we have some low-value outliers in our set (in this case, the two students who slept only 2 hours).
- The variance in this set was 10,043.22 squared minutes, and the standard deviation was 100.22 minutes.

CheckPoint

1. Indicate whether the following statements are *true* or *false*.
    a. The larger the standard deviation is, the more variable the data are. _____
    b. The variance is the square of the standard deviation. _____
    c. The standard deviation is the square of the variance. _____
    d. Standard deviation can never be less than 1.00. _____
2. Calculate the variance and standard deviation of the following set: 10, 11, 12, 15, 17, 19. _____

Answers to these questions are found on page 209.

Think About It...

THE RANGE RULE

Students of statistics often ask about shortcuts: Is there a way of finding the standard deviation of a data set that is quicker, easier, and less "math-y" than the method we've been considering here? The short answer is "no." The longer, and perhaps better, answer is, "sort of." There is a little something called the *range rule*.

The range rule says that the standard deviation of a sample can be estimated (please note the use of the word *estimated*) by simply dividing the range by 4. In math symbols, we would say that the range rule says that $SD =$ Range/4.

1. Here are three small ($n = 5$) data sets. For each, find the range, and then calculate the standard deviation (I've given you a head start in your calculations, so please use the information I've provided). Let's see if the range rule works.

Data Set:

A	A²
1	1
2	4
3	9
4	16
5	25

B	B²
1	1
3	9
6	36
15	225
23	529

C	C²
1	1
5	25
5	25
5	25
5	25

$\Sigma x = 15$ $\Sigma x = 48$ $\Sigma x = 21$

$\Sigma x^2 = 55$ $\Sigma x^2 = 800$ $\Sigma x^2 = 101$

$(\Sigma x)^2 = 225$ $(\Sigma x)^2 = 2304$ $(\Sigma x)^2 = 441$

$n = 5$ $n = 5$ $n = 5$

Mean = 3   Mean = 9.6   Mean = 4.2

Range = _____   Range = _____   Range = _____

SD = _____   SD = _____   SD = _____

2. Now apply the range rule. For each data set, is the *SD* approximately the range/4? Check the appropriate box. How different are each of these two measures for each set?

Data Set:

A	B	C
☐ Yes	☐ Yes	☐ Yes
☐ No	☐ No	☐ No

Range rule = _____   Range rule = _____   Range rule = _____

Difference _____   Difference _____   Difference _____

Answers to this exercise are found on page 231.

STANDARD DEVIATION IN CONTEXT

When first confronted with the formula for standard deviation, students often ask a very interesting question: Why do we have to *square* the deviations from the mean? Why can't we just use the absolute value of the deviations from the mean?

This is a question that statisticians and scientists have been arguing about ever since Karl Pearson first proposed the current method for calculating the standard deviation in 1893.

In 1914, an astronomer named Arthur Eddington asked this very same question as well. Eddington proposed that using the "mean absolute deviation" (usually abbreviated *MD* for mean deviation) actually worked *better* than using the standard deviation where variability is measured using a formula that squares deviations from the mean to get rid of negative deviations. Mathematically, Eddington was suggesting that variability in the data be calculated as shown in Box 5.12.

BOX 5.12

Calculating standard deviation (SD) and mean deviation (MD)

Calculating SD

$$SD = \sqrt{\frac{\sum(x - \bar{X})^2}{N}}$$

Calculating MD

$$MD = \frac{\sum(|x - \bar{X}|)}{N}$$

Example:

x	$x - \bar{X}$	$(x - \bar{X})^2$	$|x - \bar{X}|$
9	9.0 − 6.8 = 2.2	4.84	2.2
6	6.0 − 6.8 = −0.8	0.64	0.8
3	3.0 − 6.8 = −3.8	14.44	3.8
4	4.0 − 6.8 = −2.8	7.84	2.8
8	8.0 − 6.8 = 1.2	1.44	1.2
5	5.0 − 6.8 = −1.8	3.24	1.8
7	7.0 − 6.8 = 0.2	0.04	0.2
7	7.0 − 6.8 = 0.2	0.04	0.2
9	9.0 − 6.8 = 2.2	4.84	2.2
10	10.0 − 6.8 = 3.2	10.24	3.2
		$\sum = 47.6$	$\sum = 18.4$

$$SD = \sqrt{\frac{\sum(x - \bar{X})^2}{n}} = \sqrt{\frac{47.6}{10}} = \sqrt{4.76} = 2.18$$

$$MD = \frac{\sum(|x - \bar{X}|)}{n} = \frac{18.4}{10} = 1.84$$

Notice that we get two different measures of variability with these two methods. The standard deviation is larger than the mean deviation by 0.34 of our units (whatever they might be in this example). This is because squaring the deviations from the mean tends to emphasize the deviations of the outliers in the set. Both measures

> **CheckPoint**
> *Answers to questions on page 206*
>
> 1. a. TRUE ☐
>   b. TRUE ☐
>   c. FALSE ☐ The standard deviation is the square root of the variance.
>   d. FALSE ☐ The standard deviation can never be less than zero.
>
> 2. variance = 10.67 $\left(\sigma^2 = \dfrac{1,240 - 84^2/6}{6} = 10.67\right)$
>
>   SD = 3.27 $\left(\sigma = \sqrt{\sigma^2} = 3.27\right)$
>
>   **SCORE:** /5

are similar to one another in value. You might be asking yourself which one is the "right value," and this is a very good question. It turns out that both of these values are correct.

Stephen Gorard (2005) presents an interesting review of the debate that raged between Eddington and R. A. Fisher, a statistician, over which formula should be used. Eddington claimed that the *MD* worked better with real-life data than the *SD* did. Fisher responded with a careful mathematical analysis of the benefits of using the *SD* rather than the *MD*. Fisher argued that a good statistic was:

- *Consistent:* The same formula could be used to calculate the statistic in samples and in populations.
- *Sufficient:* It summarized all of the information that could be taken from a sample to estimate a population parameter.
- *Efficient:* It did the best job possible of estimating the population parameter.

According to Fisher, the two methods met the first two criteria equally well, but the *SD* method was more *efficient* than the *MD* method: The standard deviation of a sample was a more consistent and accurate estimate of the variability in the population the sample came from. This characteristic, more than the other benefit Fisher discussed (that the *SD* was easier to handle algebraically than the *MD*) has helped make the SD method traditional in modern statistics. We'll encounter this characteristic of the *SD* again in Chapter 8.

Consider also the debate about the "true" measure of anything, which we briefly discussed at the beginning of this chapter. Will we ever know the true amount of variability in a sample or a population? And if we can't know this, then how can we say one method of calculating variability is really "better" than another?

DESCRIPTIVE STATISTICS IN CONTEXT

In Chapters 1 through 5, we've discussed a number of statistics that describe data: the mean, median, mode, range, interquartile range, variance, and standard deviation. You may be asking yourself what to do with all of these statistics now that we have them. Most often, descriptive statistics wind up in presentations or publications, where they describe the basic patterns in data.

If you would like to see how researchers talk about statistics in the articles they write, several resources are available. The easiest to get to, and probably the most valuable, is the library. Find a journal that publishes articles in the field you're interested in, and read the results section. Here, the researcher will often begin by describing the data, probably using the descriptive statistics we've been talking about up to this point. The researcher will likely report the average or typical value in the data, as well as the standard deviation (these two statistics are almost always presented together). After the mean and standard deviation, the specific descriptive statistics you will come across can vary tremendously—it all depends on the questions being asked and the data collected. You might see percentages, median, mode, variance, range, or even a descriptive statistic created by the researcher just to make a particular point about the data. You will see graphs of all varieties and often tables of descriptive statistics as well.

I advise you to use published articles as templates for how to talk effectively and efficiently about statistics. After all, these articles got published, so model your own writing after success. DO NOT simply transplant your statistics into their sentences—that's plagiarism. DO pay attention to the ways in which descriptive statistics are used in scientific writing, keeping in mind the goal of writing an article about an experiment that you've spent time and effort on: You want to tell your reader what you found and what it means in language that is clear and easy to understand.

Another valuable resource is the *Publication Manual of the American Psychological Association*. This style manual is usually available in campus bookstores; if not, try the APA website (www.apa.org),), where the current style manual is also available for purchase. The APA style guide offers examples, suggestions, lists of dos and don'ts, and any number of other tips for writing for publication in psychology. Please note that the APA style manual will not tell you *what* to write (that, as always, is up to the individual author). It will, however, tell you a great deal about *how* to present what you want to say in standard APA Style.

Let's use our sleep data as an example and construct an APA Style description of the basic results. I'm going to assume that you've already written your methods section, where you describe the procedure that you used to apply your independent variable to your participants, the data that you collected from your participants, and how you measured your dependent variable. When you're ready to present your descriptive statistics to your reader, the APA style manual will provide a format for presenting that information.

SAMPLE PARAGRAPH IN APA STYLE

A sample of 33 undergraduate psychology students, enrolled in an introductory psychology class, volunteered to participate in a study of sleep patterns. Three students did not complete the entire study and so their data were removed from the data set. In all, data from 30 students were collected. The students kept track of their total sleep time (in minutes) for seven consecutive nights. On average, the students in the sample slept slightly less than the 8 hours (480 minutes) of sleep expected per night (M = 430.65 min). The typical student in the sample slept 50 minutes less than the expected 480 minutes. Sleep durations varied a great deal (SD = 104.20 min). The mean, median (Mdn = 435 min), and mode (mode = 480 min) were close together in value, indicating a roughly normal distribution of sleep times. Figure 1 shows the distribution of sleep durations in the sample. The distribution is reasonably normal in shape.

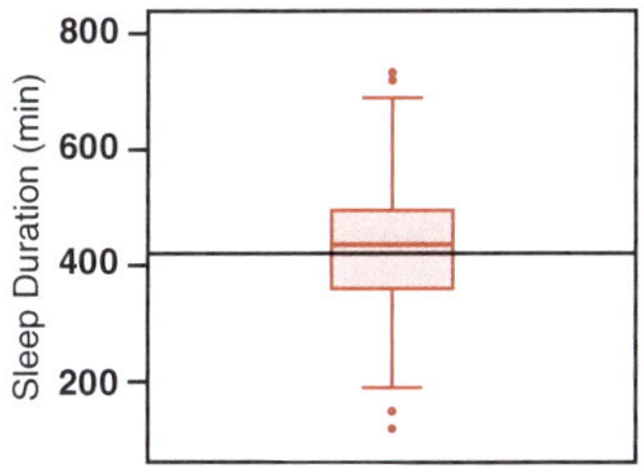

Figure 1. Distribution of sleep times on seven consecutive nights for 30 undergraduate students. The distribution is approximately normal in shape. There were two high outliers (720 and 730 minutes of sleep) and two low outliers (120 and 150 minutes of sleep) in the set. The horizontal line indicates the mean sleep time for the group (M = 430.65 min). The IQR for this data was relatively narrow, indicating that the majority of the observations were clustered around the mean and median.

Numbers less than 10 are spelled out.

See Table 5.2 below for a list of the symbols used in APA Style to refer to the descriptive statistics we've discussed so far. Refer to the APA style manual for a complete list.

Some units of time are abbreviated, while others are not. When abbreviations are used, periods are not used at the end. Table 5.2 lists the time units that are, and are not, abbreviated.

An author may create a statistic to make a point. Here, I've calculated the difference between the mean sleep duration in the sample (430 minutes) and a "typical" 8-hour night of sleep (480 minutes). The difference (50 minutes) helps delineate the difference between the sleep times in the sample and those in the population. It is descriptive, but this particular statistic has no "official" name.

Abbreviations for standard descriptive statistics are in italics. See Table 5.2 for details.

Although numbers greater than 11 are written using numerals, spell out any number that begins a sentence.

Notice that not all abbreviations are in all-capital letters.

Lesser-used descriptive statistics have no APA-approved abbreviation; how best to present these statistics is a matter left to the author. Whatever style you choose, be sure to use it consistently.

When you include a graph (called a figure in APA Style), you should refer to it first in the text, describing its main message for the reader. Any figures you include should augment rather than duplicate the text, convey essential facts, and be easy to read. Figures and tables may be placed in the text, near where they are discussed, or at the end of the paper, on separate and individual pages. Placement of figures and tables is often up to a journal's editorial staff, so they might not end up where you placed them.

The figure caption includes the title of the figure and a brief, descriptive explanation of the figure.

TABLE 5.2 APA Style symbols for descriptive statistics

Descriptive Statistic	APA Style Symbols	APA Abbreviations for Time	
Mean (sample)	M	hour	hr
Mean (population)	Μ	minute	min
Number of cases (sample)	n	second	s
Total number of cases (or population)	N	millisecond	ms
Variance (sample)	s^2	nanosecond	ns
Variance (population)	σ^2	Day, week, month, year	Not abbreviated
Standard deviation (sample)	SD		
Standard deviation (population)	Σ		

Remember that your job as the author is to explain to your reader what you saw. Make use of the descriptive statistics, and be as clear and as concise as possible.

SUMMARY

Variability in a set of measurements can be described in several ways. First, we can use the range, or the difference between the largest and the smallest measurement in the set. The range is an insensitive measure of variability because it reflects only two of the measurements in the set. An improvement on the range is the interquartile range, which measures the range of the center 50% of the data.

The variance, or the average squared deviation of a measurement from the mean, tells us how far (on average) any score is from the mean, but the deviations have all been squared, introducing a problem into this method of measuring variability. The standard deviation (the average deviation of a measurement from the mean) solves the problem of squaring the deviations and brings the measurement of variability back to the original units of measure. For this reason, the standard deviation is the measure of variability that is used most often by statisticians.

TERMS YOU SHOULD KNOW

deviation, p. 198
deviation score, p. 198
interquartile range (IQR), p. 195

range, p. 194
standard deviation, p. 203
variance, p. 200

STATISTICS IN CONTEXT

GLOSSARY OF EQUATIONS

Formula	Name	Symbols
range = maximum − minimum	The range	
IQR = 3rd quartile − 1st quartile	The interquartile range	IQR
deviation = $x - \bar{X}$	A deviation score (also a residual)	
$\sigma^2 = \dfrac{\sum(x - \bar{X})^2}{N}$	The variance definition formula	σ^2 ("sigma squared"; in the population) s^2 ("s squared"; in the sample)
$\sigma^2 = \dfrac{\sum x^2 - \dfrac{(\sum x)^2}{N}}{N}$	The variance calculation formula	
$\sigma = \sqrt{\dfrac{\sum(x - \bar{X})^2}{N}}$	The standard deviation definition formula	σ ("sigma"; in the population) s (in the sample)
$\sigma = \sqrt{\dfrac{\sum x^2 - \dfrac{(\sum x)^2}{N}}{N}}$	The standard deviation calculation formula	

WRITING ASSIGNMENT

You now have a number of descriptive statistics that will help tell the story of your data. Your job in this assignment is to describe the results of an experiment on time perception using APA Style. Examine the data shown in Table 5.3, and determine what you think the results of the experiment are telling you about age and accuracy in the perception of the passage of time. Whatever conclusion you draw should be backed up by the data. If you think there is a meaningful difference between the two age groups tested, then present the descriptive statistic (or statistics) that support(s) that conclusion. Feel free to include one or more graphs if you think it will help the reader understand what you've written.

Table 5.3 Time-estimation data

ID	Age Group	Age	Error (in ms)	ID	Age Group	Age	Error (in ms)
1	1	11	9.00	11	2	30	−3.00
2	1	6	13.00	12	2	35	3.00
3	1	8	13.00	13	2	34	7.00
4	1	15	1.00	14	2	50	4.00
5	1	15	5.00	15	2	44	5.00
6	1	5	−11.00	16	2	47	5.00
7	1	7	11.00	17	2	32	10.00
8	1	14	1.00	18	2	60	1.00
9	1	14	−1.00	19	2	30	11.00
10	1	10	−16.00	20	2	65	14.00

The Experiment: The effect of age on the perception of a short time interval.

Two groups of participants ($N = 10$ in each group) were asked to estimate the duration of an interval without reference to a watch or other timing device. Group 1 consisted of participants aged 15 years or younger. Participants in Group 2 were 30 years of age or older. All the participants were seated in front of a computer monitor and were asked to begin their estimation of the length of an 11-second interval as soon as a start signal appeared on the monitor. When they felt that 11 seconds had passed, they were to press the space bar on the computer. Each participant judged the duration of an 11-second interval three times, and a mean error measure was then calculated for each participant. Error was measured as the difference between the number of seconds they judged the interval to be and the target of 11 seconds. Negative errors indicate an underestimation of the interval (i.e., their estimation of an 11-second interval was less than 11 seconds), and positive errors indicate overestimation of the test interval.

Some things to consider as you write:

1. Describe the performance of these two groups. What was the average (for both age and error), and what were the standard deviations? What were the age ranges for the two groups? What were the ranges in the error scores for the two groups?
2. When you're calculating the mean error for each age group, it helps to use the absolute value of the error measures in your calculations. Explain why this will help your reader.
3. Did age have an effect on accuracy in this task? Was one age group more accurate than the other? Which age group was the most accurate, on average?
4. Did these two age groups differ in variability? What does variability tell you about performance on this task?

PRACTICE PROBLEMS

1. Evaluate the data sets shown in Table 5.4, and make a rough guess about the amount of variability (a great deal of variability, a moderate amount, or very little) in each set. Then, calculate the actual range, variance, and standard deviation of the data. Were your estimates of the amount of variability accurate?

Table 5.4 Estimating variability in three data sets

Set A ID	x	Set B ID	x	Set C ID	x
1	5.36	1	7.99	1	13.13
2	8.20	2	7.97	2	2.00
3	6.99	3	7.07	3	4.87
4	5.29	4	8.01	4	7.43
5	6.15	5	8.03	5	1.59
6	6.54	6	7.99	6	14.39
7	8.43	7	8.00	7	10.70
8	8.00	8	8.00	8	8.00
9	9.68	9	7.99	9	−0.34
10	8.50	10	8.00	10	10.84
11	6.40	11	7.06	11	6.10
12	9.75	12	8.00	12	15.24
13	5.15	13	8.09	13	18.22
14	8.87	14	7.92	14	3.93
15	7.55	15	8.12	15	5.22

2. Body size (including body weight) has been linked to reproductive success and to levels of aggression in feral cats (Natoli, Schmid, Say, & Pontier, 2007). The authors of that study report that a large body size leads to a high rank in the dominance hierarchy and so greater access to reproductive partners. Male cats were described by Natoli et al. as weighing approximately 20% more than females, and female domestic shorthair cats usually weigh between 7 and 9 pounds. Find the mean, median, mode, range, IQR, variance, and standard deviation for the data shown below. Based on the descriptive statistics you've calculated, how many male animals are likely in this sample?

Table 5.5 Body weight of the domestic cat (in pounds)

ID	Weight
1	9.9
2	7
3	10.2
4	10.8
5	6.3
6	9.1
7	9
8	6.3
9	9.1
10	9.1
11	11.9
12	11
13	10.2
14	12.7
15	11.1
16	7
17	8.2
18	7
19	9.1
20	8

3. According to Wikipedia.org, the smallest cat ever officially weighed and measured is a domestic shorthair named "Mr. Peebles," who weighs only 3 pounds and stands only 6 inches high (Mr. Peebles apparently has an unspecified genetic defect that has caused him to stay so small). Compare Mr. Peebles' body weight to the mean you obtained from our set of 20 cats in question 2. Where is Mr. Peebles' weight in the distribution in Table 5.5? How far from the mean is his body weight?

4. The largest cat ever officially weighed and measured was a Tabby named "Himmie," who died of respiratory failure at the age of 10 (Himmie was apparently just plain overweight). Himmie weighed an astonishing 46 pounds, 15.2 ounces. Compare Himmie's weight to the distribution you've described in question 2. How far above average weight was he? (Just FYI: *Guinness World Records* is no longer accepting nominations for heaviest pet in an effort to keep pet owners from overfeeding their animals.)

STATISTICS IN CONTEXT

5. In a study about the effects of olfactory system stimulation on the ability to concentrate at a visual search task, subjects are asked to complete a visual search task while smelling either an unpleasant or a pleasant odor. The time needed to complete the visual search task (in seconds) is recorded. Table 5.6 shows this hypothetical data. Calculate the mean, median, mode, and standard deviation of the data. Then, calculate the IQR for the data, and create a box-and-whisker plot.

Table 5.6 Time needed (in seconds) to complete visual search task while smelling an unpleasant or a pleasant odor

Unpleasant Odor	Pleasant Odor
65	80
55	70
82	63
42	77
48	75
55	71
71	58
93	80
83	71
41	72
88	85
78	155
38	75
48	66
91	71
56	122
43	69
60	84
40	95
84	70
57	86
81	120
50	96
68	72

6. Several studies (Keinan, 2002, for example) have found evidence that belief in magic and luck increased as stress levels increased. The more stressed people were, the more likely they were to engage in magical thinking. Suppose you did your own study examining the relationship between test anxiety and belief in the power of a lucky charm. You ask one group of students to bring their own personal lucky charm to their statistics final exam. Another group is the control group and has to take their statistics exam without any lucky objects present during the test.

Table 5.7 Scores on statistics final exam with and without a lucky charm

Without Lucky Charm	With Lucky Charm
82	64
95	38
68	100
75	42
70	77
70	100
79	97
79	99
76	95
76	48

a. Calculate the mean for each group.
b. Calculate the standard deviation for each group.
c. In your opinion, does having a lucky charm result in better performance on the exam? How do the two groups differ? How are they similar?

7. Match the histograms of the distributions on the next page with their descriptive statistics below.

	A	B	C	D
Mean	7.1579	5.4878	5.9535	3.3
Median	8.00	5.00	6.00	3.00
Mode	10.00	5.00	5.00	1.00
Standard deviation	2.47	2.29	2.44	2.34
Skewness	−0.68	+0.20	−0.016	+0.95

What would the skewness measurement be for a distribution that is perfectly normal in shape?

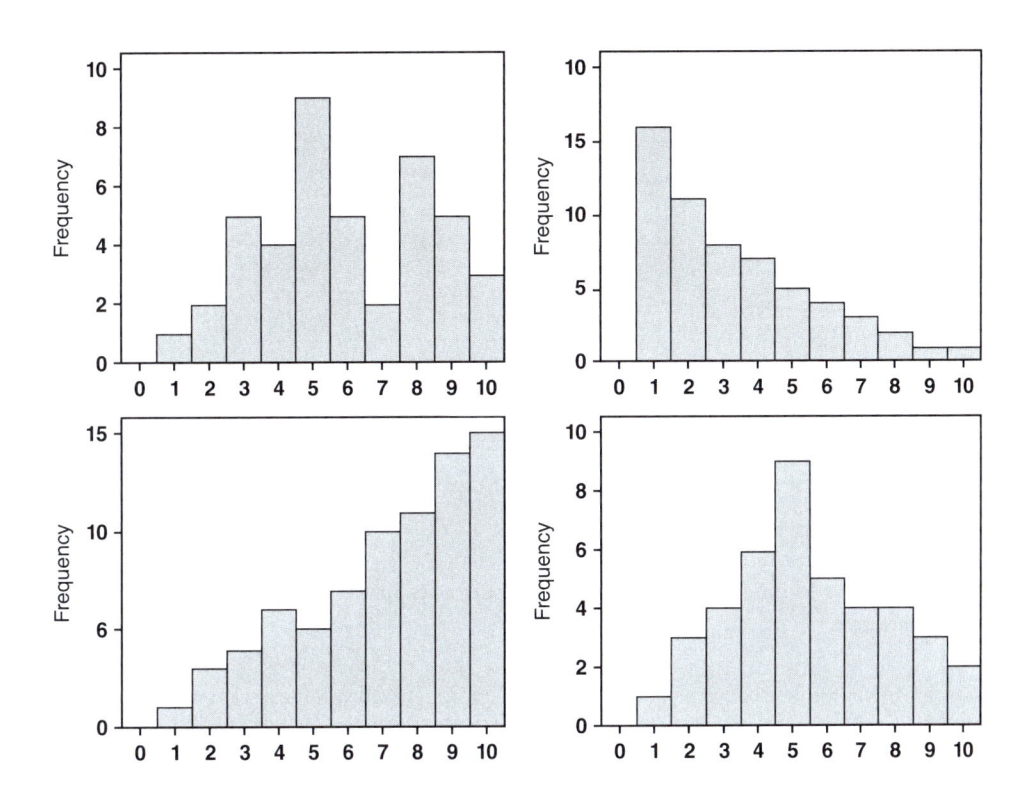

8. Suppose you were a veterinarian responsible for the health of the animals in a laboratory colony. One very important clue to health in any animal (including humans) is body weight. To help you maintain the health of the animals in your care, you need to know what a typical laboratory animal weighs (in this example, the albino Sprague-Dawley lab rat, which is very commonly used for lab research because they are calm and easy to handle). The body weights for a set of 30 adult, albino Sprague-Dawley laboratory rats (15 males and 15 females) are shown in Table 5.8. Find the mean, median, mode, range, variance, and standard deviation for this set of animals.

Table 5.8 Body weight of Sprague-Dawley rats (in grams)

Females	Males
229	510
300	485
310	476
248	528
304	448
286	498

CHAPTER 5 Variability: The "Law of Life" 219

Table 5.8 *Continued*

291	468
284	575
291	489
279	529
289	411
285	453
247	518
282	441
248	466

Once you've found these statistics (our measure of typical), answer the following questions:

a. In your newest shipment of rats, they've forgotten to tell you the sex of each animal, and all you have to go on are the body weights shown in the chart below. Using only the body weights provided, determine if the animals are probably male or probably female.

b. Were there any rats that you were unable to make a prediction for? If so, why did you have difficulty?

c. Now that we know what "typical" is, speculate how you might determine what *atypical* would be. How far away from the mean would an unhealthy animal (too fat *or* too thin) be?

Body Weight (in Grams)	Check Here if Rat is Probably FEMALE	Check Here if Rat is Probably MALE	Check Here if you can't make a Prediction About the Sex of the Animal Based on Body Weight Alone
249			
500			
390			
400			
290			

9. A relatively new area of research in personality theory focuses on how brain function is related to personality. For example, Bickart, Wright, Dautoff, Dickerson, and Barrett (2011) examined the size of the amygdala and related it to the size of a person's social network. The amygdala is a structure in the medial (toward the center of your brain) temporal lobe (the part of cortex that is under your temples). It is part of the limbic system, which controls emotional response. Bickart et al. found that people with larger

amygdalae (plural) tended to also have a larger social network (measured as the total number of social contacts a person maintains). The data in Table 5.9 are representative of what Bickart et al. found. Using the descriptive statistics we've discussed so far, describe the volume of the amygdala and the size of the social network for each of the 20 participants in this hypothetical study.

Table 5.9 Volume of the amygdala and size of social network

ID	Volume of Amygdala (in mm³)	Size of Social Network (Social Contacts Maintained)
1	2.3	26
2	4.0	44
3	2.2	35
4	4.1	42
5	3.3	29
6	3.0	29
7	3.7	22
8	2.4	11
9	3.6	46
10	3.2	5
11	4.3	44
12	3.3	32
13	4.1	11
14	4.5	40
15	2.3	42
16	2.9	27
17	2.3	16
18	4.0	35
19	3.3	9
20	3.3	44

10. The "beauty is good" effect says that physically attractive individuals are perceived as having positive qualities. In a study of this effect, Swami et al. (2010) asked college-educated men to rate the relative attractiveness of a series of photographs of women. The women in the photographs were classified by Body Mass Index (BMI) into the following categories: emaciated (BMI rating of 1 or 2), underweight (BMI rating of 3 or 4), normal weight (BMI rating of 5 or 6), overweight (BMI rating of 7 or 8), or obese

CHAPTER 5 Variability: The "Law of Life" 221

(BMI rating of 9 or 10). The men were given vignettes to read that allegedly described the personalities of the women they were being asked to rate. Some of the participants read vignettes that described positive personality traits, some read stories that described negative personalities, and a final group (the control group) did not receive any information about the personalities of the women they were rating. All of the participants were asked to rate the most attractive photograph on a 10-point scale where $1 =$ *lowest BMI* and $10 =$ *highest BMI*. The data in Table 5.10 are similar to that obtained by Swami et al. Calculate the mean, median, mode, and standard deviation for these data. Did the apparent personality of the women pictured affect the ratings of attractiveness offered by the male participants in the study?

Table 5.10 Attractiveness ratings

Negative Vignette	Positive Vignette	Control Group
3.94	3.39	5.97
2.00	3.62	4.23
3.56	3.75	3.75
2.35	3.63	3.75
3.34	4.23	3.74
3.17	3.79	5.20
2.61	5.00	4.74
4.42	4.21	3.85
4.74	3.22	2.66
2.63	3.59	3.73
2.88	4.67	4.75
2.73	3.92	3.33
3.40	3.22	4.50
4.07	4.71	4.72
3.88	4.75	4.35

11. Five groups of adolescents took the Stanford–Binet IQ test. The distribution of test scores for each group ($n = 25$ in each group, $N = 125$) are shown in Figure 5.5. The solid horizontal line shows the average score on the test ($\mu = 100$), and the two dashed horizontal lines indicate the standard deviation of scores on the test ($\sigma = 15$). The dashed line above the mean indicates a score of 115 (the mean plus one standard deviation), and the dashed line below the mean indicates a score of 75 (the mean minus one standard deviation). Describe the IQ scores in the five groups.

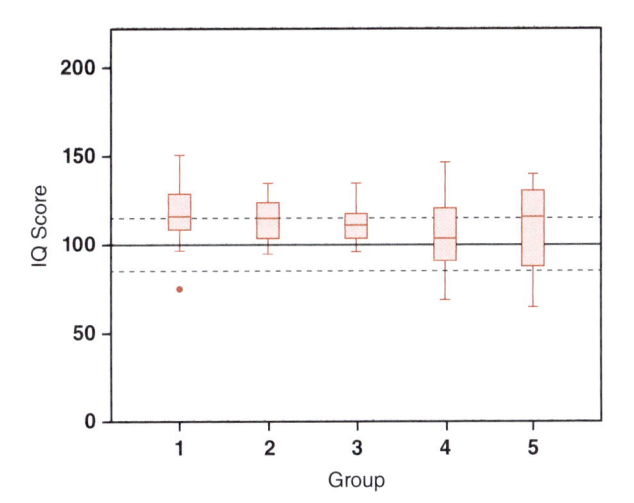

FIGURE 5.5
Box-and-whisker plot of IQ scores for five groups of adolescents.

12. Find the answer to the riddle below.

    Riddle: One day there was a fire in a wastebasket in the office of the Dean of Students. In rushed a physicist, a chemist, and a statistician. The physicist immediately starts to work on how much energy would have to be removed from the fire to stop the combustion. The chemist works on which reagent would have to be added to the fire to prevent oxidation. While they are doing this, the statistician is setting fires to all the other wastebaskets in the office. "What are you doing?" the others demand. The statistician replies, "Well, to solve the problem, _____ _____ _____ _____ _____ _____ _____."
    Solution to the riddle: Notice that there are seven (7) words in the solution to the riddle. You'll find the words in the list below. Each of the words in the solution section is associated with a number. To determine which word you need, solve the seven problems in the following problem section. The word next to the answer to the first problem is the first word in the solution to the riddle. The word next to the answer to the second problem is the second word in the solution, and so on. Round all answers to two decimal places (e.g., 2.645 rounds to 2.65).

    Solution to the riddle: Words

(3.00) we	(7.00) must	(12.00) you	(626) deviate	(7.67) obviously
(8) wonder	(69.0) a	(623) larger	(57) standard	(1.02) skewed
(100) titrate	(9) need	(10.44) sample	(3.04) population	(3.23) size

    Problems section: Solve each problem to find a number. Match the number to a word in the word section to finish the riddle.

    Data Set: 3   5   6   7   7   7   9   10   15

1. The range of the data set is _____.
2. The mean of the data set is _____.
3. $n =$ _____.
4. The sum of all the values in the data set (Σx) is _____.
5. The sum of all the values squared (Σx^2) is _____.
6. The variance is _____.
7. The standard deviation is _____.

13. Consider the following three data sets:

A	B	C
9	10	1
10	10	1
11	10	10
7	10	19
13	10	20

a. Calculate the mean of each data set.
b. Calculate the range and the IQR for each data set.
c. Calculate the variance and the standard deviation of each data set.
d. Which data set has the smallest standard deviation?
e. Use the range rule to determine the standard deviation of each data set.
f. Compare the standard deviation derived using the range rule with the standard deviation you computed using the formula. How different are they?

14. Is it possible for a data set to have negative standard deviation? Why or why not?

15. In the *Standard Deviation in Context* section, we discussed the mean absolute deviation from the mean (abbreviated *MD*), which proposed an alternate method of calculating the variability in a sample:

$$MD = \frac{\Sigma(|x - \bar{X}|)}{n}$$

Let's play with this a bit. The mean, the median, and the mode are all "averages," and they are all measures of center. Can you use another measure of central tendency as your anchor point when you're calculating the standard deviation? Suppose we calculated the mean absolute deviation from the *median* instead of the *mean*. What would happen to our measure of variability? Use the data shown in Table 5.10a to answer the following questions:

a. Calculate both the standard deviation (using the "traditional" formula) and the mean absolute deviation (using the formula for *MD* shown above) using the *mean* of the data set. Record your answers in Table 5.10b.
b. Calculate both the standard deviation and mean absolute deviation using the *median* of the data set. Record your answers in Table 5.10b.

c. How different are the measures of variability calculated using the mean from those calculated using the median?
d. Now try calculating both *SD* and *MD* using the *mode* as your center point. Record your answers in Table 5.10b.
e. Compare the results you got using the three measures of center to calculate *SD* and *MD*. Which measure of center gave you the smallest measure of variability?

Table 5.10a A small data set

x
2
2
3
4
14

Table 5.10b Results

Using the mean	
SD =	MD =
Using the median	
SD =	MD =
Using the mode	
SD =	MD =

16. The distribution in Table 5.10a is skewed (in fact, it's positively skewed, with most of the observations at the low end of the distribution and one very high outlier). Given what you know about how skew in the distribution affects the measures of center, do you think you would get different values for *SD* and *MD* using the three different measures of center (as you did in question 15) if the distribution we were working with was perfectly normal in shape? Why or why not?

17. Consider the data shown in Table 5.11.

Table 5.11 Another small data set

x	f	x + 10
25	1	
33	1	
41	1	
42	2	
43	4	

Table 5.11 *Continued*

45	2	
52	1	
54	1	
65	1	

    a. Calculate the mean and standard deviation of the data shown in Table 5.11.
    b. Now add a constant of 10 to each observation in the data set, and record your answers in the column labeled "$x + 10$."
    c. WITHOUT RECALCULATING, what will the mean of the data in column "$x + 10$" be? (Check the box next to the best answer.)
       ☐ The same mean as before a constant was added to each data point.
       ☐ The original mean $+ 10$.
       ☐ Unable to determine without recalculation.
    d. WITHOUT RECALCULATING, what will the standard deviation of the data in column "$x + 10$" be? (Check the box next to the best answer.)
       ☐ The same standard deviation as before a constant was added to each data point.
       ☐ The standard deviation before the constant was added $+ 10$.
       ☐ Unable to determine without recalculation.
    e. In your own words, define the mean and the standard deviation of a set of observations. Using these definitions, describe the effect of adding a constant to the mean and standard deviation. Did adding a constant affect one of these descriptive statistics and not the other? Why?

18. You and your roommate are embroiled in a spirited discussion about statistics. Your roommate says that the size of the standard deviation depends on the value (size) of the mean. Select a statement from the list of possible answers below, and defend it to your roomie.
    a. Yes! The higher the mean, the higher the standard deviation will be.
    b. Yes! You have to know the mean to calculate the standard deviation.
    c. No! Standard deviation measures the average deviation of scores from the mean no matter what the mean is.
    d. No! The standard deviation measures only how values differ from each other, not how they differ from the mean.

19. A paperboy wants to know all about the dogs in the neighborhoods on his newspaper delivery route—after all, forewarned is forearmed. He asks the dog owners for the weights of the dogs on his route. He then calculates the mean and standard deviation for the sample of 10 dogs he successfully obtained weights for. His standard deviation is extremely high. What do you know about the distribution of weights in this sample? Indicate whether the following statements are true (T) or false (F):
    T F a. All of the dogs are very big (they weigh a lot). Their high weights are pulling the standard deviation upward.

    T F b. The weights of the dogs may vary a great deal. Some are small dogs, and some are very large, making the standard deviation quite large.

    T F c. A few dogs in the sample could be quite large. These outliers might be creating positive skew in the data.

    T F d. A few dogs in the sample could be quite small. These outliers might be creating positive skew in the data.

20. Motor control researchers were interested in seeing how the performance of a well-learned task was altered by small perturbations. *Perturbations* are deviations from normal movement caused by an external source. In the case of this experiment, five participants were asked to point at a target. They performed six repetitions of the task. During three of their repetitions, a mechanical device bumped their arm. The distance from the target (error) was measured in millimeters. The data each participant's unperturbed (control) and perturbed repetitions are shown in Table 5.12.

Table 5.12 Error (in millimeters) during a motor-control test performed under unperturbed and perturbed conditions

ID	Condition	Error	ID	Condition	Error
1	Control	2.8	1	Perturbed	4.3
1	Control	1.7	1	Perturbed	4.0
1	Control	3.8	1	Perturbed	3.7
2	Control	4.0	2	Perturbed	4.5
2	Control	3.4	2	Perturbed	3.0
2	Control	2.3	2	Perturbed	2.7
3	Control	3.8	3	Perturbed	3.9
3	Control	1.5	3	Perturbed	2.7
3	Control	3.4	3	Perturbed	4.5
4	Control	2.5	4	Perturbed	2.2
4	Control	1.8	4	Perturbed	3.7
4	Control	3.0	4	Perturbed	1.7
5	Control	2.7	5	Perturbed	4.4
5	Control	3.1	5	Perturbed	1.7
5	Control	1.6	5	Perturbed	2.9

    a. Why did the researchers have the participants perform both tasks three times each?

    b. Calculate the mean, standard deviation, and variance for the tasks performed under the control and perturbed conditions.

21. You've been hired by a local professional baseball team to help the manager select the pitchers for an upcoming series. The team has provided you with some useful data:

the earned run averages (ERAs) of three different pitchers against the 10 teams in the league. Earned run average is the average number of runs a pitcher gives up in a nine-inning game, so lower is better.

Table 5.13   Earned run averages (ERAs) of three pitchers against 10 different opponents

Pitcher	Opponent	ERA	Pitcher	Opponent	ERA	Pitcher	Opponent	ERA
1	Team A	3.87	2	Team A	2.63	3	Team A	1.45
1	Team B	4.11	2	Team B	4.59	3	Team B	1.86
1	Team C	4.08	2	Team C	2.68	3	Team C	1.9
1	Team D	3.9	2	Team D	6.19	3	Team D	1.82
1	Team E	4.08	2	Team E	4.49	3	Team E	1.55
1	Team F	3.92	2	Team F	2.11	3	Team F	1.35
1	Team G	3.70	2	Team G	6.51	3	Team G	9.02
1	Team H	4.21	2	Team H	2.69	3	Team H	2.08
1	Team I	4.28	2	Team I	1.71	3	Team I	2.02
1	Team J	4.28	2	Team J	7.40	3	Team J	1.93

a. Find the mean ERA and standard deviation for each pitcher. Which pitcher is best, based on ERA?
b. Which pitcher is the most consistent?
c. Which pitcher is the least reliable?
d. If your team were playing the final game of the championship series against Team G, which pitcher would you NOT start?

22. A cereal company runs periodic checks on how well the machines in its factory package it's product. The company want to make sure the machines aren't putting too much or too little cereal (based on weight) in each box. Table 5.14 shows the weight in kilograms of 30 sampled packages. The company will reject any box that is either two standard deviations too heavy or two standard deviations too light, based on the mean of the sample. Are there any boxes in the sample that need to be rejected?

Table 5.14   Weights of 30 cereal boxes (in kilograms)

ID	Weight	ID	Weight	ID	Weight
1	0.355	11	0.361	21	0.352
2	0.347	12	0.355	22	0.353
3	0.347	13	0.349	23	0.351
4	0.354	14	0.353	24	0.346
5	0.351	15	0.352	25	0.354

6	0.349	16	0.349	26	0.348
7	0.349	17	0.351	27	0.345
8	0.348	18	0.354	28	0.351
9	0.351	19	0.353	29	0.347
10	0.348	20	0.348	30	0.341

23. Consider the following data set:

$$6.8, 7.3, 1.5, 2.3, 7.6, 1.2, 7.9, 8.4, 9.7, 6.9$$

Now calculate:

a. The mean.
b. The standard deviation.
c. The mean absolute deviation.

Why is the standard deviation preferred over the mean absolute deviation?

24. You've been tasked by the International Olympic Committee to detect if any of the sprinters in the 100-meter event cheated in the final race by anticipating the start gun. You know that the mean reaction time for sprinters in this event is 215 milliseconds (ms), and that the variance is 115 ms^2. If a sprinter is more than one standard deviation faster than the mean reaction time, they should be disqualified under rules of the International Association of Athletics Federation. Based on the reaction times listed below, which athlete(s), if any, should be disqualified?

Table 5.15 Starting reaction times (RT; in milliseconds) of finalists in the 100-meter sprint

Lane	Start RT
1	211
2	205
3	208
4	221
5	207
6	206
7	208
8	207

25. Table 5.16 shows the data for a small sample of patients investigated by a researcher specializing in bone health and osteoporosis. All of the patients had presented with a low-trauma fracture at a regional fracture clinic. The T-score variable is a measure of

bone density given as the standard deviation away from that of a healthy male adult. Calculate the mean, median, mode, range, IQR, variance, and standard deviation for each of the variables in the data set.

Table 5.16 T-scores and education status of 30 patients at a fracture clinic

ID	Gender	Age	T-score	Education Status
1	Male	65	−1.8	No HS
2	Female	77	−1.6	Postsecondary degree
3	Male	66	−1.2	Some postsecondary
4	Female	77	−2.3	HS grad
5	Male	71	−2.7	Some postsecondary
6	Female	77	−2.3	HS grad
7	Female	74	−1.0	Some postsecondary
8	Female	71	−2.7	Some HS
9	Female	87	−1.9	No HS
10	Male	88	−2.0	Postsecondary degree
11	Male	86	−2.7	HS grad
12	Male	68	−1.5	No HS
13	Male	72	−1.2	HS grad
14	Female	73	−1.3	Some HS
15	Female	78	−2.3	Some HS
16	Male	67	−2.7	HS grad
17	Male	83	−2.1	No HS
18	Female	67	−1.5	Postsecondary degree
19	Female	80	−1.6	Some HS
20	Female	95	−2.0	HS grad
21	Female	66	−2.7	Some postsecondary
22	Female	77	−2.1	No HS
23	Male	96	−2.3	Some postsecondary
24	Male	67	−2.7	Some HS
25	Male	73	−1.8	HS grad
26	Female	79	−1.6	No HS
27	Female	67	−2.9	Some HS

230 STATISTICS IN CONTEXT

28	Male	95	−2.4	No HS
29	Male	85	−2.2	Postsecondary degree
30	Male	74	−1.5	Some HS

26. Using the same bone health data, calculate the mean and standard deviation for the different education levels. From these descriptive statistics, how would you describe the groups relative to each other.

27. From the bone health data, construct a box-and-whisker plot for both female and male patients. How do the genders compare?

Think About It . . .

THE RANGE RULE

SOLUTIONS

1. Data Set:

A	B	C
Range = 5 − 1 = **4**	Range = 23 − 1 = **22**	Range = 5 − 1 = **4**
SD = **1.41**	SD = **8.24**	SD = **1.60**

2. Data Set:

    Apply the range rule (RR):

A	B	C
RR = **4/4 = 1**	RR = **22/4 = 5.5**	RR = **4/4 = 1**

    For each data set, is the SD approximately the range/4?

A	B	C
☑ Yes	☐ Yes	☑ Yes
☐ No	☑ No	☐ No

    How different are each of these two measures for each set?

A	B	C
1 − 1.41 = −0.41	**8.24 − 5.5 = 2.74**	**1 − 1.60 = −0.60**

REFERENCES

Brickart, K. C., Wright, C. I., Dautoff, R. J., Dickerson, B. C., & Barrett, L. F. (2011). Amygdala volume and social network size in humans. *Nature Neuroscience, 14*(2), 163–164.

Gorard, S. (2005). Revisiting a 90-year-old debate: The advantages of the mean deviation. *British Journal of Educational Studies, 53*(4), 417–430.

John F. Kennedy, 35th President of the United State: 1961-1963. 266-Address in the assembly hall at the Paulskirche in Frankfurt. June 25, 1963. Retrieved from http://www.presidency.ucsb.edu/ws/?pid=9303.

Keinan, G. (2002). The effects of stress and desire for control on superstitious behavior. *Personality and Social Psychology Bulletin, 28*(1), 102–108.

Natoli, E., Schmid, M., Say, L., & Pontier, D. (2007). Male reproductive success in a social group of urban feral cats (*Felis catus* L.). *Ethology, 113*, 283–289.

Plackett, R. L. (1958). Studies in the history of probability and statistics: VII. The principle of the arithmetic mean. *Biometrika, 45*(1/2), 130–135.

Stigler, S. M. (1986). *The history of statistics: The measurement of uncertainty before 1900*. Cambridge, MA: Harvard University Press.

Swami, V., Furnham, A., Chamorro-Premuzic, T., Akbar, K., Gordon, N., Harris, T., Finch, J., & Tovée, M. J. (2010). More than just skin deep? Personality influences men's ratings of attractiveness of women's body sizes. *Journal of Social Psychology, 150*(6), 628–647.

DESCRIPTIVE STATISTICS WITH SPSS AND R

Descriptive Statistics with SPSS

Study Time Data Set

The data set we will be using for this tutorial contains information from a fictitious study examining the study habits of 160 undergraduate students. Data collected on the study participants include sex (female, male), class (freshman, sophomore, junior, senior), and average weekly time spent studying (duration in minutes). [You can download the data set from the book's companion website: www.oup.com/us/blatchley.]

Descriptive Analyses

SPSS makes it very simple to quickly calculate descriptive statistics. In this example, we are going to determine the mean, standard deviation, minimum, maximum, range, and variance for our dependent variable: duration.

STEP 1: Open the data set in SPSS.

STEP 2: Click **Analyze**, then **Descriptive Statistics**, and then **Descriptives**.

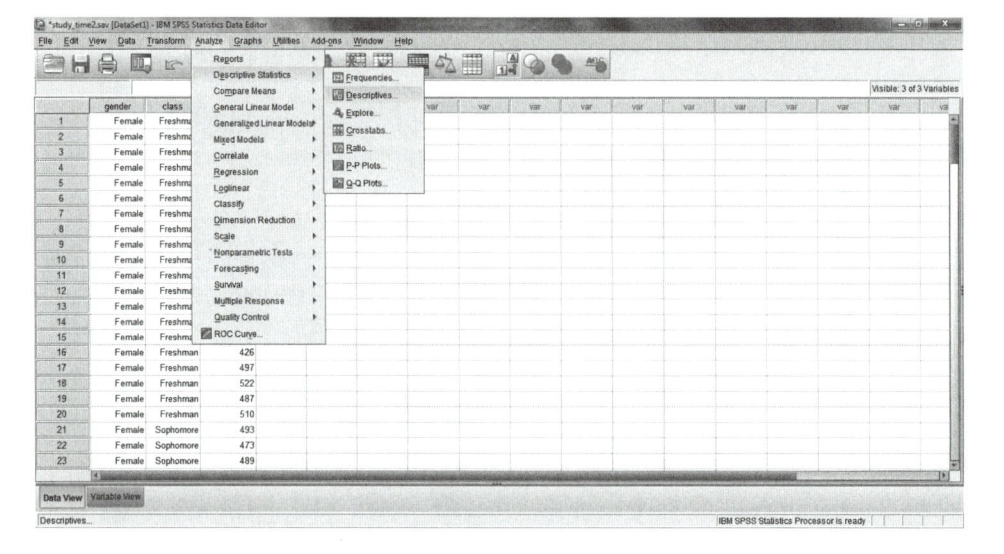

Step 3: Select the variable of interest (**duration**), and click the arrow to move it to the **Variable(s)** list. (You can also click-and-drag the variable across to the **Variable(s)** list.)

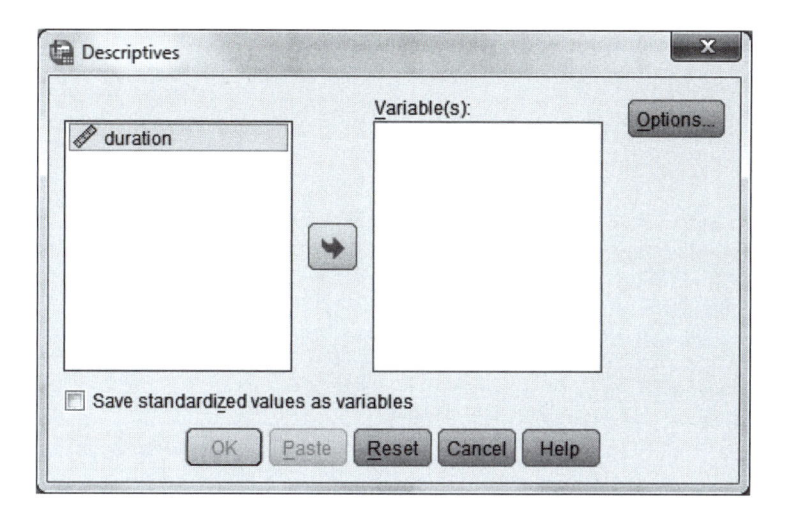

Step 4: To specify the descriptive statistics you would like SPSS to output, click the **Options** button.

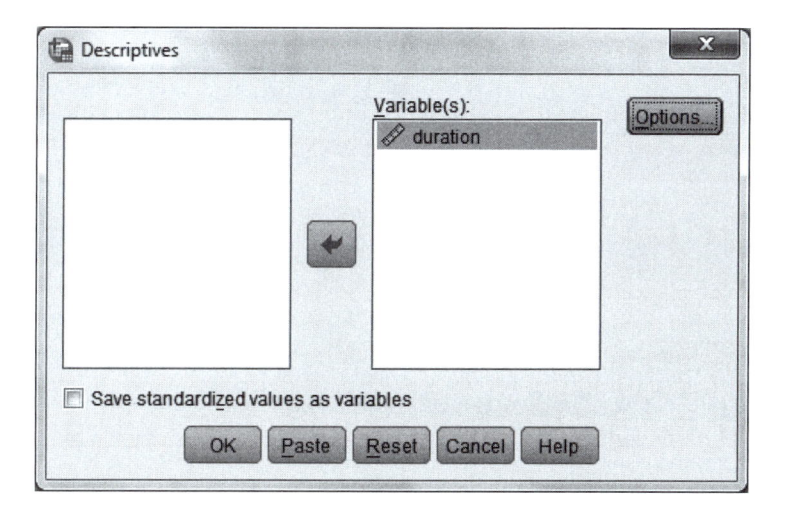

Step 5: Select the descriptive statistics you want. In this case, we'll be asking SPSS to give us the mean, standard deviation, minimum, maximum, range, and the variance. Click **Continue** when you've made your selections.

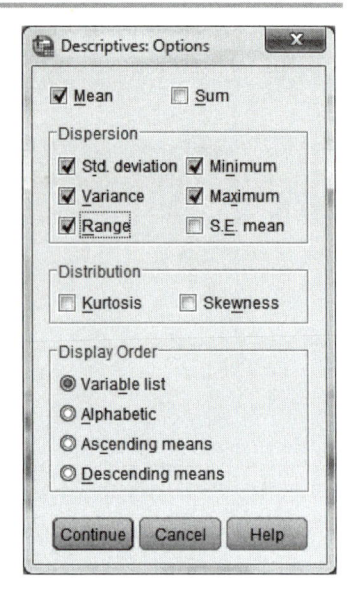

Step 6: Back at the **Descriptives** selection box, click **OK** to start SPSS working.

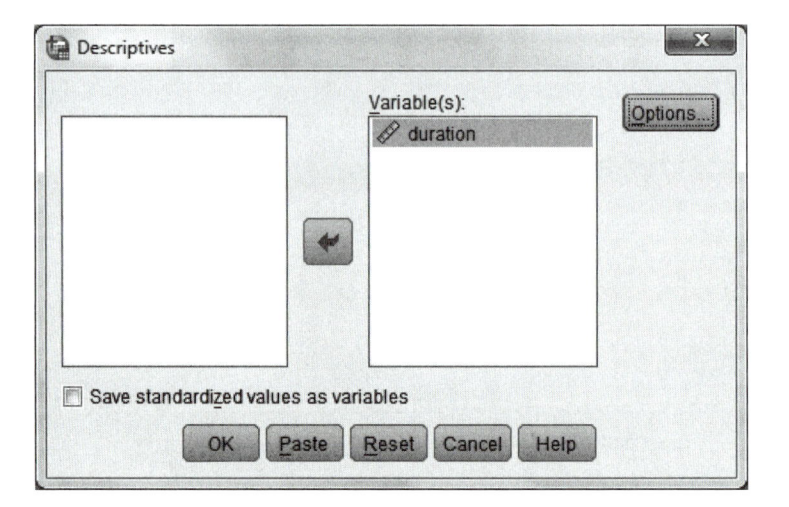

234 **STATISTICS IN CONTEXT**

Step 7: From the output, we can see that in the data set are 160 durations collected from our study participants. The study durations have a minimum weekly average of 201 minutes and a maximum of 684 minutes (range: 483 minutes). The mean duration was 452.9 minutes, with a standard deviation of 87.5 minutes (variance: 7,656.8 min^2).

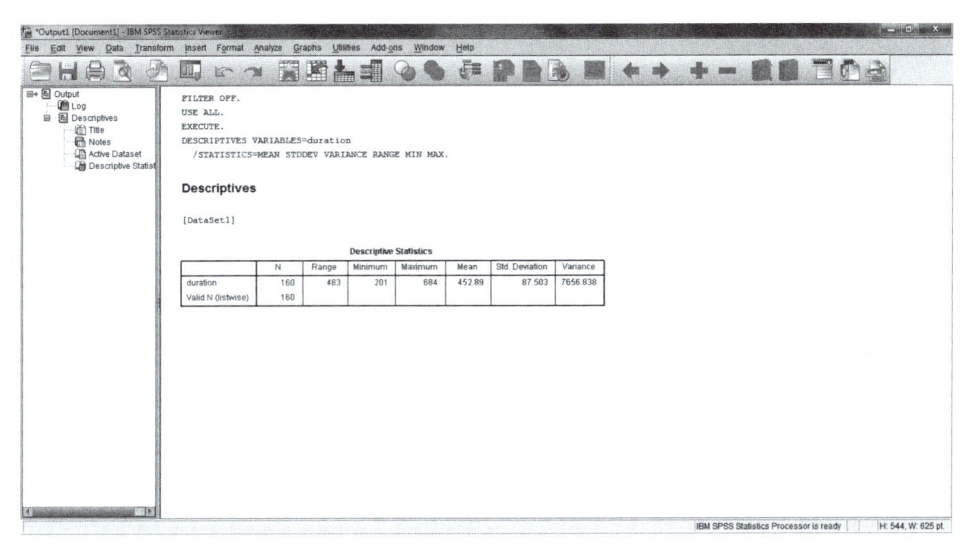

Descriptive Analyses by Group

We now have a picture of what our dependent variable looks like over the entire study population. However, that's not particularly interesting. The researchers also collected information on two independent variables: gender and class. Do these two factors influence study time duration? That's a more intriguing question. Let's take a peek at what the descriptives have to say.

Step 1: Once the data set is open in SPSS, select **Analyze** from the main menu bar. Then, select **Descriptive Statistics** and then **Explore**.

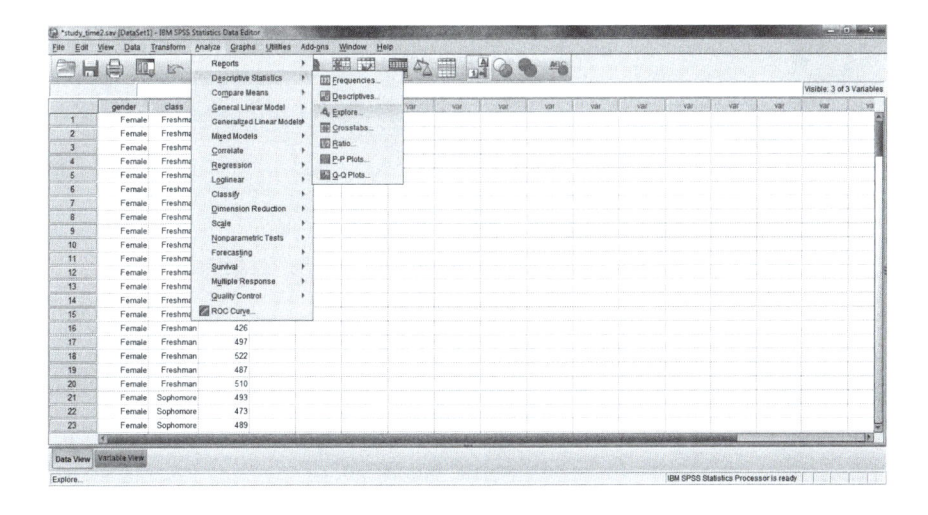

CHAPTER 5 Variability: The "Law of Life" 235

Step 2: Select your dependent variable (**duration**), and move it to the dependent list.

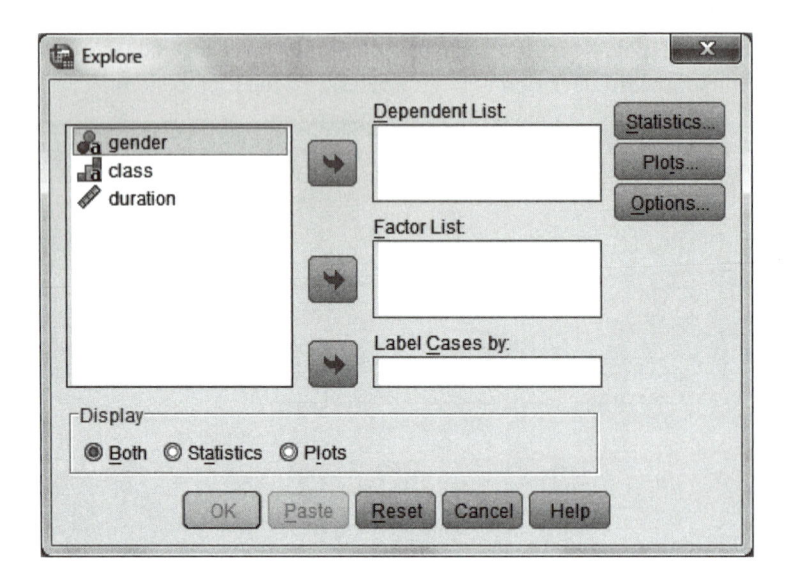

Step 3: Select the independent variables, and add them to the **Factor** list.

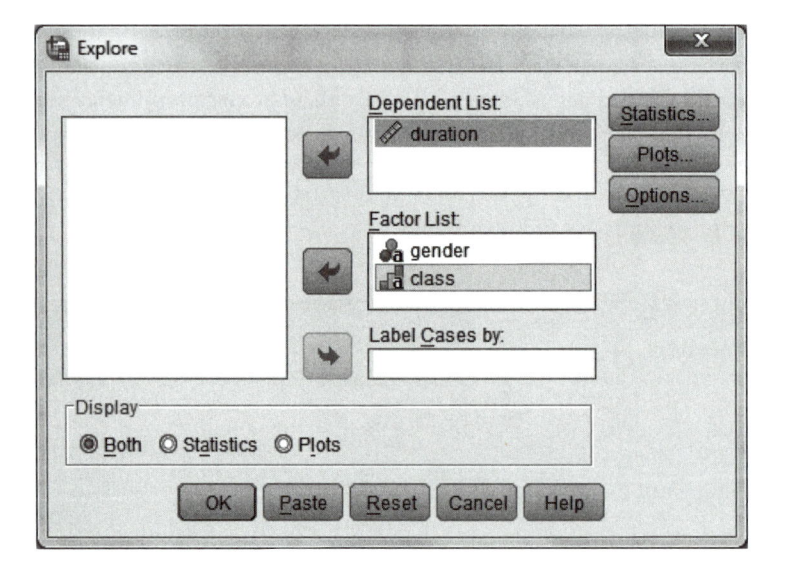

STATISTICS IN CONTEXT

Step 4: Click **Plots,** and specify the plots that you would like. Let's select **Stem-and-leaf** and **Histogram** so that we can visualize the shape of our data set. Click **Continue** to return to the **Explore** options box.

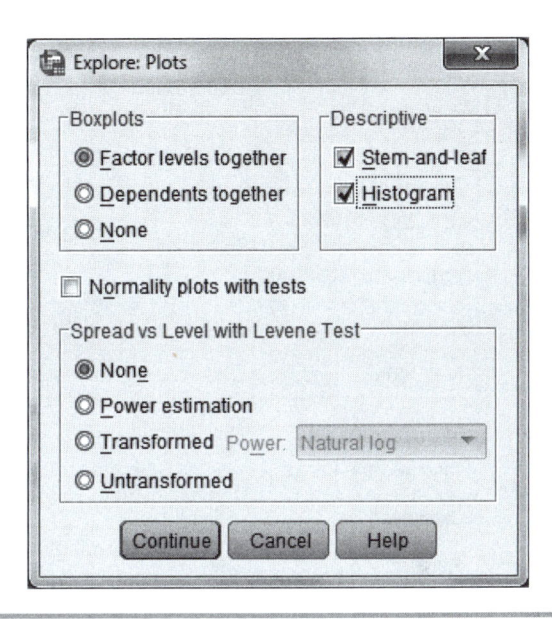

Step 5: At the **Explore** options box, under **Display**, select **Both** (Statistics and Plots). Click **OK** to start the analysis.

Step 6: Interpret the output. Since we asked SPSS to look at our dependent variable for each level of the two independent variables, it produces a lot of output (2 × gender and 4 × class). I've included just a sample focusing on gender.

CHAPTER 5   Variability: The "Law of Life"

Descriptives

Gender			Statistic	Std. Error
Duration	Female	Mean	481.79	4.203
		95% Confidence Interval for Mean — Lower Bound	473.42	
		95% Confidence Interval for Mean — Upper Bound	490.15	
		5% Trimmed Mean	480.94	
		Median	484.00	
		Variance	1413.030	
		Std. Deviation	37.590	
		Minimum	393	
		Maximum	591	
		Range	198	
		Interquartile Range	42	
		Skewness	.177	.269
		Kurtosis	.924	.532
	Male	Mean	424.00	12.403
		95% Confidence Interval for Mean — Lower Bound	399.31	
		95% Confidence Interval for Mean — Upper Bound	448.69	
		5% Trimmed Mean	423.24	
		Median	433.50	
		Variance	12306.734	
		Std. Deviation	110.936	
		Minimum	201	
		Maximum	684	
		Range	483	
		Interquartile Range	191	
		Skewness	.051	.269
		Kurtosis	−.804	.532

From the output, we can see that female participants spent more time studying across all four classes than did their male colleagues. Male participants were much more variable too.

```
Duration Stem-and-Leaf Plot for gender = Female

 Frequency          Stem & Leaf

     1.00 Extremes
(=<393)
     3.00          4 .   001
     8.00          4 .   22223333
     7.00          4 .   4444555
    12.00          4 .   666666777777
    25.00          4 .   8888888888888888999999999
    16.00          5 .   0000000000000111
     4.00          5 .   2233
     1.00          5 .   5
     3.00 Extremes
(>=570)

 Stem width:       100
 Each leaf:        1 case(s)
```

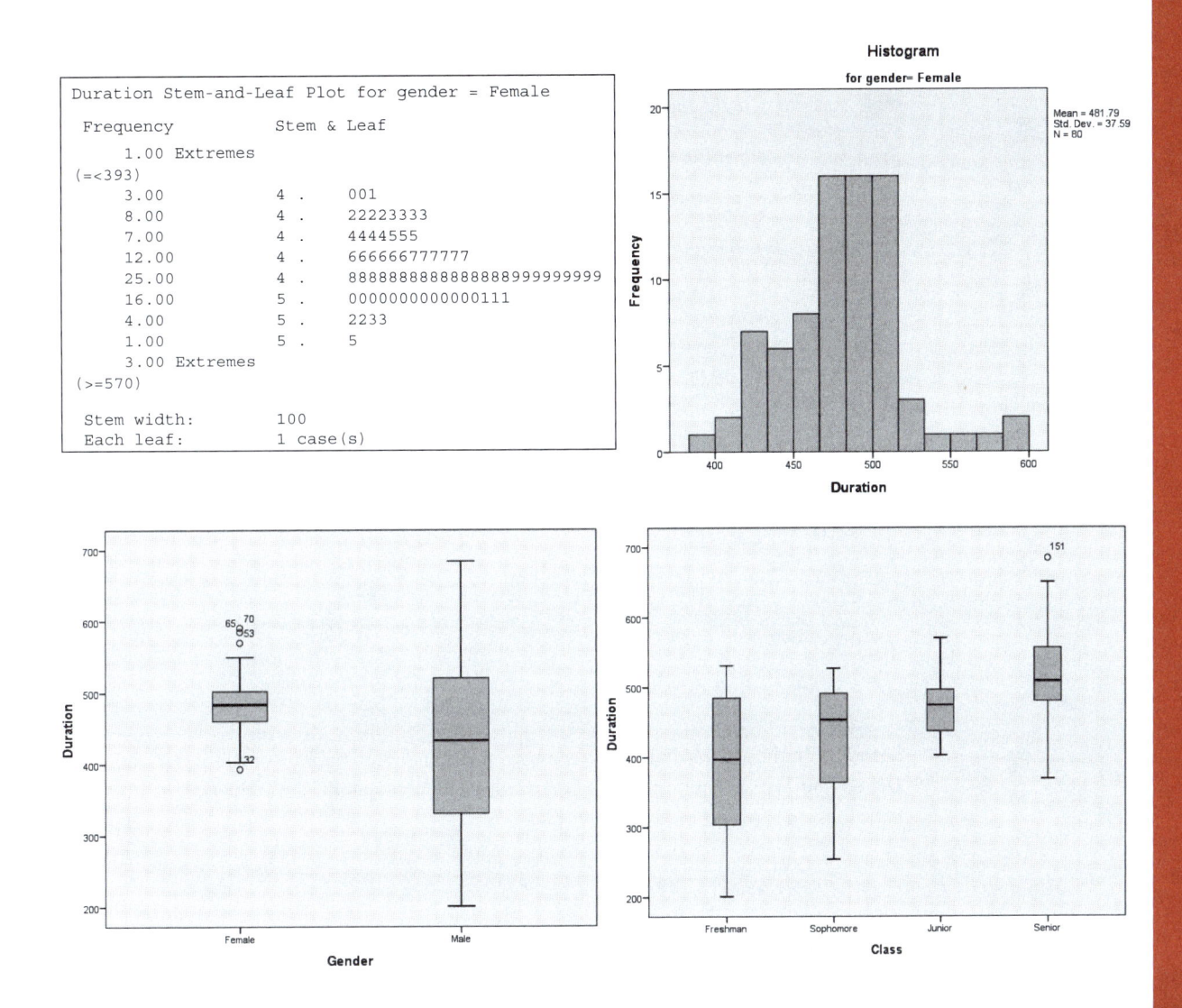

Descriptive Analyses: Interactions

Thus far, we've looked at our data as a whole and checked out how the dependent variable changed for each of our two factors. We can say that the female participants spent more time studying than the male participants. We can also say that more time was spent studying the further along in college participants got. But that may not be the whole story. In this section, we're going to drill down even more and look at the data as an interaction between the two independent variables.

CHAPTER 5 Variability: The "Law of Life" 239

At this level of analysis, we are combining the two factors (gender and class) to produce eight unique groups (female-freshman, female-sophomore, female-junior, female-senior, male-freshman, male-sophomore, male-junior, male-senior). Let's get started.

Step 1: After opening your data set in SPSS, click **Analyze**, then **Compare Means**, and then **Means**.

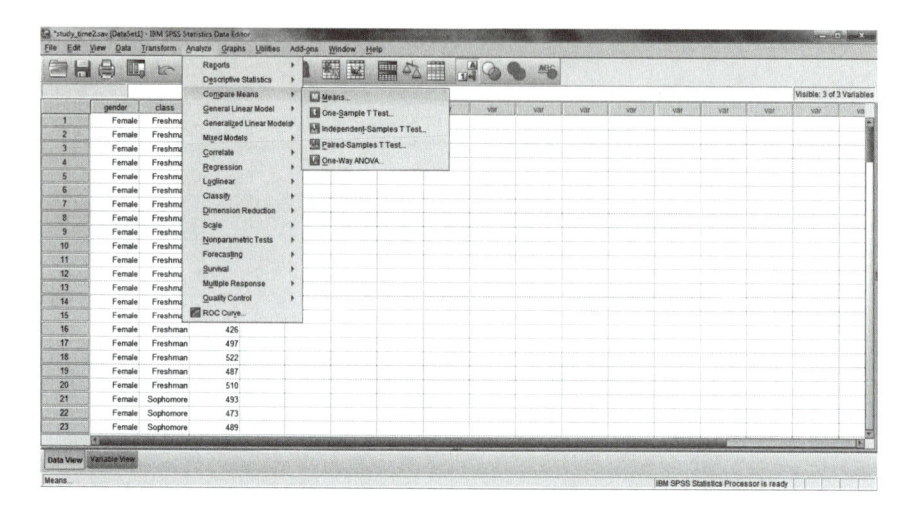

Step 2: In the **Means** dialog box, move your dependent variable (**duration**) into the **Dependent List**, and put your first independent variable (**gender**) into the **Independent list**. Click on the **Next** button.

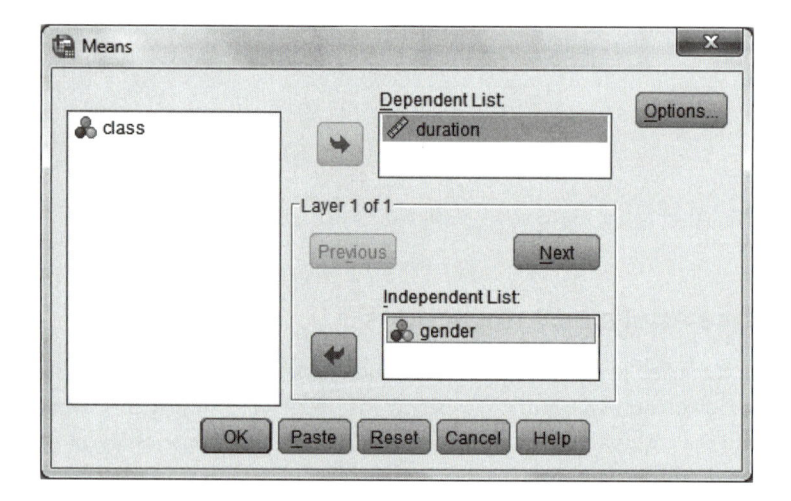

Step 3: Add the **class** variable to the **Independent List**. Notice that we created another "layer" in the dialogue box. In this situation, class is nested within gender. If we placed class in the Independent List with gender, our analysis would yield the same information as in our previous example using the **Explore** function.

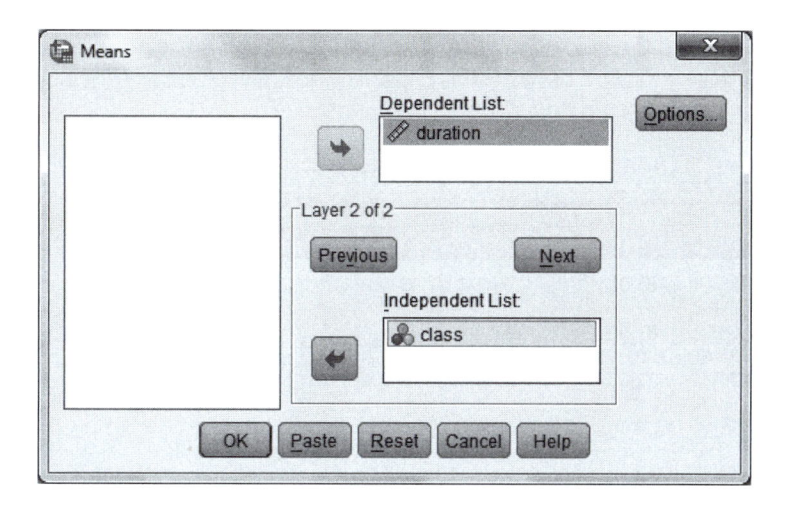

Step 4: Click on the **Options** button in the **Means** dialogue box, and move the statistics you want to appear in the output. SPSS defaults to a small subset of statistics. If you want to add several, you can use your mouse and **CTRL** key to select multiple items at a time. You can also reorder the statistics by dragging within the **Cell Statistics** list. Click **Continue** to return to the **Means** dialogue box.

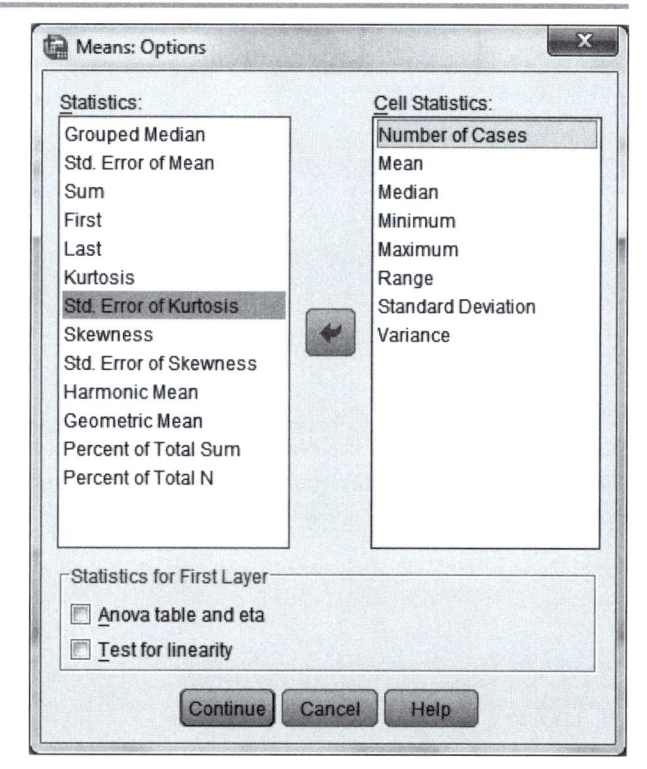

CHAPTER 5 Variability: The "Law of Life"   241

Step 5: Click **OK** in the **Means** dialogue box to run your analysis.

Step 6: Interpret your results. SPSS outputs a summary report containing the breakdown for each of the eight unique groups on all the requested descriptive statistics. The report also includes totals collapsed across the other factor. Check and confirm that the totals in this analysis match our earlier analyses.

Report

Duration

Gender	Class	N	Mean	Median	Minimum	Maximum	Range	Std. Deviation	Variance
Female	Freshman	20	481.25	485.00	426	531	105	31.492	991.776
	Sophomore	20	477.65	487.50	393	508	115	31.421	987.292
	Junior	20	468.70	477.00	403	570	167	45.781	2095.905
	Senior	20	499.55	500.50	446	591	145	35.630	1269.524
	Total	80	481.79	484.00	393	591	198	37.590	1413.030
Male	Freshman	20	296.35	304.00	201	368	167	47.557	2261.713
	Sophomore	20	375.40	364.00	254	527	273	64.292	4133.516
	Junior	20	482.15	474.50	415	560	145	40.951	1676.976
	Senior	20	542.10	550.00	369	684	315	72.239	5218.411
	Total	80	424.00	433.50	201	684	483	110.936	12306.734
Total	Freshman	40	388.80	397.00	201	531	330	101.741	10351.190
	Sophomore	40	426.53	454.00	254	527	273	71.941	5175.538
	Junior	40	475.43	474.50	403	570	167	43.410	1884.456
	Senior	40	520.82	508.50	369	684	315	60.208	3625.020
	Total	160	452.89	475.50	201	684	483	87.503	7656.838

Looking at the means for each of the groups, you can see that across the four classes, the female students' average duration of study time is relatively consistent. For males, however, there is a trend of increasing amount of time studied from class to class.

Descriptive Statistics with R

R is a bit different from SPSS. It is primarily a text-based interface (although you can install some graphic user interfaces), and it typically gives you very concise information. When we ran our descriptive analyses using SPSS, with very few clicks of the mouse we were able to generate lots of output. Sometimes with SPSS, you have to hunt for the important information as it's hidden among lots of unimportant information. R is usually the opposite: You need to tell it everything you want. This can be a challenge at times, since you may not necessarily know what you need.

To get started, we first need to load the necessary libraries (also called a package). R comes with a core set of functions, and as required, libraries are loaded to increase functionality. In this situation, we are adding the **pander** library to help with generating nicely formatted tables, the **ggplot2** library for enhanced plotting capabilities, and **dplyr** to make data manipulation in R easier to follow.

```
library(pander)
library(ggplot2)
library(dplyr)
```

Now that we have the libraries loaded, we will read in the data. We'll be using the same Study Time Data Set as we did with the SPSS tutorial. The **tbl_df** function is used to make the **study_time** data set easier to read when printed to the console.

R likes to sort categorical variables alphabetically, so after reading in the data we will reorder the class variable so that it is sorted chronologically (freshman, sophomore, junior, senior).

```
    # read in data
study_time2 <- tbl_df(read.csv("study_time.csv"))
    # reorder class factor
    study_time2$class <- factor(study_time2$class,levels(study_
    time2$class)[c(1,4,2,3)])
```

Let's check out how the data looks in R. The **glimpse** function gives a quick summary of the structure of the data set. From the output, we see that there are 160 observations and three variables, that gender and class are factors (categorical variables—nominal or ordinal data), and that duration is an integer (interval or ratio data).

```
    glimpse(study_time2)
    ## Observations: 160
    ## Variables: 3
    ## $ gender   (fctr) Female, Female, Female, Female, Female,
    Female, Fema...
    ## $ class    (fctr) Freshman, Freshman, Freshman, Freshman,
    Freshman, Fr...
    ## $ duration (int) 505, 468, 463, 437, 465, 434, 529, 483, 493,
    442, 474...
```

We will repeat our analyses that we ran in the SPSS tutorial. In our first analysis, we want to focus on our dependent variable: duration.

To calculate the descriptive statistics using the **dplyr** package (loaded earlier), we take our data set, **study_time2**, and use the pipe operator to pass the data to the **summarize** function.

In the **summarize** function, we define our summary variable names (to the left of the equals sign) and the formula to calculate it (to the right of the equals sign). Most common stats calculations are already built into R. For example, to calculate the mean of the variable duration, you use **mean(duration)**.

In the code that follows, the assignment operator <- is used to store the result (in this case, a data frame with the requested descriptive statistics) as an R object called **j**. This allows us to reuse the results without having to redo all the calculations any time we want to reuse them and makes it easier to read and follow. The last call, **pander(j)**, prints the descriptive stats in a formatted table suitable for publication. The default print of **j** can sometimes be difficult to read and is not suitable to use for publication. (Note the # symbol is used to make comments in R.)

```
j <- study_time2 %>% # data
  summarize(n = n(), # frequency count
            mean = mean(duration), # mean
            trimmed = mean(duration, trim = .05), # trimmed mean
            se = sqrt(var(duration)/length(duration)), # standard error
            median = median(duration), # median
            minimum = min(duration), # minimum value
            maximum = max(duration), # maximum value
            range = maximum - minimum, # range
            IQR = IQR(duration, type = 6), # interquartile range
            uCI = t.test(duration, conf.level = .95)$conf.int[2], # upper confidence limit
            lCI = t.test(duration, conf.level = .95)$conf.int[1], # lower confidence limit
            std_dev = sd(duration), # standard deviation
            variance = var(duration)) %>% # variance
  mutate_each(funs(round(.,2))) # round values to 2 decimal places
```

Default R Printing

```
j
## Source: local data frame [1 x 13]
##
##       n    mean  trimmed    se median minimum maximum range    IQR      uCI
##   (dbl)   (dbl)    (dbl) (dbl)  (dbl)   (dbl)   (dbl) (dbl)  (dbl)    (dbl)
## 1   160  452.89   455.71  6.92  475.5     201     684   483     94   466.56
## Variables not shown: lCI (dbl), std_dev (dbl), variance (dbl)
```

Formatted Printing

```
pander(j)
```

n	mean	trimmed	se	median	minimum	maximum	range	IQR	uCI	lCI	std_dev	variance
160	452.89	455.71	6.92	475.5	201	684	483	94	466.56	439.23	87.5	7656.84

As we did in the SPSS tutorial, we will next look at the study duration variable grouped by each level of the two categorical variables (or factors). We want to see if belonging to a group has an impact on our dependent variable. First, we'll analyze gender.

The code is nearly identical to our first look at the descriptives with the addition of the **group_by** function. This tells R to calculate the descriptives separately for each level of the gender variable. And because gender is now part of our data frame, we have to tell the rounding function not to round gender as trying to round a categorical variable doesn't work.

```
k <- study_time2 %>%
    group_by(gender) %>% # specify grouping factor
    summarize(n = n(),
              mean = mean(duration),
              trimmed = mean(duration, trim = .05),
              se = sqrt(var(duration)/length(duration)),
              median = median(duration),
              minimum = min(duration),
              maximum = max(duration),
              range = maximum - minimum,
              IQR = IQR(duration, type = 6),
              uCI = t.test(duration, conf.level = .95)$conf.int[2],
              lCI = t.test(duration, conf.level = .95)$conf.int[1],
              std_dev = sd(duration),
              variance = var(duration)) %>%
    mutate_each(funs(round(.,2)), -gender) # don't round categorical variable

pander(k)
```

gender	n	mean	trimmed	Se	median	minimum	maximum	range	IQR	uCI	lCI	std_dev	variance
Female	80	481.79	480.94	4.2	484.0	393	591	198	42.25	490.15	473.42	37.59	1413.03
Male	80	424.00	423.24	12.4	433.5	201	684	483	191.25	448.69	399.31	110.94	12306.73

And now very quickly we can look at class.

```
b <- study_time2 %>%
    group_by(class) %>% # specify new grouping variable
    summarize(n = n(),
              mean = mean(duration),
              trimmed = mean(duration, trim = .05),
              se = sqrt(var(duration)/length(duration)),
              median = median(duration),
              minimum = min(duration),
```

```
                        maximum = max(duration),
                        range = maximum - minimum,
                        IQR = IQR(duration, type = 6),
                        uCI = t.test(duration, conf.level = .95)$conf.int[2],
                        lCI = t.test(duration, conf.level = .95)$conf.int[1],
                        std_dev = sd(duration),
                        variance = var(duration)) %>%
            mutate_each(funs(round(.,2)), -class) # update to match
grouping variable

            pander(b)
```

												std_	
class	n	mean	trimmed	se	median	minimum	maximum	range	IQR	uCI	lCI	dev	variance
Freshman	40	388.80	391.22	16.09	397.0	201	531	330	183.5	421.34	356.26	101.74	10351.19
Sophomore	40	426.52	429.92	11.37	454.0	254	527	273	128.5	449.53	403.52	71.94	5175.54
Junior	40	475.43	474.14	6.86	474.5	403	570	167	63.0	489.31	461.54	43.41	1884.46
Senior	40	520.83	519.11	9.52	508.5	369	684	315	77.0	540.08	501.57	60.21	3625.02

With SPSS, when we used **Explore** to look at the data, it generated a few useful plots we could use to visualize our data. In R, we have to ask it explicitly to generate those plots. However, it's very easy to do.

We will generate a histogram and a stem-and-leaf plot for the female students' data. The **ggplot2** library provides many more options to control the plot output compared to the native R plotting capabilities and produces publication-quality figures. We'll use **ggplot2** (loaded earlier) to generate our histogram.

Default ggplot2 Histogram

```
study_time2 %>% # specify data to use
    filter(gender == 'Female') %>% # only use data from female
students
    ggplot(aes(x=duration)) + # specify the continuous variable
'duration'
    geom_histogram() # specify plot as a histogram
```

Formatted ggplot2 Histogram

With the addition of a few more lines we can make our histogram look a lot better.

```
study_time2 %>%

    filter(gender == 'Female') %>%

    ggplot(aes(x=duration)) +

    # set the width for grouping data values and make bars red
    geom_histogram(binwidth = 10, fill = "#AA1111",
```

STATISTICS IN CONTEXT

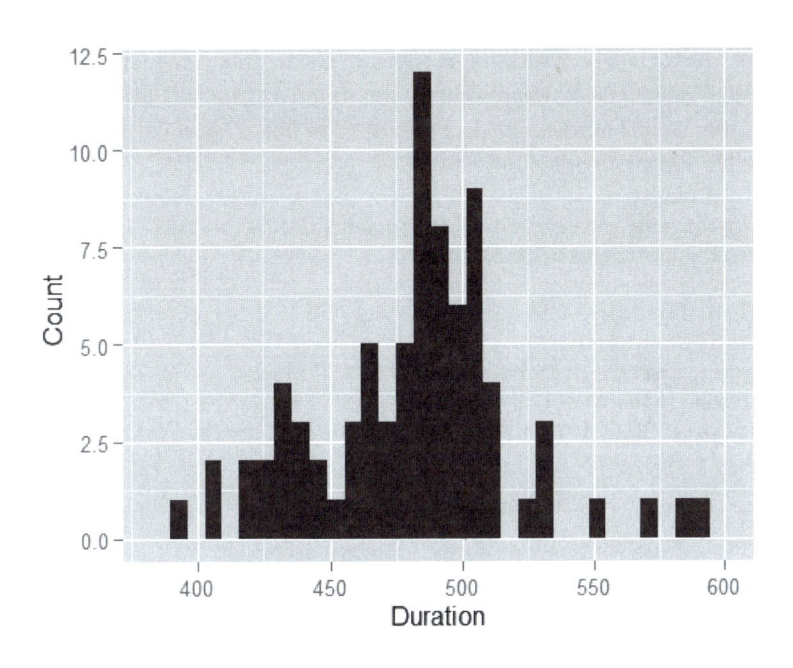

![Histogram with Count on y-axis and Duration on x-axis]

```
    color = "#AA1111") +
    xlab('Duration (in minutes)') + # x axis label
    ylab('Frequency') + # y axis label
    theme_bw() + # simple format of plot (gridlines etc.)
    ggtitle('Histogram of Variable Duration \n -Female Students-') #
Plot title, \n = new line
```

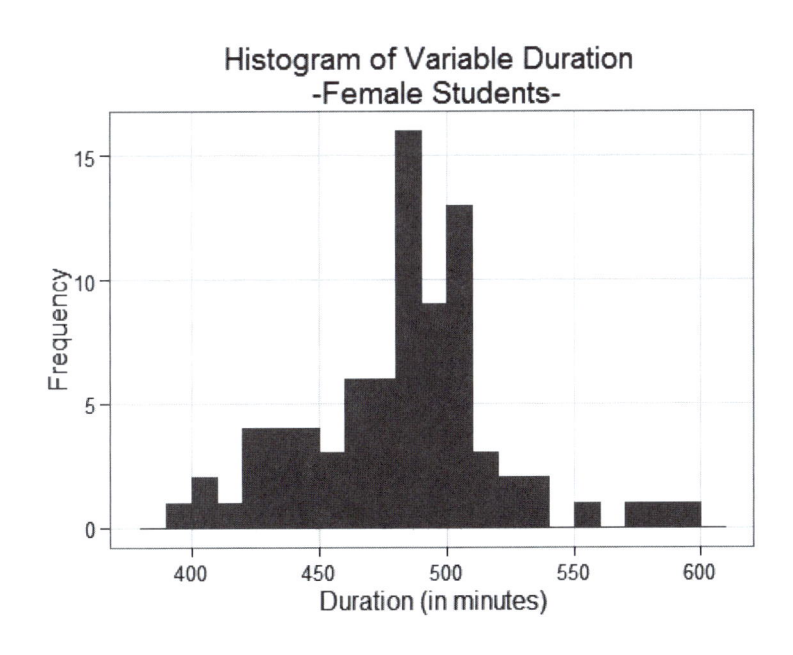

![Histogram of Variable Duration -Female Students-]

CHAPTER 5 Variability: The "Law of Life" 247

If we want to drill down and see the distributions for all eight categories of students, we tell R to fill the histogram bars with colors based on gender (**fill = gender**) and split the data onto different plots with **facet_wrap(~ class)**. The arguments **alpha** (adds transparency) and **position** (how to handle overlapping data points) allow both genders to display on the same plot without blocking each other out.

```
study_time2 %>%
    ggplot(aes(x=duration, fill=gender)) + # fill bars by gender
    geom_histogram(binwidth = 10, alpha = .5, position = "identity") +
    xlab('Duration (in minutes)') +
    ylab('Frequency') +
    facet_wrap(~ class) + # plot by class
    theme_bw() +
    ggtitle('Histogram of Variable Duration \n -Gender x Class-')
```

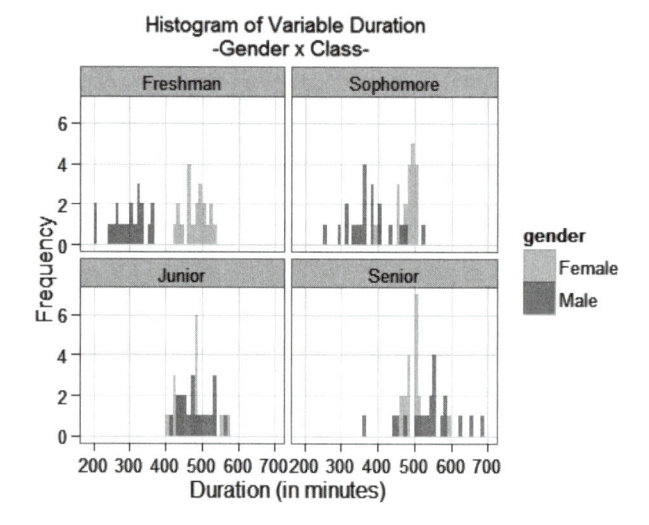

Stem-and-leaf plots were purposely designed to quickly generate a visualization of data without the benefit of graphic output. R has a built-in function that will generate the desired plot quickly.

First, we'll subset the data so only female students are included and save it to a data frame called **women**. Next, we call the stem function referencing the new dataframe and the duration variable **stem(women$duration)**. Simple.

STATISTICS IN CONTEXT

```
women <- filter(study_time2, gender == 'Female') # select only
female students
stem(women$duration)
    ##
##   The decimal point is 1 digit(s) to the right of the |
##
##   38 | 3
##   40 | 358
##   42 | 06693347
##   44 | 1256269
##   46 | 023458134788
##   48 | 00333333335677899233344799
##   50 | 0112333567889044
##   52 | 2914
##   54 | 0
##   56 | 0
##   58 | 51
```

Notice that R generated stem values for every second value (i.e., 38, 40, 42). This results in leaves that have 20 values represented to the right of the break. Thus, for 42, the values represented there are 420, 426, 426, 429, 433, 433, 434, 437. By adding the scale argument to the stem function, you are able to change this.

```
stem(women$duration, scale = 2)
##
##   The decimal point is 1 digit(s) to the right of the |
##
##   39 | 3
##   40 | 35
##   41 | 8
##   42 | 0669
##   43 | 3347
##   44 | 1256
##   45 | 269
##   46 | 023458
##   47 | 134788
##   48 | 0033333335677899
##   49 | 233344799
##   50 | 0112333567889
##   51 | 044
##   52 | 29
##   53 | 14
##   54 |
##   55 | 0
##   56 |
##   57 | 0
##   58 | 5
##   59 | 1
```

SPSS also generated box-and-whisker plots for duration across gender and class, so we'll do the same here with R. Again, we'll use the **ggplot2** library.

In our call to the **ggplot** function, we need to specify which variable will be on the *x*-axis (gender) and on the *y*-axis (duration), and to add some color, we can fill the boxes based on gender.

Since we are creating a box-and-whisker plot, we specify **geom_boxplot()**, use **guide(fill=FALSE)** to remove the legend (it's redundant), add some formatting, and *presto!* We have our plot.

```
study_time2 %>%
    ggplot(aes(x=gender, y=duration, fill=gender)) + # by gender
  geom_boxplot() + # box and whisker plot
  guides(fill=FALSE) + # remove legend
  xlab("Gender") +
  ylab("Duration (in minutes)") +
  ggtitle("Boxplot of Duration \n -Gender-") +
  theme_bw()
```

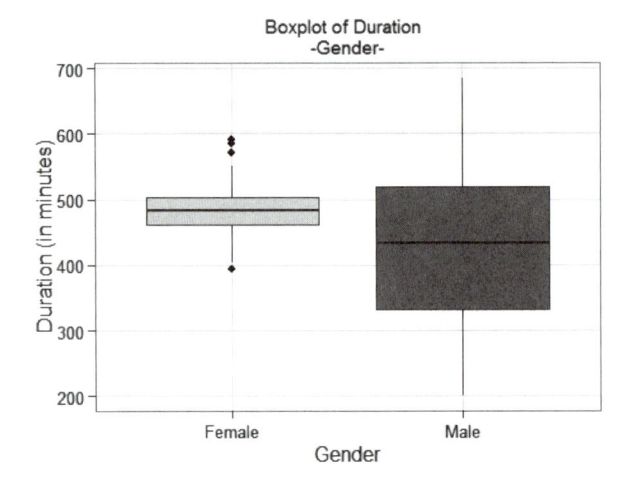

And for the box-and-whisker plot for class, we repeat the code, swapping class for gender.

```
study_time2 %>%
ggplot(aes(x=class, y=duration, fill=class)) + # by class
geom_boxplot() + # box and whisker plot
guides(fill=FALSE) + # remove legend
xlab("Class") +
ylab("Duration (in minutes)") +
ggtitle("Boxplot of Duration \n -Class-") +
theme_bw()
```

In this section, we'll be looking at the eight unique groups of study participants and determining their descriptive statistics.

Boxplot of Duration
-Class-

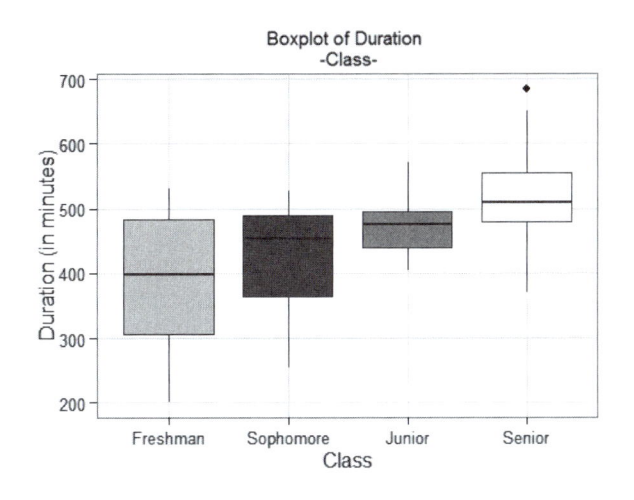

If the code looks familiar, you're right, it is. By adjusting the variables in our **group_by** function, we get the requested descriptive statistics for each of the eight student categories. For this analysis, we've requested fewer statistics.

```
x <- study_time2 %>%
group_by(gender, class) %>%
summarize(n = n(),
          mean = mean(duration),
          median = median(duration),
          minimum = min(duration),
          maximum = max(duration),
          range = maximum - minimum,
          std_dev = sd(duration),
          variance = var(duration)) %>%
   mutate_each(funs(round(.,2)), -gender, -class) # don't round categorical variables
pander(x)
```

gender	class	n	mean	median	minimum	maximum	range	std_dev	variance
Female	Freshman	20	481.25	485.0	426	531	105	31.49	991.78
Female	Sophomore	20	477.65	487.5	393	508	115	31.42	987.29
Female	Junior	20	468.70	477.0	403	570	167	45.78	2095.91
Female	Senior	20	499.55	500.5	446	591	145	35.63	1269.52
Male	Freshman	20	296.35	304.0	201	368	167	47.56	2261.71
Male	Sophomore	20	375.40	364.0	254	527	273	64.29	4133.52
Male	Junior	20	482.15	474.5	415	560	145	40.95	1676.98
Male	Senior	20	542.10	550.0	369	684	315	72.24	5218.41

CHAPTER SIX

WHERE AM I? NORMAL DISTRIBUTIONS AND STANDARD SCORES

*An unsophisticated forecaster uses statistics as a drunken man uses lamp-posts—
for support rather than illumination.*

—ANDREW LANG

Everyday Statistics

HOW "NORMAL" IS THAT CURVE?

Statisticians often say that normal distributions describe life. But what does that really mean?

Imagine you have been asked to design a line of office chairs. How big, how tall, how wide should you make them? You need to have the chair fit the buyer, and those buyers come in an amazing array of sizes and shapes. It turns out that body size—height, for example—is normally distributed, with most of us clustering around the mean and relatively few of us in the "extremely tall" or "extremely short" ends of the distribution. Engineers and designers incorporate statistical information about the distribution of body dimensions in the population into their designs to make sure that their end result will fit most people. Chapter 6 will focus on the characteristics of normal distributions and how they are used to make decisions such as these.

OVERVIEW

STATISTICS SO FAR
STANDARD SCORES
THE BENEFITS OF STANDARD SCORES
STANDARD SCORES IN CONTEXT

LEARNING OBJECTIVES

Reading this chapter will help you to . . .

- Understand how standard scores describe the position of a score within a distribution of scores. (Concept)

- Know how both IQ scores and z-scores are calculated and why they are both examples of a standard score. (Concept)

- Calculate a z-score. (Application)

- Become familiar with the 3-sigma rule and the relationship between position in a distribution (the z-score) and probability. (Concept)

- Use z-scores to estimate the probability of an observation in a normal distribution. (Application)

- Recognize how standardized tests like the SAT produce scores that use the standard normal curve to allow comparison of test performance. (Concept)

> mean (typical measurement)
> media (center of distribution)
> mode (frequently occurring observation)

STATISTICS SO FAR

In the first five chapters of this book, we have discovered a number of descriptive statistics and how they can help us describe the "take-home message" in a set of data. Measures of center tell us about the "guideposts" in our set of measurements—what a typical measurement is (the mean), where the center of the distribution is (the median), and what the most frequently occurring observation in the set is (the mode). Measures of variability tell us about how spread out the observations in our set are—how far apart the minimum and maximum values are (range and IQR) and how much a "typical" observation differs from the mean (variance and standard deviation). Frequency, relative frequency, and percentile rank begin to highlight patterns in the data, and we can display those patterns graphically using a number of different techniques. We have also talked about shapes of distributions and what they tell us about the observations in those distributions.

Researchers apply these statistics to help answer their research questions. For the beginning student of statistics, it can be quite helpful to realize that although the subject matter of the sciences may differ, the statistics scientists use are the same. The mean can describe the average dose of a drug, how often the size of a crowd prevents an individual in the crowd from calling for help during an emergency, or how much the measurements of a star's position vary between observers.

One of the most famous questions in the history of psychology is "What is intelligence?" Everyone agrees that intelligence is a quality or characteristic that humans share. But the question of what intelligence actually is, and how it can be measured, has been a bone of contention for a very long time. Historically, two explanations have been offered. One says that intelligence is innate and unitary: We are born with a certain amount of intelligence, and that amount is all we're ever going to have. The alternative explanation says that intelligence is multifaceted, malleable, and can be shaped by time and experience. The question of how to measure intelligence has been driven by which side of this philosophical fence the researcher happened to be on. French psychologist Alfred Binet stood squarely on the malleability side of that fence, and with the intelligence test he pioneered (in collaboration with Théodore Simon), he made a landmark contribution to the field of inferential statistics: the standard score (see *The Historical Context* on page 255).

STANDARD SCORES

Binet's famous test of intelligence provided a way to measure a psychological characteristic that cannot be seen or touched, or weighed or measured in the same way that the amount of a chemical in a solution or the force of a lever can be. In devising the test, Binet also created another descriptive statistic that sets the stage for the next type of statistic we'll be discussing: inferential statistics. You are probably

The Historical Context

ALFRED BINET AND INTELLIGENCE TESTING

Born in 1857 in Nice, France, Alfred Binet was a firm believer in the malleable nature of human intelligence, and the famous test that he developed reflects this perspective. Binet received his law degree in 1878 and then realized he was much more interested in science. Jean Charcot, a famous neurologist of the time (Charcot taught a young Sigmund Freud to use hypnosis), offered Binet a job at La Salpêtrière Hospital in Paris. Unfortunately, Binet published the results of some poorly designed experiments that claimed magnets could make a hypnotized patient change a phobia into its opposite (e.g., change fear of snakes into fondness for them). This study was immediately and roundly criticized by other scientists, earning Binet a reputation for careless research. As a result, he resigned from the lab and spent the next 2 years writing stories and plays and observing the mental development of his two daughters, Madeleine (age 4 at the time) and Alice (age 2½).

In 1892, Binet took a position at the Sorbonne, where he focused his research on child intellectual development. Using what he had learned from watching his daughters, he proposed that intelligence could be measured by giving children a battery of tests of their abilities, such as memory for word meaning, color, music, and more complex cognitive abilities like thinking in the abstract.

Several years earlier, mandatory education for all children in France had become the law. Binet saw that children who suffered from mental retardation but who legally had to attend school were unable to cope with the typical classroom curriculum, and he became an advocate for identifying children who needed placement in special classes. In 1904, the French government asked Binet and a graduate student, Théodore Simon, to develop a way to identify those children who would need special education outside of the French public school system.

The test Binet and Simon developed assessed "natural" or innate intelligence through a set of 30 questions of increasing complexity. Questions that 60% to 90% of children of a given age could answer easily were deemed to measure "normal" ability *for that age*. The test, called the Binet–Simon Scale, was published in 1905. Over the following years, until Binet's death in 1911, the researchers revised their scale, adding more tasks, dropping others, and developing a way to scale performance on the test, again by linking age and intelligence. A German psychologist, William Stern, added an additional refinement of this idea, saying that the ratio of mental age (MA) to chronological age (CA), multiplied by 100 to get rid of the decimal points, would provide a "mental quotient" (now called the Intelligence Quotient, or IQ score) for the child. Box 6.1 shows three examples of this calculation.

BOX 6.1 **Calculating IQ**

Normal for a child of age 6: $IQ = \left(\dfrac{MA}{CA}\right) \times 100 = \left(\dfrac{6}{6}\right) \times 100 = (1.00) \times 100 = 100$

Below normal for a child of age 6: $IQ = \left(\dfrac{MA}{CA}\right) \times 100 = \left(\dfrac{4}{6}\right) \times 100 = (0.67) \times 100 = 67$

Above normal for a child of age 6: $IQ = \left(\dfrac{MA}{CA}\right) \times 100 = \left(\dfrac{8}{6}\right) \times 100 = (1.33) \times 100 = 133$

familiar with some of the different versions of the descriptive statistic I'm referring to, although I'd be willing to bet that as often as you've probably run across them, you likely didn't know what they were. I'm talking about standard scores (note the plural here: a number of standard scores are used in everyday life).

Standard scores are descriptive statistics, or numbers that describe some aspect of a set of measurements. They are essentially measures of the *location of individual scores in a set, relative to the mean*. In fact, the definition of a **standard score** is *a score expressed relative to a known mean and standard deviation*. Think about the score that Stern developed for the Stanford–Binet test. This is a standard score: It tells you your score on the test, relative to the mean. Average on the test is always 100. If you score a 105, then you know your performance on the test is a bit above average—you know your score, relative to the mean.

Suppose I had asked you to take a new intelligence test and I reported back that you had gotten a 60 on that test. How would you know if a score of 60 was a good score? Does this score indicate that you're a genius, or average, or might be in need of some kind of remediation? Unless I provide you with some more information—some guideposts to the set of measurements that your score is a part of—you actually know very little about how you did on the test.

What else do you need to know? It would be nice to know what the mean is so that you can at least tell if you were above or below average. It would also be helpful to know how spread out all of the scores on the test were. For example, how well did you do on the test if the mean score for everyone who took the test was 50 and the standard deviation was 10? Your score is certainly above average, but if the typical person taking the test scored 50 ± 10 points, then you're really not that far above average. Now what if the mean of the distribution was 50 and the standard deviation was 1? The typical person taking the test scored 50 ± 1, and you scored a 60. Here, you score is, relatively speaking, very unusual in this narrow, tight distribution, and compared to everyone else who took the test, you did very well indeed.

THE *z*-SCORE

The first standard score we will discuss describes *the location of an individual score relative to the mean, in standard deviation units*. This particular standard score doesn't usually show up outside of statistics class (other standard scores are much more commonly used), but it is the original, definitional standard score. It's called a *z*-score, and it is calculated as shown in Box 6.2.

standard score A score expressed relative to a known mean and standard deviation.

z-score A score expressed relative to a mean of zero and a standard deviation of one; it describes the location of an observation in a distribution relative to the mean in standard deviation units.

BOX 6.2   **Calculating the *z*-score**

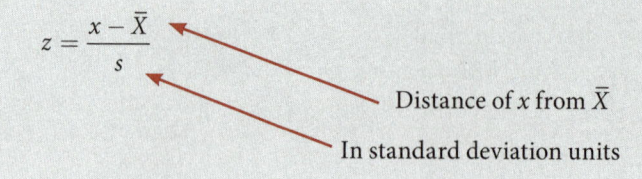

$$z = \frac{x - \bar{X}}{s}$$

Distance of *x* from $\bar{X}$

In standard deviation units

STATISTICS IN CONTEXT

Let's try calculating a few standard scores to see what they can tell us. We've got a new intelligence test that apparently has a mean of 50 and a standard deviation of 10. You scored a 60 on this test. Convert your raw score (x) into a z-score. The z you calculate will tell you the number of standard deviations away from the mean your particular test score happens to be.

BOX 6.3   **Calculating the *z*-score: Example**

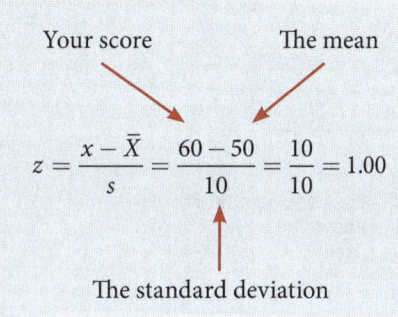

Your score →   ← The mean

$$z = \frac{x - \bar{X}}{s} = \frac{60 - 50}{10} = \frac{10}{10} = 1.00$$

↑ The standard deviation

The resulting *z*-score tells us that your score on the test is exactly one standard deviation away from (in fact, above) the mean. Now take a closer look at the calculations we've just done. The top of the formula for z ($x - \bar{X}$) should look familiar. It is the *deviation score* we used in the calculation of standard deviation in Chapter 5. The deviation score is a measure of distance from the mean. Your score of 60 is 10 points away from the mean of 50.

Think of the standard deviation in a set of observations as a ruler—in this case, a ruler that is 10 "points" long. So, 60 is 10 points away from 50, and that is the same as one ruler (or one standard deviation). The *z*-score tells us that your raw score is 1.00 standard deviation away from the mean. If I change the length of the ruler (by changing the standard deviation), then the location of a given score relative to the mean (the *z*-score) will also change (see Box 6.4). Suppose the mean of the distribution of measurements was again 50 but the standard deviation was 1. Now where is your score?

BOX 6.4   **Calculating the *z*-score: Example using a different *SD***

$$z = \frac{x - \bar{X}}{s} = \frac{60 - 50}{1} = \frac{10}{1} = 10.0$$

The same score of 60, in this now very narrow distribution, is in a different place relative to the mean of the distribution. A score of 60 in this distribution is now 10 standard deviations above average. These two distributions are shown in Figure 6.1.

Note that in the wide distribution at the top of that figure (SD = 10), 60 isn't all that far away from the mean. In the narrow distribution at the bottom (SD = 1), the same score is out in the tails of the distribution.

We can also say something about the "typicality" of the raw score of 60 in each distribution. Scores of 60 in the distribution at the top of Figure 6.1 are fairly common. They have a high frequency (note the red arrows), meaning they happen fairly often in this set of measurements, while scores of 60 in the narrow distribution at the bottom are pretty unusual (they don't happen very often in this set of measurements).

So, z-scores tell us about the relative position of an individual score in a distribution of scores, and in doing so, z-scores also describe how typical or atypical that individual score happens to be. Scores that are relatively close to the mean of a distribution are also fairly typical (meaning they occur fairly frequently within the distribution). Scores that are far away from the mean, out in the tails of the distribution, are atypical (meaning they don't happen very often).

What about someone who scored a 51 on this test? Where is that person relative to the mean in this distribution? How about someone who got a 40? What about someone whose score was the same as the mean (50)? Box 6.5 shows how these scores are converted to z-scores in this distribution.

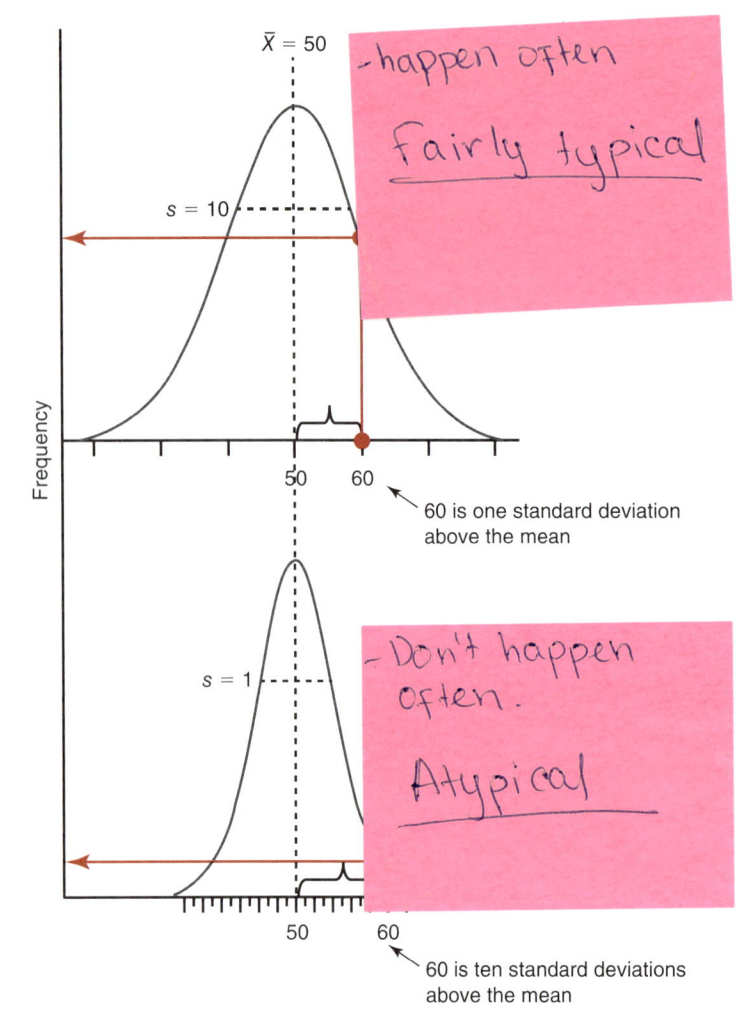

FIGURE 6.1
Position of an observation within a normal distribution.

Handwritten annotations on figure:
- happen often
- fairly typical
- 60 is one standard deviation above the mean
- Don't happen often.
- Atypical
- 60 is ten standard deviations above the mean

BOX 6.5 **Calculating the z-scores of three different results**

A	$z = \dfrac{x - \bar{X}}{s} = \dfrac{51 - 50}{10} = \dfrac{1}{10} = 0.10$	
B	$z = \dfrac{x - \bar{X}}{s} = \dfrac{40 - 50}{10} = \dfrac{-10}{10} = -1.00$	
C	$z = \dfrac{x - \bar{X}}{s} = \dfrac{50 - 50}{10} = \dfrac{0}{10} = 0$	

There are several things to notice in Box 6.5. First, when the raw score (*x*) is very close in value to the mean, the resulting *z*-value is very small (see example A). Here, a raw score of 51 is very close to the value of the mean (50 points), and that one point of difference is exactly 1/10 of a standard deviation unit (10 points = 1 standard deviation unit, so 1 point is 1/10 of a unit).

Second, when the raw score is less than the mean, the resulting *z*-score is negative (see example B). So, negative *z*-scores (just like negative deviation scores) indicate that the original raw score value was less than the value of the mean. In this case, our raw score (40) is exactly one standard deviation unit below the mean, so that raw score translates into a *z*-score of −1.00.

Lastly, when the raw score has the same value as the mean, the resulting *z*-score is zero (see example C). This should make some intuitive sense when you remember that *z*-scores tell us how far away from the mean a particular raw score is. A raw score of 50 is no different at all from a mean of 50, so it is zero standard deviations away.

So, *z*-scores tell us about the position of a score in a distribution, relative to the mean. Because position in the distribution is related to frequency, *z*-scores also tell us about the probability of a particular score within a given distribution. The farther away from the mean an individual score is (either above it or below it), the less frequently it appears in the distribution, and the rarer it is in the distribution. The reverse is also true: The closer a score is to the mean, the more probable (frequent) it is in the distribution. This relationship between position in the distribution and probability holds, generally speaking, as an axiom for any distribution. Scores that are much larger than (or much smaller than) the mean tend to be rare.

CheckPoint

Calculate the *z*-score for the following values based on the information given. Round scores to two decimal places.

1. $x = 78, \bar{X} = 45, SD = 11$ _____
2. $x = 0.012, \bar{X} = 0.014, SD = 0.001$ _____
3. $x = 1, \bar{X} = 0, SD = 2$ _____
4. $x = 1{,}000, \bar{X} = 900, SD = 245$ _____
5. $x = 3.73, \bar{X} = 5.50, SD = 1.45$ _____

Answers to this exercise are found on page 260.

THE "3-SIGMA RULE"

If the distribution you happen to be working with is normal (or even just "reasonably normal") in shape, then the relationship between distance from the mean and probability becomes very useful. This is because of another set of characteristics of

> **CheckPoint**
> *Answers to questions on page 259*
>
> 1. z-score = 3.00 ☐
> 2. z-score = −2.00 ☐
> 3. z-score = 0.50 ☐
> 4. z-score = 0.41 ☐
> 5. z-score = −1.22 ☐
>
> **SCORE:** /5

3-sigma rule A rule that describes the distribution of observations in a normal distribution. It states that in any reasonably normal distribution, 34% of the observations will be between the mean and an observation one standard deviation above the mean, 14% will be between one and two standard deviations above the mean, and 2% will be two or more standard deviations above the mean.

normal distributions called the **3-sigma rule**. Statisticians have discovered that *in any normal distribution, the following are true:*

a) Thirty-four percent of the scores in the distribution will have a value between the mean and one standard deviation above the mean.

b) Fourteen percent of the measurements in the distribution will fall between the score that is one standard deviation above the mean and the score that is two standard deviations above the mean.

c) Two percent of the observations in the distribution will be two or more standard deviations above the mean.

Basically, the 3-sigma rule says that most observations in a normal distribution will be within three standard deviations ("sigmas," or σ's) of the mean. Figure 6.2 illustrates this general rule.

No matter what the mean and the standard deviation of a distribution might be, this general rule will hold as long as the observations are normally distributed. Notice, however, that we've dealt only with the upper half of the distribution in Figure 6.2. What would the lower half of the distribution look like? Keep in mind that the distribution we're talking about is normal in shape. That means that it is, by definition, symmetrical, which means that the lower half is a mirror image of the upper half. So, 34% of the scores in this distribution will be between the mean

FIGURE 6.2
The 3-sigma rule (above the mean).

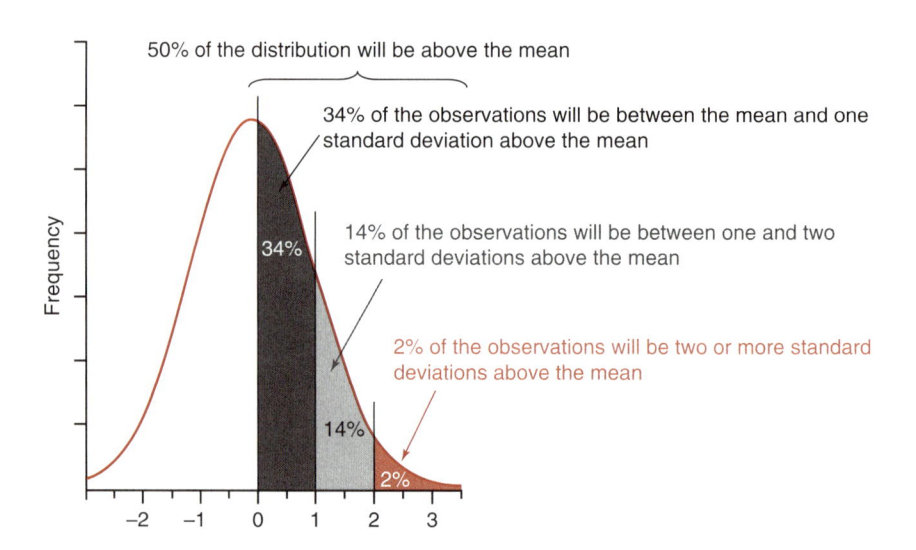

260   STATISTICS IN CONTEXT

and an observation that is one standard deviation *below* the mean, 14% will be between one and two standard deviations *below* the mean, and again, 2% will be two or more standard deviations below the mean (see Figure 6.3).

Observations that are more than two standard deviations from the mean are very rare: They make up only 4% of the observations in a normal distribution. This rule holds for *any* reasonably normal distribution. It does not matter what you are measuring, or how many things you measured. If the resulting distribution of measurements is normal (or even reasonably normal) in shape, this rule describes the frequency of observations in that set.

Given this relationship, we now not only can describe where an individual observation is in the set of observations but also can put a number to the probability of that particular score. We just convert a raw observation from the data into a *z*-score, and then use this relationship to talk about probability.

Think back to our brand-new intelligence test. If the scores on the test are normally distributed, and if the mean of the distribution of scores is 50 with a standard deviation of 10, what percentage of the scores will be above 60? Well, a score of 60 turns into a *z*-score of 1.00, and we know that in any normal distribution, 16% of the observations in that distribution will be above a *z* of 1.00. Figure 6.4 shows how I calculated the probability of a test score of 60 or better in this distribution.

What percentage of the scores on this test will be 30 or below? Converting 30 into a *z*-score gives us a *z* of −2.00. We know that 2% of the observations in a normal distribution will be two or more standard deviations either above or below the mean (because normal distributions are symmetrical), so the answer is that scores of 30 or under are pretty rare, occurring only 2% of the time in this distribution.

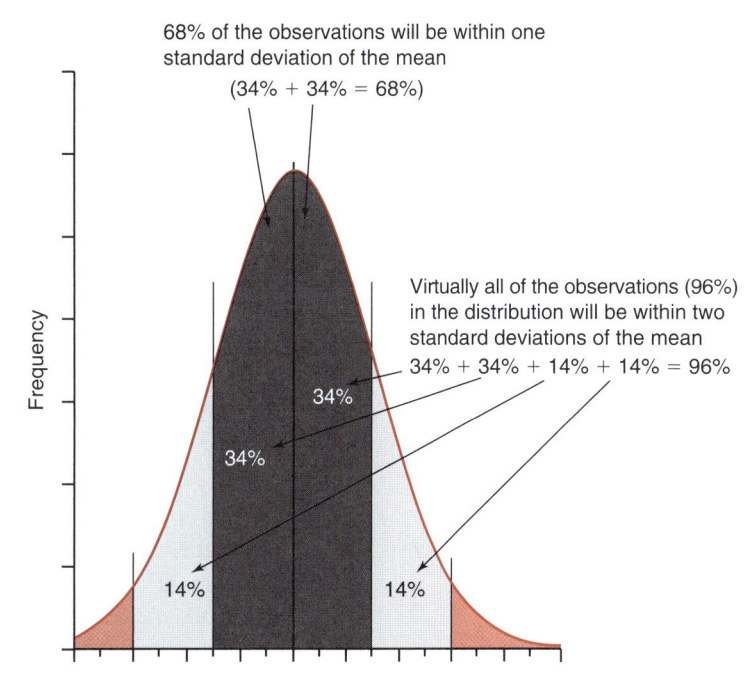

FIGURE 6.3
The 3-sigma rule (above and below the mean).

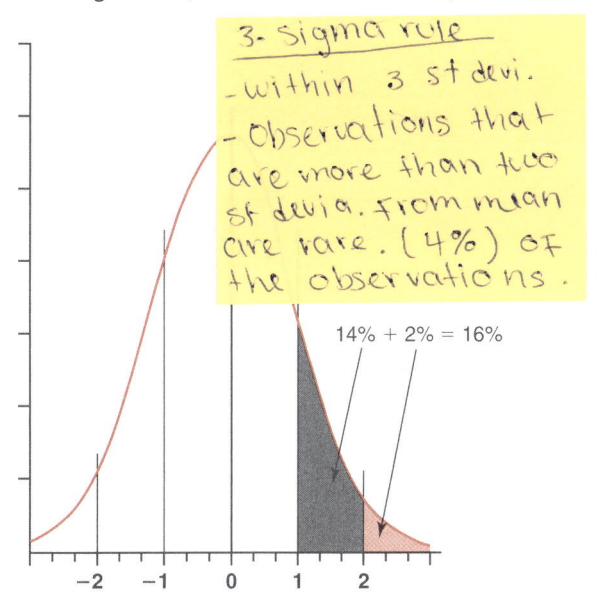

FIGURE 6.4
Calculation of proportion above a *z*-score.

CHAPTER 6 Where Am I? Normal Distributions and Standard Scores

What percentage of the scores on this test will be greater than 67? Well, a 67 turns into a *z*-score of 1.70, but how do we figure out the probability of getting a score of 67? The relationship we've talked about so far only gives us the proportions of observations at or between one, two, or three standard deviations from the mean. A *z*-score of 1.70 requires some more intricate math to figure out. Remember interpolation from your high-school math class? OK, get your calculator warmed up, roll up your sleeves, and put the coffee on—this might take a while . . . (Just kidding.)

PROPORTIONS IN THE STANDARD NORMAL CURVE

As it turns out, statisticians have already figured this out for us and helpfully provided the novice statistician with just about every possible way to slice and dice a normal distribution as well as the proportions for each slice. If you look at the table in Appendix A (beginning on page 697) you will see what is called a Table of Proportions of Area Under a Standard Normal Curve (or a *z*-table, for short).

The table is divided into three columns. In the first column (labeled "*z*"), we have *z*-scores from a *z* of 0.00 to a *z* of 4.00. In the second column, we have *the proportion* of scores between the mean and any given *z*-score. If you wanted to find the percentage of scores between the mean and a *z* of 1.70, you would look down the *z* column until you found 1.70, then read across to column B to find the *proportion* of scores in a normal distribution that would be between the mean (a *z* of 0.00) and 1.70. You should find that this value is .4554. This is a proportion, or a part out of a total of 1.00. If you want to turn it into a percentage (a part out of 100), simply multiply the proportion by 100 (or move the decimal point two places to the right) to get 45.54%. So, 45.54% of the scores in a normal distribution will be between the mean and a *z* of 1.70.

The third column of the table is also very useful. Suppose you wanted to find the percentage of scores in a normal distribution that were *at or above* a *z* of 1.70? You could recognize that 50% of the observations will be above the mean, so if 45.54% are between the mean and a *z* of 1.70, that leaves only 4.46% of the observations that can be farther out than 1.70 standard deviations.

Of course, if you were not inclined to do the math, you could just read down the *z* column until you came to 1.70 and then read across to column C, where you would find the proportion of scores that are 1.70 standard deviations or more above the mean. You should see .0446 (a proportion), which easily turns into 4.46%. So, in answer to our earlier question, the percentage of scores on this test that will be 67 or higher is 4.46%.

The table of proportions under a standard normal curve makes figuring out probabilities in normal distributions much easier, but like most things in life, practice makes perfect. So, let's practice. Keep in mind what you know about normal distributions:

- They are symmetrical.
- The mean = median = mode, so 50% of the scores will be above the mean and 50% will be below it.

STATISTICS IN CONTEXT

- Thirty-four percent of the scores will be between the mean and a score one standard deviation above the mean, 14% will be between one and two standard deviations above the mean, and 2% will be more than two standard deviations above the mean.

Suppose we wanted to know the percentage of scores in this distribution that are above a score of 80. Box 6.6 shows how we would use z-scores and the z-table to answer this question.

BOX 6.6 **Using the z-table: Example**

1) What percentage of scores will be above a score of 80?

$$z = \frac{x - \bar{X}}{x} = \frac{80 - 50}{10} = \frac{30}{10} = 3.00$$

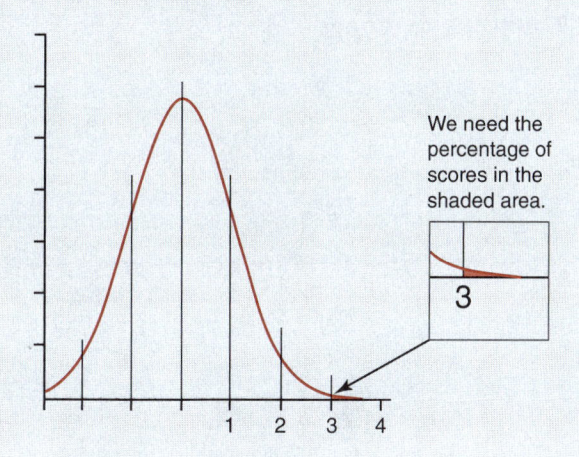

We need the percentage of scores in the shaded area.

We read down the z column until we come to a z of 3.00, then read across to column C (the proportion of scores at or above a given z value). There, we find that the proportion of scores above a z of 3.00 is .0013, which is 0.13%.

Let's return to our sleepy college students and play around with z-scores a bit more. If you remember, the mean of the distribution of sleep durations for our set of 30 college students was 408.67, and the standard deviation was 100.22 minutes. Answer the following questions about the sleep durations for this sample:

1. What percentage of the students slept at least 8 hours (480 minutes) or longer?
2. What percentage of students slept between 240 and 480 minutes?
3. What percentage of students slept 6 hours (360 minutes) or less?
4. What percentage of students slept between 2 and 3 hours?

Drawing a picture of these problems will help you visualize what is being asked. I recommend that you sketch out a normal distribution, indicate the mean on it, and then sketch in the section of that normal distribution each question is asking about.

My sketch of the first question is shown in Box 6.7. The shaded area on the picture of the normal distribution is the section of the distribution that represents sleep durations of 480 minutes or longer. We need to find the proportion of any normal distribution that would be found in this shaded area by first converting 480 minutes to a *z*-score in a distribution with a mean of 408.67 and a standard deviation of 100.22 minutes. Then, we'll use column C from the table in Appendix A to tell us how much (what proportion) of a normal distribution we would expect to see in that shaded area.

BOX 6.7 **Sketch of the normal curve showing students who slept 480 minutes or longer**

$$z = \frac{x - \bar{X}}{s} = \frac{480 - 408.67}{100.22} = \frac{71.33}{100.22} = 0.7133$$

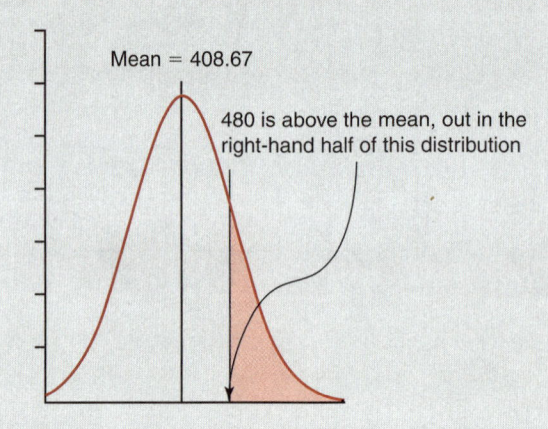

Mean = 408.67

480 is above the mean, out in the right-hand half of this distribution

---

Our duration of 480 minutes is just a bit less than one standard deviation above average in this distribution. This should make sense, given that one standard deviation in this distribution is 100.22 minutes, so a sleep duration that is exactly one standard deviation above average here would have to be 508.89 minutes—just a bit more than 480. Using column C from the *z*-table, which shows the proportion of any normal distribution at and beyond a particular *z*-score, we find that with a *z* of 0.71, we'd expect .2389, or 23.89%, of a normal distribution to be this far away from the mean or farther. The second question can be sketched as shown in Box 6.8.

BOX 6.8 **Sketch of the normal curve showing students who slept between 240 and 480 minutes**

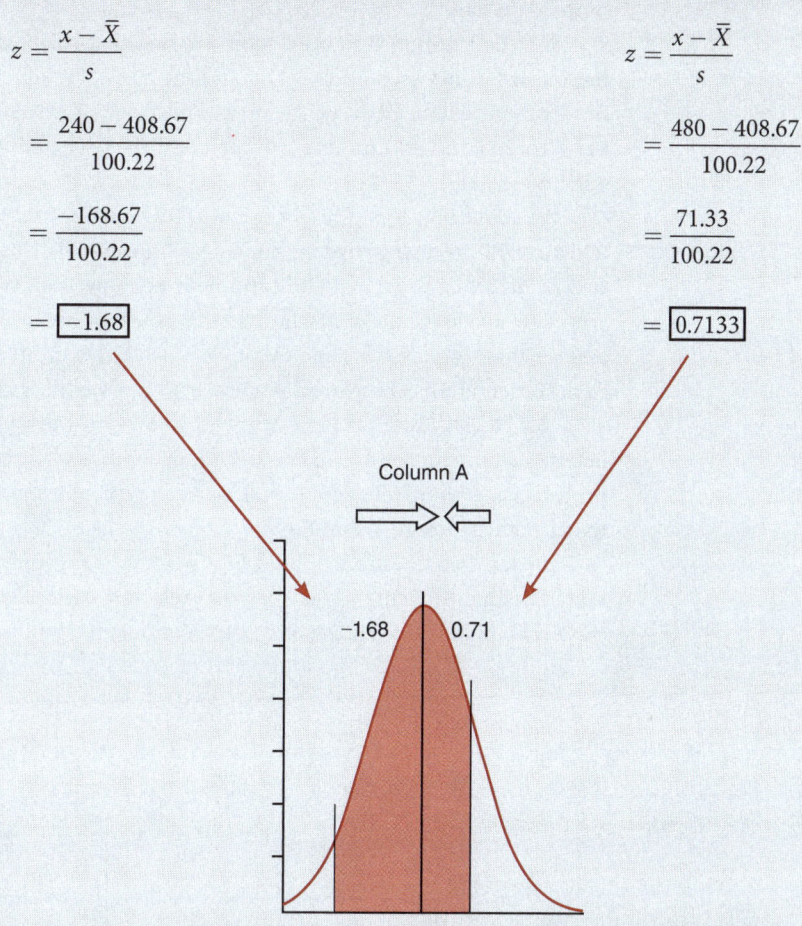

$$z = \frac{x - \bar{X}}{s}$$

$$= \frac{240 - 408.67}{100.22}$$

$$= \frac{-168.67}{100.22}$$

$$= \boxed{-1.68}$$

$$z = \frac{x - \bar{X}}{s}$$

$$= \frac{480 - 408.67}{100.22}$$

$$= \frac{71.33}{100.22}$$

$$= \boxed{0.7133}$$

HINT: Generally speaking, if the piece of the normal distribution that you're interested in crosses the mean (or includes the mean), you're going to add together the proportions you get from the table when you convert the raw scores to z-scores.

In this case, we want to find the percentage of this normal distribution between 4 hours (240 minutes) and 8 hours (480 minutes) of sleep. We need to use column B here because it shows us the proportion of the distribution between the mean and any z-score. In fact, we're going to use column B twice: once to find the proportion between the mean and whatever 240 minutes converts to as a z, and again to find the proportion between the mean and a z of 0.71 (we already know what 480 minutes converts to as a z-score from the last problem).

Notice in Box 6.8 that we have a negative z-score. This happens because the raw score of 240 is *less* than the mean. All z-scores on the left half of the distribution (below the mean) will be negative. It does not matter that the z is negative when we go to use the table. If you don't see why it doesn't matter, recall that normal distributions are symmetrical, so the left half of the distribution is a mirror image of the right half.

So, from column B in the z-table, we find that .4535, or 45.35%, of the distribution will be between a z of −1.68 and the mean. We also find that 26.11% of the observations will be between the mean and a z of 0.71. To find the total percentage between 240 and 480 minutes, we just add these percentages together, like this: 45.35% + 26.11% = 71.46%

We could also have found the proportion of observations between the mean and a z of 0.71 by using what we know about normal distributions and the information we obtained about a z of 0.71 from our first question. We know that 50% of the scores are above the mean in a normal distribution (because the mean, the median, and the mode are all the same value). We also know, from question 1, that 23.89% of the observations are above a z-score of 0.71. If we subtract 23.89% from 50.00%, we'll get 26.11% (go ahead and try it—I know you want to). I have found it very helpful to think of the normal distributions we'll be playing with in these kinds of problems as if they were jigsaw puzzles: Each piece fits together with another piece to create the whole distribution.

The last two questions should be a snap now. Just remember to make a quick sketch of the problem, convert to z-scores, and use the z-table in Appendix A. Box 6.9 shows the percentage of students who slept 6 hours or less.

BOX 6.9

Sketch of the normal curve showing students who slept 360 minutes or less

$$z = \frac{x - \bar{X}}{x} = \frac{360 - 408.67}{100.22} = \frac{-48.67}{100.22} = -0.4856 \ (\text{or} -0.49)$$

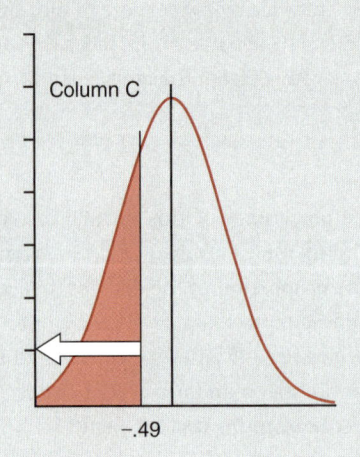

Column C

−.49

266 STATISTICS IN CONTEXT

Using column C of the z-table, we read down to a z of 0.49 and then across to find the proportion of observations 0.49 standard deviations below the mean or farther out (notice the arrow in the picture I've sketched). Column C tells us that .3121, or 31.21%, of the observations will be 360 minutes or less in this distribution. So, about 30% of the students in this distribution got 6 hours of sleep or less.

Finally, we wanted to know what percentage of students slept between 2 and 3 hours. Box 6.10 shows my sketch. Notice that 2 and 3 hours of sleep are both fairly uncommon—they are way down in the left-hand tail of this distribution. We need to find the proportion of observations in the shaded area.

BOX 6.10 **Sketch of the normal curve showing students who slept between 120 and 180 minutes**

We first need to convert 120 and 180 minutes to z-scores.

If $x = 120$, then

$$z = \frac{x - \overline{X}}{x} = \frac{120 - 408.67}{100.22} = \frac{-288.67}{100.22} = \boxed{-2.88}$$

If $x = 180$, then

$$z = \frac{x - \overline{X}}{x} = \frac{180 - 408.67}{100.22} = \frac{-228.67}{100.22} = \boxed{-2.28}$$

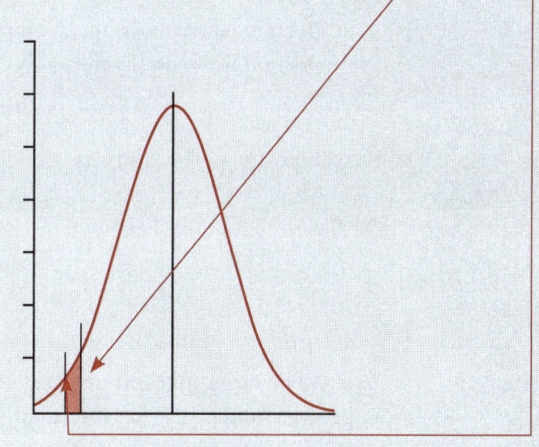

There are two ways we can go about solving for the percentage of scores in the thin section of the distribution we're interested in. First, we can use column B of the z-table, which shows us the proportion of observations between the mean and any z-score. We'll find the proportion of observations between the mean and a z of -2.88, and then we'll subtract from that the proportion of observations between the mean and a z of -2.28. This will leave us with just the shaded, gray segment we want. The result is shown in Box 6.11. (Notice that I'm showing only the left half of the distribution in that figure: We don't really need the other half here.)

BOX 6.11 **Calculating the percentage of students who slept between 120 and 180 minutes: First method**

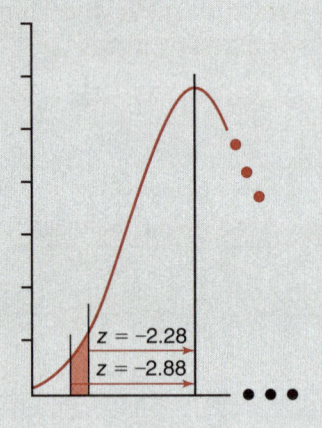

Proportion between the mean and a z of −2.88 = .4980
Proportion between the mean and a z of −2.28 = .4887
Subtracting, we get .0093

So, 0.93% (just about 1%) of the students got between 2 and 3 hours of sleep.

The other method involves column C in the z-table. Obviously, you should get just under 1% (0.93%) using column C, just as we did using column B. (If you get something different, then you know there's an error somewhere.) Column C tells us the proportion of scores that would be z standard deviations or more from the mean. We've already converted the raw scores to z-scores. Now we need to figure out what to do with that information. Take a look at Box 6.12.

BOX 6.12 Calculating the percentage of students who slept between 120 and 180 minutes: Second method

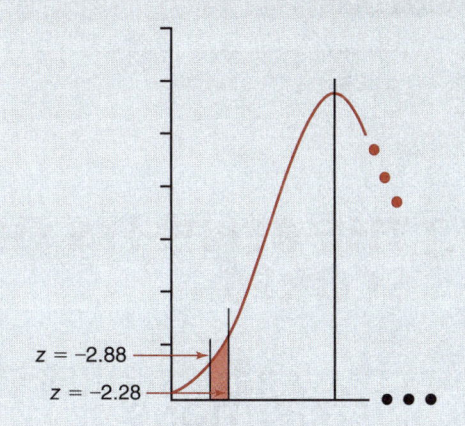

Proportion at a z of −2.28 or more =   .0113
Proportion at a z of −2.88 or more =   .0020
                                       _____
             Subtracting, we get   .0093

So, 0.93% (just about 1%) of the students got between 2 and 3 hours of sleep.

HINT: Generally speaking, if the piece of the normal distribution that you are interested in is only on one side of the mean, you will be subtracting the proportions you find in the table.

CheckPoint

Find the proportion of area under the curve to the left of the following z-scores:

1. 1.5 _____
2. −0.4 _____
3. 3.4 _____
4. 0 _____
5. −2.8 _____

Answers to this exercise are found on page 270.

> **CheckPoint**
> *Answers to questions on page 269*
>
> 1. 93.32% ☐   3. 99.97% ☐   5. 0.26% ☐
> 2. 34.46% ☐   4. 50.00% ☐
>
> **SCORE:** /5

THE BENEFITS OF STANDARD SCORES

COMPARING SCORES FROM DIFFERENT DISTRIBUTIONS

When we convert raw scores to z-scores, we are placing that score in a new distribution: the *standard normal distribution*. The z-score tells us about the position of a score in a very specific, perfectly normal, *theoretical* distribution *with a mean of zero and a standard deviation of one*. A score that in a real, empirical distribution was exactly one standard deviation away from the mean will always convert to a z-score of 1.00. A score in that real, empirical distribution that has the same value as the mean will always convert to a z-score of 0.00.

If you examine the z-table in Appendix A, you can also see that the standard normal distribution is perfectly normal in shape. Remember, we said that in a normal distribution, 50% of the scores would be greater than the mean. Column C on the z-table tells us about the proportion of scores at or above a particular z-score, and we know that a score equal to the mean will convert to a z-score of 0.00. So, the value in column C for $z = 0$ should be exactly equal to 50% (.5000)—and you'll notice that it is.

There's an additional benefit to converting scores to standard scores (e.g., z-scores), which is that this conversion will let you compare scores that come from very different distributions. Suppose we wanted to compare two different intelligence tests: the Stanford–Binet test, which has a mean of 100 and a standard deviation of 16, and our own intelligence test, which has a mean of 50 and a standard deviation of 10. Student A gets a score of 120 on the Stanford–Binet test. Student B gets a score of 67 on our intelligence test. How can we compare them? Obviously, we can't compare the raw scores because the two distributions are very different. Average performance on our intelligence test would yield a score of 50, but a score of 50 on the Stanford–Binet test would be well below average.

Look at Box 6.13, and see what happens when we convert both scores into z-scores.

STATISTICS IN CONTEXT

BOX 6.13 Using z-scores to compare results from different distributions

Student	Score on intelligence test	z-score equivalent
A	120	$z = \dfrac{x - \bar{X}}{s} = \dfrac{120 - 100}{16} = \dfrac{20}{16} = 1.25$
B	67	$z = \dfrac{x - \bar{X}}{s} = \dfrac{67 - 50}{10} = \dfrac{17}{10} = 1.70$

When we convert a score to a *z*-score, we put it on a distribution with a mean of zero and a standard deviation of one. If we convert both scores to their *z*-equivalents, then we can put them on the same distribution, which makes comparing them much easier. You can see that, relatively speaking, Student B scored higher than Student A did, even though the raw scores make it look the other way around: Student A had a score that was 1.25 standard deviations above the mean, while Student B had a score that was 1.70 standard deviations above the mean.

The defining characteristic of a standard score is that it converts a raw score from a messy, real-life empirical distribution into a score in a normal distribution with a known mean and a known standard deviation. The *z*-score distribution has a mean of zero and a standard deviation of one. Another standard score, the *T*-score (which stands for a "transformed score"), tells you the position of a score in a distribution that has a mean of 50 and a standard deviation of 10. A more familiar standardized score would be the Scholastic Aptitude Test (SAT) score, or your score on the Stanford–Binet intelligence test (your intelligence quotient, or IQ score). In fact, the world is full of standardized scores. Figure 6.5 shows you just a few.

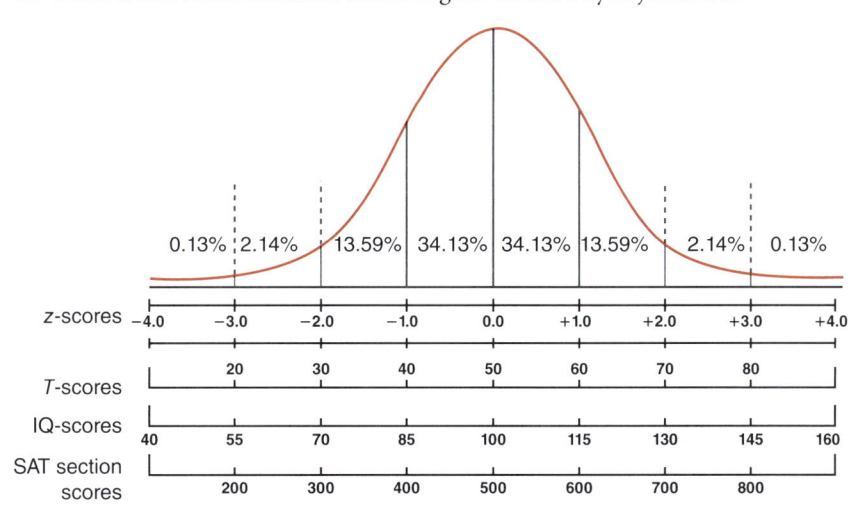

FIGURE 6.5
Types of standard scores.

	0.13%	2.14%	13.59%	34.13%	34.13%	13.59%	2.14%	0.13%
z-scores	−4.0	−3.0	−2.0	−1.0	0.0	+1.0	+2.0	+3.0	+4.0
T-scores		20	30	40	50	60	70	80	
IQ-scores	40	55	70	85	100	115	130	145	160
SAT section scores		200	300	400	500	600	700	800	

CONVERTING A z-SCORE INTO A RAW SCORE

Let's go back to our home-grown version on an intelligence test that we used as an example at the beginning of this chapter. Your score of 60 on my test turned into a z-score of 1.00 if the mean score on the test was 50 and the standard deviation was 10. If my test actually does measure intelligence, your score on another intelligence test should be in roughly the same position (your intelligence presumably hasn't changed, just the scale we're using to measure it with). So, what *should* your score be on the Stanford–Binet IQ test that has a mean of 100 and a standard deviation of 16 if, in fact, these two tests are both measuring intelligence?

There's a common-sense solution to this problem, and a more general mathematical one that will translate well to other situations. Let's go with common sense first: If your score was one standard deviation (10 points) above the mean on my test, it should also be one standard deviation (16 points) above the mean (100) on the Stanford–Binet test. So, your score should be 100 + 16 = 116 on the Stanford–Binet.

Mathematically, we can use the z-score formula backward to solve for *x* and get the same answer. We know z (1.00), and we know the mean (100) and the standard deviation (16) of the distribution of test scores we want to put our score on. All we have to do, then, is solve for *x* (the raw score on the Stanford–Binet test). Box 6.14 shows you how to do this (if you don't remember solving for *x*).

BOX 6.14

Solving for *x*

Solve for *x* in the following equation:

$$z = \frac{x - \bar{X}}{s}$$

First, multiply both sides of the equation by *s* to get rid of the denominator:

$$(s)z = \frac{x - \bar{X}}{s}(s)$$

Next, add the mean to both sides to get rid of the mean on the right side of the equation:

$$(s)z + \bar{X} = x - \bar{X} + \bar{X}$$

This leaves us with:

$$(s)z + \bar{X} = x$$

So, if z = 1.00, what would *x* be?

$$x = 16[1.00] + 100$$

$$x = 16 + 100 = 116$$

So, by reversing the formula and solving for *x*, we can convert any *z*-score to a raw score in any distribution. The raw-score equivalent of any *z* is *z* times the standard deviation plus the mean. We're going from a position in the standard normal curve back to a raw score, and we can use this solve-for-*x*-formula for any situation where we want to go from one standard score or a raw score to another scale.

CheckPoint

Determine the raw scores from the following *z* scores and distribution parameters.

1. $z = -2.34, \bar{X} = 45, SD = 7$ _____
2. $z = 0.8, \bar{X} = 10, SD = 5$ _____
3. $z = 0.1, \bar{X} = 475, SD = 35$ _____
4. $z = 0, \bar{X} = 99, SD = 18$ _____
5. $z = -1.27, \bar{X} = 0, SD = 1$ _____

Answers to this exercise are found on page 274.

CONVERTING A PERCENTILE RANK INTO A RAW SCORE

There's one more thing we can do with *z*-scores and the *z*-table. Suppose you got your grade on a 100-point statistics exam back as a percentile rank (the percentage of scores at or above your score). Let's suppose that the mean on the exam for the whole class was 85 points, the standard deviation was 8 points, and the grades were reasonably normally distributed. You can use the *z*-table and the formula for converting a *z* into an *x* to see what your grade on the exam actually was. I've put the data we need into Table 6.1.

Let's start with Student 1, whose score on the first exam was at the 10th percentile. This means that 10% of the students who took the test got a score lower than Student 1 and 90% of the class got scores higher than Student 1. Let's sketch the position of this score in a standard normal distribution. We know that it's below the mean and that only 10% of the scores were lower than this one, so it's fairly far out in the left-hand tail of the distribution. My sketch is shown in Box 6.15.

TABLE 6.1 Percentile rank

Student	Percentile Rank
1	10th
2	50th
3	60th
4	33rd
5	90th

> **CheckPoint**
> *Answers to questions on page 273*
>
> 1. 28.62 ☐
> 2. 14.00 ☐
> 3. 478.50 ☐
> 4. 99.00 ☐
> 5. −1.27 ☐
>
> **SCORE:** /5

BOX 6.15 **Sketch of the normal curve showing a score in the 10th percentile**

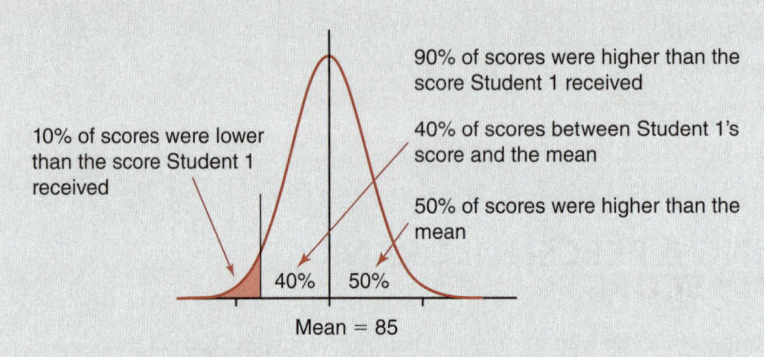

- 10% of scores were lower than the score Student 1 received
- 90% of scores were higher than the score Student 1 received
- 40% of scores between Student 1's score and the mean
- 50% of scores were higher than the mean

40% | 50%

Mean = 85

So, what score did Student 1 actually get? Remember what information the z-table in Appendix A provides. Column C will tell us the proportion of scores from any z and beyond, but we already know the proportion we need, which is 10%. We can find the z-score that cuts off 10% of the distribution by looking down column C until we come as close to 10% (.1000) as we can get, and then reading over to the z column to find the equivalent z-score that cuts off 10% of the distribution. We're using the table backward here. And when I did this, I found that a z-score of 1.28 cuts off .1003, or 10.03%, of the standard normal distribution—pretty close to 10%.

Notice that this z cuts off the top *or* the bottom 10%, and we want the bottom. So, our z-score should be negative (this will make a difference in our calculations, as you'll see in Box 6.16). Now all we need to do is convert that z of −1.28 into a raw score in a distribution with a mean of 85 and a standard deviation of 8. The score on this exam that is at the 10th percentile (90% of the scores are above it, and only 10% of the scores are below it) is 74.76 (or rounding up, 75, since most tests are not graded to the nearest hundredth of a point).

STATISTICS IN CONTEXT

BOX 6.16 Calculating a percentile rank into a raw score: Example

$$x = (s)z + \bar{X}$$
$$= 8(-1.28) + 85$$
$$= -10.24 + 85$$
$$= 74.76$$

Student 5 scored at the 90th percentile—the opposite of Student 1. Only 10% of the scores will be higher than this student's score, and 90% of them will be below it. So, Student 5 has a score that sits in the same relative position as Student 1's score, only this time, the score is above the mean. Student 1's score was 1.28 standard deviations below the mean, meaning that Student 5's score should be 1.28 standard deviations above the mean. All we have to do in order to convert Student 5's score to a raw score is multiply the mean by a positive z of 1.28. When we do that, we get a raw score of 95.24 (or rounding up, 95).

Student 2 scored at the 50th percentile, meaning that 50% of the scores on the test were higher than the one that Student 2 received and 50% were lower. No calculations are necessary at all here if you remember that the median is the score that splits the distribution in half, and in a normal distribution, the mean, median, and mode are all the same value: Student 2 got an 85 on the exam. You can use the formula we've been discussing to arrive at the same answer (see Table 6.2 below), but I hope you realize that it isn't really necessary. (Speaking of Table 6.2, I've done the rest of the calculations for you here, but do try them yourself. Remember that practice makes perfect.)

TABLE 6.2 Converting from percentile rank to raw score

Student	Percentile Rank	z-score Equivalent	Calculations $(s)z + \bar{X} = x$	Raw Score (x)
1	10th	−1.28	(8)(−1.28) + 85 = −10.24 + 85	75
2	50th	0	(8)(0) + 85 = 0 + 85	85
3	60th	+0.25	(8)(0.25) +85 = 2 + 85	87
4	33rd	−0.44	(8)(−0.44) + 85 = −3.52 + 85	81
5	90th	+1.28	(8)(1.28) + 85 = 10.24 + 85	95

Think About It...

THE RANGE RULE REVISITED

In the last chapter, we encountered the "range rule," which says that you can get a rough estimate of the standard deviation of a set of measurements by dividing the range (maximum value minus the minimum value) by four. Let's take a closer look at this rule and think about why it works. I'll give you a hint: Remember the 3-sigma rule.

The 3-sigma rule says that in a normal, bell-shaped distribution, the following will be true.

- Approximately 68% of the data will be within one standard deviation (higher or lower) of the mean.
- Approximately 95% of the data will be within two standard deviations (higher or lower) of the mean.
- Approximately 99% will be within three standard deviations (higher or lower) of the mean.

This rule is telling us that almost all of the values in our normal distribution will be between two standard deviations below the mean and two standard deviations above the mean. In other words, if we used our yardstick that was exactly one standard deviation long and laid it out along the x-axis of our distribution, we would find that roughly all of the observations in our set were within a segment of the x-axis that was four yardsticks, or four standard deviations, long.

Let's work an example of the range rule the other way around. Suppose we have a normal distribution (shown below) with a mean of 75 and a standard deviation of about 10:

$$96, 84, 84, 78, 75, 75, 66, 65, 65, 62$$

1. What score is two standard deviations above the mean? _____
2. What score is two standard deviations below the mean? _____
3. The 3-sigma rule says that 95% of the distribution should be between _____ and _____.
4. Are 95% of the scores between these two points? (Yes/No) _____
5. What is the range of the scores that are at least two standard deviations away from the mean? _____
6. We used the range rule in the last chapter to get a rough estimate of standard deviation. Some estimates were fairly close, and some were less so. Why were some of the estimates less accurate than others? (Hint: Think about the shape of the distribution.)

STATISTICS IN CONTEXT

STANDARD SCORES IN CONTEXT

I mentioned earlier that z-scores were just one kind of standard score. There are others, and you have probably come across them more often than you have z-scores. For example, you may well have taken the SAT on your way to college and this class. The score you receive from the College Board (the organization that administers and scores the SAT) is a standard score: It tells you what your score on the test was, relative to the average score on the test, and in standard deviation units.

The average SAT score on either of the two sections (math, which is worth 800 points, and critical reading and writing, which is also worth 800 points) is 500, with a standard deviation of 100 (minimum = 200 and maximum = 800 on any one section). So, perfection on the SAT would be a score of 800 on each of the sections, or 1,600 scaled (standardized) points. I'm sure that you remember the basic format of the test, and that there are, in fact, not 1,600 questions on the exam. For example, the 2016 version of the SAT had 52 reading and language questions, 44 writing questions, and 58 math questions (and one essay), for a total of 154 questions. Also, as of March 2016, there is no longer any penalty for skipping or for incorrect answers. The raw score on a given section is simply the number of questions you have answered correctly. This is then converted to a standard score on the 200–800 scales using a conversion chart. The conversion chart varies according to the date of the test. That raw score is then converted to the SAT score we're all familiar with, which tells you the position of your score in the distribution of the scores with a mean of 500 and a standard deviation of 100. You probably also got back your percentile rank (and you know what that is by now), and if you ask for it, you can also get your raw score on each section.

So, why are scores reported like this? What's the advantage of standardizing SAT scores? First, when SAT scores are standardized, we know the mean and the standard deviation, as well as the shape of the distribution that the standard scores will be a part of. In this way, comparing performance on the test across everyone who takes it can more easily be accomplished. Everyone's test result is on the same scale. Second, not only are all the results "standardized" but, in theory at least, the way the test is written, administered, and scored is also standardized, so that everyone takes the test in the same manner. Imagine trying to compare performance on a test if you knew that each test had been administered in a different way, with different questions, and with scorers who used different scoring methods.

SUMMARY

Standard scores are descriptive statistics that tell us the location of an observation in a distribution, relative to the mean and in standard deviation units. To calculate a z-score in a distribution, you subtract the value of the observation from the mean and then divide the result by the standard deviation. In normal distributions, or even reasonably normal ones, z-scores also tell us about the probability of the observation we're interested in. We can use the 3-sigma rule and the z-table to describe the relative probability of that score.

TERMS YOU SHOULD KNOW

standard score, p. 256
3-sigma rule, p. 260
z-score, p. 256

GLOSSARY OF EQUATIONS

Formula	Name	Symbol
$z = \dfrac{x - \bar{X}}{x}$	The z-score	z
$(s)z + \bar{X} = x$	Converting from z to x	

WRITING ASSIGNMENT

Research has shown that college students are generally sleep-deprived. For example, Hicks, Hernandez, and Pellegrini (2001) examined 1,585 college-age adults and found that their average sleep time (in 2001) was 6.85 hours ($SD = 1.04$ hr). Assuming that sleep durations are normally distributed in college students, compare the students in our sample of 30 college students with those in the study by Hicks et al. Feel free to use any of the descriptive statistics we've discussed so far, as well as standard scores, to help make your point.

Table 6.3 shows our growing list of the symbols used in APA Style. Notice that I've added the APA Style abbreviation for a z-score, percentage, and percentile rank to this list.

Table 6.3 APA Style symbols for descriptive statistics

Descriptive Statistics	APA Style Symbols	APA Abbreviations for Time	
Mean (sample)	M	hour	hr
Mean (population)	μ	minute	min
Number of cases (sample)	n	second	s
Total number of cases (or population)	N	millisecond	ms
Variance (sample)	s^2	nanosecond	ns
Variance (population)	σ^2		
Standard deviation (sample)	SD		
Standard deviation (population)	Σ		
z-score	z		
Percentage	% (if followed by a number) "...15% of the scores"		
Percentile rank	"percentile" "She scored in the 68th percentile"		

STATISTICS IN CONTEXT

PRACTICE PROBLEMS

1. Consider the distribution shown in Table 6.4.

    Table 6.4 Data for question 1

x	f
9.00	1
10.00	2
11.00	2
12.00	1
13.00	3
14.00	3
15.00	4
16.00	2
17.00	1
18.00	1

    a. Calculate the mean, median, mode, and standard deviation.
    b. A score of 20 has a deviation score of _____.
    c. A score of 9 has a deviation score of _____.

2. Use the data in Table 6.4 to answer the following questions:
    a. A score of 15 converts to a z-score of _____.
    b. A score of 11 converts to a z-score of _____.
    c. A score that is exactly 2 standard deviations above the mean would be _____.
    d. A score that is 1.5 standard deviations below the mean would be _____.

3. Are the data in Table 6.4 normally distributed? How can you tell?

4. For the data in Table 6.4, what percentage of the data should be at least 1.96 standard deviations above the mean?

5. Louis Terman, a professor of psychology at Stanford University in California, brought the Binet–Simon test to America, renamed it the Stanford–Binet test, and "standardized" it. The average score on the Stanford–Binet test is 100, and the distribution of IQ scores has a standard deviation of 16. In addition to standardizing Binet's test, Terman also developed a classification scheme for the scores on the Stanford–Binet test that is generally still in use today. This scheme is shown below.

    140+   ..... Genius
    120–140 ..... Very superior intelligence

110–120	.....	Superior intelligence
90–110	.....	Normal or average intelligence
80–90	.....	Dull
70–80	.....	Borderline deficiency
Below 70	.....	Feeble-minded

Convert these marker scores to *z*-scores on a distribution with a mean of 100 and a standard deviation of 16. What are the *z*-score equivalents of these marker scores? If you've taken the Stanford–Binet test and you know your own score, convert your score to a *z*-score and see where you sit on the scale of intelligence proposed by Louis Terman.

6. The performance of 20 students on their final exam on statistics in 1992 and for a second class of 20 students in 2017 is shown in Table 6.5.

Table 6.5 Scores on statistics final exam, 1992 and 2017

Class of 1992	Class of 2017
80	100
91	99
91	94
80	94
74	94
74	88
73	88
75	81
76	88
73	89
73	80
69	88
67	76
68	75
68	63
68	61
68	53
57	55
58	56
59	82

STATISTICS IN CONTEXT

a. Find the mean and standard deviation for the final exam scores in the two years, 1992 and 2017.
   b. Suppose you scored an 85 on this exam. In which distribution is your test score the farthest above the mean? How far above the mean is your score in both of these distributions?
   c. Admission to a (paid) summer internship program requires that students earn a C or better (70% or higher) on their statistics final exam. If we assume that scores (in both years) on this test are reasonably normally distributed, what percentage of students in 1992 and 2017 would qualify for this internship?

7. Graduate schools often use tests like the Graduate Record Exam (GRE) and Medical College Admissions Test (MCAT) to determine admission to a graduate program. Thousands of students take these kinds of standardized tests each year hoping for admission to graduate programs. Grades on the GRE are typically normally distributed, with a mean of 500 and a standard deviation of 100 on a given section (verbal or quantitative). Assume that the scores in your graduating class had the typical normal distribution.
   a. What percentage of your graduating class of 210 students will make it in to a graduate program that requires at least a GRE verbal score of 650?
   b. Another graduate program requires a GRE quantitative score of 766 for entry. How many of the students in your graduating class will likely have a quantitative GRE score high enough to gain entry into this program?
   c. Yet another graduate school says that students must have placed in the 75th percentile on the verbal section of the GRE to gain admission to their program. What GRE score does a student have to make on the verbal section in order to get into this program? The same school also requires a score in the 80th percentile on the quantitative section of the test. What score does a student have to earn on the quantitative section of the GRE in order to gain admission to this program?
   d. Suppose that a local manufacturer of hot dogs has a standardized test that is used to assess your hot-dog-making ability. The "How to Make a Hot Dog" test has a mean of 75 and a standard deviation of 25. Psychologists have found that the quantitative section of the GRE and the hot dog test are actually related to one another. If you score a 400 on the quantitative section of the GRE, what score will you likely make on the hot dog test?

8. Table 6.6 shows the weights of 20 domestic cats (it's the same data set that we used in Chapter 5). You may remember "Himmie" and "Mr. Peebles," the largest and the smallest cat on record, respectively. Himmie weighed in at 46 pounds, 15.2 ounces, and Mr. Peebles weighed only 3 pounds (clearly, both cats are out in the tails of the distribution).
   a. Convert Himmie's weight to a z-score in this distribution. What percentage of cats should weigh what Himmie weighs or more?
   b. Convert Mr. Peebles' weight to a z-score in this same distribution. What percentage of cats should weigh 3 pounds or less?
   c. What percentage of cats should weigh between 10 and 15 pounds?
   d. What percentage of cats should weigh 7.5 pounds or less?

e. Suppose the vet told you that your cat's weight converted to a z-score of -0.57.
   i) What does your cat weigh (in pounds)?
   ii) Does your cat have to gain weight or lose weight in order to be at the average for domestic cats?
   iii) How much weight does your cat have to gain or lose in order to be of average weight?

Table 6.6 Weights of 20 domestic cats (in pounds)

ID	Weight	ID	Weight
1	9	11	11
2	7	12	11
3	10	13	10
4	10	14	12
5	6	15	11
6	9	16	7
7	9	17	8
8	6	18	7
9	9	19	9
10	9	20	8

9. Bone mineral density (BMD) is a common measure of bone health, risk of fracture, and osteoporosis. It's also a measure that routinely reports results as standard scores. The data in Table 6.7 come from the National Health and Nutrition Survey (NHANES) and the Centers for Disease Control and Prevention (CDC).

Table 6.7 Average BMD measures for women (in mg/cm^2) by age and ethnic group

	Mean BMD (SD)		
Age	Mexican-American	White (Non-Hispanic)	Black (Non-Hispanic)
8–11	657 (120)	651 (111)	713 (134)
12–15	913 (118)	920 (138)	986 (135)
16–19	983 (112)	1004 (107)	1,077 (115)
20–29	1024 (110)	1064 (106)	1,118 (131)
30–39	1056 (133)	1065 (110)	1,130 (119)
40–49	1036 (110)	1056 (134)	1,108 (150)
50–59	960 (171)	993 (141)	1,039 (132)
60–69	895 (138)	952 (142)	1,029 (163)
70–79	825 (150)	902 (167)	997 (143)
80 and above	791 (218)	932 (141)	948 (222)

Using the data in Table 6.7, answer the following questions:

a. What is the general pattern of change in BMD across age? Is it the same pattern for all ethnic groups listed on the table?

b. Osteoporosis is defined as a BMD measure that is 2.5 standard deviations below average for a healthy 30-year-old ($z = -2.5$). What percentage of the population would you expect to be diagnosed with osteoporosis?

c. Convert a BMD measure of 720 mg/cm² to a z-score in each ethnic group and each age listed in Table 6.7. Convert a BMD measure of 720 mg/cm² to a z-score in each ethnic group listed at the following ages. Convert a BMD measure of 720 mg/cm² to a z-score in each ethnic group and each age group listed in Table 6.7.

d. Who would be diagnosed with osteoporosis in this data set?

Ethnic group	Age	z-score Equivalent (BMD = 720)	Diagnosis of Osteoporosis? (yes or no)
Mexican-American	35		
	60		
	83		
White (non-Hispanic)	35		
	60		
	83		
Black (non-Hispanic)	35		
	60		
	83		

10. Answer the questions below to solve the riddle. Each numeric answer you calculate is linked to a word on the table shown below. Round your answers to two decimal places (e.g., 2.35 rounds up to 2.40).

Riddle: How many psychologists does it take to change a light bulb?
Answer: _____ _____ _____ _____ _____ _____ _____ _____ _____ _____ _____ .

Solution section: Words

3.36) one	14.78) light	−0.71) to	6.97) change
5.00) two	6.24) that's	1.15) but	8.01) wisdom
−0.005) time	58.00) bulb	0.02) lambs	−1.23) the
1.30) apples	11.58) will	0.09) has	1.46) slowly
0.94) to	−0.09) have	1.92) was	0.02) be
343.00) western	−0.71) ready	23.01) Freudian	0.01) difference
6.14) just	3.84) want	14.78) over	107) all

Data Set A	Data Set B	Data Set C
5	48	−7
4	54	−4
8	52	−2
9	53	0
12	84	1
3	64	1
2	51	6

Questions	Word
DATA SET A	
The mean of data set A is _____.	_____
The standard deviation of data set A is _____.	_____
A score of 10 in data set A converts to a z-score of _____.	_____
The lowest score in data set A converts to a z-score of _____.	_____
DATA SET B	
The mean of data set B is _____.	_____
The standard deviation of data set B is _____.	_____
A score of 57 in data set B converts to a z-score of _____.	_____
DATA SET C	
The mean of data set C is _____.	_____
The standard deviation in data set C is _____.	_____
The raw score equivalent of a z-score of 0.43 is _____.	_____
The raw score equivalent of a z-score of 2 is _____.	_____

11. The SAT is, as you already know, a standardized test. That means that when you take the SAT, your result is on a scale with a known mean and a known standard deviation. In 2013, the national average scores on each subtest of the SAT (and the *SD*) were as shown in Table 6.8 (the maximum score possible on any one section in the 2013 test was 800, and the minimum score possible was 200).

Table 6.8 National averages on the SAT in 2013 (with *SD*)*

Statistics	Reading	Mathematics	Writing
Mean	496	514	488
SD	115	118	114

*According to the College Board Publication "2013 College Bound Seniors. Total Group Profile Report" posted at http://media.collegeboard.com/digitalServices/pdf/research/2013/TotalGroup-2013.pdf. Retrieved on May 29, 2014.

Suppose that you want to attend the very upper-crust "Blue Stocking University," which requires a score of at least 700 on the Reading test, at least 710 on the Mathematics test, and at least 680 on the Writing test, for entry into their hallowed halls.

a. What proportion of the population of students who took the test in 2013 would score 700 or higher on the Reading test?
b. What proportion of the population of students who took the test in 2013 would score 710 or higher on the Mathematics test?
c. What proportion of the population of students who took the test in 2013 would score 680 or higher on the Writing test?
d. A total of 1,660,045 high school students took the test in 2013. How many of them would be eligible for admission to Blue Stocking University on the basis of their:
   i) Reading test scores
   ii) Mathematics test scores
   iii) Writing test scores
e. What percentage of students scored between 500 and 600 on their:
   i) Reading test
   ii) Mathematics test
   iii) Writing test
f. What percentage of students taking the test would not have scored high enough on each of these tests for entry into good old Blue U on the basis of their:
   i) Reading test
   ii) Mathematics test
   iii) Writing test
g. What score would correspond to the 75th percentile on the:
   i) Reading test
   ii) Mathematics test
   iii) Writing test

12. The Stanford–Binet IQ test is another example of a standardized test. The average score on the Stanford–Binet test is 100 with a standard deviation of 16.
    a. IQ scores are normally distributed. That being the case, the 3-sigma rule would tell us that 68% of the population would have an IQ score within one standard deviation of the mean. Given this:
       i) What z-score would be exactly one standard deviation below the mean?
       ii) What IQ score would this z-score convert to?
       iii) What z-score would be exactly one standard deviation above the mean?
       iv) What IQ score would this z-score convert to?
    b. What IQ score would define the 90th percentile?

13. Suppose you took our "new" intelligence test that had a mean of 500 and a standard deviation of 100. If both this test and the Stanford–Binet test are measuring the same underlying psychological characteristic (intelligence), your score on the new test should predict your score on the Stanford–Binet test.
    a. Suppose you scored a 750 on the "new" intelligence test. What score would you predict you would receive on the Stanford–Binet test?

b. Suppose your partner scored a 525 on the "new" intelligence test. What score would you predict that your partner would receive on the Stanford–Binet test?

c. You should be able to predict scores in both directions, so what score on the "new" test would you predict someone who scored a 78 on the Stanford–Binet test would receive?

d. Suppose we gave the "new" intelligence test to 4,000 people. How many people would score at or above a 490?

14. Answer the questions shown below to complete the crossword puzzle, spelling out all numbers.

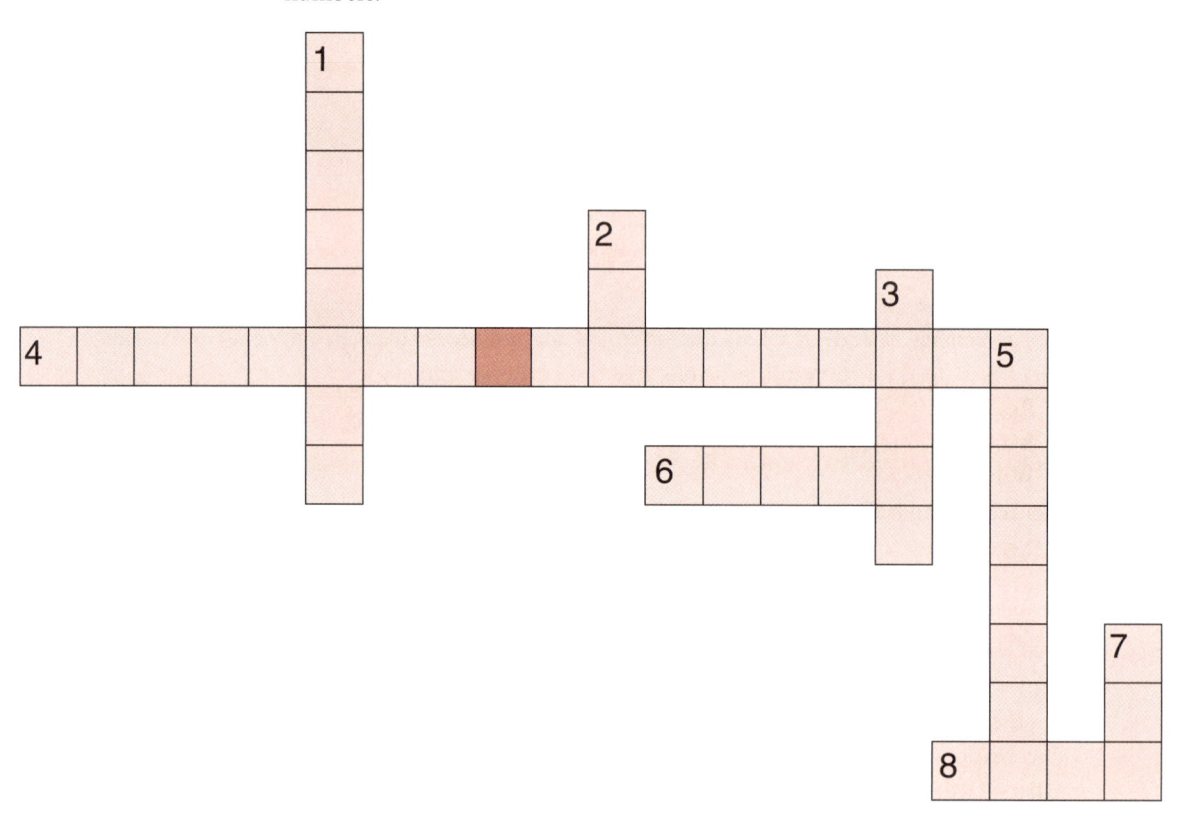

ACROSS

4. (2 words) In a z-score, the deviation of an observation from the mean is divided by _____.

6. The mean of a normal distribution is 10 and s = 0.50, a score of 14 becomes a z of _____.

8. If the mean of a normal distribution is 12.5 and s = 5, a z-score of −2.5 would correspond to a raw score of _____.

DOWN

1. The top of the z-score formula says we _____ the observation from the mean

2. The 3-sigma rule says 68% of observations in a normal distribution will be within _____ standard deviation of the mean

3. _____ % of the scores in a normal distribution are above the mean

5. An observation that is less than the mean will convert to a _____ z-score

7. The 3-sigma rule says that _____% of the observations in a normal distribution will be two or more standard deviations above the mean

STATISTICS IN CONTEXT

15. Use the z-table of Appendix A to answer the following questions:
    a. What proportion of scores should be between the mean and a z of 0.56?
    b. What proportion of scores should be between the mean and a z of 1.24?
    c. What proportion of scores should be between a z of 1.00 and 1.34?
    d. What proportion of scores should be between a z of -0.57 and -1.03?

16. Many theories about the cause of schizophrenia involve changes in the activity of a neurotransmitter chemical called dopamine in the brain. In a 1982 study, Sternberg, Van Kammen, Lerner, and Bunney examined 25 hospitalized schizophrenic patients who were treated with an antipsychotic medication. After a period of time, these same patients were classified as psychotic and nonpsychotic by hospital staff. Samples of cerebrospinal fluid (CSF) were taken from each patient and assayed for dopamine β-hydroxylase (DBH) activity. DBH is an enzyme used in the brain to break dopamine down into its inactive component parts. Levels of DBH are used as an indirect measure of dopamine levels. The more DBH present in the CSF, the more dopamine present in the brain.

    The data in Table 6.9 show the average level of DBH in two groups of schizophrenic patients. Group 1 included 15 male patients who were diagnosed with schizophrenia but did not display psychotic features of the disorder. Group 2 included 10 male patients who were diagnosed with schizophrenia and did display psychotic features. The larger the number, the more DBH was present in the CSF (units of measure are nmol/ml/mg of protein).

Table 6.9 Average levels of DBH in two groups of schizophrenic males

	Mean DBH Level	Standard Deviation
Group 1 (no psychotic features)	0.0164	.0046
Group 2 (psychotic features)	0.518	.0894

   a. Describe the difference in DBH levels between the two groups. Which group had the higher mean? Which group had more variability in their DBH levels?
   b. Let's compare the mean DBH level in Group 2 with the distribution of DBH in the nonpsychotic group (Group 1). Convert the mean DBH level in Group 2 to a z-score in the Group 1 distribution.
   c. What does the z-score you just calculated tell you about the possible relationship between DBH level and psychotic features in schizophrenia?

17. Suppose we know that on average, students are absent from class 38 times in a one-year, two-semester period, with a standard deviation of five absences (two semesters equals roughly 160 days).
    a. What percentage of students will miss between 45 and 50 days in a two-semester period?
    b. What percentage of students will miss half of the possible class days?
    c. What percentage of students will miss only 15% of their classes?

18. Convert the following z-scores to raw scores:
    a. A z-score of −1.34 in a distribution with a mean of 50 and a standard deviation of 10.
    b. A z-score of +1.96 in a distribution with a mean of 24 and a standard deviation of 3.5.
    c. A z-score of 1.72 in a distribution with a mean of 14 and a standard deviation of 3.

19. Court records in a small resort community show that the number of miles over the speed limit for convicted speeders is normally distributed with a mean of 15 mph and a standard deviation of 4 mph. Use this information to answer the following questions:
    a. What percentage of convicted speeders were going between 14 and 18 mph over the posted speed limit?
    b. If 200 speeders are arrested, about how many of them would be expected to have exceeded the speed limit by more than 20 mph?
    c. What speed is exceeded by only 5% of the speeders if the posted limit is 45 mph?

20. A student nurse reads that neonates (newborn babies) sleep an average of 18 hours on their first day in the "outside" world. Wondering about the variation in the nursery where she works, the nurse obtains the data shown in Table 6.10. Use the data to answer the following questions:
    a. What is the mean number of hours slept by infants in the nursery?
    b. What is the standard deviation of the data in Table 6.10?
    c. What percentage of infants sleep more than 17 hours a day?
    d. What percentage of infants sleep less than 18 hours a day?
    e. What percentage of infants sleep between 16 and 19 hours a day?
    f. The sleep times for "difficult" babies is three standard deviations less than the mean. How many hours per day do difficult babies tend to sleep?

Table 6.10 Number of hours slept by infants in neonatal nursery

x	f
20	3
19	4
18	10
17	5
16	1
15	2

21. A junior professor hands you a sheet of paper and tells you that on the sheet is a list of normalized grades (z-scores) from one of his colleague's classes. Being the competitive type, he wants to make sure the grades his colleague's students are receiving are similar to those of his own students. However, the evaluations differ between the two classes. Convert the z-scores in Table 6.11 into meaningful values for the junior professor's class. The junior professor's students' evaluations have a mean of 45 and a variance of 4.

Table 6.11 Class grades for five students

Student ID	Normalized Score
201454746	−0.59
201419939	2.14
201462630	1.05
201491160	−2.81
201492113	0.75

22. Two old friends were swapping stories about fishing on their local lake when they got into a disagreement over who had caught the biggest fish. Bob said his fish was biggest: It was 16.25 inches (41.23 cm) long. Doug did not agree. He was positive that his fish was the biggest: It weighed 4.5 pounds (2.04 kg). A website tracking fish populations in the local county's lakes notes that fish from the lake in question have a mean weight of 3.42 pounds ($SD = 0.4$) and a mean length of 14.09 inches ($SD = 0.8$).
    a. How can the friends make the comparison between the two fish?
    b. Who caught the bigger fish, Bob or Doug?
    c. Clearly, the biggest fish was big. If there were 10,000 fish in the lake, how many fish might you expect to be smaller than the biggest fish?

23. Sarah is a mechanical engineer working in the shoe industry. One of her clients has asked her to design a machine that can manufacture footwear that will fit 92% of the shoe-buying public. The accounting wizards from the client's firm performed a cost–benefit analysis and determined that the cost of making footwear for those unlucky 8% brings little benefit.
    The mean shoe size of the footwear-buying public is nine units with a standard deviation of three units. The measurements of foot size are normally distributed.
    a. If the footwear company is concerned with offending the tiny feet lobby by not accommodating those with small feet (i.e., small shoe sizes are made), what is the largest shoe size that Sarah's machine should make (rounding to the nearest whole size)?
    b. An athletic shoe company specializing in basketball shoes wants Sarah to design a similar machine to accommodate the larger feet of basketball players; however, they are still selling to the general public. In this case, the company still wants to have shoes that fit 92% of the population, but they don't want to miss any of the larger sizes. What is the smallest shoe size that Sarah's machine should make (rounding to the nearest whole size)?
    c. Lastly, after feeling a bit guilty that only the large-footed or small-footed individuals were getting neglected, Sarah designed a machine that neglected both. What are the largest and smallest shoe sizes Sarah's machine can make to fit the middle 92% (rounding to the nearest whole size)?

24. An elementary school district is running a spelling bee at three of its schools, and the top 10% from each school will be moving on to the state championship. The school board has asked each school to submit the names of only those students who are in the top 10% of their school. Given the information about each school's scores in

their respective spelling contests (see Table 6.12), determine the cutoff scores for each school that should be used to determine who gets to move on.

Table 6.12 Mean spelling bee scores at three district elementary schools

School	Mean	SD
Pine Street	98	10
Oak Drive	64	12
Maple Crescent	26	6

25. Determine the percentage of data points falling between the given values x_1 and x_2 for each normal distribution described in Table 6.13.

Table 6.13 Normal distributions for question 25

Mean	SD	x_1	x_2
6	2	5.5	6.3
243	17	253	265
75.4	21.2	62.8	76.1
500	100	250	750
0.453	0.014	0.449	0.453

26. A car manufacturer has determined that a small part inside the front brake assembly has a tendency to fail. This part is responsible for engaging the antilock braking system. While the brakes still work, the driver is more likely to lose control in emergency braking situations when this part is not working correctly. Engineers at the car manufacturer have been busy collecting data, and they have determined that 50% of the parts fail by the time the car reaches 10,500 miles on the odometer, and that 90% of the parts fail by the time the car reaches 12,000 miles.

The manufacturer wants to minimize any issues the failing part might cause and has said that anything greater than a 2% failure rate is unacceptable. To meet this goal, at what mileage should the cars be brought in to have the parts replaced? Assume the distance to failure measure is normally distributed.

27. McKenzie takes the commuter train into the city for work every day. She's also a stickler for details and arrives every morning at exactly 8:01 a.m. for the 8:03 a.m. train. The trains into the city, however, don't run quite as precisely as McKenzie does. While the average (mean) departure time is 8:03 a.m., the variance is 4 minutes. Assuming a normal distribution of departure times, how many times would you expect McKenzie to miss her train over the next 1,000 trips into the city?

28. Ethan is just as precise as McKenzie in question 27, but he sleeps in just a bit more and likes to live dangerously. The next train after the 8:03 a.m. train is at 8:17 a.m. He arrives at exactly 8:04 a.m. Over the next 100 trips into the city, how many times does Ethan have to wait for the second train because he missed the first?

29. Three colleges give three different general science exams to 15 first-year science students. Their scores are listed in Table 6.14.

Table 6.14 General science exam scores at three colleges

College A	College B	College C
92.6	16.8	228.9
80.9	16.0	239.8
81.0	15.2	257.3
88.5	16.6	241.0
117.2	18.1	223.2
107.9	16.3	227.5
85.2	16.4	238.3
84.8	14.5	252.0
89.8	15.3	238.2
101.1	17.1	227.6
78.8	15.9	230.0
112.5	15.3	233.7
98.5	15.1	218.5
99.9	14.8	235.8
102.3	21.3	238.8

a. Graph the data using a histogram to check if they appear to be normally distributed.
b. Calculate the mean and standard deviation for each university.
c. Calculate the z-score for each science student in each school.
d. What are the means and standard deviations for each school using the z-scores instead of the raw scores?
e. How many students does each school have at or above the 90th percentile based on this general science exam?

30. A new tool to measure depression in young adults has just been developed, and scores are conveniently distributed normally. Individuals scoring more than two standard deviations away from the mean score on the tool are classified as "high risk." Those who score between one and two standard deviations away from the mean score are classified as "moderate risk," and individuals who are within one standard deviation of the mean score are classified as "low risk."
a. What percentage of young adults would you expect to be classified as "moderate risk"?

b. The mean score of the tool is 33, and the standard deviation is 6. Classify the cases described in Table 6.15 based on their raw scores from the new depression tool.

Table 6.15 Depression scores for five young adults

Case ID	Score
1	36
2	28
3	47
4	20
5	22
6	31

31. The risk of negative outcomes for children is strongly associated with family income. Generally, those families that fall below a certain level of income have significantly greater number of poor outcomes (e.g., increased mental illness, increased school dropout, and increased crime). The yearly family income threshold often used to identify poverty for a family of four is $17,000. Given the following information about different counties within one state and assuming that family income is normally distributed:
    a. Determine the expected percentage of below-income-threshold families for each county.
    b. Calculate the family income for the 1% (i.e., 99th percentile) for each county.

Table 6.16 Mean family incomes in five counties

County	Mean	SD
A	$34,495.00	$12,632.00
B	$45,667.00	$9,812.00
C	$58,634.00	$11,723.00
D	$29,187.00	$4,521.00
E	$69,455.00	$23,931.00

Think About It . . .

THE RANGE RULE REVISITED

SOLUTIONS

1. What score is two standard deviations above the mean?   **95.**
2. What score is two standard deviations below the mean?   **55.**

3. The 3-sigma rule says that 95% of the distribution should be between **55 and 95.**
4. Are 95% of the scores between these two points? (Yes/No) **Yes, 9 of the 10 scores are between 95 and 55.**
5. What is the range of the scores that are at least two standard deviations away from the mean? **95 − 55 = 40. The range rule says that the standard deviation should be 40/4 = 10, and it is.**
6. Why were some of the estimates less accurate than others? **The range rule applies when the distribution is normal in shape. The estimates made using the range rule will be precise (perfect) only if the distribution is perfectly normal. Estimates made on distributions that are skewed or less than perfectly normal will be less accurate. However, since they are only estimates of the standard deviation, the fact that the estimates are less than perfect likely won't matter to most users of the range rule. In fact, if you graph the distribution I've given you above, you'll see that it is not normal. The actual standard deviation of this set of numbers is 10.3. The range rule estimates the standard deviation to be 10.**

REFERENCES

Hicks, R. A., Fernandez, C., & Pellegrini, R. J. (2001). Self-reported sleep durations of college students: Normative data for 1978–79, 1988–89, and 2000–01. *Perceptual and Motor Skills, 93*, 139–140.

Sternberg, D. E., van Kammen, D. P., Lerner, D., & Bunney, W. E. (1982). Schizophrenia: Dopamine ß-hydroxylase activity and treatment response. *Science, 216*(4553), 1423–1425.

CHAPTER SEVEN

BASIC PROBABILITY THEORY

The 50-50-90 rule: Anytime you have a 50-50 chance of getting something right, there's a 90% probability you'll get it wrong.

—ANDY ROONEY

Everyday Statistics

HOW TO WIN BIG MONEY . . . MAYBE

They come in every color of the rainbow, decorated with icons for luck, love, and money and offering players everything from a measly buck to millions. Lottery scratch cards are seen by some as a one-way ticket to easy street and by others as a nonstop trip to the poor house. Indeed, sometimes gamblers do beat the odds, winning multiple times and confounding the experts.

Statisticians can calculate the probability that you will scratch your way to easy money. Recently, a woman from North Carolina reportedly won nine times in a span of 4 months. And between 2008 and 2010, Dr. Joan Ginther—a retired professor of statistics, touted as the "Luckiest Woman in the World"—won big cash prizes four times, walking away with a cool $20 million. The odds that a person buying just one scratch card at a time will win the lottery four times has been calculated at 1 in 18 septillion—that's one chance in 18,000,000,000,000,000,000,000,000 (24 zeros). Were these women just lucky, or did they find a way to beat the system? Chapter 7 will discuss the basics of probability so that you can decide for yourself.

OVERVIEW

PROBABILITY
PROBABILITY AND FREQUENCY
USING PROBABILITY
PROBABILITY IN CONTEXT

LEARNING OBJECTIVES

Reading this chapter will help you to . . .

- Explain the gambler's fallacy. (Concept)

- Describe how probability and frequency are related to one another. (Concept)

- Understand basic set theory, including conditional probability and the difference between independent and dependent events. (Concept)

- Calculate the probability of an event. (Application)

- Calculate combined probability using the AND and OR rules. (Application)

- Know the difference between theoretical and empirical sets. (Concept)

- Evaluate the ways that probability is used in modern science. (Concept)

PROBABILITY

For many students new to statistics, the prospect of studying probability is daunting. I sympathize completely with the hesitation I see in many of them. I remember delaying the decision to take my own first course in statistics because I was 100 percent certain that I would never be able to grasp the mathematical intricacies of probability. I was almost disappointed to discover that understanding probability wasn't all that hard. It turns out that the language and symbols used in probability theory are often the most difficult bit. But if you have ever learned a foreign language, you will have faced this vocabulary problem before. My advice is to make a list of new terminology and treat these new words the same way you would any new vocabulary in any course. You probably (no pun intended) think I'm just saying this right now, but I hope you will soon see that I'm telling the truth: You already know quite a bit about the probability theory we will discuss here; you just don't know that you know it.

We have already encountered the idea of probability, first when we considered frequency and again when we examined graphs of frequency distributions. You can think about probability as being the same thing as frequency. Things that happen very frequently—the sun coming up in the morning, for example—are highly probable events. Things that happen very infrequently—like picking the winning numbers in the lottery—are improbable.

PROBABILITY AND FREQUENCY

The story of the "Gambling Nobleman" (see *The Historical Context* beginning on page 297) points to a pattern in the history of human beings and probability. Gamblers have always wanted to increase their chances of winning, and they soon recognized that being able to measure the likelihood of achieving a specific outcome when rolling dice, flipping cards, or spinning a wheel provided an enormous advantage over less knowledgeable competitors. It should come as no surprise, then, that some of the first attempts to measure probability were focused on counting how many times something happened. You will see this focus on frequency in set theory, the foundation of probability studies. We will start our discussion with a word that gets used a lot in probability: *set*.

BASIC SET THEORY

Sets, Subsets, and Events

set theory The mathematical study of the infinite, or the mathematical study of sets.

set A well-defined collection of things, objects, or events.

In **set theory**, a **set** is a well-defined collection of things, objects, or events. In a deck of cards (the set), where cards are numbered from 1 (ace) to 10 and face cards are labeled from jack to king, there are 52 *events* in the set. Sets are designated by capital letters (A, B, C, and so on), so let's call the set of all red cards in a fair deck of cards "set A." So, A then equals {ace ♦ ♥, 2 ♦ ♥, 3 ♦ ♥, 4 ♦ ♥, 5 ♦ ♥, · · ·, Q ♦ ♥, K ♦ ♥} for a total of 26 elements or events (half the cards in a deck of 52 are red, so 26 of them are red and 26 are black). Typically, the sets that we deal with in statistics are collections of events, objects, measurements, or outcomes that we are trying to study.

Set theory also includes other types of sets. The two that we really need to talk about are the following:

- **Subsets**: portions of a larger set. If every element in A is also in B, then A is a subset of B. If A = {1, 2, 3} and B = {1, 2, 3, 4, 5, 6, 7, 8, 9}, then A is a subset of B. In symbols, the relationship would look like this: A ⊆ B (this mathematical "sentence" reads "A is a subset of B").
- **Universal set**: includes all possible events, outcomes, measurements, objects, etc. The symbol for the universal set is the letter S.

subset A portion of a larger set.

universal set The set of all possible events, outcomes, measurements, objects, etc.

gambler's fallacy The erroneous belief that an event is less likely to happen if it follows a series of similar events, or that past events can change the probability of a random future event.

The Historical Context

THE GAMBLER'S FALLACY

The age-old quest to win at games of chance has been one of the driving forces in the development of our understanding of probability. History and literature are rife with examples of the rich suddenly becoming the poor because they were convinced they had a system that would guarantee them a big win at the casino. In his short story "The Case of the Gambling Nobleman," Colin Bruce uses the familiar characters created by Sir Arthur Conan Doyle to involve Dr. Watson and Sherlock Holmes with one such hapless aristocrat.

In this story, a young woman comes to Sherlock Holmes begging for help because her fiancé is burning through the family fortune in a desperate attempt to replenish his bank account by playing roulette. The fiancé has a system, one based on his understanding of the laws of probability. After he mortgages his entire estate and dashes off for the casino, his soon-to-be bride runs just as fast to Sherlock Holmes.

The roulette wheel consists of a grid of numbers, 1 through 36, on either black or red squares. Even numbers are traditionally on red slots of the wheel, with odd numbers on black. On a French wheel (which this particular nobleman would likely be using), there is one green box: This is the zero slot, and it belongs to the house (see Figure 7.1).

FIGURE 7.1
The French roulette wheel.

When you play roulette, you can bet on a single number, a range of numbers, the colors black and red, or whether the number will be odd or even. The croupier spins a marble ball in one direction and the roulette wheel in the opposite direction. When gravity takes over, the ball loses its momentum and falls onto the wheel and into one of the numbered slots. If the ball falls into your slot, you win. If the ball falls into the zero slot, the house wins everything (all the players lose).

CHAPTER 7 Basic Probability Theory   297

So, our desperate fiancé has a system. He says that

> the laws of probability always assert themselves in the long term, but not in the short. Thus... if you toss a coin two or three times, it may well turn up the same way each time. But if you toss it a few hundred times, you are very likely to get a more even division between heads and tails. As the number of times you toss it tends towards infinity, the ratio of heads to tails is guaranteed to become ever closer to unity (p.27).

Sherlock Holmes agrees that the fiancé is, so far, correct. The young man continues, saying he would wait until an

> uninterrupted sequence of one color—either red or black—had continued for some time. Then he would place a modest bet on the opposite color. For he knew that in order for the laws of chance to be obeyed, a preponderance of the other color would become inevitable. (p.27).

Sherlock Holmes sighs, shaking his head, and asks the young woman if her fiancé has put this plan into action. Holmes recognizes that the young man has fallen victim to the **gambler's fallacy**: the erroneous belief that an event is less likely to happen if it follows a series of similar events, or that past events can change the probability of a random future event.

So, what is the problem with the fiancé's system? First, think about the odds of winning at roulette to begin with. Suppose you bet, as the young man in the story does, on a color—let's say black. What are the odds that on the first spin of the wheel, you'll win? There are 18 red and 18 black slots on the wheel, plus the zero slot, for a total of 37 slots. The odds of winning if you bet black are 18/37, or .4864. In other words, slightly less than half the time the ball will land on a black slot, giving you about one chance of winning in every 2.06 bets.

The young man is correct when he says that eventually, the number of times the ball lands on red will come to equal the number of times it lands on black—that the ratio will approach unity (1.00). However, there are two gigantic problems with his system.

The first is that each spin of the roulette wheel is an independent event: The occurrence of one event has no effect on the occurrence of the other. The assumption that if red has come up 25 times in a row, black is more likely to come up on the 26th spin of the wheel requires that the history of events on the wheel has somehow been kept track of. Yet as far as anyone has been able to tell, neither the cosmos nor the roulette wheel keeps track of the number of times a particular color wins. The likelihood that the ball will land on a black slot is exactly the same whether the ball has landed on red 5, 15, or 25 previous spins in a row: It's still .4826.

The second problem has to do with the fiancé's rather carefree statement that as the number of events he's betting on (coin tosses or spins of the roulette wheel) approaches *infinity* (my emphasis), the ratio of heads to tails will even out. This is an excellent example of the *Law of Large Numbers*, also known as the *Law of Averages*. Eventually, heads will equal tails, and red will equal black, but it takes a very, very, *very* large number of trials for this to happen (think about how many times you'd have to toss a coin or spin the roulette wheel to "approach infinity" and you'll get a feel for how large "large" actually is in this case).

The gambler's fallacy is also known as the *Monte Carlo fallacy*, after a famous example of this misunderstanding of the rules of probability. On August 18, 1913, at the infamous Casino de Monte-Carlo, the ball landed on black a record 26 times in a row. By the 15th time black won, players began to bet heavily on red, believing that chances that the next spin would come up black were astronomically low and getting lower. As the players doubled and tripled their bets, the casino raked in the francs. Eventually, the casino won millions, and the players walked away very much poorer, but wiser, for their efforts.

We have talked about universal sets and subsets before; we've just used different names for them. You can think of the *population* as the universal set and a *sample* taken from the population as a subset. In symbols, A ⊆ S means that sample A is a subset of the population, S. In fact, any sample can be considered a subset of the population it came from because samples consist of observations or measurements that are part of the larger, parent population.

The basic equation in probability theory is that the probability of A (our subset or sample) is equal to the number of elements in A divided by the number of elements in S (the universal set or the population). Notice that probability here is defined in terms of *frequency*—the number of elements in A and in S. In mathematical terms, it is written the way it is presented in Box 7.1.

BOX 7.1   **The probability equation**

$$p(A) = \frac{\#(A)}{\#(S)}$$

Examples of Set Theory

Let's go back to our deck of cards. The universal set here would be S = {ace ♣ ♦ ♥ ♠, 2 ♣ ♦ ♥ ♠, 3 ♣ ♦ ♥ ♠, 4 ♣ ♦ ♥ ♠, 5 ♣ ♦ ♥ ♠, ···, Q ♣ ♦ ♥ ♠, K ♣ ♦ ♥ ♠}, for a total of 52 events. Suppose we want to figure out the probability of drawing from the deck a very small subset that consists of only the three of clubs (3♣). Our subset is A = {3♣} (there's only one 3♣ in a fair deck of cards; if there's more than one in your deck, I'm not playing cards with you). So, what is the probability that, when drawing a card at random from the deck, it will be the three of clubs? Take a look at Box 7.2.

BOX 7.2   **The probability equation: Example**

$$p(A) = \frac{1}{52} = .019 = .02$$

We have one chance out of a total of 52 events of pulling out a three of clubs. That's the same as a 2% chance (or 2 chances out of 100). Probabilities are often expressed as percentages, or chances out of 100, for the same reason that frequencies are often expressed as percentages. Converting to percentages puts all sets on the same footing and allows comparisons across sets that might have very different numbers of elements.

The probability of an event can never be less than 0.00 or more than 1.00. In other words, it cannot happen more often than always (a probability of 1.00, or 100%) or less often than never (a probability of 0.00). If we're calculating probability and wind up with a negative number or one greater than 1.00, we will know immediately that we have made an error.

What is the probability of picking *any* three out of the deck? For that matter, what are the chances of picking one of any card out of the deck? Start with picking any three, and you should quickly see that our odds of picking a three out of a deck of cards is 4 out of 52 because subset A = {3♣, 3♥, 3♠, 3♦} and our population S = {all 52 cards}, so $p(A) = 4/52$, or .0769, or .08 rounding up. So, we have about an 8% chance of picking any three out of the deck. In fact, we have about an 8% chance of picking one of any card out of the deck.

What is the probability that we will draw a face card (J, Q, or K) out of the deck? Well, there are four jacks, four queens, and four kings in the deck, for a total of 12 face cards. There are 52 cards in total in the deck, so the probability of drawing a face card is 12/52, or 23% (23 chances in 100 draws).

Now, what is the probability of drawing a king—any king—out of the set of face cards? We've changed our definition of the population (the universal set) so that S = {J♣♥♠♦, Q♣♥♠♦, K♣♥♠♦}, for a total of 12 face cards. Four of them are kings, so the probability of drawing a king out of the set of face cards is 4/12, or 33 chances out of 100.

CheckPoint

Calculate the probabilities of the following draws assuming a normal 52-card, four-suit deck with values that range from 2 (low) through ace (high). Express your answer as a decimal to three significant digits.

1. Drawing a black card _____
2. Drawing an ace _____
3. Drawing a red jack, queen, or king _____
4. Drawing a black card with a value of 7 or higher _____
5. Drawing a card with a value of 5 or lower _____

Answers to these questions are found on page 302.

CONDITIONAL PROBABILITY

We know that the probability of drawing a king from a set made up of all face cards is 4/12, or 33%. What happens to the probability of this event (drawing a king from the set of face cards) "without replacement"—in other words, if we *do not* put back the card we just drew before we draw again? On the first draw, our population S = {4J, 4Q, 4K}, for a total of 12 cards. If the first card drawn was a king, then the probability that the next card we draw will also be a king is $p(K) = 3/11$. There are now only three kings left in S, and S has decreased in size by one card. If the first card drawn was not a king, then $p(K) = 4/11$ (all four kings are still in S, but the size of S has still decreased by one).

This is an example of **conditional probability**, the probability of a given event that *depends on* (or is *conditional upon*) the occurrence or nonoccurrence of another event. In symbols, this situation is usually written as $p(B|A)$, meaning the probability (p) of event B happening given that event A has happened. In our example, the probability of drawing a king on the second draw is conditional upon (or depends on) the outcome of the first draw.

We need to consider two additional issues in probability before we get any more complex. The first is the question of *independence*: Are the events in question independent of one another or dependent on one another?

- **Independent events**: the occurrence of one event does not change the probability of the other event; written as $p(B|A) = p(B)$.
- **Dependent events**: the occurrence of one event alters the probability of the occurrence the other event.

The probability of drawing a king on the second draw is an example of a dependent event because the probability depends on whether or not a king was drawn the first time. Now suppose we put the first card back in the deck after the first draw—in other words, we sampled "with replacement." What is the probability of drawing a king on the second draw?

You are correct if you reasoned that it is the same probability as that of drawing a king on the first draw. If we put the first card back, we have the same chances of drawing a king the second time as we did the first time we reached into the set to pick a card. This is an example of independent events—the probability of drawing a king on the second draw when we replace the first card is independent of what that first card was. There are still 4 chances out of 12 because we put the first card back.

In addition to the question of independence and dependence, we need to consider whether or not the events in question are **mutually exclusive**. Two events are mutually exclusive if the occurrence of one means the others cannot occur at the same time. If we toss a coin, for example, we can get either heads or tails, but not both. But if we toss the coin and get heads, then the probability of getting tails is now automatically zero. So, the probability of getting heads does affect the probability of getting tails. Two mutually exclusive events cannot be independent because the occurrence of one *does* affect the occurrence of the other. In set theory, two mutually exclusive sets are considered **disjoint**, meaning that they do not overlap at all.

COMBINING PROBABILITIES

Suppose we want to know the probability that two events will *both* happen. For example, what is the probability of drawing two kings in a row from our set of face

---

conditional probability The probability of a given event that depends on (or is conditional upon) the occurrence or nonoccurrence of another event.

independent events Two events, each of which occurs without having any effect on the probability of the occurrence of the other.

dependent events Two events in which the occurrence of one alters the probability of the occurrence the other.

mutually exclusive (or disjoint) Denoting two sets that do not share *any* elements.

> **CheckPoint**
> *Answers to questions on page 300*
>
> 1. .500 ☐
> 2. .077 ☐
> 3. .115 ☐
> 4. .308 ☐
> 5. .308 ☐
>
> SCORE: /5

cards if we sample without replacement (making the two events dependent)? How do we figure this out?

The AND Rule

AND rule The probability of event A and event B occurring together.

The situation I've just described calls for us to use the **AND rule**. It is written symbolically as $p(A \cap B)$, which translates into English as "the probability of event A *and* event B." The upside-down U is the symbol that represents the intersection of two sets or events: We use it to show that we are interested in cases where these two events, A and B, coincide.

To find the answer to our question, we will need to combine the probability of each event in a particular way. Finding the probability of each event by itself is easy. Take a look at Box 7.3.

BOX 7.3 **Finding the probability of two events separately**

$$p(A, \text{ or K on the first draw}) = 4/12, \text{ or } .33$$

$$p(B, \text{ or K on the second draw}) = 3/11, \text{ or } .27$$

So, how do we combine them? The AND rule states that when we want to know the probability of event A *and* event B, we *multiply* the individual probabilities together. In symbols, it looks like what is shown in Box 7.4.

BOX 7.4 **Using the AND rule to find the probability of dependent events**

$$p(A \cap B) = p(A) \times p(B|A)$$

$$p(A \cap B) = .33 \times .27 = .09, \text{ or 9 chances out of 100}$$

Notice that the probability of both event A and event B happening (9%) is less than the probability of either event A by itself (33%) or event B by itself (27%). This should make some inherent sense. Isn't it fairly unlikely that we will draw first a king and then another king on two sequential draws? Both events have to happen to meet our criteria for success, and that's a fairly unusual sequence of occurrences.

The OR Rule

What is the probability of drawing *either* a king *or* a queen on the first draw? In this case, we're looking for event A *or* event B—either one will work. This is written symbolically as written symbolically as $p(A \cup B)$, where the U-shaped symbol represents the union of the two sets or events (in this case, kings and queens together). Notice that the two events are independent. The probability of drawing a king is *independent* of the probability of drawing a queen because they are two entirely separate, or mutually exclusive, events: Drawing a king on the first draw does not change the number of queens still in the set, and vice versa. From a common sense at the poker table point of view, the probability of drawing either a king or a queen should be more likely than the probability of drawing two kings in a row.

Again, we will need to find the individual probabilities of each event and then combine them somehow. And again, finding the individual probabilities is fairly simple. When we want to find the probability of one event *or* another, we apply what is creatively called the **OR rule**, and we *add* their individual probabilities. In symbols, we say $p(A \cup B) = p(A) + p(B)$.

OR rule The probability of either event A or event B occurring.

BOX 7.5 **Using the OR rule to find the probability of independent events**

$$p(A, \text{ or drawing a king the first time}) = 4/12, \text{ or } .33$$
$$p(B, \text{ or drawing a queen the first time}) = 4/12, \text{ or } .33$$
$$p(A \cup B) = p(A) + P(B)$$
$$p(A \cup B) = .33 + .33 = .66$$

So, in this case, it's actually fairly likely (a better than 50/50 shot) that we will draw either a king or a queen out of the set of face cards. Again, this should make some inherent sense: There are only 12 cards in S, and we want to know the probability of 8 of those 12 happening on the first draw. That's 66% no matter how you slice it (8/12 = .66).

Combining the AND & OR Rules

Let's get really creative. What is the probability of drawing *either* a jack *or* a red card on the first draw? The solution to this question is shown in Box 7.6, but think

CHAPTER 7 Basic Probability Theory

the steps through before you take a look at the answer. I'll bet you can figure this out for yourself.

First, we have to figure out the individual probabilities: the probability of drawing a jack and the probability of drawing a red card. That should be fairly straightforward. But then we have to think about this: Some of the jacks are also red cards. These events (being a jack and being a red card) are *not* mutually exclusive. We want to add together the probabilities of A and B, but we also want to remove the probability that A (a jack) will be an example of B (a red card). In other words, we want to remove the probability that event A (drawing a jack) is dependent on event B (A ∩ B) (drawing a jack or a red card). How can we do that?

The answer is that we need to subtract the probability that A and B intersect (A ∩ B) from the combined probability of A or B (A ∪ B). In symbols, we would say that we're doing this:

BOX 7.6   **Combining the AND & OR rules**

$$p(A \cup B) = p(A) + p(B) - p(A \cap B)$$

Once we fill in the appropriate numbers, we will have our answer. Remember to do the addition first, then the subtraction (thinking back to your first math classes and the mnemonic device for remembering the order of operations in an equation: "Please Excuse My Dear Aunt Sally," or parentheses, exponents, multiplication, division, addition, and then subtraction).

BOX 7.7   **Working out the solution to our AND & OR rule equation**

$p(\text{jackJ} \cap \text{red card}) = 2/12$ (there are 2 red jacks in the set of 12 face cards)

$p(\text{jack} \cap \text{red card}) = 2/12 = .166667$, or .17

$p(\text{jack}) = 4/12 = .33$ (there are 4 jacks in the set of 12 face cards)

$p(\text{red card}) = 6/12 = .50$ (6 of the 12 face cards are red cards)

$p(\text{jack} \cup \text{red card}) = p(J) + p(\text{red card}) - p(J \cap \text{red card})$

$p(\text{jack} \cup \text{red card}) = .33 + (.50) - .17$

$p(\text{jack} \cup \text{red card}) = .83 - .17 = .66$, or a 66% chance

---

STATISTICS IN CONTEXT

There's a good chance (66%) that the first card we draw out of the deck will be either a jack or some kind of red card.

Upping the Ante: Finding the Probability of More Than Two Dependent Events

Now let's play poker. What is the probability that the first four cards you're dealt will all be aces? We will assume you're playing with only one deck of cards and with three friends (a total of four people at the table), and that you are dealt to first. Obviously, there is no replacement of the cards after each card is dealt, so we are sampling without replacement and are trying to find the probability of dependent events. We have also changed our definition of the universal set again. Now we're dealing with a fair deck of 52 cards, not just the 12 face cards.

On the first draw, the probability of your being dealt an ace is 4/52: There are four aces in the deck, and you get the first card of the deal. So, you have 4 chances in 52 of that first card being an ace. On the second deal, things have changed. First, the total number of cards in the deck has been reduced by four, so now there are 48 cards in the deck. And if you got an ace on the first draw, then there are only three aces left in the deck, assuming no one else got one (and to make our lives simpler, we'll assume you are the only person who got dealt an ace). So, let's figure out the remaining probabilities, as shown in Box 7.8.

BOX 7.8 **Calculating the probability of being dealt four aces**

$$p(\text{ace on the first deal}) = 4/52 = .08$$

$$p(\text{ace on the second deal}) = 3/48 = .06$$

$$p(\text{ace on the third deal}) = 2/44 = .05$$

$$p(\text{ace on the fourth deal}) = 1/40 = .025$$

Now, should we add or multiply the individual probabilities? Since we need to know the probability that you will be dealt an ace *and then* another ace, *and then* another, *and then* a fourth, this is an AND situation, and we should multiple the individual probabilities.

So, the probability that you will be dealt four aces in a row = .08 × .06 × .05 × .025 = .000006, or 6 chances out of every 100,000 times you sit down to play and all of these circumstances apply. Now consider that this is a very artificial situation: We set it up so that no one else was dealt an ace, we're playing with only one deck, and you happen to be the person who gets dealt to first. The real odds of being dealt four of a kind are considerably less than what we just calculated and also happen to be why I don't play poker—at least, not for money.

CheckPoint

Calculate the conditional probabilities of the following draws, assuming a normal 52-card, four-suit deck with values that range from 2 (low) through ace (high) and no replacement after draws. Express your answer as a decimal to three significant digits.

1. Drawing two aces in a row _____
2. Drawing a 7 or an 8 on the first draw _____
3. Drawing a jack or a queen on the second draw, assuming the first card drawn was a queen _____
4. Drawing a black card on the first draw and a red jack on the second draw _____
5. Drawing a card with a value of 3 or lower on the first draw and a card with a value of 6 or lower on the second draw _____
6. Drawing a card with a value of 3 or lower on the fourth draw, assuming the first three draws produced a jack, a 2, and a 10 _____
7. Drawing a red jack on the first draw and a black jack on the second draw or a 7 on the first draw and another 7 on the second draw _____
8. Drawing at least one king in five draws _____

Answers to these questions are found on page 308.

USING PROBABILITY

So, why are we talking about poker and roulette and kings and aces and hearts and clubs? It's because inferential statistics, our next topic, is based on probability theory. We're going to make use of the rules of probability to determine the probability of getting a particular outcome (like a particular sample mean) in an experiment. To do that, however, we need to talk about one more bit of jargon. We need to make a distinction between an *empirical* set of observations and a *theoretical* one.

The difference between a theoretical set of anything and an empirical set is fairly straightforward. A **theoretical set** of observations is one based on theory: It is the set of observations we expect to get if the theory is true—in other words, the set of observations that ought to happen. An **empirical set** of observations is one based on experience: It is the set of observations that we actually get when measuring something. We can again compare these two ideas to the ideas of the population and the sample.

Most scientists setting up an experiment have a pretty good understanding of what they ought to see in their data. They've read the literature, they know what other researchers have found, and they know what models (mathematical and otherwise) describe the data they are interested in. The population of measurements is, for most scientists doing experiments, what theoretically ought to happen. But this population of all possible measurements is, generally speaking, theoretical, or as yet unmeasured. The sample of measurements that a scientist actually collects when running an experiment is empirical. It is based on personal experience with the variables, and it is "real" in a way that a theoretical distribution is not.

To illustrate the point, I will ask you to consider a population of playing cards you are probably not familiar with: the Mah Jong deck. There are 144 cards in a

theoretical set A set of observations based on theory.

empirical set A set of observations based on experience.

Mah Jong deck, and the systems of suits and values on those cards are quite different from the more familiar Western deck. Table 7.1 shows you the distribution of cards in a Mah Jong deck.

Theoretically, what is the probability that we could reach into the deck of 144 cards and draw out a wheel card? There are 36 wheel cards in the deck of 144, so the probability would be 36/144, or 25%. What is the probability that we would draw out a number card? Or a bamboo stick? I hope you see that *theoretically*, we are just as likely to draw out a bamboo stick as we are a wheel or a number (according to the information in Table 7.1). When it comes to the other cards, the probabilities change. Theoretically, what is the probability of drawing out a flower card? We have 4 chances out of 144, so the chances are significantly lower that we will draw a flower card at random (4/144 = 2.8% chance). How about drawing out a dragon card? These odds are a bit better: There are 12 dragon cards in the deck of 144 cards, so we have an 8.3% chance of picking one dragon out at random. If we drew a picture of these probabilities, we would have something that looked like the bar graph shown in Figure 7.2.

Notice that all three of the "suits" are, *theoretically*, equally likely. The probability of drawing out a flower or a season card is also equal. For any card, the *actual* hand that we are dealt when we play cards is an empirical distribution and a subset

TABLE 7.1 Distribution of cards in Mah Jong deck

Type	Value	Total Number	
Bamboo sticks	• 4 of each, numbered 1–9	36	
Wheels	• 4 of each, numbered 1–9	36	"Suit" cards
Numbers	• 4 of each, numbered 1–9	36	
Wind	• 4 each of East • 4 each of South • 4 each of West • 4 each of North	16	
Dragon	• 4 each of Red • 4 each of Green • 4 each of White	12	
Flower	• Plum • Orchid • Chrysanthemum • Bamboo	4	"Honor" cards
Season	• Spring • Summer • Autumn • Winter	4	
		TOTAL 144 cards	

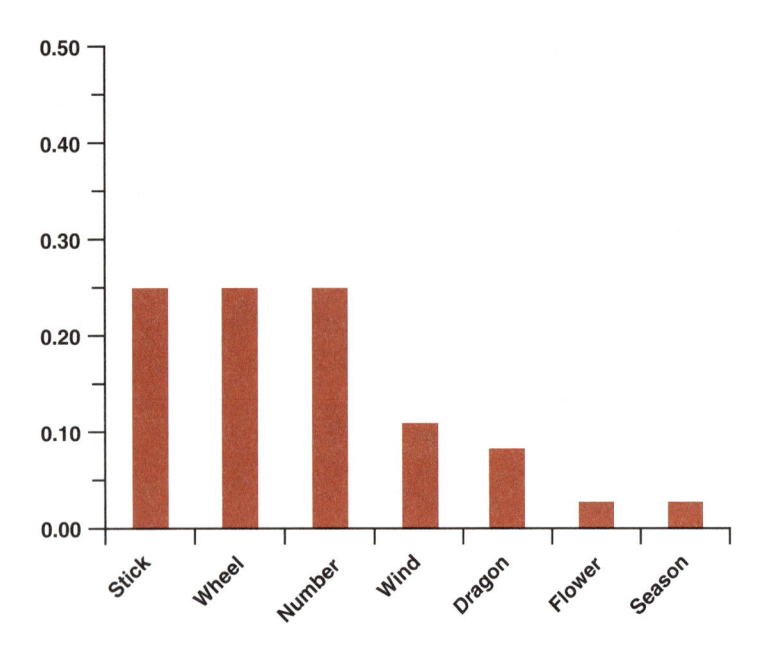

FIGURE 7.2
Probability distribution of Mah Jong cards.

of the population. All of the cards in our hand are part of the population: They are all representative of the cards in the population. But the hand that we are dealt (the empirical sample from this population) will probably not look like a miniature of the population. We may get all wheels, or only number cards, or a dragon, a flower, and two seasons.

Although our actual hand may look very different from the population, this is not a problem as long as we didn't do anything to the deck that would bias the hand that we were dealt, making it more or less likely that we would draw any particular subset of cards. The sample of observations that we draw out of the population is empirical. As long as nothing has happened that biased the observations drawn for the sample, that sample should, theoretically at least, "represent" the population. We should be able to see at least some of the population in the hand that we are dealt.

CheckPoint Answers
Answers to questions on page 306

1. .005 ☐ $p(\text{ace}) \times p(\text{ace}) = \dfrac{4}{52} \times \dfrac{3}{51} = .077 \times .059 = .005$

2. .154 ☐ $p(7) + p(8) = \dfrac{4}{52} + \dfrac{4}{52} = .154$

3. .137 ☐ $p(J) + p(Q) = \dfrac{4}{51} + \dfrac{3}{51} = .137$

4. .020 ☐ $\dfrac{26}{52} \times \dfrac{2}{51} = .020$

5. .057 ☐ $\dfrac{8}{52} + \dfrac{19}{51} = .057$

6. .143 ☐ $\dfrac{7}{49} = .143$

7. .006 ☐ $\left(\dfrac{2}{52} \times \dfrac{2}{51}\right) + \left(\dfrac{4}{52} \times \dfrac{3}{51}\right) = .006$

8. .400 ☐ $\dfrac{4}{52} + \dfrac{4}{51} + \dfrac{4}{50} + \dfrac{4}{49} + \dfrac{4}{48} = .400$

SCORE: /8

308   STATISTICS IN CONTEXT

Think About It...

PROBABILITY THEORY AND CARD GAMES

Gambling, often with decks of cards, has been a part of human culture the world over. There are almost as many games of chance as there are cultures on our planet. And there are a number of very different kinds of cards these games are played with.

The traditional Western deck is one of the smaller sets. We're used to seeing 52 cards, ranging in value from an ace to a king and divided into four suits. The idea of suits and varying values assigned to each card is fairly universal, but the number of suits and the number of cards within a suit vary considerably. For example, in India, "Ganjifa" cards look quite different from the ones we're accustomed to in the United States. First, the cards are round rather than rectangular. The number of suits in a deck also varies from region to region in India, with 8 suited decks in some places and up to 12 suited decks in others.

My personal deck of Indian playing cards is a 10-suit deck, with 12 cards in each suit. Two cards in each suit are "court cards" (kings and queens), and the rest are number cards. Each of the 10 suits represents a different incarnation of the Hindu god Vishnu, and each suit has its own sign:

- Suit 1 (Matsya, the first incarnation): a fish
- Suit 2 (Kurma, the second incarnation): a turtle
- Suit 3 (Varaha, the third incarnation): a boar
- Suit 4 (Narasimha, the fourth incarnation): a lion
- Suit 5 (Vamana, the fifth incarnation): a drinking vessel or an umbrella
- Suit 6 (Parashurama, the sixth incarnation): an ax
- Suit 7 (Ramachandra, the seventh incarnation): an ax or monkeys
- Suit 8 (Krishna, the eighth incarnation): a cow or a "Chakra quoit," a disc-shaped weapon
- Suit 9 (Buddha, the ninth incarnation): a lotus
- Suit 10 (Kalkin, the tenth incarnation): a horse or a sword

1. How many cards are there in my deck of cards? _____

2. Let's compare the odds of being dealt the following cards in the Ganjifa and traditional Western decks. Each row represents a card being dealt to you, without replacement.

	Western Deck	Ganjifa Deck
First card dealt	A club _____	A turtle _____
Second card	A club _____	A turtle _____
Third card	A club _____	A turtle _____

Think About It... continued

	Western Deck	Ganjifa Deck
Fourth card	A club _____	A turtle _____
p(4 of a kind)	_____	_____

3. With which deck is being dealt four of a kind less likely? _____

Answers to this exercise can be found on page 321.

PROBABILITY IN CONTEXT

The heart of descriptive statistics could probably be called *demographics*: using numbers to describe the people in a society. Likewise, the heart of inferential statistics is *probability*: using numbers to predict the likelihood of events in a population. The rules of probability were first described by people trying to win at games of chance. But while probability may have started out as part of a game, it soon became part of science. Today, it is used to determine if a conclusion drawn about a sample is a good conclusion or a bad one.

We will be treating samples drawn from a population as good representations (meaning they are "unbiased representations") of that population (we will see how to draw a random sample in Chapter 8). The statistical tests that we use with these samples will allow us to determine if a given sample is probably representative of the population or, alternatively, if it is so unusual that it cannot be considered representative of the population. If it is really unusual, even though it is unbiased, we get to start asking some interesting questions about why it is so unusual.

We will be using the basic rules of probability when we discuss inferential statistics. What you need to know to move on to inferential statistics is just what we have discussed here:

- The idea of samples being subsets of the universal set or population
- The AND rule
- The OR rule
- The relationship between frequency and probability

Let's consider sleep and college students one more time in an example of how science uses of probability. How would a scientist approach the question of whether or not college students are sleep-deprived? Brick, Seely, and Palermo (2010) looked at a group of students with a reputation for not getting enough sleep: medical students. These researchers wanted to know if medical students were actually sleep-deprived. To find out, they surveyed 341 medical students between the ages of 21 and 43 and asked them to complete the Pittsburgh Sleep Quality Index (PSQI). Developed by Buysse, Reynolds, Monk, Berman, and Kupfer (1989), the PSQI is a questionnaire designed to measure both sleep quantity and quality. Scores on the PSQI can range from 0 to 21 points. High scores indicate poor sleep, and scores higher than 5 indicate a possible clinical sleep problem. Good sleepers (people getting enough good-quality sleep) score about 2.67 on the PSQI, with a standard deviation of 1.70.

Brick et al. (2010) hypothesized that if medical students were not getting enough sleep, their individual PSQI scores (for all 341 students) should be higher than "normal." The mean for the whole sample should, in turn, be higher than the average PSQI score for the population of normal sleepers (2.67 ± 1.70).

Everyone needs sleep, but we don't all need the same amount. If we were to look at the distribution of PSQI scores in the population (for *everyone* who sleeps, not just medical students), we would find something that resembles the distribution shown in Box 7.9. The arrow at the center of the distribution points to the hypothetical average PSQI score in this population. There's some variability in the distribution, meaning that the PSQI score varies from person to person. Notice also that the distribution is normal in shape. Why did I assume the distribution of PSQI scores would be normal? Remember what Quetelet said about the distribution of most biological characteristics: Typically (I could say *theoretically*), they are normally distributed, with lots of scores around the mean and relatively few at the extreme high and low ends of the distribution. The distribution of PSQI scores in the entire adult population is our *theoretical* set. PSQI scores are measures of a biological characteristic, and if we could measure them in the entire population, they should behave like statistical theory suggests—in other words, they should be normally distributed.

BOX 7.9 **The distribution of PSQI scores in the population**

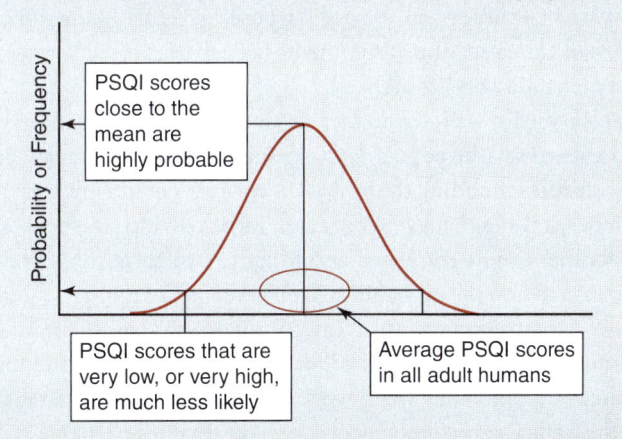

The set of 341 PSQI scores in our sample is our *empirical* set. These students actually took the PSQI, and the researchers calculated an empirical mean from their data. If medical students are no more sleep-deprived than anyone else, there should be a very good chance (a high probability) that the average PSQI score in the sample would be close to the mean level in the population. Do you see why we can make this prediction?

Most of the PSQI scores in the population are close to the population mean in value. It would be unusual, although not impossible, to pick out a sample of people and find a PSQI score that was extremely high unless all or most of the people in our sample didn't get enough sleep. If we found a sample mean that was way out in the upper tail of this distribution, what would we assume about the people in our sample?

You can think of a *frequency distribution* as being the same as a *probability distribution*. Frequent and highly probable PSQI scores would be near the average score in the population. Very high PSQI scores, or very low ones, would be very rare—unless there happened to be something odd or different about our sample. Scientists just love it when samples look odd or different—in other words, when there is a big difference between the theoretical and the empirical. When they find an odd sample mean, they start asking all kinds of interesting questions about *why* the sample mean is odd or different. Was the PSQI score in the sample really high because all of the people in the sample are sleep-deprived medical students? Is something else going on in the sample making the scores look different? Have we made an error of some kind? In subsequent chapters, we'll use basic probability theory and logic like this to help us determine the probability of sample means in order to make conclusions about specific hypotheses when we use inferential statistics.

But wait, you say: What did Brick and colleagues find out about the supposedly sleep-deprived medical students? To find out, you will need to complete the *Practice Problems* at the end of this chapter.

SUMMARY

Probability and frequency are, for our purposes, essentially the same thing. The more frequent an event is in a set of measurements (a deck of cards, a sample, or a population), the higher the probability of that event.

Sets are defined as well-defined collections of objects or events. In statistics, we often deal with the universal set (the population) and subsets of that universal set that we call samples.

The basic rule for finding the probability of an event in a given set is to work out the ratio of the number of times that event occurs in the set divided by the total number of events.

Probabilities can be combined according to two basic rules: The AND rule states that when you want to find the probability of one event *and* another, you multiply the individual probabilities of each event together. The OR rule states that when you want to find the probability of one event *or* another, you add the individual probabilities together.

Researchers commonly work with both theoretical and empirical sets of measurements. A theoretical set is the set of observations that researchers expect because, in theory, those observations ought to occur. An empirical set is the set of events or observations the researcher actually observes.

TERMS YOU SHOULD KNOW

AND rule, p. 302
conditional probability, p. 301
dependent events, p. 301
disjoint, p. 301
empirical set, p. 306
gambler's fallacy, p. 297
independent events, p. 301

mutually exclusive, p. 301
OR rule, p. 303
set, p. 296
set theory, p. 296
subsets, p. 297
theoretical set, p. 306
universal set, p. 297

GLOSSARY OF EQUATIONS

Formula	Name	Symbol
$p(A) = \frac{\#(A)}{\#(S)}$	Probability	$p(A)$
$p(A \cap B) = p(A) \times p(B \mid A)$	The AND rule	$p(A \cap B)$
$p(A \cup B) = p(A) + p(B)$	The OR rule	$p(A \cup B)$

PRACTICE PROBLEMS

1. In the *Probability in Context* section of this chapter, you read that probability can tell the researcher about the likeliness that a particular sample mean will occur within a given population. Why do researchers use the sample mean, rather than some other descriptive statistic calculated from the sample?

2. In the Mah Jong deck described in Table 7.1 (p. 307), find the probability of the following:
   a. Drawing an honor card
   b. Drawing any dragon card
   c. Drawing a 9 of bamboo card
   d. Drawing a 1 of bamboo, a 1 of wheels, a 1 of numbers

3. The Ganjifa deck of cards, described in the *Think About It . . .* feature on page 309, varies from region to region in India. One version has 96 cards, divided equally into eight suits: white or silver coin, crown, sword, slave, red or gold coin, harp, merchandise, and bill of exchange. Use this information to answer the following questions:
   a. How many cards are in each suit?
   b. What is the probability of drawing first a crown card and then a harp card?
   c. What is the probability of drawing either a sword card or a red coin card?

4. Today, you are washing socks. Everyone knows that the dryer eats socks, so let's figure out the following probabilities, which may help us determine if it's time to find a new dryer:
   a. You put the following nine pairs of socks into the dryer: three pairs of black socks, two pairs of blue socks, one pair of red socks, and three pairs of white socks. What is

the probability that the first sock you take out of the dryer (removing them one at a time) is a blue sock?
b. If you take your socks out of the dryer one at a time, how many socks will you have to remove to guarantee a pair of socks of the same color?
c. What is the probability that you will remove all of the white socks in a row (first a white sock, then another white sock, then another white sock, etc.)?

5. Table 7.2 shows the results of a study of probability by Wolf conducted in 1882 (cited in Hand, Daly, Lunn, McConway, and Ostrowski, 1994). Wolf tossed one die 20,000 times and recorded the number of times each of the six possible faces showed up.

Table 7.2 Results of 20,000 rolls of a die

Face	1	2	3	4	5	6
Frequency	3,407	3,631	3,176	2,916	3,448	3,422

Answer the following questions about the data:
a. What is the probability of rolling a:

1_____   4_____
2_____   5_____
3_____   6_____

b. Do you think the die was "fair" or "biased"? Why?

6. Norton and Dunn (1985, cited in Hand, Daly, Lunn, McConway, and Ostrowski, 1994) examined the relationship between snoring and risk of heart disease. They surveyed their participants and classified them according to the amount of snoring reported by their spouses. The participants were also categorized according to whether they had been diagnosed with heart disease. Their data are shown in Table 7.3.

Table 7.3 Snoring and heart disease in a sample of 2,484 adult males

Heart Disease	No Snoring	Occasional Snoring	Snoring Nearly Every Night	Snoring Every Night
Yes	24	35	21	30
No	1355	603	192	224

Use these data to answer the following questions.
a. How many participants suffer from heart disease?
b. What is the probability that a participant in this study suffers from heart disease?
c. How many participants snore nearly every night?
d. How many participants do not snore at all?
e. What is the probability that a participant in this study does not snore?
f. What is the probability that a participant in this study snores at least to some degree?
g. How many participants snore nearly every night or every night?
h. What is the probability of a participant snoring every night or nearly every night?
i. What is the probability of snoring and suffering from heart disease?

j. Norton and Dunn hypothesized that participants who snore would be more likely than participants who do not snore to suffer from heart disease. Do you think that the results of their study supported their hypothesis?

7. In a 1985 study, Breslow (cited in Hand, Daly, Lunn, McConway) and Ostrowski, 1994) examined the death rates from coronary heart disease in a set of British male physicians. He also classified the doctors according to whether or not they smoked cigarettes. Table 7.4 shows the number of physicians who died from coronary heart disease.

Table 7.4 Age, smoking, and heart disease in a sample of male physicians

Age (in years)	Nonsmokers	Smokers
35–44	2	32
45–54	12	104
55–64	28	206
65–74	28	186
75–84	31	102

Use the data to answer the following questions:
a. How many doctors were surveyed?
b. How many of these doctors reported that they smoked? How many reported that they did not smoke?
c. What is the probability that a person between the ages of 35 and 44 will die of coronary disease? What is the risk that a smoker in this age group will die of heart disease?
d. What is the probability that a smoker between the ages of 75 and 84 will die of coronary disease?
e. At what age is the probability of dying of coronary disease the highest for smokers?

8. Let's return to the Brick et al. (2010) study of sleep quality and quantity in a very large sample of medical students. The data shown in Table 7.5 are similar to what Brick et al. found, although with a more manageable n.

Table 7.5 PSQI scores and sleep durations for 20 medical students

Student ID	PSQI Score	Sleep Duration (in hours)
1	7.15	7.46
2	3.40	9.29
3	6.42	6.84
4	10.65	7.82
5	2.71	6.15
6	5.07	5.53
7	6.93	6.31
8	1.00	6.93

Table 7.5 *Continued*

9	9.28	8.40
10	9.60	5.86
11	7.74	4.92
12	5.42	6.84
13	11.91	6.37
14	7.63	5.81
15	6.51	6.52
16	3.94	6.43
17	7.17	7.01
18	7.35	7.43
19	2.63	8.47
20	5.27	8.06

Use the data to answer the following questions:
a. What is the average PSQI score in the sample?
b. What is the standard deviation of the PSQI scores in the sample?
c. What is the average sleep duration in the sample?
d. What is the standard deviation of the sleep durations in the sample?

9. Figure 7.3a shows the theoretical normal distribution of PSQI scores in the population of adult humans, and Figure 7.3b shows the theoretical distribution of sleep durations in the population (obtained from the National Sleep Foundation's "Sleep in America Survey," 2006).
   a. Convert the mean PSQI score from the sample (in question 8) into a *z*-score. If the population is normal in shape, then we should be able to use the *z*-table in Appendix A to predict the probability of an average PSQI score, like the one in the sample, in the population.
   b. Indicate the approximate position of the sample mean PSQI score on the appropriate population distribution. What is the probability of getting a PSQI score equal to the sample mean in this distribution?

FIGURE 7.3
Distributions of (a) PSQI scores and (b) sleep durations in the population.

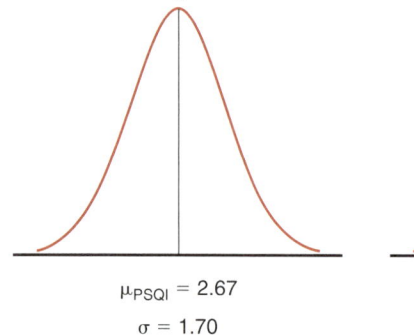

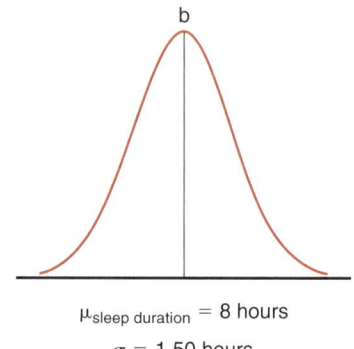

a
$\mu_{PSQI} = 2.67$
$\sigma = 1.70$

b
$\mu_{sleep\ duration} = 8$ hours
$\sigma = 1.50$ hours

c. Convert the mean sleep duration in the sample into a z-score in the population of sleep durations. What is the probability of getting a sleep duration equal to or higher than the sample mean in this population?
d. Indicate the approximate position of the sleep duration sample mean on the appropriate population distribution.
e. In your opinion, is there any evidence that the medical students were meaningfully different in either sleep duration or PSQI score from the rest of the population?

10. In 2011, the Centers for Disease Control and Prevention (CDC) report the following distribution of a selected set of symptoms of emotional disturbance in the population of US adults 18 years or older. The symptoms the CDC asked about were:
    - Feelings of sadness (all of the time or some of the time)
    - Feelings of hopelessness (all of the time or some of the time)
    - Feelings of worthlessness (all of the time or some of the time)
    - Feeling that everything was an effort (all of the time or some of the time)

Table 7.6 Emotional disturbance in the population of US adults, 18 years or older

Sex	All Adults	Sadness All or Most of the Time	Sadness Some of the Time	Hopeless All or Most of the Time	Hopeless Some of the Time	Worthless All or Most of the Time	Worthless Some of the Time	Effort All or Most of the Time	Effort Some of the Time
Male	113,034	2,287	6,608	1,689	4,204	1,792	3,199	5,162	7,730
Female	121,850	3,901	10,995	2,983	5,478	2,387	4,558	7,319	10,914
Totals									

a. How many people in total were surveyed by the CDC?
b. How many women reported any of the symptoms of mental illness in this survey?
c. How many men reported any of the symptoms of mental illness in this survey?
d. What is the probability that a woman will report a symptom of mental illness?
e. What is the probability that a man will report a symptom of mental illness?
f. What is the probability that a man will report feeling sadness of some type?
g. What is the probability that a woman will report feeling helpless some of the time?
h. What is the probability that either a man or a woman will report that things are an effort all or most of the time?

11. Suppose a small college has 500 freshmen, 400 sophomores, 300 juniors, and 200 seniors. You randomly select one student from the school. What is the probability that the one student you select at random will be:
    a. A freshman
    b. A senior
    c. A sophomore or a junior
    d. Not a senior

12. Pygmies live in central Africa in a region called Ituri (after the river by that name). Colin Turnbull lived with a group of pygmies during the early 1950s and told of his experiences in a book called *The Forest People* (1962). Turnbull reports that pygmies,

on average, are less than 4 feet 6 inches tall. For each of the questions below, assume that the height of pygmies is normally distributed with a mean of 4 feet 3 inches and a standard deviation of 2 inches.
   a. Pygmies live in huts made up of a framework of branches covered with leaves. Materials can be gathered and the hut constructed in an afternoon. If the opening is 4 feet high, what proportion of the pygmies will have to duck to enter?
   b. Pygmy culture has few rules. I will make one up in order to have a question. Suppose only those pygmies who were between 4 feet 2 inches and 4 feet 6 inches tall are allowed to sing in a coming-of-age ceremony. What is the probability that a pygmy selected at random would be allowed to sing at the ceremony?
   c. In the group Turnbull lived with, there was a main group of about 20 families (65 people) plus a subsidiary group of about 4 families (15 people). The subsidiary group was led by Cephu. If one person from the camp were chosen at random, what is the probability that the person would be from Cephu's subgroup?

13. One hundred slips of paper bearing the numbers from 1 to 100 are placed into a large hat and thoroughly mixed. What is the probability of drawing:
    a. The number 17 _____
    b. The number 92 _____
    c. Either a 2 or a 4 _____
    d. An even number, and then a number from 3 to 19 inclusive (the number drawn first is replaced before the second draw) _____
    e. A number from 96 to 100 or a number from 70 to 97 _____

14. The following questions refer to the throw of one die:
    a. What is the probability of obtaining an odd number on one throw if the die is "fair"?
    b. What is the probability of obtaining seven odd numbers in seven throws if the die is fair?

15. The following questions refer to the throw of two dice:
    a. What is the probability of rolling a 1 on each die?
    b. What is the probability that the sum of the two dice will be a number larger than 8 if the number on the first die is 6?
    c. What is the probability of rolling a 6 on one die and either a 2 or a 3 on the other?

16. Eighty-one Americans (all adults) are asked to name their favorite team sport. The results are shown in Table 7.7.

Table 7.7 Favorite team sport named by 81 adult Americans

Sport	Number Favoring the Sport
American football	39
Basketball	15
Baseball/softball	15
Soccer	8
Ice hockey	4

A person from this sample is selected at random.
   a. What is the probability of selecting someone whose favorite sport is American football?
   b. What is the probability of selecting someone whose favorite sport is ice hockey?
   c. What is the probability of selecting someone whose favorite sport is either American football or soccer?

17. The exam scores of a group of 200 students are normally distributed with a mean of 75 and a standard deviation of 5. A student from this group is selected randomly.
    a. Find the probability that the student's grade is greater than 80. (Hint: Convert 80 to a z-score first.)
    b. Find the probability that the student's grade is less than 50.
    c. Find the probability that the student's grade is between 50 and 80.
    d. Approximately how many students have grades of greater than 80?

18. Suppose you had a normal distribution with a mean of 50 and a standard deviation of 10.
    a. What percentage of scores are at least 69.6 or more?
    b. What percentage of scores are 66.4 or less?
    c. What is the probability of randomly selecting a score in this distribution that is 2.33 standard deviations or more above the mean?

19. The average litter of border collie puppies has a mean of six puppies and a standard deviation of two puppies. Agueda, who breeds and trains border collies as search-and-rescue dogs, delivered a litter of 10 pups. Two of the pups are white and brown, one is white and black, and the remaining seven are white, brown, and black.
    a. What is the probability of randomly selecting a puppy with brown fur but no black fur?
    b. What is the probability of randomly selecting a puppy with brown or black fur but not brown and black fur?
    c. What is the probability of randomly selecting a puppy with both brown and black fur?
    d. What is the probability of randomly selecting two puppies that are white and brown only, on the first two selections?
    e. What proportion of border collie litters yield 10 puppies or more?

20. Professor Lee's Introduction to Social Statistics class has 200 registered students. The class average for the midterm is 72% with a standard deviation of 8%. During the term, 70% of Professor Lee's students attend class regularly. However, on the last day of class, when Professor Lee provides exam advice, 90% of students attend (i.e., the regular 70% and an additional 20%, who are all new faces). During the last class, Professor Lee distributes course evaluation forms and randomly selects 10 students to complete extended course evaluations. What is the probability that:
    a. The first student Professor Lee selects will have regularly attended the lectures?
    b. The first student Professor Lee selects will have attended only one lecture?
    c. All 10 students Professor Lee selects will have regularly attended the lectures?
    d. The first student Professor Lee selects will have earned a grade of 85% or higher on the midterm?
    e. The first student Professor Lee selects will have earned a grade of either 80% or higher or 50% or lower?

21. Complete the crossword puzzle shown below. (Note: ' is an apostrophe—no letter is needed here.)

ACROSS

4. Events in the tails of a normal distribution are improbable or _____
7. (2 words) Belief that a past event can change the probability of a future event
9. (2 words) All possible events, outcomes, measurements, objects, etc
11. When the occurrence of one event alters the probability of another event, the two events are _____
13. (2 words) $P(A \cup B) = P(A) + P(B)$
14. A collection of things, objects, or events

DOWN

1. Probability that depends on the occurrence of another event is called _____
2. (2 words) When nothing in set A is also in set B
3. _____ and probability are the same thing
5. An example of a universal set in statistics
6. When $P(A|B) = P(B)$, A and B are _____
8. (2 words) $P(A \cap B) = P(A) \times P(B|A)$
10. An example of a subset of the universal set
12. A portion of the larger set

320   STATISTICS IN CONTEXT

Think About It...

PROBABILITY THEORY AND CARD GAMES

SOLUTIONS

1. How many cards are there in my deck of cards? $12 \times 10 = 120$ **cards in total**
2. Let's compare the odds of being dealt the following cards in the Ganjifa and traditional Western decks. Each row represents a card being dealt to you, without replacement.

		Western Deck	Ganjifa Deck
First card dealt		A club $13/52 = $ **0.25**	A turtle $10/120 = $ **0.083**
Second card		A club $12/51 = $ **0.235**	A turtle $9/119 = $ **0.076**
Third card		A club $11/50 = $ **0.22**	A turtle $8/118 = $ **0.068**
Fourth card		A club $10/49 = $ **0.204**	A turtle $7/117 = $ **0.060**
p(4 of a kind)	$.25 \times .235 \times .22 \times .204 = $ **.0026**	$.083 \times .076 = .068 \times .06 = $ **.000020**	

3. With which deck is being dealt four of a kind less likely?   **It would be much less likely to be dealt four of a kind in the Ganjifa deck. In the traditional Western deck, 25% of the cards are clubs (our target suit), while in the Ganjifa deck, the target suit makes up only 8.3% of the total.**

REFERENCES

Brick, C. A., Seely, D. L., & Palermo, T. M. (2010). Association between sleep hygiene and sleep quality in medical students. *Behavioral Sleep Medicine, 8,* 113–121.

Bruce, C. (2001). *Conned again, Watson: Cautionary tales of logic, math and probability.* Cambridge, MA: Perseus Publishing.

Buysse, D. J., Reynolds, C. F., Monk, T. H., Berman, S. R., & Kupfer, D. J. (1989). The Pittsburgh Sleep Quality Index (PSQI): A new instrument for psychiatric research and practice. *Psychiatry Research, 28*(2), 193–213.

Centers for Disease Control and Prevention. (2011). *National Health and Nutrition Examination Survey, Mental Health Depression Screener (DPQ_F).* Retrieved from https://wwwn.cdc.gov/nchs/nhanes/2009-2010/DPQ_F.htm#DPQ010

Hand, D. J., Daly, F., Lunn, D., McConway, K., & Ostrowski, L. (Eds.). (1994). *Handbook of small data sets.* London: Chapman and Hall.

National Sleep Foundation. (2006). *Sleep in America Poll.* Retrieved from http://www.sleepfoundation.org.

Turnbull, C. (1962). *The forest people.* New York: Simon and Schuster.

CHAPTER EIGHT

THE CENTRAL LIMIT THEOREM AND HYPOTHESIS TESTING

Chance is the pseudonym God uses when He'd rather not sign His own name.
—ANATOLE FRANCE

Everyday Statistics

EVER BEEN PART OF A POLL?

Another election season has come and gone, and the pollsters were out in force, trying to predict which candidate was going to win. Have you ever been called for this kind of survey? If so, then you've been a part of a random sample drawn from a population.

Statistics tells us that pollsters can't interview everyone; it just isn't feasible. Instead, they ask a set of randomly selected voters, all part of the voting population, and let these folks stand in for the rest of us. Typically, survey makers buy a Random Digit Dial (RDD) sample of cellphone and landline numbers. The interviewer then randomly selects a member of the household to be interviewed—the adult with the most recent birthday, for instance, or the youngest male in the home (young men frequently decline to participate in these surveys, so they're often targeted in efforts to increase their usually low representation in the sample). When news anchors tell us who leads the polls, they're relying on the results of random samples like these. We will learn more about random samples and how they're measured in Chapter 8.

OVERVIEW

INTRODUCTION: ERROR IN STATISTICS

THE CENTRAL LIMIT THEOREM

DRAWING SAMPLES FROM POPULATIONS

USING THE CENTRAL LIMIT THEOREM

ESTIMATING PARAMETERS AND HYPOTHESIS TESTING

HYPOTHESIS TESTING IN CONTEXT

LEARNING OBJECTIVES

Reading this chapter will help you to . . .

- Understand Kinnebrook's famous error, and explain how error is defined in statistics. (Concept)

- Explain the differences between descriptive statistics and inferential statistics. (Concept)

- Recite and interpret the three statements that make up the central limit theorem. (Concept)

- Understand the null and alternative hypotheses and directional and nondirectional hypotheses. (Concept)

- Write an appropriate null and alternative hypothesis. (Application)

- Know how to select a simple random sample from a population. (Concept)

- Describe how the central limit theorem is used in inferential statistics. (Concept)

INTRODUCTION: ERROR IN STATISTICS

What is an error? Does the size of the error determine whether it is important or not? How big an error can you make before you're in danger of being accused of being incompetent or—worse—a liar?

Most people think of an error as a simple mistake. Instead of sugar, you added salt to your coffee: That's an error. Statisticians, however, have a very specific meaning for the word and, as we will see in this chapter, have spent years working out ways to measure that error.

Dutch scientist Friedrich Bessel was struck by how two astronomers observing the same phenomenon could come up with different measurements of what they saw. He chalked this up to what he called the "personal equation" (see *The Historical Context* on page 325 for the full account). Researchers following Bessel discovered that a number of variables could affect our perception of time and create error in the measurements we make. These investigators also developed a definition of the term *error* for use in statistics. An **error** in statistics is the amount by which an observation differs from its expected value. Forget substituting salt for sugar: A statistical error would be expecting that a sugar cube contained one teaspoon of sugar and finding out that it actually contains 1.34 teaspoons, or 0.996 teaspoons.

error The amount by which an observation differs from its expected value.

INFERENTIAL STATISTICS

We ended the last chapter with a discussion of the difference between an empirical and a theoretical distribution. For our purposes, a *theoretical* distribution represented what ought to happen in, for example, an experiment, according to the rules of mathematics or logic. An *empirical* distribution represented the "actual" results of an experiment, regardless of how illogical or unlikely they might appear to be. **Inferential statistics**, defined as the branch of statistics dealing with conclusions, generalizations, predictions, and estimations based on data from samples, can help us in our quest to compare empirical and theoretical distributions.

inferential statistics The branch of statistics dealing with conclusions, generalizations, predictions, and estimations based on data from samples.

THE SCIENTIFIC METHOD

Science, as a method of finding out how the world and the people and things in it work, has adopted inferential statistics as one of its tools of choice. When we use a traditional "scientific method," we observe or study an aspect of the natural world, usually in a sample that we have drawn from the larger population. We measure this subset of the population and base our conclusions about how things work in that population on what happened in that subset.

On the surface, assuming something about the entire population based on just a sample of it seems a fairly foolish thing to do. After all, samples are generally very small sets of measurements, while the populations they come from are generally enormous. One might easily, and accidentally, miss something important in the population by selecting only a tiny piece of it to study. We need a link between samples, sample means, and the population. So, what exactly is the relationship between a sample and the population it was drawn from?

The Historical Context

KINNEBROOK'S ERROR AND STATISTICS

Let's look at the case of poor David Kinnebrook, who made one of the more famous errors in the history of psychology.

In 1796, Mr. Kinnebrook was employed as an assistant to Britain's Royal Astronomer, Sir Nevil Maskelyne. Several years earlier, Maskelyne had been asked to examine time itself as part of an international race to discover a viable method for finding the longitude of ships at sea (a process that required the development of very precise clocks).

In his attempt to develop a means of measuring longitude, Maskelyne asked Kinnebrook to measure what is known as the "transit" of a star. Kinnebrook was to record "the moment a given star passed the meridian wire in the Greenwich telescope" (Mollon & Perkins, 1996, p. 101). Unfortunately for Kinnebrook, his measures of this transit were not the same as Maskelyne's. In fact, Maskelyne accused Kinnebrook of making repeated and serious errors in his judgment of the time it took the star to pass the wire. Maskelyne even fired him as a result.

Kinnebrook's observation differed from Maskelyne's by 0.8 of a second, which translates to roughly a quarter of a mile along the equator—a deviation of an observation from the expected value that could have dire consequences to a sailor lost at sea.

What does an astronomer's mistake in measuring the movement of a star have to do with the history of psychology and the study of errors by statisticians? About 20 years later, Maskelyne was approached by a Dutch astronomer named Friedrich Bessel. Bessel was more interested in the errors that observers made than in the transit of the stars. Bessel became intrigued by the differences in measurement made by Kinnebrook and Maskelyne, and he conducted a series of experiments on "interobserver" differences, using himself and several friends as subjects. He found that these differences were the norm rather than the exception, and eventually developed a way to measure these differences that he called the "personal equation."

Kinnebrook's error may have been small, but it was incredibly influential. In their article, published on the 100th anniversary of the founding of psychology, Mollon and Perkins suggest that "[h]istorians have taken Kinnebrook's dismissal to be the event that gave birth to experimental psychology" (1996, p. 101).

THE CENTRAL LIMIT THEOREM
MEASURING THE DISTRIBUTION OF LARGE SETS OF EVENTS

I've mentioned before that the history of statistics is, in large part, the history of human attempts to measure uncertainty (a.k.a. variability or error). Take, for example, the French mathematician Abraham De Moivre, who was interested in estimating error in (what else?) games of chance. He described what he found through his investigations in one of the first books on probability theory, entitled *The Doctrine of Chance: A Method of Calculating the Probabilities of Events in Play* (published in Latin in 1711 and in English in 1718) and said to be highly appreciated by gamblers.

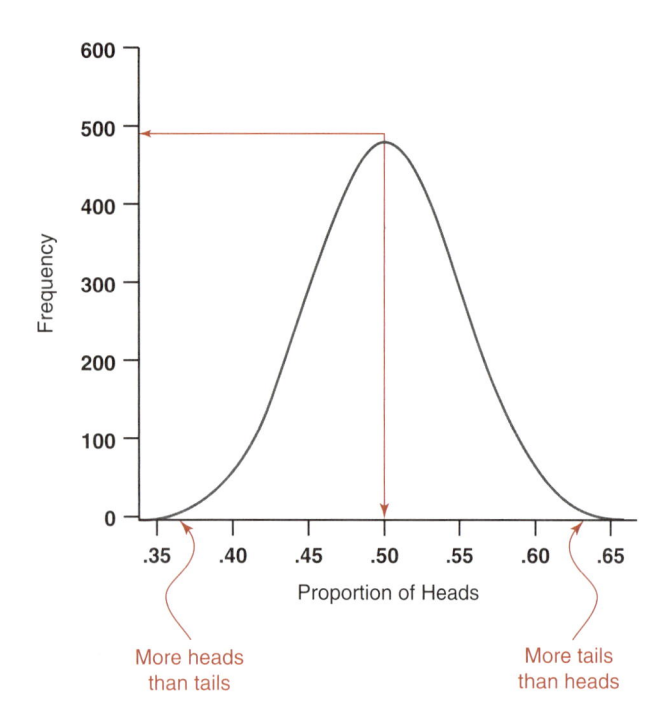

FIGURE 8.1
Representation of the results of De Moivre's experiment throwing sets of coins.

In one such investigation, De Moivre took a large number of coins and tossed them many, many times, recording the number of heads that came up as the result of each toss. But rather than performing the test once on all of the coins, he divided the coins into sets and performed the experiment with each set, so that he had a number of results rather than just one.

What De Moivre found was that the proportion of heads that appeared, after a massive number of tosses of these sets of coins, was normally distributed with a mean of 0.50. Figure 8.1 illustrates the results of De Moivre's coin-tossing experiment. On the *x*-axis, we have proportion of heads per set of coins tossed. On the *y*-axis, we have frequency, which, you will recall, can also be read as probability.

What De Moivre found was that most of the time, the set of coins up came about 50% heads and 50% tails. This is indicated by the red arrows, showing that the proportion of heads = .50. This outcome happened more often than any other: 50% heads and 50% tails was the most probable outcome for each set. Within some sets, however, there were more heads than tails, and within others, there were more tails than heads. It was this uncertainty about the outcome that De Moivre was interested in.

After examining the outcomes of many studies like this one, De Moivre concluded that generally, *sums of a large number of events will be normally distributed*. In this case, an event is either of the two possible results (heads or tails) of flipping a coin. The sum of events is the total of either result divided by the number of times the coin was flipped. The sets of coins that De Moivre tossed can be thought of as a series of samples (each is a subset of all possible coins), and by repeatedly tossing sets of coins and reducing each set to a single number that represents the sum of the number of heads in the set, De Moivre effectively reduced each sample to a single statistic.

Another French mathematician, Pierre Simon Laplace, took De Moivre's idea and generalized it, including other examples of sums—the mean, for instance. Laplace said that "any sum or mean . . . will, if the number of terms is large, be approximately normally distributed" (Stiegler, 1986, p. 137). This statement suggests that (theoretically) if you have a very large number of sums of events (a very large number of samples, and a distribution created out of the means of each of these samples), those *means* will be normally distributed. Remember Sherlock Holmes and the gambling nobleman in Chapter 7? This is the foundation of the nobleman's misunderstanding of probability. He just didn't understand how big a "very large number" of things actually was.

Notice that Laplace, like De Moivre, describes how large sets of *statistics* (numbers that describe some aspect of a set of observations of measurements) are distributed, not the distribution of the individual observations themselves. The *theoretical distribution* of all possible sample means will be normal in shape. There will be an average sample mean (the mean of this theoretical distribution) and a degree of error or variability.

THE LAW OF LARGE NUMBERS AND THE CENTRAL LIMIT THEOREM

Jacob Bernoulli, a Swiss mathematician, was working on another characteristic of large sets of measurements. Bernoulli proposed what has come to be called **the law of large numbers** (**LLN**), which predicts the results of doing the same experiment a large number of times. Basically, the LLN says that the mean of the results of a large number of repetitions of an experiment will be close to the mean of the population. The more you repeat the trials, the closer the mean result will be to the population mean. In essence, the LLN says that sample means, especially means of large samples, will almost always *be representative of* (i.e., be close to the value of) the mean of the population. If we don't know the population mean, we can assume that our empirical sample mean will be a good predictor of the theoretical population mean.

law of large numbers (LLN) The mean of the results of an experiment repeated a large number of times will be close to the mean of the population.

A number of other mathematicians (S. D. Poisson, Carl Friedrich Gauss, Adolphe Quetelet, Jerzy Neyman, Egon Pearson, just to name a few) refined the ideas that De Moivre, Laplace, and Bernoulli were playing with, and out of this work has come a theorem (an idea that has been demonstrated to be true or is assumed to be demonstrable) that is the foundation of inferential statistics. This theorem is called the central limit theorem (here, the word "central" means the same thing as "foundational").

The **central limit theorem** (**CLT**) is a series of statements that explain how statistics from any one empirical sample are related to the theoretical population parameters the sample came from. The CLT incorporates the conclusions that De Moivre, Laplace, and Bernoulli came to—that *a large set of sample means will be normally distributed*, and that *the means of the samples that make up this large set will approximate the mean of the population*. When we use inferential statistics to draw a conclusion about the population based on a sample mean, we are assuming that the central limit theorem is true and that we can justifiably use something discovered in a sample to draw an inference about what will be seen in the population as a whole. De Moivre, Laplace, and others laid the foundation for telephone pollsters everywhere: The average response to a question about voting preference made by members of a sample should be close to the average response in the whole population of voters, especially if the sample is very large.

central limit theorem (CLT) A series of statements that explain how any empirical sample mean is related to the theoretical population it came from.

The central limit theorem allows us to compare events in a sample with events in a population using an intermediary distribution—a distribution that you can think of as being between the population and a sample that came from the population. This distribution, called the sampling distribution of the means, is the distribution of sample means that Laplace and the others were describing.

CHAPTER 8 The Central Limit Theorem and Hypothesis Testing

Think About It...

THE LAW OF LARGE NUMBERS AND DICE GAMES

The illustration below shows what Bernoulli was talking about in his law of large numbers. Imagine rolling a single, six-sided die. Each side of the die has an equal probability to land facing up. You can roll a 1, 2, 3, 4, 5, or 6. The average roll of a single die, in theory, will be a 3.5.

1. How did I determine that the average roll of a single die is 3.5? _____
_____

If you roll a set of 10 dice, each die will land with a single face showing. On average, you should still roll a 3.5, even though empirically, you might get a sample of 10 numbers like this:

(6, 6, 3, 2, 5, 1, 3, 5, 3, 5; mean = 3.9)

Roll them again and you might get something like this:

(4, 3, 2, 4, 2, 5, 1, 6, 2, 1; mean = 3.0)

These two sample means are quite different, and neither is exactly 3.5 (the theoretical average we expect). Right now, we have made just a few rolls of the die, so we're in the boxed portion of the graph in Figure 8.2, which shows theoretical average value (on the *y*-axis) of the roll of just one die (the dark gray line) across many, many rolls of that die (the *x*-axis). Within this boxed portion, we can expect to experience a lot of fluctuation (error) in our estimates of the population mean. Notice that the theoretical average of just one die is always the same.

2. Mark the location of the two sample means calculated above on the *y*-axis of the graph.

3. How many rolls of the set of 10 dice (approximately) would you have to conduct before the mean roll was equal to 3.5 (the population mean)? _____
_____

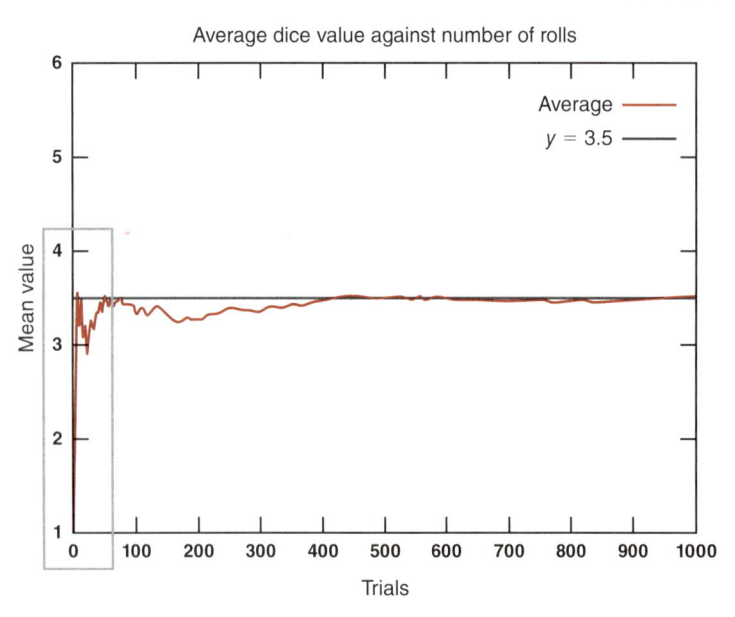

FIGURE 8.2
The average value of the roll of one die against the number of rolls made.

Answers to this exercise can be found on page 352.

328 **STATISTICS IN CONTEXT**

DRAWING SAMPLES FROM POPULATIONS

THE SAMPLING DISTRIBUTION OF THE MEANS

Most of the distributions we've been talking about so far have consisted of a set of measurements and a list of how often each measurement occurred in the set. We have displayed these distributions as tables (frequency distributions), where one column contains the individual measurements and the adjacent column lists the frequency of each measurement. We have also drawn pictures of these distributions, where the *x*-axis of the graph lists each individual measurement and the *y*-axis shows the frequency.

The **sampling distribution of the means** (**SDM**) is somewhat different. It is a *distribution of sample statistics*, usually the sample mean, rather than a distribution of individual measurements. Think again of our sleeping students. Rather than drawing a sample from one college and then tracing the distribution of those scores, we prepare an SDM by drawing samples from many colleges, calculating the mean of each sample, and graphing the distribution of those means. If you drew a picture of the SDM, as I have done in Figure 8.3, you would display a set of statistics (again, in a sampling distribution of the *means*, the statistics displayed are sample means) on the *x*-axis, with frequency (as always) on the *y*-axis.

The SDM is created by repeatedly *sampling* from the population. **Sampling** refers to *the procedure of selecting samples from the population*. There are two caveats in creating a sampling distribution of the means. First, all of the samples we draw must be the *same size*. Second, we must use a *random selection* of elements for each sample (more about this shortly). Figure 8.4 illustrates the steps you would use in creating a sampling distribution of the means.

sampling distribution of the means (SDM) A distribution of the means of all possible samples of a given size, selected at random from a given population.

sampling The procedure of selecting samples from the population.

FIGURE 8.3
Comparison of the elements in a population and in a sampling distribution of the means.

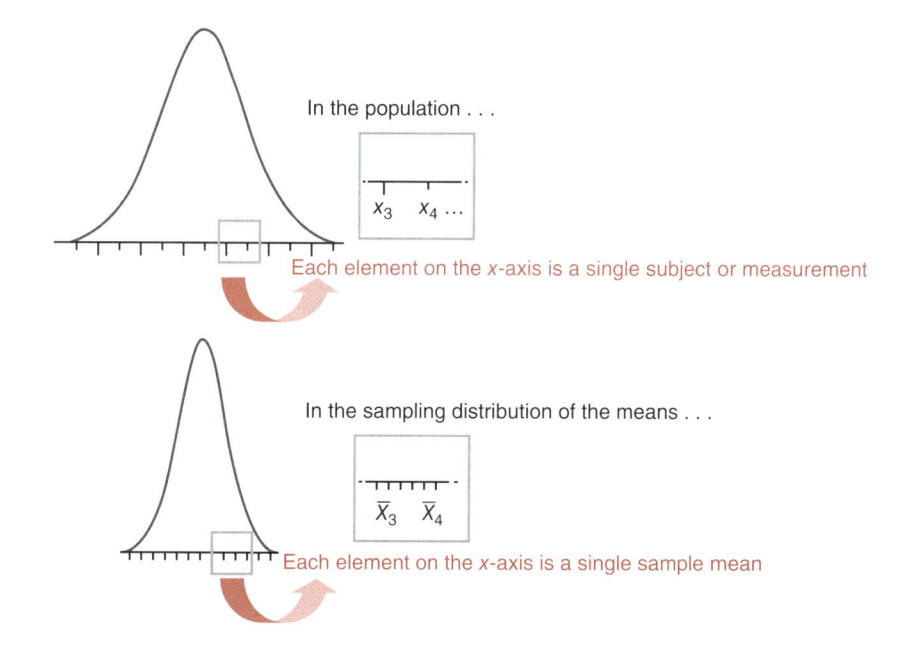

In the population . . .
x_3 x_4 ...
Each element on the *x*-axis is a single subject or measurement

In the sampling distribution of the means . . .
$\bar{X}_3$ $\bar{X}_4$
Each element on the *x*-axis is a single sample mean

CHAPTER 8 The Central Limit Theorem and Hypothesis Testing

Step 1. Collect *every possible sample of a given size* from the population.

↓

Step 2. Calculate the mean of each and every sample.

↓

Step 3. Determine the frequency of each sample mean.

↓

Step 4. Display each sample mean, in order from smallest to largest, on the *x*-axis of a line graph.

↓

Step 5. Display the frequency of each sample mean on the *y*-axis.

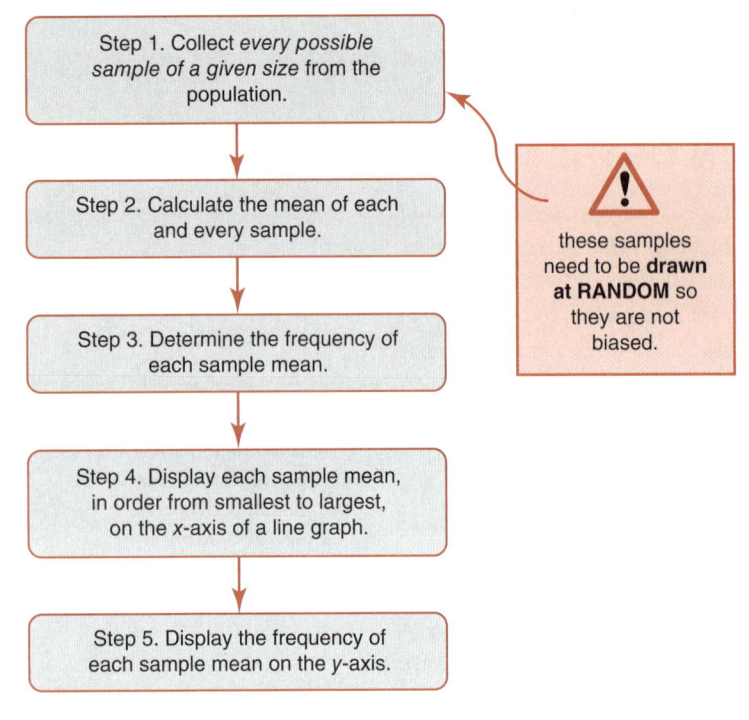

⚠ these samples need to be **drawn at RANDOM** so they are not biased.

FIGURE 8.4
The five steps involved in creating a sampling distribution of the means.

Both the population and the SDM typically are *theoretical* distributions. Scientists rarely, if ever, have the time or the money necessary to collect data from the entire population. Nor do they have the time to collect every possible sample from a population so as to create the entire SDM. The easiest way to solve the time crunch is to determine the relationship between, on the one hand, the population of all values and the SDM, composed of all possible sample means from all possible samples, and on the other hand, a single, potentially empirical sample. If the mean of your easy-to-collect empirical sample, which is one of the samples in the SDM, ultimately can tell us about the population the sample came from, then we can leave both of these theoretical samples just as they are—theoretical—and work with our empirical data. Letting random chance decide which observation, person, event, etc. will be in the sample reduces potential bias in the sample, making it a better representation of the entire population.

As you can see in Figure 8.4, the empirical sample—the set of measurements that we actually study—is just *one* of the samples in the SDM. This is because the single sample of measurements we obtain would be *one* of the samples we could have theoretically drawn from the population. The empirical sample that we study is related to the SDM, and it is also probably evident that the SDM is related to the population. After all, if you collected each and every sample from the population, you've essentially reproduced the population, but in what can be thought of as "bite-size" pieces.

THE THREE STATEMENTS THAT MAKE UP THE CENTRAL LIMIT THEOREM

From a common-sense perspective, then, the sample that we study should reflect the population it came from, especially if that sample is large. Statisticians examined these three related distributions (two theoretical and one empirical) and developed the central limit theorem to describe, from a mathematical perspective, how all three are related to one another. The CLT comprises three statements. Here's what the first statement says:

1. The distribution of sample means (the SDM) will "approach" a normal distribution as *n* (the sample size) increases, and the SDM will have a mean and a standard deviation.

In other words, if you created a sampling distribution of the means using reasonably large samples, the distribution of sample means that resulted when you graphed the frequency of each sample mean would be normal in shape. Notice that this statement is very similar to what De Moivre, LaPlace, and Bernoulli were saying about distributions of very large sets of measurements.

This distribution of sample means would have a mean (the mean of all the sample means, or a "mean sample mean") and a standard deviation, just like any distribution we've discussed so far. The symbols are used to refer to these statistics are:

- $\mu_{\bar{X}}$ for the mean sample mean
- $\sigma_{\bar{X}}$ for the standard deviation of the sampling distribution of the means

These symbols remind you of what these statistics represent—think of $\mu_{\bar{X}}$ as a reminder that this is a mean (μ) of a set of sample means ($\bar{X}$) and of $\sigma_{\bar{X}}$ as a reminder that the variability in the SDM is the variability (σ) among the sample means ($\bar{X}$). The standard deviation of the sampling distribution of the means ($\sigma_{\bar{X}}$) is also known as the **standard error** (thankfully). Figure 8.5 illustrates how samples are drawn from a population and how the mean of each sample makes up the SDM.

standard error The standard deviation of the sampling distribution of the means.

The second statement in the central limit theorem says:

2. The mean of the SDM (the mean sample mean) will always be equal to the mean of the population ($\mu = \mu_{\bar{X}}$).

This is a useful statement that will (I hope) make sense if you stop and think about it. If you pulled *each and every possible sample* from the population (thereby basically recreating the population, but in small chunks that are all the same size) and then used just the mean to describe each of those samples, the mean of all of those sample means will be exactly the same as the population mean. This is because we have included every possible element from the population in our SDM.

It is highly likely that *any* sample we draw at random from the population will have "come from" the center of the population. Remember, the 3-sigma rule tells us that 68% of the observations in the population are within one standard deviation of the mean of the population, so there is a good chance that an observation close to the population mean in value will be selected for any sample we draw, just because there are so many of them.

If most of the observations that make up any given sample are close to the mean of the population, it stands to reason that most of the samples will have means that are also close in value to the mean of the population. We can say that the mean of the empirical sample we study is probably close in value to the mean of the theoretical population the sample came from, just by random chance. Look at Figure 8.5, and notice the relationship between each sample mean and the population mean. Each sample mean is an estimate of the population mean, and bigger samples provide us with better estimates.

FIGURE 8.5
The central limit theorem (CLT) illustrated.

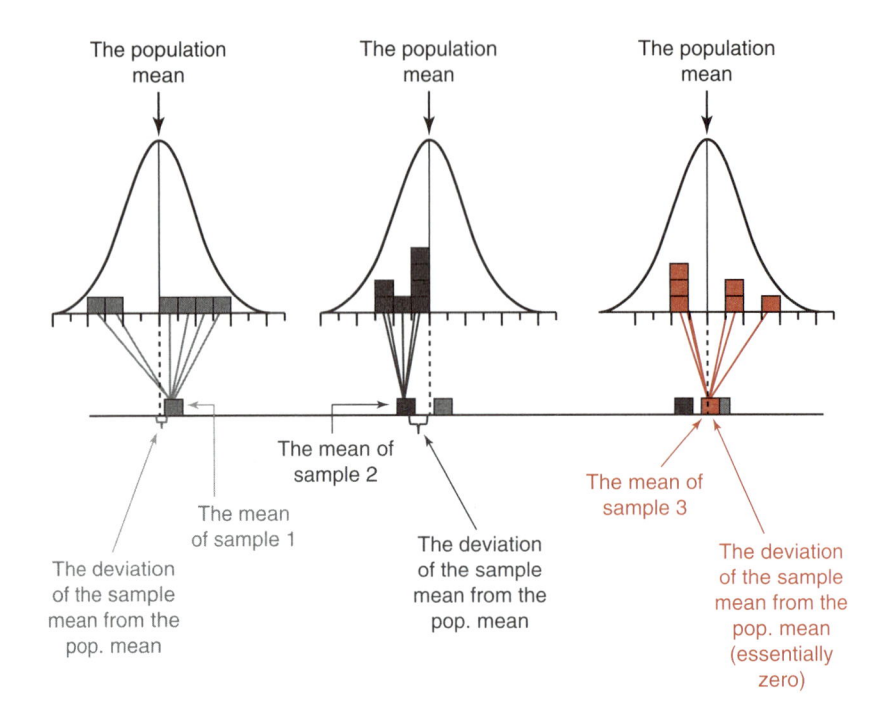

Finally, the third statement in the CLT says:

3. The standard deviation of the SDM (the standard error) will always be equal to the population standard deviation divided by the square root of the sample size $\sigma_{\bar{X}} = \dfrac{\sigma}{\sqrt{n}}$.

This means that the variability in the sampling distribution of the means (the variability among the means of each of the samples) is also related to the variability in the population the samples came from, just not in quite as straightforward a relationship as are $\mu_{\bar{X}}$ and μ. This statement tells us that *there will be less variability in the SDM than there was in the population*. We are dividing the population standard deviation by the square root of the sample size, so the result of this operation ($\sigma_{\bar{X}}$) will have to be smaller than σ.

Again, this should make some intuitive sense. We created the SDM by dividing the population up into small chunks of equal size. When we then take the next step and reduce each sample down to a single number (the sample mean, or $\bar{X}$), there should be less variability between each of those sample means than there was in the population.

The amount of "narrowing" in the SDM variability (compared to the amount of spread in the population) depends on how big each bite we took happens to be. If we take very large samples out of the population, then the means of each of the samples we took should be fairly close together ($\sigma_{\bar{X}}$ will probably be much smaller than σ because n was a large number). If we take tiny samples from the population, then our resulting SDM will have *a lot* of samples in it (remember, we have to take every possible sample of a given size from the parent population, and there will be more of them if they are small samples). The standard error won't be a whole lot smaller than σ because n was a small number. Take a look at Figure 8.5 once again, and see if you can visualize this last statement of the CLT.

CheckPoint

Calculate the standard error ($\sigma_{\bar{X}}$) using the information provided in the questions below. Round your answers to two decimal places.

1. $\sigma = 14.2$, $n = 40$ _____
2. $\sigma = 3.43$, $n = 30$ _____
3. $\sigma = 154$, $n = 50$ _____
4. $\sigma = 7.4$, $n = 16$ _____
5. $\sigma = 14.38$, $n = 100$ _____

Answers to these questions are found on page 334.

RANDOM SAMPLING

Before we go any further, we need to say a word about the warning sign (⚠) back in Figure 8.4. We draw empirical samples from the theoretical population because populations are usually too big and too changeable to measure directly. Good sampling procedures will give us good empirical samples to test by generating samples that are representative of the population. Generally speaking (and you will probably hear more about this in a research methodology class), good sampling of a population means using random sampling techniques.

The principle of **random sampling** is fairly straightforward. A random sample is one in which each element of the population has an equal chance of being selected. Putting random chance in charge of the selection process reduces bias in who or what gets to be in the sample. The experimenter, along with any expectations, wants, needs, and desires for a specific result, is removed from the process and the sample is less likely to reflect his or her bias. Random sampling is a *method* of sampling, not an *outcome* of that sampling procedure.

The easiest way to make sure that all of the elements of a population have an equal chance of being selected for membership in a given sample is the "lottery" technique. If your state has a lottery, you may have seen this technique in action. Think of the sample as the winning lottery number. It's composed of a set of elements (numbers) drawn from a larger population (a ridiculously large population of possible numbers, in the case of a lottery). To make sure the drawing is random (i.e., that all numbers have an equal chance of being selected for inclusion in the winning set), the numbers are painted on ping-pong balls, and the balls are then dropped into a hopper, mixed thoroughly, and "drawn" at random (an air current determines which ping-pong ball makes it up the chute and into the winning number set). It's like putting the procedure for drawing numbers from a hat on steroids.

Let's try creating a sampling distribution of the means with a fairly unusual population that I made up (I drew the numbers that make up the population at random from a random number table). This population, shown in Box 8.1, has only 20 observations in it (like I said, it's a very small and unusual population).

random sampling A procedure for generating a sample in which each element of the population has an equal chance of being selected.

> **CheckPoint**
> *Answers to questions on page 333*
>
> 1. $\sigma_{\bar{X}} = 2.25$ ☐
> 2. $\sigma_{\bar{X}} = 0.63$ ☐
> 3. $\sigma_{\bar{X}} = 21.78$ ☐
> 4. $\sigma_{\bar{X}} = 1.85$ ☐
> 5. $\sigma_{\bar{X}} = 1.44$ ☐
>
> **SCORE:** /5

BOX 8.1 — A tiny population

9	7	13	10
8	10	8	8
10	12	10	5
8	9	10	11

We are going to "sample" from this population, making sure that each sample has an *n* of 4. Next, we'll calculate the mean of each sample, and then we'll calculate both $\mu_{\bar{X}}$ and $\sigma_{\bar{X}}$ so that we can compare them to the population parameters (μ and σ). According to the central limit theorem, μ should be equal to $\mu_{\bar{X}}$, and $\sigma_{\bar{X}}$ should be equal to $\frac{\sigma}{\sqrt{n}}$.

First, we'll need to know what μ and σ are. Box 8.2 shows the calculations of the population mean and standard deviation. Try it on your own first, and then check your calculations.

BOX 8.2 — Calculating the mean and standard deviation of our tiny population

$$\sigma = \frac{\sum x}{n} = \frac{180}{20} = 9.00$$

$$\sigma = \sqrt{\frac{\sum x^2 - \frac{(\sum x)^2}{n}}{n}} = \sqrt{\frac{1{,}700 - \frac{(180)^2}{20}}{20}} = \sqrt{\frac{1{,}700 - 1{,}620}{20}} = \sqrt{\frac{80}{20}} = \sqrt{4} = 2.00$$

STATISTICS IN CONTEXT

Now we need to start drawing out samples of four elements each from the population. We'll sample "with replacement" using the lottery technique. We write the numbers that are in the population onto small pieces of paper, place the pieces of paper into a receptacle (I'm using a hat because I'm a stickler for tradition), mix them thoroughly, and then draw out 10 samples of four numbers each. Obviously, we will get a different result each time we do this; for the moment, let's use the results shown in Box 8.3, based on my own draw from the hat.

BOX 8.3   **10 samples of $n = 4$ each**

Samples	1	2	3	4	5	6	7	8	9	10
	11	11	13	8	10	8	13	6	10	6
	8	10	8	12	8	6	8	8	11	10
	8	11	11	10	8	9	11	9	6	7
	8	6	8	13	5	10	9	10	8	9

Sums and means for each sample:

	1	2	3	4	5	6	7	8	9	10
Sum	35	38	40	43	31	33	41	33	35	32

Divide by $n = 4$ for each sample:

	1	2	3	4	5	6	7	8	9	10
Mean	8.75	9.50	10.00	10.75	7.75	8.25	10.25	8.25	8.75	8.00

Notice that not all of the means are the same—there is variability among them. However, it should be relatively easy to find the mean of all of the sample means and the variability between sample means. Box 8.4 shows you these calculations.

With only 10 small samples ($n = 4$ in each) drawn from this population (and nowhere near all possible samples of $n = 4$), we're not doing too badly. The mean sample mean of our set of 10 sums is not exactly equal to 9.0 (it's 9.25), but we are only a little bit off. According to the central limit theorem, the standard error (the standard deviation of these sample means) should theoretically be equal to the population standard deviation divided by the square root of n $\sigma_{\bar{X}} = \frac{\sigma}{\sqrt{n}}$. So, the standard error should be

$$\sigma_{\bar{X}} = \frac{2}{\sqrt{4}}$$
$$= \frac{2}{2}, \text{ or } 1.00$$

CHAPTER 8   The Central Limit Theorem and Hypothesis Testing

BOX 8.4

Calculating the mean and standard deviation of the sample means

Mean of all the Sample Means	Standard Deviation of the Sample Means
$\dfrac{\sum \bar{X}}{n} = \dfrac{90.25}{10} = 9.25$	$\sqrt{\dfrac{\sum \bar{X}^2 - \dfrac{(\sum \bar{X})^2}{10}}{10}} = \sqrt{\dfrac{824.17 - \dfrac{90.25^2}{10}}{10}}$
	$= \sqrt{\dfrac{824.17 - 814.51}{10}}$
	$= \sqrt{\dfrac{9.66}{10}} = \sqrt{0.966}$
	$= 0.98$

The standard deviation of the sample means, based on just 10 samples of four taken from the population (again, nowhere near the total number of all possible samples that we could have taken), is 0.98, almost what we predicted it would be. We're off, but not by too much.

Even with just a few of the thousands of potential samples of four that we could have drawn from the population, we still come very close to the values the CLT predicts. If we continued to draw samples from the population (and feel free to do so), we would eventually find that $\mu_{\bar{X}} = \mu$ and $\sigma_{\bar{X}} = \dfrac{\sigma}{\sqrt{n}}$. In Box 8.5, I've added 20 more sample means (for a total of 30) to our growing SDM. Notice what happened to the mean sample mean and the standard error.

Adding 20 more samples to our sampling distribution of the means resulted in the mean sample mean (9.025) and the standard error (0.99) moving toward the value predicted by the CLT (9.00 and 1.00 respectively).

BOX 8.5

Samples and their means

Sample Number	Mean	Sample Number	Mean	Sample Number	Mean
1	8.75	11	9.00	21	11.00
2	9.50	12	9.50	22	8.00
3	10.00	13	8.75	23	9.25
4	10.75	14	9.50	24	10.00

5	7.75	15	10.50	25	8.75
6	8.25	16	7.50	26	9.25
7	10.25	17	9.75	27	8.75
8	8.25	18	7.25	28	9.50
9	8.75	19	9.75	29	8.75
10	8.00	20	7.00	30	8.75

Samples 1–10 are from Box 8.3.

$$\frac{\sum \bar{X}}{n} = \frac{270.75}{30} = 9.025$$

and

$$\sqrt{\frac{\sum \bar{X}^2 - \frac{(\sum \bar{X})^2}{n}}{n}} = \sqrt{\frac{2{,}473.02 - \frac{270.75^2}{30}}{30}}$$

$$= \sqrt{\frac{2{,}473.02 - 2{,}443.52}{30}}$$

$$= \sqrt{\frac{29.5}{30}} = \sqrt{0.9833} = 0.99$$

USING THE CENTRAL LIMIT THEOREM

The central limit theorem tells us that samples are related to the population that they were drawn from in two very specific ways:

1. In terms of the mean sample mean in the sampling distribution of the means.
2. In terms of the variability of the SDM, or how spread out individual sample means in the SDM are.

According to the central limit theorem, the mean of a single sample is usually a good, unbiased estimate of the mean of the SDM (especially when random selection has been used to create the sample). The mean of the SDM and the standard deviation of the SDM (the standard error) are, in turn, good estimates of the population mean and the population standard deviation, respectively. If we can describe the characteristics of our sample (and by now we certainly can), then we can make an educated guess about the population parameters that sample came from. This, as we'll soon see, makes testing hypotheses about what effects the independent variable might be having on a dependent variable possible.

ESTIMATING PARAMETERS AND HYPOTHESIS TESTING

So about now you may be asking, what exactly is **hypothesis testing**?

Typically, when you decide to set up a study, you have first developed a theoretical explanation of how the event or condition you are studying works. This theoretical explanation is usually based on previous or similar research and is called a **hypothesis**, from the Greek *hypotithenai*, meaning "to suppose."

For example, I might hypothesize that student sleep durations during finals are very short compared to their sleep times during other periods of time because students are stressed out, pulling all-nighters, and studying much more than they usually do—all factors that prevent sleep. Or, it could be true that students sleep the clock around when stressed, so our mean sleep time in stressed students could hypothetically be closer to 24 hours.

Suppose I hypothesize, based on my own experience and a review of the literature on sleep in college students, that stress makes college students sleep less than the typical, presumably unstressed student does. I would then set up my experiment and get ready to estimate parameters of theoretical populations.

Scientific hypotheses are usually formatted as "if . . . then" statements that refer to the effect that an independent variable (IV) has on a dependent variable (DV), like this:

> *If* the IV has an effect, *then* the mean value of the DV in the empirical sample will be some value quite different from the average sleep time in the population.

In terms of the experiment we have been considering, I might state our hypothesis as follows:

> *If* high stress levels decrease sleep duration, *then* students will spend less time sleeping when they're stressed than is average in the (unstressed) population.

The study that I conduct will look for evidence that supports my hypothesis that stress level affects time spent sleeping.

What if stress really does shorten sleep duration, making our sample mean very different from the population mean? If the sample mean is way out in the tails of the SDM, then we can conclude that the independent variable had a big effect on the dependent variable—that the IV caused a much larger effect on the DV than did just plain old random chance.

Notice that random chance and error are essentially the same thing in this situation. We need to be able to tell if the difference between our sample mean and the theoretical population mean is meaningful, or if it's just the result of random chance.

THE NULL AND ALTERNATIVE HYPOTHESES

Generally, scientists propose two very specific and mutually exclusive hypotheses. Remember from Chapter 7 that two events are mutually exclusive if the two sets do

---

hypothesis testing The procedures used to determine if the null hypothesis can be rejected or accepted in favor of the alternative hypothesis.

hypothesis A theoretical explanation made on the basis of limited evidence as a starting point for further research.

not share any elements. In this case, *mutually exclusive* means that the hypotheses cannot both simultaneously be true (for that matter, they cannot both simultaneously be false, either).

The first hypothesis is called the **null hypothesis**, (the symbol for the null hypothesis is H_0, or "*H* sub-zero"), where the word "null" refers to "nothing." The null hypothesizes about a specific population called the *null population*. Specifically, H_0 proposes that the mean of the null population (from which our empirical sample will be drawn) is exactly what other researchers have found it to be—and what we expect it to be. In the context of our sleep study, the null hypothesis would state that the average sleep time in the population our sample comes from is 8 hours. If the mean of our empirical sample deviates from 8 hours, that deviation is the result of random chance and random chance alone, not because stress at finals is exerting an effect on the subjects. You can see why the null hypothesis is also called the hypothesis of no effect.

Note that in the null hypothesis, we are estimating what we think the parameters of the population that our sample comes from might be. If the independent variable (stress) does nothing, then our sample represents, and should look the same as, the population of unstressed sleeping college students. And if the independent variable has no effect, then any variation that we see between our sample mean and our null hypothesis population mean represents error and is due to random chance alone.

The second hypothesis is called the **alternative hypothesis** (the symbol for the alternative hypothesis is H_1, or "*H* sub-one"). It refers to an explanation of the behavior or event we're studying that is an alternative to the null—in fact, it refers to an alternative theoretical population where our independent variable (stress) actually has an effect on sleep time. Since both hypotheses cannot simultaneously be true, the logical alternative to the null is that the independent variable *does* have an effect: Stress, and not just random chance, is at least part of the reason for our sample mean to deviate from what we theoretically expected to see in the population.

The alternative hypothesis suggests that the independent variable has had an effect, that it did change our sample, and that our sample therefore came from some other population with some other average sleep time—a *theoretical* alternative population made up of stressed college students that has a mean and standard deviation of something other than 8 hours per night. Figure 8.6 illustrates the thinking behind our experiment.

DIRECTIONAL HYPOTHESES

Because the null hypothesis always states that the independent variable has no effect, it falls to our alternative hypothesis to state

null hypothesis The hypothesis of no effect, which claims that the independent variable has no effect on the population. Symbol: H_0.

alternative hypothesis The hypothesis that makes the claim that the independent variable does have an effect on the population. Symbol: H_1.

The actual mean sleep time in the general population of college students = μ_0.

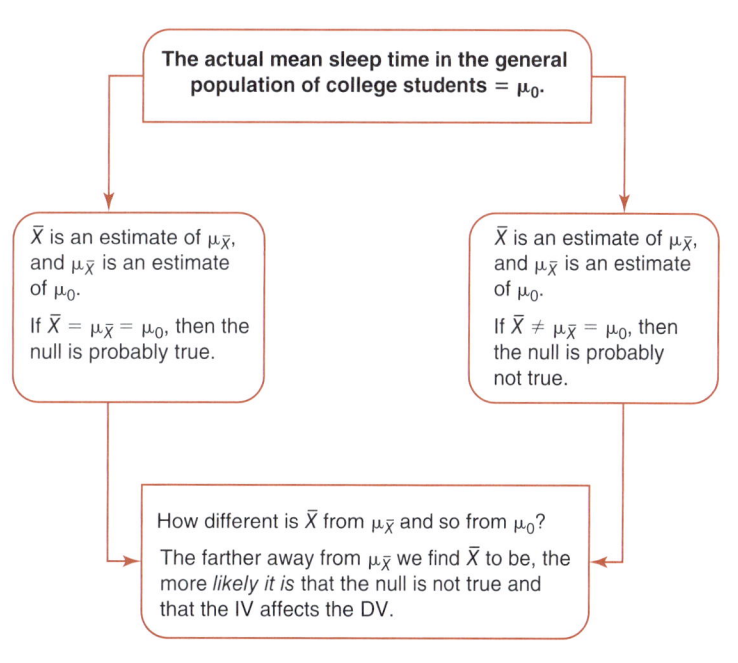

$\bar{X}$ is an estimate of $\mu_{\bar{X}}$, and $\mu_{\bar{X}}$ is an estimate of μ_0.

If $\bar{X} = \mu_{\bar{X}} = \mu_0$, then the null is probably true.

$\bar{X}$ is an estimate of $\mu_{\bar{X}}$, and $\mu_{\bar{X}}$ is an estimate of μ_0.

If $\bar{X} \neq \mu_{\bar{X}} = \mu_0$, then the null is probably not true.

How different is $\bar{X}$ from $\mu_{\bar{X}}$ and so from μ_0?

The farther away from $\mu_{\bar{X}}$ we find $\bar{X}$ to be, the more *likely it is* that the null is not true and that the IV affects the DV.

FIGURE 8.6
The logic of hypothesis testing.

specifically what kind of difference we expect the independent variable to make or how our two theoretical populations will differ. For example, suppose we had reason to expect that stressed-out students would sleep less than nonstressed students.

In our proposed experiment, the alternative hypothesis would state specifically that we think the mean sleep duration of the sample we studied will be *less than* the mean sleep duration hypothesized to exist in the null population. Our sample of stressed-out students no longer represents the null population. Instead, they represent some (theoretical) alternative population of stressed-out students who all have very short sleep durations. This is known as a **directional hypothesis** because we are specifying a direction for the difference between our sample mean and the hypothesized population mean.

On the other hand, we could set out to do our experiment with no idea about how the null and alternative populations might differ, and we would want our alternative hypothesis to reflect this. So, we would hypothesize that the sample of stressed-out students would have a mean sleep duration that was different from the mean in the population in some way, either longer or shorter. This would be a **nondirectional hypothesis**, stating that the null and alternative populations *are* different, but not suggesting *how* these means are different. Box 8.6 shows how we would write out our two hypotheses.

Both hypotheses refer to the relationship between the means of two theoretical populations. If the independent variable has no effect, then μ_0 (the mean of the population referred to in the null hypothesis) and μ_1 (the mean of the population hypothesized about in the alternative hypothesis) overlap and have essentially the same means. If the null is false, then the means of the two distributions will be different.

directional hypothesis A form of the alternative hypothesis claiming that the effect of the independent variable will be to change the value of μ in a specific direction, either upward or downward, relative to the mean of the null population.

nondirectional hypothesis A form of the alternative hypothesis claiming that the effect of the independent variable will be to change the value of μ without specifying whether the value is expected to increase or decrease.

BOX 8.6 — **Null, directional alternative, and nondirectional alternative hypotheses**

$H_0: \mu_0 = \mu_1$ ⟶	The null hypothesis always proposes no difference between the two populations, that these two theoretical populations essentially overlap.
$H_1: \mu_0 < \mu_1$ or $H_1: \mu_0 > \mu_1$ ⟶	Directional alternative hypotheses propose a difference in a particular direction (above or below the mean).
$H_0: \mu_0 \neq \mu_1$ ⟶	Nondirectional alternative hypothesis proposes simply that the null is false and the means will not be equal in one direction or the other.

Figure 8.7 illustrates these populations and their relationships. Remember that populations, and the SDMs that are created by sampling from them, are theoretical. Only the sample of observations we actually obtain and study is empirical.

If the null is true, then the distributions described in the alternative and null hypotheses overlap one another and have identical means (dark pink and dark gray circles, respectively, in Figure 8.7). The relatively small difference between the means of the two populations is probably due to random chance rather than some real difference caused by the independent variable. However, if the samples in the SDM have been changed by the independent variable, then the mean of the SDM will reflect this. The SDM would then reflect the population of stressed-out college students in the alternative hypothesis. And the mean of the SDM would now reflect the mean of the population hypothesized about in the alternative (dark gray circle in Figure 8.7). This will happen if the null is not true.

In statistical terms, we set up these two mutually exclusive hypotheses, then we look for evidence that we can *reject the null hypothesis*. It sounds somewhat backward, but if you reject the null, you can say that you've found no evidence that the independent variable made no difference. If you *fail to reject the null*, then you do not have evidence that there is always no difference, and it's back to the drawing board for you and your hypothesis.

Generally speaking, the farther away from the population mean our sample mean is, the more likely that something other than just random chance is influencing our data, that our null hypothesis is false, and that our independent variable actually has had a meaningful effect on the subjects in our sample.

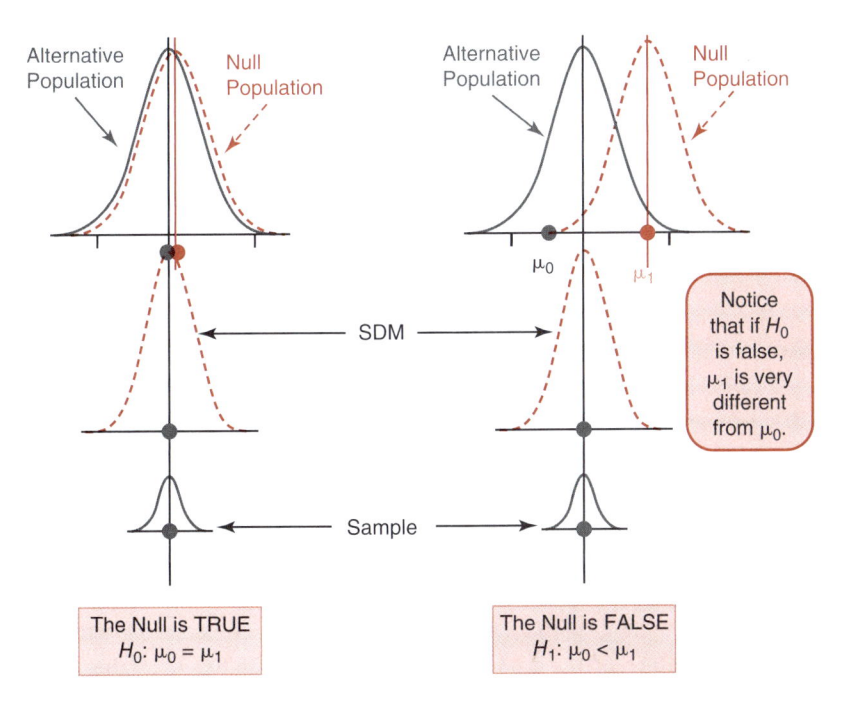

FIGURE 8.7
Comparing null and alternative hypotheses.

CHAPTER 8 The Central Limit Theorem and Hypothesis Testing   341

CheckPoint

Symbolize the alternative hypotheses that are stated here. Should they be *directional* or *nondirectional* hypotheses?

1. Ambitious persons score lower on measures of subjective well-being than do people in general. _____
2. The mean IQ score of college students in the psychology department is higher than the mean IQ score for the population of college students. _____
3. Persons with depression produce less of the neurotransmitter GABA than the average person does. _____
4. The mean sleep time of an American college students is different from the mean sleep time of Americans. _____
5. The compensation rate of white, male workers is higher than that of workers in general. _____
6. Men are less likely to have anxiety disorders than the average person. _____

Answers to these questions are found on page 344.

HYPOTHESIS TESTING IN CONTEXT

In science, hypothesis testing takes place within the context of research. When scientists do research, they have formulated a *model* of the natural world and are asking a question about that model. So, what's a model? Generally speaking, models are *formal statements of the relationship between variables that are expressed in the form of mathematical equations.* Usually, these models describe how one or more random variables are related to one or more other random variables. The model that we develop in hypothesis testing asks how an empirical sample is related to a theoretical population.

Let's use an example. Suppose I want to test the effectiveness of meditation in reducing stress. I will select a sample of 20 students (using random selection, of course) and teach them to meditate. I will then measure stress level in my sample and compare it to the stress level in the population. Since stress tends to decrease blood flow to the periphery of the body, I will use hand temperature as my measure of stress: The more stressed someone is, the lower that person's hand temperature should be.

I expect that stress levels, and so hand temperatures, will vary in the null population. Not everyone has exactly the same hand temperature—some people have cold hands without being stressed, others have warm hands without being relaxed. Most people will have hand temperatures close to the population average, however, and variation in hand temperature away from average will be the result of random chance (error). So, my null hypothesis is that the average hand temperature in my sample of meditators will be the same as the average hand temperature in the null population (because meditation has no effect on the stress response).

On the other hand, if meditation really does reduce stress, I should see that reflected in the hand temperatures of my sample. I should see that the average hand temperature of the students in my sample is higher than the average in the nonmediating null population. My sample represents the conditions in the alternative population—the population of unstressed, warm-handed meditators. The process that I would follow in carrying out my hypothesis test would look like this:

- First, I would identify the research question, starting with the independent variable. In this case, the independent variable is meditation (I want to see the effect of meditation on stress).
- Second, I would identify how I am going to measure the dependent variable. In this case, I have decided to measure hand temperature in order to see if meditation has an effect on stress. This is the parameter that I will be making hypotheses about.
- Third, I would write my hypotheses, both of which will be about hand temperature in a theoretical population:
  - The null hypothesis makes a prediction about the mean of the null population. The null hypothesizes that the IV has no effect at all: $H_0: \mu_0 = \mu_1$ (the mean of the null population is equal to the mean of the alternative population).
  - The alternative hypothesis makes a prediction about the mean of the alternative population. This hypothesis specifies the type of effect I think the IV has. In this case, I would write a directional hypothesis specifying that I think the independent variable (meditation) should increase the values of the dependent variable (hand temperature) $H_1: \mu_0 < \mu_1$ (the mean of the null population is less than the mean of the alternative population).
- Then, I collect the data (the hand temperatures of the 20 students in the sample) and compare them to the hypothesized mean of the null population.

I want to determine which of the two hypotheses I've written out is supported by the evidence. Think about how to determine if something is true—it's not as easy as it might seem at first glance. The evidence I have is a sample mean. And if the null is true, then the mean of the sample should be close to the mean of the null population—but how close? If the alternative is true, then the mean of the sample should be less than the mean in the population—but how much lower? What I really need to know is the answer to a fairly specific question: What is the probability that the average hand temperature of the meditators would be higher than the theoretical average in the null population just by random chance alone? If this probability is low, then it would be logical to assume that something else is contributing to the difference in sample means—something in addition to random chance. What is that "something"? Could it be meditation?

It's not easy to prove something is true. But what if we can show that one of the hypotheses is *not* true? In hypothesis testing, we look for evidence that the null hypothesis is not true, and that we can therefore *reject* it. If the evidence shows that we can reject the null, then we can logically accept the alternative. As we progress through the next several chapters, we will return to this idea of hypothesis testing and rejecting/accepting hypotheses.

CheckPoint Answers
Answers to questions on page 342

1. $H_1: \mu_0 > \mu_1$ (a one-tailed test) ☐
2. $H_1: \mu_0 < \mu_1$ (a one-tailed test) ☐
3. $H_1: \mu_0 > \mu_1$ (a one-tailed test) ☐
4. $H_0: \mu_0 \neq \mu_1$ (a two-tailed test) ☐
5. $H_1: \mu_0 < \mu_1$ (a one-tailed test) ☐
6. $H_1: \mu_0 > \mu_1$ (a one-tailed test) ☐

SCORE: /6

SUMMARY

We've covered quite a bit in this chapter, much of it essential to the way that inferential statistics are used in science.

We started with a discussion of the meaning of the term *error* in statistics. Errors are differences between the empirical evidence and the theoretical, predicted value. We then discussed De Moivre, Laplace and Bernoulli, who told us that very large sets of measurements generally are normally distributed, and that the mean of a sample will reflect the mean of the population it came from.

We then discussed a related theorem that deals with how big sets of numbers or statistics behave, called the central limit theorem (CLT). The CLT states that if we took every possible sample of a given size from the population using random sampling, calculated the mean of each of those randomly selected samples, and then plotted the mean of each sample and the frequency of each sample mean, we would produce a sampling distribution of the means, or SDM. This distribution would be normal in shape, and it would have a mean ($\mu_{\bar{X}}$, or mean sample mean) and a standard deviation ($\sigma_{\bar{X}}$, or standard error) of its own. The mean sample mean of this distribution would always be equal to the value of the mean of the population each sample came from ($\mu_{\bar{X}} = \mu$). The amount of variability in the SDM would always be equal to the amount of variability in the population divided by the square root of the sample size $\sigma_{\bar{X}} = \dfrac{\sigma}{\sqrt{n}}$.

Lastly, we discussed how the CLT is used to see if a given independent variable has an effect on a particular dependent variable. To do so, we create a null hypothesis in which we assume that the IV has no effect on the DV, as well as an alternative hypothesis that assumes the IV does have an effect on the DV. We then apply the IV to see if the mean value of the DV ($\bar{X}$) has been changed from what we originally thought it would be. We also introduced the idea that distance from our estimates of the population mean ($\mu_{\bar{X}}$ and) would be a relatively simple way to decide if our IV had an effect on our DV ($\bar{X}$). In the next chapter, we will explore what "different from" means in this situation and introduce our first two inferential statistical tests, based on the CLT.

STATISTICS IN CONTEXT

TERMS YOU SHOULD KNOW

alternative hypothesis, p. 339
central limit theorem (CLT), p. 327
directional hypothesis, p. 340
error, p. 324
hypothesis, p. 338
hypothesis testing, p. 338
inferential statistics, p. 324
Law of Large Numbers (LLN), p. 327
nondirectional hypothesis, p. 341
null hypothesis, p. 339
random sampling, p. 333
sampling, p. 329
sampling distribution of the means (SDM), p. 329
standard error, p. 331

GLOSSARY OF EQUATIONS

Formula	Name	Symbol
$\sigma_{\bar{X}} = \dfrac{\sigma}{\sqrt{n}}$	The standard error of the SDM	$\sigma_{\bar{X}}$
$\mu_{\bar{X}} = \mu$	The mean of the SDM	$\mu_{\bar{X}}$

PRACTICE PROBLEMS

1. How should you go about creating a sampling distribution of the means (an SDM)?

2. The central limit theorem (CLT) predicts that sampling from a normally distributed population results in a normally distributed SDM. What would the shape of the SDM be if the population is skewed? Why?

3. What would a distribution that has no variability in it at all look like? Explain your answer.

4. Typically, levels of cortisol (a hormone secreted into the bloodstream in response to stress) are highest just after a person wakes up, then fall slowly throughout the rest of the day—a pattern known as a circadian rhythm. Polk, Cohen, Doyle, Skoner, and Kirschbaum (2005) wanted to know if certain aspects of our personalities could predict levels of cortisol. They measured the personalities of a randomly selected group of 344 people, using a standardized personality test, and classified the participants into either a negative attitude (NA) group or a positive attitude (PA) group. They also measured levels of cortisol, at the same time of day, for each participant.
    a. The null hypothesis (in words) would be: _____.
    b. The alternative hypothesis (in words) would be: _____.
    c. The null and alternative (in symbols) would be:
        Null = _____. Alternative = _____.

5. Johnston, Tuomisto, and Patching (2011) wanted to know if stress created and measured in the lab was higher than stress created and measured in a natural setting. They compared a measure of physiological stress (heart rate) in people who had had their stress created and measured in the lab with people whose stress was created and measured in a natural setting.
   a. The null hypothesis (in words) would be: _____.
   b. The alternative hypothesis (in words) would be: _____.
   c. The null and alternative (in symbols) would be:
      Null = _____.   Alternative = _____ .

6. In a classic study of human memory, Elizabeth Loftus (1975) asked two groups of subjects to watch a very short film of a traffic accident (the film itself lasts less than a minute, and the accident only a few seconds). In the film, a car runs a stop sign into oncoming traffic, ultimately causing a five-car, chain-reaction accident. The two groups were then given a questionnaire with 10 questions on it. For each group, only two questions were targets; the rest were just filler. For Group 1, the first target question was "How fast was car A (the car that ran the stop sign) going when it ran the stop sign?" Group 2 was asked "How fast was car A going when it turned right?" Both groups received the same final question on the form, which was "Did you see the stop sign for car A?"
   For the target question about the speed of the car:
   a. The null hypothesis (in words) would be: _____.
   b. The alternative hypothesis (in words) would be: _____.
   c. The null and alternative (in symbols) would be:
      Null = _____.   Alternative = _____.

7. For each of the following samples, use the available information to calculate (where possible) or predict the following statistics:
   - $\bar{X}$
   - SD (standard deviation)
   - $\mu_{\bar{X}}$
   - $\sigma_{\bar{X}}$

   For all three samples, $\mu = 20.00$ and $\sigma = 15.00$.
   a. Sample 1:
      2
      3
      2
      5
      3
      3
      4
      3
      4
      4

b. Sample 2:

Stem	Leaf
0	7
1	2, 3, 7, 7
2	8, 9, 9
3	
4	0, 1

c. Sample 3:
$n = 50$
$\Sigma x = 1{,}270.533$
$\Sigma x^2 = 33{,}706$

8. Draw the probable SDMs for the sets of data in question 7, and indicate the approximate location of the sample mean on each. Do you think that the difference between $\mu_{\bar{X}}$ and $\bar{X}$ will be a meaningful, significant difference in each of these data sets?

9. Table 8.1 shows the results of a study of probability by Wolf (cited in Hand, Daly, Lunn, McConway, & Ostrowski, 1994).

Table 8.1 Results of 20,000 rolls of a die

Face	1	2	3	4	5	6
Frequency	3,407	3,631	3,176	2,916	3,448	3,422

Wolf tossed one die 20,000 times and recorded the number of times each of the six possible faces showed up (we've already seen these data in Chapter 7).
   a. What is the theoretical average roll of the die?
   b. What is the empirical average roll of the die?
   c. De Moivre, LaPlace, and the CLT predict that these two means should be the same. Are they? If not, why do they differ?

10. Table 8.2 shows how four samples were collected. For each sample, indicate whether it was randomly selected.

Table 8.2 Selection process for four samples

Sample	Selection process
1	A researcher wants to compare stress levels in first-year and fourth-year college students. She goes to a fast-food restaurant known to be frequented by college students and positions herself outside the door between the hours of 11:00 a.m. and noon, asking every other college-age person entering the restaurant to be a participant in her study.

Table 8.2 Continued

2	Another researcher wants to study stress as well, but he decides to follow tradition and use students enrolled in Introductory Psychology. He posts a sign-up sheet outside of each of the classrooms used for the course asking for volunteers.
3	A third researcher decides to use student ID numbers and a table of random numbers to select her two samples. She obtains a list of the ID numbers for all first- and fourth-year students enrolled in her school. She then selects numbers from each list at random using a random number table.
4	Yet another researcher decides to bypass the paperwork needed to gain access to student ID numbers (usually confidential and not easily given out). She decides to use the campus phone book and a random number table. She assigns a number to each phone number listed in the book and then selects phone numbers from the list using a random number table.

11. Indicate whether each of the following statements is true (T) or false (F).
    a. T F According to the CLT, as *n* increases, $\sigma_{\bar{X}}$ also increases.
    b. T F According to the CLT, the standard deviation (*s*) of a single sample always equals the standard deviation of the population (σ).
    c. T F According to the CLT, the mean of a single sample ($\bar{X}$) always equals the mean of the population (μ).
    d. T F Statisticians typically use μ to estimate $\bar{X}$.
    e. T F As σ increases, $\sigma_{\bar{X}}$ also increases.
    f. T F According to the CLT, the sampling distribution of the means will be normal in shape.
    g. T F The CLT says that $\mu_{\bar{X}} = \mu$.
    h. T F The CLT says that there will be less variability in the SDM than in the population from which all the samples are drawn.

12. Which of the following are null hypotheses? Circle all that apply.
    a. $\mu_{\bar{X}} = \mu$
    b. $\mu_{\bar{X}} \neq \mu$
    c. $\mu_{\bar{X}} > \mu$
    d. $\mu_{\bar{X}} < \mu$

13. Which of the following are directional hypotheses? Circle all that apply.
    a. $\mu_{\bar{X}} = \mu$
    b. $\mu_{\bar{X}} \neq \mu$
    c. $\mu_{\bar{X}} > \mu$
    d. $\mu_{\bar{X}} < \mu$

14. For each of the following, indicate whether the question being asked is directional (D) or nondirectional (ND).
    a. Does offering free samples of food in the grocery store make people more likely to buy the food being offered?
    b. Does meditation change the perception of the passage of time?
    c. Does belief in being lucky change performance on the Stroop task?

d. Does academic examination stress reduce immune system function?
e. Does providing observers with advance knowledge of the location of an object in space affect their ability to detect that object?

15. Suppose you had a sampling distribution of the means and you selected a series of sample means from it. Use the 3-sigma rule and the central limit theorem to answer the following questions about the SDM and those sample means:
    a. If the mean of the population from which all of the samples were drawn was 300 with a standard deviation of 10, what would the mean sample mean in the SDM ($\mu_{\bar{x}}$) be?
    b. If all of the samples drawn from the population had $n = 20$, what would the standard error of the SDM ($\sigma_{\bar{x}}$) be?
    c. What percentage of the sample means in the SDM would be within one standard error above $\mu_{\bar{x}}$ and one standard error below $\mu_{\bar{x}}$?
    d. What percentage of the sample means in the SDM would be between $\mu_{\bar{x}}$ and one standard error above the mean?
    e. What percentage of the sample means in the SDM would be at least two standard errors or more below $\mu_{\bar{x}}$?

16. Suppose you wanted to know how far above $\mu_{\bar{x}}$ a particular sample mean was in standard error units (after all, the SDM is a normal distribution). Could you use the formula for finding a z-score to find out? Think about the statistics/parameters that describe the sampling distribution of the means ($\mu_{\bar{x}}$, $\sigma_{\bar{x}}$) and the population (μ, σ). Modify the formula for z (shown below) so that it would tell you what you need to know.

$$z = \frac{x - \bar{X}}{s}$$

17. Suppose you had a sample of $n = 25$. What would the population standard deviation be if:
    a. $\sigma_{\bar{x}} = 10$
    b. $\sigma_{\bar{x}} = 0.5$
    c. $\sigma_{\bar{x}} = 2$

18. Susan rolls one fair die 25 times and gets the results shown in Table 8.3. Use these results to answer the questions that follow.

Table 8.3 Results of 25 rolls of a die

Face	1	2	3	4	5	6
Frequency	3	6	7	5	0	4

a. What is the theoretical probability of rolling a 4?
b. What is the empirical probability of rolling a 4?
c. What is the theoretical probability of rolling a 5?
d. Is the fact that Susan never rolled a 5 indicative of a problem with her die, or is it possible that the die would never land with the 5 showing in 25 rolls of the die?

19. Now let's add another die. Use the table shown below to determine the possible sums of two fair dice by adding the numbers rolled on Die 1 with Die 2 (I've started the table for you).

		Die 1					
		1	2	3	4	5	6
Die 2	1	2	3	4			
	2						
	3						
	4						
	5						
	6						

   a. What is the smallest sum of two dice possible?
   b. What is the largest sum of two dice possible?
   c. What is the theoretical probability of rolling two dice and getting a 7?

20. Stating hypotheses (general, null, and alternative) can be daunting, but like most things, it gets easier with practice. Let's try it with some examples from the literature. For each example, state the research hypothesis, the null hypothesis, and the alternative hypothesis.
    - Hays, Klemes, and Varakin (2016) were interested in how we judge the duration of a task. They knew that the perceived time can lengthen or shorten depending on aspects of the stimulus itself (e.g., brightness can affect duration judgments) and what we're doing at the time (e.g., our emotional state can affect duration judgments). They wondered if manipulating the orientation of a visual stimulus (flipping the scene upside-down) would affect the subjective perception of time. To find out, they first trained participants to judge short (400 milliseconds) and long (1,600 milliseconds) intervals. The participants were shown either inverted or upright scenes of both indoor and outdoor environments (offices, living rooms, streets, etc.) and asked to indicate if the duration of the scene was closer to the short or the long standard.
    - Gillen and Berstein (2015) were investigating the influence that having a tan might have on hiring decisions. They knew that just as a pale complexion was the marker of personal wealth and a life of leisure in the nineteenth century, having a deep tan is a marker of affluence and social status today. They presented participants with a résumé complete with a photograph of the job applicant (all participants read the same resumé and saw the same person in the photograph). Some participants saw the untanned picture of the job applicant; others saw the digitally altered tanned photo. Participants were asked whether they would recommend the person for the job and, if so, to rate the strength of their recommendation on a scale from 1 to 10 (where 10 is the strongest possible recommendation).
    - Collisson, Kellogg, and Rusbasan (2015) were interested in the perception of psychology as a science and the factors that influence that perception. They noted that biology and physics textbooks tended not to include as many in-text citations of

research as do psychology texts. They wondered if the number of citations included might influence the readers perception of the scientific nature of the material being presented. Maybe fewer citations makes the material look less controversial and more generally agreed upon? To find out, they asked participants to read an excerpt from a "textbook" in psychology or biology that contained either zero, three, or eight citations (the typical number of citations in science textbooks in general had been found to be three). The participants then rated the scientific nature of the text on a survey that asked them to assess whether the material they'd read seemed to be accepted by all members of the discipline and whether the text needed more citations.

21. Complete the crossword puzzle shown below.

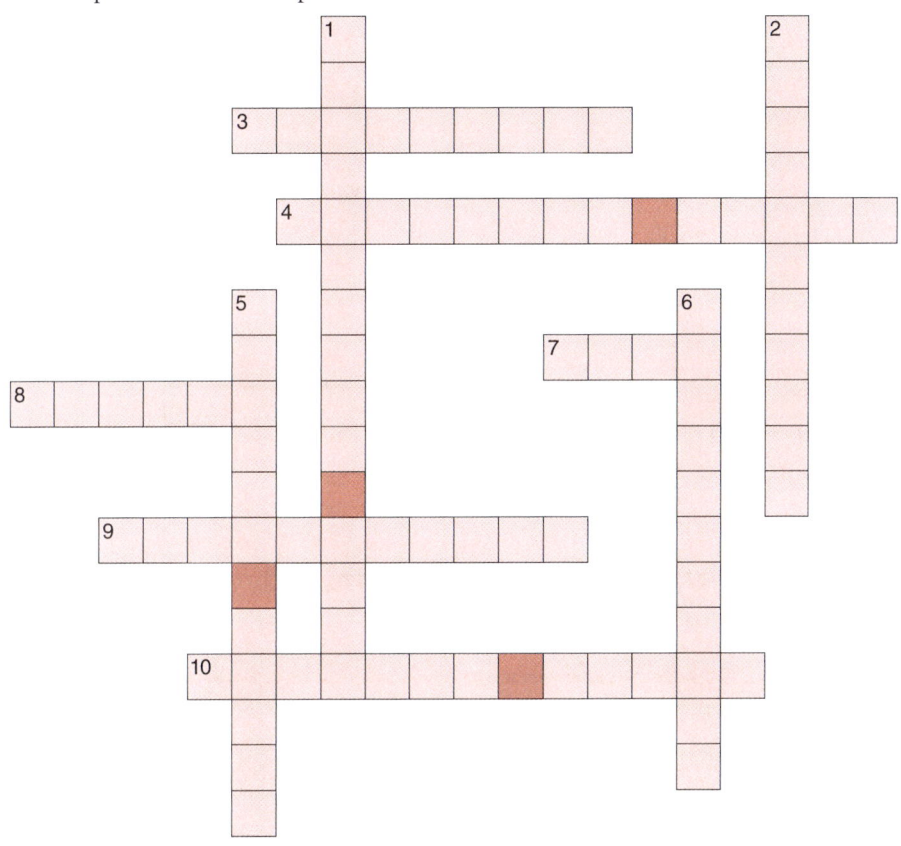

ACROSS
3. A distribution based on experience
4. (2 words) Symbol is σ_x
7. Hypothesis symbolized by H_0
8. Sampling process ensuring that all elements of pop. have an equal chance _____ of being selected
9. For example: $H_1 : \mu_0 < \mu_1$
10. (2 words) Theorem explaining how empirical samples are related to theoretical populations

DOWN
1. (2 words) Procedures to determine if null hypothesis can be rejected
2. A distribution based on theory
5. (2 words) The x-axis of SDM is composed of all possible given sizes
6. Hypothesis symbolized by H_1

CHAPTER 8 The Central Limit Theorem and Hypothesis Testing   351

Think About It...

THE LAW OF LARGE NUMBERS AND DICE GAMES

SOLUTIONS

1. How did I determine that the average roll of a single die is 3.5? **I added up the numbers possible on all of the faces of the die (1 + 2 + 3 + 4 + 5 + 6) and got 21. I then divided 16 by the number of faces (6) and got 3.5.**
2. Mark the location of the two sample means calculated above on the *y*-axis of the graph.
3. How many rolls of the set of 10 dice (approximately) would you have to conduct before the mean roll was equal to 3.5 (the population mean)? **At least 400 rolls (red arrow), and maybe as many as 425 (gray arrow) would be needed before the mean roll consistently = 3.5.**

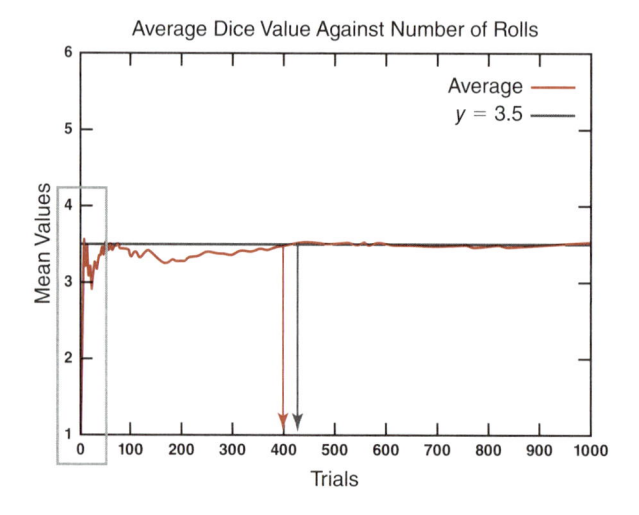

Average Dice Value Against Number of Rolls

REFERENCES

Collisson, B., Kellogg, J., & Rusbasan, D. (2015). Perceptions of psychology as a science: The effect of citations within introductory textbooks. *North American Journal of Psychology, 17*(1), 77–87.

De Moivre, A. (1718). *The doctrine of chances; Or, a method of calculating the probabilities of events in play.* London: W. Pearson.

Gillen, M., & Bernstein, M. (2015). Does tanness mean goodness? Perceptions of tan skin in hiring decisions. *North American Journal of Psychology, 17*(1), 1–15.

Hand, D. J., Daly, F., Lunn, D., McConway, K., & Ostrowski, L. (Eds.). (1994). *Handbook of small data sets.* London: Chapman and Hall.

Hays, J., Klemes, K. J., & Varakin, D. A. (2016). Effects of scene orientation and color on duration judgments. *North American Journal of Psychology, 18*(1), 27–44.

Johnston, D. W., Tuomisto, M. T., & Patching, G. R. (2011). The relationship between cardiac reactivity in the laboratory and in real life. *Health Psychology, 27*(1), 34–42.

Loftus, E. (1975). Leading questions and the eyewitness report. *Cognitive Psychology, 7*, 560–572.

STATISTICS IN CONTEXT

Mollon, J. D., & Perkins, A. J. (1996, March 14). Errors of judgment at Greenwich in 1796. *Nature, 380*, 101–102.

Polk, D. E., Cohen, S., Doyle, W. J., Skoner, D. P., & Kirschbaum, C. (2005). State and trait affect as predictors of salivary cortisol in healthy adults. *Psychoneuroendocrinology, 30*(3), 261–272.

Stiegler, S. M. (1986). *The history of statistics: The measurement of uncertainty before 1900*. Cambridge, MA: Harvard University Press.

CHAPTER NINE

THE z-TEST

Law of Probability Dispersal: Whatever it is that hits the fan will not be evenly distributed.

—ANONYMOUS

Everyday Statistics

FEAST OR FAMINE?

Children should spend most of their early lives engaged in the vitally important task of simply growing. Between birth and the age of about 18, children grow bigger, stronger, taller, and heavier.

That is, unless they don't.

Have you ever wondered why your pediatrician seemed to be obsessed with how much you weighed and how tall you were? Doctors monitor the ratio of height and weight in children because, statistically speaking, this ratio is a very strong indicator not only for the health of the child but also for the health of the population as a whole.

One of the best measures of malnutrition is the z-score: The distance a child's weight is from the median weight for children of the same height can predict the health of that child. Malnutrition is defined as a z-score of −2.00 (two standard deviations below the median) or more. Distance from the mean was never more important than it is here. Chapter 9 will discuss the z-score and what it can tell us.

OVERVIEW

ERROR REVISITED

HOW DIFFERENT IS DIFFERENT ENOUGH? CRITICAL VALUES AND p

THE z-TEST

ON BEING RIGHT: TYPE I AND TYPE II ERRORS

INFERENTIAL STATISTICS IN CONTEXT: GALTON AND THE QUINCUNX

LEARNING OBJECTIVES

Reading this chapter will help you to . . .

- Appreciate how the Trial of the Pyx is related to hypothesis testing. (Concept)

- Learn how critical values, p-values, and the alpha level are related to one another and how they are used in hypothesis testing. (Concept)

- Understand the z-test. (Concept)

- Calculate a z-value, and use it to determine whether the null hypothesis should be rejected or retained. (Application)

- Understand Type I and Type II errors and how they help statisticians to interpret the results of inferential tests. (Concept)

ERROR REVISITED

You may have heard the old expression "a difference that makes no difference is no difference at all." I don't know the origin of this sentiment, but it describes what we are doing with inferential statistics very well. Using the central limit theorem (CLT) that we discussed in Chapter 8, we can estimate parameters of the population we are interested in and then use empirical data to decide if any differences we see between our estimation and the parameter are probably the result of chance. Is it a difference that makes a difference, or is it a difference that makes no difference at all?

BOX 9.1 **The inferential *z*-formula**

The original *z*-score

$$z = \frac{x - \bar{X}}{s}$$

← Difference between two means →
← Variability →

Where is a single score (*x*) in a normal distribution relative to a mean ($\bar{X}$) in standard deviation units (*s*)?

Within a sample

The inferential *z*-formula

$$z = \frac{\bar{X} - \mu_{\bar{X}}}{\sigma_{\bar{X}}}$$

Where is a single mean ($\bar{X}$) in a normal distribution relative to a mean ($\mu_{\bar{X}}$) in standard units ($\sigma_{\bar{X}}$)?

Within the SDM

In Chapter 8, we wanted to know if sleep durations during stressful periods of time were different from sleep durations during less stressful times. To find out, we used the CLT, which proposes that the mean of the sample we collect will be one of the samples in a theoretical sampling distribution of the means (SDM). The CLT also says that the position of the sample mean within the SDM will tell us the probability that this difference is, in statistical language, *significant*. A result is called significant if the probability that it would happen by random chance alone is very low. If the difference we see between our sample mean and the population mean is statistically significant, it means that something more than random chance is contributing to the difference. And we can assume that the independent variable, the IV, is that "something more."

What we need now is a way to tell where, relative to the mean, a particular element in the SDM happens to be. We've already discussed a method for measuring location in a distribution called the *z*-score (see Chapter 6). The *z*-score was a statistic that described where a single element of a sample is, relative to the mean of the sample, in standard deviation units. The *z*-score works no matter what distribution we happen to be in, as long as that distribution is normal in shape. Since the central limit theorem states that the sampling distribution of the means will always be normally distributed, we can use the z-score formula, with some minor alterations, to find out where in the SDM our sample mean is. Box 9.1 illustrates these minor changes in the formula for z.

If you think of the z-score formula as a process or a series of instructions to follow, you should see that these two formulas are the same. At the top of each formula, we are asking *how different are two measurements?* At the bottom, we are *dividing that difference by a measure of variability.* The resulting z-score will tell the location of an element in a distribution, relative to the mean, in standard deviation units.

The differences between the two formulas are in the symbols each formula uses. The symbols used differ because the distributions we are describing differ. Notice that the symbols refer to specific single elements within the distribution and the mean of that distribution. In the formula on the left-hand side of Box 9.1, the specific single elements in the distribution (the values on the x-axis) are individual measurements. In the formula on the right-hand side, the individual elements (again, values on the x-axis) are sample means. The formula on the left tells us the location of a single element in a distribution of individual measures, while the formula on the right tells us the location of a single sample mean in a distribution of sample means. You guessed it: We are in the SDM now.

This pattern of looking for a difference between two means in standard deviation units will show up again and again as we discuss inferential statistics, so get used to seeing it, or something that looks like it. There's a pattern here: Find the difference on the top, and divide that by error (or variability) on the bottom.

CheckPoint

Calculate the z-statistic using the sample means, mean sample means, and standard errors provided. Round your answers to two decimal places.

1. $\bar{X} = 7{,}500$; $\mu_{\bar{X}} = 3{,}650$; $\sigma_{\bar{X}} = 1{,}200$ _____
2. $\bar{X} = 20$; $\mu_{\bar{X}} = 26$; $\sigma_{\bar{X}} = 4$ _____
3. $\bar{X} = 74$; $\mu_{\bar{X}} = 68$; $\sigma_{\bar{X}} = 6$ _____
4. $\bar{X} = 141$; $\mu_{\bar{X}} = 100$; $\sigma_{\bar{X}} = 15$ _____
5. $\bar{X} = 19.2$; $\mu_{\bar{X}} = 21.7$; $\sigma_{\bar{X}} = 3.25$ _____
6. $\bar{X} = 15.5$; $\mu_{\bar{X}} = 16.2$; $\sigma_{\bar{X}} = 1.3$ _____
7. $\bar{X} = 225$; $\mu_{\bar{X}} = 135$; $\sigma_{\bar{X}} = 45$ _____
8. $\bar{X} = 29$; $\mu_{\bar{X}} = 30.5$; $\sigma_{\bar{X}} = 1$ _____

Answers to these questions are found on page 359.

The Historical Context

THE TRIAL OF THE PYX

FIGURE 9.1
Commemorative coins undergo conformity testing during the Trial of the Pyx at Goldsmith's Hall in London, England, in January 2017.

The money we use today has no intrinsic value. Dollar bills, dimes, and bitcoins have value because we have agreed that they have value and that we can exchange them for goods and services.

It has not always been like this. Coins used to be minted of real gold or silver, and the value of the coin was determined by the amount of gold or silver it contained. But how did people know that the coins they were using actually contained the proper amount of gold to give them the agreed-upon value? A government might want to have a means of ensuring that the coins it was distributing were actually worth what they were supposed to be worth, in order to maintain the people's faith in their money and their government. This brings us to the ancient and mysterious Trial of the Pyx.

Begun in the twelfth century in Britain and continuing even today, the Trial of the Pyx is a procedure, cloaked in ceremony, for determining the amount of gold in the coins issued by the Royal Mint. The Pyx is simply a box. Traditionally, the Master of the Mint would select coins at random from a given day's production and put them in the box. The setting aside of coins for the Pyx would continue throughout the year until the required number of coins had been selected.

At the appointed time, the "trial" would take place. A judge (officially, the "Queen's Remembrancer"—a post created in 1154, just to give you some idea of the age of this ceremony) would assemble a jury from the Worshipful Company of Goldsmiths. The jurors had 2 months to count, weigh, and assess the purity of the coins in the Pyx. Purity was measured compared to "trial plates"—metal plates of gold or silver with a known purity, also set aside in the Pyx.

There was an allowable margin of error (a standard deviation) built in to the tests so that the coins in the Pyx could deviate a bit from the standard in either weight or purity. This window of tolerance was called the "remedy." The actual size of the remedy varied from year to year.

If the coins were found to be very different from what was expected, the Master of the Mint had a problem. If the coins were found to contain less than the expected amount of gold yet were still within the remedy, the Master was allowed to keep the shortfall. However, if the coins were found to contain more than the expected amount of gold—if they were over the limits of the remedy—then the Master would be "at the prince's mercy or will in life and members" (Stigler, 2002, p. 390). Apparently, princes don't take kindly to people wasting their gold.

Can you see hints of our discussion of the basis of inferential testing in this elaborate ceremony? From a statistical point of view, there is a sample selected at random (the coins pulled out for the Pyx) and compared with a simple null hypothesis to a standard (the expected weight of a given number of coins with a given purity). In addition, the trial takes into account a window around the expected average, an allowance for a certain amount of forgivable variation in the comparison (the remedy). This comparison puts the central limit theorem (CLT) to work.

CheckPoint Answers
Answers to questions on page 357

1. 3.21 ☐
2. −1.50 ☐
3. 1.00 ☐
4. 2.73 ☐
5. −0.77 ☐
6. −0.54 ☐
7. 2.00 ☐
8. −1.50 ☐

SCORE: /8

HOW DIFFERENT IS DIFFERENT ENOUGH? CRITICAL VALUES AND *p*

Now we can compare our sample mean with our hypothesized population mean using the *z*-formula, and we can use the resulting *z*-score to decide if our sample is different enough for that difference to be significant. We also know (check with Chapter 6 to refresh your memory) that the farther out in the tail of a normal distribution an observation is, the less probable the observations is. This is illustrated in Figure 9.2.

In a normal distribution, distance from the center defines probability of occurrence. The sample mean ($\bar{X}_1$) in Figure 9.2 has a low probability of occurrence by random chance alone. The probability that a sample mean equal to the mean of the SDM, or a value very close to it, will show up just by random chance is considerably higher.

ASSUMPTIONS IN HYPOTHESIS TESTING

Using the sampling distribution of the means comes with several assumptions. Three of these assumptions come from the central limit theorem (i.e., the SDM is normal in shape, and its mean and standard deviation can be predicted by the mean and standard deviation of the population). The fourth assumption, which is also related to the CLT as a whole, is that *the SDM reflects a true null*—that all the samples in our SDM come from a population where

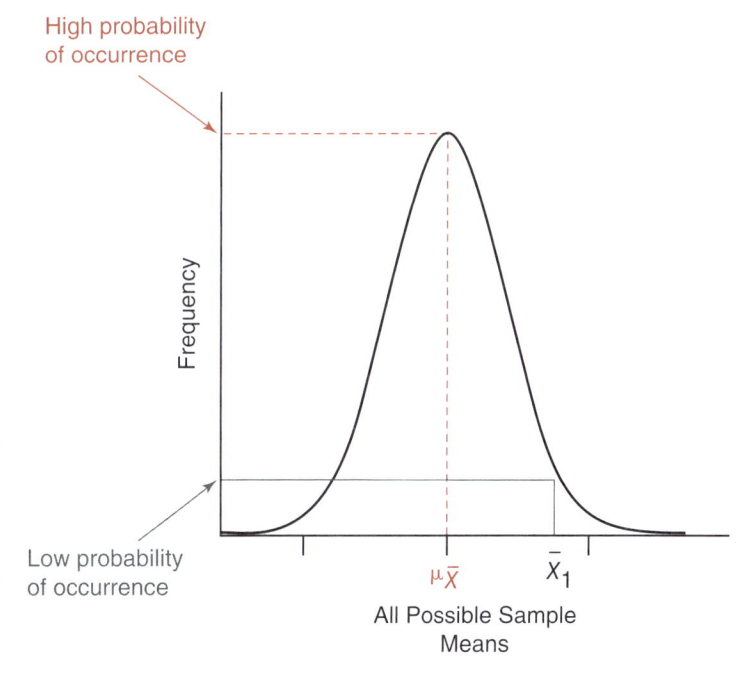

FIGURE 9.2
Probability in a normal distribution (the SDM).

the independent variable has no effect at all and where only random chance is operating. If the null is true, it means that the only factor causing variability (difference among the means in the SDM) is random chance.

So, if a sample mean is close to the mean of the SDM, that sample mean happens quite frequently. It should not be considered odd or unusual at all, because it is highly likely that the little bit of difference between it and the mean value is just caused by random chance. However, if our sample mean is found way out in the tail of the sampling distribution of the means, then it is *unlikely* that the difference between it and the mean was caused by random chance alone. And it is very, very likely that our IV is causing or contributing to the deviation of our sample mean from the population mean. Put another way: If the null is true, it is very unlikely that our sample mean would be found way out in the tail of the sampling distribution of the means.

This leads us to one more important question: How far out in the tail of the SDM does our sample have to be for us to say that the independent variable has had an effect? How different is "different enough"? Excellent question, and one that statisticians have been considering ever since the central limit theorem was first proposed. And the answer is that if your sample mean is in *the outer 5% of the distribution*, your sample mean is significantly different from $\mu_{\bar{X}}$

THE OUTER 5%: THE REJECTION REGION

We can use the *z*-table in Appendix A to find how far away from the center of the distribution a sample mean has to be in order for it to be considered meaningfully different. The *z*-score that defines difference is called the **critical value of *z* (or the critical *z*-value)**. The symbol for the critical value is *alpha* (α). Alpha tells us the *probability or position in the SDM that defines an unlikely observation if the null is true.*

Refer to *z*-table in Appendix A, and look up a *z*-score of 1.00. Check out the probability of a score on the *x*-axis being between a *z* and 1.00 and the mean. You should see that roughly 34% of all of the scores in a normal distribution (whether they are individual *x*'s or sample means) are between a $z = 1.00$ and the mean (you can use the 3-sigma rule to answer this question as well). So, a *z*-score of 1.00 defines the "inner" 34% and "cuts off" the outer (toward the tail) 16% of the distribution. If our sample mean was at least 1.00 standard errors above the mean, our α-level would be .1600 (i.e., we would be in the outer 16% of the distribution). We might say that our sample mean could be defined as different, but not different enough to be called meaningful or significant.

How far out in the tail do we need to be in order to be in the outer 5% of the distribution? We can use the table backward to discover this. We need the *z*-score that cuts off the outer 5% of a normal distribution. Read down column C until you come as close to .05000 (the outer 5%) as possible. Then, read across to the *z*-column to discover what this *z*-value has to be. You should find that a *z*-score

critical value of *z* (critical *z*-value) The *z*-value that defines a significant result; the position in the SDM that defines the rejection region in a *z*-test.

of either +1.65 or −1.65 will cut off the upper or lower 5% of the distribution. Take a look at Figure 9.3 to see how this works.

If our sample mean is more than 1.65 standard errors above, or below, $\mu_{\bar{X}}$, then the mean is in the outer 5% of the distribution, and we have a meaningful, *significant difference* for a one-tailed or directional hypothesis. This region (above a z of 1.65 *or* below a z of −1.65) is called the **rejection region**, because if our sample mean is at least 1.65 standard errors away from the mean, we can *reject* the null hypothesis and conclude that we have a significant difference caused by our independent variable (plus random chance, of course).

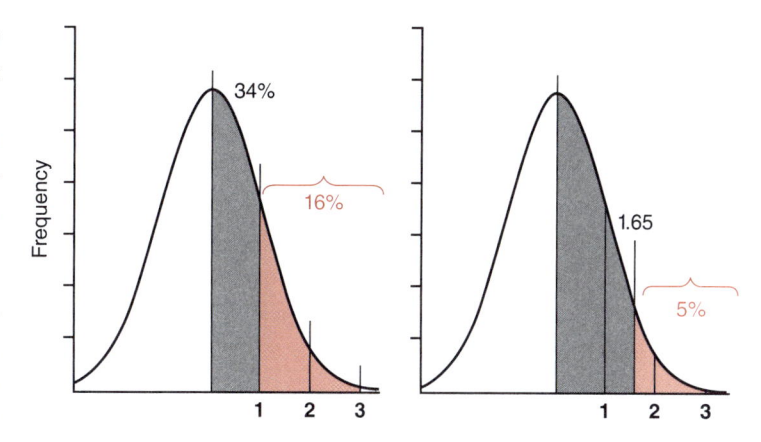

FIGURE 9.3
The outer 16% (on the left) and the outer 5% (on the right) of the sampling distribution of the means.

Let's take a brief pause to add to your statistical vocabulary. We have several new terms to define. First, as we already know, the z that cuts off the outer 5% of the distribution is called the critical z-value because it is critical to our definition of "different from $\mu_{\bar{X}}$." A z of +1.65 defines "different from the population mean" as anything in the outer 5% of the distribution. When we define different as the outer 5%, we are saying that if our sample mean is at least +1.65 standard errors above the population mean, we will decide that it is different enough from the population mean to be considered meaningful (significant). We will, as a result, reject the null hypothesis (our sample mean falls in the rejection region).

rejection region The proportion of the SDM containing the outcomes that will cause rejection of the null hypothesis.

We can define the size of the rejection region (how much of the SDM we'll consider to be the "tail") using a new term: the **alpha (α) level**. The alpha level defines the size of the rejection region and is usually set before you even begin your experiment. In our example, α = .05 (the outer 5%). By tradition, an alpha of .05 is the largest rejection region acceptable. The British statistician R. A. Fisher was the first to propose the idea of hypothesis testing and p-values, as well as the first to set the rejection region at 5%. He selected this measure as a matter of convenience, saying, "We shall not often be astray if we draw a conventional line at .05" (Fisher, 1934, p. 82). A rejection region larger than 5% of the SDM is unacceptable. However, you can always define the rejection region as *smaller* than 5%. For example, you could set your alpha to .01, or the outer 1% of the distribution. (I will have more to say about alpha levels a bit later, and then we'll talk about why you might want to set your alpha level at something smaller than 5%.)

alpha level The probability, or position in the SDM, that defines an unlikely observation if the null is true; the size of the rejection region.

There is one more new term to consider, and that is something called the **p-value**. The p-value is related to the alpha level and is technically defined as the probability of our sample mean being in the rejection region by random chance alone. Think about this for a minute. What would make our sample mean differ

p-value The probability that the sample mean is in the rejection region by random chance alone.

from the population mean? There are two factors to consider here. First, random chance might be responsible for the difference we see between these two means. Second, the IV might be responsible. It is possible that our sample mean might appear to be very different from the population mean, but that random chance is the reason why. So, we want to keep our *p*-value quite low—in fact, we'd like to make it as small as possible. Again, traditionally, the outer 5% of the distribution is considered to be a small enough risk of mistaking random chance for the effect of the independent variable. (And again, I'll have more to say about this new term later on.)

Think About It . . .

p-VALUES AND ALPHA LEVELS

The distinction between *p*-values and alpha levels is a subtle one that often gives students some difficulty. So, let's see if we can smooth those difficulties out. Alpha sets the standard for how far out in the tail of the SDM the statistic we're calculating must be (in this case, a *z*-statistic) in order to reject the null hypothesis. The *p*-value tells you how far out in the tail your statistic actually was. In order to reject H_0, the *p*-value must be less than or equal to the alpha level.

Consider the following experiment, one that you may have actually carried out in grade school. Remember growing pea plants in Dixie cups? Did you ever ask the musical question, "Does playing music to the growing pea plant make it grow bigger, faster, and with more pea pods per plant than plants raised in silence?" The experiment we would set up to answer this question is simple. We plant a set of peas and rear them with Mozart playing in the background. We let the plants mature, and then we compare the average number of pea pods from our sample of plants with the average the Burpee Seeds says we can expect.

Before we even plant our first seed, however, we need to decide on how big our rejection region should be. To keep things simple, let's use the standard, traditional outer 5% of the distribution ($\alpha = .05$). Because we are hypothesizing that music should make our plants bigger and better, we will conduct a directional test and assign only the upper 5% of the tail of the distribution as our rejection region.

We will then calculate a *z*-statistic, which will tell us how far away our sample mean is from the population average number of pea pods in standard error units. To say that our sample mean is significantly different, we need it to exceed the critical *z*-value.

1. What is the critical *z*-value for a one-tailed test, $\alpha = .05$? _____
The normal distribution shown here represents the sampling distribution of the means we're working with.

2. Draw a line showing where the mean of the null population would be.

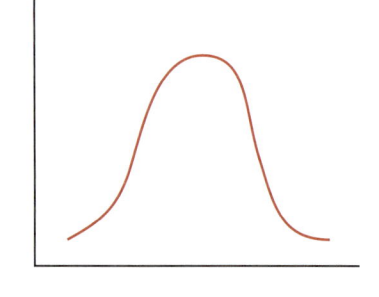

3. Draw a line showing the location of the critical z-value for α = .05 on a one-tailed test.

4. Suppose that when we ran our z-test we got an observed z-statistic of +1.93. Draw an arrow that shows (roughly) where a z of 1.93 would be.

5. On the z-table in Appendix A, find the proportion of values in a normal distribution that would be at least +1.93 or above. _____

   This is the p-value for your test. The p-value is the probability of the observed z-statistic occurring due to random chance alone in a given SDM.

Answers to this exercise can be found on page 387.

FINDING THE z-VALUE IN A NONDIRECTIONAL HYPOTHESIS

What do we do when we have a nondirectional hypothesis? We cannot use 1.65 standard errors away from the mean if we're looking at either the lower or the upper tails, because then we would be saying that what we call significantly different from the population comes from the outer 10% of the distribution (5% on each side).

With 1.65 standard errors as a cutoff point for the outer 5% of the distribution, we are actually defining the upper 5% *or* the lower 5% for a total of the outer 10%, depending on the direction we're interested in. (Remember the OR rule from Chapter 7: We need to add these two probabilities together.) Tradition defines a total of the outer 5% of the distribution as "different" for both directional and nondirectional tests. We should find the z-value that cuts off the upper 2.5%, and therefore the lower 2.5%, of the distribution so that we can add the probabilities for a total of 5%. Take a look at the z-table in Appendix A, and find the z-value that cuts off the outer 2.5% of the distribution (look for the proportion as close to .0250 as you can get). You should discover that a z of ±1.96 cuts off the upper and lower 2.5% of the distribution, keeping our α = .05 rule intact.

So, if we're using a two-tailed test, and if our sample mean is at least 1.96 standard errors above the mean *or* 1.96 standard errors below the mean, then we can conclude that our sample mean reflects the effect of the independent variable on

TABLE 9.1   Critical values for z-tests

Type of Test	Outer 5%	Outer 1%	Outer 0.01%
One-tailed test	Either + 1.65 or −1.65	Either + 2.33 or −2.33	Either + 3.10 or −3.10
Two-tailed test	±1.96	±2.36	±3.30

our subjects. We have a significant difference between our sample mean and the hypothesized population mean.

As I mentioned earlier, several other standard critical values can be used to define "different enough." Table 9.1 shows these other critical values. We will talk about why you might want to select a different critical value later on in this chapter.

CheckPoint

Using the following z-statistics, determine whether or not we can reject the corresponding null hypotheses. Express your answers as a comparison of the z-statistics and the appropriate critical values, and state whether or not you have grounds to reject the null.

1. $z = 0.25$; $H_1: \bar{X} \neq \mu_{\bar{X}}$; $\alpha = .05$ _____

2. $z = 2.94$; $H_1: \bar{X} > \mu_{\bar{X}}$; $\alpha = .0001$ _____

3. $z = 1.90$; $H_1: \bar{X} \neq \mu_{\bar{X}}$; $\alpha = .10$ _____

4. $z = -1.25$; $H_1: \bar{X} < \mu_{\bar{X}}$; $\alpha = .05$ _____

5. $z = 1.71$; $H_1: \bar{X} > \mu_{\bar{X}}$; $\alpha = .15$ _____

Answers to these questions are found on page 366.

THE z-TEST

z-test An inferential test used to compare the mean of a single sample with the mean of a population.

Believe it or not, we have been discussing our first inferential test, called the **z-test**. We use the z-test when we want to compare the mean of a single sample we have collected with the mean of a population. To use this test, we need to know the parameters of the population we are working with. Specifically, we need to know the population mean and standard deviation in order to use the formulas we've been discussing. If you take a look at the z-formula in Box 9.2, you will see why we need to know these two parameters.

Let's compare the mean sleep duration of the students in our stressed-out sample with the average sleep duration in the population. Right about now, you may be thinking, *Oh great—now I have to find the average number of hours of sleep for the typical, nonstressed college student. Where am I going to find this?*

Usually, experimenters go to the literature to see what other researchers have found. And we are in luck, because sleep (and what too little or too much of it does to academic performance) has been a topic of study in a number of labs over the years. For example, Hicks, Fernandez, and Pellegrini (2001) examined

> **BOX 9.2** **The *z*-formula**
>
> $$z = \frac{\bar{X} - \mu_{\bar{X}}}{\sigma_{\bar{X}}}$$
>
> **The *z*-formula**
> The *top*
> Subtract the sample mean from the mean of the SDM.
> The *bottom*
> Divide that by the standard error of the SDM.
>
> **Remember that**
> $$\mu = \mu_{\bar{X}}$$
> and
> $$\sigma_{\bar{X}} = \frac{\sigma}{\sqrt{n}}$$

1,585 college-age adults and found that their average sleep time was 6.85 hours ($SD = 1.04$ hr). If we then assume that 6.85 hours of sleep is typical of all college students, we know what to expect in the population *and* the sampling distribution of the means if the null is true. Both distributions should look like what is shown in Figure 9.4 (this figure might look familiar; it's the same one we used in Chapter 8).

Please take note of the relationship between the population (in pink) and the SDM which is our guess about the population generated by the central limit theorem (in gray). When the null hypothesis is true and our IV (stress) has had no effect on the students' sleeping times, these two distributions overlap one another. The mean of the sampling distribution of the means is approximately equal to the mean of the population, just as the CLT predicts.

The overlap between the two distributions isn't perfect: The two means are not exactly the same. Why is there any difference at all between these two means? If you said random chance, you were right. We can never eliminate the effects of random chance. However, we can predict the probability that our IV is the cause of any differences we see between means. We do this by assuming the null is true and then looking for evidence that it is *not* true. This sounds odd, but it is easier to show that something isn't true than it is to show that it is true, and true every time. So, considering that, you can see why statisticians chose this "backward" approach to answering the question. This is illustrated in the diagram on the right of Figure 9.4.

Notice that the difference between the two means here is quite large, and that the critical values for both our test ($z = -1.65$) and our sample mean ($M = 5.67$) are to the left of (below) the hypothesized mean of the population. If the difference

FIGURE 9.4
Relationship between null and alternative populations.

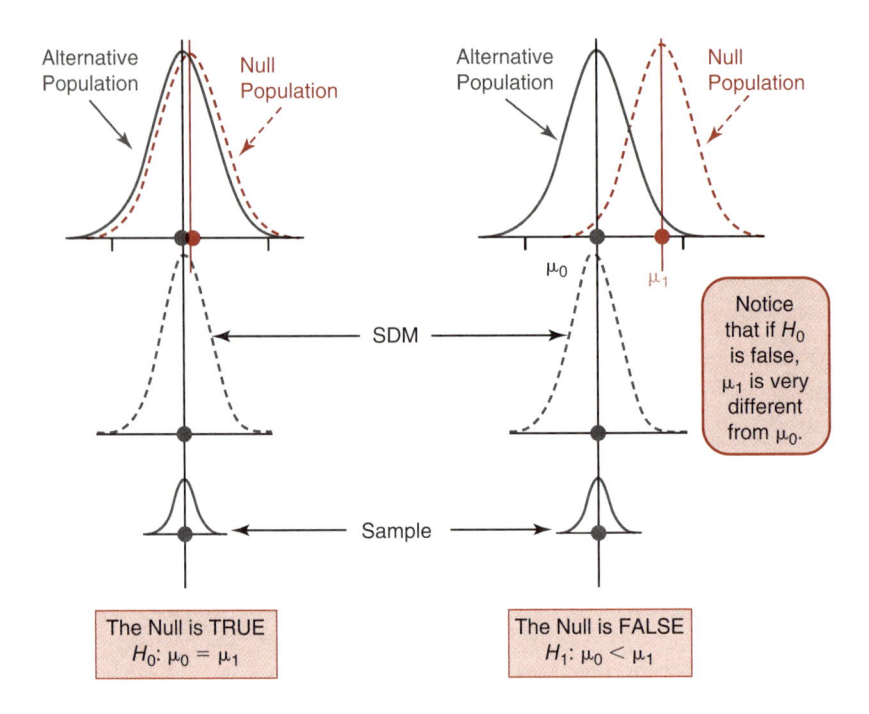

Left panel: Alternative Population and Null Population overlap; labeled SDM and Sample. **The Null is TRUE** $H_0: \mu_0 = \mu_1$

Right panel: Alternative Population centered at μ_0, Null Population centered at μ_1. Notice that if H_0 is false, μ_1 is very different from μ_0. **The Null is FALSE** $H_1: \mu_0 < \mu_1$

between the two means is far enough out in the tail to be past our critical value, we'll assume that the null hypothesis is false, that stress really does affect sleep time, and that stress might be a significant factor in how long college students sleep in general.

Now we are set up and have all of the pieces that we need to run the z-test. Box 9.3 shows what we know—the data, our research question, and our hypotheses. I have noticed that if we keep track of the things we know as we start, as well as the values for the variables we are calculating as we go along, then this process is a snap. I have used a table to keep track of what we know, what we need to calculate, and what our hypotheses are. I have found this kind of table to be quite useful. Consider using it yourself. Be careful, though—this is a big box.

CheckPoint
Answers to questions on page 364

1. ☐ $-1.96 < 0.25 < 1.96$ ∴ we cannot reject the null
2. ☐ $2.94 < 3.10$ ∴ we cannot reject the null
3. ☐ $1.90 > 1.65$ ∴ we reject the null
4. ☐ $-1.25 < -1.65$ ∴ we cannot reject the null
5. ☐ $1.71 > 1.03$ ∴ we reject the null

SCORE: /5

BOX 9.3 Research question: What is the effect of stress on sleep duration in college students?

What we Know	The Data	Hypotheses in Words	Hypotheses in Symbols
* $\mu = 6.85$ hrs * $\sigma = 1.04$ hrs ♦ $\mu_{\bar{X}} = 6.85$ hrs ♦ $\sigma_{\bar{X}} = \dfrac{\sigma}{\sqrt{n}} = \dfrac{1.04}{\sqrt{10}}$ ✓ $\sum x = 56.75$ ✓ $\bar{X} = 5.67$ ✓ $s = 1.55$ $n = 10$ Critical $z = -1.65$ $\alpha = .05$ Sources * Hicks et al. (2001) ♦ The CLT ✓ Our calculations	The Sample Hours of Sleep 5.50 5.00 4.50 6.00 7.00 8.00 2.00 7.00 6.00 5.75 $\sum x = 56.75$	**The Null** The null hypothesis says that stress has no effect at all on sleep time. Any deviation from an average of 6.85 hours of sleep during finals is caused by random chance. **The Alternative** The alternative states a *directional* difference. Stress *decreases* the amount of sleep students get during finals.	**The Null** $H_0: \mu_{\bar{X}} = \mu$ or $H_0: \mu_{\bar{X}} = 6.85$ **The Alternative** $H_1: \mu_{\bar{X}} < \mu$ or $H_1: \mu_{\bar{X}} < 6.85$

Notice the source of the information we're using in our test of the null hypothesis. The mean of the population is an estimate from the study by Hicks et al. (2001). This mean represents the average sleep time in the population if the null is true and stress does not affect sleep time. Information about the mean of the SDM comes from the central limit theorem. We calculated the standard error of the SDM using a formula for doing so that, again, comes from the CLT. We use all of these pieces to calculate z, which will tell us how many standard errors our sample mean is from the mean of the SDM. The distance from the mean tells us how likely or unlikely our sample mean is in this population if the only thing operating on the subjects in our sample is random chance. Box 9.4 shows the calculation of z with our data.

Our null hypothesis predicts that our sample mean (5.675 hours of sleep) will be close to the population mean (6.85 hours of sleep). Is it? Well, 5.675 hours of sleep is certainly less than the prediction of the mean in the population. The question is, is it different *enough*? Our alternative hypothesis states that if the mean in our sample is less than the mean of the population *by 1.65 standard errors*, then we can say that the subjects in our sample get significantly less sleep than do the unstressed people in the population.

We have already established that if the mean of our sample is in the outer 5% of the sampling distribution of the means, we have evidence that the null hypothesis is not true. And, we have our critical value of -1.65. If our observed z-value (the one we just finished calculating) is more than 1.65 standard errors *below* the mean of the

BOX 9.4 Calculation of z

$$\sum x = 56.75$$

$$\bar{X} = \frac{\sum x}{n} = \frac{56.75}{10} = 5.675$$

$$\sigma_{\bar{X}} = \frac{\sigma}{\sqrt{n}} = \frac{1.04}{\sqrt{10}} = \frac{1.04}{3.16} = 0.33$$

$$z = \frac{\bar{X} - \mu_{\bar{X}}}{\sigma_{\bar{X}}} = \frac{5.675 - 6.85}{0.33} = \frac{-1.175}{0.33} = -3.56$$

Our sample mean is 3.56 standard errors below the mean.

$z_{obs} > z_{critical}$ —3.56 is farther out in the tail than —1.65.

Decision: Reject the null hypothesis.

SDM, then we know that the difference between the sample mean and the population was significant, and that stress significantly decreased sleep time. Figure 9.5 illustrates the relationship between the sample mean and the mean of the SDM for this example.

Our sample mean turns out to be way below the population mean (in fact, our sample mean converted to a z of −3.56, more than three-and-a-half standard errors *below* the mean of the SDM) and way below our critical value of −1.65. We can tell that our mean is extremely rare in the null population, where only random chance is operating; students in our sample slept an average of 1.18 hours less than did nonstressed students. It looks as though stress has changed our subjects so much that they no longer represent the null population of unstressed students.

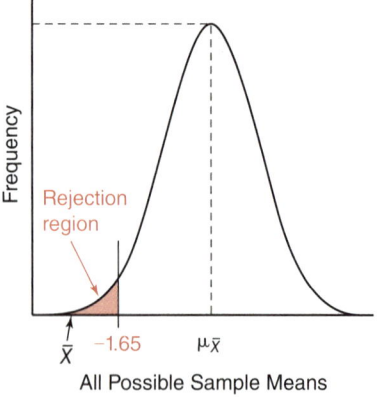

z-statistic The ratio of the difference between a sample mean and the mean of the population divided by the standard error.

The mathematics involved in finding our z-score are fairly simple—a bit of adding and subtracting and taking one square root. It is nothing your typical calculator can't handle. In fact, finding a z-score is so simple that many statistical software programs, SPSS among them, won't even calculate it for you. I find it is actually more time-consuming to use the computer to find a **z-statistic** than it is to calculate it by hand. Try it for yourself and see.

Understanding the concept behind the math (and why we're doing the math) is crucial to understanding the z-test—so crucial, in fact, that we will try another example to make sure we've all got it. The main idea to remember is that if we know something about the population our sample came from, we can predict the mean and variability in the SDM. With this information in hand, we can determine the location of our sample mean within the SDM and then use that information to come to a conclusion about the effect of our independent variable.

* The critical z-value of −1.65 defines "different enough to matter."

* Our mean is in the rejection region (farther out in the tail than 1.65 standard errors).

* In fact, our mean is −3.56 standard error below the mean.

FIGURE 9.5
Results of z-test for effects of stress on sleep duration.

ANOTHER EXAMPLE

The National Sleep Foundation (www.sleepfoundation.org) reports that sleep deprivation can produce a number of physical and psychological effects. Cognitively, sleep deprivation produces loss of concentration, sleepiness, and forgetfulness. One might expect that sleep-deprived students will have lost some of their ability to concentrate and remember the material they're learning. If this is so, we might well see a decrease in their cognitive performance.

We will use a measure of our participants' short-term memory called the digit span to test for cognitive performance. The participants in our sample will be presented with a series of lists of numbers and asked to repeat them back to the examiner. If they are correct with the first, short list (three digits long), they will then be given a longer list to remember and repeat. The longest list that a participant can remember is that person's digit span. The average digit span that a normal adult can repeat back without error is seven with a standard deviation of two digits.

We get a random sample of 15 students. Each student will undergo a sleep deprivation procedure where their sleep is monitored each night, and after 4 hours of sleep, they are woken up. Participants will be sleep-deprived nightly for a total of 2 weeks. This should allow us to see the effect of sleep deprivation (the independent variable) on cognitive performance, measured here in terms of short-term memory (the dependent variable). We will then administer the digit span test and record the average digit span of the sleep-deprived students in our sample. Our research question can be stated this way: Does sleep deprivation affect cognitive performance as measured by the digit span test? Box 9.5 shows what we know, and Box 9.6 solves for z.

BOX 9.5   **Research question: Does sleep deprivation affect digit span?**

What we Know	The Data	Hypotheses in Words	Hypotheses in Symbols
* $\mu = 7$ * $\sigma = 2$ ♦ $\mu_{\bar{X}} = 7$ ♦ $\sigma_{\bar{X}} = \frac{\sigma}{\sqrt{n}} = \frac{2.00}{\sqrt{15}} = 0.52$ ✓ $\sum x = 108$ ✓ $\bar{X} = 7.2$	Digital Span 7 6 6 8 8 10 7 8 6 7 9 7 5 10 4 $\sum x = 108.00$	**The Null** The average digit span score for sleep-deprived students is the same as the average for students in the general population.	**The Null** $H_0: \mu_{\bar{X}} = 7.0$
$n = 15$ Critical $x = \pm 1.96$ $\alpha = .05$ (2-tailed test) Sources * Previous research ♦ The CLT ✓ Our calculations		**The Alternative** The average digit span score for sleep-deprived students is NOT the same as the average for the population. (Notice that this is a nondirectional test.)	**The Alternative** $\mu_{\bar{X}} \neq 7.0$

CHAPTER 9   The z-Test

BOX 9.6 Solving for z

$$\sum x = 108.00$$

$$\bar{X} = \frac{\sum x}{n} = \frac{108}{15} = 7.20$$

$$\sigma_{\bar{X}} = \frac{\sigma}{\sqrt{n}} = \frac{2.00}{\sqrt{15}} = \frac{2.00}{3.87} = 0.52$$

$$z = \frac{\bar{X} - \mu_{\bar{X}}}{\sigma_{\bar{X}}} = \frac{7.20 - 7.00}{0.52} = \frac{0.20}{0.52} = 0.38$$

$$z_{obs}\ (0.38) < z_{critical}\ (\pm 1.96)$$

Decision: 0.38 < 1.96, so we have failed to reject the null.

Our observed z does not fall in the rejection region (our sample mean is not more than 1.96 standard errors away from the mean of the SDM), so we cannot reject the null. We do not have evidence that the null is false, so we have to assume that it is true—and that sleep deprivation does not significantly affect short-term memory span as measured by the digit span test.

We found a slight difference between the digit span scores of the sleep-deprived students and those of everyone else, but apparently not a significant one. We also found that stressed students slept significantly less than the typical (nonstressed) student in the population. Our sample mean was significantly different from the population mean, which told us that stress was probably a factor in significantly decreasing sleep duration. In both cases, our sample mean was serving as an *estimate* of the true population mean. An estimate is an educated guess about the value of something. In this case, we are estimating the value of the mean of the alternative population.

STATISTICS AS ESTIMATES

All samples are estimates, or representations, of the population they came from. Some estimate the mean of the population very well, with little error, and some do not. Sample means are estimates of population parameters as well. The difference between a sample mean and a population mean is called the error. Every time we pull a sample out of the population and calculate a mean for that sample, we're estimating the mean of the population that sample came from. If our sample mean is close to the mean of the SDM/null population (remember that they are the same), we can say that any difference between it and the population mean is not meaningful.

If our sample mean is way out in the tails of the SDM, however, then it isn't doing such a good job of representing the null population mean. If our sample mean happens to be in the outer 5% of the SDM, it is so different from the population mean that we say we can reject the null hypothesis altogether. The mean sleep duration in our sample does not represent the population of unstressed students very well at all.

ON BEING RIGHT: TYPE I AND TYPE II ERRORS

So, we can now calculate a *z*-score to tell us whether our sample mean is significantly different from a population mean. Students are often surprised that even with two statistics that emphasize how different our sample is from the population, and how much our IV has affected our sample, the conclusion we come to can still be wrong.

Because we're using probabilities and are defining things as being probably different from one another, we always run the risk of being wrong. We defined a difference as at least the outer 5% and counted any mean at least that far out as being different enough to matter. (You may have noticed that some tables list other definitions of "different" that are even farther out in the tails of a SDM; I'll have more to say on those in a minute.)

Suppose, however, that while we were selecting a sample, we pulled an outlier from the population to be in our sample? This unusual person or thing will pull the mean of the sample toward itself, and could potentially make our sample mean much bigger, or much smaller, than it would be without this oddball element. We haven't made an intentional error, but our conclusion will reflect this outlier and it will be an error nonetheless.

Or, suppose our sample is small, because of circumstances beyond our control, and underestimates the variability of the population. We believe our sample represents the population, but in reality it doesn't.

In an effort to recognize that these errors might exist in a study (they often are difficult to see until we get our result), statisticians have developed some methods for controlling these errors. Let me give you an example.

AN EXAMPLE

Suppose you work for an environmental protection group, and since you're just beginning, you get the job of going out in the field and following up on complaints. A neighborhood association calls your boss and says that the water in their neighborhood has started to taste "funny." A new processing plant just went in nearby, so they want someone out there with test tubes PDQ to test the water for toxins and poisons. Guess who gets called to do the work? That's right, you do.

So, you carefully collect a number of samples of the potentially polluted water, haul them back to the lab, and start testing. What you find are some elevated levels of some chemicals that are naturally in the water supply but are also produced as byproducts of the processing plant. What you need to find out is simple: Are they elevated enough to be trouble, or are they roughly what the government says they should be? You know what the typical levels in safe water should be (the population mean and standard deviation), so you decide to use a *z*-test to see if there's a problem.

You can be right in your conclusion, or you can be wrong, in one of two ways. Let's focus on how you could be wrong. First, suppose the *z*-value you calculate is in the rejection region (and you've checked and made sure you didn't make an error in your calculations), so you reject the null. The null says that the levels of chemical you're

Type I error The decision to reject a true null.

Type II error The decision to accept a false null.

testing for are not meaningfully different from the amounts usually present in the water. By rejecting the null, you've just decided that the levels are too high, and that the plant down the road should be shut down until you can find out what's leaking.

The owner of the company that runs the plant is not going to be happy. The neighborhood association may triumphantly announce that they "told you so" and call a press conference to excoriate big, thoughtless, and cruel industry. All because you made what is known as a **Type I error** and probably didn't even know it. A Type I error happens when you *reject a null hypothesis that is true*.

You might just as easily have made a **Type II error**, where you *accepted the null that is false*. In this case, you decided that the levels of the chemical in the water were a little high but, based on your *z*-value, not high enough to be considered significantly different than the normal levels in the drinking water. Now the owner of the big plant that really is adding pollution to the groundwater is very happy (and might even get to say "I told you so" himself). The neighborhood association, however, now thinks that the water is fine, when it actually isn't, and that can be very bad indeed.

You can never get rid of either error—they exist because of the nature of statistics and probability. You can, however, take steps to limit how often you might make them. Be careful though: If you limit the risk of a Type I error, you increase the risk of a Type II error. Which one is worse? When he proposed using $p = .05$, Fisher said, "A scientific fact should be regarded as experimentally established only if a properly designed experiment rarely fails to give this [.05] level of significance" (1926, p. 504). We should not rely on just one experiment, on just one result, when we're drawing our conclusions. Hypotheses need to be repeatedly tested before we decide that the results are unlikely to be the result of chance alone.

Table 9.2 illustrates the fix we're in with Type I and Type II errors. Across the top are the two types of errors we risk making. Down the left-hand side are the decisions we could make.

When the null is true (and the IV has no effect at all) yet we reject it (we say the null isn't true, meaning our IV has an effect), we have made a Type I error. Rejecting a null that is false is the correct decision. How could we control the risk of making a Type I error?

Since we can only make a Type I error when we reject the null, why don't we make the null harder to reject? How do we do that? We move the alpha level farther out in the tail—in other words, we make sure the difference between the sample and the mean of the SDM have to be even bigger before we can reject the null. So, instead of using $\alpha = .05$, we set our rejection region to the outer 1% ($\alpha = .01$). Our critical *z*-value (the one we must exceed to reject H_0) now moves from 1.65 to 2.33.

TABLE 9.2   Type I and Type II errors

Decision Made by Experimenter	Reality	
	The Null is True	The Null is False
Reject the null hypothesis	Type I error	Correct decision
Do not reject the null hypothesis	Correct decision	Type II error

This makes it harder to reject H_0 because the difference between the two means must be even larger in order to be in the rejection region. Setting the alpha level to a smaller sliver of the SDM decreases the chances of making a Type I error, but making the rejection region smaller now makes the "acceptance region" larger and increases the risk of making a Type II error. What should we do to control Type II errors?

When we reject H_0, we are concluding that our sample mean does not represent the mean of the population. It might represent some other population quite well, but not the one we're examining. To reduce the chances of making a Type II error, where we say the null is false when it really is true, we need to make the sample we draw from the population the best possible representation of that population. To do that, we can increase the sample size. Big samples represent the population they came from better than do small samples, so let's make sure we draw the largest possible sample we can from the population.

p-VALUES

So, to change the probability that we'll make a Type I error, we decrease the alpha level from .05 to .01 or .001. Remember when I said that an alternate name for our definition of different was *p*? The letter *p* refers to the probability of making a Type I error. To reduce Type I, set *p* at .01 or .001.

To change the probability of making a Type II error, increase the sample size. If your sample better represents your population, you'll make Type II errors less often. However, you will be more likely to reject a true null (Type I error), and there can be severe consequences to that, as we've just seen. It's a balancing act—and a tricky one. As you begin to think about your research question and how you'll answer it, consider how you can be wrong and what it means along the way. This will help you decide what your alpha level should be and how to interpret your results.

INFERENTIAL STATISTICS IN CONTEXT: GALTON AND THE QUINCUNX

The central limit theorem (CLT) and what it tells us about the population, the sampling distribution of the means, and how our empirical sample mean fits in to the SDM are foundational to inferential statistics. The theory actually explains why we can base a conclusion about an entire population of possible participants on our examination of only one sample taken from the population. Sir Francis Galton, cousin to Charles Darwin and one of the last of the "gentleman scientists" (wealthy members of the aristocracy who dabbled in science more as a hobby than as a career), invested a great deal of his time and energy into developing a model of the central limit theorem to help in his attempt to explain variability in his studies of the heritability of traits (genetics). Galton's model is a machine called the Quincunx, a name taken from the shape of the array of pins he used in it. Never the most accomplished of mathematicians, Galton used the physical model of the Quincunx as proof of the central limit theorem's truth.

Two versions of Galton's Quincunx are depicted in Figure 9.6. The device consists of an array of pins, traditionally in a pattern called a Quincunx—a symmetrical arrangement of five dots, with one in each of the four corners and one in the center—like the five dots on a die (although most modern Quincunxes don't use this traditional pattern of pins).

Shot pellets, poured through a funnel into the top of the machine, run down through the array of pins. A large piece of glass covers the entire machine to keep the pellets from bouncing anywhere other than downward. When a shot hits a pin, it has a 50-50 chance of bouncing either left or right. It will hit several pins before its journey through the Quincunx ends in one of several slotted compartments ranged across the bottom of the contraption.

Galton found that if he poured a large number of shot pellets down through the array of pins, the resulting distribution of shot pellets in the bins at the bottom of the array would be normal in shape. Most of the pellets would wind up in the center bins, and the probability of a pellet ending up in a particular bin decreased as the distance of the bin from the center of the distribution increased.

Think of the set of shot pellets in the funnel as the elements of the population. Each ball or shot pellet that drops down through the array of pins is a sample mean from a randomly selected sample. The distribution of pellets in the bins when we're done represent the sampling distribution of the means. Pellets (sample means) collecting in the center bins would have a value close to the average in the population and would happen quite frequently. Pellets that collect in the outer bins are quite different from the population average and don't happen very often.

In a Quincunx, the only variable acting on the sample means is random chance. Random chance determines if the ball will bounce to the left or the right. Random chance determines if the ball lands in the middle, to the far left, or just to the right of the center of the set of bins. One of these balls is your empirical sample mean. If only random chance is determining where your sample is, what is the probability that your empirical mean will end up in the bin as far away from the center as is possible? That probability should be fairly low if your Quincunx is fair and square.

If only random chance is operating, what is the probability that your sample mean will be close to the center bin? Again, assuming a fair Quincunx, that probability should be fairly high. We start off assuming that only random chance is operating in our null hypothesis. Then, we check the actual position of the

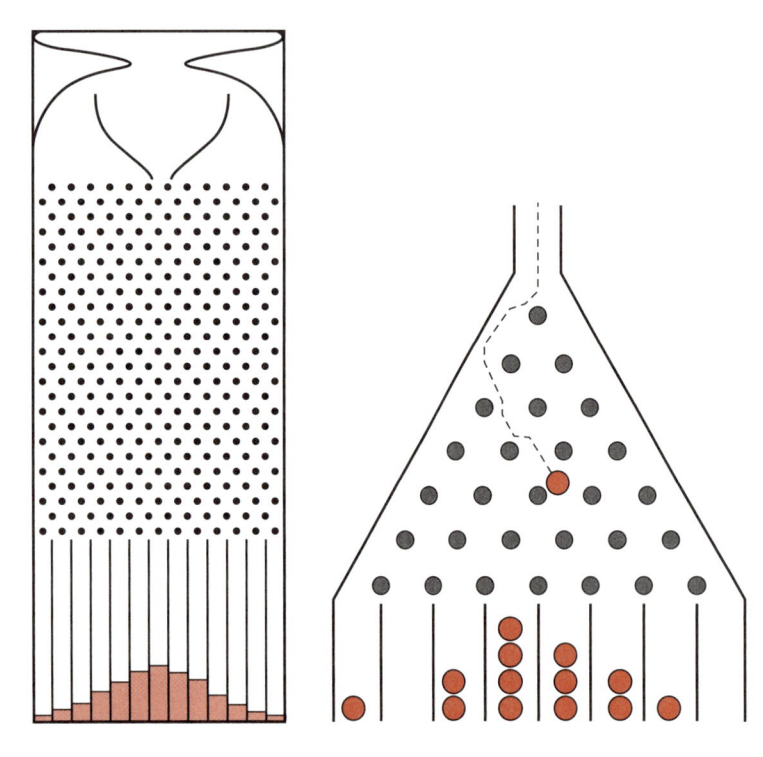

FIGURE 9.6
Examples of Galton's Quincunx.

374   STATISTICS IN CONTEXT

empirical mean to see if we have support for our hypothesis. If we don't have any evidence against it, we assume that the null is true and that our independent variable has no effect. If we add the IV into the mix and find our sample mean is now out to the left or right of the center of the bins, then we have support for our alternative hypothesis. And now we're doing science.

A number of working models of a Quincunx are available online, where you can see for yourself what happens when you play around with the parameters of the CLT. (See *Quincunx Websites* at the end of the chapter for a selection of these websites.) For example, change *n* and see what happens to the normal distribution and the number of samples you need to use to create a normally distributed SDM. On some online models, you can change the number of bins at the bottom, or the number of rows of pins within the Quincunx itself. Does the number of rows of pins the pellets must bounce off of change the shape of the SDM? What happens if you interrupt the flow of pellets as they bounce down through the rows? What happens if you change the size of the sample that you pour through the pins? You can play with this model of the central limit theorem and explore the limitations of sampling from a population to draw conclusions about that population. You can also see a modern Quincunx in a game show called "The Wall." Contestants answer questions as a ball bounces through the Quincunx, crossing their fingers that they will answer the question correctly (making them eligible to win money) and that ball will land on the big payouts at the bottom. The biggest prizes are found to the far left and far right in the bottommost array where random chance is least likely to send the ball. As Galton could tell you, winning the big money is by no means a certainty, but in the first season of the show two teams of contestants walked away with more than a million dollars (The Wall, 2017).

SUMMARY

Let's summarize. We want to know what effect an independent variable will have on the participants in a randomly selected sample, taken from a given population.

We make several assumptions before we even begin to test our hypotheses:

- First, we assume that the measurements in the population we're sampling from are normally distributed, before we add the IV.
- Second, we assume that this population has a mean (μ) and a standard deviation (σ), and that we know these parameters.
- Third, we assume that the central limit theorem describes the relationship between a theoretical SDM and the population accurately (meaning $\mu_{\bar{X}} = \mu$ and $\sigma_{\bar{X}} = \dfrac{\sigma}{\sqrt{n}}$ if the IV has no effect on our set of measurements).

We then form two mutually exclusive hypotheses about the effect of this independent variable on the participants that came from our population and created the SDM:

- The null hypothesis says that the IV has no significant effect on our participants.
- The mutually exclusive alternative hypothesis predicts that the IV does have an effect. According to the CLT, if the IV has no effect, then $\mu_{\bar{X}} = \mu$.

Next, we also estimate what the mean of the population would be if, in fact, the null hypothesis is false (i.e., if the IV *does* have an effect and $\mu_{\bar{X}} \neq \mu$). We will find $\bar{X}$ and use it to estimate $\mu_{\bar{X}}$. If $\mu_{\bar{X}} \neq \mu$, then we have some good evidence that the addition of the IV has made the alternative population differ from the null population.

We also have a decision rule defining what "different from the mean" needs to be in order for us to say that we have a meaningful difference. The critical value defines the *probability* that the difference between $\bar{X}$ and $\mu_{\bar{X}}$ is due to the IV and not just random chance. If $\bar{X}$ comes from the outer 5% of the sampling distribution of the means, then we can say that there is a 5% chance, and only a 5% chance, that random variability is the cause of this difference.

To determine the difference between our two means, we run a *z*-test. The resulting *z*-value tells us if our sample mean falls in the rejection region, and whether or not we can reject the null.

We should also consider the risk of making either a Type I or a Type II error, and adjust our experiment in order to limit these errors.

TERMS YOU SHOULD KNOW

alpha (α) level, p. 361
critical value of *z*, p. 360
p-value, p. 361
rejection region, p. 361

Type I error, p. 372
Type II error, p. 372
z-statistic, p. 368
z-test, p. 364

GLOSSARY OF EQUATIONS

Formula	Name	Symbol
$z = \dfrac{\bar{X} - \mu_{\bar{X}}}{\sigma_{\bar{X}}}$	The *z*-test	*z*

WRITING ASSIGNMENT

Select one of the studies discussed in this chapter and write an APA style results section for that study.

PRACTICE PROBLEMS

1. Consider the elements of the formula for calculating z $\left[z = \dfrac{\bar{X} - \mu_{\bar{X}}}{\sigma_{\bar{X}}} \right]$.

    a. Which distribution (population, SDM, or sample) are we calculating a value of *z* in?
    b. What does the top of the equation tell us?

c. What does the bottom of the equation tell us?
d. If the variability in the population suddenly increased, what would happen to the value of $\sigma_{\bar{X}}$?
e. If n increases, what happens to $\sigma_{\bar{X}}$?

2. There is a relationship between percentile rank and the critical z-value in a one-tailed z-test.
   a. What is percentile rank?
   b. What is the relationship between these two measures?

3. Identify the Type I and Type II errors in the descriptions below.
   a. Your lab has just developed a new cancer-fighting drug that you hope can significantly reduce tumor size. Unfortunately, the side effects of the drug can be severe, including hair loss, intestinal lesions, intestinal bleeding, and an increased risk of death. The drug is entering the testing phase.
      i) What would a Type I error be?
      ii) What would a Type II error be?
      iii) Which type of error is the most serious for the drug company developing the drug?
      iv) Which type of error is the most serious for the patients taking the drug?
      v) How can you reduce the risk of these errors?
   b. A number of studies have linked low levels of a neurotransmitter called serotonin (5-HT) with the severity of several mental illnesses—for example, depression and anxiety disorders. Suppose you develop a new treatment for depression that raises 5-HT levels. Manipulating levels of chemicals in the brain is not without risk, however, because although high levels of 5-HT are associated with reduction of the symptoms of depression, 5-HT also plays a significant role in sleep, cognitive processes, arousal, sexual behavior, and pain perception. Your drug is ready for testing.
      i) What is a Type I error in this case, and what are its consequences for you as the developer of this drug?
      ii) What is a Type II error in this case, and what are its consequences for you as the developer of this drug?
   c. Everyone knows that lowering your cholesterol is a good thing to try to do. However, there is evidence that lowering cholesterol levels can produce rather disturbing changes in behavior. Extremely low levels of cholesterol (lower than most people can achieve by changing their diet alone) have been found to be associated with an increase in risk-taking behavior, homicide, suicide, and what is referred to as "nonillness mortality." Your job is to describe the effects of a new statin drug, reputed to work better than anything else on the market, to a group of patients suffering from high blood pressure. In an effort to provide complete information, you describe the Type I and Type II errors associated with taking this drug. What are they?

d. You are interested in diffusion of responsibility—the phenomenon that can be seen when other people are present in an emergency, making any one individual less likely to take action in that situation. Research has shown that diffusion of responsibility is very rare when a person is confronted by an emergency alone, and that it is likely to occur when at least three other people are present. You want to know if the gender makeup of the crowd matters as well. To find out, you expose your 15 male research participants to an emergency and measure the amount of time it takes for the research participant to open the door to the room and notify the experimenters about the emergency. Participants sit in a windowless room, along with a group of three males (same sex) or three females (opposite sex) while smoke is gradually pumped into the room through the ventilation system.
  i) What is a Type I error in this case, and what are its consequences?
  ii) What is a Type II error in this case, and what are its consequences?

4. Since we're talking about lower cholesterol and the possible side effects of a very low cholesterol level, let's explore the research a bit. According to Goodman et al. (1988), normal total blood cholesterol levels should be less than 200 mg/dl ($SD = 44.4$ mg/dl). Levels higher than 240 mg/dl indicate high cholesterol. Dr. Michael Criqui, writing for the American Heart Association (1994), says that levels below 160 mg/dl are categorized as *hypocholesterolemia*, or abnormally low

Table 9.3   Total cholesterol in 25 patients after taking new statin drug

Cholesterol Levels (in mg/dl)				
219	191	198	214	163
264	248	182	235	209
152	148	145	189	213
230	181	180	100	219
102	249	282	188	264

levels of cholesterol. Suppose you want to test this new statin mentioned in question 3(c) to see if it really does create a decrease in cholesterol. Use the data in Table 9.3 to answer the following questions.

a. Calculate $\mu_{\bar{X}}$, $\sigma_{\bar{X}}$, $\bar{X}$, and σ.
b. Describe the effects of the statin drug on your sample. Do you see any change in the level of cholesterol after taking the drug?
c. What is your research question? What are the null and alternative hypotheses? Decide if you want to do a one-tailed or a two-tailed test.
d. Calculate the value of z. What is your critical z-value?
e. Sketch a normal distribution, and draw the critical value or values of z.
f. What is your conclusion?

5. The Stanford–Binet IQ test is an example of a standardized test. This basically means that a population mean and standard deviation have been established for the test. We know that the average score in the population of Stanford–Binet test takers is 100 with a standard deviation of 15. Since cholesterol is an essential element of cell membranes in the central nervous system (it controls the permeability of the cells in the brain— extremely low levels can lead to "leaky" cells), you decide to test the cognitive abilities of patients who suffer from hypocholesterolemia (thought be genetic in origin) using their score on the Stanford–Binet test as a measure of cognition. Your data are as presented in Table 9.4.

Table 9.4 IQ score for 15 patients with hypocholesterolemia

IQ score		
106	95	93
80	103	92
73	93	106
101	115	109
112	83	96

a. Calculate $\mu_{\bar{X}}$, $\sigma_{\bar{X}}$, $\bar{X}$, and σ.
b. Describe the effects of hypocholesterolemia on IQ score in your sample. Do you see any change in cognitive ability as we have defined it?
c. What is your research question? What are the null and alternative hypotheses? Decide if you want to do a one-tailed or a two-tailed test.
d. Calculate the value of z. What is your critical z-value?
e. Sketch a normal distribution, and draw the critical value or values of z.
f. What is your conclusion?

6. For the previous two questions, describe the Type I and Type II errors and their consequences.

7. For each of the populations defined in Table 9.5, calculate a z-score and make a decision about the difference between the sample mean and the population mean for each, using the appropriate critical value.

Table 9.5 List of population parameters

Population Mean	Population Standard Deviation	Sample Size (n)	Sample Mean	Alpha Level
100	30	50	105	.05, two-tailed
50	13	20	45	.01, one-tailed
207	25	9	200	.001, one-tailed

8. Martha Taft was the wife of Robert Taft, who served as Senator from Ohio in the 1940s and early 1950s. Mr. Taft had the reputation of being something of a cold fish. Because they lived in the public eye, Martha needed to be the social, gregarious, and amusing half of their partnership. She seems to have been well suited to the task. When asked what she thought about statistics, she is rumored to have said:

> I always find that statistics are hard to swallow and impossible to digest. The only one I can ever remember is that if all the people who go to sleep in church were laid end to end _____ _____ _____ _____ _____ _____ _____.

Find the rest of the quote by solving the problems shown below. Once again, each of the words in the solution section is associated with a number. To determine which word you need, solve the problems in the problem section. The word next to the answer to the first problem is the first word in the solution to the riddle, the word next to the answer to the second problem is the second word in the solution, and so on. Round all answers to two decimal places (e.g., 2.645 rounds to 2.65).

Problems section: Solve each problem to find a number. Match the number to a word in the word section to finish the riddle.

Words:

(12.8) they	(15.6) need	(51.76) a	(± 1.96) comfortable
(6.01) excellent	(128) would	(7.19) lot	(± 2.98) boring
(2,156) be	(1.11) more	(1.29) winsome	(52.00) less

Data set:
The sample: 4   6   7   7   9   10   20   20   20   25
The population parameters: $\mu_{\bar{x}} = 10.00$
$\sigma = 8.00$

a. The sample mean is _____.
b. The sum of all the values in the sample (Σx) is _____.
c. The sum of all the values squared (Σx^2) is _____.
d. The variance is _____.
e. The standard deviation is _____.
f. The observed z-value is _____.
g. The critical z-value if $\alpha = .05$ for a two-tailed test is _____.

9. An educational psychologist is interested in seeing if Return To College (RTC) students—that is, students who have not attended school for more than 4 years in order to work or raise a family—are more motivated to achieve higher grades than students who have not taken a break in their education. She knows that the distribution of grade point averages (GPAs) for students at her college who have not interrupted their education is normally distributed with a mean of 2.90 and a standard deviation of 0.89. She collects the following sample of GPAs for 10 RTC students. Use the data in Table 9.6 to answer the following questions.

Table 9.6 GPAs for RTC students
GPA
3.70
2.60
3.10
3.00
3.10
3.80
2.60
3.40
3.00
2.80

a. Find the mean GPA for the RTC students in the sample.
b. Determine the null and alternative hypotheses.
c. Decide if the GPAs earned by RTC students differ significantly from the population.

10. You want to compare the performance on a learning task of a group of 20 subjects who have been given Prozac with the performance of the nondrugged population. You know that the mean score on this learning task in the population is seven correct responses with a population standard deviation of two. The mean number of correct responses for the drugged sample was 6.1.
    a. Did drugged subjects perform differently than nondrugged subjects?
    b. How do you know?

11. The superintendent of a school district claims that the children in her district are brighter, on the average, than the general population. The superintendent knows that the national average for school-age children on the Stanford–Binet IQ test is 100 with a standard deviation of 15. She collects IQ scores from 10 students in her district in order to compare her sample with the rest of the nation. Her results are shown in Table 9.7.

Table 9.7 Score on the Stanford–Binet IQ test for 10 students
Test Scores
105
109

Table 9.7 Continued
115
112
124
115
103
110
125
99

a. Should you use a one-tailed or a two-tailed test?
b. What would a Type I error mean in this case?
c. Can the superintendent correctly claim that the students in her district have higher IQ scores?

12. Too much sodium in our diet can have significant health consequences. Suppose you wanted to find out if switching from a red-meat hot dog to a one made from poultry (chicken and turkey) would reduce the amount of sodium you were consuming. You go online and discover that on average, a red-meat hot dog contains 233 milligrams of sodium per ounce with a standard deviation of 30.98 milligrams. You do a quick survey of 17 poultry dogs in the supermarket and find that the average sodium content of the brands available in your local market is 256.35 milligrams per ounce. Should you switch to poultry-based hot dogs?

13. After several years of study, a psychologist interested in human performance in flight simulators knows that reaction time to a red overhead emergency indicator is normally distributed, with $\mu = 205$ milliseconds and $\sigma = 20$ milliseconds. The psychologist wants to know if the color of the emergency indicator matters. Using a yellow indicator light, she looks to see if there is an effect on reaction time.
    a. Identify the independent and dependent variables
    b. Is this a one-tailed or a two-tailed question?
    c. If the psychologist obtained an average reaction time of $\bar{X} = 195$ milliseconds for a sample of 25 people, what conclusion could she come to about the effect of the color of the indicator on reaction time?

14. A researcher is trying to assess some of the physical changes that occur in addicts undergoing alcohol withdrawal. For the population of non-alcohol-addicted adults, the average body temperature is $\mu = 98.2$ degrees Fahrenheit with $\sigma = 0.62$ degrees. The

data shown in Table 9.8 show the body temperatures of a sample of alcoholics during withdrawal.
   a. Identify the independent and dependent variables.
   b. Is this a one-tailed or a two-tailed test?
   c. Is there evidence of a significant change in body temperature during withdrawal?

Table 9.8 Body temperature for a sample of alcoholics during withdrawal

Body Temperature (in °F)	
98.6	99.9
99.0	101.0
99.4	99.6
100.1	99.5
98.7	100.3
99.3	97.2

15. Suppose that the average birth weight for the population of newborn infants in your home state is $\mu = 6.393$ pounds with $\sigma = 1.43$ pounds. An investigator would like to see if the birth weight of infants born to mothers who smoke cigarettes is significantly lower than that seen in the population. A random sample of women who smoked during their pregnancy is selected, and the birth weight of their infants is recorded. The data are shown in Table 9.9.

Table 9.9 Birth weight of babies born to mothers who smoked during pregnancy

Weight (in Pounds)	
5.07	5.29
4.41	5.29
4.85	4.63
6.17	5.07
7.05	5.73
4.85	4.41
5.51	5.07

   a. Identify the independent and dependent variables
   b. Is this a one-tailed or a two-tailed question?
   c. What should the scientist conclude about the effect of cigarette smoking on infant birth weight?

16. Wearable technology is all the rage these days. Adults of all ages are sporting activity-tracking devices that will count their steps, measure their heart rate, and even calculate the number of calories burned during exercise. Suppose you wanted to know if this kind of immediate feedback would encourage even more walking among healthy adults who already walk daily. You discover that healthy adults take an average of about 10,000 steps per day with a standard deviation of 2,300 steps. You select a sample of 14 healthy adults, give each one of them an activity tracker to wear on their wrists, and ask them to track the number of steps per day they take for a set period of time. You discover that in your sample, the mean number of steps per day taken was 13,400. Did the fitness trackers increase the number of steps taken by the people in your sample?

17. The manufacturer of a popular preservative for hot dogs states that his product will preserve the dogs for an average of 35 days (in refrigeration) with a standard deviation of 4.5 days. You want to test out his statement, so you select a random sample of 50 hot dogs that have been treated with his preservative and determine how long the dogs stay "edible" in refrigeration. Your sample mean is 32 days.
    a. State the null and alternative hypotheses.
    b. Identify the independent and dependent variables.
    c. Interpret the results of your test.
    d. Does the evidence support the manufacturer's 35-day freshness guarantee?

18. Last year, the state legislature passed a bill aimed at decreasing the mean expenditure of college-age students on liquor. At the time, the mean amount of money spent on alcohol by college students (per year) was $115.00 with a standard deviation of $29.00. This year (after the bill has been in effect for a while), you select a random sample of 85 college students and determine how much they spend on liquor per year. Your sample mean is $110.00.
    a. State the null and alternative hypotheses.
    b. Interpret the results of your test.
    c. Did the bill have an effect on how much college students spend on liquor per year?

19. A large sugar refinery states that the average American consumes 156 pounds of sugar per year with a standard deviation of 9 pounds. In response, the American Medical Association (AMA) conducts a year-long study to measure sugar consumption. They select, at random, 25 people and monitor their sugar intake. The sample mean is 148 pounds of sugar per year
    a. State the null and alternative hypotheses.
    b. Decide whether you should do a one-tailed or a two-tailed test.
    c. Interpret the results.
    d. Do the results of the AMA study refute or support the statement made by the sugar company?

20. A popular tutoring agency advertises that students who use this service see their overall academic averages increase by 15%, with a standard deviation of 3%, within 6 months. Maria, a skeptical employee of the tutoring agency, believes the agency is overstating its effectiveness and decides to test its claim. Maria randomly selects 10 of the agency's new tutees and tracks the changes in their overall academic averages over a period of 6 months. Use Maria's observations, displayed in Table 9.10, to help her decide whether the tutoring agency's claims are as dubious as she believes. Adopt an alpha level of .05.

Table 9.10 Tutees' overall academic averages

Average Before Starting Tutoring	Average After 6 Months of Tutoring
54	62
71	80
70	68
46	68
60	73
60	65
59	64
48	59
65	77
75	84
$\Sigma = 608$	$\Sigma = 700$

a. State the null and alternative hypotheses in symbols.
b. Decide whether you should do a one-tailed or a two-tailed tailed test.
c. Interpret the results.

21. It is often said that pet owners are happier people, but does owning a pet really improve a person's well-being? To answer this question, you decide to study whether Americans who own dogs have different subjective well-being scores than Americans in general. You know that the mean subjective well-being score for Americans is 6.4 with a standard deviation of 0.6. The mean subjective well-being score of American dog owners is 7.2. Adopt an alpha level of .05.
a. State the null and alternative hypotheses in symbols.
b. Do American dog owners have different subjective well-being scores than the general American population?

22. Complete the crossword shown below.

ACROSS
3. Another name for a one-tailed test
5. You make a Type II error when you _____ a false null
6. The typical rejection region consists of the outer _____ percent of the SDM
7. The number that illustrates the size of the rejection region
8. (2 words) To use the z-test, you must know both the population mean and the _____ _____
9. The value of z that must be exceeded in order to reject the null hypothesis
11. An estimated range of values in a population is this kind of estimate
12. An array of five dots in a specific pattern: ∴
13. A sample mean is this kind of estimate of the population mean

DOWN
1. (2 words) The p-value indicates the probability of an observed statistic occurring by this alone
2. (2 words) The ratio of the difference between the sample and population means divided by the standard error
4. (2 words) The proportion of the SDM that contains outcomes that will result in the rejection of the null hypothesis
10. You make a Type I error when you _____ a false null

Think About It . . .

p-VALUES AND ALPHA LEVELS

SOLUTIONS

1. What is the critical *z*-value for a one-tailed test, $\alpha = .05$?   **+1.65**
2. Draw a line showing where the mean of the null population would be.
3. Draw a line showing the location of the critical *z*-value for a $\alpha = .05$ on a one-tailed test.
4. Suppose that when we ran our *z*-test we got an observed *z* statistic of +1.93. Draw an arrow that shows (roughly) where a *z* of 1.93 would be.
5. On the *z*-table, find the proportion of values in a normal distribution that would be at least +1.93 or above.   **02680, or 2.68%. Notice that the *p*-value is less than the alpha level. Our peas have significantly more pea pods than do those raised in silence.**

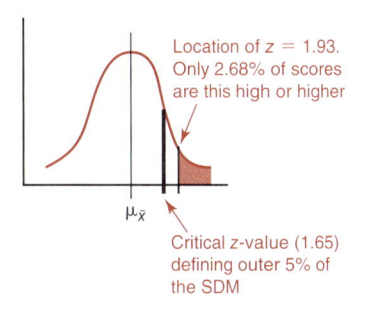

REFERENCES

Criqui, M. H. (1994). Very low cholesterol and cholesterol lowering: A statement for healthcare professionals from the American Heart Association Task Force on Cholesterol Issues. *Circulation, 90*, 2491.

Fisher, R. A. (1926)). The arrangement of field experiments. *The Journal of the Ministry of Agriculture, 33*, 503–513.

Goodman, D. S., et al. (1988). Report of the National Cholesterol Education Program Expert Panel on Detection, Evaluation, and Treatment of High Blood Cholesterol in Adults. *Archives of Internal Medicine, 148*(1), 36–69.

Hicks, R. A., Fernandez, C., & Pellegrini, R. J. (2001). Self-reported sleep durations of college students: Normative data for 1978–79, 1988–89 and 2000–01. *Perceptual and Motor Skills, 93*, 139–140.

Stigler, S. M. (2002). The Trial of the Pyx. In *Statistics on the table: The history of statistical concepts and methods* (3rd ed.) (pp. 383–402). Cambridge, MA: Harvard University Press.

The Wall (game show). (2017). Retrieved from https://en.wikipedia.org/wiki/The_Wall_(game show).

Quincunx Websites

Math Is Fun: The Quincunx: www.mathsisfun.com/data/quincunx.html

The Galton Board: http://webphysics.davidson.edu/Applets/galton4/galton_mean.html

Quincunx Demonstration: www.sixsigmastudyguide.com/quincunx/

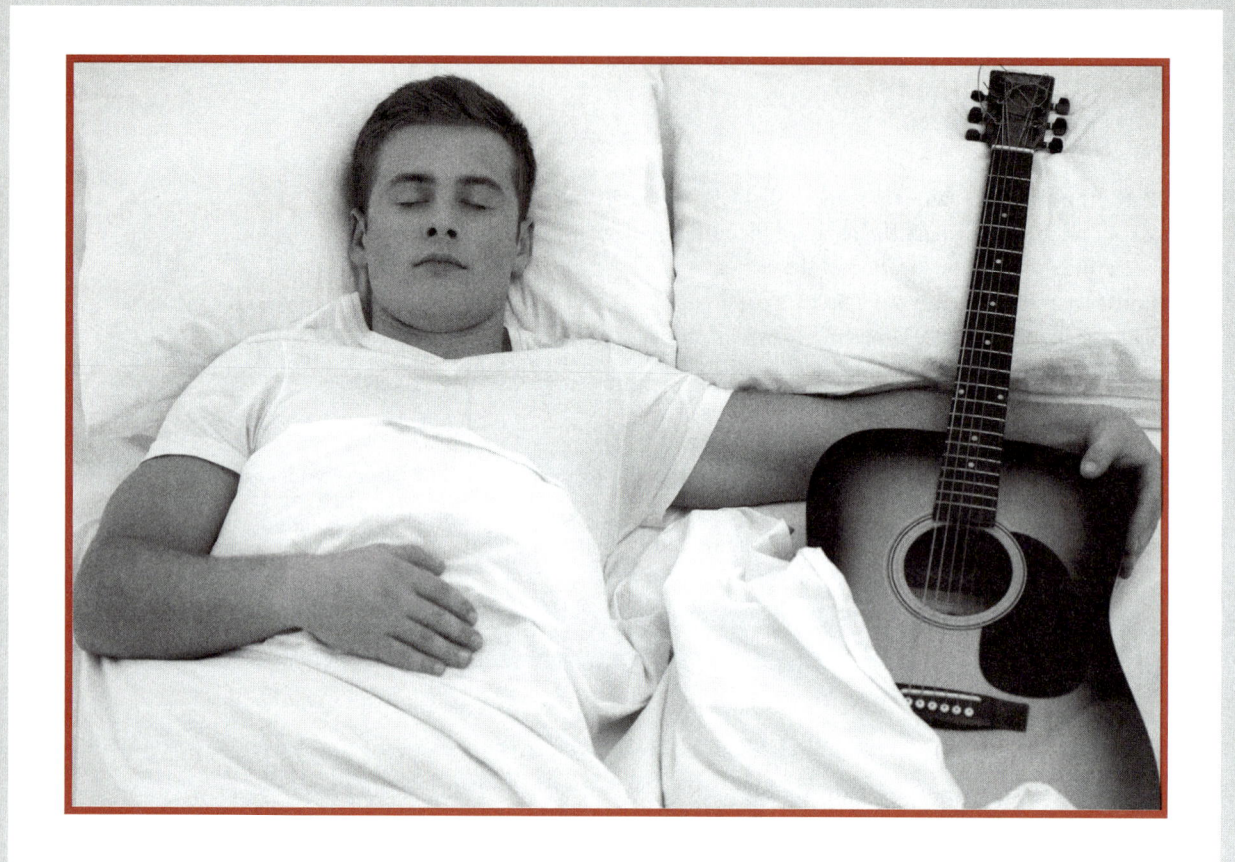

CHAPTER TEN

t-TESTS

The great tragedy of Science—the slaying of a beautiful hypothesis by an ugly fact.
—THOMAS H. HUXLEY

Everyday Statistics

THE MUSIC OF OUR DREAMS

Lots of us dream in color. Visual imagery is the most commonly reported type of sensory experience in a dream, while the experiences of a smell or a taste are very rare. Interestingly, even though dreamers often report hearing voices or other acoustic signals in dreams, hearing music or a recognizable tune is incredibly rare.

Given that the content of dreams is thought to reflect our everyday experiences, a group of researchers at the University of Florence in Italy (Uga, Lemut, Zampi, Zilli, & Salzarulo, 2006) wondered if people who spend their days deeply enmeshed in music (musicians and singers) would dream about music more often than those of us who are not so musically inclined. Would their dream content reflect their everyday experiences as well?

The researchers used an independent-samples *t*-test to compare the number of dreams containing recognizable music for a sample of musicians and a sample of nonmusicians. They found that musicians' dreams were significantly more likely to have musical content than were the dreams of the nonmusicians. (Was science fiction writer Philip K. Dick right, then, when he suggested that "androids dream of electric sheep"?)

Chapter 10 will talk about *t*-tests and how they are used to compare groups.

OVERVIEW

INFERENTIAL TESTING SO FAR

WILLIAM GOSSET AND THE DEVELOPMENT OF THE *t*-TEST

THE HISTORICAL CONTEXT: STATISTICS AND BEER

"STUDENT'S" FAMOUS TEST

WHEN BOTH σ AND μ ARE UNKNOWN

INDEPENDENT-SAMPLES *t*-TESTS ASSUMPTIONS

DEPENDENT-SAMPLES *t*-TESTS

t-TESTS IN CONTEXT: "GARBAGE IN, GARBAGE OUT"

LEARNING OBJECTIVES

Reading this chapter will help you to . . .

- Describe how Gosset's contribution to statistics improved estimations of the population standard deviation using small samples. (Concept)

- Explain how the *z*-score, the *z*-test, and the *t*-test are related. (Concept)

- Understand degrees of freedom and how they are related to sample size. (Concept)

- Identify independent and dependent samples in real data. (Concept and Application)

- Define the standard error of the difference, and explain how it is calculated from the pooled standard deviations in the two groups being compared. (Concept)

- Calculate a *t*-value for single-sample, independent-sample, and dependent-samples *t*-tests, and use the statistic to determine whether the null hypothesis should be rejected or retained (Application)

- Discuss the assumptions that underlie *t*-tests. (Concept)

INFERENTIAL TESTING SO FAR

Chapter 9 introduced the *z*-test, which allows researchers to compare the mean of a sample with the mean of a population and determine if that difference is likely or unlikely under certain conditions. A scientist just needs to follow the guidelines set up by the central limit theorem and select a random sample from a normally distributed population. Providing the sample is large (remember that the larger *n* is, the more the SDM represents the population), the mean of the SDM will be the same as the mean of the population ($\mu_{\bar{X}} = \mu$), and the variability in the SDM will equal the variability in the population divided by the square root of the sample size ($\sigma_{\bar{X}} = \frac{\sigma}{\sqrt{n}}$).

When creating a statistical test, it is quite useful to know the parameters that describe the population. In reality, however, down in the bowels of the lab, we may work toward getting the biggest sample possible, but typically our samples will be nowhere near the size the *z*-test assumes them to be. Then, we often have another problem that makes using a *z*-test difficult: We might know the mean of the population, but we often don't know the population standard deviation. Average performance in a population can be estimated ("known") fairly easily: It is often a combination of performance levels seen in the past, the design of a specific test or object, and/or our predictions about what we are likely to see. Variability in performance is more of a moving target. Differences among individual people or things in the way they respond to a stimulus can be quite variable, and can change over time as well. This makes variance harder to know before testing begins.

Think for a moment about manufacturing something like light bulbs. According to 1000Bulbs.com, a lighting blog, incandescent bulbs are designed to last an average of 1,200 hours. This means that the average lifespan in the population is "known." However, we also know that not all of the light bulbs in the population will last exactly that long, and some may last longer. The length of time that incandescent bulbs actually last is variable and unknown. What to do, what to do?

WILLIAM GOSSET AND THE DEVELOPMENT OF THE *t*-TEST

A young man named William Gosset developed the solution to the problem of both small samples and unknown standard deviations. Working for the Guinness Brewing Company at the turn of the twentieth century, he developed a method for comparing means using the basic format of the *z*-test, but with small samples. Gosset knew that the statistics describing samples—like the mean and standard deviation of a sample—could be used as estimates of the parameters of the population. He also knew that very large samples provided better estimates of the population parameters than did small samples, for fairly obvious reasons.

Working with various grains and yeasts, Gosset examined sample after sample and found a pattern in the way the standard deviation in an individual small sample estimated the variability in the population: The standard deviation of a sample tended to slightly *underestimate* the standard deviation of the population. In other words, there was slightly less variability in a sample taken from a population than there was in the original population. Gosset needed a way to make the sample standard deviation just a little bit bigger in order for it to provide a better estimation of the population. A look at Box 10.1 will remind you of the formula for the standard deviation, or SD.

BOX 10.1   Calculating the sample standard deviation

$$s = \sqrt{\frac{\sum(x - \bar{X})^2}{n}}$$

Question: What would happen to the value of *s* if we made *n* just a little bit *smaller*?

If *n* were slightly smaller, then we would be dividing the sum of all the deviations from the mean by a slightly smaller number, which would make SD just a little bit bigger and, thus, a better estimate of the population standard deviation. The formula that we have been using to find the standard deviation up to this point is actually the formula you would use to calculate the standard deviation in a population, when you don't have to use a sample to estimate average deviation from the mean. When we have access to all of the observations in a population, we can divide deviations from the mean by *n*. When we have to use a subset of the population—a sample—to estimate variability in the population, we divide by $n - 1$.

Gosset rewrote the equation for finding *z* in the *z*-test using his new estimate for the population variability (as shown in Box 10.2). Now he had a method for finding the position of a sample mean in the SDM that works when using small samples to estimate the standard error of the means. This would come to be known as **Student's *t*-test**, for reasons explained in *The Historical Context* on page 393.

Gosset also postulated the *t*-distribution. Like the distribution of *z*-scores, the *t*-distribution showed the probability of a particular *t*-score as a function of that *t*'s distance from the mean in estimated standard error units. He calculated these probabilities using his formula to estimate the variability of the SDM with $n - 1$ (a shortened version of his *t*-distribution is shown in Appendix B).

Student's *t*-test The inferential test proposed by William Gosset to allow comparison of two sample means when sample sizes are small and the parameters of the population are unknown.

BOX 10.2 The *z*-test and *t*-test compared

$$s = \sqrt{\frac{\sum(x-\bar{X})^2}{n}} \quad \text{changed to} \quad \hat{s} = \sqrt{\frac{\sum(x-\bar{X})^2}{n-1}}$$

Note the change in the symbols: *s* represents the standard deviation in a sample, while $\hat{s}$ (pronounced "*s*-hat") can be used as an estimate of the population standard deviation.

$$z = \frac{\bar{X} - \mu_{\bar{X}}}{\sigma/\sqrt{n}} \quad \text{changed to} \quad t = \frac{\bar{X} - \mu_{\bar{X}}}{\hat{s}/\sqrt{n}}$$

For a *z*-test, we calculate the standard error of the means (variability in the SDM) using the population standard deviation (σ). When we don't know the population standard deviation, we estimate it ($\hat{s}$), and then we estimate the standard error of the means with that guess about the population parameters.

"STUDENT'S" FAMOUS TEST

Our next topic is Mr. Gosset's *t*-test. At work in his laboratory, Gosset faced a number of problems, two of which we have already discussed: He needed to test different strains of barley, hops, and so on, using small samples, and he often had no idea how much variability in size, taste, and such existed in the populations of various grains used in making beer. Gosset's test allowed him to make a good guess about the standard deviation of the population, and to do so using a small sample. Let's take a closer look at the *t*-test.

First, the basic format of the test is essentially the same as that for the *z*-test. Remember (and take a look at Box 10.3 if you don't) that the top of the *z*-test formula compared two means, the mean of a population and the mean of a sample that came from that population, in order to see how different they were. On the bottom of the *z*-test formula we have the standard error (*SE*) from the SDM. At the bottom of the *t*-score, we have the **estimated standard error** ($s_{\bar{X}}$).

estimated standard error The standard deviation of a sample, calculated with Gosset's $n-1$ formula and used to estimate the population standard deviation.

The tops of both the *z*-test and the *t*-test are identical because both tests are asking the same question: How different are these two means, the sample mean and the population mean? The bottoms of the *z*- and *t*-tests are also essentially the same. The only real difference is that with the *t*-test, the standard error of the mean (or *SEM*) is based on two *estimates*. The first estimate is called $\hat{s}$ ("*s*-hat"). This estimate uses Gosset's formula for making the sample standard deviation a better guess about the population standard deviation. Since $\hat{s}$ is our estimate of the population standard deviation (σ), we just slide $\hat{s}$ into the formula to find the standard error.

The Historical Context

STATISTICS AND BEER

Born in 1876 in Canterbury, England, William Sealy Gosset attended New College at Oxford and graduated with degrees in both mathematics and chemistry. In 1899, shortly after graduating, young William took a job working for the Guinness Brewing Company in Dublin, Ireland. You may be wondering what a man now famous for his work as a statistician was doing in a brewery, and you wouldn't be alone. Some scholars have suggested that he was a statistician acting as some sort of adviser to the brewery, while others say he was a brewer who liked to play with statistics on the side. In truth, his job with the brewery often involved working with small samples of various grains and yeasts to improve the brewing process and the taste of the beer. Gosset himself recognized that although statistical tests were available for use in making comparisons between two samples (e.g., the z-test), these tests assumed that you were using very large samples. Large samples were a problem in this case, though, because the process of making beer was (and is) very sensitive to temperature and timing—both processes that are easily controlled in the small scale but are more difficult to manage on a large scale.

Gosset went on to publish the data from his work at the brewery, calling his test a t-test to distinguish it from the z-test. However, don't look for Gosset's work under his own name. The Guinness company had been burned by a former employee who had published some trade secrets. As a result, all employees of the brewery were banned from publishing anything. Gosset got special permission from the head of the company to publish under a pseudonym after he convinced the powers-that-be that what he was writing about had nothing to do with making beer and everything to do with statistics. The pseudonym Gosset chose was "A. Student," a name that stuck with him for the rest of his publishing life. In Gosset's honor, this test he devised is called Student's t-test.

And if you ever feel overwhelmed when faced with a statistical formula, just remember what Gosset wrote about his own relationship with mathematics. R. A. Fisher (another famous mathematician and great friend of Gosset's) apparently had the habit of referring to some mathematical solutions as "self-evident," implying that these mathematical solutions were so simple everyone could see them. Gosset reportedly said that "self-evident translated for him as 'two hours of hard work before I can see why'" (Gosset, cited in Johnson & Kotz, 1997, p. 328).

FIGURE 10.1
William Sealy Gosset (1908).

Because it's based on an estimate, we say that $\dfrac{\hat{s}}{\sqrt{n}}$ is the *estimated* standard error.

Just as we used the format for the z-score, which finds the *position* of a single score in a sample, in order to find the *position* of a single sample mean in the SDM, we use the t-test to find the *position* of a single sample mean in the SDM when we've

BOX 10.3   The *z*-score, *z*-test, and *t*-test compared

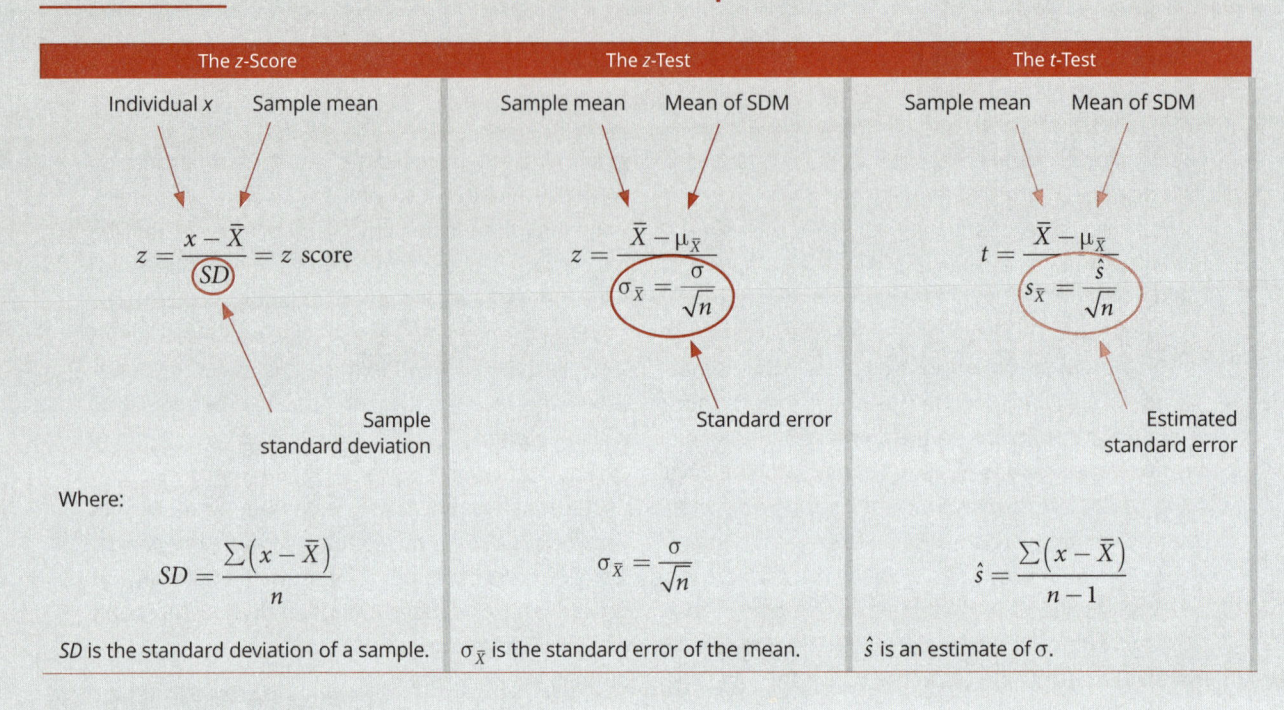

The *z*-Score	The *z*-Test	The *t*-Test
Individual *x*, Sample mean	Sample mean, Mean of SDM	Sample mean, Mean of SDM
$z = \dfrac{x - \bar{X}}{SD} = z\text{ score}$	$z = \dfrac{\bar{X} - \mu_{\bar{X}}}{\sigma_{\bar{X}} = \dfrac{\sigma}{\sqrt{n}}}$	$t = \dfrac{\bar{X} - \mu_{\bar{X}}}{s_{\bar{X}} = \dfrac{\hat{s}}{\sqrt{n}}}$
Sample standard deviation	Standard error	Estimated standard error

Where:

$$SD = \dfrac{\sum (x - \bar{X})}{n}$$

$$\sigma_{\bar{X}} = \dfrac{\sigma}{\sqrt{n}}$$

$$\hat{s} = \dfrac{\sum (x - \bar{X})}{n - 1}$$

SD is the standard deviation of a sample. $\sigma_{\bar{X}}$ is the standard error of the mean. $\hat{s}$ is an estimate of σ.

used an estimate of the population standard deviation. The single-sample *t*-test is shown in Box 10.3, so let's try an example using that test.

THE SINGLE-SAMPLE *t*-TEST

Let's return to the sleepy college students and see how they are doing. While we've been gone, an enterprising researcher has decided to see what effect a daily program of exercise might have on the number of hours of sleep these students are getting. So, we now have a sample, selected at random from the population of currently enrolled undergraduate college students. We want to compare their average sleep duration after 3 months of this daily program of exercise with the mean sleep duration in the population. We know the population mean sleep time (we will use the standard recommendation of 8 hours of sleep per night), but this time around, we don't know how much variability there is in the population. Box 10.4 shows us what we know and what we'll need to determine in order to make this comparison.

We have more blanks in what we know than we did when we were using the *z*-test in Chapter 9. That's because we don't know quite as much going into the *t*-test. The steps involved are as follows:

BOX 10.4 **Research question: Does regular daily exercise change sleep time?**

What We Know	The Sample	Hypotheses in Words	Hypotheses in Symbols
$\mu = 8$ hours	9	The null	The null
$\sigma = $ UNKNOWN	6		
$\mu_{\bar{X}} = 8$ hours	10	The average sleep duration for students who exercised *is the same as* the average sleep duration for students in the general population.	$\mu_{\bar{X}} = 8.00$
$s_{\bar{X}} = \dfrac{\hat{s}}{\sqrt{n}}$	9		
	10		
	12		
	9		
The sample	11		
$\checkmark \Sigma x = ?$	6		
$\checkmark \bar{X} = ?$	10		
	12		
	6		
$\hat{s} = \sqrt{\dfrac{\Sigma x^2 - \dfrac{(\Sigma x)^2}{n}}{n-1}}$	10	The alternative	The alternative
	8		
	11	The average sleep duration for students who exercised is something other than the average sleep duration for students in the general population	$\mu_{\bar{X}} \neq 8.00$
$n = 15$			
$t = \dfrac{\bar{X} - \mu_{\bar{X}}}{\hat{s}/\sqrt{n}} = ?$			
Critical $t = ?$			
$\alpha = .05$ (1-tailed test)		(Notice this is a non-directional test.)	

1. We need to estimate the standard deviation of the population, which we will do by calculating $\hat{s}$ in the sample.
2. Next, we will use $\hat{s}$ as a stand-in for the population standard deviation and calculate the estimated standard error ($s_{\bar{X}}$).
3. Then, we calculate t.

We also apply the same set of assumptions that we used when performing the *z*-test—specifically, that our sample came from a population that was normally distributed. This allows us to use the central limit theorem. Get your calculator warmed up because here we go. Box 10.5 shows the calculation of *t*.

Now we have a value for *t* (2.42). If you remember, when we were doing a *z*-test, we took our "observed" *z*-value and compared it to a critical value. If the observed *z* was farther out in the tail(s) of the distribution, and if it exceeded a critical *z*-value,

BOX 10.5 Does daily exercise improve sleep time? Calculating t

What We Know	Estimating σ	Calculating t
$\sum x = 139$	$\hat{s} = \sqrt{\dfrac{\sum x^2 - \dfrac{(\sum x)^2}{n}}{n-1}}$	$t = \dfrac{\bar{X} - \mu_{\bar{X}}}{\hat{s}/\sqrt{n}} = \dfrac{9.26 - 8.00}{2.02/3.87}$
$\bar{X} = \dfrac{139}{15} = 9.27$	$= \sqrt{\dfrac{1{,}345 - \left(\dfrac{19{,}321}{15}\right)}{14}}$	$\dfrac{2.02}{3.87} = 0.52$
$\sum x^2 = 1{,}345$	$= \sqrt{\dfrac{1{,}345 - (1{,}288.07)}{14}}$	$t = \dfrac{1.26}{0.52} = 2.42$
$(\sum x)^2 = (139)^2 = 19{,}321$	$= \sqrt{\dfrac{56.93}{14}}$	
	$= \sqrt{4.07} = 2.02$	

then we could say that we could reject the null hypothesis and that we had found support for the alternative hypothesis.

Because we knew the value of the population standard deviation (σ) and, therefore, the value of the standard error of the mean (*SEM*) in the SDM ($\sigma_{\bar{X}}$), we could precisely locate the *z*-score that cut off the top 5% (or the top and bottom 1%, etc.) of the distribution of the probability of all sample *z*-scores. If we did a *z*-test using the outer 5% of the right-hand tail of the distribution as our acceptable measure of "different enough," our critical *z*-value would always be +1.64, no matter how big our sample size.

Gosset recognized that with a *t*-test, this is no longer possible. Why? Because we no longer know the value of either σ or $\sigma_{\bar{X}}$. We need to *estimate* these measures, and because we're estimating, we are using potentially faulty information and creating error in our estimate. Gosset also knew that the bigger the sample, the better it estimated the population parameter σ.

In fact, Gosset hypothesized that there would be an entire family of possible *t*-distributions, each providing an estimate of the population and each dependent on how big the sample was. Gosset distinguished each potential distribution by assigning it a sort of identification code, using a number he called the *degrees of freedom* (*df*).

DEGREES OF FREEDOM

The term **degrees of freedom** refers to the number of values in the calculation of a statistic that are free to *vary* (i.e., free to take on any value). Take a look at Box 10.6 to see an example of what Gosset meant by degrees of freedom.

degrees of freedom The number of observations in a set that are free to vary given one or more mathematical restrictions on the set.

BOX 10.6 Degrees of freedom: An example

Suppose that n = 6 and the mean of these six subjects is 36. Here are the first five elements in the sample. Determine the sixth element.

1) 45
2) 30
3) 40   The first five values can be any number.
4) 41
5) 32
6) x Once they're set, the sixth value = x.

$$\frac{\sum x}{n} = \bar{X}$$

$$\frac{\sum x}{n}(n) = \bar{X}(n)$$

$$\sum x = \bar{X}(n)$$

$$\sum x = 36(6) = 216$$

so ...

$$x = 216 - 188 = 28$$

As you can see, once we discover what the first five measurements are, the last one—the sixth element—is no longer free to vary. There is only one value that the sixth element in the set can take on. If it is any other value, the mean no longer equals 36. There are $n - 1$, or 5, degrees of freedom in this set.

The distribution of the probability of all possible *t*-scores depended on the *size of the sample* used. Take a look at Figure 10.2 to see why the size of the sample is so important.

Because the distribution of *t*-values only *estimates* the population (we are using guesses rather than the true data), we need to know which *t*-distribution we are dealing with in order to determine the critical *t*-value that will define "different enough" for us. Unlike the critical *z*-value, the critical *t*-value will change when the size of the sample changes. This is because the critical *t*-value is dependent on how good our estimate of the population standard deviation is, and that depends on sample size.

Each of the curves in Figure 10.2 represents a distribution of *t*-values based on samples of

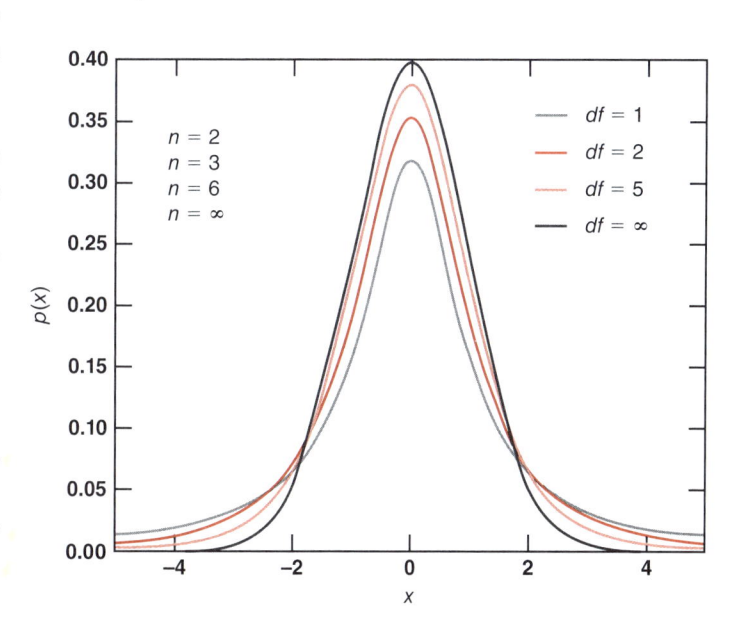

FIGURE 10.2
Relationship between sample size and shape of the *t*-distribution.

different sizes. The mean *t*-value does not change as sample size changes, but the shape of the distribution does. And this change follows a pattern: As *n* gets larger, the distribution becomes more and more normal in shape. In fact, our largest sample is based on an infinitely large sample ($n = \infty$). Basically, a sample size of infinity would be found in a sample that included the entire population. This distribution is perfectly normal in shape. Now take a look at the distribution when $n = 2$: The tails of this distribution are "fat" compared to the normal distribution when $n = \infty$. So, the probability of obtaining a *t*-value out in the tails of this distribution is higher than if we had used larger samples.

Each *t*-distribution varies in its shape according to the size of the sample used minus one—in other words, by its degrees of freedom. So, each *t*-distribution can be identified by its *df*. We're now ready to go find whether our observed *t*-value of 2.42 is significant.

In the null hypothesis, we said that the difference between our sample mean came from a population that looked exactly like the population with a mean of 8.00 hours of sleep. We found that *t* was 2.42, so we now need to find out if our observed *t* exceeds a critical *t*-value. Table 10.1 below shows the table of critical *t*-values, separated by

TABLE 10.1 Abbreviated table of critical *t*-values

1-TAIL α =	.1	.05	.025	.01	.005
2-TAILS α =	.2	.1	.05	.02	.01
df = 1	3.078	6.314	12.706	31.821	63.656
2	1.886	2.920	4.303	6.965	9.925
3	1.638	2.353	3.182	4.541	5.841
4	1.533	2.132	2.776	3.747	4.604
5	1.476	2.015	2.571	3.365	4.032
6	1.440	1.943	2.447	3.143	3.707
7	1.415	1.895	2.365	2.998	3.499
8	1.397	1.860	2.306	2.896	3.355
9	1.383	1.833	2.262	2.821	3.250
10	1.372	1.812	2.228	2.764	3.169
11	1.363	1.796	2.201	2.718	3.106
12	1.356	1.782	2.179	2.681	3.055
13	1.350	1.771	2.160	2.650	3.012
14	1.345	1.761	2.145	2.624	2.977
15	1.341	1.753	2.131	2.602	2.947
16	1.337	1.746	2.120	2.583	2.921

both *df* (down the column on the far left) and alpha (α) level (across the top). Notice that this is a partial table of critical *t*-values. The full table is in Appendix B.

Let's use an alpha level of .05 because the consequences of making a Type I error are not very severe. Take a look at Table 10.1, and find the column that will provide us with a critical *t*-value and $\alpha = .05$ for a two-tailed test. Our sample consisted of 15 students. So, our degrees of freedom is $15 - 1 = 14$, and we read down the column until we come to a *df* of 14. Where the row for *df* = 14 and the column for $\alpha = .05$ intersect, we'll find our critical *t*-value. For our experiment, *t*-critical = 2.145. Our conclusion is shown in Box 10.7.

BOX 10.7 **Checking our observed *t*-value against the critical *t*-value**

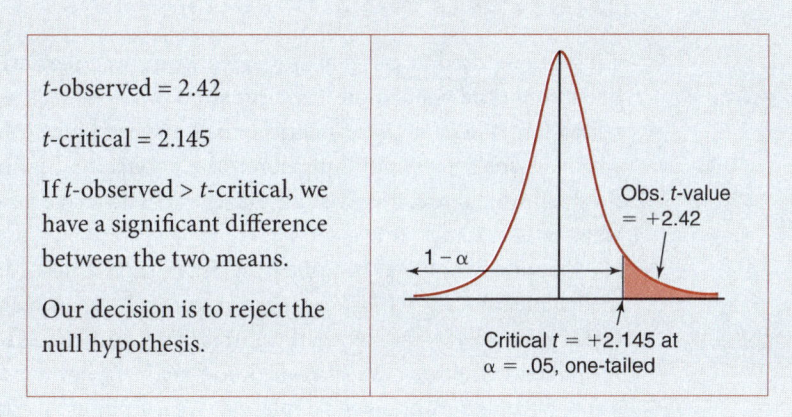

t-observed = 2.42

t-critical = 2.145

If *t*-observed > *t*-critical, we have a significant difference between the two means.

Our decision is to reject the null hypothesis.

Obs. *t*-value = +2.42

Critical *t* = +2.145 at α = .05, one-tailed

Because our observed *t*-value is in the rejection region, we can reject the null hypothesis and say that there is a significant difference between the mean number of hours of sleep for participants who got daily exercise and the average amount of sleep in the null population. In fact, looking at the value of our sample mean, we can conclude that the participants in our sample got significantly more sleep than did people who did not exercise daily.

CheckPoint

Samantha suspects the eight students in her engineering study group earned higher exam grades, on average, than did the students in their class as a whole (i.e., $H_1: \bar{X} > \mu_{\bar{X}}$). The class average for the engineering exam is 70. The exam grades for the members of Samantha's study group are 74, 68, 86, 61, 75, 71, 77, and 72. Answer the following questions to help Samantha discover whether or not her study group's average grade is higher than the class average by a statistically significant amount. Use an alpha level of .05, and round your answers to two decimal places, where appropriate.

1. Calculate $\hat{s}$.
2. Calculate the estimated standard error (ESE).

3. Calculate *t* using a single-sample *t*-test.
4. Calculate *df*.
5. Identify the critical *t*-value, compare it to the *t*-observed, and state whether you should accept or reject the null hypothesis.

Answers to these questions are found on page 402.

WHEN BOTH σ AND μ ARE UNKNOWN

The logic we used to go from situations where we knew everything there was to know about the population (the *z* situation, where we knew both σ and μ) to the *t*-situation (where we have small samples and we know μ but not σ) also works when we don't know anything about the population at all. For a single-sample *t*-test, all we did was use what we did know ($\hat{s}$) to make an educated guess about what we didn't know (σ).

About now you may be wondering, isn't a sample mean an estimate of the population mean? Can't we just use what we know ($\bar{X}$) to estimate what we don't know (μ)? If your mind is taking you down this path, you are correct.

One solution we have when we know neither μ nor σ is to collect two random samples from this unknown population. We use one as an estimate of the null population (the population that has not received the independent variable or IV—or in this case, the *treatment variable*): We will call this group the **control group**. Then, we'll use the other sample, called the **experimental group**, as an estimate of the alternative population, the one that has received the IV, or the treatment variable. We can then compare these two estimates of the population mean to see if the independent variable has had an effect.

This sort of test, where both the population mean and standard deviation are unknown and so we use samples to estimate them, is called a *two-sample t-test*. If these two-sample means are the same, did the treatment variable have an effect? How different do they have to be for us to conclude that the treatment or IV had an effect?

The *independent-samples t-test* and its cousin, the *dependent-samples t-test*, are two-sample *t*-tests. Before we begin our discussion of these two tests, we need to discuss the terms *independent samples* and *dependent samples*.

To really understand what would make two samples we're collecting dependent on one another, it may be easier to think about what would make them independent. **Independent samples** are not linked to one another in any way. The elements in the first sample (the participants, or the plants, or the animals you want

control group A sample, typically selected at random from the population that does not receive the independent variable. This group represents the population in the null hypothesis.

experimental group A sample, typically selected at random, that represents the population of people who have been treated or have received the independent variable. This group represents the effect of the IV.

independent samples Two or more samples that have been drawn by random selection and random assignment of subjects to groups, so the samples are not linked together in any way.

to study) are selected at random, as are the elements in the second sample: There are no deliberate links between these two samples.

It should be fairly obvious that if independent samples are not linked to one another, then **dependent samples** are. In fact, researchers can deliberately link two samples in several ways, and these methods are shown in Table 10.2.

The first method is a classic. It is called the *before-and-after* or *repeated-measures* technique. If you obtain two measurements from each participant, one before you've administered the treatment and the other after you've administered it, then you have dependent samples. The way the subject reacts to the treatment (Group 2) is dependent on the way the subject reacted without the treatment (Group 1) because it's the same person in both conditions.

For example, suppose two people take part in a study of an antianxiety drug. Person 1 is jumpy and anxious, talks way too fast, can't sit still, and so on. Person 2 is phlegmatic, stolid, calm, and appears to barely even notice any change in the environment around him. The effect of the antianxiety drug, if it works, should be linked to the characteristics they displayed before each subject took the drug. The effect the drug has depends on what we might call the personality of the person taking it.

The second method shares some characteristics with the repeated-measures design. In the *matched-samples* method, the researcher starts with an idea of what variables might confound the effect of her independent variable—in other words, which variables might mask the effect of the IV, which ones might mimic it, and the like. So, to eliminate the effect of these other variables as much as possible, the researcher matches each subject on those variables before the experiment begins. If you are testing the same antianxiety drug, you might match all of your subjects on their level of anxiety before you give them the drug—make sure they're all equally anxious at the beginning of the study. Then, randomly assign these anxious participants to either the control group or the experimental group and compare the responses of the experimental group (who received the drug) with the responses of the control group (who did not).

dependent samples Two or more samples that share variability. Individual members of the samples are linked together, dependent on each other, in some way.

TABLE 10.2   Matched-pairs techniques

	Description
Repeated measures	The same subjects are tested repeatedly—for example, before and after administration of the IV (pretest/posttest design).
Matched pairs	Subjects are given a test before the experiment begins and are matched according to their scores on the pretest. All subjects begin the experiment sharing at least one factor: their score on the pretest.
Natural pairs	Pairing of subjects occurs naturally, before the experiment begins—for example, parent–child or sibling connections link the two subjects.

> **CheckPoint**
> *Answers to questions on page 400*
>
> 1. $\hat{s} = 7.21$ ☐
> 2. ESE $= 2.69$ ☐
> 3. $t = 1.12$ ☐
> 4. $df = 7$ ☐
> 5. $1.12 < 1.895 \therefore$ accept ☐
>
> **SCORE:** /5

The third method can be considered a special case of the matched-pairs technique. When you use the *natural-pairs technique*, you are taking advantage of links between subjects that are inborn or genetic as opposed to links created by the experimenter. The best example of this technique is probably a "twin study," where researchers take advantage of the natural, genetic links between twins. Identical twins share the same genetic code, so if a researcher presents an independent variable to one twin and not to the other, we can compare the effect of the IV without worrying about any difference an individual's genetic code might have on the response to the IV.

Whether you are using independent-samples or dependent-samples *t*-tests, the question you're asking with these tests is essentially the same: Are these two guesses about population means the same or different? The form of the test should also look familiar. Remember, we have the difference on the top and the error on the bottom. Let's start with the independent-samples *t*-test.

INDEPENDENT-SAMPLES *t*-TEST

Suppose we want to find out if drinking a glass of warm milk before bedtime actually helps people go to sleep more quickly. Let's further suppose that we don't know the average time it takes for people to fall asleep without warm milk, let alone with it. So, we go to our population and draw out a random sample of 40 people. Then, we *randomly assign* each person to one of two groups—remember, we're trying to eliminate links between the people in our two groups.

Group 1 ($n = 20$) will be our control group: They will not get any warm milk. Group 2 ($n = 20$) is the experimental group: They will get 6 ounces of warm milk 20 minutes before going to bed. Both groups sleep in a sleep lab for the night so that they can be monitored. Using an electroencephalograph (or EEG), researchers note the number of minutes needed for each participant to fall asleep, measured as the time it takes for Sleep Stage 1 EEG patterns to appear in a participant's EEG. Box 10.8 shows you the formula for calculating this *t*.

Take a look at the top of the equation. Once again, we're looking at the difference between the mean of the control group ($\bar{X}_C$), our estimate of the mean time needed to fall asleep in the population of people who do not drink warm milk, and the mean of the experimental group ($\bar{X}_E$), our estimate of the mean time needed to fall sleep in the population of people who drink warm milk, just as we've done in both the *z*-test

BOX 10.8   Formula for calculating *t*

$$t = \frac{\bar{X}_C - \bar{X}_E}{s_{\bar{X}_C - \bar{X}_E}} \qquad \text{where} \qquad s_{\bar{X}_C - \bar{X}_E} = \sqrt{\left(\frac{\hat{s}_C}{\sqrt{n_C}}\right)^2 + \left(\frac{\hat{s}_E}{\sqrt{n_E}}\right)^2}$$

and the single-sample *t*-test. Now take a look at the bottom of the formula. This is our error term. The symbol ($s_{\bar{X}_C - \bar{X}_E}$) refers to a measure of variability in a distribution called the *sampling distribution of the difference between the means*, or *SDDM*.

THE STANDARD ERROR OF THE DIFFERENCE

Previously, we wanted to know where the mean from our empirical sample was in the theoretical sampling distribution of the means (SDM). If our sample mean was unlikely in the SDM, that meant our IV was unlikely to have had any effect and that our null hypothesis could not be rejected. With a single-sample *t*-test, we were looking for the location of a single sample mean in a distribution of all possible means of a given size.

Now we have a different distribution. We want to know how likely a *given difference between two sample means* might be if the null is true, and we need a different kind of sampling distribution. Why not create a distribution of all possible $\bar{X}_C - \bar{X}_E$? One by one, we will select a sample in the control population, subtract from it a sample of the same size from the experimental population, and then put that difference between means aside and do it all again.

We're creating a sampling distribution, but this time, it's a distribution of *differences between means*—the **sampling distribution of the difference between means**, or SDDM. We'll do the *t*-test to see where in this SDDM our empirical difference between the two sample means we've got in front of us actually is. If the null hypothesis is true and warm milk makes no difference at all, then the difference in mean time to fall asleep from each group should be zero (the average difference between the two means if the null is true). If the difference in mean time to fall asleep is other than zero, we need to find out how far away from zero it is, just as we do with a single-sample *t*-test.

The term at the bottom of the independent-samples *t*-test formula is called the estimated standard error of the difference. Taking a closer look at it, we can see that it's really composed of two measurements that ought to look familiar. Box 10.9 compares the error terms of the single-sample and the independent-samples *t*-tests.

When we have two samples to compare, our error term becomes the *pooled variability of both samples, added together*, which serves as an estimate of the standard error of the difference. Because both of our samples are the same size, they are equally good estimates of variability in the population they come from. We're adding these estimates to get a measure of the variability in the SDDM. Easy, as they say, as pie.

sampling distribution of the difference between means
A distribution of the difference between the means of all possible samples of a given size, selected at random from a given population.

BOX 10.9 Error terms in single-sample and independent-samples *t*-tests

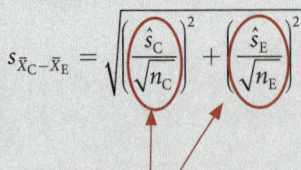

$$s_{\bar{X}_C - \bar{X}_E} = \sqrt{\left(\frac{\hat{s}_C}{\sqrt{n_C}}\right)^2 + \left(\frac{\hat{s}_E}{\sqrt{n_E}}\right)^2}$$

Notice we have *two* estimations of the population standard deviation, one from each sample.

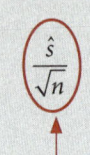

Error term for the independent-samples *t*-test

$$\frac{\hat{s}}{\sqrt{n}}$$

When a single sample is used to estimate standard error.

Error term for the single-sample *t*-test

FINDING THE DIFFERENCE BETWEEN TWO MEANS: AN EXAMPLE

We're ready to try our new *t*-test to see if warm milk makes it easier to fall asleep. We've selected two samples, warmed the milk, and given it to our experimental group. We have hooked everyone up to the EEG, and now we are ready to go. Box 10.10 shows you our setup.

Notice that because we are using two samples, we need to take both of these groups into consideration when we calculate *t* and when we determine the degrees of freedom. So, the calculation of our overall *df* includes the *df* in the control group *and* the *df* in the experimental group. The formula for finding *df* becomes $df = (n_C + n_E) - 2$. Notice also that this is the same as finding the *df* for each group using $n - 1$ and then adding those two values together: $df_{total} = (n_C - 1) + (n_E - 1)$.

Once again, we're comparing two sample means on the top and dividing by a measure of the combined error of both samples on the bottom. We have what we need, so let's get started. The calculations for *t* are shown in Box 10.11.

Our *t*-value of 1.38 tells us that our sample mean is 1.38 estimated standard errors of the difference above the mean of zero. We have a difference in the mean number of minutes it took for the control group (who did not receive any milk before bedtime) and the experimental group (who did receive warm milk) to fall asleep. It looks as though it took the experimental group a bit less time to fall asleep than it took the control group. But we don't know yet if this difference is big enough to be meaningful, or if it is the result of random chance. And we won't know until we compare our observed *t*-value with the critical *t*-value from the table in Appendix B, so let's do that now. Box 10.12 illustrates this comparison.

BOX 10.10 **Research question: Does warm milk decrease the time needed to fall asleep?**

What We Know	The Sample		Hypotheses
Control group	Control	Experimental	The null
$\sum_C = 237.82$	10.65	10.04	
	9.54	21.99	$H_0: \mu_C - \mu_E = 0$
	8.00	10.13	
$n = 20$	12.37	15.72	
	13.14	8.69	
$\bar{X}_C = 11.89$ min	20.00	8.52	
	10.15	10.16	The alternative
Experimental group	10.56	10.00	
	11.88	10.32	$H_1: \mu_C - \mu_E \neq 0$
$\sum_E = 205.00$ min	8.30	7.48	
	7.58	11.90	Two-tailed test
$n = 20$	16.52	7.07	
	16.81	5.17	or
$\bar{X}_E = 10.25$	12.00	7.74	
	16.91	5.33	$H_1: \mu_C - \mu_E > 0$
$df = (n_C + n_E) - 2$	7.60	13.03	
	8.59	9.36	One-tailed test
$t = \dfrac{\bar{X}_C - \bar{X}_E}{s_{\bar{X}_C - \bar{X}_E}}$	10.22	11.26	
	9.73	9.82	
	17.27	11.27	
	$\sum_C = 237.82$	$\sum_E = 205.00$	

Notice that the experimental group (represented by the gray dashed line) does in fact have a mean that is above zero (the red control group)—but is it different enough for us to call it significant? To determine this, we need to find our critical *t*-value, and to do that, we need to know our degrees of freedom. To find our *df*, we sum or pool the *df* for each group (*df* = 19 + 19 = 38). Then, we go to the table of critical *t*-values.

We have failed to reject the null hypothesis, which said that warm milk would have no significant effect on the number of minutes needed to fall asleep. Warm milk may make you feel better, but it has no meaningful effect on our dependent

BOX 10.11 Calculating *t* for our soporific milk study

$$t = \frac{\bar{X}_C - \bar{X}_E}{s_{\bar{X}_C - \bar{X}_E}} \quad \text{where} \quad s_{\bar{X}_C - \bar{X}_E} = \sqrt{\left(\frac{\hat{s}_C}{\sqrt{n_C}}\right)^2 + \left(\frac{\hat{s}_E}{\sqrt{n_E}}\right)^2}$$

which can be rewritten as ...

$$s_{\bar{X}_C - \bar{X}_E} = \sqrt{\frac{\sum x_C^2 - \frac{(\sum x_C)^2}{n_C} + \sum x_E^2 - \frac{(\sum x_E)^2}{n_E}}{n_C(n_E - 1)}}$$

$$t = \frac{\bar{X}_C - \bar{X}_E}{\sqrt{\frac{\sum x_C^2 - \frac{(\sum x_C)^2}{n_C} + \sum x_E^2 - \frac{(\sum x_E)^2}{n_E}}{n_C(n_E - 1)}}} = \frac{11.87 - 10.25}{\sqrt{\frac{3,090.48 - \frac{(237.82)^2}{20} + 2,362.14 - \frac{(205.00)^2}{20}}{20(20 - 1)}}}$$

$$= \frac{1.62}{\sqrt{\frac{\left(3,090.48 - \frac{56,558.35}{20}\right) + \left(2,362.14 - \frac{42,025}{20}\right)}{380}}} = \frac{1.62}{\sqrt{\frac{(3,090.48 - 2,827.92) + (2,362.14 - 2,101.25)}{380}}}$$

$$= \frac{1.62}{\sqrt{\frac{(262.56 + 260.89)}{380}}} = \frac{1.62}{\sqrt{\frac{523.45}{380}}} = \frac{1.62}{\sqrt{1.38}} = \frac{1.62}{1.17} = 1.38$$

---

variable. The mean number of minutes needed to fall asleep was slightly greater in the control group than in the experimental group, but that difference is not significant.

FINDING THE DIFFERENCE BETWEEN MEANS WITH UNEQUAL SAMPLES

In the example that we just finished working, we had two samples with the same number of people in each group. Ideally, we'd like for our groups to have equal numbers all the time, but very often, we wind up with unequal *n*'s. There are many reasons for this, but perhaps the most common is something we call **subject mortality**. No, it doesn't mean that someone in your study has died; this

subject mortality The voluntary withdrawal of a participant from a study.

BOX 10.12 Comparing the observed and critical *t*-values for our sleep study

The sampling distribution of the difference between the means.

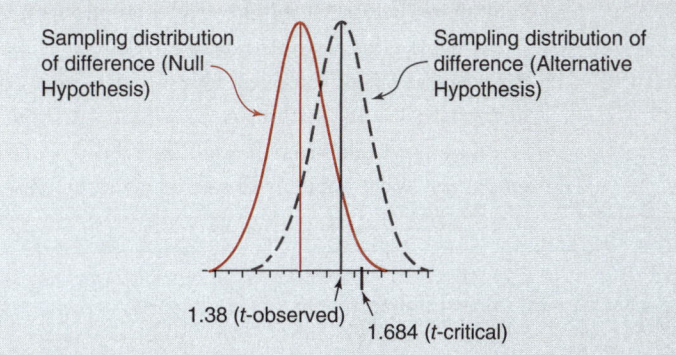

Sampling distribution of difference (Null Hypothesis)

Sampling distribution of difference (Alternative Hypothesis)

1.38 (*t*-observed)   1.684 (*t*-critical)

is a different, and far more serious, problem that cannot be handled by adjusting a formula. Instead, it means that we have had a participant or two drop out of the study and refuse to participate any further. It isn't what we want to happen, but participants always have the right to refuse to participate at any point in the study. So, it's a problem we need to be ready for.

If one group is larger than the other, then when we use our *t*-test, we will have one estimate that is better than the other. We are thus more likely to have a *t*-value that is biased toward the larger group. We're going to need to adjust the formula to correct for this difference. What should we adjust? Well, the top of the formula probably shouldn't be changed. Since we're using the standard deviations in each sample as guesses about the standard deviation in each population (control and experimental), how about a change here? You guessed it: We'll have to adjust the error term on the bottom to correct this problem. Box 10.13 shows the formula with the adjustment for unequal *n*'s.

This should look somewhat familiar—you've seen all but the very last term here before. We're still summing the variability in the two samples, and you can see the expanded form of the equations for both in the formula. In the red box, we have the formula for standard deviation in the control group, and in the gray

BOX 10.13 Adjusting the *t*-formula for unequal *n*'s

$$t = \frac{\bar{X}_C - \bar{X}_E}{s_{\bar{X}_C - \bar{X}_E}} = \frac{\bar{X}_C - \bar{X}_E}{\sqrt{\frac{\left[\sum x_{C^2} - \frac{(\sum x_C)^2}{n_C}\right] + \left[\sum x_{E^2} - \frac{(\sum x_E)^2}{n_E}\right]}{(n_C + n_E) - 2} \left(\frac{1}{n_C} + \frac{1}{n_E}\right)}}$$

box, we have the same for the experimental group. After we've summed these two boxes together, we divide by $df = (n_C + n_E) - 2$. We're just pooling these two measures of variability. The only term that is new is the last one: $(1/n_C + 1/n_E)$. This is a "weighting term," which you can think of as a way to make the larger sample count a bit more (because it should be a better guess) than the smaller sample.

Let's try this: Using the data we've already collected from our warm milk study (just to make things simpler), let's suppose that three people from the control group drop out (sleeping with electrodes pasted to your head is not as much fun as it sounds, and these three decided they didn't want to continue). Now we have unequal n's. What happens to our conclusions? Box 10.14 shows the data from the

---

BOX 10.14 **Our research question remains the same**

What We Know	The Samples		Hypotheses
Control group	Control	Experimental	The null
$\sum_C = 211.66$	10.65	10.04	$H_0: \mu_C = \mu_E$
	9.54	21.99	
	8.00	10.13	or
$n = 17$	12.37	15.72	
	13.14	8.69	$H_0: \mu_C = \mu_E = 0$
$\bar{X}_C = 12.45$ min	20.00	8.52	
	10.15	10.16	
Experimental group	10.56	10.00	The alternative
	11.88	10.32	
$\sum_E = 205.00$ min	8.30	7.48	$H_1: \mu_C > \mu_E$
	7.58	11.90	
	16.52	7.07	
$n = 20$	16.81	5.17	or
	12.00	7.74	
$\bar{X}_E = 10.25$	16.91	5.33	$H_1: \mu_C - \mu_E > 0$
	7.60	13.03	
$df = (n_C + n_E) - 2$	8.59	9.36	One-tailed test.
	10.22	11.26	
$t = \dfrac{\bar{X}_C - \bar{X}_E}{s_{\bar{X}_C - \bar{X}_E}}$	9.73	9.82	
	17.27	11.27	
	$\sum_C = 211.66$	$\sum_E = 205.00$	
	$n = 17$	$n = 20$	

warm milk study with three observations "grayed out." These observations have been eliminated from the data set and the resulting calculations of t. Box 10.15 shows the calculation of t using the appropriate formula and our new data.

BOX 10.15 **The test**

$$t = \frac{\bar{X}_C - \bar{X}_E}{s_{\bar{X}_C - \bar{X}_E}} = \frac{\bar{X}_C - \bar{X}_E}{\sqrt{\dfrac{\sum x_C^2 - \dfrac{(\sum x_C)^2}{n_C} + \sum x_E^2 - \dfrac{(\sum x_E)^2}{n_E}}{(n_C + n_E) - 2}\left(\dfrac{1}{n_C} + \dfrac{1}{n_E}\right)}}$$

$$= \frac{12.45 - 10.25}{\sqrt{\dfrac{2{,}857.21 - \dfrac{(211.66)^2}{17} + 2{,}362.14 - \dfrac{(205)^2}{20}}{(17 + 20) - 2}\left(\dfrac{1}{17} + \dfrac{1}{20}\right)}}$$

$$= \frac{2.2}{\sqrt{\dfrac{2{,}857.21 - \dfrac{44{,}799.96}{17} + 2{,}362.14 - \dfrac{42{,}025}{20}}{35}(0.06 + 0.05)}}$$

$$= \frac{2.2}{\sqrt{\dfrac{(2{,}857.21 - 2{,}635.29) + (2{,}362.14 - 2{,}101.25)}{35}(0.11)}}$$

$$= \frac{2.2}{\sqrt{\dfrac{221.92 \_ 260.89}{35}(0.11)}}$$

$$= \frac{2.2}{\sqrt{\dfrac{482.81}{35}(0.11)}} = \frac{2.2}{\sqrt{13.79(0.11)}} = \frac{2.2}{\sqrt{1.52}} = \frac{2.2}{1.23} = \boxed{1.79}$$

Our observed *t*-value is 1.79. When we compare it with the tabled critical value when *df* = 35 (see Appendix B), we find ourselves with the same problem we had last time we used the table. We have a critical value for *df* = 30, or for *df* = 40, but not for *df* = 35. I would again use the critical value for *df* = 30 (*t*-critical = 1.697) because it is farther out in the tail than is the one for *df* = 40 (*t*-critical = 1.684).

Our observed *t*-value is 1.79, and our critical *t*-value is 1.697. Our *t*-observed is larger than the *t*-critical—in other words, *t*-observed is farther out in the tail of the distribution than is *t*-critical—so we can *reject the null hypothesis*. We can now conclude that a glass of warm milk does make it easier to fall asleep. If you take a look at the data, you can see why we were more likely to see a significant result when the three participants dropped out. Eliminating three data points from the control group changed the mean number of minutes needed for the people in this group to fall asleep. The mean decreased, which made the difference between the means of the control and experimental group larger, which resulted in a different conclusion about the effects of warm milk.

By the way, you can use the formula for unequal n's even when the n's are equal. It will give you exactly the same answer for the standard error of the difference. Try it yourself using the data from the equal n's example and see. I've included the calculations at the end of the chapter to check yourself (see *Using the Formula for Unequal n's with Equal n's* on page 433).

CheckPoint

There is a large body of psychological research devoted to social priming, which is the idea that the presentation of certain stimuli can change people's attitudes or behavior. Working within this research field, Ruwayda devised an experiment to test her hypothesis that social priming with money will increase people's competitiveness (i.e., $H_1: \mu_C < \mu_E$). She asked a control group and an experimental group, each comprising 10 randomly selected participants, to complete a survey that measured their competitiveness. The control group completed the survey in an unadorned room. The experimental group completed the survey in the same room after a poster depicting paper money had been tacked to the wall in clear view of the participants. The observed competitiveness scores are displayed in the table here.

Using the data in the table, answer the following questions to help Ruwayda determine whether the money-primed group's competitiveness scores are higher than those of the experimental group by a statistically significant amount at an alpha level of .05. Round your answers to two decimal places, where appropriate.

Competitiveness Scores

Control Group	Experimental Group
6.4	5.8
5.2	6.3
7.8	5.9
6.1	8.8
2.3	7.3
4.7	5.6
8.1	6.8
4.6	3.6
6.8	4.7
5.9	6.5
?$_C$ = 57.9	?$_E$ = 61.3
n = 10	n = 10

1. Calculate the estimated standard error of the difference.
2. Calculate t using an independent samples t-test.
3. Calculate df.
4. Identify the critical t-value, compare the result to t-observed, and state whether you should accept or reject the null hypothesis.

Answers to these questions are found on page 412.

ASSUMPTIONS

When we use a *t*-test to analyze our data, we are making a series of **assumptions**. If we violate these assumptions, then the results of our *t*-test may be compromised. First, we are assuming that the dependent variable distributions are reasonably *normally distributed* and indicate that the two populations that we're comparing are also *normally distributed*. Second, we are assuming that both groups (control and experimental) are *approximately equal in variability*. And third, we are assuming that the groups (control and experimental) are *independent* of one another. This last assumption is appropriate only for the independent-samples *t*-test.

If we violate any of these assumptions, the results of the *t*-test may be misleading. But not to worry, there are solutions to violations of each of these assumptions. For example, if the assumption of normality has been violated (i.e., if one or both of our samples have serious outliers in them), we can use a nonparametric test to compare the two groups (we will talk about what a nonparametric test is and why it solves the problem of skewed data in Chapter 14). Alternatively, if the population distribution is reasonably normal (i.e., not severely skewed), we can go ahead and use the *t*-test because statisticians have discovered that the *t*-test is *robust*. A **robust** test is one that is resistant to deviations from the assumptions (including the assumption of normality) that underlie all hypothesis testing. A *t*-test will provide us with reasonably accurate results even when our population is not perfectly normal. This is particularly true when the sample sizes are equal for both groups, and when our samples are large.

If the second assumption of equal variability has been violated and the standard deviations of these two groups are really very different, there are some modifications we can make to the formula used to calculate *t*. These modifications are a bit beyond the scope of this book, but if you use a software package like SPSS to analyze your data, the program will automatically check to make sure you haven't violated this assumption—and if you have, it will recalculate *t* using an alternative formula.

Lastly, if we've violated the assumption of the independence of the two groups (control and experimental), we can use what is known as a *paired-samples t-test*, also known as the *dependent-samples t-test*, which is our next topic.

> **assumptions** The Conditions or requirements for carrying out a particular hypothesis test.

> **robust** Describing a test or statistic that is resistant to errors that are the result of violations of an assumption (e.g., the assumption of normality).

Think About It . . .

t-TESTS AND SAMPLE SIZE

Mr. Gossett, slaving away over a hot batch of hops, wanted to find a way to run inferential tests when the sample size was very small. He didn't want to use a *z*-test because it required that you know the population standard deviation. He knew that large sample sizes would be beneficial when the population standard deviation was unknown, because large samples would give a better estimate of the population

Think About It... continued

parameters than small ones would. But Gossett needed to use small samples, so he set about developing a method that would allow him to do so effectively.

Just what effect does sample size have on a *t*-test? Let's start with the formula for calculating our *t*-statistic for a single-sample *t*-test and "reverse engineer" the *t*-equation to solve for the difference (*D*) between $\bar{X}_E$ and $\mu_{\bar{X}}$ in this situation:

$$t = \frac{\bar{X} - \mu_{\bar{X}}}{\frac{\hat{s}}{\sqrt{n}}} = \frac{D}{\frac{\hat{s}}{\sqrt{n}}}$$

Minimal *D* required for significance can be found using $D = (t_{crit})\left(\frac{\hat{s}}{\sqrt{n}}\right)$

1. What happens to the value of *t* as *n* increases? _____

2. What happens to the critical value of *t* as *n* increases? _____

3. Suppose $\hat{s} = 10$ and $n = 2$. Use a one-tailed test and $\alpha = .05$ for all critical values.
    a. If $n = 2$, $df = $ _____, and the critical *t*-value would be _____.
    b. If $n = 2$, the difference between the two means must be _____ for *t* to be significant.

4. Now let's increase *n*.
    a. If $n = 20$, *df* would be _____, and the critical *t*-value would be _____.
    b. Now how different must the two means be for *t* to be significant?
    _____
    c. If $n = 100$, the difference between the two means must be _____ for *t* to be significant.

Is it possible to have too small a sample size? J. C. F. de Winter (2013) at the Delft University of Technology set up his computer to run 100,000 repetitions of *t*-tests (all three types that we have considered here) with an extremely small sample size of $n = 2$. He found that even when the sample consisted of only two measurements, the *t*-test still worked well. The *t*-test is indeed quite robust.

CheckPoint
Answers to questions on page 410

1. $s_{\bar{X}_C - \bar{X}_E} = 0.70$ ☐
2. $t = -0.49$ ☐
3. $df = 18$ ☐
4. $0.49 < 1.734 \therefore$ we accept the null ☐

SCORE: /4

DEPENDENT-SAMPLES *t*-TESTS

INTRODUCTION

With an independent-samples test, we can compare two sample means that represent two different populations: the control group and the experimental group. When our samples are independent of one another—as they were in our milk sleep study—it means the scores in one sample do not depend on the scores in the other. The fact that the two sets of measurements are independent of one another also means that the amount of variability in the control group is independent of the amount of variability in the experimental group. Since there is no overlap between the two measures of variability, we can add them together without having to be concerned that one measure is related to the other.

At least one of you is out there saying, "*A-ha!* That must mean that if the two groups are *dependent* on one another, then our measures of variability are also dependent, and so we will have to change the way we calculate the error term in the *t*-test!" Well, congratulations, you're correct!

When we want to compare two sample means that are dependent on one another, we will need to compensate for the fact that the two samples are not independent in our calculation of the error term. Meanwhile, the top of the *t*-test formula, indicating the difference between the two means, stays the same. (I should tell you that the mathematics required are very simple—much simpler than what was required for the problem we just worked through.)

USING A DEPENDENT-SAMPLES *t*-TEST: AN EXAMPLE

Here is our example: There's an old saying that in the spring, a young man's fancy lightly turns to thoughts of love. We'll update this and remove the inherent sexism, rephrasing it to say that young people of both sexes are influenced by the season, and that in the spring, they are more likely to be distracted by thoughts of romance than they are in the fall.

We will select a random sample of 12 young people enrolled in college, and we'll measure their distractedness, once in the fall and again in the spring, using a repeated-measures design. We're testing each subject twice, once before they should be distracted by love and again after the spring has had its effect on everyone.

To measure distractedness, we will ask the participants to take a reaction time test developed to measure distraction (Lavie & Forster, 2007). On this test, participants are asked to look for a target stimulus that is embedded in an array of distracter stimuli. As soon as they've found the target, they press the key on the computer keyboard that matches the target stimulus. The time needed to press the key is recorded, with shorter reaction times indicating less distractibility. We will also measure distraction at the same relative point in each semester (let's say midterm) to try to equalize the stress levels created by a potential host of variables other than the season.

We are using a dependent-samples *t*-test to assess our hypothesis. We could have selected a random sample of people in the fall, measured their distractedness, and put that data aside so that we could get a different sample in the spring to compare them with. But a repeated-measures design will give us cleaner and more powerful results. Why? Because most of the variables that might affect sleep time other than the time of year will be consistent in the life of the participant (or at least we hope so) across time. This way, we can look just at the effects of the season on their distractibility.

CALCULATIONS AND RESULTS

Box 10.16 shows the formula for calculating *t* in a dependent-samples case. Notice that we don't have to worry about differing sample sizes with this kind of design: We need *pairs* of measurements (two from each person), and anyone without a matched pair of measurements cannot be included in the data set.

BOX 10.16 **Formula for a dependent-samples *t*-test**

$$t = \frac{\bar{X}_{fall} - \bar{X}_{spring}}{s_{\bar{D}}} \quad \text{where} \ldots \quad s_{\bar{D}} = \frac{\hat{s}_D}{\sqrt{n}} = \frac{\sqrt{\frac{\sum D^2 - \frac{(\sum D)^2}{n}}{n-1}}}{\sqrt{n}}$$

and ...

$$D = x_{1_{fall}} - x_{1_{spring}}$$

At least some of this will look familiar. Of course, the top of the *t*-formula looks familiar because, after all, we're comparing two means again. But the error term on the bottom has once again changed. Even so, the basic format of all of the *t*-tests we'll discuss is the same. We're looking for another version of the standard error of the difference and the equation for estimating it, and except for the fact that we're dealing with "D's" instead of "*x*'s," it should also look really familiar.

Basically, we're going to take two scores (fall and spring) from each participant and subtract them, one from the other. For subject 1, we'll subtract his spring distractedness score ($x_{1_{spring}}$) from his fall distractedness score ($x_{1_{fall}}$) to find the difference, for that one person, between spring and fall. We will repeat this for each person in our sample and, essentially, squish the two sets of measurements we got from each participant into a single number: the difference score, or D. We will wind up with one D score per person.

Next, we will find the standard deviation of the D scores using the $n - 1$ formula. Then, we will divide that by the square root of n to get the estimated standard error of the difference scores. (By the way, have you noticed that we're using the term *standard error* when we're talking about sampling distributions, and the term *standard deviation* when we're talking about populations or samples? This will help make the distinction between distributions clear.) In Box 10.17, we can see the whole setup for our experiment. Box 10.18 shows the calculation of t.

BOX 10.17 **Are college students more distracted in the spring than in the fall?**

What We Know	The Rata (Reaction Time in Milliseconds)				The Hypotheses
	Fall	Spring	D	D²	
$t = \dfrac{\bar{X}_{fall} - \bar{X}_{spring}}{s_{\bar{D}}}$	539	715	−176	30,976	The null
	548	710	−162	26,242	
	534	776	−242	58,564	$\mu_f - \mu_s = 0$
$s_{\bar{D}} = \dfrac{\hat{s}_D}{\sqrt{n}}$	433	357	76	5,776	or
	548	734	−186	34,596	$\mu_D = 0$
	514	727	−213	45,369	
$= \dfrac{\sqrt{\dfrac{\sum D^2 - \dfrac{(\sum D)^2}{n}}{n-1}}}{\sqrt{n}}$	336	734	−388	150,544	The alternative
	541	706	−165	27,225	
	543	717	−174	30,276	$\mu_f - \mu_s \neq 0$
	646	738	−92	8,464	or
$D = X - Y$	556	526	30	900	$\mu_D \neq 0$
	549	640	−91	8,281	
	$\Sigma = 6{,}297$	$\Sigma = 8{,}080$	$\Sigma = -1{,}783$	$\Sigma = 427{,}215$	This is a two-tailed test.

The difference between distractedness scores in the fall and the spring (the top of the *t*-equation) has the same values as the average or mean difference ($\bar{D}$) score. And by converting our two sets of measurements (fall and spring distractedness scores) into a single column of D scores, we're now essentially doing a single-sample *t*-test. Our single sample of measurements consists of the single column of D scores. The error term reflects this. We're using the sample of D scores to estimate the variability in the population, $\hat{s}_D$, and then using that estimate of the population variability to estimate the standard error, $\dfrac{\hat{s}_D}{\sqrt{n}}$.

We've now got an observed *t*-value of −4.23. Is it significant? We'll need to compare this value with a critical value from the *t*-table in Appendix B. Given how

BOX 10.18 Calculating t

What We Know	Calculations
$(\sum D) = -1{,}783$	$t = \dfrac{\bar{X}_{fall} - \bar{X}_{spring}}{s_{\bar{D}}} = \dfrac{\bar{D}}{s_{\bar{D}}}$
$\sum D^2 = 427{,}215$	
$n = 12$	$t = \dfrac{-148.58}{35.11} = -4.23$
$\sum_{fall} = 6{,}297$	
$\sum_{spring} = 8{,}080$	$s_{\bar{D}} = \dfrac{\hat{s}_D}{\sqrt{n}} = \dfrac{\sqrt{\dfrac{\sum D^2 - \dfrac{(\sum D)^2}{n}}{n-1}}}{\sqrt{n}} = \dfrac{\sqrt{\dfrac{427{,}215 - \dfrac{(-1{,}783)^2}{12}}{12-1}}}{\sqrt{12}}$
$\bar{X}_{fall} = 524.75$	
$\bar{X}_{spring} = 673.33$	$= \dfrac{\sqrt{\dfrac{427{,}215 - \dfrac{3{,}179{,}089}{12}}{12-1}}}{\sqrt{12}} = \dfrac{\sqrt{\dfrac{427{,}215 - 264{,}924.08}{12-1}}}{3.46}$
$\bar{X}_{fall} - \bar{X}_{spring} = -148.58$	
$\hat{D} = \dfrac{\sum D}{n} = \dfrac{-1{,}783}{12} = -148.58$	$= \dfrac{\sqrt{\dfrac{162{,}290.92}{11}}}{3.46} = \dfrac{\sqrt{14{,}753.72}}{3.46} = \dfrac{121.46}{3.46} = 35.11$

we've calculated t, what would the df be? If you said the number of pairs minus one, you're correct. Our df is $12 - 1$, or 11. We use the same t-table that we've been using right along, and we'll use an alpha of .05, again because the consequences of a Type I error are not serious.

You should find that for a two-tailed test, the critical t-value is -2.201. Can we reject the null hypothesis here? Yes, because our observed t-value is farther out in the tail than the critical t-value. Distractedness scores change significantly from fall to spring. Our data show that the participants in our sample were significantly more distracted in the spring than they were in the fall (the reaction times tended to be 148 milliseconds longer in the spring than in the fall, indicating more distractibility).

CheckPoint

Ashton wonders if dog owners in her hometown of Chicago are less likely to walk their dogs outdoors in the winter, when it is cold and more uncomfortable to be outside, than in the summer (i.e., $H_1: \mu_S > \mu_W$). To test her hypothesis, Ashton records the number of times per week a sample of 10 randomly selected dog owners walk their dogs in August. Six months later, she contacts the same dog owners again to find out the number of times per week they walk their dogs in February. The observed frequencies of dog walks are displayed in the table below.

Number of Dog walks in a week

Summer	Winter
5	5
2	3
5	3
4	4
6	4
3	2
7	4
5	4
5	3
6	4
$\Sigma = 48$	$\Sigma = 36$
$n = 10$	$n = 10$

Using the data in the table, answer the following questions to help Ashton discover whether dog owners are less likely to walk their dogs in the winter than in the summer by a statistically significant amount at an alpha level of .05. Round your answers to two decimal places, where appropriate.

1. Calculate ΣD^2.
2. Calculate $(\Sigma D)^2$.
3. Calculate SD_D.
4. Calculate t using a dependent-samples t-test.
5. Calculate df.
6. Identify the critical t-value, compare it to t-observed, and state whether you should accept or reject the null hypothesis.

Answers to these questions are found on page 419.

Think About It...

t-TESTS AND VARIABILITY

Researchers are generally much more excited about a statistically significant difference between group means than they are about a difference that is not significant. As a result, researchers are often very interested in finding the experimental design most likely to reveal a statistically significant difference when there actually is a difference to see.

The formula for finding *t* can be generalized like this:

$$t = \frac{\text{difference between groups caused by IV}}{\text{variability}}$$

$$= \frac{\text{difference between groups caused by IV}}{\text{individual differences} + \text{random chance}}$$

We cannot manipulate the difference between the groups because that's cheating: The difference caused by the IV *is what it is* (to use a popular expression). If the top of the *t*-ratio is off limits, however, what about the bottom? Can we reduce variability and make our observed *t*-value stand out?

Variability here is the result of two influences. First and foremost is the fact that just about everything in the natural world—people, plants, animals—varies. We may all be the same, generally speaking, but individually, we are all quite different. We all have unique personal histories, experiences, beliefs, and expectations. As a result, we respond to the experiments we participate in differently.

The second influence on variability is random chance, a.k.a. measurement error. We cannot remove the influence of random chance, but we can reduce the "noise" of individual differences by making the two sets of people we're examining as identical as we possibly can. And what is more identical than using the same set of people in both the experimental and the control group?

Here are the data from two studies of the effects of caffeine on heart rate. Study A compared an experimental group, who received 300 milligrams of caffeine and then had their heart rates measured, with a control group, who received no caffeine prior to heart rate measurement. In Study B, one group of individuals had their heart rates measured twice, before and after drinking a caffeinated beverage.

Calculate the observed value of *t* for both situations. In both studies, heart rate was measured in beats per minute (BPM).

Study A	
Control	Experimental
71	70
71	78
71	80
73	81
70	75
$\bar{X} = 71.2$	$\bar{X} = 76.8$
$SD = 1.10$	$SD = 4.44$
$n = 5$	$n = 5$

Study B	
Before	After
71	70
71	78
71	80
73	81
70	75
$\bar{X} = 71.2$	$\bar{X} = 76.8$
$SD = 1.10$	$SD = 4.44$
$n = 5$	$n = 5$

1. How different are the two *t*-tests? _____

2. Can you think of a problem associated with using a repeated-measures design? _____

Answers to this exercise can be found on page 436.

> **CheckPoint**
> *Answers to questions on page 416*
>
> 1. $\Sigma D^2 = 28$ ☐
> 2. $(\Sigma D)^2 = 144$ ☐
> 3. $SD_D = 0.39$ ☐
> 4. $t = 3.09$ ☐
> 5. $df = 9$ ☐
> 6. $3.09 > 1.833 \therefore$ reject ☐
>
> **SCORE:** /6

t-TESTS IN CONTEXT: "GARBAGE IN, GARBAGE OUT"

We now have four inferential tests (one *z* and three different kinds of *t*'s) that we can use to answer different kinds of questions. Each of these tests is appropriate in a particular type of situation, and it is primarily the responsibility of the user to determine if a given situation is the appropriate one for the test they're using. Table 10.3 recaps the tests and the situations each one is appropriate for.

We need to be aware of which situations call for each test, because if we use the wrong test, we will reach an incorrect conclusion about our data. As we rely more and more on computers to do the heavy lifting (in the case of statistics, the

TABLE 10.3 Which test should I use? Four situations

Test	Number of Samples	When to Use it
z-test	1	To compare a single sample mean with a known population mean where the population standard deviation is also known.
single-sample *t*-test	1	To compare a single sample mean with a known population mean where the population standard deviation is unknown.
independent-samples *t*-test	2	To compare two independent-sample means (i.e., sample participants are selected and assigned to groups at random) where nothing is known about the population (neither mean nor standard deviation).
dependent-samples *t*-test	2	To compare two dependent-sample means (i.e., participants in sample 1 are deliberately linked in some way to participants in sample 2) where nothing is known about the population (neither mean nor standard deviation).

mathematics of each test), we run an increasing risk of using the wrong test to analyze our data. This is because most statistical analysis software will not indicate whether the test we are attempting to use is the wrong one for the situation: It will simply analyze the data that we've entered into the program. Crunching the numbers as it was designed to do, the program will give us an incorrect answer, and we will reach an incorrect conclusion—often without even knowing it. On the other hand, if we're aware of when a particular test is appropriate, we will produce accurate results. Remember the old acronym GIGO, "garbage in, garbage out": It always applies when we're using computers.

Once we have finished with our data analysis, we will want to report the results, and as we've already discussed, we will likely use APA Style to report them. Previously, we used the results of a z-test to illustrate the format for an APA Style report. However, since z-tests are only very rarely used (we almost never know both the population mean and standard deviation, the two criteria for a z-test), let's use the results of one of our t-tests here.

When we tested the question of whether college students were more distracted in the spring than in the fall, we found that students were significantly more distracted in the spring. We might say something like this in our results section:

> In order to determine if students were more distracted in the spring than in the fall, we measured the reaction times in a computerized distraction task (Lavie & Forster, 2007) of a sample of 12 students at two points during the academic year. The first measurements were obtained at the midpoint of the fall semester and the second set at the midpoint of the spring semester. Generally, students were more distracted (had longer reaction times) in the spring ($M = 524.75$ ms, $SD = 73.06$ ms) than they did in the fall ($M = 673.33$ ms, $SD = 113.25$ ms). A dependent-samples t-test revealed that this difference was statistically significant: $t_{(11)} = -4.23, p < .05$.

Notice how we would refer to the mean and standard deviation in the text: A capital M, italicized, is used for the mean, and since the mean should always be accompanied by the standard deviation, capital SD, also italicized, is used to indicate the standard deviation. Notice also the APA Style format for reporting the results of a statistical test, outlined in Figure 10.3.

The formatting used to report statistics is simply a way of making sure that statistics are reported in a consistent manner so that readers of scientific reports can easily find and interpret them. The sixth edition of the *Publication Manual of the American Psychological Association* (2010) asks that authors always report the results for inferential tests so that they "include the obtained magnitude or value of the test statistic, the degrees of freedom, [and] the probability of

$$t_{(11)} = -4.23, p < .05$$

- Degrees of freedom
- The probability of obtaining the statistic by chance alone
- Abbreviation for the test used
- Observed value of the statistic

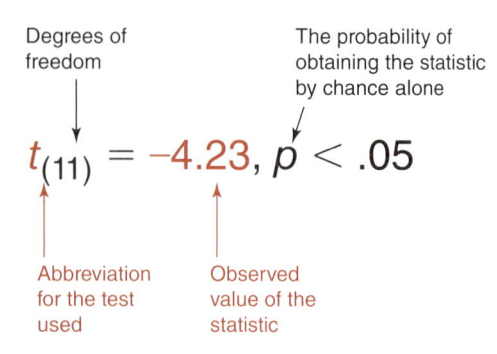

FIGURE 10.3
APA Style formatting for reporting statistical test results.

420   STATISTICS IN CONTEXT

obtaining a value as extreme as or more extreme than the one obtained (the exact *p* value) . . ." (p. 34).

Typically, presentations of the results of hypothesis tests begin with the author telling the reader precisely which test was used. We also need to present the APA Style–formatted "statement" (as shown in Figure 10.3) of the results of that test, beginning with the italicized abbreviation for the specific test that was used. We used a *t*-test, so the statement starts with an italicized *t* (obviously, if we had used a *z*-test, the symbol for the type of test used would be *z*). Our overall job is to present the reader with enough information about the statistical analysis we did that the reader can interpret and understand our analysis. Remember how much the significance of the *t*-value we calculated depended on the size of the sample or samples that we used? For this reason, we must also present the degrees of freedom.

Lastly, we want to tell our reader whether or not our results were statistically significant, so we present the *p*-value. Remember that "*p*" tells the reader the probability that we would have gotten the observed statistic that we calculated just by random chance alone. If there is less than a 5% chance that we would have gotten that observed statistic by random chance alone, then we say that the statistic is statistically significant. If our results had not been statistically significant, we would have had a "greater than" sign instead of the "less than" sign we used here. We should always report the results of a non–statistically significant test, because readers will want to see them. A nonsignificant result is nothing to be ashamed of—it is still a result, and it still tells the reader something about the data.

SUMMARY

The material we have covered in this chapter will be much easier to remember and to process if you look for links between the individual inferential tests we've discussed. All three tests (single-sample *t*-test, independent-samples *t*-test, and dependent-samples *t*-test) are asking the same question: *How different are these two means if we take variability in the distribution of all those possible differences into account?*

The single-sample *t*-test compares the mean of a single sample with the known mean of the population the sample came from. If the independent variable does not affect the participants in the sample at all, then the sample mean should be the same as the population mean.

The independent-samples and dependent-samples *t*-tests are also asking how different two means are, but now the means are both from samples taken from populations. One sample stands in for the null population, where the IV has no effect; the other stands in for the alternative population, where the IV has changed things.

All of the formulas for the *t*-test reflect this basic similarity in the question being asked. The structure of each formula consists of a measure of difference on the top and a measure of error on the bottom. The way we calculate error changes as the kind of error we're looking for changes.

A student once told me that she remembered that *t*-tests are appropriate when you want to compare two means by thinking of the title to a very old song, "Tea for Two" (by Irving Caesar and Vincent Youmans). You might rephrase that "*t* for two," perhaps. Comparing more than two means requires a different test, which we'll be looking at in the next chapter.

TERMS YOU SHOULD KNOW

assumptions, p. 411
control group, p. 400
degrees of freedom, p. 396
dependent samples, p. 401
estimated standard error, p. 393
experimental group, p. 400

independent samples, p. 400
robust, p. 411
sampling distribution of the difference between means, p. 403
Student's *t*-test, p. 391
subject mortality, p. 406

GLOSSARY OF EQUATIONS

Formula	Name	Symbol
$\sigma = \sqrt{\dfrac{\sum(x - \bar{X})^2}{n}}$	**Sigma** The population standard deviation	σ
$\hat{s} = \sqrt{\dfrac{\sum(x - \bar{X})^2}{n-1}}$	**s-hat** The population standard deviation estimated from a sample	$\hat{s}$
$s_{\bar{X}} = \dfrac{\hat{s}}{\sqrt{n}}$	**s-sub-X-bar** The estimated standard error (variability in the SDM)	$s_{\bar{X}}$
$t = \dfrac{\bar{X} - \mu_{\bar{X}}}{S_{\bar{X}}}$	**t** The estimated position of the sample mean in the SDM; an "estimated *z*"	t single-sample
$s_{\bar{X}_C - \bar{X}_E} = \sqrt{\dfrac{\sum x_C^2 - \dfrac{(\sum x_C)^2}{n_C} + \sum x_E^2 - \dfrac{(\sum x_E)^2}{n_E}}{n_C(n_E - 1)}}$	The estimated standard error of the difference when $n_1 = n_2$	$s_{\bar{X}_C - \bar{X}_E}$
$s_{\bar{X}_C - \bar{X}_E} = \sqrt{\dfrac{\sum x_C^2 - \dfrac{(\sum x_C)^2}{n_C} + \sum x_E^2 - \dfrac{(\sum x_E)^2}{n_E}}{(n_C + n_E) - 2}\left(\dfrac{1}{n_C} + \dfrac{1}{n_E}\right)}$	The estimated standard error of the difference when $n_1 \neq n_2$	$s_{\bar{X}_C - \bar{X}_E}$

Formula	Description	Symbol
$s_{\bar{D}} = \dfrac{\hat{s}_D}{\sqrt{n}} = \dfrac{\sqrt{\dfrac{\sum D^2 - \dfrac{(\sum D)^2}{n}}{n-1}}}{\sqrt{n}}$	The estimated standard error of the difference for a dependent-samples *t*-test	$s_{\bar{D}}$
$\bar{D} = \dfrac{\sum D}{n}$	The average D score	$\bar{D}$
$t = \dfrac{\bar{X}_1 - \bar{X}_2}{s_{\bar{D}}}$	*t*	*t* dependent samples
$D = X - Y$	The difference score	D

WRITING ASSIGNMENT

Using the data set from Table 10.6 in the *Practice Problems*, write a report in APA Style on the data and the results of the inferential test you ran. Describe your participants, justify your alpha level, include the results of your inferential test in APA Style, and make note of any problems you see in the design of this study.

PRACTICE PROBLEMS

1. The formulas for the *z*- and for all three *t*-tests share some common features. What do the equations have in common? How are they different?

2. Take a look at the last row on the *t*-table in Appendix B. Do any of these values look familiar? What does it mean to say that $df = \infty$? Why would you have $df = \pm 1.96$ for a *t*-test when $\alpha = .05$ and $n = \infty$?

3. Take a look back at Box 10.13 (p. 407). Notice that $df = 30$ gives us a critical *t*-value that is slightly larger than when $df = 40$. As the sample size gets bigger, what happens to the critical *t*-values on the table? Do they get larger or smaller, and why?

4. There is a shortcut you can use when you're interpreting the results of your *t*-test, no matter what type of *t*-test you run. That is, if the *t*-observed value is 1.00 or less, you automatically know that you do not have a significant result. Explain why this is true.

5. Cognitive psychologists know that humans find it easier to remember items at the beginning and at the end of a list of items, while having more difficulty with items in the middle. This is an example of the "serial position effect" (also known as the "primacy/recency effect"). Suppose we think that the level of education one has might make a difference to the tendency to remember items in the middle of the list. We select two groups of subjects with $n = 10$ in each group. Members of Group 1 have 12 years of

formal education, and members of Group 2 have 20 or more years of formal education. Both groups are given a list of 60 words to remember in a given amount of time. Later, the number of words correctly recalled from the middle third of the list is measured (Table 10.4).

Table 10.4 Number of words recalled from middle 20 words on the list

Group 1	Group 2
16	16
14	17
15	9
16	20
11	3
14	4
13	14
12	19
14	13
13	7

a. What test (z-test, or t-test for single samples, independent samples, or dependent samples) should you use to analyze these results, and why? Will you use a one-tailed or a two-tailed test?
b. What significance level will you use, and why?
c. What is the result of your analysis? Did your independent variable have a significant effect on your dependent variable?
d. Interpret your results in an APA Style sentence using the correct format for presenting the results of your inferential test.

6. Most sleeping pills reduce the period of dreaming sleep, also known as rapid eye movement (REM) sleep. You want to validate the claims of a new sleeping pill, which is advertised as not affecting REM sleep at all. You randomly select two groups of subjects. Group 1 receives a placebo, and Group 2 receives the recommended dosage of the new sleeping pill. REM sleep is monitored in a sleep lab during the course of one night's sleep, and the total number of minutes spent in REM are counted for each subject. Table 10.5 summarizes the results.

a. What test should you use to analyze these results, and why? Will you use a one-tailed or a two-tailed test?
b. What significance level will you use, and why?
c. What is the result of your analysis? Did your IV have a significant effect on your DV?
d. Interpret your results in an APA Style sentence using the correct format for presenting the results of your inferential test.

Table 10.5 Number of minutes of REM sleep during an 8-hour sleep period

Placebo	Sleeping Pill
90	65
92	58
76	52
82	60
90	57
75	63
86	59
78	77
106	
84	

7. Suppose another researcher was interested in the effectiveness of the sample sleeping pill from question 6. This researcher asked patients suffering from insomnia to serve as subjects in his experiment testing out the new pill. Subjects were matched in terms of the severity of their insomnia and then separated into the same two groups (placebo and drug) to have their REM sleep measured in the same manner. The results are summarized in Table 10.6.
   a. What test should the researcher use to analyze the results, and why? Should he use a one-tailed or a two-tailed test?
   b. What significance level should he use, and why?
   c. Provide an analysis of the researcher's results. Did the IV have a significant effect on the DV?
   d. Interpret the results in an APA Style sentence using the correct format for presenting the results of an inferential test.

Table 10.6 Number of minutes of REM sleep in patients matched for insomnia

Placebo	Sleeping Pill
110	99
89	84
86	94
87	87
85	87
87	106
74	102
85	91

8. Depression is a serious mental disorder that has been shown to respond very well to cognitive therapy. You want to test the effectiveness of cognitive therapy on a group of people suffering from anxiety. You select a random sample of 20 people diagnosed with anxiety disorder and treat them with cognitive therapy. Then, you administer a standardized test of anxiety that has a mean of 50 and a standard deviation of 10. Higher scores indicate more anxiety. Use the data in Table 10.7 to answer the following questions:
   a. What test should you use to analyze these results and why? Will you use a one- or a two-tailed test?
   b. What significance level will you use and why?
   c. What is the result of your analysis? Did your IV have a significant effect on your DV?
   d. Interpret your results in an APA Style sentence using the correct format for presenting the results of your inferential test.

Table 10.7   Scores on standardized test of anxiety

Subject ID	Test Score	Subject ID	Test Score
1	46	11	57
2	38	12	48
3	34	13	63
4	38	14	46
5	60	15	59
6	60	16	53
7	45	17	56
8	68	18	74
9	53	19	39
10	55	20	48

9. Suppose you wanted to redo the study described in question 8, but now using your own test of anxiety that has not been standardized. You know from your own testing that the mean score for a person without anxiety should be 50, but you have not collected enough data on the test to determine the standard deviation in the population. Test the subjects from question 8 using your anxiety test. Compare the two means, but this time, estimate the variability in the population and the SDM.
   a. What is the result of your analysis? Did your IV have a significant effect on your DV?
   b. Calculate the value of D (even if your results are not statistically significant).
   c. Interpret your results in an APA Style sentence using the correct format for presenting the results of your inferential test.

10. The Stanford–Binet IQ test is a *standardized* test of intelligence. This means that average intelligence is already known (average IQ = 100) and that the test has been assessed

across many samples and thousands of subjects, so the standard deviation of scores on this test is also known ($SD = 15$). Let's assume that you've developed a drug that you believe promotes above-average IQ in children. Use the following data to assess the effectiveness of your medication.

$\Sigma x = 5{,}250$
$n = 50$

a. What test should you use to analyze these results, and why? Will you use a one-tailed or a two-tailed test?
b. What significance level will you use, and why?
c. What is the result of your analysis? Did your IV have a significant effect on your DV?
d. Interpret your results in an APA Style sentence using the correct format for presenting the results of your inferential test.

11. Suppose you had the following data:

$\Sigma x_C = 278$ $\Sigma x_E = 344$
$\Sigma x_C^2 = 11{,}478$ $\Sigma x_E^2 = 20{,}520$
$n_C = 7$ $n_e = 6$

Calculate the appropriate t-value for these data, and interpret your results.

12. Suppose you had the following data:

$\Sigma x_C = 147$ $\Sigma x_E = 75$
$\Sigma x_C^2 = 7{,}515$ $\Sigma x_E^2 = 1{,}937$
$n_C = 4$ $n_e = 4$

Calculate the appropriate t-value for these data, and interpret your results.

13. In the spaces provided, identify the symbols shown below.

Symbol	
$s_{\bar{X}_C - \bar{X}_E}$	
df	
α	
σ	
SD	
μ	
$s_{\bar{X}}$	
$\mu_{\bar{X}}$	
n_c	
D	

14. In the spaces provided, describe the following situations:

Repeated measures	
Matched pairs	
Natural pairs	

15. Using the following observed data, determine the critical *t*-value for each observed *t*-value, and interpret the results of each test.

Observed *t*-Value	Group 1 (*n* =)	Group 2 (*n* =)	Type of Test	Alpha Level	Critical *t*-Value	Interpretation (Reject or Retain H_o)
1.533	9	9	two-tailed	.05		
−2.069	10	6	one-tailed	.01		
3.820	5	5	two-tailed	.01		
3.000	9	9	one-tailed	.05		
−5.990	29	29	two-tailed	.05		
1.960	61	61	one-tailed	.05		

16. Take another look at the table of critical *t*-values in Appendix B. In general, is it easier to get a significant result if you ask a two-tailed or a one-tailed question. Why? Be sure to include a definition of what you mean by the term "easier."

17. For each of the tests in question 15, write out the results of the test using the APA format for reporting the results of a *t*-test.

18. In an early study of the effects of frustration on feelings of hostility, Miller and Bugelski (1948) had a group of boys at a camp rate their attitudes toward two minority groups (Mexicans and Japanese). The campers then participated in a long, difficult, and frustrating testing session that kept them away from their weekly movie. Finally, the boys again rated their attitudes toward the minority groups. The scores in Table 10.8 represent the number of unfavorable traits attributed to minorities and are similar to those of Miller and Bugelski. Does the intervening unpleasant task alter attitudes toward minorities?

Table 10.8 Number of unfavorable traits attributed to minorities

ID	Before Testing	After Testing
1	5	6
2	4	4
3	3	5

4	3	4
5	2	4
6	2	3
7	1	3
8	0	2

19. A number of studies have explored the relationship between neuroticism and alcoholism using an animal model. Here is a typical study. From each of seven litters, two cats were randomly selected. One cat from each pair was assigned to an experimental group, which was subjected to a procedure that induced temporary neurosis; the other cat in each pair was placed in a control group. Then, all cats were offered milk spiked with 5% alcohol. The amount consumed in 3 minutes was measured in milliliters and is shown in Table 10.9. Does neurosis increase the consumption of alcohol?

Table 10.9 Amount of alcohol-spiked milk consumed (in milliliters)

Littermate Pair	No Experimental Neurosis	Experimental Neurosis
1	63	88
2	59	90
3	52	74
4	51	78
5	46	78
6	44	61
7	38	54

20. To test compliance with authority, Stanley Milgram carried out a classic experiment in social psychology. Milgram's study required that subjects administer increasingly painful shocks to seemingly helpless victims in an adjacent room (the "victims" in the next room were actors and never received any actual shocks). Suppose we want to determine if willingness to comply with authority (in this case, the experimenter who orders the increasingly intense shocks) can be affected by having someone else present to help a participant make the decision about whether or not to follow orders. Six randomly selected volunteers are assigned to a "committee"; each committee consists of the volunteer participant and two other people, both confederates of the experimenter, whose job is to agree with whatever decision the real participant in the study makes. Six other participants are assigned to perform this task alone. Compliance is measured using a score between 0 and 25, depending on the point at which the subject refuses to comply with authority: A score of 0 indicates unwillingness to comply at the very outset of the experiment, while a score of 25 indicates willingness to comply with orders completely. Table 10.10 shows the compliance scores. Does a committee affect compliance with authority?

Table 10.10 Compliance scores (0–25)

Committee	Solitary
2	3
5	8
20	7
15	10
4	14
10	0

21. An investigator wishes to determine whether alcohol consumption causes a deterioration in the performance of drivers. Before a driving test, volunteers are matched for body weight and then asked to drink a glass of orange juice. Participants in Group 1 drink plain orange juice; participants in Group 2 drink orange juice that has been laced with vodka. Performance is measured by the number of errors made on a driving simulator. A total of 20 participants are randomly assigned in equal numbers to the control (orange juice only) and experimental (orange juice and vodka) groups. Table 10.11 shows the data. Does vodka cause performance to deteriorate?

Table 10.11 Number of errors made on a driving simulator

Control Group	Experimental Group
19	27
16	28
16	28
19	25
23	28
21	29
19	28
18	23
14	27
17	20

22. A psychologist wants to determine the effect of instruction on the time required to solve a mechanical puzzle. Ten randomly selected volunteers are given the mechanical puzzle to solve as rapidly as possible. Before they start, the participants are told that the puzzle is extremely difficult to solve and that they should expect to have difficulty. The psychologist knows that the average time needed to solve the puzzle when participants have not been primed to expect a specific outcome is 7 minutes. Table 10.12 shows the results. Did the instructions given prior to the task affect the performance of the task?

Table 10.12 Number of minutes needed to solve a mechanical puzzle

Participant ID	Minutes Needed
1	5
2	20
3	7
4	23
5	30
6	24
7	9
8	20
9	6
10	10

23. Age and obesity are the two strongest risk factors for treatment-resistant hypertension (TRH), prompting many doctors to predict that as our population ages and gets heavier, the incidence of this unhealthy condition will increase. Suppose you have a new drug that you think will reduce TRH. Normal blood pressure is a reading of 120 mm Hg (the systolic pressure when your heart contracts to push blood through your circulatory system) over 80 (the diastolic pressure when your heart relaxes between beats). Moderate to severe high blood pressure is indicated by a systolic pressure of 160 mm Hg or above, or a diastolic pressure of 100 mm Hg or more. Seven patients who have been diagnosed with high blood pressure volunteer for your study. In a supine position, they have their diastolic pressure measured before they take the drug. Their blood pressure is taken again after the drug has been administered and has had time to have an effect. Table 10.13 shows the results of your study. Does your new drug reduce blood pressure?

Table 10.13 Supine diastolic pressure (in mm Hg)

Patient	Before Drug	After Drug
1	98	82
2	96	72
3	140	90
4	120	108
5	130	72
6	125	80
7	110	98

24. You believe that the daily consumption of chocolate speeds up the learning of statistics. The mean number of weeks it takes statistics students (ones not eating chocolate, poor souls) to master their subject is 13 weeks. You select a random sample of 16 students, feed them a daily dose of chocolate, and then send them to statistics class. Later, you calculate the number of weeks it took them to complete their course work. Your sample mean is 11.0 weeks with $\hat{s} = 2.7$ weeks. Does chocolate decrease the amount of time it takes to master statistics?

25. Krithiga's research focuses on strategies that make workplace policies more palatable to employees. She hypothesizes that employees will rate the fairness of new workplace policies more favorably if they are given some control in the process of creating these policies. To test her hypothesis, Krithiga generates a control group of employees randomly selected from Firm A and an experimental group of employees randomly selected from Firm B. At Firm B, Krithiga implements a process in which workers and management collaboratively draft a policy to deal with romantic relationships between colleagues; she subsequently administers a survey to the experimental group to measure the perceived fairness of the policy. Krithiga then takes Firm B's newly drafted policy to Firm A, where it is unilaterally implemented by the management team. Krithiga surveys the control group from Firm A to measure their perceptions of the fairness of the policy. The perceived fairness scores for the participants in the control and experimental groups are presented in Table 10.14; higher scores represent a greater degree of perceived fairness. Does giving employees some control over the process of creating new workplace policies increase the perceived fairness of the policies?

Table 10.14   Perceived fairness scores

Control Group	Experimental Group
4.5	3.5
3.25	5.75
6	8
5	6.5
5.75	7
2.5	7.5
8	4.5
3.5	2.75
4	7.25
3.5	7.25
$\Sigma_C = 46$	$\Sigma_E = 60$
$n = 10$	$n = 10$

26. Do you ever get the feeling you're being watched? It has been postulated that the feeling of being watched often compels people to bring their behavior into conformity with the normative expectations they associate with their social environment. Tikvah wondered if creating the feeling of being watched might compel her coworkers to be more charitable. She placed a collection jar in the breakroom at her office to raise funds for an ill colleague with high medical expenses. Every other day, Tikvah would tack a poster of a pair of eyes to the breakroom bulletin board, to give those passing by the tip jar the experience of feeling they were being watched. Table 10.15 displays the total donations made on days with and without the poster on the bulletin board over a 20-day period. Does the feeling of being watched may people more inclined toward charitable giving?

Table 10.15 Daily donation totals (in dollars)

No Poster	Poster
17.84	62.50
65.14	15.40
24.52	28.00
30.00	40.00
15.25	33.50
40.00	27.25
60.00	55.75
18.65	12.78
22.30	31.64
24.00	60.00
$\Sigma_C = 317.70$	$\Sigma_E = 366.82$
$n = 10$	$n = 10$

USING THE FORMULA FOR UNEQUAL n'S WITH EQUAL n'S

Did you try the formula for unequal n's when the n's were actually equal (see p. 405)? If you did, you should have gotten something like this:

$$s_{\bar{X}_C - \bar{X}_E} = \sqrt{\frac{\Sigma x_C^2 - \frac{(\Sigma x_C)^2}{n_C} + \Sigma x_e^2 - \frac{(\Sigma x_E)^2}{n_{Ee}}}{(n_C + n_{eE}) - 2}\left(\frac{1}{n_{cC}} + \frac{1}{n_E}\right)}$$

$$= \sqrt{\frac{3{,}090.48 - \frac{(237.82)^2}{20} + 2{,}362.14 - \frac{(205)^2}{20}}{(20 + 20) - 2}\left(\frac{1}{20} + \frac{1}{20}\right)}$$

$$= \sqrt{\frac{(3{,}090.48 - 2{,}827.89) + (2{,}362.14 - 2{,}101.25)}{38}(0.05 + 0.05)}$$

$$= \sqrt{\frac{(262.56)+(260.89)}{38}(0.10)}$$

$$= \sqrt{\frac{523.45}{38}(0.10)}$$

$$= \sqrt{13.775(0.10)}$$

$$= \sqrt{1.3775}$$

So, the standard error of the difference, when $n_1 = n_2 = 20$, is $\sqrt{1.3775} = 1.17$, which is the same value we found when we used the formula for equal n's (Box 10.19).

BOX 10.19

Data when $n_1 = n_2 = 20$

What We Know	The Samples		Hypotheses
Control group	Control	Experimental	The null
$\sum_C = 237.82$	10.65	10.04	$H_0: \mu_C = \mu_E = 0$
	9.54	21.99	
	8.00	10.13	
$n = 20$	12.37	15.72	
	13.14	8.69	
$\bar{X}_C = 11.89$ min	20.00	8.52	
	10.15	10.16	The alternative
Experimental group	10.56	10.00	
	11.88	10.32	
$\sum_E = 205.00$ min	8.30	7.48	$H_1: \mu_C - \mu_E \neq 0$
	7.58	11.90	Two-tailed test.
$n = 20$	16.52	7.07	
	16.81	5.17	or
$\bar{X}_E = 10.25$	12.00	7.74	
	16.91	5.33	
$df = (n_C + n_E) - 2$	7.60	13.03	$H_1: \mu_C - \mu_E > 0$
	8.59	9.36	
	10.22	11.26	One-tailed test.
	9.73	9.82	
$t = \dfrac{\bar{X}_C - \bar{X}_E}{s_{\bar{X}_C - \bar{X}_E}}$	17.27	11.27	
	$\sum_C = 237.82$	$\sum_E = 205.00$	

Think About It...

t-TESTS AND SAMPLE SIZE

SOLUTIONS

1. What happens to the value of *t* as *n* increases?   **As *n* gets larger, the denominator of the *t*-ratio gets smaller—we're dividing $\hat{s}$ by a larger and larger number, so $\dfrac{\hat{s}}{\sqrt{n}}$ gets smaller and smaller. This means that the *t*-value we calculate will increase. The numerator (D) gets divided by a smaller number as *n* increases, so *t* gets bigger.**

2. What happens to the critical value of *t* as *n* increases?   ***t*-critical will get smaller as *n* increases. If you go down the *df* column in the table of Appendix B (remembering that *df* = *n* − 1), you should see the critical values decrease.**

3. Suppose $\hat{s}$ = 10 and *n* = 2. Use a one-tailed test and an α = .05 for all critical values.
   a. If *n* = 2, *df* = 2 − 1, or 1, and the critical *t*-value would be **6.314**.
   b. If *n* = 2, the difference between the two means must be **44.78** for *t* to be significant.

4. Now let's increase *n*.
   a. If *n* = 20, *df* would be **20 − 1, or 19**, and the critical *t*-value would be **1.729**.
   b. Now how different must the two means be for *t* to be significant? **3.96**

$$D = (t_{crit})\left(\dfrac{\hat{s}}{\sqrt{n}}\right)$$
$$= 1.729\left(\dfrac{10}{\sqrt{19}}\right)$$
$$= 1.729\left(\dfrac{10}{4.36}\right)$$
$$= 1.729\,(2.29)$$
$$= 3.96$$

   c. If *n* = 100, the difference between the two means must be **1.658** for *t* to be significant.

$$D = (t_{crit})\left(\dfrac{\hat{s}}{\sqrt{n}}\right)$$
$$= 1.658\left(\dfrac{10}{\sqrt{100}}\right)$$
$$= 1.658\left(\dfrac{10}{10}\right)$$
$$= 1.658\,(1)$$
$$= 1.658$$

As sample size increases, the degree of difference between means needed to reject H_0 gets smaller.

t-TESTS AND VARIABILITY

SOLUTIONS

1. How different are the two *t*-tests?

   Study A: $\quad t = \dfrac{\bar{X}_c - \bar{X}_e}{s_{\bar{X}_c - \bar{X}_e}} = \dfrac{71.2 - 76.8}{2.05} = \dfrac{-5.6}{2.05} = \boxed{-2.73}$

   Difference between the two group means $= -5.60$

   $$s_{\bar{X}_c - \bar{X}_e} = \sqrt{\left(\dfrac{\hat{s}_c}{\sqrt{n}}\right)^2 + \left(\dfrac{\hat{s}_e}{\sqrt{n}}\right)^2} = \sqrt{\left(\dfrac{1.10}{\sqrt{5}}\right)^2 + \left(\dfrac{4.44}{\sqrt{5}}\right)^2}$$
   $$= \sqrt{0.49^2 + 1.99^2} = \sqrt{4.2002}$$
   $$= \boxed{2.05}$$

   Study B: $\quad t = \dfrac{\bar{X}_{\text{before}} - \bar{X}_{\text{after}}}{s_{\bar{D}}} = \dfrac{-5.60}{1.78} = \boxed{-3.15}$

   Difference between before and after group means $= -5.60$

   $$s_{\bar{D}} = \dfrac{\hat{s}_D}{\sqrt{n}} = \dfrac{3.97}{\sqrt{5}} = \boxed{1.78}$$

2. Can you think of a problem associated with using a repeated-measures design? **There is less variability (pink circles in the answer to question 1 above) in the dependent-samples *t*-test than in the independent-samples *t*-test. As a result, the *t*-value (red boxes) that we calculate is larger (farther away from zero) in the dependent test than in the independent test. (FYI: Both are statistically significant.) One potential problem with the repeated-measures design is something called a "carry-over" or "fatigue effect." These problems arise when performance in one condition in the experiment *carries over* and affects performance in the other condition, or when the participant becomes tired (or bored with the procedure altogether) during the first condition, which then affects his or her performance in the second condition. In the example cited here, a carry-over effect would occur if an aspect of heart rate in the "before caffeine" group carried over to affect heart rate in the "after caffeine" group.**

REFERENCES

American Psychological Association. (2010). *Publication manual of the American Psychological Association* (6th ed.). Washington, DC: American Psychological Association.

de Winter, J. C. F. (2013). Using the Student's *t*-test with extremely small sample sizes. *Practical Assessment, Research & Evaluation, 18*(10). Retrieved from http://pareonline.net/getvn.asp?v=18&n=10

Johnson, N. L., & Kotz, S. (1997). Section 6: Applications in medicine and agriculture. In *Leading personalities in statistical sciences: From the seventeenth century to the present* (pp. 327–329). New York, NY: Wiley Interscience.

Lavie, N., & Forster, S. (2007). High perceptual load makes everybody equal: Eliminating individual differences in distractibility with load. *Psychological Science, 18*(5), 377–381.

Miller, N. E., and Bugelski, R. (1948). Minor studies of aggression: II. The influence of frustration imposed by the in-group on attitudes expressed toward out-groups. *Journal of Psychology, 25*(2), 437–442.

Uga, V., Lemut, M. C., Zampi, C., Zilli, I., & Salzarulo, P. (2006). Music in dreams. *Consciousness and Cognition, 15*(2), 351–357.

Conducting *t*-Tests with SPSS and R

The data set we will be using for this tutorial is drawn from the Behavioral Risk Factor Surveillance System (BRFSS), a nationwide health survey carried out by the Centers for Disease Control and Prevention. The set we will be using is based on results of the 2008 questionnaire, completed by 414,509 individuals in the United States and featuring a number of questions about general health, sleep, and anxiety, among other things. The version we will be using is a trimmed-down version of the full BRFSS data set; You can find it on the companion site for this textbook, www.oup.com/us/blatchley

t-Tests with SPSS

We will be conducting an independent *t*-test to determine if there is a statistically significant difference in the quality of sleep of men and women interviewed for the BRFSS study. This is an independent *t*-test because, by definition, our observations are independent of one another, since an individual cannot be both male and female (at least not in this survey). Note that I have removed any men and women who indicated they had had zero nights of poor sleep in the previous 30 days. Therefore, this test looks at a sample of men and women who had between 1 and 30 days of poor sleep in the previous 30 days.

The null and alternative hypothesis for this test are:

H_0: There is no difference in the means.

H_1: There is a difference in the means.

or, phrased another way:

H_0: There is no statistical difference in the average number of nights of poor sleep for men and for women.

H_1: There is a statistical difference in the average number of nights of poor sleep for men and for women.

STEP 1: Load the data set into SPSS.

STEP 2: Set up the test. Click **Analyze**, then **Compare Means**, and then **Independent-Samples T Test**.

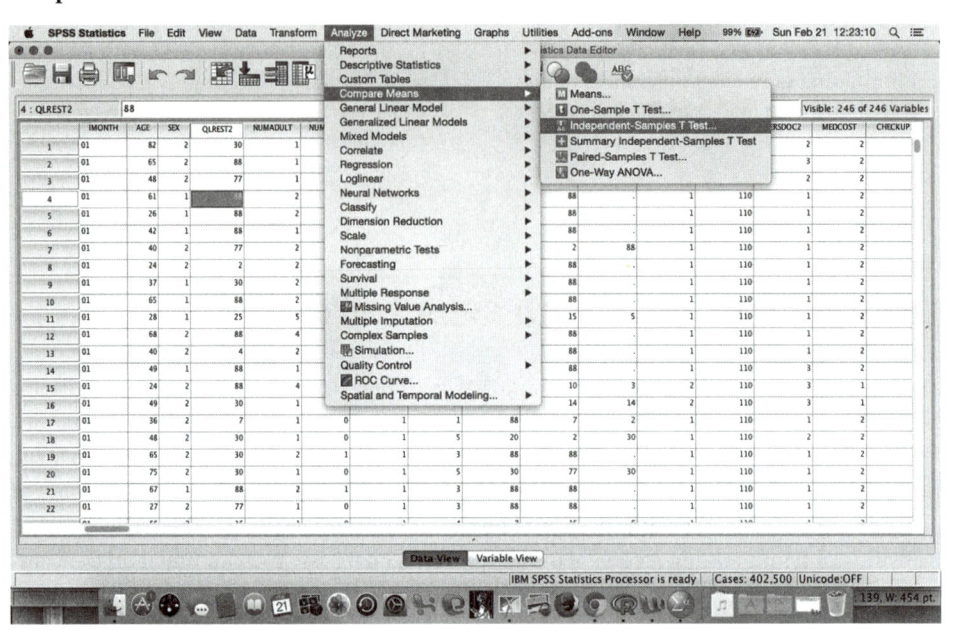

STEP 3: Select your dependent variable. Transfer the dependent variable "How Many Days Did You Get Enough Sleep" (QLREST2) to the **Test Variable(s)** box by selecting it and then clicking on the top arrow button.

STEP 4: Select your independent variable. Transfer the independent variable "Respondents Sex" (Sex) to the **Grouping Variable** box by selecting it and then clicking on the bottom arrow button.

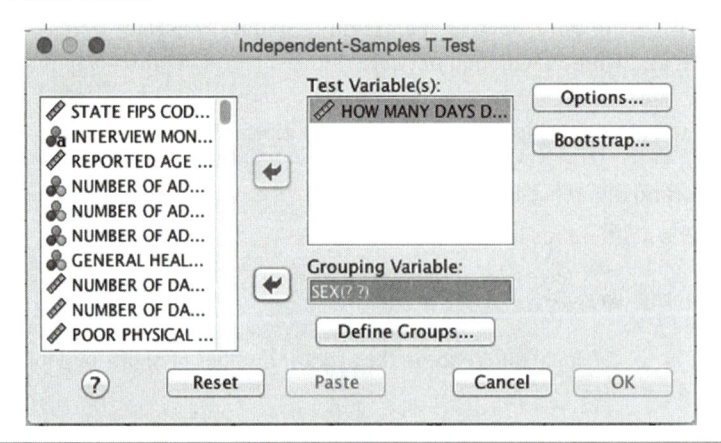

STEP 5: Define your groups. You will need to define the groups that the independent variable applies to. In this case, the variable uses 1 and 2 to denote males and females, respectively.

Click on the **Define Groups** button underneath the **Grouping Variable** box. Set the appropriate values for each group. Click **Continue**.

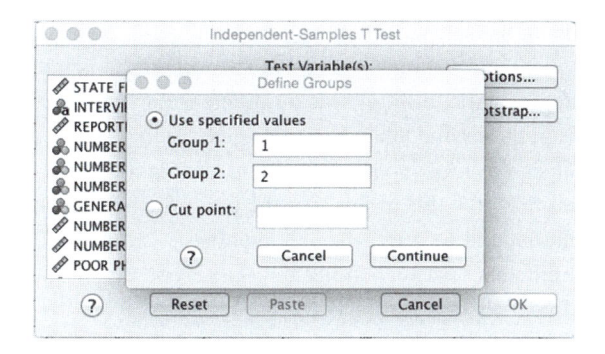

Note that in order to do this, you will need to know how the data have been defined and which variables you want to compare. In this case, we are examining a dichotomous variable (sex), and we only have two choices. So, the exercise may seem redundant. However, this is part of SPSS's flexibility: In the event that you have more than two categories, you will want to know which groups you are attempting to compare and what their numeric categorization is.

STEP 6: Select the confidence interval. Click on the **Options...** button. This will allow you to select the confidence interval (CI) for our *t*-test (remember: $1 - \alpha =$ the CI). SPSS defaults to 95%. In other words, we are saying that we have statistical significance at $p < .05$.

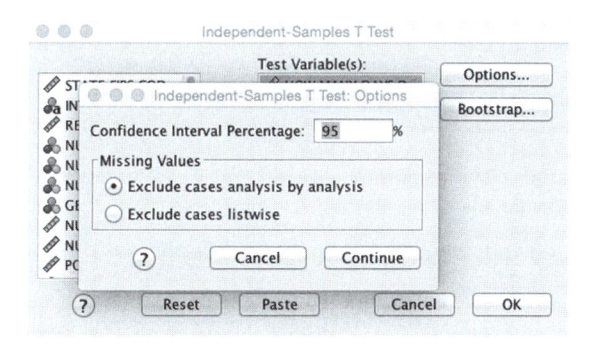

If you would like to increase or decrease the confidence interval, you can do so here. As a rule of thumb, we have a minimum confidence interval of 95% (or an α of 5%).

Click **Continue**, and then click **OK**.

STEP 7: Interpret your results. The results will be generated in the **Output** window in SPSS.

T-Test

Group Statistics

	RESPONDENTS SEX	N	Mean	Std. Deviation	Std. Error Mean
HOW MANY DAYS DID YOU GET ENOUGH SLEEP I	Male	93050	11.99	10.113	.033
	Female	167805	12.54	10.244	.025

Independent Samples Test

		Levene's Test for Equality of Variances		t-test for Equality of Means						
		F	Sig.	t	df	Sig. (2-tailed)	Mean Difference	Std. Error Difference	95% Confidence Interval of the Difference Lower	Upper
HOW MANY DAYS DID YOU GET ENOUGH SLEEP I	Equal variances assumed	100.847	.000	-13.279	260853	.000	-.553	.042	-.635	-.472
	Equal variances not assumed			-13.329	194204.308	.000	-.553	.042	-.635	-.472

CHAPTER 10 *t*-Tests 439

We can see from the statistics that the mean number of poor nights' sleep is higher for women than for men, but we want to know if that difference is statistically significant. To determine if there is a statistically significant difference in the means, we must consult the middle three cells in the second table. This table gives us the *t*-value, the degrees of freedom, and the statistical significance (*p*-value) shown as the Sig. (2-tailed) column. If the Sig. (2-tailed) value is less than .05, this means that the mean difference between the two groups is statistically significant, and we reject the null hypothesis. If the Sig. (2-tailed) value is greater than .05, then the difference between the mean values is not statistically significant, and we fail to reject the null hypothesis.

You can see in the table above that the *p*-value is less than .005 and is therefore significant at the .05 level. We can reject the null hypothesis that the mean number of nights of poor sleep are the same for men and women.

t-Tests with R

To carry out an independent two-variable *t*-test in R, we just need to use a few lines of code. We will also use code to load the data and then remove the observations we are not interested in using. The formula for a *t*-test can be easily manipulated to complete other *t*-tests, including paired *t*-tests and independent *t*-tests.

To load the data and work with it, we must enable the following packages:

- pander
- foreign

STEP 1: Load the data set.

```
PoorSleep<-read.csv("Desktop/Poor Sleep.csv")
```

"PoorSleep" is what we have chosen to call the data set. When we want to reference it or elements of it, we will use PoorSleep in our formulas. You can choose any name you wish, though you should use a name that you can recognize or remember easily. To reference a specific variable within the dataset we would type PoorSleep$variablename.

STEP 2: Trim the data set to include observations with data, and observations with greater than zero nights of poor sleep.

Remember that in the SPSS example, we compared individuals who had reported at least one night of poor sleep in the last 30 days. We are not concerned with individuals who reported having zero nights of poor sleep in the last 30 days, nor are we concerned with individuals who did not answer the question.

The survey codes individuals who opted not to answer as 99; individuals who were unsure of their answer are coded as 77. Anyone who reported having had zero nights of poor sleep in the previous 30 days was coded as 88. Since we are only concerned with people who had more than zero nights of poor sleep, we will need to recode these three values:

```
#recode 99 to missing for QLREST2
PoorSleep$QLREST2[PoorSleep$QLREST2==99] <-NA

#recode 77 to missing for QLREST2
PoorSleep$QLREST2[PoorSleep$QLREST2==77] <-NA

#recode 88 to missing for QLREST2
PoorSleep$QLREST2[PoorSleep$QLREST2==88] <-NA
```

STEP 3: Drop missing variables from the data set.

```
#DATA SET WITHOUT N/A VARIABLES
FullPoorSleep <- na.omit(PoorSleep)
```

STEP 4: Conduct a two-sided independent *t*-test.

```
t.test(FullPoorSleep$QLREST2~FullPoorSleep$Sex, mu=0,
alt="two.sided", conf=0.95, var.eq=T, paired=F)
```

What does this actually mean?

The code outside of the brackets, **t.test**, indicates what type of significance test we are conducting. Inside the brackets, we are referencing the specific data that we are interested in comparing. In this case, we are analyzing the effect an individual's sex has on the number of nights of poor sleep they have.

We use the ~ between the two variables to indicate that our *y* variable, or DV (QLREST2), is numeric and that our *x* variable, or IV (sex), is a binary factor.

mu=0 indicates that our null hypothesis is the difference between the mean, which is equal to zero.

This is a two-sided *t*-test, so we specify that **alt="two.sided"**.

We are using a confidence interval of 95%, so we indicate that **conf=0.95**.

R assumes the variance in each group is not equal; therefore, we must specify **var.eq=T**.

We are not conducting a paired *t*-test, so we indicate **paired=F**.

As we've already seen, there are several options for *t*-tests, including one-sample and paired *t*-tests. To run these tests, we need to manipulate the syntax slightly.

When both *x* and *y* are numeric, replace the ~ with a so that we have **t.test(y1,y2)**

If we were conducting a paired *t*-test, we would simply indicate **paired=T**.

STEP 5: Analyze the results.

```
    Two Sample t-test
data:  FullPoorSleep$QLREST2 by FullPoorSleep$Sex
t = -13.279, df = 260850, p-value < 2.2e-16
alternative hypothesis: true difference in means is not equal to 0
95 percent confidence interval:
 -0.6351654 -0.4717833
sample estimates:
mean in group 1 mean in group 2
       11.98933        12.54280
```

The results show us that with a *t*-statistic of −13.279 and a *p*-value of .000, we reject the null hypothesis that the true difference in the means is equal to zero.

CHAPTER ELEVEN

ANALYSIS OF VARIANCE

If we knew what we were doing, it wouldn't be called research, would it?
—ALBERT EINSTEIN

Everyday Statistics

DOES SIGHT AFFECT OUR NIGHTMARES?

In dreams, we can do impossible things—fly, walk on water, play miniature golf with aliens. But not all dreams are happy and upbeat, and some can be downright scary.

Meaidi, Jennum, Prito, and Kupers (2014) at the University of Copenhagen, Denmark, decided to study these downright scary dreams to learn whether there were differences in the way blind and sighted people experience nightmares. Using a technique called analysis of variance, or ANOVA, they compared the dream content of congenitally blind people, the so-called "late blind" (people who lost their sight either in childhood or as adults), and a group of "sighted controls." Among the variables they measured was the number of nightmares experienced by these three groups.

The researchers found that the congenitally blind had significantly more nightmares than did either the late blind or the controls. Their research suggests that all of us, sighted or not, use nightmares as "threat simulations," or a way to practice coping with the threats we experience in real life. Perhaps being blind from birth results in more perceived threats and a stronger need to practice coping with them?

Chapter 11 will discuss ANOVA and how this statistical technique can help us compare more than two groups at a time.

OVERVIEW

COMPARING MORE THAN TWO GROUPS

ANALYSIS OF VARIANCE: WHAT DOES IT MEAN?

ANOVA TERMINOLOGY

POST-HOC TESTING

MODELS OF F

ONE-WAY ANOVA ASSUMPTIONS

FACTORIAL DESIGNS OR TWO-WAY ANOVAS

ANOVA IN CONTEXT: INTERPRETATION AND MISINTERPRETATION

LEARNING OBJECTIVES

Reading this chapter will help you to . . .

- Explain Fisher's approach of using comparisons of variability to determine differences in group means. (Concept)

- Understand between-group and within-group variability. (Concept)

- Define the terms *sum of squares* and *mean square*. (Concept)

- Calculate between-group and within-group sum of squares and mean square. (Application)

- Calculate an F-value, and use the statistic to determine whether the null hypothesis should be rejected or retained. (Application)

- Calculate a Tukey HSD statistic, and apply it to an interpretation of the results of a significant F-test. (Concept and Application)

- Recognize the assumptions that underlie F-tests. (Concept)

COMPARING MORE THAN TWO GROUPS

In Chapter 10, we introduced three different *t*-tests, each based on using the variability in a sample to estimate the variability in a sampling distribution and the population. We used the single-sample *t*-test when we knew the population mean but not the population standard deviation. We used the independent-samples and dependent-samples *t*-tests when we didn't know anything at all about the parameters in the population. All three *t*-tests can be used to compare, at most, two sample means. In this chapter, we will explore what we can do when we want to compare more than just two means.

Most people think of statistics and mathematics as abstract, theoretical, and all too often just plain incomprehensible. Many of the men and women who have shaped statistics into what we use today, however, turned to statistics to solve very specific problems. In Chapter 10, we met William Gosset, one of these greats, who needed a way to use very small samples to estimate very large population parameters. His friend and fellow statistician Ronald Aylmer (R. A.) Fisher (who, by the way, developed the degrees of freedom that Gosset used in his famous *t*-test) faced a slightly different problem: how to determine which fertilizer will work the best for farmers (see *The Historical Context* on page 445). First statistics and beer, now statistics and fertilizer—what's next?

ANALYSIS OF VARIANCE: WHAT DOES IT MEAN?

Both *z*-tests and *t*-tests use variability to calculate the effect of an independent variable, or IV, on a particular dependent variable, or DV. If two means are very different from one another, and if we estimate that the variability in the distribution of differences is small, then our average difference between the means will be unusual, out in the tail of the distribution, and we will judge it likely to be the result of more than random chance.

An *analysis of variance*—or **ANOVA**, for short—allows us to compare means in more than just two groups, again in terms of their variability. An example will help here. Let's consider blood doping, a topic that has been in the news quite a lot lately.

Blood doping is a technique believed to provide an athlete with extra endurance during a race. There are three techniques used to "dope" an athlete, all of them intended to increase the competitor's red blood cell count. More red blood cells means more oxygen going to the muscles, and that translates into more aerobic capacity and (at least in theory) better endurance. The three techniques are:

ANOVA An inferential test used to test hypotheses about two or more means.

STATISTICS IN CONTEXT

The Historical Context

FERTILIZER, POTATOES, AND FISHER'S ANALYSIS OF VARIANCE

As a child, R. A. Fisher was fascinated by mathematics, statistics, and astronomy. His vision was severely impaired, but this was no impediment to his scholarship: He had tutors who would read mathematics to him, and young Fisher would picture the problems in his head. He eventually became so good at visualizing the solutions that he would frequently present his answers with absolutely no proof, saying that the solution was so obvious it didn't need to be proven. Gosset remarked that these "obvious" solutions to problems were generally not obvious to anyone else until they'd slogged through the math and could see on paper what Fisher saw in his head.

After graduating from Cambridge with his MA in mathematics, Fisher tried his hand at a number of jobs—statistician for an investment company, farmer in Canada, math teacher (which he hated, perhaps because his mathematics were not obvious to his students either). Feeling snubbed by the academic community after a paper he had submitted for publication was reduced to a footnote in a larger paper by Karl Pearson, he took a job as a statistician at the Rothamsted Agricultural Experimental Station in Harpenden, England. There, he was given access to a warehouse of data, collected over 90 years, all carefully bound in leather volumes and full of information on everything having to do with growing crops. Fisher was in his element. Supplementing this massive database with findings from his own experiments, he made major contributions to science, including the analysis of variance, or ANOVA.

In a 1923 paper ("Studies in Crop Variation II"), Fisher describes a series of experiments he performed on the effect of different variables on the growth of potatoes. He compared "three types of fertilizer, ten varieties of potatoes, and four [types] of soil" (Salsburg, 2001, p. 48). Anyone who has tried to grow veggies in a home garden knows that there's variability in the crop you get: Even with great care, some plants yield more than others do. Fisher developed the mathematics needed to determine how much of the overall variability in the growth of a particular crop came from a specific variable (amount of rain, type of fertilizer, characteristics of the soil, etc.). He "analyzed the variability" to develop his method of splitting the variability up, hence the ANOVA test.

By the way, the Rothamsted Agricultural Experimental Station is still in existence and actively engaged in cutting-edge agricultural research.

FIGURE 11.1 Ronald Aylmer Fisher.

- Homologous transfusion, or HT, in which blood from a compatible donor is stored and then injected into the athlete just before the athletic event.
- Autologous transfusion, or AT, where blood from the athlete herself or himself is removed several months prior to the event, stored, and then transfused back into the athlete just before the competition.

- Use of the drug erythropoietin, or EPO, a naturally created hormone that controls the production of red blood cells production in humans.

Blood doping can be difficult to detect because it involves injecting human blood products rather than an identifiable foreign substance. However, with the potential rewards come several risks. Boosting the number of red blood cells makes the blood thicker and more likely to clot, thereby increasing the risk of stroke and heart attack. Accepting blood from a matched donor can invite transmission of infection, and taking medication to increase red blood cell count can have several adverse health effects, including liver damage and high cholesterol levels.

A HYPOTHETICAL STUDY OF BLOOD DOPING

Let's suppose we want to find out if there is any difference in how well each of these doping methods works. We will select our sample (using random selection) from the population. Participants will be assigned (randomly) to one of four groups:

- Group 1 will undergo blood doping using the HT method.
- Group 2 will undergo AT doping.
- Group 3 will use EPO.
- Group 4 will serve as a control group and will not receive blood doping of any type.

All of the participants will be given a test of endurance where scores will vary from 0 to 25 and higher scores indicate greater endurance.

BOX 11.1

Hypothetical effect of different methods of blood doping*

HT	AT	EPO	Control
20	22	15	5
24	16	19	9
19	20	23	11
25	22	20	10
15	19	18	9
$\Sigma = 103$	$\Sigma = 99$	$\Sigma = 95$	$\Sigma = 44$
$n = 5$	$n = 5$	$n = 5$	$n = 5$
$\bar{X} = 20.6$	$\bar{X} = 19.8$	$\bar{X} = 19.0$	$\bar{X} = 8.8$

*Maximum endurance score possible = 25

Box 11.1 shows the hypothetical endurance scores for each group at the end of our experiment. Before we start, take a minute to predict the outcome. Do you think that all three methods of blood doping will work for participants in all three groups in the same way? If not, which group should see the greatest benefit? Which group should see the least benefit?

ASSESSING BETWEEN-GROUP AND WITHIN-GROUP VARIABILITY IN OUR HYPOTHETICAL RESULTS

Take a look at the data in Box 11.1, and answer our first question: Without doing any statistical analysis for the moment, did the doping methods have the same effect on the endurance scores of these athletes? How can you tell? Do you think that blood doping made a difference in endurance scores at all? What would you look at to see if blood doping mattered?

Fisher (if he were available to help us with this study) would advise that we look at what he called *between-group variability* in order to answer this question. Between-group variability is just what it sounds like: variability or difference across group means. If you look at the data, you can see that the means describing the data in the four groups are all different. The participants in the HT group had the highest endurance score, followed by those in the AT and the EPO groups; participants in the control group bring up the rear with the lowest average endurance scores.

What would make these four means differ from one another? Well, some of the difference is probably due to random chance: Everyone is different, and maybe we happened to get some people with very low or very high endurance scores in one of our groups. Some of this difference, though, might be caused by our independent variable. Maybe the fact that the groups experienced different doping methods (or no doping at all) is making a difference in the endurance scores.

Speaking of variability, take a look within any one group, and ask yourself whether or not everyone in this group performing in exactly the same way. Do all five people in a group get the same score on the endurance test? Or, do scores within any one group vary, suggesting that performance on the endurance test varies even within a group of participants who shared the same experience?

We are looking now at what Fisher called *within-group variability*, caused by random chance or the fact that no matter how many variables we might have in common with one another, we are all still different. To address this situation, Fisher devised a test to measure the influence of random chance on within-group variability. He saw that if we compare the variability between groups (caused by our IV and by random chance) with the variability within groups (caused by random chance), we can "average out" random chance and be left with the effect of the independent variable. Box 11.2 illustrates the logic of the ANOVA procedure.

BOX 11.2 **Averaging out the effects of random chance between and within groups**

Type of doping method used (our IV)			
HT	AT	EPO	Control
20	22	15	5
24	16	19	9
19	20	23	11
25	22	20	10
15	19	18	9
$\Sigma = 103$	$\Sigma = 99$	$\Sigma = 95$	$\Sigma = 44$
$n = 5$	$n = 5$	$n = 5$	$n = 5$
$\bar{X} = 20.6$	$\bar{X} = 19.8$	$\bar{X} = 19.0$	$\bar{X} = 8.8$

Variability within groups

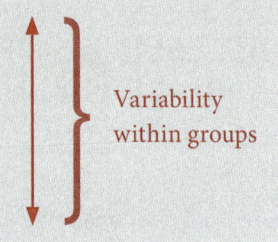

Variability across or between groups

The ANOVA test compares:

$$\frac{\text{variability between}}{\text{variability within}} = \frac{\text{variability caused by IV} + \text{variability caused by random chance}}{\text{variability caused by random chance}}$$

Random chance cancels out, leaving us with just the effect of the IV.

---

one-way ANOVA The inferential test used when there is one factor (or independent variable) with more than two levels (or groups being compared).

factor The independent variable under investigation in an ANOVA.

levels The values that an independent variable can take on. In an experiment, the number of levels equals the number of groups compared (e.g., an independent variable with four levels creates four groups for comparison).

ANOVA TERMINOLOGY

Our blood doping experiment is an example of a **one-way ANOVA**, which refers to the *one* IV (usually called a **factor** in analysis of variance) with more than two **levels** (in ANOVA, the number of treatments or groups being compared). In our experiment, we have one factor (the type of blood doping used) that has four levels (the types of blood doping—including no doping at all—that we're comparing).

The fact that we have a *one-way* test should suggest something to you. Indeed, there is also a **two-way ANOVA**, used to compare two factors (IVs), each with at least two levels. Suppose I think that the sex of the athlete is having an effect on endurance scores and might interact with the method of blood doping used. Now I have two factors—doping method used (with four levels) and sex (with two levels)—and I want to see what effect they both have on our DV (endurance score).

448 STATISTICS IN CONTEXT

In this chapter, we will discuss both one-way and two-way ANOVAs, beginning with the simplest case: our blood doping study, which requires a one-way ANOVA.

THE ONE-WAY ANOVA PROCEDURE

Mathematically, what we need to do is calculate the amount of variability *between the groups* and the amount of variability *within the groups*. We then compare the two using the ratio shown in Box 11.2. Fisher and his assistants worked out the mathematics that allow these sorts of comparisons, and they're all based on analyzing the variability in an experiment. Instead of trying to limit or eliminate variability, these tests use that variability to compare groups.

Fisher actually wanted to find the *mean* variance between groups and the *mean* variance within them. To do so, he first had to calculate the overall variability between and then within groups. Fisher called the overall variability between groups the **sum of squares between (SS_b)** and the overall variability within groups the **sum of squares within (SS_w)**.

SUMS OF SQUARES

You might be wondering where the term *sum of squares* came from and what it has to do with variability. It actually is a sort of nickname for variability, and it comes from the formula used to calculate it. Let's take a look at the equation in Box 11.3 to see what squares we might be summing. When we calculate variance, we are *summing the squared deviations around the mean*—hence **sum of squares**.

two-way ANOVA The inferential test used when there are two factors (or independent variables), each with two or more levels (or groups being compared).

sum of squares between groups (SS_b) A measure of the total variability between groups.

sum of squares within groups (SS_w) A measure of the total variability within groups.

sum of squares The sum of the squared deviations of an event or a sample mean from the mean. The quantity of difference between two means, squared.

BOX 11.3 **Definition formula for variance: The sum of squares**

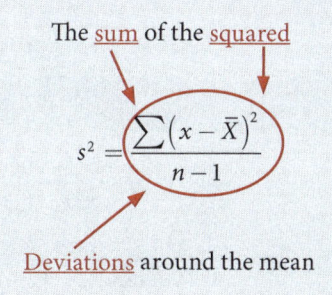

The sum of the squared

$$s^2 = \frac{\sum(x - \bar{X})^2}{n-1}$$

Deviations around the mean

Fisher then divided the overall variability between and within the groups (SS_b and SS_w) essentially by n (actually by a version of the *df*, or $n - 1$) in order to find what he called the **mean squares between groups (MS_b)** or the **mean squares within groups (MS_w)**. The ratio of the measure of the average variability between the groups (reflecting the effect of the factor and random chance) to the measure of average variability within the groups (reflecting the effect of random

mean squares between groups (MS_b) A measure of the average effect of the independent variable on the dependent variable; the ratio of the total variability between groups (SS_b) to the degrees of freedom between groups (df_b).

CHAPTER 11 Analysis of Variance   449

mean squares within groups (MS_w) A measure of the average effect of random variability within groups; the ratio of the total variability within groups (SS_w) to the degrees of freedom within groups (df_w).

F-statistic The statistic generated by an ANOVA, calculated as the ratio of variability between groups (MS_b) to variability within groups (MS_w).

chance) yields a statistic called F (in honor of Fisher). The size of the **F-statistic** tells us about how much of an effect the IV, or factor, has had: The bigger F is, the greater the effect of the IV. Box 11.4 illustrates the general procedure for calculating F.

Before we take a look at the formulas, suppose the IV had no effect at all. What would the value of F be? Box 11.5 shows what would happen in this case.

The effect of the IV would be zero, and F would be close to 1.00. (Random chance between the groups might be less than random chance within the groups, so F can actually be less than 1.00 but cannot be less than zero.) There is no upper limit to the value of F because there is no limit to the effect of the IV. Thus, the closer F is to 1.00, the less likely it is that our IV has had a significant effect, and the larger the value of F, the more likely it is to be statistically significant.

BOX 11.4   **General procedure for calculating F**

$$F = \frac{\text{effect of the IV} + \text{effect of random chance}}{\text{effect of random chance}}$$

or

$$F = \frac{MS_b}{MS_w}$$

BOX 11.5   **If the IV has no effect at all**

$$F = \frac{\cancel{\text{effect of the IV}} + \text{effect of random chance}}{\text{effect of random chance}}$$

$$= \frac{\text{effect of random chance}}{\text{effect of random chance}}$$

Finding the Sum of Squares: An Example

The equations for calculating the variability between and within groups are shown in Box 11.6 and Box 11.7. You should see some familiar formulas here; after all, we're calculating variance again.

BOX 11.6   Finding the sum of squares between and within groups

$$SS_b = \sum \left[\frac{(\sum x_i)^2}{n_i} \right] - \frac{(\sum x_{total})^2}{N_{total}} \qquad SS_w = \sum \left[\sum x_i^2 - \frac{(\sum x_i)^2}{n_i} \right]$$

$$SS_{total} = \sum x_{total}^2 - \frac{(\sum x_{total})^2}{N_{total}} \qquad SS_{total} = SS_b + SS_w$$

BOX 11.7   Finding the mean sum of squares between and within groups

$$MS_b = \frac{SS_b}{df_b} \qquad \text{and} \qquad MS_w = \frac{SS_w}{df_w}$$

df_b (degrees of freedom between groups) = number of groups (K) $-$ 1

df_w (degrees of freedom within groups) = total number of participants (N) minus the number of groups (K), or $N_{total} - K$

$$MS_{total} = \frac{SS_{total}}{df_{total}}$$

where $df_{total} = N - 1$

Note that $df_b + df_w = df_{total}$

There are some caveats to discuss here. First, although you can easily calculate the MS_{total}, it is not necessary for finding the value of F, so it is often left out of the description of an ANOVA in a publication. However, SS_{total} (overall total variability) is very often calculated and used as a check on the correct calculation of SS_b and SS_w. That is because the total sum of squares (total overall variability) is made up of the variability between groups and the variability within groups. You can't have unaccounted-for variability just floating around your design. If $SS_b + SS_w \neq SS_{total}$, you have done something wrong, somewhere, and need to go back and double-check your calculations. Let's work through finding the SS_b (in Box 11.8) and SS_w (in Box 11.9) to demonstrate how these formulas work. Box 11.10 shows the calculation of the SS_{total}, which we will use to double-check our calculations of both SS_b and SS_w.

CHAPTER 11 Analysis of Variance

BOX 11.8 Calculating the total variability between groups (SS_b) in our doping study

HT	AT	EPO	Control
20	22	15	5
24	16	19	9
19	20	23	11
25	22	20	10
15	19	18	9
$\Sigma = 103$	$\Sigma = 99$	$\Sigma = 95$	$\Sigma = 44$
$n = 5$	$n = 5$	$n = 5$	$n = 5$

The Data

The formula for SS_b

$$SS_b = \underbrace{\sum \left[\frac{(\sum x_i)^2}{n_i}\right]}_{\text{Part I}} - \underbrace{\frac{(\sum x_{total})^2}{N_{total}}}_{\text{Part II}}$$

Instructions for Part I

1. For each group (*i*), add up the scores within the group and square the total.
2. Divide that total squared by the number of scores in the group. Repeat Steps 1 and 2 for all the groups.

$\sum \left[\frac{(\sum x_i)^2}{n_i}\right]$ Part I

$$\sum \left[\frac{(\sum x_i)^2}{n_i}\right] = \left[\frac{(103)^2}{5}\right] + \left[\frac{(99)^2}{5}\right] + \left[\frac{(95)^2}{5}\right] + \left[\frac{(44)^2}{5}\right]$$

$$= \left[\frac{10,609}{5}\right] + \left[\frac{9,801}{5}\right] + \left[\frac{9,025}{5}\right] + \left[\frac{1,936}{5}\right]$$

$$= 2,121.8 + 1,960.2 + 1,805.0 + 387.2 = 6,274.2$$

Instructions for Part II

1. Add up all of the scores in all groups, then square the total.
2. Divide Step 1 by the total number of participants in the study (*N*).

$\frac{(\sum x_{total})^2}{N_{total}}$ Part II

$$\frac{(\sum x_{total})^2}{N_{total}} = \frac{(103 + 99 + 95 + 44)^2}{20} = \frac{341^2}{20} = \frac{116,281.0}{20} = \boxed{5,814.05}$$

Now we subtract Part II from Part I to find the SS_b.

$$SS_b = \sum \left[\frac{(\sum x_i)^2}{n_i}\right] - \frac{(\sum x_{total})^2}{N_{total}}$$

$$= 6,274.2 - 5,814.05$$

$$= \boxed{460.15}$$

STATISTICS IN CONTEXT

So, our SS_b—the total variability between the groups—is 460.15. Did you see parts of the formula for SS_b that were familiar? Fisher used the formula for variance that we're all familiar with to break out, or *partition*, the variability into *between-group* and *within-group* variability. You should see pieces of the formula for variance (in particular, the calculation formula) in the calculation of the overall variance between groups (SS_b).

Now that we have the SS_b, we need to determine the overall variability within groups (Box 11.9). Some of the pieces we need to calculate SS_w are also part of the formula for SS_b, so we're about halfway there.

BOX 11.9 **Calculating the overall variability within groups (SS_w) in our doping study**

HT (x^2)	AT (x^2)	EPO (x^2)	Control (x^2)
20 (400)	22 (484)	15 (225)	5 (25)
24 (576)	16 (256)	19 (361)	9 (81)
19 (361)	20 (400)	23 (529)	11 (121)
25 (625)	22 (484)	20 (400)	10 (100)
15 (225)	19 (361)	18 (324)	9 (81)
$\Sigma = 103$	$\Sigma = 99$	$\Sigma = 95$	$\Sigma = 44$
$\Sigma^2 = 2,187$	$\Sigma^2 = 1,985$	$\Sigma^2 = 1,839$	$\Sigma^2 = 408$
$n = 5$	$n = 5$	$n = 5$	$n = 5$

$$SS_w = \sum \left[\sum x_i^2 - \frac{(\sum x_i)^2}{n_i} \right]$$

The formula for SS_w

Instructions for finding SS_w

1. For each group, sum the scores in the group, square that total, and divide that total by the number of scores within the group. (Hint: We've already done this part.)
2. Subtract Part I from the sum of all the scores squared in a group, set that piece aside, repeat the brackets, and then sum all the individual pieces.
3. Add these group statistics together.

$$SS_w = \sum \left[\sum x_i^2 - \frac{(\sum x_i)^2}{n_i} \right]$$

$$= \left[2,187 - \frac{(103)^2}{5} \right] + \left[1,985 - \frac{(99)^2}{5} \right] + \left[1,839 - \frac{(95)^2}{5} \right] + \left[408 - \frac{(44)^2}{5} \right]$$

$$= [2,187.0 - 2,121.8] + [1,985.0 - 1960.2] + [1,839 - 1,805] + [408.0 - 387.2]$$

$$= 65.2 + 24.8 + 34.0 + 20.8$$

$$= \boxed{144.8}$$

One final piece that we should calculate is the total sum of squares. Box 11.10 shows these calculations.

BOX 11.10 Finding the total sum of squares

$$SS_{total} = \underbrace{\sum x_{total}^2}_{\text{Part I}} - \underbrace{\frac{(\sum x_{total})^2}{N_{total}}}_{\text{Part II}}$$

Notice that we already have these two parts. Part I comes from the formula for SS_w, and Part II comes from the formula for SS_b.

$$= \underbrace{[(2{,}187 + 1{,}985 + 1{,}839 + 408)]}_{\text{Part I}} - \underbrace{\left[\frac{(103 + 99 + 95 + 44)^2}{20}\right]}_{\text{Part II}}$$

$$= 6{,}419 - \frac{341^2}{20} = 6{,}419.00 - 5814.05 = \boxed{604.95}$$

We now have the values for all of the sums of squares, which we will need to calculate MS_b and MS_w so that we can then determine F. But right now, a lot of these elements are sort of flying around. To keep track of the sums of squares, the mean squares, the degrees of freedom, and so on, we use what we call a **source table**, as shown in Box 11.11. Most computer programs that perform ANOVAs will present the results of their calculations in a source table similar to the one shown here.

source table A table used to record the sums of squares, the mean squares, and the degrees of freedom required to find an *F*-statistic using ANOVA.

BOX 11.11 Source table*

Source	SS	df	MS	F
Between groups	SS_b	$K - 1$	$\dfrac{SS_b}{df_b}$	$\dfrac{MS_b}{MS_w}$
Within groups	SS_w	$N - K$		
Total	SS_{total}	$N - 1$		

Note: $SS_b + SS_w = SS_{total}$
$df_b + df_w = df_{total}$ } Use these to check your calculations.

*The word "source" refers to the source of variability for each component of *F*.

Notice that we have the calculations we will need to find our F-value. As we fill in the source table for our experiment, also notice that once we have calculated the SS_b and SS_w, we're just about ready to calculate F. Box 11.12 shows the calculations for MS_b, MS_w, and our F-statistic.

BOX 11.12 — Calculating MS and F

$$MS_b = \frac{SS_b}{df_b} = \frac{460.15}{3} = 153.38$$

$$F = \frac{MS_b}{MS_w} = \frac{153.38}{9.05} = 16.95$$

$$MS_w = \frac{SS_w}{df_w} = \frac{144.8}{16} = 9.05$$

And, finally, Box 11.13 shows the completed source table. Our observed F-statistic is 16.95 with $df_b = 3$ and $df_w = 16$.

BOX 11.13 — Completed source table*

Source	SS	df	MS	F
Between groups	460.15	3	153.38	16.95
Within groups	144.80	16	9.05	
Total	604.95*	19		

* Notice that $SS_b + SS_w = SS_{total}$.

Is our observed F significant? To determine this, we need to once again compare our observed statistic with one from a table of critical statistics, this time called an F-table (see Appendix C). And just as with the z-tests and t-tests we have already seen, the table shows you F-values that cut off the outer tails of a distribution of F-statistics. We need to be farther out in the tail of the F-distribution than these critical values in order to say that our observed F is significant. Figure 11.2 shows the distribution of F-values.

CHAPTER 11 Analysis of Variance

FIGURE 11.2
The *t*-distribution and the *F*-distribution.

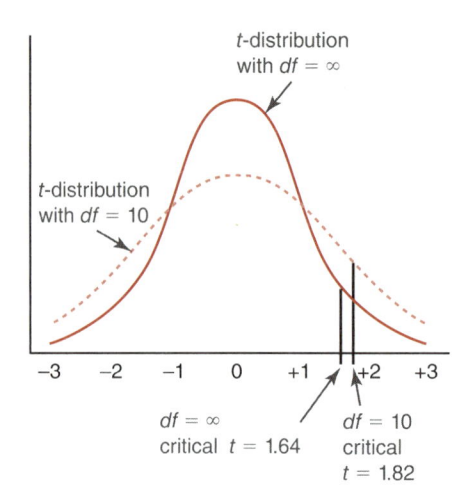

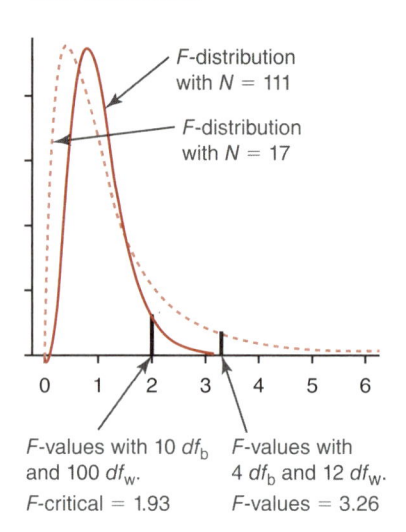

The *t*-distribution

t-distribution with $df = \infty$
t-distribution with $df = 10$

$df = \infty$ critical $t = 1.64$
$df = 10$ critical $t = 1.82$

The *F*-distribution

F-distribution with $N = 111$
F-distribution with $N = 17$

F-values with 10 df_b and 100 df_w.
F-critical = 1.93

F-values with 4 df_b and 12 df_w.
F-values = 3.26

Think About It...

t FOR TWO AND *F* FOR MANY

It is possible to use the ANOVA test to compare two group means. (It's kind of like swatting flies with a sledgehammer: You can do it, but you're using more tool than you really need.) If you do calculate *F* when you have 1 *df* between groups, you will find that $F = t^2$.

Let's try it. Here are scores for two groups. Calculate both *F* and an independent-samples *t*. Does $F = t^2$?

Group 1: 5   6   9   10   ($\Sigma x = 30$)   ($\Sigma x^2 = 242$) $n = 4$

Group 2: 1   2   2   0   ($\Sigma x = 5$)   ($\Sigma x^2 = 9$) $n = 4$

Now compare the critical *t*- and *F*-values. You should also find that:

$$F_{critical} = (t_{critical})^2$$

and vice versa:

$$t_{critical} = \sqrt{F_{critical}}$$

Let's think about this conceptually, rather than mathematically:

1. What does the top of the *t*-test ask you to compare?
2. What does MS_b represent if you have only two groups?
3. What does the term at the bottom of the *t*-test represent?

STATISTICS IN CONTEXT

4. Is there any similarity between the calculation of MS_w and the calculation of the pooled variability used in the *t*-test?
5. How are the calculations of the variability in these two tests different? (Hint: Think about what measure of variability is being used in both—standard deviation and variance.)

Answers to this exercise can be found on page 495.

Comparing the *F*-Distribution and the *t*-Distribution

Compare the *F*-distribution with the *t*-distribution that we have been using so far, and notice that they differ from one another. The most obvious difference between the two distributions is in their shape. The distribution of *t*-values is normal in shape, while the *F*-distribution is positively skewed: The right-hand tail of the distribution is a great deal longer than the left-hand tail. This is because there are no negative *F*-values. Think about this conceptually for a minute: Does it make sense that there could be less than zero variability in an ANOVA design? What would negative variability look like?

The second thing to notice is that both distributions are affected by sample size. For the *t*-distribution, as df ($n-1$) gets larger, the shape of the distribution becomes more normal (the distribution "pulls in its skirts"), which means that the critical *t*-value gets smaller as the samples get bigger. The same is true for the *F*-distribution. Notice that as the total number of participants in the study gets bigger, and as the number of groups in your model gets larger, the critical *F*-value gets smaller. The *F*-distribution pulls in its skirts with larger *N*'s as well.

So, to the back of the book we go, looking at Appendix C to see the table of critical *F-value*s. I've reproduced a part of the *F* table in Table 11.1 (shown below) so that we can talk about how to read the table.

There are several things to notice about this table. First, it's very large. The *F*-table takes up several pages, so when you use it, make sure you are at the right spot. You do that by reading across the top of each page of the table until you find the correct degrees of freedom between groups (df_b). So, if you used five groups in your study, you would have four degrees of freedom between groups ($df_b = k - 1$, or $5 - 1 = 4$), and you would read over to the column headed "4" df_b. (The reason for the large gaps in the headings here is that the table in Appendix C has been shortened to make it viable in a textbook with a limited number of pages for tables.)

We used four groups in our study, so we need the column in Table 11.1 marked "3" at the top because $df_b = k - 1$ (number of groups minus one) $= 4 - 1 = 3$. Next, we need to read down the left-hand side of the table (and this is what makes the table so large) until we come to the df_w ($N - k$, or total number of subjects minus the number of groups) in our study. We had 20 participants in our study and four groups, so $df_w = 20 - 4 = 16$. We read down the column marked $df_b = 3$ until we come to the row marked $df_w = 16$: There, we converge on a critical *F*-value of 3.24. This particular *F*-table shows significant values when $\alpha = .05$. We would move on to the next table to find the significant *F*-values for other alpha levels. Look up the critical *F*-value from the table marked $\alpha = .01$, and you should find that it is 5.29.

TABLE 11.1 Table of significant F-values, α = .05

df within \ df Between	1	2	3	4	5	6
1	161.4	199.50	215.71	224.58	230.16	233.99
2	18.51	19.00	19.16	19.25	19.30	19.33
3	10.13	9.55	9.28	9.12	9.01	8.94
4	7.71	6.94	6.59	6.39	6.26	6.16
5	6.61	5.79	5.41	5.19	5.05	4.95
6	5.99	5.14	4.76	4.53	4.39	4.28
7	5.59	4.74	4.35	4.12	3.97	3.87
8	5.32	4.46	4.07	3.84	3.69	3.58
9	5.12	4.26	3.86	3.63	3.48	3.37
10	4.96	4.10	3.71	3.48	3.33	3.22
11	4.84	3.98	3.59	3.36	3.20	3.09
12	4.75	3.89	3.49	3.26	3.11	3.00
13	4.67	3.81	3.41	3.18	3.03	2.92
14	4.60	3.74	3.34	3.11	2.96	2.85
15	4.54	3.68	3.29	3.06	2.90	2.79
16	4.49	3.63	3.24	3.01	2.85	2.74
17	4.45	3.59	3.20	2.96	2.81	2.70
18	4.41	3.55	3.16	2.93	2.77	2.66
19	4.38	3.52	3.13	2.90	2.74	2.63
20	4.35	3.49	3.10	2.87	2.71	2.60

Once again, in order for our observed F-statistic to be significant, it must exceed the critical F-value from the table. The result we obtained ($F = 16.95$) is statistically significant at both $\alpha = .05$ and $\alpha = .01$. Endurance scores differ significantly across the three blood doping methods (and a control) that we compared. So, we might say something like this in our write-up:

> We compared the effects of three methods of blood doping on the endurance scores of athletes. Twenty randomly selected participants were randomly assigned to one of four treatment groups, such that each group had five participants. The first group received homologous doping (HT), the second group received autologous doping (AT), and the third group received the drug EPO,

which increases red blood cell count. A fourth group served as the control group and was not blood-doped. After doping, each participant took an endurance test, where higher scores (maximum = 25) indicated higher endurance levels. A one-way ANOVA revealed a statistically significant difference in endurance scores as a function of method of doping administered: $F_{(3,16)} = 16.95, p < .01$.

This statement tells the reader that we used an *F*-test (ANOVA), that we had 3 *df* between groups and 16 *df* within groups (the values are, by tradition, shown with the df_b listed first and then the df_w), and that the value of *F* was 16.95, which was significant at least at the .01 level.

CheckPoint

Tanya specializes in pain management. In an effort to better understand the effectiveness of competing strategies for reducing pain, she devises an experiment in which four groups, each comprising 10 randomly assigned participants who suffer from chronic back pain. Group 1 receives a placebo, Group 2 receives cortisone injections, Group 3 is guided through a meditation exercise, and Group 4 serves as a control. After administering treatments, Tanya collects self-reported pain scores from each participant. The self-reported pain scores range from 0 (*no pain*) to 10 (*unspeakable pain*). Use the pain scores, reported in the table below, to answer the following questions and help Tanya determine whether the differences between the mean pain scores for each treatment group are statistically significant at an alpha level of 0.05. Round your answers to two decimal places, where appropriate.

Self-reported pain scores

Placebo	Cortisone	Meditation	Control
4	3	5	8
6	2	4	4
3	6	7	5
3	5	6	7
6	3	2	9
5	4	7	6
7	6	5	4
2	1	3	4
5	2	4	5
3	4	6	7
Σ = 44	Σ = 36	Σ = 49	Σ = 59
n = 10	n = 10	n = 10	n = 10

1. Calculate SS_b. _____
2. Calculate SS_w. _____
3. Calculate df_b. _____
4. Calculate df_w. _____
5. Calculate MS_b. _____
6. Calculate MS_w. _____
7. Calculate F. _____
8. Identify the critical *F*-value, compare it to *F*-observed, and state whether you should accept or reject the null hypothesis. _____

Answers to these questions are found on page 462.

Think About It...

THE *F*-STATISTIC

The *F*-statistic is the ratio of the average variability between our groups and the average variability within our groups. The larger our *F*-statistic is, the larger the difference between group means in our design. Let's take a closer look at the *F*-statistic. Answer the following questions, referring to Figure 11.2 for help if you need it.

1. Can *F* ever be a negative number? _____
2. The *F*-distribution is skewed. Is the skew positive or negative? _____
3. Suppose that the null hypothesis is true and there is no meaningful difference between the groups. What should the value of *F* be? _____
4. Keeping in mind that MS_w is a measure of random chance, can the value of *F* be between 0 and 1? _____
5. Suppose *F* was 0.40. What would that tell you about amount of variability within groups compared to the amount of variability between groups? _____
6. Are *F*-tests ever directional? _____
7. Think about the way that the null and alternative hypotheses are written for an *F*-test. If you have more than two groups to compare, does it make sense to write a directional alternative hypothesis? _____

Answers to this exercise can be found on page 495.

POST-HOC TESTING

We still have some unanswered questions here, even with our significant *F*-value. The *F*-test tells us that the type of doping used did have an effect on endurance score. But the *F*-test tells us only that. It does not tell us if any one type of doping was better than, or worse than, any other type, or which type of doping caused the overall difference that the *F*-test is detecting.

If we go back to the means of each group, we can get a feel for where the significant *F* is coming from, so let's do that. The mean endurance score was 20.6 for athletes undergoing HT doping, 19.8 for those who experienced AT doping, 19.0 for those taking EPA, and 8.8 for the control group. The HT doping group had the highest endurance scores, the AT doping and EPO doping groups had slightly lower endurance scores, and the control group had the lowest endurance scores. But are these differences meaningful? Is the difference between the AT and EPO groups (19.8 − 19.0, or 0.8) really important? To answer this question, we need to use a **post-hoc test** (literally, an "after the fact" test) of all possible pairs of comparisons. The post-hoc test compares each pair of group means. If we have three groups, we can make several "pairwise" comparisons (see Box 11.14).

post-hoc test A test used to compare pairs of group means.

BOX 11.14 Possible pairwise comparisons*

Groups	1 (HT Doping)	2 (AT Doping)	3 (EPO Doping)	4 (Control)
1 (HT doping)	HT vs. HT 1 vs. 1 = 0	HT vs. AT 1 vs. 2	HT vs. EPO 1 vs. 3	HT vs. control 1 vs. 4
2 (AT doping)	AT vs. HT 2 vs. 1 (same as 1 vs. 2)	AT vs. AT 2 vs. 2 = 0	AT vs. EPO 2 vs. 3	AT vs. control 2 v. 4
3 (EPO doping)	EPO vs. HT 3 vs. 1 (same as 1 vs. 3)	EPO vs. AT 3 vs. 2 (same as 2 vs. 3)	EPO vs. EPO 2 vs. 2 = 0	EPO vs. control 3 vs. 4
4 (control)	Control vs. HT 4 vs. 1 (same as 1 vs. 4)	Control vs. AT 4 vs. 2 (same as 2 vs. 4)	Control vs. EPO 4 vs. 3 (same as 3 vs. 4)	Control vs. control 4 vs. 4 = 0

*Notice that the values on the diagonal are always zero (one group minus itself = zero). Values below the diagonal are mirror images of the values above the diagonal.

There are six pairs of groups that we can compare. Let's start with the first row in this table. Here, we're comparing HT doping with AT doping (1 vs. 2), HT doping with EPO doping (1 vs. 3), and HT doping with our control group (1 vs. 4). Notice that we're ignoring the very first comparison—HT doping vs. HT doping—because it doesn't make sense.

Moving to the second row, we can ignore the first comparison because we have already considered it: Comparing HT and AT doping is the same as comparing AT and HT doping (the group comparisons shaded in light pink in Box 11.14 are "mirror images" of these three comparisons, so we don't need to make these comparisons.) On the second row, we continue comparing group means, AT vs. EPO and AT vs. controls. That leaves the final comparison, EPO doping vs. controls.

We have already discussed an inferential test that allows comparison of two group means, and someone reading this right about now is surely shouting, "Just do a t-test!" That's a great idea in principle, but there's a problem with repeatedly doing any inferential test on a given set of data: Basically, each time you use a t-test to compare two group means, you risk making a Type I error (rejecting a null hypothesis that you shouldn't). We've already decided that the minimal acceptable risk for a Type I error is .05 (5 times out of every 100 times we use the t). And we

> **CheckPoint Answers**
> *Answers to questions on page 459*
>
> 1. $SS_b = 27.8$
> 2. $SS_w = 104.6$
> 3. $df_b = 3$
> 4. $df_w = 36$
> 5. $MS_b = 9.27$
> 6. $MS_w = 2.91$
> 7. $F = 3.19$
> 8. $F_{obs} = 3.19 > F_{crit} = 2.87 \therefore$ reject
>
> **SCORE:** /8

would need to do three *t*-tests to compare all possible pairs of samples here. That means our risk of a Type I error would increase every time we ran another *t*-test. Running two tests increases the risk of a Type I error to 10%. Running three increases the risk to roughly 15% (calculating the risk isn't quite equivalent to multiplying .05 by the number of times the test is run, but it will give you a value that is close enough for our argument). At this rate, the risk rises past our limit of 5 times out of 100 and quickly becomes unacceptable.

What we really need is to do all three pairwise comparisons at once, rather than repeating the process. Statisticians have developed a number of these **multiple-comparison tests** (also known as *post-hoc tests*) that allow us to do this.

multiple-comparison tests A series of tests developed to explore differences between means when the *F*-test is significant. The Tukey honestly significant difference Test is an example of a multiple-comparison test.

THE TUKEY HSD TEST WITH EQUAL *n*'S

Multiple-comparison tests differ depending on the kind of post-hoc question we want to answer. We'll discuss one particular kind of post-hoc test called the **Tukey honestly significant difference (or HSD) test.** The test is named after John Tukey, a famous statistician we met in Chapter 3, when we discussed another of his inventions, the stem-and-leaf plot.

Tukey honestly significant difference (or HSD) test A specific type of multiple-comparison test used to assess pairwise comparisons among means.

This test is designed to compare all possible pairs of sample means, all simultaneously, and to minimize the probability of making a Type I error. You will see that the Tukey HSD test looks very much like the *t*-test we've already discussed, but with one small change. The difference, as you may have guessed, is in how error or variability is calculated. Box 11.15 shows the formula for calculating a Tukey HSD statistic for our blood doping example. We will be calculating a Tukey statistic for each pair of means being compared, so we will calculate six Tukey HSD statistics in this example.

BOX 11.15 — Calculating Tukey HSD when *n*'s are equal for each sample*

$$\text{Tukey HSD} = \frac{\bar{X}_1 - \bar{X}_2}{s_{\bar{X}}}, \text{ where } s_{\bar{X}} = \sqrt{\frac{MS_w}{N_{total}}} \text{ and } N_{total} = n \text{ common to each sample}$$

HT vs. AT	$\text{Tukey HSD} = \dfrac{\bar{X}_1 - \bar{X}_2}{s_{\bar{X}}} = \dfrac{20.6 - 19.8}{\sqrt{9.05/5}} = \dfrac{0.8}{\sqrt{1.81}} = \dfrac{0.8}{1.35} = \boxed{0.59}$
HT vs. EPO	$\text{Tukey HSD} = \dfrac{\bar{X}_1 - \bar{X}_2}{s_{\bar{X}}} = \dfrac{20.6 - 19.0}{\sqrt{9.05/5}} = \dfrac{1.6}{\sqrt{1.81}} = \dfrac{1.6}{1.35} = \boxed{1.19}$
HT vs. control	$\text{Tukey HSD} = \dfrac{\bar{X}_1 - \bar{X}_2}{s_{\bar{X}}} = \dfrac{20.6 - 8.8}{\sqrt{9.05/5}} = \dfrac{11.8}{\sqrt{1.81}} = \dfrac{11.8}{1.35} = \boxed{8.74}$
AT vs. EPO	$\text{Tukey HSD}^* = \dfrac{\bar{X}_1 - \bar{X}_2}{s_{\bar{X}}} = \dfrac{19.8 - 19.0}{1.35} = \dfrac{0.8}{1.35} = \boxed{0.59}$
AT vs. control	$\text{Tukey HSD}^* = \dfrac{\bar{X}_1 - \bar{X}_2}{s_{\bar{X}}} = \dfrac{19.8 - 19.0}{1.35} = \dfrac{11.0}{1.35} = \boxed{8.15}$
EP vs. control	$\text{Tukey HSD}^* = \dfrac{\bar{X}_1 - \bar{X}_2}{s_{\bar{X}}} = \dfrac{19.0 - 8.8}{1.35} = \dfrac{10.2}{1.35} = \boxed{7.56}$

*NOTE: I've stopped calculating the error term because that term is the same for all comparisons when *n*'s are equal.

BOX 11.16 — Possible pairwise comparisons

Groups	1 (HT Doping)	2 (AT Doping)	3 (EPO Doping)	4 (Control)
1 (HT doping)		(20.6 − 19.8) Tukey = **0.59**	(20.6 − 19) Tukey = **1.19**	(20.6 − 8.8) Tukey = **8.74**
2 (AT doping)			(19.8 − 19) Tukey = **0.59**	(19.8 − 8.8) Tukey = **8.15**
3 (EPO doping)				(19 − 8.8) Tukey = **7.56**
4 (control)				

CHAPTER 11 Analysis of Variance

Box 11.16 shows all six the comparisons we're making in the traditional matrix format for this post-hoc test. Once we have the Tukey HSD post-hoc statistics, though, what should we do with them? If you said we should compare them to a table of significant Tukey statistics, you're correct.

There are one or two things to note in Box 11.15. First, because we have the same number of subjects in each group, we only need to calculate our error term (measure of variability) once. The value of $s_{\bar{x}}$ stays the same for each comparison. Variability and the size of the sample are related to one another—as long as the sample sizes are the same, our estimate of variability can stay the same. If our sample sizes are not equal, we will have to change how we calculate the HSD statistic for each comparison (more about this a little later).

We have calculated six HSD statistics, one for each comparison that we're making. The larger the HSD statistic, the more likely it is that this comparison, this difference between means, is significant. We can get a feel for each comparison by just looking at the HSD statistics we calculated. There was a relatively small difference between the mean for Group 1 and the mean for Group 2, and our HSD statistic is correspondingly relatively small. The biggest difference between pairs of means is between Groups 1 and 4 (AT doping and the control group), and we have a correspondingly big value for the HSD statistic. But we need more than just who is bigger than whom here—we need to know if the difference between the two means is significant or not.

We will determine the significance of our statistic in the usual way, by comparing it to a critical value obtained from a table of critical values. You will find said table of critical Tukey HSD values in Appendix D. Table 11.2 shows a partial table of Tukey HSD statistics. If our observed Tukey statistics are larger than the tabled valued, then our pairwise comparison is statistically significant—that is, those two means differ significantly.

To find the critical Tukey value (we need to find only one, as you'll see in a minute), we read across the top of the table, showing the number of levels (or groups) of the independent variable. We have four levels of type of doping used in our experiment, so we'll read across to the fourth column. We then read down the left-hand side, showing the number of degrees of freedom within groups (df_w). We have 16 df_w, so we read down to the sixteenth row. We will find our critical value where the column and the row intersect. The top value in each box of the table shows the critical values if our alpha is set at .05; the bottom value represents the critical values if $\alpha = .01$. Our critical Tukey HSD value at $\alpha = .05$ is 4.05, and at $\alpha = .01$, it's 5.19. If our observed Tukey HSD value is greater than 4.05 or 5.19, then the difference between the values of the two groups in our pair is statistically significant. Box 11.17 shows the results of this post-hoc comparison.

We can now add to our original conclusion. Not only did we get a significant overall effect from our independent variable (as shown by our *F*-test), but we also can say that the real difference between our groups seems to be coming from the

TABLE 11.2 Partial table of critical Tukey HSD values

| df for Error Term | α | \multicolumn{7}{c}{Number of Levels} |
		2	3	4	5	6	7	8
5	.05	3.64	4.60	5.22	5.67	6.03	6.33	6.58
	.01	5.70	6.98	7.80	8.42	8.91	9.32	9.67
6	.05	3.46	4.34	4.90	5.30	5.63	5.90	6.12
	.01	5.24	6.33	7.03	7.56	7.97	8.32	8.61
7	.05	3.34	4.16	4.68	5.06	5.36	5.61	5.82
	.01	4.95	5.92	6.54	7.01	7.37	7.68	7.94
8	.05	3.26	4.04	4.53	4.89	5.17	5.40	5.60
	.01	4.75	5.64	6.20	6.62	6.96	7.24	7.47
9	.05	3.20	3.95	4.41	4.76	5.02	5.24	5.43
	.01	4.60	5.43	5.96	6.35	6.66	6.91	7.13
10	.05	3.15	3.88	4.33	4.65	4.91	5.12	5.30
	.01	4.48	5.27	5.77	6.14	6.43	6.67	6.87
11	.05	3.11	3.82	4.26	4.57	4.82	5.03	5.20
	.01	4.39	5.15	5.62	5.97	6.25	6.48	6.67
12	.05	3.08	3.77	4.20	4.51	4.75	4.95	5.12
	.01	4.32	5.05	5.50	5.84	6.10	6.32	6.51
13	.05	3.06	3.73	4.15	4.45	4.69	4.88	5.05
	.01	4.26	4.96	5.40	5.73	5.98	6.19	6.37
14	.05	3.03	3.70	4.11	4.41	4.64	4.83	4.99
	.01	4.21	4.89	5.32	5.63	5.88	6.08	6.26
15	.05	3.01	3.67	4.08	4.37	4.59	4.78	4.94
	.01	4.17	4.84	5.25	5.56	5.80	5.99	6.16
16	.05	3.00	3.65	4.05	4.33	4.56	4.74	4.90
	.01	4.13	4.79	5.19	5.49	5.72	5.92	6.08
17	.05	2.98	3.63	4.02	4.30	4.52	4.70	4.86
	.01	4.10	4.74	5.14	5.43	5.66	5.85	6.01
18	.05	2.97	3.61	4.00	4.28	4.49	4.67	4.82
	.01	4.07	4.70	5.09	5.38	5.60	5.79	5.94
19	.05	2.96	3.59	3.98	4.25	4.47	4.65	4.79
	.01	4.05	4.67	5.05	5.33	5.55	5.73	5.89
20	.05	2.95	3.58	3.96	4.23	4.45	4.62	4.77
	.01	4.02	4.64	5.02	5.29	5.51	5.69	5.84

BOX 11.17　Tukey HSD statistics

Comparison	Critical Tukey
HT vs AT	
HT vs EPO	
HT vs control	Tukey = 8.74**
AT vs EPO	
AT vs control	Tukey = 8.59**
EPO vs control	Tukey = 7.56**

*Critical Tukey value at $\alpha = .05 = 4.05$
**Critical Tukey value at $\alpha = .01 = 5.19$

difference between the athletes who experienced blood doping of any kind (Groups 1, 2, and 3) and the athletes in the control group who did not receive doping of any kind. We might say something like this:

> We compared the effects of three methods of blood doping on the endurance scores of athletes. Twenty randomly selected participants were randomly assigned to one of four groups, such that each group had five participants. The first group received homologous doping (HT), the second group received autologous doping (AT), and the third group received the drug EPO, which increases red blood cell count. The fourth group served as the control group and was not blood-doped. After doping, each participant took an endurance test, where higher scores (maximum = 25) indicated higher endurance levels. A one-way ANOVA revealed a significant difference in endurance scores as a function of method of doping administered: $F_{(3,16)} = 16.95$, $p < .01$. Tukey's HSD test was used to determine the nature of the differences between groups. The Tukey HSD test revealed that mean endurance scores were significantly higher in the HT ($M = 20.6$), AT ($M = 19.8$), and EPO groups ($M = 19.0$) than in the control group ($M = 8.8$): $p < .01$.

THE TUKEY HSD TEST WITH UNEQUAL N'S

We need to do a post-hoc test only if we have already shown that there is a significant effect of our independent variable (i.e., a significant F-statistic). If we find that our IV has had no effect, then we don't need to compare means to see where an effect is coming from. As I mentioned earlier, the calculation of the error term for the Tukey HSD statistic depends on sample size. If the *n*'s are not equal—in other words, if there are different numbers of people within each group—then we have to change how we calculate the HSD. Box 11.18 shows this change.

BOX 11.18 **Performing a Tukey HSD test when *n*'s are unequal**

$$\text{Tukey HSD} = \frac{\bar{X}_1 - \bar{X}_2}{s_{\bar{X}}},$$

$$\text{where } S_{\bar{X}} = \sqrt{\frac{MS_w}{2}\left(\frac{1}{n_1} + \frac{1}{n_2}\right)}$$

Notice that a "weighting factor" shows up again to do exactly the same thing it was doing last time (see the *t*-test for independent samples with unequal *n*'s in Chapter 10): It is giving greater weight to the larger of the two samples or groups being compared because large samples are better representatives of the population they came from than are small samples.

Because our sample sizes are unequal, our measure of the variability in the experiment ($s_{\bar{x}}$) will not be the same for each comparison—it will change as the sample size changes. Let's run an example with this formula just for practice. Suppose, when we were doing this experiment, that one of our EPO doping participants decided to drop out before the end of the study. Now our sample sizes are unequal. This means that our measures of center and spread for that sample change, and our *F*-statistic, change. Box 11.19 shows what happens to the data when this one person drops out.

Our *F* changed (from 16.95 to 17.74), but it is still significant at the .01 level. Now, if we want to compare all possible means to see where this significant *F* is coming from, we will need to use the formula that corrects for the unequal sample sizes. Box 11.20 shows this comparison. The difference between the two methods of finding Tukey HSD values is that here we need to account for the differences in sample size (akin to what we did when using *t*-tests with unequal sample sizes). Yes, once again, our error term changes to give the larger sample more weight.

Our Tukey HSD values have indeed changed, but the conclusion we draw at the end has not. We still have a significant difference in endurance scores across the groups (i.e., our *F*-statistic is significant). And our Tukey HSD test still shows that any kind of blood doping seems to increase endurance scores above the level of our baseline control group. The variability between groups that drives the significant *F*-test seems to be coming from the difference between the doped athletes and the control group.

BOX 11.19 **Changes to the experiment when one subject drops out**

Group 1 (HT)	Group 2 (AT)	Group 3 (EPO)	Group 4 (Control)
20	22	15	5
24	16	19	9
19	20	23	11
25	22	20	10
15	19	18	9
$\Sigma x = 103$	$\Sigma x = 99$	$\Sigma x = 72$	$\Sigma x = 44$
$\Sigma x^2 = 2{,}187$	$\Sigma x^2 = 1{,}985$	$\Sigma x^2 = 1{,}310$	$\Sigma x^2 = 408$
$\bar{X} = 20.6$	$\bar{X} = 19.8$	$\bar{X} = 18.0$	$\bar{X} = 8.8$
$n = 5$	$n = 5$	$n = 4$	$n = 5$

$$SS_b = \sum \left[\frac{(\sum x_i)^2}{n_i}\right] - \frac{(\sum x_{total})^2}{N_{total}}$$

$$= \left[\frac{(103)^2}{5}\right] + \left[\frac{(99)^2}{5}\right] + \left[\frac{(72)^2}{4}\right] + \left[\frac{(44)^2}{5}\right] - \frac{(318)^2}{19}$$

$$= [2{,}121.80] + [1{,}960.20] + [1{,}296.00] + [387.20] - 5{,}322.32$$

$$= 5{,}765.20 - 5{,}322.32$$

$$= 442.88$$

$$SS_w = \sum \left[\sum x_i^2 - \frac{(\sum x_i)^2}{n_i}\right]$$

$$= \left[2{,}187 - \frac{(103)^2}{5}\right] + \left[1{,}985 - \frac{(99)^2}{5}\right] + \left[1{,}310 - \frac{(72)^2}{4}\right] + \left[408 - \frac{(44)^2}{5}\right]$$

$$= [(65.2) + (24.8) + (14.0) + (20.8)]$$

$$= 124.8$$

Source	SS	df	MS	F	p
Between	442.88	3	147.63	**17.74**	**.01**
Within	124.80	15	8.32		
Total	567.68	18			

BOX 11.20 **Tukey HSD comparison with unequal sample size**

$$\text{Tukey HSD} = \frac{\bar{X}_1 - \bar{X}_2}{s_{\bar{X}}},$$

$$\text{where } s_{\bar{X}} = \sqrt{\frac{MS_w}{2}\left(\frac{1}{n_1} + \frac{1}{n_2}\right)}$$

Group 1 vs. Group 2

$$\frac{\bar{X}_1 - \bar{X}_2}{\sqrt{\frac{MS_w}{2}\left(\frac{1}{n_1} + \frac{1}{n_2}\right)}} = \frac{20.6 - 19.8}{\sqrt{\frac{8.32}{2}\left(\frac{1}{5} + \frac{1}{5}\right)}}$$

$$= \frac{0.8}{\sqrt{4.16(0.4)}} = \frac{0.8}{\sqrt{1.664}} = \frac{0.8}{1.36} = 0.59$$

Group 1 vs. Group 3 (remember that Group 3 now has only four participants in it)

$$\frac{\bar{X}_1 - \bar{X}_3}{\sqrt{\frac{MS_w}{2}\left(\frac{1}{n_1} + \frac{1}{n_2}\right)}} = \frac{20.6 - 18.0}{\sqrt{\frac{8.32}{2}\left(\frac{1}{5} + \frac{1}{4}\right)}}$$

$$= \frac{2.6}{\sqrt{4.16(0.45)}} = \frac{2.6}{\sqrt{1.872}} = \frac{2.6}{1.37} = 1.90$$

Now we have the error terms for the comparisons we are going to make for the remaining groups. If both groups have five participants in them, then the error term is 1.36. If one group has five participants and the other has four, our error term will be 1.37. Let's make the rest of our comparisons.

Group 1 vs. Group 4

$$\frac{\bar{X}_1 - \bar{X}_4}{\sqrt{\frac{MS_w}{2}\left(\frac{1}{n_1} + \frac{1}{n_2}\right)}} = \frac{20.6 - 8.8}{1.36} = 8.68$$

Group 3 vs. Group 4

$$\frac{\bar{X}_3 - \bar{X}_4}{\sqrt{\frac{MS_w}{2}\left(\frac{1}{n_1} + \frac{1}{n_2}\right)}} = \frac{18.0 - 8.8}{1.37} = 6.72$$

Group 2 vs. Group 3

$$\frac{\bar{X}_2 - \bar{X}_3}{\sqrt{\frac{MS_w}{2}\left(\frac{1}{n_1} + \frac{1}{n_2}\right)}} = \frac{19.8 - 18.0}{1.37} = 1.31$$

Group 2 vs. Group 4

$$\frac{\bar{X}_2 - \bar{X}_4}{\sqrt{\frac{MS_w}{2}\left(\frac{1}{n_1} + \frac{1}{n_2}\right)}} = \frac{19.8 - 8.8}{1.36} = 8.09$$

CheckPoint

Calculate the Tukey HSD statistics using the information provided in the following questions. Round your answers to two decimal places.

1. $\bar{X}_1 = 14.1; \bar{X}_2 = 13; MS_w = 3.23; n_1 = 6; n_2 = 6$ _____
2. $\bar{X}_1 = 23.4; \bar{X}_2 = 17.9; MS_w = 8.82; n_1 = 7; n_2 = 9$ _____
3. $\bar{X}_1 = 4.5; \bar{X}_2 = 7; MS_w = 1.74; n_1 = 5; n_2 = 5$ _____
4. $\bar{X}_1 = 34; \bar{X}_2 = 31.6; MS_w = 4.11; n_1 = 5; n_2 = 5$ _____
5. $\bar{X}_1 = 17.6; \bar{X}_2 = 14.34; MS_w = 6.73; n_1 = 8; n_2 = 11$ _____

Answers to these questions are found on page 472.

MODELS OF *F*

Like most statistical tests, ANOVA has its own jargon—that is, its own language to specify the type of ANOVA being used. Table 11.3 reviews the language that we have used so far and introduces some new jargon into the equation. The hope of people who coin these terms is that they will make the concepts discussed here a bit more understandable.

Notice that the terms *factor* and *independent variable* (or *IV*) are interchangeable. Factors are described as having *levels*, where a level refers to the amount or type of the factor that is used to classify subjects (to create groups). Our blood-doping experiment consisted of one factor (the type of doping examined) with four levels (HT, AT, EPO, and control). Notice also that all of the ANOVAs described in Table 11.3 test the effects of the factor or factors on a *single* dependent variable (DV). Our experiment examined the effects of our one factor on our one DV, which

TABLE 11.3 ANOVA terminology

Term	Definition	One-Way ANOVA Example	Two-Way ANOVA Example
Factor	The independent variable (IV); the variable being manipulated or tested	Type of doping (1) with four levels	Type of doping (1) Sex (2)
Level	The amount or type of IV being administered; the classification of an IV	HT doping (1), AT doping (2), EPO doping (3), control (4)	IV1 (type of doping) – 3 levels IV2 (sex) – 2 levels
Dependent variable	The variable being measured to see the effects of the IV	Mean endurance score by type of doping Single *F-statistic* generated	Mean endurance score by type of doping (main effect F_1), sex (main effect F_2), and the interaction between sex and type of doping (interaction effect F_3).
ANOVA model	The formalization of relationships between variables in the form of mathematical equations	Fixed-effects model	3 x 2 factorial model

was endurance score after blood doping. You can examine the effects on more than one DV using a *multivariate ANOVA*, or *MANOVA*, but we're not going to be discussing this type of test in this textbook.

ONE-WAY ANOVA ASSUMPTIONS

Using the terminology presented in Table 11.3, we could describe the model of ANOVA that we have used so far in this chapter as *fixed-factor* or *fixed-effect ANOVA*. This is because a fixed number of levels of the IV are applied to each level in the same way. The researchers have deliberately selected and planned for the levels of the independent variable to test—in this case, the three types of blood doping and a control group. If we have a fixed-effect model, we are assuming that our groups of subjects come from normally distributed populations that differ only in terms of their individual means. Just as we've done with the *z*- and *t*-tests, we start with our assumption that the null hypothesis is true and then look for evidence that our factor had a significant effect on the dependent variable. We also assume (again, just as we've done for the *z*-tests and independent-samples *t*-tests) that the observations are independent, that the populations our samples came from are reasonably normally distributed, and that these populations have approximately equal variances. Again, statisticians make these assumptions to simplify the statistical analysis and make it easier to draw conclusions. Violations of these assumptions don't mean that we cannot do a statistical analysis, but they probably do mean that we will need to use a somewhat different test that can handle violations. These situations are beyond the scope of an introductory statistics class, but you will likely encounter them in your next statistics course.

FACTORIAL DESIGNS OR TWO-WAY ANOVAS

What if we had more than one factor that we wanted to examine? Suppose there were two factors we thought might have an effect on the dependent variable—type of doping and the sex of the athlete, for example. When we use more than one IV to categorize our samples, we would be using what's known as a **factorial model**, or **factorial design**

Using a one-way ANOVA, we generated a single *F*-statistic that told us whether or not the variability between groups resulted from our independent variable or from random chance. When we add another IV to the mix, we can generate three *F*-statistics. Two of the *F*-statistics we will generate are called **tests of main effects**—the effect of one IV alone, disregarding the other IV. Main effect 1 (and the F_1 statistic that we'll generate) will tell us about the effect of the first IV alone, disregarding the effects of any other variable in the design. The second main effect (F_2) tells us about the effects of the second IV alone, again disregarding the rest of the variables. The last *F*-statistic is called the **interaction effect**, and as the name suggests, it tells us about the joint effects of the IVs on the DV together (F_3 or F_I).

factorial model (or factorial design) A two-way ANOVA design used to compare means of all possible combinations of the independent variables being tested.

test of main effects In a two-way ANOVA, an *F*-statistic that indicates the effect of one independent variable alone, without accounting for the other independent variable.

interaction effect In a two-way ANOVA, the joint effect of two or more independent variables on the dependent variable.

CheckPoint
Answers to questions on page 470

1. 1.51 ☐
2. 5.24 ☐
3. −4.24 ☐
4. 2.64 ☐
5. 3.88 ☐

SCORE: /5

AN EXAMPLE: ALBERT BANDURA'S STUDY OF IMITATING VIOLENCE

Let's try an example. Do you think we humans need to be rewarded or punished in order to learn effectively, or can just watching what happens to other people teach us how to behave? Albert Bandura, along with two colleagues, set up a now famous experiment to determine just that (Bandura, Ross, & Ross, 1963). They asked 40 boys and 40 girls, selected at random, to watch one of two movies. Both movies showed adults hitting, pounding on, pushing, and assaulting a balloon doll called a Bobo doll. Bobo, pictured in Figure 11.3, has a weighted bottom so that he always bounces back for more punishment.

In half of the movies, the adult assaulting the doll was the same sex as the observer child, and in the other half, the adult beating up the doll was the opposite sex of the observer child. In the end, Bandura et al. had four groups:

FIGURE 11.3
Bandura's famous Bobo doll experiment. The female model is shown in the top row; children from the experimental group are shown in the middle and bottom rows.

STATISTICS IN CONTEXT

1. Male children who saw the male adult model.
2. Male children who saw the female adult model.
3. Female children who saw the male model.
4. Female children who saw the female model.

The dependent variable was the child's willingness to imitate the behavior seen on film. After seeing the movies, the children were allowed a 10-minute play period in a room full of age-appropriate toys, one of which was the Bobo doll. The number of imitations of the aggressive behavior directed toward the Bobo doll was counted for each child. The results shown in Box 11.21 are similar (but not identical) to those reported by Bandura et al.

BOX 11.21 **Results of a Bobo doll experiment: Number of aggressive imitations**

Sex of Child Observer

		Male	Female	Row totals
Sex of Adult Model	Male	106 117 108 101 97 $\Sigma x = 529$ $N = 5$ $\bar{x} = 105.8$	41 40 34 38 42 $\Sigma x = 195$ $N = 5$ $\bar{x} = 39.0$	$\Sigma x = 724$ $N = 10$ $\bar{x} = 72.4$
	Female	51 50 49 49 45 $\Sigma x = 244$ $N = 5$ $\bar{x} = 48.8$	58 51 60 56 62 $\Sigma x = 287$ $N = 5$ $\bar{x} = 57.4$	$\Sigma x = 531$ $N = 10$ $\bar{x} = 53.1$
	Column totals	$\Sigma x = 773$ $N = 10$ $\bar{x} = 77.3$	$\Sigma x = 482$ $N = 10$ $\bar{x} = 48.2$	Overall: $\Sigma x = 1,255$ $N = 20$ $\bar{x} = 62.75$

Right brace: Main effect of the sex of the adult model

Bottom brace: Main effect of the sex of the child observer

The two main effects that we're looking for are (1) the effect of the sex of the adult model and (2) the effect of the sex of the child observer on the number of aggressive imitations performed against the Bobo doll. The top row of the table in Box 11.21 shows the number of aggressive imitations by boys and girls when the model was male. The overall mean number of aggressive actions when the model was male was 72.4, compared to a mean number of aggressive actions of 53.1 when the model was female. It looks as though we have a difference in willingness to imitate that depends on the sex of the model. Children, on balance, seem to have been more willing to imitate when the model was male than they were when the model was female.

The second main effect is the effect of the sex of the observer, disregarding (for the moment) the sex of the model. Compare the averages of column 1 (male children) and column 2 (female children) to get an idea of what effect this factor had. Male children seem to be more willing to imitate the actions of a model than were female children (an average of 77.3 aggressive actions for boys versus 48.2 for girls).

The main effects (possibly significant—we don't know yet) are relatively easy to see when we look at the data. Often, the most interesting effect is the interaction between the two main factors. Take a look at the data, and see if you think there might be an interaction between the two factors in our version of Bandura's experiment. Does the sex of the adult model interact with the sex of the child observer? Do male children change in their willingness to imitate when the sex of the model changes? Are female children affected by the sex of the model in the same way that males are, or they affected in a different way?

GRAPHING THE MAIN EFFECTS

The easiest way to visualize both the main effects and the interaction is to construct graphs of them. Figure 11.4 shows graphs of the two main effects and of the interaction.

Let's take a look at the main effects (A and B) first. The top graph (A) shows the effect of the sex of the adult model in the film on the number of aggressive acts the child (regardless of the sex of the child) performs against the Bobo doll. The children in the study seem to be more willing to imitate the actions of a male model than to imitate those of a female model. The number of aggressive acts for all children when the model was male (the F-test for this main effect) will tell us if the difference in number of aggressive acts the children perform differs significantly as a function of the sex of the model.

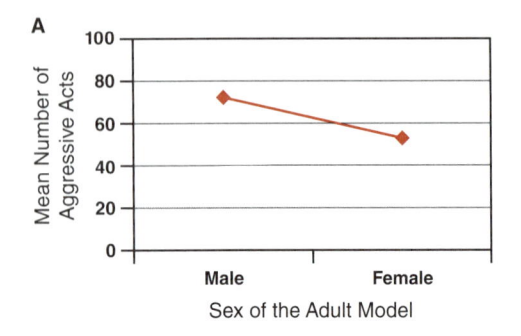

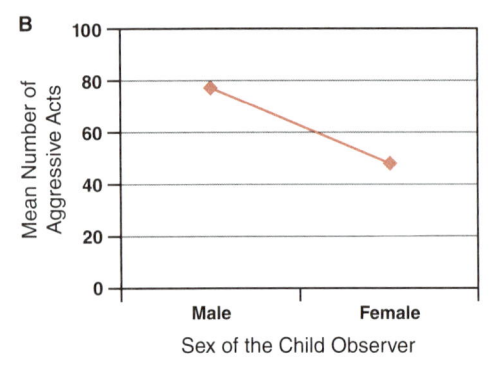

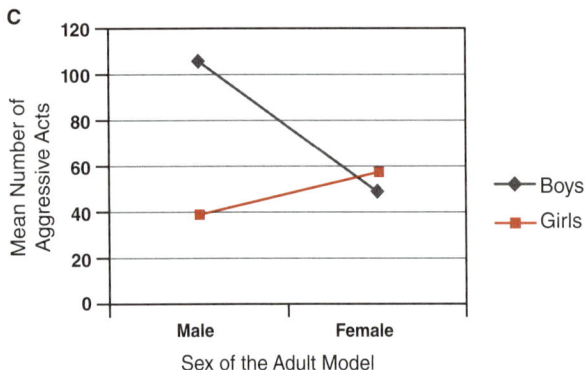

FIGURE 11.4
Graphic representation of data from Bandura's Bobo doll study.

Graph B shows the effect of the second IV, the sex of the child observer, again on the number of aggressive actions performed against poor Bobo. When you examine this graph, you can see that male children performed more aggressive actions against the doll than did female children. The *F*-test for this main effect will tell us if this is a statistically significant difference.

Now take a look at graph C—the graph of the interaction between the two independent variables. To create this graph, I selected one IV (in this case, it happened to be the sex of the adult model) and asked the computer to draw a separate line for each level of the second IV (red for female child observers and dark gray for male child observers). It does not matter which IV we select to be scaled on the *x*-axis (if you try reversing the two IVs on this graph, you will see what I mean). I generally prefer to put the independent variable with the highest number of levels on the *x*-axis and then draw separate lines for each level of the other IV, just for the sake of simplicity. Again, with two levels in each of our IVs here, we can set either variable as the *x*-axis.

What does this graph show us? Does it look as though there's a significant interaction between these two variables? Think about this question one variable at a time, and take a look at Box 11.22.

BOX 11.22 **Graphs showing (a) significant and (b) nonsignificant main effects of the IV**

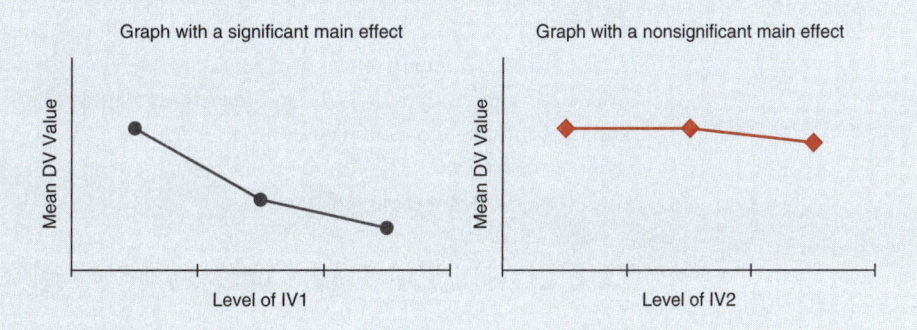

In the left-hand graph in Box 11.22, notice that there are three levels to the IV (whatever it might be), and that the average DV "score" decreases from Level 1 to Level 2, to Level 3. The graph on the right shows what happens when there is no effect from this IV: The mean DV score does not change from Level 1 to Level 2 to Level 3. A sloping line indicates a possible effect, and a flat, horizontal line indicates no effect by the IV. Now take a look at a graph of significant and non-significant interactions in Box 11.23.

BOX 11.23 **Graphs showing (a) significant and (b) nonsignificant interactions of two IVs**

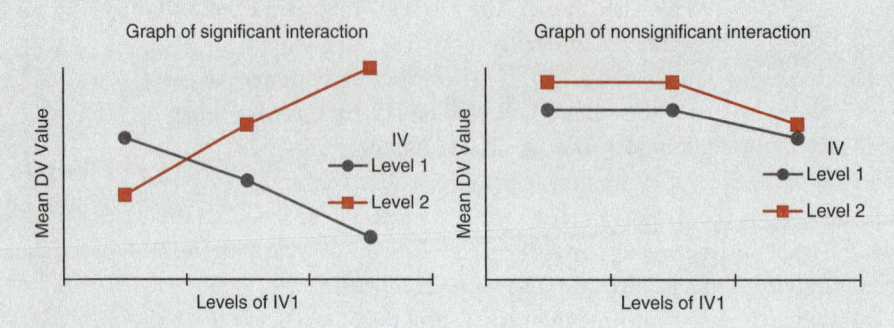

Graph of significant interaction — IV: Level 1, Level 2; axes: Mean DV Value vs. Levels of IV1

Graph of nonsignificant interaction — IV: Level 1, Level 2; axes: Mean DV Value vs. Levels of IV1

In the graph of the significant interaction (on the left) in Box 11.23, we can see that the two independent variables have two different effects on the DV. Variable 1 tends to increase the average DV value across levels, while IV2 has the opposite effect—it tends to decrease the average DV score across levels. In the graph of the nonsignificant interaction on the right, both IVs have the same effect on the DV. Across Levels 1 and 2, there's no change in the DV, but both IV's decrease between Levels 2 and 3.

In the graph of the Bandura study data, note that we probably have a significant interaction (the *F*-test for the interaction will tell us if this is true). It is possible to see both the two main effects and how they interact with one another here. Boys imitate more often than do girls, and boys imitate most often if the model is male, while girls imitate most often if the model is female. The sex of the observer and the sex of the model interact: Children imitate the model of the same sex more than they do the model of the opposite sex.

THE LOGIC OF THE TWO-WAY ANOVA

Before we discuss the results of the two-way ANOVA for the Bandura data (I hope you haven't read ahead—it's supposed to be a surprise), let's talk about the logic of the portioning of the variability in a two-way situation. Box 11.24 shows the logic behind the three *F*-tests we do with a two-way ANOVA.

In a two-way ANOVA, we partition the total variability into sources as we do for a one-way ANOVA. The only difference is that we have more independent variables, and so more sources of variability. The first IV contributes variability between its levels, as does the second IV. Finally, some variability is contributed to the total variability in the model by the interaction between the two IVs. Our assumptions about the populations that our samples have come from remain the same in a two-way ANOVA: We will assume that the null hypothesis about the effects of our IVs (we now have two) is true and then look for evidence so that we can reject

BOX 11.24 **The logic of two-way ANOVA**

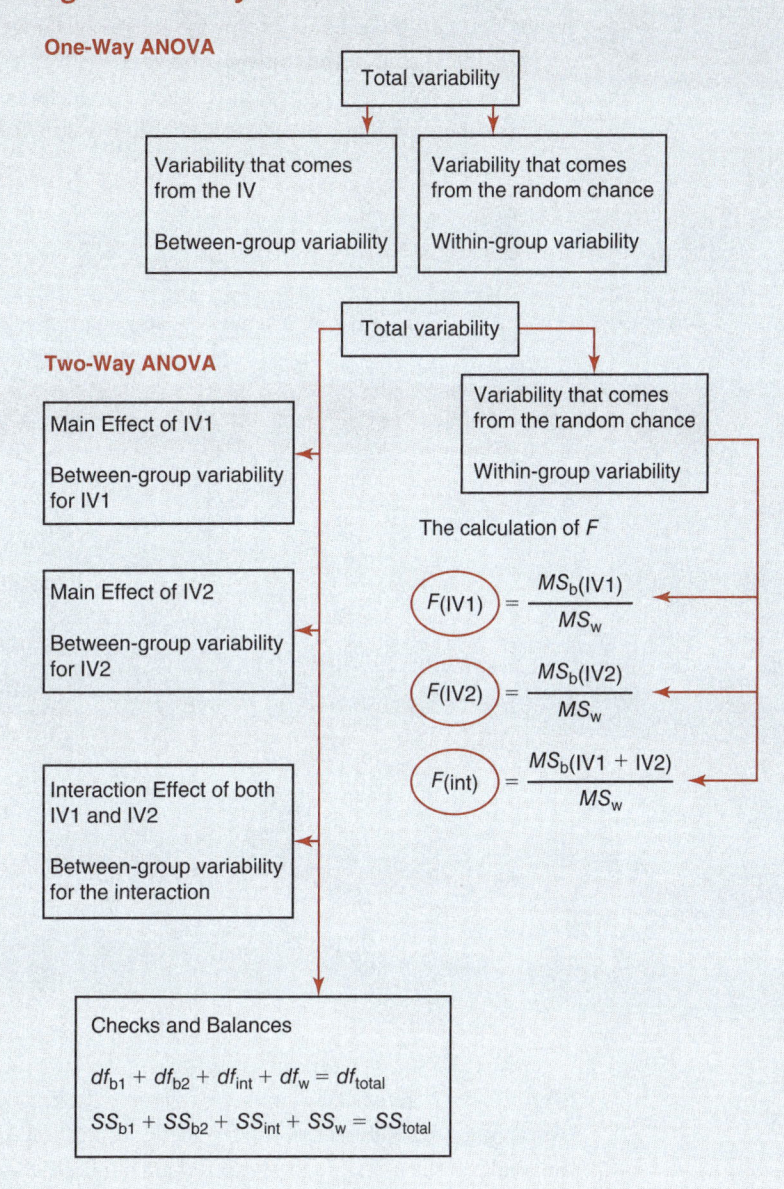

One-Way ANOVA

Total variability
- Variability that comes from the IV / Between-group variability
- Variability that comes from the random chance / Within-group variability

Two-Way ANOVA

Total variability

- Main Effect of IV1 / Between-group variability for IV1
- Main Effect of IV2 / Between-group variability for IV2
- Interaction Effect of both IV1 and IV2 / Between-group variability for the interaction
- Variability that comes from the random chance / Within-group variability

The calculation of F

$$F(IV1) = \frac{MS_b(IV1)}{MS_w}$$

$$F(IV2) = \frac{MS_b(IV2)}{MS_w}$$

$$F(int) = \frac{MS_b(IV1 + IV2)}{MS_w}$$

Checks and Balances

$df_{b1} + df_{b2} + df_{int} + df_w = df_{total}$

$SS_{b1} + SS_{b2} + SS_{int} + SS_w = SS_{total}$

the null. We will also assume that the populations our samples have come from are normally distributed and have roughly equal variance.

USING THE TWO-WAY ANOVA SOURCE TABLE

We won't be covering the mathematical formulas for calculating the sums of squares, and so on, for a two-way ANOVA in this textbook. If you understand the logic behind the two-way ANOVA and have access to software that will perform the calculations for you, what you really need to know is how to interpret the graphs that can be generated from the means of the groups you use and how to interpret the output of the software you've used. We will be discussing the output of SPSS as we have been right along in this text. So, let me show you what you'll get if you put these data into SPSS and ask for a two-way factorial ANOVA.

BOX 11.25 **SPSS source table for a two-way ANOVA***

Tests of Between-Subjects Effects DV: # of aggressive acts against Bobo

Source	Type III Sum of Squares	df	Mean Square	F	Sig.
Corrected Model	13202.950(a)	3	4400.983	194.090	.000
Intercept	78751.250	1	78751.250	3473.043	.000
IV1 (sex of model)	1862.450	1	1862.450	82.137	.000
IV2 (sex of child)	4234.050	1	4234.050	186.728	.000
IV1 * IV2 (interaction)	7106.450	1	7106.450	313.405	.000
Error	362.800	16	22.675		
Total	92317.000	20			
Corrected Total	13565.750	19			

*I've added the description of each IV.

Box 11.25 shows a typical source table for a two-way ANOVA. The data come from our 2 × 2 factorial design. You can ignore the lines labeled "Corrected Model," "Intercept," and "Corrected Total": These are not important in our discussion of the results of the F-test, so I've grayed them out here. The rows labeled "IV1," "IV2," and "IV1 * IV2" show the two main effects in our design and the interaction (IV1 and IV2 together). These are important. A "Type III Sum of Squares" is a measure of overall variability using the equation we have been discussing in this chapter (the fact that it's labeled Type III suggests that there might be other types, and there are). The sums of squares indicate the overall variability that comes from IV1 and IV2 and from the interaction of these two (IV1 and IV2 together).

The column labeled "*df*" indicates the degrees of freedom for each IV and for the interaction between the two variables. Notice that each IV has just one *df*—why do you suppose that's the case? Remember that we calculated the *df* for our one

IV in the one-way ANOVA example by subtracting one from the number of levels for that independent variable. We'll do the same thing here. We have two levels to IV1, so our calculation of df for this IV is $df = 2 - 1 = 1$. Because IV2 also has two levels, its df is 1 as well. The df for the interaction is calculated by multiplying the df for the first IV by the df for the second IV ($df_{IV1} * df_{IV2}$), which for us means that the $df_{int} = 1 * 1 = 1$.

When we determined the mean sums of squares (*MS*) for the one IV in a one-way ANOVA, we divided the overall sum of squares between and the sum of squares within by their respective df's. We'll do the same thing here to get the MS_{IV1}, MS_{IV2}, and $MS_{int.}$ And finally, all three *F*-values are calculated as you might expect, given how we calculated *F* for the one-way ANOVA. We will divide each *MS* by the MS_w to find an *F*-value. To determine if the *F*-value is significant, we'll read across to the column labeled "Sig." and then read down that column for the two main-effect *F*'s and the interaction *F*. If the number in the column labeled "Sig." is .05 or less, then we will call that *F* significant.

There's one caveat here: SPSS reports only the first three digits after the decimal point in this column. So, a "Sig." value of .000 would be less than .05, even if we don't know what the value of the number in the fourth decimal place is. When you report a *p*-value like .000, you go with what you know. Here, we know that $p < .001$, so we would report our results like that.

INTERPRETING THE RESULTS

So, what does the source table in Box 11.25 tell us? First, the sex of the model in the film is a significant factor in our experiment. Regardless of the sex of the child observing the movie, male models were imitated more often than were female models: $F_{(1,16)} = 82.137$, $p < .001$. Second, the sex of the observer was also significant. Regardless of the sex of the model in the film, male children imitated the aggression they saw more often than did female children: $F_{(1,16)} = 186.73$, $p < .001$. Lastly, we had a significant interaction between the sex of the model and the sex of the child. Children were more likely to imitate the behavior of a model of the same sex as the child than they were to imitate a model of the opposite sex: $F_{(1,16)} = 313.405$, $p < .001$.

When we were doing a one-way ANOVA and had a significant *F*-statistic, we said that we needed to do a post-hoc test in order to determine where the variability in the *F* was coming from. Post-hoc tests are available for two-way ANOVAs, but the one most frequently used in this situation is a *t*-test. I know, I know, I said that you shouldn't do multiple *t*-tests, but the procedure here is to do *t*-tests only for the significant factors. For example, suppose that we had three levels to a given IV and that the *F*-statistic for that IV was not significant. We would not need to do any post-hoc comparisons because the *F* tells us there are no significant differences between groups here. If we had a significant *F* for the other IV, we could go back to the means for that factor. If we saw a large difference between, for example, the mean for Group 1 and the mean for Group 3, but very little difference between Group 2 and Groups 1 and 3, then we could run a *t*-test comparing Group 1 versus Group 3 and only do one *t*-test.

Because our situation in the Bandura test involves two IVs, each with only two levels, we don't need to do any post-hoc testing at all. The *F*-statistic tells us there is significant variability between two means for IV1, which indicates that the difference between the means of the two levels is automatically the source for that significant *F*.

ANOVA IN CONTEXT: INTERPRETATION AND MISINTERPRETATION

Statistics were formally introduced into psychology labs at almost the same time that psychology came into its own as a discipline. Adolphe Quetelet introduced the idea of the "average man" in 1853. The first scientific laboratory in psychology was created by a German physiologist named Wilhelm Wundt in 1875. In London (1877–1878), Sir Francis Galton worked away on trying to find a way to produce a statistic that would describe the degree to which two variables were related to one another, or "correlated." In 1908, Gossett introduced the *t*-test designed to compare the means of two groups, and in 1919, Fisher came out with the ANOVA that allowed multiple variables to be compared. Like them or not, statistics quickly became a fundamental part of the formal study of human behavior. Today, almost all students of psychology must take a basic statistics course as a part of their core curriculum.

The two problems students seem to encounter most often involve (1) selecting the appropriate test for the question they're asking and then (2) interpreting the results of that test. Notice, please, that Chapter 16 is entitled "What Test Should I Use, and Why?" In that chapter, we will discuss this question in some depth, but it's never too early to start thinking about how the tests we've already covered can be (appropriately) used. Most statistics texts (and this one is no exception) will provide you with some kind of decision tree or flow chart to help you with this determination. I've found that if students themselves make this kind of chart, they tend to remember it better, and make better use of it, than they do when the chart is designed by someone else. There is a decision tree in the back of this text, but I'd like to encourage you to make your own.

Let's start by making a list of the kind of information you need to know before you select the test you'd like to use. Box 11.26 shows you the beginnings of my list, and we will add to it as we introduce more tests.

The last question on this list, the one about your own context, relates to the other stumbling block students seem to have with statistics after they've left the classroom: What does all this output stuff mean? I have mentioned before that we tend to rely more and more on computers to do the mathematical part of statistics, but you can provide a computer with almost any kind of data, ask it to calculate almost any statistic, and it will quickly provide you with output. Understanding that output is the "hard" part, and something that requires a human brain. A computer just spits out the numbers; it cannot tell you what that output means.

> **BOX 11.26** **Questions to ask before selecting an inferential test**
>
> 1. What kind of data do I have? Qualitative (nominal or ordinal) or quantitative (interval or ratio)?
> 2. How many independent variables do I have? One? More than one?
> 3. What do I know about the population? The mean and standard deviation? Just the mean? Nothing?
> 4. What are my options? What test should I do? So far, we've covered the *z*-test, three kinds of *t*-tests, and two kinds of ANOVAs. We will be adding to this list.
> 5. What is my point? What is my context for this test?

Interpreting the output of a statistical test gets easier as you gain more experience in asking research questions and in using statistical tests to answer them. However, we all know that there's some truth in another old saying: "Experience is a hard teacher because she gives the test first and the lesson afterward." Those of us who teach statistics, and those of us who use them, probably come across one problem in the interpretation of a statistical test more often than any other—the confusion over the meaning of the word "significant." Researchers and students often confuse the word "significant" used in a statistical sense with the more practical interpretation of this word as an indicator of importance.

If you're reading a mystery and the grizzled, cigar-smoking lead detective says that he's stumbled upon a "significant clue," you know that the clue is very important. In fact, that clue is more important than any other clue. Researchers often mistakenly assume that "significant" means important in this same way when it is used in a statistical sense. It does not.

Statistics users often say a result that is significant at an alpha level of .01 is "more significant" than that is one significant at an alpha level of .05—in this case, they are using the word "significant" like the detective in the mystery does. However, significance in statistics just means that we've probably obtained a meaningful result. Meaningful at $\alpha = .05$ "means" the same thing as does meaningful at $\alpha = .01$.

A significant result obtained at the end of an inferential analysis is "really as much a function of the sample size and experimental design as it is a function of strength of relationship" (Helberg, 1996, p. 7). In both cases, the *p*-value tells you the risk of making a Type I error, and the risk of making a Type I error depends on your sample size, on your methodology, on the effect the IV has on the DV, and even on random chance. The result of an inferential test really says nothing about how big a deviation from the null hypothesis we've obtained. Remember that a significant difference is *probably* meaningful and, like many conclusions we humans draw, should be taken with a grain of salt.

In the next chapter, we will discuss two statistics that have been developed at least partly in response to this common misinterpretation of what a significant result can actually tell you. Confidence intervals and effect size measurement will help researchers address the problem of what "significant" actually means. My general words of advice in using the inferential statistics we've discussed so far is to remember your research question and direct your conclusions about the results of the test you ran back to that question. Don't try to make the test say more than it can.

SUMMARY

The ANOVA procedure was developed to investigate differences between several (i.e., more than two) groups. A one-way ANOVA compares variability across three or more levels of one independent variable. A two-way ANOVA compares variability across two independent variables, each with two or more levels. The two-way ANOVA also provides the opportunity to investigate how these two IVs interact or influence one another, which is often more interesting than the effects of the two IVs considered separately.

The statistic that is calculated in the ANOVA test is called an F-statistic (in honor of its creator, R. A. Fisher). The F-statistic is calculated by dividing the average variability *between* groups (caused by both random chance and the effect of the IV) by the average variability *within* groups (caused by random chance). The F-statistic that is generated describes the effects of the IV over and above the effect of random chance. An F of one would indicate that the IV would have no effect at all on the DV. There is no top or maximum value of F.

If a significant effect of the IV on the DV is found, several post-hoc tests, or tests of multiple comparisons, should be done to determine which group or groups are being affected. We discussed one of these post-hoc tests, called the Tukey honestly significant difference (or HSD) test. This test will compare all possible pairs of group means.

A two-way ANOVA (also known as a factorial ANOVA) allows the researcher to examine the effects of two IVs (also known as the main effects of each IV alone) as well as the effect of the interaction between these two IVs on a single DV. We discussed how to interpret the graphs of the main effects and interaction and how to interpret the source table for the two-way ANOVA.

TERMS YOU SHOULD KNOW

ANOVA, p. 445
factor, p. 448
factorial design, p. 471
factorial model, p. 471
F-statistic, p. 450
interaction effect, p. 471
levels, p. 448
mean squares between groups (MS_b), p. 449
mean squares within groups (MS_w), p. 450
multiple-comparison test, p. 462
one-way ANOVA, p, 448

post-hoc test, p. 460
source table, p. 454
sums of squares, p. 449
sums of squares between groups (SS_b), p. 449
sums of squares within groups (SS_w), p. 449
tests of main effects, p. 471
Tukey honestly significant difference (or HSD) test, p. 462
two-way ANOVA, p. 448

GLOSSARY OF EQUATIONS

Formula	Name	Symbol
$SS_b = \sum \left[\dfrac{(\sum x_i)^2}{n_i} \right] - \dfrac{(\sum x_{total})^2}{N_{total}}$	Sums of squares between groups	SS_b
$SS_w = \sum \left[\sum x_i^2 - \dfrac{(\sum x_i)^2}{n_i} \right]$	Sums of squares within groups	SS_w
$MS_b = \dfrac{SS_b}{df_b}$	Mean squares between groups	MS_b
$MS_w = \dfrac{SS_w}{df_w}$	Mean squares within groups	MS_w
$F = \dfrac{MS_b}{MS_w}$	F-statistic	F
$HSD = \dfrac{\bar{X}_1 - \bar{X}_2}{s_{\bar{X}}}$	Tukey HSD when *n*'s are equal	Tukey HSD
$s_{\bar{X}} = \sqrt{\dfrac{MS_w}{N_{total}}}$	Tukey HSD error term when *n*'s are equal	$s_{\bar{X}}$
$HSD = \dfrac{\bar{X}_1 - \bar{X}_2}{s_{\bar{X}}}$	Tukey HSD statistic when *n*'s are unequal	Tukey HSD
$s_{\bar{X}} = \sqrt{\dfrac{MS_w}{2}\left(\dfrac{1}{n_1} + \dfrac{1}{n_2}\right)}$	Error term for Tukey HSD when *n*'s are unequal	$s_{\bar{X}}$
$SS_{total} = \sum x_{total}^2 - \dfrac{(\sum x_{total})^2}{N_{total}}$	Sum of squares total	SS_{total}
$K - 1$ where K = number of levels	Degrees of freedom between groups	df_b
$N - K$ where N = total number of participants	Degrees of freedom within groups	df_w
$N - 1$	Total degrees of freedom	df_{total}

WRITING ASSIGNMENT

Here is a chance for you to practice interpreting the results of a statistical test. As you're working through the problem, keep in mind what your research question is and what the test results actually can tell you about that question.

Consumption of alcohol is associated with increased levels of aggressive behavior in adult humans as well as other mammals. Alcohol use and misuse is frequently seen in college students (Lyvers, Czerczyk, Follent, & Lodge, 2009). Let's see what happens when we measure willingness to imitate the aggressive behavior of a model in three different groups of college-age drinkers.

A group of male college students of legal drinking age were given the Alcohol Use Disorders Identification Test (AUDIT), which provides a score from 0 to 40. Scores between 0 and 8 indicate a low-risk drinker, scores from 9 to 15 indicate a level of drinking that is hazardous to the person taking the test, and scores of 16 or higher indicate a harmful level of drinking. All subjects were asked to watch a film featuring a person they admired acting in an aggressive manner toward a Bobo doll. The students were then taken to a room, empty except for a Bobo doll, and told they were to wait in the room for 30 minutes for a school counselor to arrive. The counselor would be giving them a lecture on the dangers of alcohol consumption. The number of aggressive actions performed toward the doll during a 30-minute exposure were counted.

Use the one-way ANOVA test to determine if aggressive behavior depends on the degree of risk in drinking behavior (based on the AUDIT score). Then, write an APA Style summary of the data (shown in Table 11.4) and your conclusion. Remember to state your research question in your summary. If you find a statistically significant result, do the appropriate post-hoc test to determine where that significant difference is coming from. Finally, if you get a significant result, speculate a bit: What aspect of the data shown here would you look at to see if the IV is having a strong or weak effect on the DV?

Table 11.4 Number of aggressive acts toward Bobo as a function of AUDIT score

Low Risk	Hazardous Risk	Harmful Risk
0	5	8
0	2	7
1	1	10
0	3	10
2	1	15
1	2	20

PRACTICE PROBLEMS

1. What is the general formula for *F*? Can an *F*-value ever be negative? Why or why not?

2. If *F* can be either negative or positive, what would the shape of the *F*-distribution be? Would it be normal or skewed? If *F* cannot be negative, what would the shape of the *F*-distribution be?

3. Suppose your *F*-value for a single IV was 0.00. Given that value, what do you know about the MS_b and MS_w?

4. Type III sums of squares (the type that we've been dealing with in this chapter) assume that the number of subjects in each group being compared is equal. If the number of subjects in each group is radically different, then the researcher should use a Type I sum of squares formula. Think about the formula for variance (in fact, write it down). Why does the number of subjects in each group matter so much when we're analyzing variance? What would happen to the estimate of the amount of variability within a group or between groups as the number of subjects in that group or groups goes up or down?

5. Using the terms *factors* and *levels*, describe an independent-samples *t*-test.

6. Fill in the blanks in the following source tables using the information available to you.
    a. Experiment 1: A new diet pill is being tested for effectiveness. Four groups of women are selected. Women in Group 1 ($n = 10$) are given 5 milligrams of the drug, those in Group 2 ($n = 10$) are given 10 milligrams, Group 3 ($n = 10$) gets 15 milligrams, and Group 4 ($n = 10$) receives a placebo (control). The women are asked to record the percentage of weight lost after taking the pill daily for 6 months. The source table shown below indicates their results. *Reject Ho*

Source	SS	df	MS	F
Between	263.2	3	87.73	8.65
Within	365.2	36	10.14	
Total	628.4	39		

    Is *F* significant? If so, at what alpha level?
    b. Experiment 2: Remember growing beans in pots in grade school? Suppose we ask children in a second-grade class ($n = 30$ students) to see if music affects plant growth. The children are asked to plant bean seeds in pots. One-third of the class plays classical music to their growing bean plants, one-third plays rock-and-roll to their plants, and the final third plays rap/hip-hop music to their plants. When the first bean pod appears on the plant, the plants are carefully removed and rinsed of any soil. Then, the weight of the now dry plant (in ounces) is measured.

Source	SS	df	MS	F
Between			196	2.05
Within				
Total		29		

Is F significant? If so, at what alpha level?

c. Experiment 3: Suppose we decide to improve on the experiment in part b by introducing another IV. Our new IV is watering frequency, and it has three levels: twice a day (Level 1), once a day (Level 2), and once every other day (Level 3). We will also add a control group to the mix (a group of plants that grow in silence). When the first bean pod appears, indicating plant maturity, the plants are again removed from their pots, rinsed, dried, and then weighed.

Source	SS	df	MS	F
Music	536.34			
Water	34.32			
Music × water				
Within	1,220.75	48		
Total	1,813.48	59		

Are the two main effects significant? Is the interaction F significant?

7. A table showing data from an experiment is presented below (Table 11.5). Use the appropriate descriptive statistic to summarize the data found in the highlighted boxes on the table. Construct line graphs of the two main effects and of the interaction.

Table 11.5 Data for question 7

	1	2	3	4	Row Totals
1	$\Sigma = 15$ $n = 5$	$\Sigma = 30$ $n = 5$	$\Sigma = 37$ $n = 5$	$\Sigma = 40$ $n = 5$	$\Sigma = 122$ $N_{(R1)} = 20$
2	$\Sigma = 19$ $n = 5$	$\Sigma = 18$ $n = 5$	$\Sigma = 11$ $n = 5$	$\Sigma = 15$ $n = 5$	$\Sigma = 63$ $N_{(R2)} = 20$
3	$\Sigma = 22$ $n = 5$	$\Sigma = 35$ $n = 5$	$\Sigma = 11$ $n = 5$	$\Sigma = 15$ $n = 5$	$\Sigma = 83$ $N_{(R3)} = 20$
Column totals	$\Sigma = 56$ $N_{(C1)} = 15$	$\Sigma = 83$ $N_{(C2)} = 15$	$\Sigma = 59$ $N_{(C3)} = 15$	$\Sigma = 70$ $N_{(C4)} = 15$	$\Sigma = 268$ $N_{(tot)} = 60$

8. Use the information provided in the following source tables to determine if the F-statistic shown is significant. Conduct a Tukey HSD test for any significant F-values, and interpret the result of that test (assume equal n's).

a.

Source	SS	df	MS	F
Between	579.5	2	289.75	
Within	1,277.1	18	70.95	
Total	1,856.6	20		

b.

Source	SS	df	MS	F
Between	240.67	2		
Within	1,354.00	18		
Total	1,594.67	20		

c. Refer to the two source tables in parts a and b above. In one table, the F is significant; in the other, it is not significant. If we assume that these two tables represent the same experiment, done twice with different subjects the second time, what changed when F was nonsignificant?

9. For each of the four graphs shown in Figure 11.5, determine if there is evidence of a main effect for one or both of the IVs, and if a significant interaction between the two IVs used in the experiment is likely.

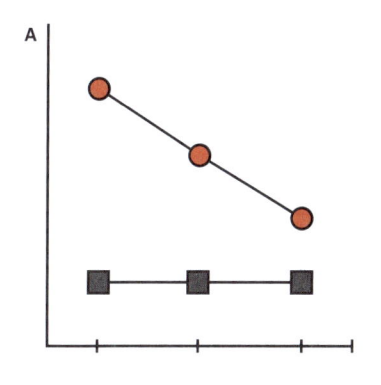

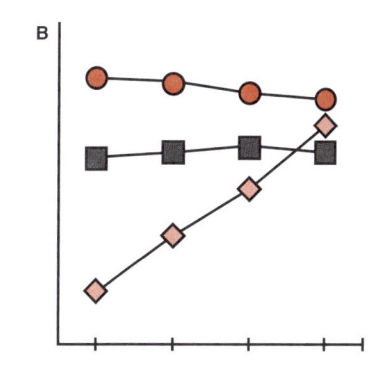

FIGURE 11.5
Graphs for question 9.

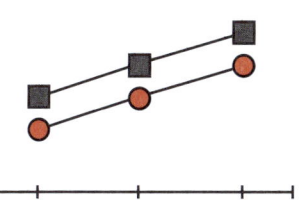

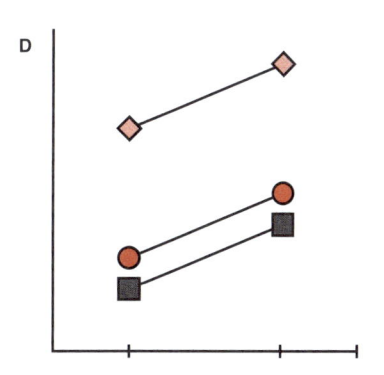

CHAPTER 11 Analysis of Variance

10. For each of the graphs shown in question 9, determine the number of levels of each of the IVs used. IV1 is shown on the *x*-axis; IV2 levels are shown in separate lines.
    a. Graph 1_____
    b. Graph 2_____
    c. Graph 3_____
    d. Graph 4_____

11. You are interested in discovering if alcohol impairs one's ability to perform a test of manual dexterity. You randomly assign 21 subjects to one of three groups. Subjects in Group 1 ($n = 6$) consume a drink with 3 ounces of alcohol (the equivalent of about three beers), subjects in Group 2 ($n = 9$) consume a drink with 1 ounce of alcohol (equivalent to one beer), and subjects in Group 3 ($n = 6$) consume a drink with no alcohol in it at all. After waiting 30 minutes, all subjects are tested on a manual dexterity task. High scores on the manual dexterity test indicate better performance. Conduct an ANOVA using the data in Table 11.6 to determine if alcohol has an effect on performance. If the *F*-statistic is significant, conduct a Tukey HSD post-hoc test to compare means. Describe the effect of alcohol on manual dexterity.

Table 11.6 Scores on manual dexterity test

Group 1	Group 2	Group 3
31	42	69
44	56	57
31	31	67
44	47	62
56	44	72
47	48	65
	46	
	39	
	41	

12. You are interested in determining if there is a significant difference in the number of colds experienced by children at four different elementary schools in your area. You randomly select five children from each school and record the number of colds each child experienced between September 1st and June 1st of the previous year. The data are shown in Table 11.7. Does the number of colds vary according to the school attended?

Table 11.7 Number of colds experienced in four schools

School 1	School 2	School 3	School 4
3	3	2	2
2	4	2	1

2	4	3	1
2	5	2	2
1	4	1	1

13. The association value of a stimulus is a measure of its meaningfulness. An experiment was performed to see the association value of nonsense syllables on learning. Stimuli with high association value are easily associated with another stimulus or easy to remember in a memory test. In an experiment like the one described here, nonsense syllables (consonant–vowel–consonant, or CVC, trios) are used. A CVC syllable with an association value of zero would be one that had no meaning for the participants (e.g., KAX). A syllable with a very high association value (100%) would be one that could be found at the beginning of an English word (e.g., BEC). Four treatment levels were randomly assigned to 32 subjects, with eight subjects in each level. The association values of the lists were 25% for the first list, 50% for the second list, 75% for the third list, and 100% for the fourth list. The dependent variable was time (in seconds) needed to learn the list well enough to correctly recite it twice. The data are shown in Table 11.8. Does association value affect the time needed to learn a list of nonsense syllables?

Table 11.8   Time needed (in seconds) to learn the list of criteria for four levels of association

25%	50%	75%	100%
62	42	28	18
61	40	20	17
50	48	27	20
61	31	36	18
72	50	28	14
54	49	19	15
42	31	28	16
33	39	17	21

14. A human factors psychologist studied three computer keyboard designs. Three samples of individuals were given material to type on a particular keyboard, and the number of errors committed by each subject was recorded. The data are shown in Table 11.9. Does typing performance differ significantly among the three types of keyboards?

Table 11.9   Number of errors made on three different keyboards

Keyboard A	Keyboard B	Keyboard C
0	6	6
4	8	5

Table 11.9 *Continued*

0	5	9
1	4	4
0	2	6

15. For each of the *F*-tests shown in Table 11.10, conduct the Tukey HSD post-hoc test, and describe which group means differ.

   a. A new antidepressant medication is tested at low, medium, and high doses. Dependent variable is score on the Beck Depression Inventory for each participant. High scores indicate high levels of depression.

   Table 11.10 Depression scores for three trial groups

Group	n	Mean	SD
1	5	38.2	8.17
2	5	20.6	8.32
3	5	14.8	7.85
Total	15	24.533	12.75
MS_w	65.9		

   b. Rats are randomly assigned to be reared in three different environments: impoverished (alone in a small wire-mesh cage), a standard environment (in a large cage with other rats of the same sex), or an enriched environment (in a very large cage, with many other rats and toys that are changed on a regular basis). After 2 months, rats are tested in a maze, and the number of trials needed to learn the maze with no errors is recorded (Table 11.11).

   Table 11.11 Number of trials required to learn a maze for three groups of rats

Group	n	Mean	SD
1	6	19.17	3.31
2	6	16.33	3.33
3	5	9.20	1.92
Total	17	15.24	5.04
MS_w	8.926		

   c. Three groups of volunteers are examined. Group 1 studies a list of 10 CVC nonsense syllables in a silent room, Group 2 studies the same list in a room with a sound playing at constant volume, and Group 3 studies the list in a room where the sound volume changes unpredictably. After 30 minutes of studying, the number of nonsense syllables each participant is able to remember is measured (maximum possible score = 10). The number of syllables recalled differs significantly across sound groups (Table 11.12).

Table 11.12 Memory scores for three groups

Group	n	Mean
1	8	6
2	8	4
3	8	3
Total	24	15.24
MS_w	4.18	

d. The effectiveness of pet therapy versus traditional therapy is examined. Volunteers are randomly selected, assigned to one of four groups, and asked to perform a stressful task. Participants in Group 1 perform the task alone (control group), those in Group 2 perform the task with a good friend present, Group 3 performs the task with their dog present, and Group 4 performs the task after having a stress-reduction therapy session with an accredited and acclaimed therapist. Heart rate (in beats per minute) is measured during the task (Table 11.13). A one-way ANOVA finds a significant difference in heart rate across the four groups.

Table 11.13 Mean heart rates recorded during a stressful task for four groups

Group	n	Mean
1	10	80
2	10	85
3	11	46
4	9	71
Total	40	70.0
MS_w	26.986	

16. Using the following observed information, determine the critical F-value for each observed F-test, and interpret the results of each test.

Observed F-Value	Number of Groups Compared	Number of Subjects Per Group	Alpha Level	Degrees of Freedom (b, w)	Critical F-Value	Interpretation (Reject H_o or Retain H_o)
5.06	3	9, 9, 9	.05			
4.00	4	6, 5, 6, 8	.01			

CHAPTER 11 Analysis of Variance

Continued						
3.82	5	10, 10, 10, 10, 10	.01			
3.00	6	9, 9, 5, 8, 9, 9	.05			
16.2	3	15, 15, 15	.05			
1.96	5	20, 20, 18, 16, 15	.05			

17. Take a look at the table of critical F-values in Appendix C, and answer the following questions:
    a. What is the largest critical F-value in the table?
        i. What is the alpha level associated with this critical F-value?
        ii. What are the degrees of freedom between and within for this critical F-value?
        iii. How many groups are being compared if you use this critical F-value?
        iv. How many participants would you have if you used this critical F-value?
    b. What is the smallest critical F-value in the table (for df_w other than infinity)?
        i. What is the alpha level associated with this critical F-value?
        ii. What are the degrees of freedom between and within for this critical F-value?
        iii. How many groups are being compared if you use this critical F-value?
        iv. How many participants would you have if you used this critical F-value?
    c. Find F-critical $= 1.96$.
        i. What are the degrees of freedom between and within for this critical F-value?
        ii. What alpha level would you need to use with the critical F-value?
        iii. How many groups are being compared if you use this critical F-value?
        iv. How many participants would you have if you used this critical F-value?

18. Using the following information, determine the critical Tukey value for each test (assume equal number of participants per group).

Observed Tukey Value	Number of Groups Compared	Number of Participants in the Study	Alpha Level	Degrees of Freedom (b, w)	Critical Tukey Value
3.50	3	27	.05		
6.72	4	24	.01		
3.82	5	35	.01		
3.00	6	66	.05		
16.2	3	33	.05		

19. You are interested in knowing whether a significant difference exists between the academic standing of students who sit closer to and who sit farther away from the front of the classroom. You randomly select 10 children who sit at the front of their respective classrooms, 10 children who sit in the middle of their classrooms, and 10 children who sit at the back of their classrooms and then record their final grades at the end of the academic year. The data are displayed in the Table 11.14. Does academic standing vary according to where students sit in relation to the front of their classrooms?

Table 11.14 Final grades of 30 students

Front	Middle	Back
72	84	63
64	40	71
81	77	71
90	73	76
68	64	44
84	71	86
53	52	67
60	68	75
77	78	76
79	60	58
$\Sigma = 728$	$\Sigma = 667$	$\Sigma = 687$
$n = 10$	$n = 10$	$n = 10$

a. Calculate SS_b and SS_w.
b. Calculate df_b and df_w.
c. Calculate MS_b and MS_w.
d. Calculate F.
e. Identify the critical F-value, compare it to F-observed, and state whether you should accept or reject the null hypothesis.

20. Use the information provided in the questions below to calculate F.
    a. $MS_b = 332; MS_w = 218$
    b. $SS_b = 1,247; df_b = 4; MS_w = 214.4$
    c. $SS_b = 39.8; df_b = 3; SS_w = 129.44; df_w = 16$
    d. $SS_b = 3,648; SS_w = 50,085; k = 5; N = 50$

21. What causes the variance in the numerator of the F-statistic? What cause the variance in the denominator of the F-statistic?

22. Complete the crossword shown below.

ACROSS

1. The *F*-statistic is named after him
4. The IV that has more than two levels in a one-way ANOVA is called this
5. The values that an IV can take on
8. The variability in an ANOVA that reflects the average effect of the IV on the DV
9. This type of table lists the various types Variability in an ANOVA
10. (2 words) The ratio of MS_b and MS_w
11. Creator of the post-hoc HSD test
12. The "V" in ANOVA stands for this

DOWN

2. (3 words) The sum of the squared deviations from the mean
3. The variability in an ANOVA that reflects the effect of random chance
6. A two-way ANOVA design that compares all possible combinations of means is called this
7. When one level of an IV influences other IVs

494   STATISTICS IN CONTEXT

Think About It . . .

t FOR TWO AND *F* FOR MANY

SOLUTIONS

I used SPSS to speed things along. Here's what SPSS found:

- $t = -2.014$, $df = 6$, $p = .091$. The difference between the two means ($M_1 = 7.5$, $SD = 2.3$, and $M_2 = 37.00$, $SD = 29.2$) is not statistically significant.
- $F = 4.056$, $df_b = 1$, $df_w = 6$, $p = .091$. Once again, the difference between the two means is not statistically significant.
- Does $F = t^2$? Yes: $t^2 = -2.014^2 = 4.056$.
- The critical *t*-value ($df = 6$, two-tailed test at $\alpha = .05$) is 2.447.
- The critical *F*-value ($df_b = 1$ and $df_w = 6$, $\alpha = .05$) is 5.99.
- $2.447^2 = 5.987809$, or 5.99.

1. What does the top of the *t*-test ask you to compare? **The difference between two group means.**
2. What does MS_b represent if you have only two groups? **The difference between two group means.**
3. What does the term at the bottom of the *t*-test represent? **The pooled variability of both groups.**
4. Is there any similarity between the calculation of MS_w and the calculation of the pooled variability used in the *t*-test? **Yes. MS_w is a measure of the pooled variability of all of the groups. If you only have two groups, MS_w should be very closely related to the pooled variability in the *t*.**
5. How are the calculations of the variability in these two tests different? **The *t*-test uses the pooled standard deviation of the groups. The ANOVA uses pooled variance. Variance = standard deviation squared.**

THE *F*-STATISTIC

SOLUTIONS

1. Can *F* ever be a negative number? **No. *F* is the ratio of variability between and variability within groups. Variability can never be less than zero, so a negative *F* is impossible.**
2. The *F*-distribution is skewed. Is the skew positive or negative? **F is positively skewed (a long right-hand tail). Most values of *F* are low, with some very high outliers.**
3. Suppose the null hypothesis is true and there is no meaningful difference between the groups. What should the value of *F* be? **In a perfect world, *F* would equal 1.00 in this situation (the same amount of variability with groups as between groups). However, in the real world, *F* can be less than 1.00 (but never less than 0.00).**
4. Keeping in mind that MS_w is a measure of random chance, can the value of *F* be between 0 and 1? **Yes, when there is more variability within groups than there is between them.**

5. Suppose *F* was 0.40. What would that tell you about amount of variability within groups compared to the amount of variability between groups?   **It would tell me that we have more variability within our groups (people within any given group are inconsistent in their responses) than we have between groups (groups differ from one another, but not as much as individuals within groups do).**
6. Are *F*-tests ever directional?   **No.**
7. Think about the way that the null and alternative hypotheses are written for an *F*-test. If you have more than two groups to compare, does it make sense to write a directional alternative hypothesis?   **We can't write a directional alternative because the *F*-test is designed to look for any difference between groups, not a specific difference. Groups can, and often do, differ in either direction.**

REFERENCES

Bandura, A., Ross, D., & Ross, S. A. (1963). Imitation of film-mediated aggressive models. *Journal of Abnormal and Social Psychology, 66*, 3–11.

Fisher, R. A., & Mackenzie, W. A. (1923). Studies in crop variation II: The manurial response of different potato varieties. *Journal of Agricultural Science, 13*(3), 311–320.

Helberg, C. (1996). Pitfalls of data analysis (or how to avoid lies and damned lies). *Practical Assessment, Research & Evaluation, 5*(5). Retrieved from http://PAREonline.net/getvn.asp?v=5&n=5

Lyvers, M., Czerczyk, C., Follent, A., & Lodge, P. (2009). Disinhibition and reward sensitivity in relation to alcohol consumption by university undergraduates. *Addiction Research and Theory, 17*(6), 668–677.

Meaidi, A., Jennum, P., Prito, M., & Kupers, R. (2014). The sensory construction of dreams and nightmare frequency in congenitally blind and late blind individuals. *Sleep Medicine, 15*, 586–595.

Salsburg, D. (2001). *The lady tasting tea: How statistics revolutionized science in the twentieth century*. New York: A. W. H. Freeman/Owl Book.

Using SPSS and R for ANOVA

The data set we will be using for this tutorial is drawn from the Behavioral Risk Factor Surveillance System (BRFSS), a nationwide health survey carried out by the Centers for Disease Control and Prevention. The set we will be using is based on results of the 2008 questionnaire, completed by 414,509 individuals in the United States and featuring a number of questions about general health, sleep, and anxiety, among other things. The version we will be using is a trimmed-down version of the full BRFSS data set, which you can find on the companion website for this textbook (www.oup.com/us/blatchley).

One-Way ANOVA in SPSS

We will be conducting a one-way ANOVA test to determine if there is a statistically significant difference in sleep quality for individuals with different levels of overall life satisfaction. The BRFSS asks respondents to provide their level of personal life satisfaction on a scale of 1 (*very satisfied*) to 4 (*very dissatisfied*). Respondents also have the option of answering "I don't know" or not responding at all.

We are exploring the difference between individuals who had at least one poor night's sleep in the 30 days before completing the questionnaire and who answered the life satisfaction question.

The null and alternative hypotheses are:

H_0: All group population means are identical ($\mu_1 = \mu_2 = \mu_3 = \mu_4$).

H_1: At least one group population mean is different.

or

H_o: The average number of poor night's sleep does not differ by the level of life satisfaction ($\mu_1 = \mu_2 = \mu_3 = \mu_4$).

H_1: The average number of poor night's sleep will differ depending on life satisfaction.

STEP 1: Load the data into SPSS.

STEP 2: Set up the test. Click **Analyze**, then **Compare Means**, and then **One-Way ANOVA**.

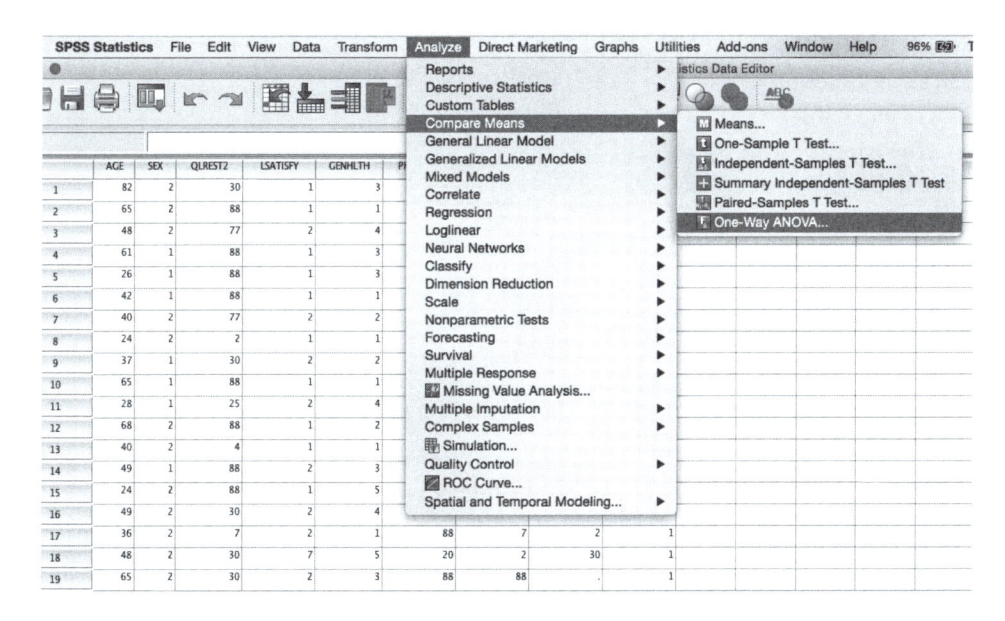

STEP 3: Select your dependent variable. Transfer the variable **How Many Days Did You Get Enough Sleep (QLREST2)** to the **Test Variable(s)** box by selecting it and then clicking on the top arrow button.

STEP 4: Select your factor. Transfer the variable **Satisfaction with Life (LFSATISFY)** to the **Factor** box by selecting it and then clicking on the bottom arrow button.

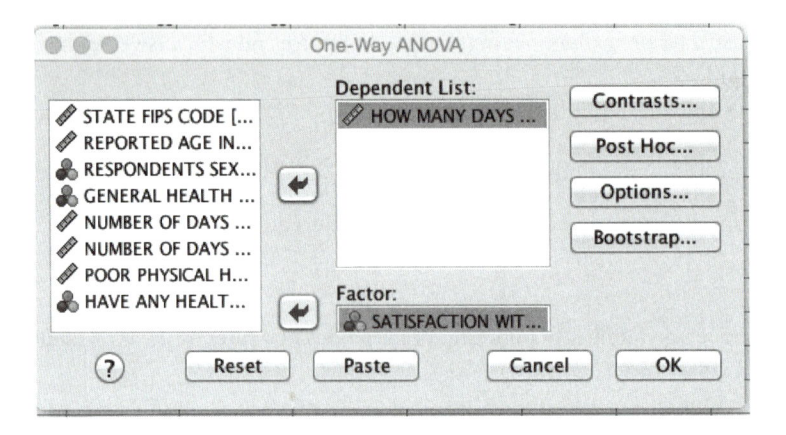

Remember, in an ANOVA test, the "Factor" is the independent variable—the variable being manipulated or tested.

STEP 5: Include Descriptive Statistics. Click **Options**, and select **Descriptive** and **Means plot**.

Descriptive will provide general descriptive statistics.

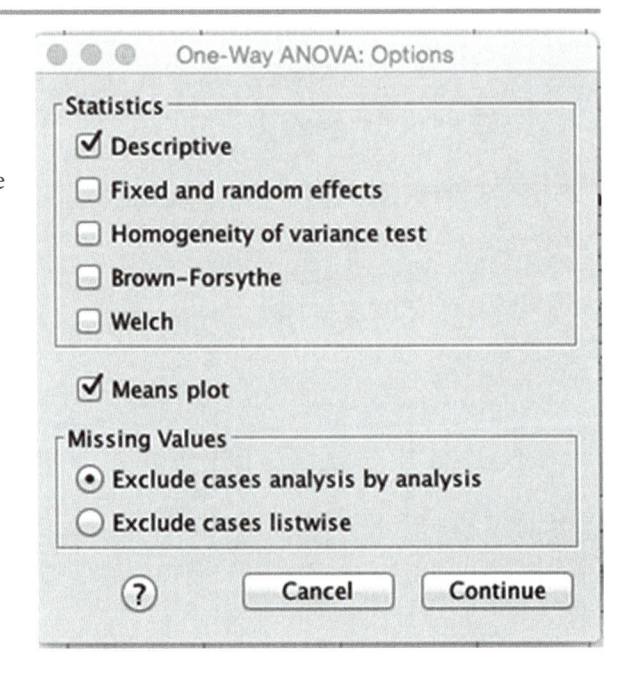

498 STATISTICS IN CONTEXT

STEP 6: Set up the post-hoc test. Click **Post-hoc…**, and select **Tukey** under the **Equal Variances Assumed** heading. This will automatically conduct post-hoc testing using Tukey's HSD.

Remember that a one-way ANOVA only tells us if there is a statistical difference between groups; it does not tell us what the difference is OR what the size of that difference is. If we want this information, we need to conduct post-hoc testing: It's a good idea to select this option if you think there will be a difference in the means of your test groups.

Click **Continue** and then **OK**.

STEP 7: Analyze the results.

The **Descriptives** table is shown first. This gives us an indication of some of the key attributes of the data. You will get a rough idea of what the data look like and what you can expect from the ANOVA results.

Descriptives

HOW MANY DAYS DID YOU GET ENOUGH SLEEP I

	N	Mean	Std. Deviation	Std. Error	95% Confidence Interval for Mean Lower Bound	95% Confidence Interval for Mean Upper Bound	Minimum	Maximum
Very Satisfied	104078	10.32	9.516	.029	10.26	10.37	1	30
Satisfied	131642	12.96	10.184	.028	12.90	13.01	1	30
Dissatisfied	13650	18.66	10.267	.088	18.48	18.83	1	30
Very Dissastisfied	3444	21.72	9.790	.167	21.40	22.05	1	30
Total	252814	12.30	10.174	.020	12.26	12.34	1	30

You can see from the values above that there appear to be large differences in the average number of poor night's sleep for each of our groups. Note that the number increases as life satisfaction decreases. However, even individuals who are very satisfied or satisfied with their life quality may have experienced up to 30 days of poor sleep.

Notice, too, the number of observations in each category. In this case, we do not have equal sample sizes, which means that our measures of variability are going to be different for each comparison in the experiment.

CHAPTER 11 Analysis of Variance 499

The mean plot shows results similar to those of the Descriptives table.

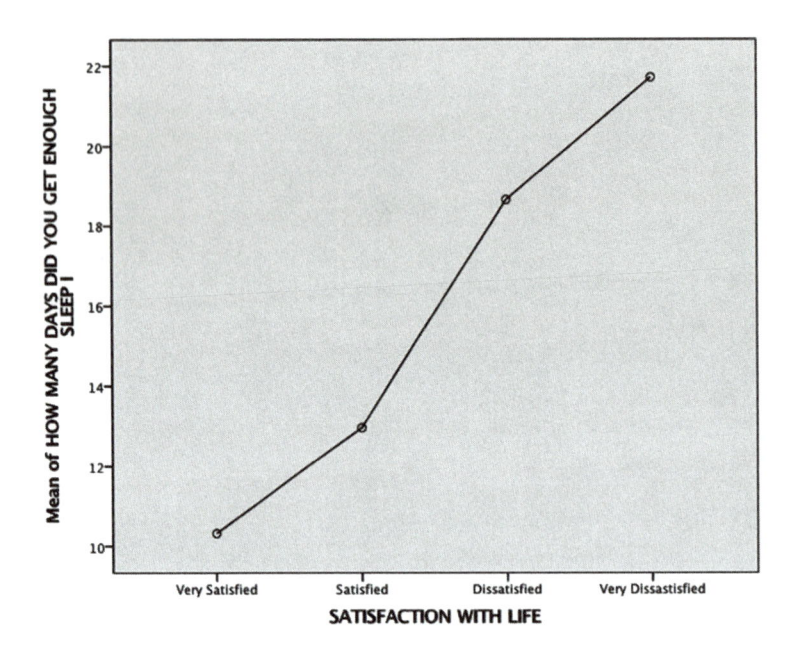

The ANOVA results are displayed next. The final column of the table shows the statistical significance value, a.k.a. the *p*-value. If this value is less than .05, we can reject the null hypothesis that all of the sample means are equal and conclude that at least one of the sample means is different from the rest at the 95% confidence interval. If the *p*-value is greater than .05, then we cannot reject the null hypothesis, and we conclude that there is no statistically significant difference in the population means.

ANOVA

HOW MANY DAYS DID YOU GET ENOUGH SLEEP I

	Sum of Squares	df	Mean Square	F	Sig.
Between Groups	1323633.13	3	441211.045	4489.208	.000
Within Groups	24846825.5	252810	98.283		
Total	26170458.7	252813			

In this case, the *p*-value is < .0005, and we reject H_0.

If we had found statistically significant results, then we would proceed with Tukey's HSD post-hoc test to determine which categories had significant differences and what the magnitude of the differences were. The first table, Multiple Comparisons, shows us the difference between each of the average groups.

500 STATISTICS IN CONTEXT

Multiple Comparisons

Dependent Variable: HOW MANY DAYS DID YOU GET ENOUGH SLEEP I
Tukey HSD

(I) SATISFACTION WITH LIFE	(J) SATISFACTION WITH LIFE	Mean Difference (I–J)	Std. Error	Sig.	95% Confidence Interval Lower Bound	Upper Bound
Very Satisfied	Satisfied	−2.642*	.041	.000	−2.75	−2.54
	Dissatisfied	−8.339*	.090	.000	−8.57	−8.11
	Very Dissastified	−11.406*	.172	.000	−11.85	−10.97
Satisfied	Very Satisfied	2.642*	.041	.000	2.54	2.75
	Dissatisfied	−5.697*	.089	.000	−5.93	−5.47
	Very Dissastified	−8.765*	.171	.000	−9.20	−8.33
Dissatisfied	Very Satisfied	8.339*	.090	.000	8.11	8.57
	Satisfied	5.697*	.089	.000	5.47	5.93
	Very Dissastified	−3.068*	.189	.000	−3.55	−2.58
Very Dissastified	Very Satisfied	11.406*	.172	.000	10.97	11.85
	Satisfied	8.765*	.171	.000	8.33	9.20
	Dissatisfied	3.068*	.189	.000	2.58	3.55

*. The mean difference is significant at the 0.05 level.

The table shows the difference between the mean for our focus category (I) and the other categories in the survey (J). We will see each value twice, once as a positive value and once as a negative value. This is because SPSS conducts the mean difference calculation for each group as the focus category (I) and the remaining groups as the other category (J). The difference between the values is the same, but the sign will differ depending on which value comes first in the equation.

HOW MANY DAYS DID YOU GET ENOUGH SLEEP I

Tukey HSD[a,b]

SATISFACTION WITH LIFE	N	Subset for alpha = 0.05 1	2	3	4
Very Satisfied	104078	10.32			
Satisfied	131642		12.96		
Dissatisfied	13650			18.66	
Very Dissastified	3444				21.72
Sig.		1.000	1.000	1.000	1.000

Means for groups in homogeneous subsets are displayed.
a. Uses Harmonic Mean Sample Size = 10503.522.
b. The group sizes are unequal. The harmonic mean of the group sizes is used. Type I error levels are not guaranteed.

This table shows us what the mean values for each of these groups is and how many observations were in each category.

Two-Way ANOVA in SPSS

In the previous example, we saw a statistically significant difference in the number of poor night's sleep for individuals with different levels of overall life satisfaction. People who were more satisfied tended to have had fewer nights of poor sleep over the preceding 30 days.

What if we also wanted to see whether the results differed for men and women? To study the combined effect of sex and life satisfaction on sleep quality over a 30-day period, we need to complete a two-way ANOVA.

STEP 1: Load the data into SPSS.

STEP 2: Set up the test. Click on **Analyze,** then **General Linear Model,** and then **Univariate**

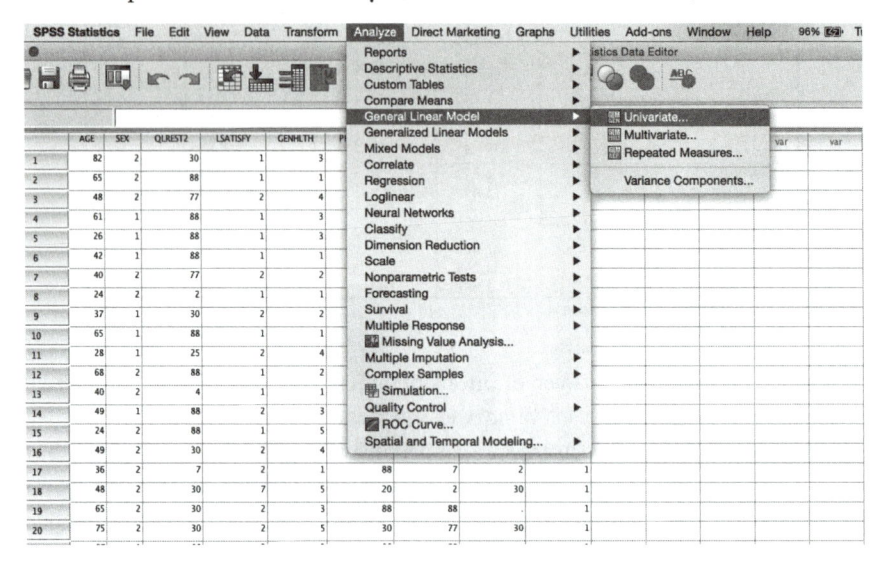

STEP 3: Select your dependent variable. Transfer the variable **How Many Days Did You Get Enough Sleep (QLREST2)** to the **Test Variable(s)** box by selecting it and then clicking on the top arrow button.

STEP 4: Select your fixed factors. Transfer the variable **Satisfaction with Life (LFSTISFY)** and **Respondents Sex (Sex)** to the **Factor** box by selecting them and then clicking on the second arrow button. You can select them one at a time, or by clicking while pressing the **Control** button your keyboard, you can select them both at the same time.

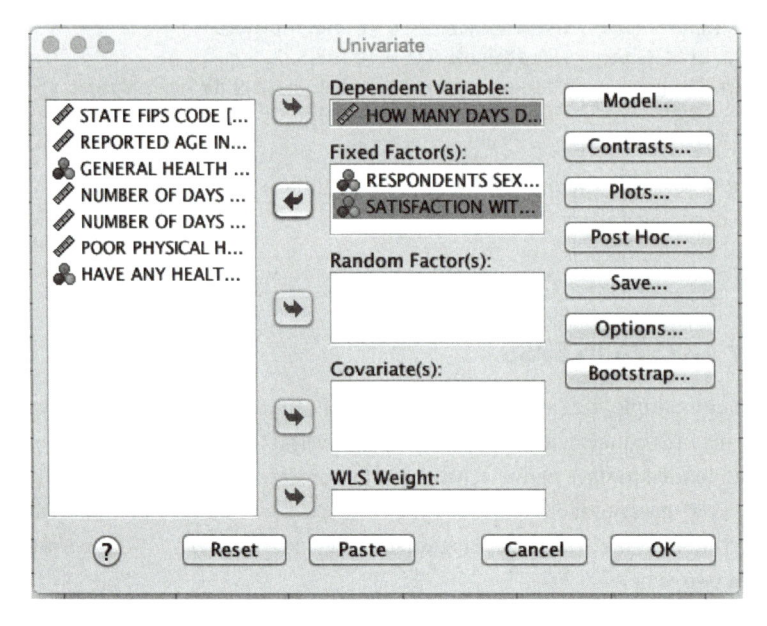

502 **STATISTICS IN CONTEXT**

STEP 5: Plot the relationships. Click **Plots…**, then transfer the variable **SEX** to the **Horizontal Ax**is box by selecting it and clicking the first arrow. Transfer the variable **LSATISFY** to the **Separate Lines** box by selecting it and clicking the second arrow. Click **Add**. This creates a term **SEX*LSATISFY**.

Transfer the variable **LSATISFY** to the **Horizontal Axis** box by selecting it and clicking the first arrow. Transfer the variable **SEX** to the **Separate Lines** box by selecting it and clicking the second arrow. Click **Add**. This creates a term **LSATISFY*SEX**.

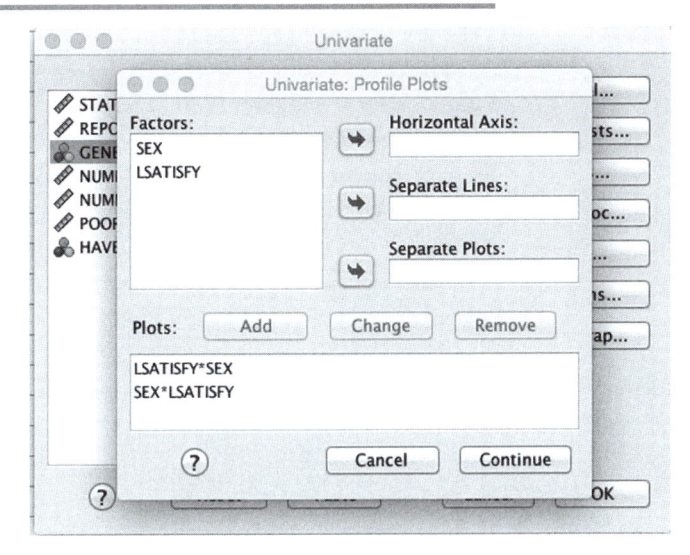

This will give you a visual interpretation and an initial impression of the relationship between the variables. These plots will help you interpret the data. If the lines are parallel, it is unlikely that the interaction between the variables is statistically significant. If the lines are not parallel, or if they cross, it is more likely that the interaction is statistically significant.

STEP 6: Select your post-hoc and Descriptive Stats. Click **Options…**, which will list all of your Univariate model options. Select the interaction term that we created **SEX*LSATISFY**.

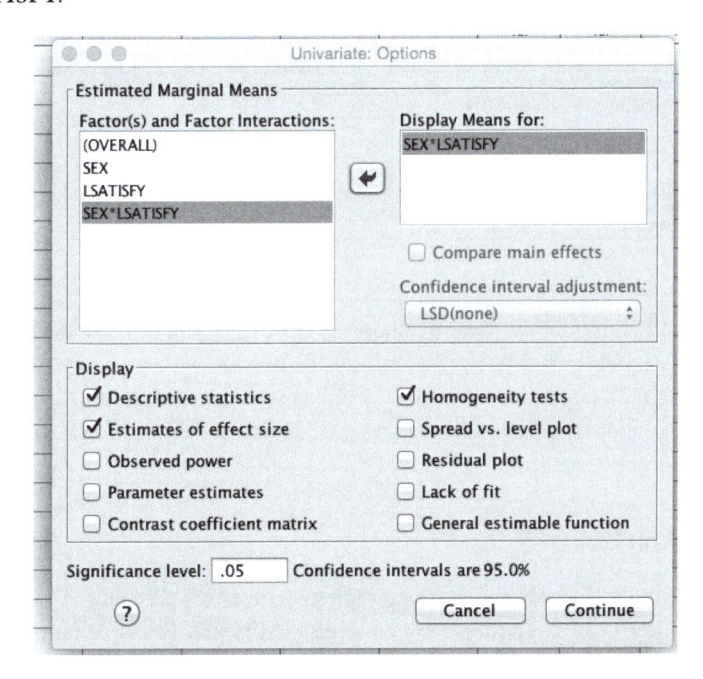

Note that we are once again conducting the test with a 95% confidence interval.

CHAPTER 11 Analysis of Variance   503

STEP 7: Analyze your results.

Looking at the plots we created in Step 5, we can see that the plots of Satisfaction with Life by Sex cross one another. The plot on the right shows increasing numbers of poor night's sleep as life satisfaction decreases; however, the magnitude of the increase becomes greater for men than for women in the "Very Dissatisfied" category. This indicates that the interaction of these variables may have statistical significance.

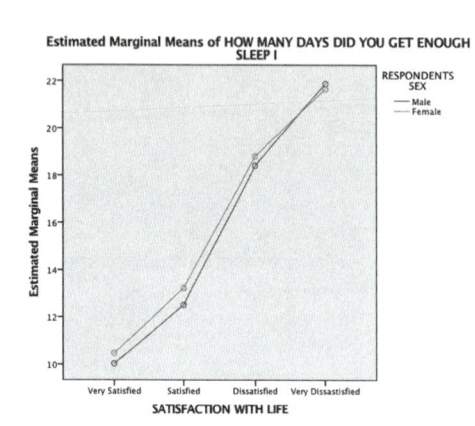

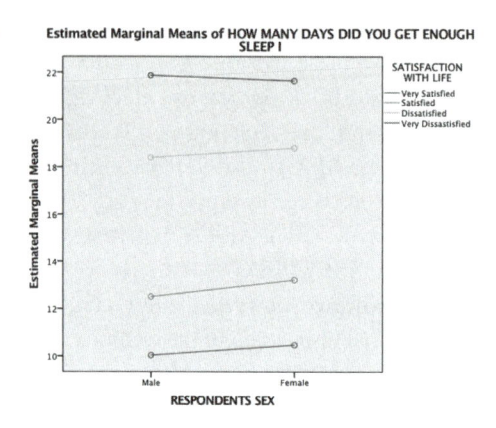

The Tests of Between-Subjects Effects table will tell us if the interaction is statistically significant.

Tests of Between-Subjects Effects

Dependent Variable: HOW MANY DAYS DID YOU GET ENOUGH SLEEP I

Source	Type III Sum of Squares	df	Mean Square	F	Sig.	Partial Eta Squared
Corrected Model	1344276.8[a]	7	192039.550	1955.546	.000	.051
Intercept	9608922.01	1	9608922.01	97848.036	.000	.279
SEX	1062.514	1	1062.514	10.820	.001	.000
LSATISFY	1208587.47	3	402862.489	4102.365	.000	.046
SEX * LSATISFY	1621.818	3	540.606	5.505	.001	.000
Error	24826181.8	252806	98.203			
Total	64404122.0	252814				
Corrected Total	26170458.7	252813				

a. R Squared = .051 (Adjusted R Squared = .051)

We are specifically interested in the interaction between Sex and Life Satisfaction, shown in the table as SEX*LSATISFY. The Sig. column (or the *p*-value) tells us at what level the interaction term is significant. We can see above that it is significant at the .001 level, or the 0.1% level. We can conclude from these results that gender and life satisfaction have a statistically significant impact on the number of poor night's sleep for respondents in the BFRSS.

One-Way ANOVA in R

Let's now try using R to study the same examples we investigated using SPSS. To recap, we want to see if there is a correlation between sleep quality and life satisfaction. To conduct one-way or two-way ANOVAs in R, however, you need to pay careful attention to letter case: Uppercase and lowercase letters can mean different things in R.

For this procedure, you will need to make sure you have the foreign package loaded and installed. To complete these steps, go to **Packages and Data**, and select **Package Installer.**

In the Package installer window, click **Get List**, and then search for "foreign." Click **Install Selected**; make sure that the **Install Dependencies** check box is selected. Exit the window.

Go to "Packages and Data," and click **Package Manager**.

In the Package Manager window scroll through to "foreign," and select the check box beside it. Make sure the status column changes to "loaded." This will load the foreign package into R and allow you to access other types of data files.

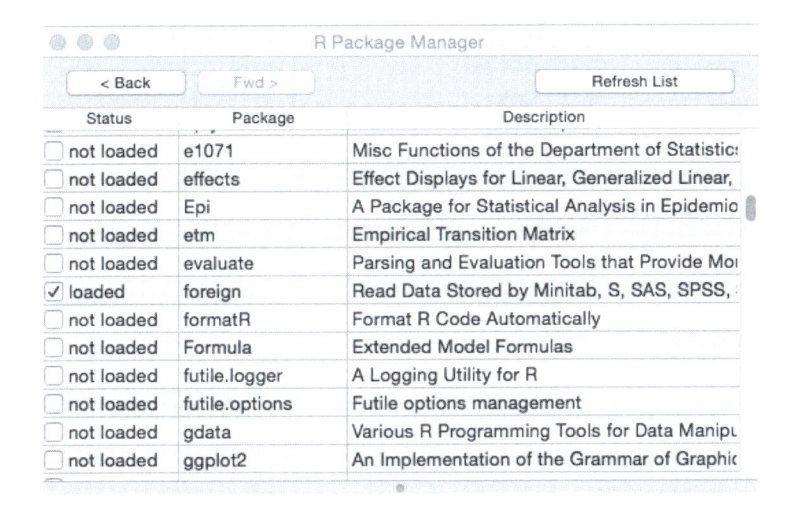

Status	Package	Description
not loaded	e1071	Misc Functions of the Department of Statistics
not loaded	effects	Effect Displays for Linear, Generalized Linear,
not loaded	Epi	A Package for Statistical Analysis in Epidemio
not loaded	etm	Empirical Transition Matrix
not loaded	evaluate	Parsing and Evaluation Tools that Provide Mor
✓ loaded	foreign	Read Data Stored by Minitab, S, SAS, SPSS,
not loaded	formatR	Format R Code Automatically
not loaded	Formula	Extended Model Formulas
not loaded	futile.logger	A Logging Utility for R
not loaded	futile.options	Futile options management
not loaded	gdata	Various R Programming Tools for Data Manipu
not loaded	ggplot2	An Implementation of the Grammar of Graphic

Alternatively, you can type **library(foreign)** into the consol.

Now we're ready to start our ANOVA.

STEP 1: Load the data into R.

```
#Load the data
LifeSatisfaction - read.csv("example.csv")
```

STEP 2: What do the data look like? Examine the data to make sure they look the way you think they should and were loaded correctly.

```
> str(LifeSatisfaction)

'data.frame':   414509 obs. of 15 variables:
$ QLREST2  : int 30 88 77 88 88 88 77 2 30 88 ...
$ LSATISFY : int 1 1 2 1 1 1 2 1 2 1 ...

> summary(LifeSatisfaction)
```

QLREST2
Min.:   1.00
1st Qu.:   5.00
Median: 25.00

LSATISFY
Min.:   1.00
1st Qu.:   1.000
Median: 2.000

```
Mean:     40.24              Mean:      1.667
3rd Qu.:  88.00              3rd Qu.:   2.000
Max.:     99.00              Max.:      9.000
NA's:         3              NA's:      10624
```

STEP 3: Recode missing variables for QLREST2. Remember that we are only interested in individuals who had between 1 and 30 nights of poor sleep in the previous 30 days. The BRFSS allows respondents to opt out of questions (coded as 99) or answer "I Don't Know" (coded as 77). Of course, some respondents may have had no trouble sleeping (coded as 88).

```
> LifeSatisfaction$QLREST2[LifeSatisfaction$QLREST2==88]<-NA
> LifeSatisfaction$QLREST2[LifeSatisfaction$QLREST2==77]<-NA
> LifeSatisfaction$QLREST2[LifeSatisfaction$QLREST2==99]<-NA
```

STEP 4: Recode missing variables for LSATISFY. Again, respondents could opt out of answering the question (coded as 9), or they could indicate that they were unsure (coded as 7). We will recode these variables as missing so that they are not taken into consideration by R when conducting the ANOVA.

```
> LifeSatisfaction$LSATISFY[LifeSatisfaction$LSATISFY==7]<-NA
> LifeSatisfaction$LSATISFY[LifeSatisfaction$LSATISFY==9]<-NA
```

STEP 5: Review the statistics after dropping variables. Has our recoding worked?

```
>summary(LifeSatisfaction)
```

```
QLREST2              LSATISFY              SEX
Min.:  1.00          Min.:  1.00           Min.:      1.000
1st Qu.:  4.00       1st Qu.:  1.000       1st Qu.:   1.000
Median: 10.00        Median:   2.000       Median:    2.000
Mean:   12.35        Mean:     1.607       Mean:      1.624
3rd Qu.: 20.0        3rd Qu.:  2.000       3rd Qu.:   2.000
Max:    30.00        Max:      4.000       Max.:      2.000
NA's:   153,654      NA's     14,465
```

Our numbers appear to be different. We no longer see a max value of 99 under the QLREST2 column or a max number of 9 for LSATISFY, and our mean for both categories is within the range of the data we are interested in.

STEP 6: Code LSATISFY as a categorical variable. We want to make sure that R is treating our LSATISFY variable as four unique groups and not one group with different values. To do this, we must change the data type from an integer to an ordinal factor.

Check the structure of the variable.

```
str(LifeSatisfaction$LSATISFY)
int [1:414509]  1 1 2 1 1 1 2 1 2 1 ...
```

Right now, it is being viewed by R as an integer and not unique categories. This will impact how our ANOVA is run, the number of degrees of freedom in our *F*-test, and the overall results. To ensure we have the same results in R as we do in SPSS, we must recode the variable as a factor.

```
> LifeSatisfaction$LSATISFY<-ordered(LifeSatisfaction$LSATISFY)
> str(LifeSatisfaction$LSATISFY)
 Ord.factor w/ 4 levels "1"<"2"<"3"<"4": 1 1 2 1 1 1 2 1 2 1 ...
```

STEP 7: Run the test. Conduct the ANOVA to see if participants with different Life Satisfaction ratings have different mean nights of poor sleep in the past 30 days

```
> aov.ex1 = aov(QLREST2~LSATISFY,data=LifeSatisfaction)
```

When setting up the ANOVA, it is important to note the order of your arguments. The first variable in the code will always be your DV, and the second variable will always be your IV.

STEP 8: Review the results. Unlike SPSS, the results of the ANOVA do not automatically show up as output. You must request to see the results.

```
> #view the summary table
> summary(aov.ex1)
              Df      Sum Sq       Mean Sq      F value      Pr(>F)
LSATISFY      3       1323633      441211       4489         <2e-16 ***
Residuals  252810     24846826     98
---
Signif. codes:  0 '***' 0.001 '**' 0.01 '*' 0.05 '.' 0.1 ' ' 1
161695 observations deleted due to missingness
```

Show the means so that we are able to see the value for each factor.

```
> print(model.tables(aov.ex1,"means"),digits=3)
Tables of means
Grand mean

12.29766

LSATISFY
            1       2       3       4
         10.3     13    18.7    21.7
rep  104078.0 131642 13650.0 3444.0

view graphical output
> boxplot(QLREST2~LSATISFY,data=LifeSatisfaction)
```

From our results, we see that there is a statistical difference in the means for at least one of our Life Satisfaction groups. To determine which categories had statistically significant differences and what the magnitude of those differences is, we must proceed with Tukey's HSD test.

STEP 9: **Run the** post-hoc testing. Again, for this code, R will do the calculation but you must request to see the data in a separate line of code. The first line below is the Tukey HSD test, and the second line is calling up the results based on what we have named them (in this case, posthoc).

```
> posthoc<-TukeyHSD(x=aov.ex1)
> posthoc
```

Tukey multiple comparisons of means
    95% family-wise confidence level

Fit: aov(formula = QLREST2 ~ LSATISFY, data = LifeSatisfaction)

$LSATISFY

	diff	lwr	upr	p adj
2-1	2.641622	2.535982	2.747262	0
3-1	8.338936	8.107088	8.570783	0
4-1	11.406474	10.965366	11.847583	0
3-2	5.697314	5.468298	5.926329	0
4-2	8.764852	8.325226	9.204479	0
4-3	3.067539	2.581879	3.553198	0

Unlike SPSS, R will show us the results for each group mean difference only once (remember that the output chart in SPSS shows us the result for each factor compared to the other factors, i.e., 1–2 and 2–1, 1–3 and 3–1, etc.).

The first column in the output above shows us the comparison, the second column shows us the value of the difference, the third and fourth columns show us the confidence interval, and the last column is the *p*-value, or the Sig. value. In this example, all of the differences are statistically significant at the .005 level.

Two-Way ANOVA in R

Suppose we believe there is a difference between the way that men's and women's life satisfaction impacts their quality of sleep, and we want to test whether this is the case. To do so, we will have to conduct a two-way ANOVA.

The code and process are very similar to what we would use in a one-way ANOVA. The differences are noted here.

STEP 1: Load your data into R.

```
#Load the data
LifeSatisfaction - read.csv("example.csv")
```

STEP 2: What do the data look like? Examine the data to make sure they look the way they are supposed to and were loaded correctly.

```
> str(LifeSatisfaction)

'data.frame': 414509 obs. of 15 variables:
$ QLREST2  : int  30 88 77 88 88 88 77 2 30 88 ...
$ LSATISFY : int  1 1 2 1 1 1 2 1 2 1 ...
$ SEX      : int  2 2 2 1 1 1 2 2 1 1 ...
```

508 STATISTICS IN CONTEXT

```
> summary(LifeSatisfaction)
```

QLREST2	LSATISFY	SEX
Min.: 1.00	Min.: 1.00	Min.: 1.000
1st Qu.: 5.00	1st Qu.: 1.000	1st Qu.: 1.000
Median: 25.00	Median: 2.000	Median: 2.000
Mean: 40.24	Mean: 1.667	Mean: 1.624
3rd Qu.: 88.00	3rd Qu.: 2.000	3rd Qu.: 2.000
Max.: 99.00	Max.: 9.000	Max.: 2.000
NA's: 3	NA's: 10624	

STEP 3: Recode missing variables for QLREST2. Remember that we are only interested in individuals who had between 1 and 30 poor night's sleep in the last 30 days. The BRFSS allows respondents to opt out of questions (coded as 99) or answer "I don't know" (coded as 77). Respondents who had no trouble sleeping are coded as 88.

```
> LifeSatisfaction$QLREST2[LifeSatisfaction$QLREST2==88]<--NA
> LifeSatisfaction$QLREST2[LifeSatisfaction$QLREST2==77]<-NA
> LifeSatisfaction$QLREST2[LifeSatisfaction$QLREST2==99]<-NA
```

STEP 4: Recode missing variables for LSATISFY. Again, respondents could opt out of answering the question (coded as 9), or they could indicate that they were unsure (coded as 7). We will recode these variables as missing so they are not taken into consideration by R when conducting the ANOVA.

```
> LifeSatisfaction$LSATISFY[LifeSatisfaction$LSATISFY==7]<-NA
> LifeSatisfaction$LSATISFY[LifeSatisfaction$LSATISFY==9]<-NA
```

STEP 5: Review the statistics after dropping variables. Has our recoding worked?

```
>summary(LifeSatisfaction)
```

QLREST2	LSATISFY	SEX
Min.: 1.00	Min.: 1.00	Min.: 1.000
1st Qu.: 4.00	1st Qu.: 1.000	1st Qu.: 1.000
Median: 10.00	Median: 2.000	Median: 2.000
Mean: 12.35	Mean: 1.607	Mean: 1.624
3rd Qu.: 20.0	3rd Qu.: 2.000	3rd Qu.: 2.000
Max: 30.00	Max: 4.000	Max.: 2.000
NA's: 153,654	NA's 14,465	

Our numbers appear to be different. We no longer see a max value of 99 under the QLREST2 column or a max number of 9 for LSATISFY, and our mean for both categories is within the range of the data we are interested in.

Note that Sex does not have any missing variables and has a min of 1 and max of 2. No observations fall outside of the binary male/female categories, which is why we did not have to recode any of the data.

STEP 6: Code LSATISFY and SEX as categorical variables. We want to make sure that R is treating our LSATISFY as variable as four unique groups and SEX as two unique groups. To do this, we must change the data type from an integer to an ordinal factor.

Check the structure of the variable.

```
str(LifeSatisfaction$LSATISFY)
int [1:414509] 1 1 2 1 1 1 2 1 2 1 ...
str(LifeSatisfaction$SEX)
int [1:414509] 2 2 2 1 1 1 2 2 1 1 ...
```

Right now, it is being viewed by R as an integer and not unique categories. This will impact how our ANOVA is run, the number of degrees of freedom in our *F*-test, and the overall results. To ensure we have the same results in R as we do in SPSS, we must recode the variable as a factor.

```
> LifeSatisfaction$LSATISFY<--ordered(LifeSatisfaction$LSATIFSY)
> str(LifeSatisfaction$LSATISFY_

> LifeSatisfaction$SEX<-ordered(LifeSatisfaction$SEX)
> str(LifeSatisfaction$SEX)
 Ord.factor w/ 2 levels "1"<"2": 2 2 2 1 1 1 2 2 1 1 ...
```

STEP 7: Run the test. Conduct the two-way ANOVA to see if men and women with different Life Satisfaction ratings have different mean poor night's sleep in the past 30 days.

You can conduct the test in two ways, using very similar code. It really just depends on how much you want to type and how you want the output table to look.

OPTION 1

```
-aov(QLREST2~LSATISFY + SEX + LSATISFY:SEX,
data=LifeSatisfaction)
```

OPTION 2

```
-aov(QLREST2~LSATISFY*SEX, data=LifeSatisfaction)
```

When setting up the two-way ANOVA, it is important to note the order of your arguments. The first variable in the code will always be your DV and the variables after the ~ will always be your IV(s).

STEP 8: Review the results. Unlike SPSS, the results of the ANOVA do not automatically show up as output. You must request to see the results.

```
> summary(aov.ex2)
              Df   Sum Sq   Mean Sq   F value    Pr(>F)
LSATISFY       3  1323633    441211  4492.870   < 2e-16 ***
SEX            1    19022     19022   193.701   < 2e-16 ***
LSATISFY:SEX   3     1622       541     5.505  0.000889 ***
Residuals         252806  24826182        98
---
Signif. codes:  0 '***' 0.001 '**' 0.01 '*' 0.05 '.' 0.1 ' ' 1
161695 observations deleted due to missingness
```

You will notice the results are identical for both methods of typing in the formula:

```
> summary(aov.ex2b)
              Df   Sum Sq   Mean Sq   F value    Pr(>F)
LSATISFY       3  1323633    441211  4492.870   < 2e-16 ***
SEX            1    19022     19022   193.701   < 2e-16 ***
LSATISFY:SEX   3     1622       541     5.505  0.000889 ***
Residuals         252806  24826182        98
---
Signif. codes:  0 '***' 0.001 '**' 0.01 '*' 0.05 '.' 0.1 ' ' 1
161695 observations deleted due to missingness
```

You will also notice that the results look slightly different from what we obtained in SPSS. That is because there is a statistical nuance in R that will cause slightly different values in your two-way ANOVA results. To obtain the correct *F* and Sig. values, we do the following:

```
> drop1(aov.ex2,~.,test="F")

Single term deletions

Model:
QLREST2 ~ LSATISFY + SEX + LSATISFY:SEX
              Df  Sum of Sq       RSS      AIC   F value     Pr(>F)
                             24826182  1159674
LSATISFY       3    1208587  26034769  1171685  4102.365  < 2.2e-16 ***
SEX            1       1063  24827244  1159683    10.820  0.0010044 **
LSATISFY:
SEX            3       1622  24827804  1159684     5.505  0.0008893 ***
---
Signif. codes:  0 '***' 0.001 '**' 0.01 '*' 0.05 '.' 0.1 ' ' 1
```

Above, we see the same results to the ones we obtained in SPSS. Our *F*-values match and are statistically significant at the .05 level. From these results, we can conclude that there is a statistically significant difference in the average number of poor night's sleep for men and women with different levels of life satisfaction.

CHAPTER TWELVE

CONFIDENCE INTERVALS AND EFFECT SIZE: BUILDING A BETTER MOUSETRAP

Data is not information, Information is not knowledge, Knowledge is not understanding, Understanding is not wisdom.

—CLIFF STOLL AND GARY SCHUBERT

Everyday Statistics

CIGARETTE SMOKING AND THE MOVIES

We humans are creatures of habit, even when that habit is dangerous to our lives and physical well-being. Cigarette smoking is a perfect example. Many people—even smokers themselves—are aware of the risks, but smoking is a notoriously hard habit to break. So, why start in the first place?

Many of our habits begin when we see people we admire displaying that behavior. This kind of modeling has frequently been linked to smoking. Do you think, for example, that seeing a favorite movie star smoking in a film influences smoking behavior? A group of researchers used a measure of effect size (a statistic called an odds ratio) and confidence intervals to find out (Distefan, Pierce, & Gilpin, 2004).

The researchers asked more than 2,000 nonsmoking teenagers in California to identify their favorite movie stars. The teens were invited to watch a selection of movie scenes featuring the named stars and then asked to categorize the actors into two groups, comprising those who smoked on screen and those who did not.

Three years later, the researchers asked the same group of teens if they had picked up the smoking habit. Those who had were likely to have identified actors who smoked on screen as their favorite stars. It turns out we do imitate our Hollywood heroes. This chapter will show you what the confidence intervals used in this study are all about.

OVERVIEW

USING ESTIMATIONS
ESTIMATES AND CONFIDENCE INTERVALS
EFFECT SIZE: HOW DIFFERENT ARE THESE MEANS, REALLY?
EFFECT SIZE AND ANOVA
STATISTICS IN CONTEXT: THE CI VERSUS THE INFERENTIAL TEST

LEARNING OBJECTIVES

Reading this chapter will help you to . . .

- Explain what the Neyman–Pearson lemma says about confidence intervals. (Concept)

- Know the difference between precision and confidence. (Concept)

- Distinguish between point estimates and interval estimates. (Concept)

- Calculate a confidence interval around a sample mean in a *z*-test and single-sample *t*-test, as well as for differences between sample means in independent- and dependent-samples *t*-tests and for an *F*-test. (Application)

- Calculate a confidence interval for differences between sample means in independent- and dependent-samples *t*-tests and for an *F*-test. (Application)

- Understand the Cohen's *d* effect size statistic. (Concept)

- Calculate effect size for *z*-, *t*-, and *F*-tests. (Application)

- Describe how inferential tests and confidence intervals are related to one another. (Concept)

USING ESTIMATIONS

Named for Polish mathematician Jerzy Neyman—introduced in *The Historical Context* on page 515—and British statistician Egon Pearson, the Neyman–Pearson lemma says that if we think of the sample mean as an estimate of a population mean, we can make that mean a better estimate of that population mean by building an interval around it called the *confidence interval* (CI).

In a 1933 article, Neyman wrote that most statisticians are concerned with making a confident decision about what the mean of a particular population might be. Since all measurements are really estimates (think back to David Kinnebrook and his estimates of time in the observatory), what we ought to focus on is *confidence*, not *precision*. Here's what Neyman had to say:

> It is impossible to assume that any particular T [the estimate] is exactly equal to θ [the parameter being estimated]. What a statistician needs is a way to measure the accuracy of T, usually solved by finding the variability in T and then writing the results in the form $T \pm S_T$. . . . [T]he smaller S_T is, the more accurate is the estimate of T of θ. (Neyman, 1937, p. 346)

Neyman is arguing that if we do an inferential test comparing the mean of a sample and the mean of a population, and if that test produces a significant result, we can use that sample mean as an *estimate* of the *true* or *alternative* population mean. Our confidence in that estimate depends on two characteristics of the data: our definition of different (i.e., the alpha level we're using) and the variability in the null population.

Why would building an interval or a window around a sample mean make that mean a better estimator of the population mean? Let's start with an understanding of the difference between a *point estimate* and an *interval estimate* and between *precision* and *confidence*.

When we use a single number as our estimate of some other value, we are making a **point estimate**. If that estimate is identical or very close to the value of the measure we are trying to estimate, then that estimate is **precise**.

If we use a range of values instead of a single measurement to make our estimate, then we are making an **interval estimate**. This is precisely what Neyman and Pearson are advocating we do in creating a confidence *interval*. If our estimate is accurate, then we can call it a precise estimate. **Confidence** refers to how certain we are that our estimate is a good one. An example will help.

Suppose we were in Las Vegas and wanted to bet on what the value of a single card drawn from a set of cards might be. To make this example a bit simpler, let's say that the set of cards consisted of only one suit (diamonds) between 10 and ace. The casino allows us to make either a point estimate or an interval estimate of the value of a single card drawn at random from the set. Take a look at Box 12.1 to see an illustration of point versus interval estimates and precision versus confidence in this situation.

point estimate An estimation of a population parameter (a single point in the population) made using a sample statistic (a single point in the sample).

precise Denoting a measurement of quantity that is identical or very close to that quantity's true value.

interval estimate An interval (range) of probable values of an unknown population parameter, often the population mean, based on data from a sample; an estimated interval in a population.

confidence The degree to which an estimate is reliable.

> **BOX 12.1** **Point versus interval estimates and confidence versus precision**

Point Estimate vs. Interval Estimate	Confidence vs. Precision
Card chosen is the queen of diamonds.	**Confidence:** The degree of surety you have in your estimate. **Precision:** The accuracy of your estimate.
Point estimate: *You say that the card chosen will be the 10 of diamonds.* ➡ Result: You lose.	↑**High precision** usually means **low confidence** ↓.
Interval estimate: *You say that the card chosen will be between the 10 and the ace of diamonds.* ➡ Result: You win!	↓**Low precision** usually means **high confidence** ↑.

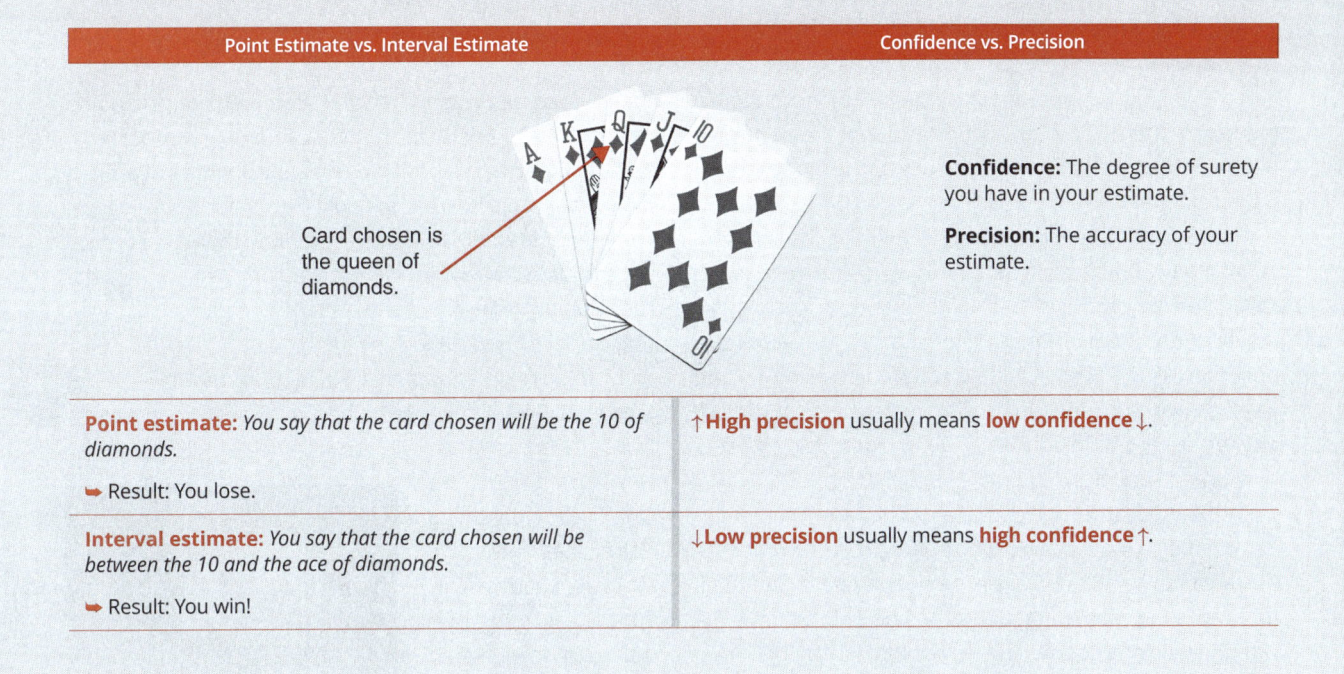

If we make a point estimate about which card will be selected at random, then whatever our estimate is, it will have a high level of precision (we are specifying only one card). But since the odds that we will be right are fairly low (the probability would be 1 out of 5, or 0.20), our confidence in that estimate is probably fairly low as well. On the other hand, if we make an interval estimate, then our level of precision is fairly low (we're specifying a range of possible values), but our confidence is fairly high (the card must be between the upper and lower limits of the interval we have specified).

ESTIMATES AND CONFIDENCE INTERVALS

We have discussed five different inferential tests so far. In each, we begin with a known or a hypothesized population mean—the average amount of time spent sleeping by adult humans, for example. We know that most adults sleep about 8 hours per night (this "known mean" is a guess or an estimate as well). However, people in the population will vary in how much time they sleep, and that variability is being caused by random chance. We start with the **null population** (see Figure 12.2), treat the sample we drew from it, and look for an effect of that

null population The population hypothesized about in the null hypothesis.

The Historical Context

JERZY NEYMAN: WHAT'S A LEMMA?

Let me introduce you to Jerzy Neyman, a congenial man who spent most of his professional life as a professor of mathematics and statistics at the University of California, Berkeley.

Professor Neyman was born in imperial Russia in April 1894. He won prizes at school as a child and went on to study first physics and then mathematics at college. He survived the Russian Revolution (with just two arrests—once for being Polish at a time when Russia and Poland were at war and once for selling matches on the black market) as well as World War I, all while studying mathematics and statistics.

In 1920, he earned the equivalent of his MA in mathematics and took a job teaching. Once word got around that he was about to be arrested again, he and his family left Russia for Poland. Neyman eventually earned his PhD, and in 1925, with the help of a good friend, he left Poland for England to work with Karl Pearson (considered the founder of modern statistics), William Gossett (whom we met in Chapter 10), and Pearson's son Egon.

Jerzy and Egon were interested in the problem of hypothesis testing using small samples. How do you take a relatively small sample from a huge population and make the mean of that small sample a good estimator of the population mean? In a series of 10 papers between 1928 and 1933, the two men developed an answer to this question that is now known as the *Neyman–Pearson theory*, or the *Neyman–Pearson lemma*. In mathematics, a *lemma* (from a Greek word meaning "premise") is a proven statement that is used as proof of another statement. Neyman and Pearson's lemma states that we can make a better estimate of a population mean by building an interval around it. Neyman and Pearson not only developed the idea of confidence intervals, they also described the ways in which we can be wrong when we draw a conclusion about the results of an inferential test. You're already familiar with these errors (Type I and Type II).

In 1937, Neyman was offered a job at the University of California, Berkeley, where he began a statistics laboratory. During World War II, Neyman turned his talents to designing bomb sites and trying to make them as accurate as possible. Although he signed the loyalty oath, swearing that he was not involved in any activity to overthrow the U.S. government, he actively supported his colleagues in the faculty at Berkeley who had refused to sign. He died in Oakland, California, in August 1981.

FIGURE 12.1
Jerzy Neyman.

---

treatment. In our example, our independent variable, or IV, is stress, and sleep duration is our dependent variable, or DV. Before we stress the sample, I would say that average sleep duration in our sample mean is probably close to 8 hours, the average sleep duration in the population. After we stress the people in the sample, what will the average amount of sleep in the sample be?

If the IV (stress) has no effect, the mean sleep duration in the sample should be about 8 hours. If stress changes how much time we spend asleep, then the sample should have a mean sleep time of less than, or more than,

8 hours. And if our sample mean has been changed by the IV so much that it is out in the outer 5% of the population distribution (the standard definition of "different" in statistics), then it no longer does a good job of estimating the known population mean of 8 hours. If the sample mean does not estimate the null population of people who sleep about 8 hours a night, what population does it represent?

It's that last question that we haven't been able to address, at least so far. An inferential statistical test will tell us if our IV had a meaningful effect on the sample, but it won't tell us about this other population—in this case, the population of stressed people who sleep some other amount of time per night. This other population (the true population or the **alternative population**) consists of the sleep times of stressed adult humans, and it would be quite beneficial to the research we're doing if we could estimate sleep time for the people is this other population.

The topic of this chapter, the **confidence interval (CI)**, is a method of estimating the mean of the true population, once we've found a significant inferential test result. The CI describes a *range of values that, with a particular degree of certainty, will include an unknown population parameter*—in this case, a population mean. We will add this descriptive statistic to our statistical toolbox to help make our conclusions stronger. Let's start with the first test we discussed: the *z*-test.

alternative population The population hypothesized about in the alternative hypothesis.

confidence interval (CI) An interval estimate of a population parameter, usually the population mean.

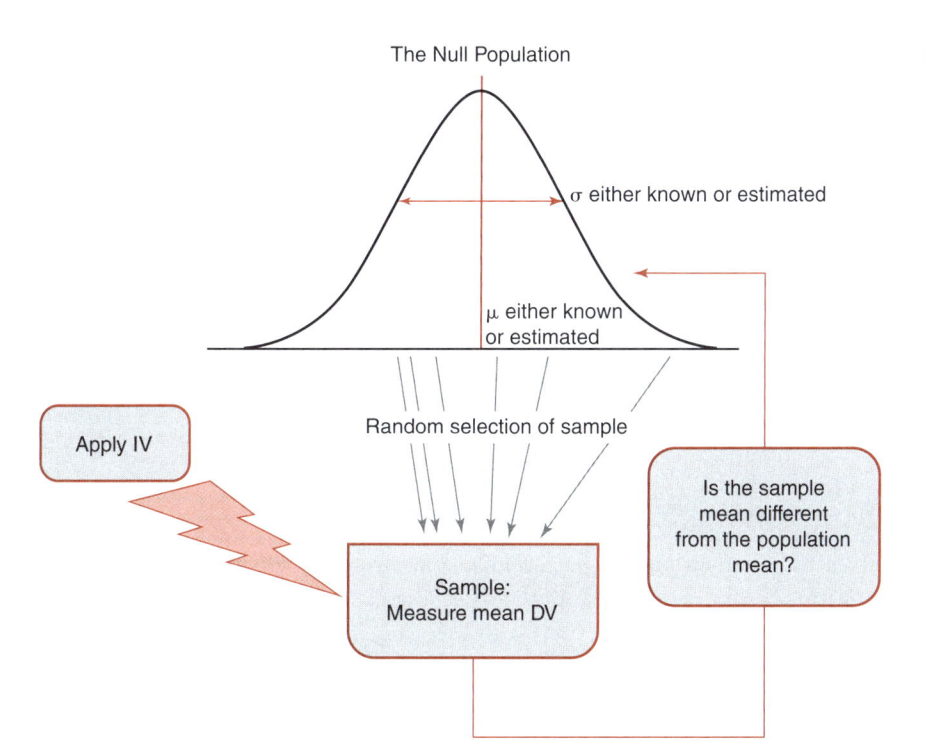

FIGURE 12.2
The null population.

CHAPTER 12 Confidence Intervals and Effect Size: Building a Better Mousetrap

CIS AND THE z-TEST

In Chapter 9, we learned that Hicks, Fernandez, and Pellegrini (2001) had found that the average sleep time of college students was 6.85 hours ($SD = 1.04$ hr), and we wanted to know if stressed students sleep fewer hours than nonstressed students. So, we compared the mean sleep time of our sample, which presumably came from this null population and so ought to have a mean sleep time of 6.85 hours *if the IV has no effect*, with this known population mean. To do so, we used a z-test (we knew μ and σ, allowing us to use z). Box 12.2 shows the data, the test we conducted, and the conclusion that we came to.

BOX 12.2 — Results of our stressed-out students study

Descriptive Statistics from the Sample	Our Hypotheses	The Test and the Conclusion
* $\mu = 6.85$ hours * $\sigma = 1.04$ hours ♦ $\mu_{\bar{X}} = 6.85$ hours ♦ $\sigma_{\bar{X}} = \dfrac{\sigma}{\sqrt{n}} = \dfrac{1.04}{\sqrt{10}}$ ✓ $\sum x = 56.75$ ✓ $\bar{X} = 5.67$ ✓ $s = 1.55$	**The Null** $H_0: \mu_{\bar{X}} = \mu$ or $H_1: \mu_{\bar{X}} < 6.85$ Stress does nothing to sleep time.	$z = \dfrac{\bar{X} - \mu_{\bar{X}}}{\sigma_{\bar{X}}} = \dfrac{5.675 - 6.85}{0.33} = \dfrac{-1.175}{0.33} = -3.56$ $z_{\text{critical}} > z_{\text{observed}}$ The value of -3.56 is farther out in the left-hand tail than is the critical z-value.
$n = 10$ Critical $z = -1.65$ $\alpha = .05$ Sources * Hicks et al. (2001) ♦ The CLT ✓ Our calculations	**The Alternative** $H_1: \mu_{\bar{X}} < \mu$ or $H_1: \mu_{\bar{X}} < 6.85$ Stress decreases sleep time.	**Conclusion:** Reject the null hypothesis. The mean sleep time in our sample of 10 stressed-out college students is significantly lower than the mean sleep time seen in the null population.

We found that stress significantly decreased the sleep time of the 10 college students in our sample. So, we now know that stressed students are different from nonstressed students, at least in their sleeping patterns. But what is that average amount of time we could expect a *stressed* student to sleep? What is the mean of the alternative population? To find out, we will use the mean of our sample (5.675 hours) as an estimate of the mean of the alternative population. To make our estimate one we can be confident of, we'll calculate a confidence interval around that mean. In fact, we'll calculate a 95% CI around our sample mean.

How did I specify the "width" of our confidence level? Good question. Take a look at the two populations in Figure 12.3, and think about the decision we're making. The distribution on the right represents the null population. According to Hicks et al. (2001), it has a mean of 6.85 hours and a standard deviation of 1.04 hours. The dashed gray line in the distribution shown on the left represents the mean sleep time in our sample (5.67 hours). Notice that our sample mean is in the rejection region (the red region in the null distribution), so we concluded that a mean of 5.67 hours of sleep was significantly lower than the population mean of 6.85 hours.

We will use our sample mean of 5.67 hours of sleep per night as an estimate of the mean sleep time in the alternative population of stressed people, and we will use the variability in the null population to create our confidence interval. So, we will be 95% confident that the true mean sleep time for stressed adults in the alternative population is within the window that we calculate.

We were using a one-tailed test, so we defined "different" as the outer 5% (in our case, the outer 5% of the left-hand tail of the null population). We discovered that the lower 5% of this distribution would be defined by a sleep time that is 1.65 standard errors or more below the mean of the null population (our critical z-value for our one-tailed test was -1.65). At the same time, "not different" (i.e., the same) can be defined as $1 - \alpha$. So, *different* is a sample mean in the lower 5% of the distribution (a z of -1.65 or more) and *not different* should be $1 - .05$, or the remaining 95% of the distribution (see the arrows in Figure 12.3).

If we use this critical z-value to define the upper and lower limits of our interval, however, we will be defining a 90% CI. We'll be adding the critical z that defines

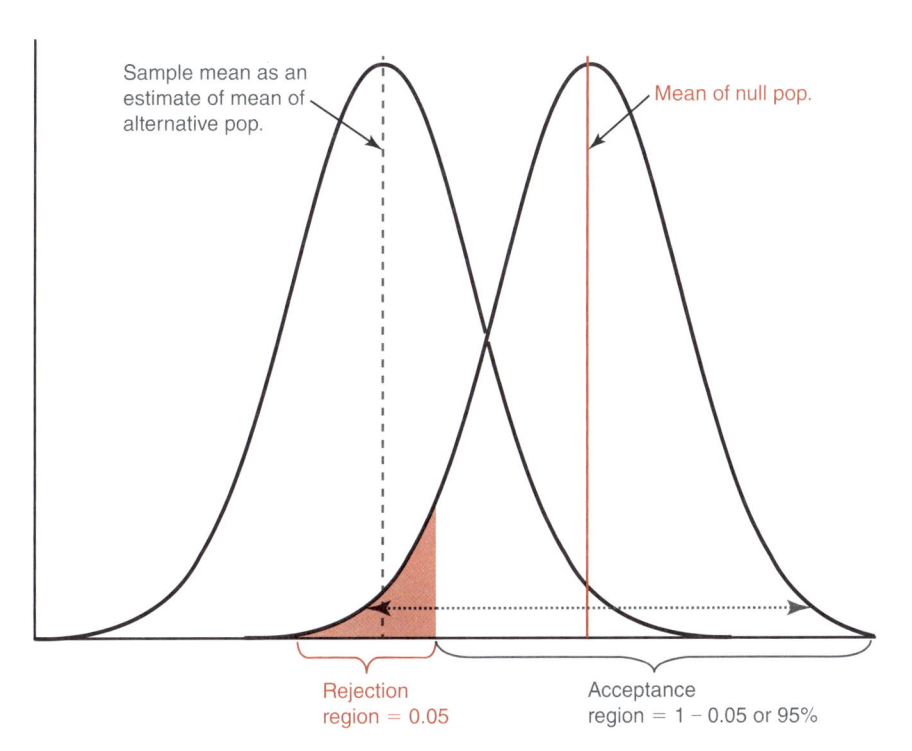

FIGURE 12.3
Distributions for the null and alternative populations in our study of stressed-out sleeping students.

CHAPTER 12 Confidence Intervals and Effect Size: Building a Better Mousetrap

5% of the distribution to the mean to find the upper limit and subtracting that same value from the mean to find the lower limit. How can we fix this to find the 95% CI? We use a critical z-value of ±1.96 to define the outer 5% and the inner 95%.

The width of the interval that we calculate, as Neyman suggested, is based on the alpha level we use in our inferential test and on the variability in the data. The critical value we use to determine the limits of this interval is always two-tailed because we're defining one limit above the sample mean and another limit below it. The more variability there is in the null population and in our sample from it, the wider our interval around the mean will have to be. Box 12.3 shows the formula for calculating the CI around our sample mean.

BOX 12.3 **Calculating the confidence interval around our sample mean**

$$\text{Upper limit } (UL) \text{ of the CI} = \bar{X} + (z_{\text{critical}})(\sigma_{\bar{X}})$$

$$UL = 5.67 + (1.96)(.33) = 5.67 + 0.65 = 6.32$$

$$\text{Lower limit } (LL) \text{ of the CI} = \bar{X} - (z_{\text{critical}})(\sigma_{\bar{X}})$$

$$LL = 5.67 - (1.96)(.33) = 5.67 - 0.65 = 5.02$$

We are 95% sure that the mean of the true population (representing stressed students) is between 5.02 hours and 6.32 hours of sleep.

The two-tailed critical z-value tells us what "different from the mean of the SDM" is by defining it as the outer 5%. Therefore, if our mean is not in that outer 5%, it must be in the inner 95%, making us 95% sure that the population mean we're looking for is in the window, somewhere.

Think About It . . .

WHAT DOES A CONFIDENCE INTERVAL REALLY MEAN? (PART 1)

The sample mean serves as a point estimate of the population mean. The sample mean sits at the center of the interval we construct because it is the best point estimate we have of an unknown population mean. We add a "margin of error" to the sample mean to get the upper and lower limits of the CI. The margin of error is found by multiplying a constant z-value—the critical z-value that defines the level of confidence we want to have (95% in the example shown here)—by the standard error ($\sigma_{\bar{X}}$).

$$\bar{X} \pm 1.96(\sigma_{\bar{X}})$$

When we're done, we say that we are, for example, 95% confident that the population mean we are interested in is between the upper and lower limits of the CI we have constructed. We're confident, but are we accurate?

Suppose we had the following situation: We know that μ = 50 and σ = 10. We draw a sample of 25 from the population and find $\bar{X} = 48$. Let's construct a 95% CI around this estimate of the population mean:

$$\text{CI} = \bar{X} \pm 1.96\left(\frac{\sigma}{\sqrt{n}}\right) = 48 \pm 1.96\left(\frac{10}{\sqrt{25}}\right) = 48 \pm 1.96(2) = 48 \pm 3.92$$

So, the upper limit of the CI is 51.92, and the lower limit is 44.08. The population mean of 50 is within this interval, just as we were confident it would be.

1. Can we reject H_0? _____

2. What does rejecting H_0 tell us about the relationship between $\bar{X}$ and μ? _____

Since we have only drawn one sample from the population, however, it is entirely possible that our CI might *not* contain μ—we're not 100% confident, after all. What we are really confident of is that *in the long run and in theory*, a certain percentage of the intervals (95% in this example) would contain μ. That means that if we repeated this process many times, 95% of all of the intervals we constructed would contain the true value of μ.

Suppose we were to repeat this process a total of 20 times. Now we would have 20 CIs and 20 estimates of μ.

3. How many of the CIs we construct should contain the population mean? _____

4. How many of the CIs we construct should NOT contain the population mean? _____

5. How often can we reject H_0? _____

Let's try it.

6. Finish the table shown below.

$\bar{X}$	CI	$\bar{X}$	CI	$\bar{X}$	CI	$\bar{X}$	CI
48	44.08–51.92	49		47		51	
47	43.08–50.92	53		50		51	
50	46.08–53.92	43		53		47	
53	49.08–56.92	50		49		50	
58		52		51		49	

7. Do these sample means seem like reasonable estimates of µ (if only random chance is operating)? _____

8. How many intervals contained µ? _____

9. Why did we not see what we were confident of seeing? (Think about the meaning of "in the long run" as you answer this question.) _____

Answers to this exercise can be found on page 547.

CIS AND THE SINGLE-SAMPLE *t*-TEST

Let's go back to our question in Chapter 10 about the effect of daily exercise on the amount of time college students spend sleeping to see how a confidence interval can add to the conclusion that we draw when we use *t*-tests. Box 12.4 shows the data, the test, and the conclusion that we made in this example.

BOX 12.4 **Results of our exercise and sleep study**

The Data	Single-Sample *t*-Test
$H_0: \bar{X} - \mu_{\bar{X}} = 0.00$ $H_1: \bar{X} - \mu_{\bar{X}} > 0.00$ $\bar{X} = \dfrac{139}{15} = 9.26$ $\mu_{\bar{X}} = \mu = 8.00$ $df = 15 - 1 = 14$ $\alpha = .05; t_{\text{critical}} = 2.145$	$t = \dfrac{\bar{X} - \mu_{\bar{X}}}{\dfrac{\hat{s}}{\sqrt{n}}} = \dfrac{9.26 - 8.00}{\dfrac{2.02}{3.87}}$ $\dfrac{2.02}{3.87} = 0.52$ $t = \dfrac{1.26}{0.52} = 2.42$

$t_{\text{observed}} = 2.42$

$t_{\text{critical}} = 2.145$

If $t_{\text{observed}} > t_{\text{critical}}$, we have a significant difference between the two means. Our decision is to reject the null hypothesis.

We knew that the average amount of nightly sleep in the population was 8 hours. Our null hypothesis was that the mean of the sample of people who exercise regularly would be the same as the mean of the null population (again, 8 hours of sleep). The alternative was that the average sleep time of regular exercisers would be longer than that of people in the null population (a one-tailed test). We drew a sample of people at random from this population, had them exercise daily for a set period of time, and then measured their sleep time. Our conclusion was that our sample mean was sufficiently different from the null population mean for us to conclude that daily exercise causes a change in the sleep time of these subjects (notice that daily exercise increased the amount of time spent sleeping).

What else can we determine from examining these data? Well, the mean of our sample is a poor estimate of the mean of the null population. However, it would probably be a good estimate of the mean of the alternative population—the population of adult sleepers who exercise daily. We will use a confidence interval to make our estimate of the mean of the alternative population.

The CI that we will create around our sample mean will allow us to be 95% confident that the true mean of the alternative population will be within the interval. Box 12.5 shows the formula for calculating the CI around our sample mean. Remember, we need a two-tailed critical t-value to find our upper and lower CI limits. With 14 df, the two-tailed critical t-value when $\alpha = .05$ is ± 2.145.

BOX 12.5 **Calculating a confidence interval around our sample mean**

Upper limit (UL) of the CI $= \bar{X} + ($two-tailed critical t-value$)($estimated standard error or $s_{\bar{X}})$

$UL = 9.26 + (2.145)(0.52) = 9.26 + 1.115 = 10.37$

Lower limit (LL) of the CI $= \bar{X} - ($two-tailed critical t-value$)($estimated standard error or $s_{\bar{X}})$

$LL = 9.26 - (2.145)(0.52) = 9.26 - 1.115 = 8.14$

We can be 95% sure that the mean of the true population is between 8.14 and 10.37 hours of sleep.

CheckPoint

Calculate confidence intervals for the following means. Use the information provided in each question to identify the appropriate critical value for the confidence interval equation.

1. $\bar{X} = 6.2$; $\alpha = .01$; $s_{\bar{X}} = 0.4$; $df = 6$ _____
2. $\bar{X} = 331.7$; $\alpha = .01$; $\sigma_{\bar{X}} = 13$ _____
3. $\bar{X} = 49.8$; $\alpha = .05$; $s_{\bar{X}} = 3.2$; $df = 4$ _____
4. $\bar{X} = 0.13$; $\alpha = .05$; $\sigma_{\bar{X}} = 0.012$ _____
5. $\bar{X} = 68.5$; $\alpha = .10$; $\sigma_{\bar{X}} = 3.8$ _____

Answers to these questions are found on page 525.

CIS AND INDEPENDENT- AND DEPENDENT-SAMPLES *t*-TESTS

Let's review. In both the *z*-test and the single-sample *t*-test, we were comparing a single-sample mean with a known population mean. If the single-sample mean was sufficiently different from the mean of the null population, then we said that it was more likely to be a representative of the alternative population. We used a *z*-test when the mean and the standard deviation of the null population were known, and a single-sample *t*-test when the mean of the null population was known, but we needed to estimate the standard deviation of the null population.

When both the mean and the standard deviation were unknown, we used estimates of both the null and alternative populations. Since we don't know a thing about the null population, the question we're asking about these two estimates has to do with the difference between them. One of the means represents the null population, which does not receive the IV; the other mean represents the alternative population, which receives the IV. In the *t*-test that we conduct to compare these two means, we assume that the difference will equal zero (meaning the two means are the same and the IV has no effect). A difference of other than zero indicates that the alternative population has been affected by treatment with the IV.

When the two means that we compare are independent of one another, the test we use to compare them is an independent-samples *t*-test. And if the two means are dependent on one another, we use a dependent-samples *t*-test. The measure of variability that we use to calculate *t* in either test is based on standard error, just as it was in a *z*-test or single-sample *t*-test. The only difference is that our standard error is the standard deviation in the sampling distribution of the differences between the means. Box 12.6 shows you these characteristics of an independent samples *t*-test.

BOX 12.6 **Estimates of variability in an independent-samples *t*-test**

$$t = \frac{\bar{X}_C - \bar{X}_E}{s_{\bar{X}_C - \bar{X}_E}}$$

where the error term is a combination of the variability in each estimate (sample):

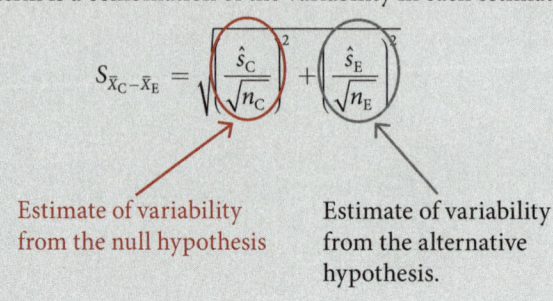

$$S_{\bar{X}_C - \bar{X}_E} = \sqrt{\left(\frac{\hat{s}_C}{\sqrt{n_C}}\right)^2 + \left(\frac{\hat{s}_E}{\sqrt{n_E}}\right)^2}$$

Estimate of variability from the null hypothesis

Estimate of variability from the alternative hypothesis.

In Chapter 10, we used an independent-samples *t*-test to compare the time it took to fall asleep in the null population, comprising people who had received no treatment, with the sleep onset time in the alternative population, comprising people who had been given some warm milk (the treatment, or IV). In this particular case, we found no significant difference between the sample of people who had received warm milk (the experimental group) and those who had not (the control group).

We hypothesized that the two means would differ by 0 minutes, but they actually differed by 1.62 minutes. When we turned that difference into a *t*-score, we saw that our observed *t*-value of 1.36 was less than the critical *t*-value of 1.68, so we could not say that warm milk had a significant effect on sleep onset time. We actually don't need to construct a confidence interval here because the inferential test has shown us that the two means are approximately equal, but let's calculate a CI just to see what happens.

What would the formula for the CI be in this case? You have probably noticed a pattern in the formulas so far, so let's go back and review. Take a look at Box 12.7 to see how we created our intervals for the *z*- and single-sample *t*-tests.

BOX 12.7   **Calculating CIs for the *z*-test and single-sample *t*-test**

Calculating a confidence interval for a *z*-test

$$CI = \bar{X} \pm (z_{critical})(\sigma_{\bar{X}})$$

Calculating a confidence interval for a single-sample *t*-test

$$CI = \bar{X} \pm (t_{critical})(s_{\bar{X}})$$

We essentially multiply the variability in the null population by the critical value that represents our definition of "different" in a two-tailed version of the test we're using. (Notice that if you are using a two-tailed test, then you already know what your critical value for the CI will be.) We then add that product to the sample mean we are building our interval around in order to get the upper limit for

---

CheckPoint Answers
Answers to questions on page 523

1. 6.2 ± 1.48 (4.72, 7.68) ☐
2. 331.7 ± 33.54 (298.16, 365.24) ☐
3. 49.8 ± 8.88 (40.92, 58.68) ☐
4. 0.13 ± 0.024 (0.106, 0.154) ☐
5. 68.5 ± 6.23 (62.27, 74.73) ☐

SCORE: /5

the interval, and then subtract that same product from the sample mean in order to find the lower limit. We can apply that same logic to the *t*-test for independent samples, as shown in Box 12.8.

BOX 12.8 — Calculating a CI for an independent-samples *t*-test

Calculating a confidence interval for an independent-samples *t*-test

$$CI = (\bar{X}_1 - \bar{X}_2) \pm (t_{critical})(s_{\bar{X}_1 - \bar{X}_2})$$

There's only one real difference in this calculation compared to what we did for the *z*- and single-sample *t*-tests. Once again, we are multiplying the critical value, which determines our level of confidence, by the variability in the null population. However, we are now building the CI around the *difference* between the two sample means. In our null hypothesis, we proposed that the difference between the two sample means would be zero. In our alternative hypothesis, we proposed that the difference between these two sample means would be something other than zero. And our measure of variability is called the standard error of the difference (between the two means). It should make sense, then, that our CI surrounds this difference. Because our inferential test showed us that the two means are not different from one another, the mean of the null population (zero) should be within the interval that we create. Box 12.9 shows our calculations.

BOX 12.9 — Calculating a confidence interval for our sleepy milk study

$$CI = (\bar{X}_1 - \bar{X}_2) \pm (t_{critical})(s_{\bar{X}_1 - \bar{X}_2})$$

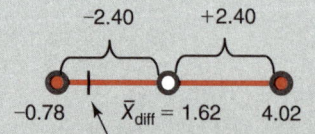

$LL = 1.62 - (2.02)(1.19)$ $UL = 1.62 + (2.02)(1.19)$
$ = 1.62 + (2.40)$ $ = 1.62 - 2.40$
$ = -0.78$ $ = 4.02$

−2.40   +2.40

−0.78 $\bar{X}_{diff} = 1.62$ 4.02

Null hypothesis (difference between sample means = 0) is contained within the interval because our *t*-value was not significant.

A confidence interval gives the researcher a measure of what the true population mean would be, given that we have shown our sample is actually likely in the null population. If we use the pattern we've discussed for CIs, it should be easy to figure out the formula for finding the upper and lower limits for the CI around a mean in a dependent-samples *t*-test. We also discussed an example of the dependent-samples *t*-test in Chapter 10, so let's go back and see what we have.

We asked a question about whether college students are more distracted in the spring semester than they are in the fall. We used a dependent-samples *t*-test because we had a repeated-measures design: Each subject served as his or her own control, because we tested each subject twice—once in the spring and again in the fall. Our dependent variable was a score on a reaction time test thought to be a measure of distraction. The null hypothesis was that there would be no difference between distraction in the spring and in the fall, and so no difference in reaction time $\left(\mu_{\bar{X}_{fall}} - \mu_{\bar{X}_{spring}} = 0\right)$. The alternative hypothesis was that the difference between the fall and the spring reaction times would be something other than zero. Box 12.10 will remind you of what we found.

BOX 12.10   **Results of our spring–fall distractedness study**

What we Found	Dependent-Samples *t*-Test
$\bar{X}_{fall} = 524.75$ ms	$t = \dfrac{\bar{X}_1 - \bar{X}_2}{S_{\bar{D}}}$
$\bar{X}_{spring} = 673.33$ ms	
$\bar{D} = -148.58$ ms	
$t_{observed} = -4.2$	$S_{\bar{D}} = \dfrac{\hat{s}_D}{\sqrt{n}} = \dfrac{\sqrt{\dfrac{\sum D^2 - \dfrac{(\sum D)^2}{n}}{n-1}}}{\sqrt{n}}$
$t_{critical} = 2.201$	
$df = 12 - 1 = 11$	

We concluded that we could reject H_0; we can say that students are significantly more distracted in the spring than they are in the fall. You should be able to predict whether the null population mean will be inside or outside the interval we create, based on the significant *t*-test. You should also be able to predict what the formula for calculating the *UL* and *LL* of the CI looks like. Check with Box. 12.11 to make sure you're correct.

Is the null hypothesis mean (zero) within our confidence interval? No, and we could predict that it would not be there because our inferential *t*-test told us there was a meaningful difference between the mean reaction time in the fall and the mean reaction time in the spring.

BOX 12.11   Calculating a confidence interval for a dependent-samples *t*-test

$$\text{CI} = \bar{D} \pm (t_{\text{critical}})(s_{\bar{D}}) = -148.58 \pm (2.201)(35.11)$$

$$UL = 148.58 + (2.201)(35.11) = -71.30$$
$$LL = -148.58 - (2.201)(35.11) = -226.86$$

Think About It...

WHAT DOES A CONFIDENCE INTERVAL REALLY MEAN? (PART 2)

The first time we paused to think about what confidence intervals really mean, we constructed CIs when the population mean and standard deviation were known. What is different about the CIs we construct when we have to estimate σ and the standard error?

The basic format of the CI remains the same: The sample mean, our estimate of μ, still sits in the middle of an interval. And the width of the interval is still dependent on our desired level of confidence (we'll use 95% in this example, again) plus or minus the margin of error (critical value times the standard error or the estimated standard error, as the case may be).

Compare the two formulas shown below:

$$\text{CI} = \bar{X} \pm (z_{\text{critical}})(\sigma_{\bar{X}}) \qquad \text{CI} = \bar{X} \pm (t_{\text{critical}})(s_{\bar{X}})$$
$$n = 10 \; (\sigma = 10) \qquad\qquad n = 10 \; (\hat{s} = 10)$$

1. What is the z_{critical} for a 95% CI? _____

2. What is the t_{critical} for a 95% CI? _____

Suppose I choose a larger sample size ($n = 30$) and recalculated the margin of error for both tests.

3. What is the z_{critical} for the 95% CI when $n = 30$? _____

4. What is the t_{critical} for the 95% CI when $n = 30$? _____

5. What is the z_{critical} for the 95% CI when $n = 61$? _____

6. What about when $n = 121$? _____

7. What is the t_{critical} for the 95% CI when $n = 61$? _____

8. What about when $n = 121$? _____

9. What happens to the width of the CI for the *z*-test as *n* increases? _____

10. What happens to the width of the CI for the *t*-test as *n* increases? _____

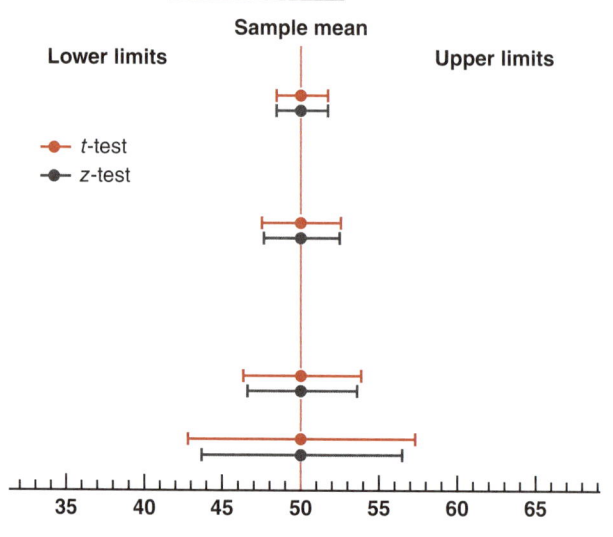

11. Why are the CIs for the *t*-test larger than the CIs for the *z*-test when *n* is small (e.g., *n* = 10 or 30)? _____

12. Why are the CIs for both tests approximately the same size when *n* is large (e.g., *n* = 61 or 121)? _____

Answers to this exercise can be found on page 548.

CheckPoint

Calculate confidence intervals for the following differences between means:

1. $\bar{X}_1 = 642$; $\bar{X}_2 = 589$; $s_{\bar{x}_1-\bar{x}_2} = 4.6$; $df = 8$; $\alpha = .05$ _____
2. $\bar{X}_1 = 17.9$; $\bar{X}_2 = 24.1$; $s_{\bar{D}} = 0.9$; $n_1 = 20$; $N = 40$; $\alpha = .01$ _____
3. $\bar{X}_1 = 6.6$; $\bar{X}_2 = 6.1$; $s_{\bar{x}_1-\bar{x}_2} = 0.11$; $n_C = 10$; $n_E = 10$; $\alpha = .10$ _____
4. $\bar{D} = 42$; $s_{\bar{D}} = 5.8$; $n_2 = 6$; $N = 12$; $\alpha = .05$ _____
5. $\bar{X}_1 = 102$; $\bar{X}_2 = 98.6$; $s_{\bar{x}_1-\bar{x}_2} = 0.38$; $df = 16$; $\alpha = .05$ _____

Answers to these questions are found on page 530.

EFFECT SIZE: HOW DIFFERENT ARE THESE MEANS, REALLY?

Suppose that we got a significant inferential test result. We would then know that the two means we are comparing are meaningfully different and that our IV has had an effect on our DV. We could then calculate a CI around our mean and tell our interested readers about the likely value of the mean of the alternative distribution. Both of these results are interesting, but we're still missing some information. The inferential test that we use to examine the data tells us the probability that the result we got was due to random chance or caused by the IV. The CI tells us what the mean of the true population would probably be. And a new statistic, called **Cohen's *d***, tells us about how much of an effect our IV actually had on the DV—in other words, the **effect size**.

A common error in interpreting the result of a statistical test is to assume the alpha level that defines "different" in our test tells us how big or strong the effect of the IV on the DV actually is. For example, suppose we find that the difference between the means in our two samples is significant at an alpha of .05. Another research team finds that the difference is significant at an alpha of .01. Many students and researchers will describe the result that is significant at the .01 level as indicative of a stronger or "larger" effect than the one that is significant at an alpha of .05. Yet the inferential test can only tell us about the probability that random chance is responsible for any difference we see in the data. It makes no sense to equate that probability with the strength or size of the effect the IV has.

Cohen's *d* A measure of effect size.

effect size A measure of the size of the effect an IV has on the DV. Generally, the ratio of the difference between the estimate of the true population mean and the mean of the null population divided by the variability (or an estimate of the variability) in the null population.

CheckPoint
Answers to questions on page 530

1. 53 ± 10.61 (42.39, 63.61) ☐
2. −6.2 ± 2.57 (−8.77, −3.63) ☐
3. 0.50 ± 0.19 (0.31, 0.69) ☐
4. 42 ± 14.91 (27.09, 56.91) ☐
5. 3.4 ± 0.81 (2.59, 4.21) ☐

SCORE: /5

It is entirely possible that we might have a very small effect that happens to be statistically significant in a very tight, consistent population. At the same time, we might have an IV that causes a very large and strong effect on the DV but is not statistically significant because the population has a great deal of variability. In fact, the strength or the size of the effect that a given IV has on a given DV is dependent on the variability in the data (again!). Let's take a look at the formula for calculating Cohen's *d* in Box 12.12.

BOX 12.12 Basic format of Cohen's *d*

$$\text{Cohen's } d = \frac{\mu_1 - \mu_2}{\sigma_{\bar{X}}}$$

where μ_1 represents the null mean and μ_2 represents the alternative mean

$$d = \frac{\mu_1 - \mu_2}{\sigma} \quad \text{versus} \quad t = \frac{\mu_1 - \mu_2}{\sigma_{\bar{X}}}$$

Many of you may be saying, "Hmmm, this looks really familiar." If so, you are correct in thinking so. The general format for calculating *d* should look familiar; in fact, it should remind you of the basic format for finding *z* or *t*. The difference is in the denominator of the ratio. To find *d*, we are dividing the difference between the two means (or our estimates of these two means from the samples that we took) by the variability in the *null* population.

In the calculation of *t*, we divide that same difference by the standard error—or the variability in the sampling distribution of the means. Cohen's *d* is telling us about the size of the effect the IV had (the difference between the mean assuming the IV had no effect and the mean we actually found after applying the IV). The *t*-statistic is telling us about the relative location of the sample mean in the sampling distribution of the means. The relative location (e.g., close to the hypothesized mean or far away from it) tells us about the probability that the mean is caused by random chance.

The actual format of the formula for effect size changes as a function of the kind of inferential test we're doing. These formulas are listed in Box 12.13. See if you can predict what changes in each formula—I'll bet you already know.

STATISTICS IN CONTEXT

> **BOX 12.13** **Formulas for calculating effect size**
>
> **Single-sample mean**
>
> $$z\text{-test:} \quad d = \frac{\bar{X} - \mu_0}{\sigma}$$
>
> $$t\text{-test:} \quad d = \frac{\bar{X} - \mu_0}{\hat{s}}$$
>
> **Two-sample means**
>
> $$\text{Independent-samples } t\text{-test:} \quad d = \frac{\bar{X}_1 - \bar{X}_2}{\hat{s}}$$
>
> when $n_1 = n_2$ when $n_1 \neq n_2$
>
> $$\hat{s} = \sqrt{n}\left(s_{\bar{X}_1 - \bar{X}_2}\right) \qquad \hat{s} = \sqrt{\frac{\hat{s}_1^2(df_1) + \hat{s}_2^2(df_2)}{df_1 + df_2}}$$
>
> $$\text{Dependent-samples } t\text{-test:} \quad d = \frac{\bar{X}_1 - \bar{X}_2}{\hat{s}}, \text{ where } \hat{s} = \sqrt{N}\left(s_{\bar{D}}\right)$$

Notice that we're using $\hat{s}$ to estimate the population variability. We used $\hat{s}/\sqrt{n}$ to estimate the standard error, so now we're going backward (multiplying by $\sqrt{n}$) to get back to our estimate of population variability.

We found a significant result when we compared the mean of a sample of stressed college students with the population mean (proposed by Hicks et al., 2001) using a z-test. We calculated a CI around our estimate of what the mean sleep time in the stressed population would be, and now we can calculate a d-statistic to see how large the effect of stress was on sleep time. Take a look at Box 12.14 to see the calculation of d.

> **BOX 12.14** **Calculating the size of the effect of stress on sleep time**
>
> $$d = \left|\frac{\bar{X} - \mu_{\bar{X}}}{\sigma}\right| - \left|\frac{5.67 - 6.85}{1.04}\right| = \left|\frac{-1.18}{1.04}\right| = 1.13$$

Notice that we are taking the *absolute value* of the effect size. This should make some sense because a negative effect size would mean that the effect of the variable was less than zero. A variable can have an effect (i.e., the means can differ by some amount) or not (i.e., the means are the same), but that effect cannot be negative. Effect size is an estimate of the amount of variance in a design that is explained, and negative variance is an impossibility.

Now we need to know what this means—does a d of 1.13 mean we have a small, a medium, or a large effect? Jacob Cohen (1969), the author of the d-statistic, proposed the following conventions for interpreting d:

TABLE 12.1 Interpretation of Cohen's *d*

Value of *d*	Interpretation
0.20	Small effect
0.50	Medium effect
0.80 or larger	Large effect

In our study, stress has a large effect on sleep durations. In fact, we can now say quite a bit about the effect of stress on sleep. For example:

In 2001, Hicks et al., proposed that the mean number of hours of sleep seen in the population of college-age students was 6.85 hours with a standard deviation of 1.04 hours. In the present study, we compared the sleep durations of 10 college-age students who had been subjected to stress with the population mean proposed by Hicks et al. We hypothesized that stress would significantly alter mean sleep durations in the stressed sample. A z-test revealed that sleep durations were significantly decreased in the stressed sample ($z(10) = -3.56$, $p < .01$) and that stress produced a large effect ($d = 1.13$). A 95% CI suggested that the mean sleep duration in the population of stressed students would be between 5.31 and 6.21 hours of sleep.

CheckPoint

Use the information provided in the questions below to calculate Cohen's *d* for the following means and differences between the means, and provide interpretations of whether the effect size of the corresponding IVs is "small," "medium," or "large."

1. z-test: $\bar{X} = 18$; $\mu_0 = 17.6$; $\sigma = 1.6$ _____
2. Single-sample t-test: $\bar{X} = 6.4$; $\mu_0 = 5.2$; $\hat{s} = 1.4$ _____
3. Independent-samples t-test: $\bar{X}_1 = 112$; $\bar{X}_2 = 108$; $\hat{s} = 2.6$ _____
4. Dependent-samples t-test: $\bar{X}_1 = 4.2$; $\bar{X}_2 = 3.6$; $\hat{s} = 1.1$ _____

Answers to these questions are found on page 534.

EFFECT SIZE AND ANOVA

We have also discussed the inferential test called an ANOVA, which allows us to compare more than two means simultaneously. The resulting *F*-statistic tells us again whether it is more probable that the results we see in the data are due to random chance or to the effect of the IV. We can also calculate a descriptive statistic that will tell us about the strength of that effect. For an ANOVA, this effect size statistic is called an *eta-squared statistic* (the symbol for eta is η, so the symbol for the eta-squared statistic is $η^2$). The general formula for $η^2$ is shown in Box 12.15. The resulting number is interpreted as the proportion of variability in the dependent variable that can be explained or accounted for by variability in the independent variable.

BOX 12.15 **Eta squared: The effect size statistic**

$$η^2 = \frac{SS_{between}}{SS_{total}}$$

Let's look at an example to help explain. In Chapter 11, we looked at athletes who used different blood-doping techniques to achieve better performance. We discovered a correlation between athletes' endurance scores and the type of blood doping the athletes used. Box 12.16 shows the results of that ANOVA and the calculation of effect size for that comparison.

BOX 12.16 **Blood doping study: ANOVA and effect size source table**

Source	SS	df	MS	F
Between groups	460.15	3	153.38	16.95*
Within groups	144.80	16	9.05	
Total	604.95	19		

*significant at α = .01 level

Calculation of effect size (*f*) Statistic

$$η^2 = \frac{SS_b}{SS_{total}} = \frac{460.15}{604.95} = 0.76$$

A Rule of Thumb for Interpreting $η^2$

0.02 = small effect
0.13 = medium effect
0.26 or larger = large effect

The effect size is fairly large: 76% of the variability in endurance scores (our dependent variable) can be explained by variability in the effects of blood doping. The closer the value of $η^2$ is to 1.00, the stronger the effect of the independent variable.

> **CheckPoint**
> *Answers to questions on page 532*
>
> 1. $d = 0.25$; small ☐
> 2. $d = 0.86$; large ☐
> 3. $d = 1.54$; large ☐
> 4. $d = 0.55$; medium ☐
>
> **SCORE:** /4

So, the type of blood doping used not only was the probable reason for differences in endurance scores but also had a very large effect on the dependent variable. Eta squared is usually reported for simple "balanced" one-way ANOVA designs. The term *balanced* means that you have the same number of participants in each group tested. The advantage of eta squared is that it is very simple to calculate and very easy to interpret. It is, however, a biased measure of effect size because its value depends on the number of effects being compared. If we were doing a two-way ANOVA and divided total variability into more than just one between-group term, the value of eta squared would change. To get around this problem, SPSS calculates *partial eta squared*, or ηp^2. Fortunately, partial eta squared is also very easy to calculate and to interpret, and the formula is shown in Box 12.17.

BOX 12.17 **Partial eta squared**

$$\eta p^2 = \frac{SS_{effect}}{SS_{effect} + SS_{within}}$$

Two things need to be noted here. First, we've switched from SS_b (sum of squares between) to a more general term—SS_{effect}—and you may be wondering why. The reason has to do with the second thing we should notice: When we're doing a one-way ANOVA, there is only one IV. This means that the between-group variability, caused by our one IV, is all represented in one term, SS_b. When we do a two-way ANOVA, we have two IVs, or two "effects," that we want to assess. So, we need to calculate an effect size (ηp^2) for each one. Again, an example will help here. Let's look back at the results of the two-way ANOVA we did in Chapter 11 to examine the data in the Bobo doll study (see Box 12.18).

Suppose we had only the effect of the model as our IV. What would η^2 be? Well, since we have only one "effect," the value of eta squared would be exactly the same as the value of partial eta squared. But when we have more than one IV, and more than one way to divide up the between-group variability, our partial eta squared reflects the changes in between-group sum of squares. The partial eta squared is interpreted in the same way that eta squared is. Notice also that the measures of partial eta squared *do not* sum to 100%.

> **BOX 12.18** **Calculating the effect size for the two IVs in the Bobo doll study**
>
> Effect of the sex of the model
>
> $$\eta p^2 = \frac{SS_{effect}}{SS_{effect} + SS_{within}} = \frac{1,862.45}{1,862.45 + 362.80} = \frac{1,862.45}{2,225.25} = 0.837$$
>
> Effect of the sex of the child
>
> $$\eta p^2 = \frac{SS_{effect}}{SS_{effect} + SS_{within}} = \frac{4,234.05}{4,234.05 + 362.80} = \frac{4,234.05}{4,596.85} = 0.921$$
>
> Effect of the interaction between the two IVs
>
> $$\eta p^2 = \frac{SS_{effect}}{SS_{effect} + SS_{within}} = \frac{7,106.45}{7,106.45 + 362.80} = \frac{7,106.45}{7,469.25} = 0.951$$
>
> | Source | Type III Sum of Squares | df | Mean Square | F | Sig. |
> |---|---|---|---|---|---|
> | IV1 (sex of model) | 1,862.450 | 1 | 1,862.450 | 82.137 | .000 |
> | IV2 (sex of child) | 4,234.050 | 1 | 4,234.050 | 186.728 | .000 |
> | IV1 * IV2 (interaction) | 7,106.450 | 1 | 7,106.450 | 313.405 | .000 |
> | Error | 362.800 | 16 | 22.675 | | |
> | Total | 92,317.000 | 20 | | | |

STATISTICS IN CONTEXT: THE CI VERSUS THE INFERENTIAL TEST

Inferential statistical tests have been in use for a very long time. Whenever scientists wanted to know if two or three means were meaningfully different, out came the appropriate test, and with a few calculations, the researchers had their answer. If the results of the test were significant, the researchers might add to the discussion of the results of an experiment by showing the confidence interval, too. The really zealous researchers began to talk about the size of the effect that a particular IV had as well, but mention of both CIs and *d* were few and far between—until recently.

What changed? Recently, statisticians have begun asking questions about whether we need *both* the CI and the inferential test. Think back to our test of

$\mu = 6.57$

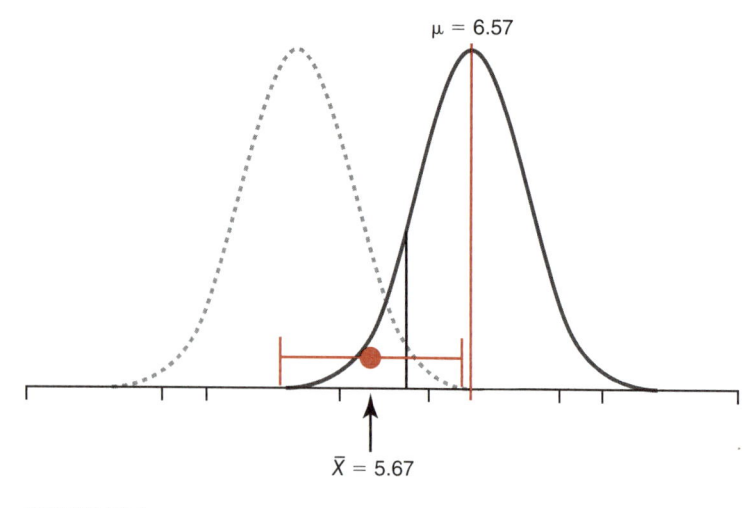

$\bar{X} = 5.67$

FIGURE 12.4
Confidence interval versus the z-test.

stress in sleeping college students. Don't the CI and the inferential test both tell us about the difference between the two means we're comparing? Let's use our study of stressed students as an example.

Suppose we do a side-by-side comparison of the two methods (CI and z, in this example) with our sleeping students. Take a look at Figure 12.4 to see what I'm talking about.

When we converted the difference between the sample mean and the population mean to a z-score, the resulting z was very far out in the left-hand tail of the SDM. In fact, it was much more than 1.65 standard errors below the mean (the dashed gray line)—it was 3.65 standard errors below the mean, firmly in the rejection region. We can say that the sample mean is different from the population mean and reject H_0.

When we built a CI around that mean, we essentially concluded the same thing. Notice that the mean of the known population is not within the CI. In fact we can say that we're 95% confident that the mean of the sample is different from the mean of the population (just what the z-test told us). We are also 95% sure that all those possible sample means between the mean and the one we observed are similar to the population mean. We have the difference between the two point estimates and its significance, along with an interval estimate, available for use.

Some researchers favor the significance test; others prefer the CI. Both tests are testing the null hypothesis, one by determining how often a sample mean would show up in the outer tails of the distribution by random chance alone and the other by determining how often the population mean would show up in the window around the sample mean.

More and more psychology journals are requiring researchers submitting publications to include the CI with the results of their inferential tests. Researchers submitting the results of a study without including the CI will likely be asked to resubmit that manuscript with the CI information included—or simply be flat-out rejected for publication.

Effect size is also being required for publication by more and more APA Style journals. Since the three statistics, when considered together, give a much clearer picture of the results of an experiment, it just makes sense to include them all. Cover all your bases, and get into the habit of including all three sources of information in your results sections.

SUMMARY

In this chapter, we discussed two additional statistics that can be used to help the researcher interpret the results of a study.

The first statistic, the confidence interval (CI), describes a range of values that should encompass the true mean of the alternative hypothesis once the null hypothesis has been rejected.

The second statistic, the effect size (usually measured by Cohen's *d* statistic), describes the size or strength of the effect that the IV has on the DV.

The inferential test statistic, the CI, and *d* all provide the researcher with information concerning the probability that a given result was caused by a hypothesized IV (inferential test statistic), the location of the true population mean (CI), and the estimated size of the effect of the IV (effect size).

TERMS YOU SHOULD KNOW

alternative population, p. 517
confidence, p. 529
confidence interval (CI), p. 517
Cohen's *d*, p. 514
effect size, p. 529

interval estimate, p. 514
null population, p. 516
precise, p. 514
point estimate, p. 514

GLOSSARY OF EQUATIONS

Formula	Name	Symbol
$CI = \bar{X} \pm (z_{critical})(\sigma_{\bar{X}})$	Confidence interval for use with a *z*-test	CI
$CI = \bar{X} \pm (t_{critical})(s_{\bar{X}})$	Confidence interval for use with a single-sample *t*-test	
$CI = \bar{X} \pm (t_{critical})(s_{\bar{X}_1 - \bar{X}_2})$	Confidence interval for use with an independent-samples *t*-test	
$CI = \bar{D} \pm (t_{critical})(s_{\bar{D}})$	Confidence interval for use with a dependent-samples *t*-test	
$d = \dfrac{\mu_1 - \mu_2}{\sigma_{\bar{X}}}$	General formula for effect size measure	*d*
$d = \dfrac{\bar{X} - \mu_0}{\sigma}$	Effect size for a *z*-test	
$d = \dfrac{\bar{X} - \mu_0}{\hat{s}}$	Effect size for a single-sample *t*-test	

Formula	Name	Symbol
$d = \dfrac{\bar{X}_1 - \bar{X}_2}{\sqrt{n}\left(s_{\bar{X}_1 - \bar{X}_2}\right)}$	Effect size for an independent-samples *t*-test (equal *n*'s)	
$d = \dfrac{\bar{X}_1 - \bar{X}_2}{\sqrt{\dfrac{\hat{s}_1^2\,(df_1) + \hat{s}_2^2\,(df_2)}{df_1 + df_2}}}$	Effect size for an independent-samples *t*-test (unequal *n*'s)	
$d = \dfrac{\bar{X}_1 - \bar{X}_2}{\sqrt{N}\,(s_{\bar{D}})}$	Effect size for a dependent-samples *t*-test	
$f = \dfrac{\sqrt{\dfrac{K-1}{N_{total}}(MS_{between} - MS_{within})}}{\sqrt{MS_{within}}}$	Effect size for a one-way ANOVA	f
$\eta^2 = \dfrac{SS_{between}}{SS_{total}}$	Eta squared: Effect size for a one-way ANOVA	η^2
$\eta p^2 = \dfrac{SS_{effect}}{SS_{effect} + SS_{within}}$	Partial eta squared: Effect size for a two-way ANOVA	ηp^2

WRITING ASSIGNMENT

In question 5 of the *Practice Problems* below, which concerns osteoprotegerin levels in individuals with schizophrenia, you will be asked to conduct an inferential test, calculate a CI, and measure the effect size. Write a short sentence or two about the differences between the CI and the inferential test. Do they tell you the same thing? Do you think we need to do both? What are the benefits and drawbacks of calculating both a CI and conducting an inferential hypothesis test?

PRACTICE PROBLEMS

1. To calculate a CI, you must use a two-tailed critical value appropriate to the test (*z* or *t*) that you are using. Explain why a two-tailed value must be used. Feel free to use an example.

2. Briefly, what is the difference between the null population and the alternative population?

3. Suppose the mean of a population is 65 and the standard deviation is 12. Calculate the following CIs around a sample mean of 50. (Use the table of *z*-values in Appendix A for your calculations, and think carefully about how you might find a 68% CI—we discussed a rule that will help you here.)
   a. 95%
   b. 99%

c. 68%
d. 50%
e. 34%
f. 60%
g. What happens to the width of the CI as the critical z-value changes?

4. Developmental psychologists are interested in finding out if we are more affected by events that occur in early childhood than by those that occur in adolescence or adulthood. Suppose we conducted a study of the effects of experiencing a traumatic event (loss of a parent or caretaker) as a child (younger than 12 years old) and as an adult (30 years or older). The effects will be assessed based on levels of serum cortisol, a hormone released by people experiencing stress (high levels of cortisol can have serious physical consequences).

Table 12.2 Cortisol levels (in mg/dl) in children and adults after a traumatic event

Children	Adults
5.89	6.23
4.55	6.21
5.10	23.24
4.64	10.09
3.37	4.02
6.14	9.42
3.99	21.91
5.99	19.31
4.44	6.63
6.23	3.61

Answer the following questions about the data:
a. What test should we do, and why? Choose from the tests we've discussed so far: z-test, single-sample t-test, independent-samples t-test, dependent-samples t-test, and one-way ANOVA.
b. What does a typical participant look like? Do children differ from adults in their experience of a traumatic event?
c. What conclusions can we draw from the results of the inferential test?
d. If appropriate, calculate a 99% CI and determine the effect size for our data.

5. Several studies have found that mental disorders are associated with elevated levels of inflammatory markers. Hope et al. (2010) found that serum levels of an inflammatory marker called osteoprotegerin (OPG) was elevated in patients who had been

diagnosed with severe mental disorders. The OPG levels were measured in individuals diagnosed with different types of mental disorder. To make things a bit easier, we will compare individuals with schizophrenia and healthy volunteers. We'll further suppose that we know what the average level of OPG should be in a healthy person with no mental illness. The data are shown in Table 12.3 below.

Table 12.3 Levels of OPG (in ng/ml) in healthy controls and individuals with schizophrenia

Healthy Controls	Individuals with Schizophrenia
$\mu = 2.52$	3.00
$\sigma = ?$	3.90
$n = 10$	2.81
	2.37
	4.93
	3.75
	4.94
	3.73
	2.06
	3.14

Answer the following questions about the data:
  a. What test should we do, and why? Choose from the tests we've discussed so far: z-test, single-sample t-test, independent-samples t-test, dependent-samples t-test, and one-way ANOVA.
  b. What conclusions can we draw from the results of the inferential test?
  c. If appropriate, calculate a 95% CI and determine the effect size for the data.

6. Studies of the level of cholesterol in patients with depression have suggested that those who attempt suicide tend to have lower-than-normal levels of cholesterol (measured in spinal fluid). The data in Table 12.4 illustrate the results seen in a study of the

Table 12.4 Cholesterol levels (in ng/ml) in suicide attempters and healthy controls

	Suicide	Control
n	12	10
Mean	138.14	144.40
Median	135.56	133.32
Mode	102.57	108.63
Standard deviation $\left(s_{\bar{X}_1 - \bar{X}_2}\right) = 11.95$	26.37	29.70

relationship between cholesterol level and suicidality in adolescents done by Plana et al. (2010). Total cholesterol levels in two groups of adolescents were compared.

Answer the following questions about the data:
a. What test should we do, and why? Choose from the tests we've discussed so far: z-test, single-sample t-test, independent-samples t-test, dependent-samples t-test, and one-way ANOVA.
b. What conclusions can we draw from the results of the inferential test?
c. If appropriate, calculate a 95% CI and determine the effect size for our data.

7. The image here depicts the 95% confidence intervals for 50 samples drawn at random from the population. The dashed red line illustrates the mean of the population.
   a. What percentage of the samples will include the mean of the population?
   b. What percentage of the samples will indicate that the mean of the sample is significantly different from the mean of the population?
   c. Draw an arrow pointing to the samples that would most likely be in the rejection region if the distribution is alpha = .05.

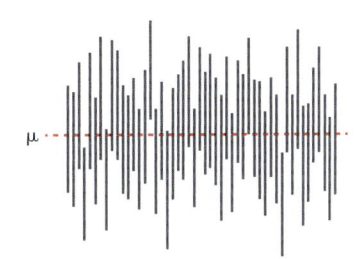

8. Did you know that chimpanzees have a measurable personality? Several different measures are designed for use by animal caretakers to assess and describe personality in the chimp. Lilienfeld, Gershon, Duke, Marino, and de Waal (1999) developed a measure of abnormal personality, or psychopathology, for use with chimpanzees. They called the measure the Chimpanzee Psychopathy Measure, or CPM. In their study of 34 chimpanzees, they found that CPM scores ranged from a low of 47.96 to a high of 90.72, with higher scores indicating more psychopathology. The mean score on the CPM was 68.17 ($SD = 11.06$). Let's replicate part of their study and compare CPM scores for male and female chimpanzees. Table 12.5 shows a summary of the data collected by the researchers in this study.

Table 12.5 Male versus female chimpanzee CPM scores

Males	Females	t-Test Result	Effect Size
Mean CPM = 72.98	Mean CPM = 65.87	$t(32) = 1.81, p < .08$	.66

Answer the following questions about the data:
a. Write a short interpretation of the data.
b. What does the result of the t-test tell you?
c. What does the result of the effect size measure tell you?
d. Do these two tests agree or disagree?

9. Construct a 95% CI and write a short sentence of interpretation for the data in sample A shown below.

Sample A
1
2
3
4
5
6
7

10. Construct a 99% CI and write a sentence of interpretation for the data in sample B shown below.

Sample B
1
3
5
7
9

11. Given the sample data

$$\Sigma X^2 = 1{,}980;\ \Sigma X = 156;\ n = 13$$

construct a 95% confidence interval and write a sentence of interpretation.

12. A random sample of 64 first graders had mean of 14 on a test of mathematical comprehension ($SD = 2.00$). Construct a 95% confidence interval around the mean, and explain in a sentence what it means in terms of first graders and their mathematical comprehension.

13. Many manufacturers of light bulbs advertise their 75-watt bulbs as having a life of 750 hours (about a month). Two students wired up 12 bulbs and recorded the elapsed time until they burned out. The students obtained the following statistics:

$$\Sigma X = 8{,}820;\ \Sigma X^2 = 6{,}487{,}100;\ n = 12$$

Perform a *t*-test, and write a conclusion about the manufacturers' claim. Calculate the 95% CI around the sample mean, and write a sentence of interpretation.

14. The "Personal Control" (PC) scale is a test that measures the degree to which people feel in control of their personal lives (the higher the score, the more control people feel). The mean PC scale score for the population of all college students is 51. What about students

542   STATISTICS IN CONTEXT

who are in academic difficulty? How do they feel about the control of their personal lives? The hypothetical data shown in Table 12.6 are PC scores for students in academic difficulty. Compare the mean of this sample with the average PC score in the population using the appropriate statistical test. Write a conclusion about the feelings of personal control among students in academic difficulty. Calculate a 95% CI and effect size, and interpret the results of your tests.

Table 12.6 PC scores for 12 students in academic difficulty

ID	PC Score
1	48
2	44
3	53
4	35
5	58
6	42
7	55
8	37
9	50
10	46
11	40
12	32

15. Humans can learn by watching and imitating a "model," but researchers have argued about whether animals can do the same. For example, Thorndike (1911) concluded that animals were incapable of learning by imitation, but Hobhouse (1901) reported that cats, dogs, otters, elephants, monkeys, and chimpanzees could all learn by imitation. Suppose the following study was conducted to investigate this question. One group of hungry cats was shown food being obtained from underneath a vase. Another group was not shown this scene, although there was food under the vase. Shortly afterward, the time (in seconds) required to upset the vase and find the food was recorded for each animal. The data are shown in Table 12.7. Compare the times for the two samples using the appropriate inferential test. Calculate a 95% CI and effect size. Which of these two theorists does the data support?

Table 12.7 Time (in seconds) needed to find food

Shown	Not Shown
18	25
15	22

Table 12.7 Continued

15	21
12	19
11	16
9	15

16. For samples a and b below, calculate the value of *t* (using a single-sample *t*-test) and determine whether the difference between the sample mean and the population mean is statistically significant (using a one-tailed test). For which sample is the value of *t* larger? Calculate effect size for both a and b, and interpret this number.

Sample a	Sample b
$\bar{X} = 3.725$	$\bar{X} = 5.64$
$\mu = 2.00$	$\mu = 2.00$
$\hat{s} = 15$	$\hat{s} = 15$
$n = 399$	$n = 89$

17. For samples a and b, calculate the value of *z* and determine whether the difference between the sample mean and the population mean is statistically significant. For which sample is the value of *z* larger? Calculate the effect size for both a and b, and interpret this number.

Sample a	Sample b
$\bar{X} = 62$	$\bar{X} = 62$
$\mu = 70$	$\mu = 70$
$\sigma = 8$	$\sigma = 8$
$n = 16$	$n = 400$

18. Use the data in the source table shown below to calculate effect size.

Source	SS	df	MS	F
Between	392	2	196	4.00
Within	833	17	49	
Total	1,225	19		

19. Using the data in the source table shown below, calculate *F* for the main effects and the interaction. Then, calculate effect size for both main effects and the interaction.

544 STATISTICS IN CONTEXT

Source	SS	df	MS	F
Music	536.34	3	178.78	
Water	34.32	2	17.16	
Music × water	22.07	6	3.68	
Within	1,220.75	48	25.43	
Total	1,813.48	59		

20. Ji received a grade of 57 on his paper. Ji knows the grades for the other students in his tutorial on this paper are 82, 71, 59, 92, 67, 72, 70, 68, 65, 76, 40, and 65.
    a. Construct a 95% confidence interval around the average paper grade for the class as a whole.
    b. State the upper and lower limits of the 95% confidence interval. Does Ji's grade fall within the interval estimate?

21. How damaging are negative stereotypes surrounding old age? A researcher studying the effect of negative subliminal messaging on older adults devises a study on the effects of age priming on walking speed. Age priming is a technique in which experiment participants are "primed" with stereotypes surrounding old age. The control group and the experimental group each comprise 10 randomly selected participants. The alternative hypothesis states that the walking speed of people who receive negative age priming will be slower than the walking speed of those who are not primed. When the participants arrive, they are invited into a room under the pretense that they will be briefed about a reaction-time study they will supposedly be participating in. However, the reaction-time study is a diversion, and the briefing the participants receive is a fiction. The true purpose of the briefing is to prime the experimental group by uttering a myriad of words negatively associated with old age. After receiving their fictitious briefings, participants are asked to walk down the hallway to another room where their reaction time will be measured. What is really measured is the time it takes participants to walk down the hallway. The participants' walk times are reported in Table 12.8 below.

Table 12.8 Walking times (in seconds) for primed and unprimed participants

Control group	Experimental group
24	21
20	32
15	26
22	41
19	18
31	30
23	22

Table 12.8 Continued

16	23
38	37
14	30
$\Sigma = 222$	$\Sigma = 280$
$n = 10$	$n = 10$

a. Is there a statistically significant difference between the mean walk times of the control and experimental groups at the .05 level?
b. Is there a statistically significant difference between the mean walk times of the control and experimental groups at the .01 level?
c. Construct 95% and 99% confidence intervals around the difference between the mean walk times of the control and experimental groups.

22. Complete the crossword shown below.

546 **STATISTICS IN CONTEXT**

ACROSS	DOWN
2. Modified eta squared effect size measure is called this	1. A proven statement that is used as proof of another statement
5. First name of Neyman's collaborator	3. CIs can be used instead of this kind of statistical test
7. The population mean hypothesized about in H_1	6. Statistician who developed the idea of CI
8. (2 words) Interval estimate of a population parameter, usually the population mean	6. (2 words) $d = \dfrac{\mu_1 - \mu_2}{\sigma}$
9. (2 words) Name of ANOVA effect size measure	10. Type of estimate with high precision but low confidence
10. The accuracy of your estimate is called called this	11. Developer of the *d*-statistic that measures effect size

Think About It...

WHAT DOES A CONFIDENCE INTERVAL REALLY MEAN? (PART 1)

SOLUTIONS

1. Can we reject H_0? **No. H_0 can be rejected only when μ falls outside the CI (and therefore in the rejection region); we cannot reject it here.**
2. What does rejecting H_0 tell us about the relationship between $\bar{X}$ and μ? **They are very far apart—so far apart that we have to assume that more than just random chance is operating in our distributions.**
3. How many of the CIs we construct should contain the population mean? **95% of 20 = 19.**
4. How many of the CIs we construct should NOT contain the population mean? **5% of 20 = 1.**
5. How often can we reject H_0? **Defined by α, so 5% of the time, or about once in these 20 samples.**
6. Finish the table shown below. (I've highlighted the cases where μ is outside the CI.)

$\bar{X}$	CI	$\bar{X}$	CI	$\bar{X}$	CI	$\bar{X}$	CI
48	44.08–51.92	49	45.08–52.92	47	43.08–50.92	51	47.08–54.92
47	43.08–50.92	53	49.08–56.92	50	46.08–53.92	51	47.08–54.92
50	46.08–53.92	43	**39.08–46.92**	53	49.08–56.92	47	43.08–50.92
53	49.08–56.92	50	46.08–53.92	49	45.08–52.92	50	46.08–53.92
58	**54.08–61.92**	52	48.08–55.92	51	47.08–54.92	49	45.08–52.92

7. Do these sample means seem like reasonable estimates of μ (if only random chance is operating)? **Yes. Most are fairly close to μ, which we would expect if the population were normally distributed.**
8. How many intervals contained μ? **18, which is one less than we predicted.**

CHAPTER 12 Confidence Intervals and Effect Size: Building a Better Mousetrap

9. **Why did we not see what we were confident of seeing?** Two reasons. The first is that we have only 20 samples here, which isn't a very long run (think about the Law of Large Numbers and our discussion of how long the long run needs to be), and the second is random chance. We've set up our intervals so that we will probably see one μ outside of the CI in 20 CIs, but again, only probably. If we took many more samples (a really long run), we'd see that μ was contained within 95% of the CIs built.

WHAT DOES A CONFIDENCE INTERVAL REALLY MEAN? (PART 2)

SOLUTIONS

1. What is the $z_{critical}$ for a 95% CI? **1.96.**
2. What is the $t_{critical}$ for a 95% CI? $df = 9$ so $t_{crit} = 2.62$.
3. What is the $z_{critical}$ for the 95% CI when $n = 30$? **1.96.**
4. What is the $t_{critical}$ for the 95% CI when $n = 30$? $df = 29$, so $t_{crit} = 2.045$.
5. What is the $z_{critical}$ for the 95% CI when $n = 61$? **1.96.**
6. What about when $n = 121$? **1.96.**
7. What is the $t_{critical}$ for the 95% CI when $n = 61$? **2.00.**
8. What about when $n = 121$? **1.98.**
9. What happens to the width of the CI for the z-test as n increases? **The larger the sample, the narrower the CI.**
10. What happens to the width of the CI for the t-test as n increases? **The larger the sample, the narrower the CI.**
11. Why are the CIs for the t-test larger than the CIs for the z-test when n is small (e.g., $n = 10$ or 30)? **Because the margin of error for the z-test is based on a constant critical z-value while the CIs for the t-test are based on an estimate of σ. And, small samples create less precise estimates of μ.**
12. Why are the CIs for both tests approximately the same size when n is large (e.g., $n = 61$ or 121)? **Again, sample size matters a lot here. Large samples provide better estimates of the population parameters than do small samples. In fact, the critical value of t will exactly equal the critical value of z when n is very, very large (compare the critical values of t and z when n = infinity).**

REFERENCES

Cohen, J. (1992). A power primer. *Psychological Bulletin, 112*, 155–159.

Distefan, J. M., Pierce, J. P., & Gilpin, E. A. (2004). Do favorite movie stars influence adolescent smoking initiation? *American Journal of Public Health, 94*(7), 1239–1244.

Hicks, R. A., Fernandez, C., & Pellegrini, R. J. (2001). Self-reported sleep durations of college students: Normative data for 1978–79, 1988–89 and 2000–01. *Perceptual and Motor Skills, 93*, 139–140.

Hobhouse, L.T. (1901). *Mind in evolution*. New York: Macmillan.

Hope, S., Melle, I., Aukrust, P., Agartz, I., Lorentzen, S., Eiel, N., Steen, S. D., Ueland, T., & Andreassen, O. A. (2010). Osteoprotegerin levels in patients with severe mental disorders. *Journal* of *Psychiatry and Neuroscience, 35*(5), 304–310.

Lilienfeld, S. O., Gershon, J., Duke, M., Marino, L., & de Waal, F. B. M. (1999). A preliminary investigation of the construct of psychopathic personality (psychopathy) in chimpanzees (*Pan troglodytes*). *Journal of Comparative Psychology, 113*(4), 365–375.

Neyman, J. (1937).Outline of a theory of statistical estimation based on the classical theory of probability. *Philosophical Transactions of Royal Society of London, Series A, 236*(767), 333–380.

Plana, T., Gracia, R., Mendez, I., Pintor, L., Lazaro, L., & Castro-Fornieles, J. (2010). Total serum cholesterol levels and suicide attempts in child and adolescent psychiatric inpatients. *European Journal of Child and Adolescent Psychiatry, 19*, 615–619.

Thorndike, E. L. (1911). *Animal intelligence*. New York: Macmillan.

CHAPTER THIRTEEN

CORRELATION AND REGRESSION: ARE WE RELATED?

Statistics can be made to prove anything—even the truth.
—AUTHOR UNKNOWN

Everyday Statistics

ANTS ARE PREDICTING THE FUTURE!

Well, maybe not the future exactly, but they might be able to predict earthquakes.

Researchers in Germany discovered a correlation between the behavior of red wood ants and earthquake activity (Berberich, Berberich, Grumpe, Wohler, & Schreiber, 2013). The researchers set up cameras to monitor the nests, or "mounds," of two colonies of ants, 24/7, over a three-year period between 2009 and 2012. The mounds were in a region of Germany that is "seismically active," experiencing roughly 100 earthquakes per year.

Red wood ants have a predictable pattern in their daily behavior that, if you were to chart it, would be roughly M-shaped: Their activity rises between dawn and midday, then falls; it rises again in the late afternoon, then falls again at night. Just before an earthquake, however, ant activity fell precipitously, and this suppression of activity lasted until the day after the quake.

Seismologists have been searching for years for a marker of earthquake activity. Maybe the lowly red wood ant is the solution to this age-old problem. Chapter 13 will discuss correlations and regression and how, like the red wood ant (maybe), these statistics can let us predict the future.

OVERVIEW

THE CORRELATION COEFFICIENT
TESTING HYPOTHESES ABOUT *R*
THE LEAST SQUARES REGRESSION LINE, A.K.A. THE LINE OF BEST FIT
SOME CAUTIONARY NOTES
COEFFICIENT OF DETERMINATION
STATISTICS IN CONTEXT: CORRELATIONS AND CAUSATION

LEARNING OBJECTIVES

Reading this chapter will help you to . . .

- Describe the results of Galton's study of how "mother" and "daughter" sweet pea sizes are co-related (correlated). (Concept)

- Understand the least squares regression line and regression to the mean. (Concept)

- Recognize positive and negative correlations and strong and weak correlations. (Concept)

- Calculate the Pearson product–moment correlation coefficient (r) and the coefficient of determination (r^2). (Application)

- Calculate the regression equation to predict a *Y*-value from an *X*-value. (Application)

- Explain the differences between linear and curvilinear relationships. (Concept)

THE CORRELATION COEFFICIENT

The person responsible for seeing a useful statistic in vegetables and their offspring was Sir Francis Galton. His goal was to bring careful measurement and statistics to the biological sciences, in particular to help scientists describe the characteristics of heredity. One of his more enduring contributions to statistics—accomplished through the study of sweet pea plants (see *The Historical Context*)—was the idea that it should be possible to calculate a statistic that measures the degree to which two variables are *correlated*, or dependent on one another. His work, in collaboration with Karl Pearson, led to the description of the *Pearson product–moment correlation coefficient*, which is commonly known by its simpler name, the **correlation coefficient (r)**.

Correlation coefficients have become quite popular and are used in just about every discipline you can name, from psychology to literature and from economics to history. For example, in oncology (the study of cancer), several researchers have shown a strong positive correlation between height and the risk of developing cancer for both men and women (see Green et al., 2011). Let's examine this a bit more closely. First, what does the statement that height and risk of cancer are related to one another, both strongly and positively, mean? If you're tall, should you be worried?

Pearson product-moment correlation coefficient [or correlation coefficient(r)] A statistic that measures the degree of relationship or dependence between two variables, X and Y. The coefficient can vary between −1.00 (a perfect negative correlation) and +1.00 (a perfect positive correlation). An r value of zero (0) indicates that X and Y are independent of one another, and are not correlated.

The Historical Context

STATISTICS AND SWEET PEAS

Sir Francis Galton (see Figure 13.1) was a cousin to Charles Darwin, and he shared Darwin's interest in biology, particularly in heredity. He was wealthy enough not to need a paying job, and this enabled him to follow his fancy almost anywhere he wanted to go.

Galton was fascinated by heredity and what we could, or could not, inherit from our parents, grandparents, and so on. He went looking for a model organism to study and settled on sweet pea plants—first, because they were abundant and easy to grow from seeds, and second, because they could self-fertilize, meaning that each daughter plant had only one parent. With sweet pea plants, Galton didn't have to figure out how much of a genetic contribution to the characteristics of the plant each parent had made.

In 1875, Galton sent packages of sweet pea seeds to seven friends. The seeds varied in size from package to package. His friends planted the seeds, collected the seeds from these second-generation plants, and mailed them back to Galton. Galton then made a plot (called a *scatterplot*) showing the mean size of the mother (first-generation) peas on the *x*-axis and the mean size of the daughter seeds from a given mother on the *y*-axis (both in 1/100ths of an inch). The resulting graph is shown in Figure 13.2.

Each dot on the graph represents the mean size of a mother seed and the mean size of the resulting daughter seeds grown from that mother. Since all of the mother seeds in a given bag were the same size, calculating the mean and standard deviation of the mother seeds was very easy.

Galton immediately recognized that his scatterplot contained a great deal of information. First, he saw a general pattern in the data: As the size of the mother seed increases, the mean size of the daughter seed also increases. It isn't a perfect relationship—notice that if the mother seed was 0.16 inch wide, the average size of the daughter seed was about 0.16 inch wide as well, but when you move up in mother size to 0.17 inch, the average size of the daughter seeds *decreased*. Even so, Galton could describe a general trend upward in the set of means. To do so, he connected the means with a dashed line. This line wavers up and down as the means change, but generally, it rises from lower left to upper right.

Then, Galton drew a straight line that he believed did the best job of describing the relationship between the two variables (the red line [my emphasis] on the graph). Galton determined that the size of the mother pea and the size of her daughters were *co-related* (correlated). In other words, they were related to one another in a very specific way that could be described with a single, straight line that he called the *regression line*.

Galton also plotted the overall mean parent size (the vertical dotted line rising from the *x*-axis) and the overall daughter size (the horizontal dotted line extending from the *y*-axis). When he did this, he noticed something else about the data. The mothers at either extreme (either very small mother seeds or the very large ones) had offspring that were less variable than did the mother seeds in the middle range of the *x*-axis. Figure 13.3 illustrates what Galton saw with a simpler data set (taken from Stanton, 2001).

This graph shows the size of two daughter seeds for each mother (the squares in the graph). The black dots connected by the red line are once again the means. The dashed line rising from the *x*-axis shows the overall average size of a mother seed (0.10 inch), and the dashed line extending horizontally from the *y*-axis shows the overall average size of an offspring pea (about 0.09 inch). Notice that the variability of the offspring of mothers at either extreme of the data is less than the variability of the offspring of mothers from the middle. The individual offspring at the extreme edges of the distribution are closer in size to their overall mean seed size than are the offspring of mother seeds that are in the middle. Galton described the tendency for variability at the extreme ends of a distribution to resemble the mean measurement as

FIGURE 13.1
Sir Francis Galton.

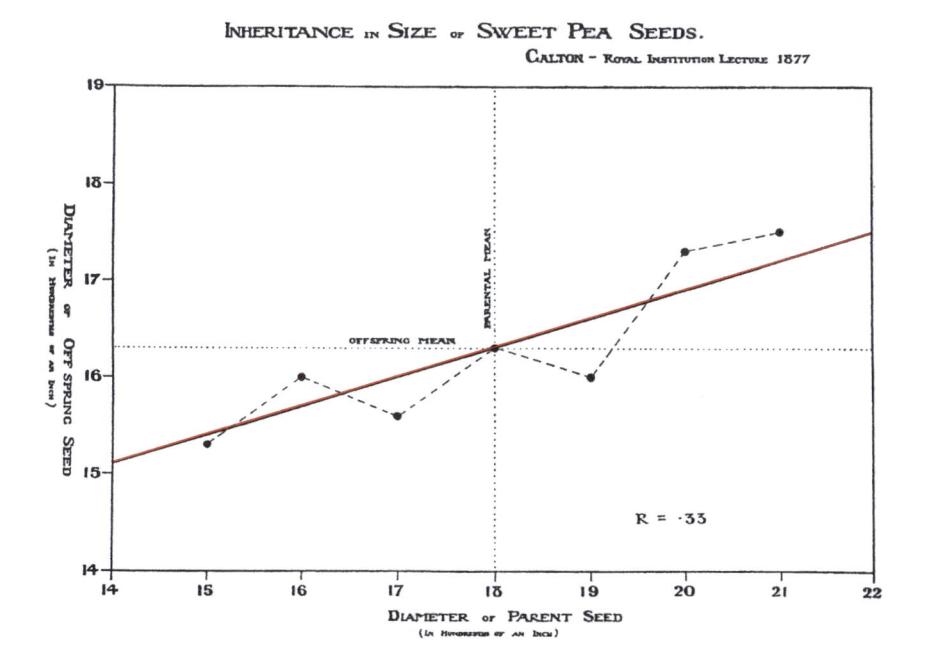

FIGURE 13.2
Galton's original scatterplot (to the nearest 0.01 inch)

CHAPTER 13 Correlation and Regression: Are We Related? 553

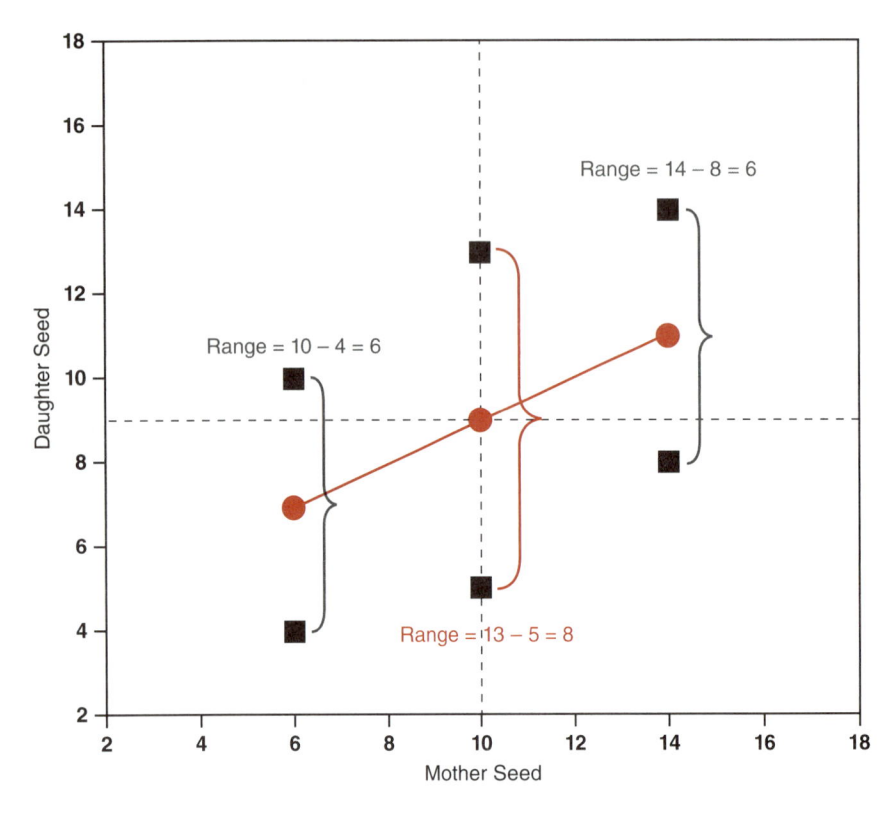

FIGURE 13.3
Regression to the mean.

regression to the mean, and he also determined that it was based on variability in the data.

Galton was not a mathematician, and although he saw these relationships in the data and recognized that a statistic describing the correlation between two variables could be calculated, he left the actual development of the mathematics behind the correlation statistic to his colleague, Karl Pearson. As a result, the statistic Galton envisioned isn't named after Galton but after Pearson. The complete and formal name for the statistic is the Pearson product–moment correlation coefficient, or simply the correlation coefficient, symbolized by the italic letter r.

POSITIVE AND NEGATIVE CORRELATIONS

correlation The degree to which two variables (X and Y) are dependent on one another; the degree to which two variables (X and Y) vary together.

positive correlation A relationship between variables where an increase in the value of X tends to be paired with an increase in the value of Y.

A **correlation** can be described as either *positive* (*direct*) or *negative* (*indirect*). Figure 13.4 shows positive and negative correlations—how would you describe the difference between the two?

When you have a **positive correlation**, or a positive relationship between the two variables, high values on the *x*-axis variable (height, for example) tend to be paired with high values on the *y*-axis variable (risk of cancer); at the same time, low *X*-values tend to be paired with low *Y*-values (check out the dashed lines on Figure 13.4A). If you were to draw a straight line that describes the overall relationship between the *X* and *Y* variables, the line would start in the lower left-hand corner of the graph and rise to the upper right-hand side.

A classic example of a positive correlation comes from Galton's *anthropometric* studies (literally, "measuring humans"). Galton noticed that the heights of parents and their children tend to be positively related to one another. The taller the parents are, the taller their offspring tend to be. This trend does not simply increase upward

554 STATISTICS IN CONTEXT

or downward to infinity (where tall parents tend to have tall children who in turn have even taller children, who have even taller children, etc.) because of **regression to the mean**. At some point, very tall parents will have offspring that are shorter than they are, as the average height of the next generation tends to regress toward the mean height of adults.

Other examples of positive correlations, to name just a few, include the following:

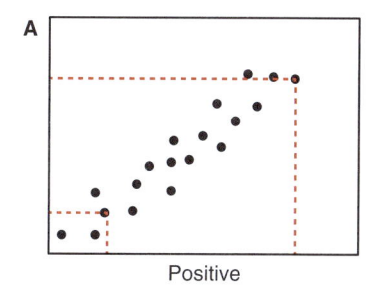

FIGURE 13.4
Two directions for correlations.

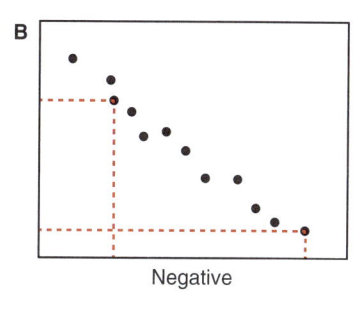

- Yearly income and number of years of education (the more education one has, the higher one's yearly income tends to be).
- Number of calories ingested during the day and body weight (the more you eat, the more you weigh).
- Demand for electricity and the heat of the day (demand for electricity tends to go up as the temperature outside goes up).
- Cancer and height: the taller a person is, the higher their risk of developing cancer.

regression to the mean The phenomenon, first described by Galton, that variables at the extreme first measured will tend to be closer to the mean value of that variable at the second measurement.

In contrast, in a **negative correlation**, low values on the *x*-axis variable tend to be paired with high values on the *y*-axis variable, and vice versa (shown in Figure 13.4B). The straight line drawn to describe this relationship would fall from the upper left-hand corner to the lower right-hand corner. I'm sure you can think of a few examples of negative correlations on your own, but here are some to get you started:

negative correlation A relationship between variables where an increase in the value of *X* tends to be paired with a decrease in the value of *Y*.

- The age of a car and its value at resale (the older your car is, the less you tend to get for it at resale).
- Performance on a statistics test and the number of hours of sleep the night before (scores on the test tend to be higher for students who got a good night's sleep).
- The heat outside and the number of layers people wear (the hotter it is, the fewer articles of clothing we tend to wear).

WEAK AND STRONG CORRELATIONS

Correlations are also described in terms of the *strength* of that relationship—strong, moderate, weak, or nonexistent. Figure 13.5 shows examples of positive and negative relationships of various strengths.

How would you describe the difference between a strong correlation and a weak one? Mathematically, Pearson's correlation coefficient can take on any value between −1.00 and +1.00 (an *r*-value of zero indicates no correlation at all between the two variables). An *r*-value of either −1.00 or +1.00 indicates a perfect relationship between *X* and *Y* (see Figure 13.5, G and H).

It's very easy to find a straight line that describes a perfect relationship because the points in the scatterplot of this relationship fall in a straight line. What this means in terms of the data is that for every unit of increase in the value of the X variable, there is a constant increase (or decrease) in the value of the Y variable. As the relationship decreases in strength, the value of r moves away from perfection, towards zero. The points on the scatterplot also change: They spread out. There is more variability in the data, and less consistent change in Y as X changes when the relationship is less than perfect. Traditionally, correlations of −.80 or +.80 or more are defined as indicating a strong relationship (negative or positive), while correlations of less than −.50 or +.50 indicate moderate to weak relationships.

When the r-value gets very close to zero, it is very difficult to find a single line that will describe the relationship between X and Y. In fact, a perfectly horizontal line and a perfectly vertical line describe the relationship in Figure 13.5E equally well. The horizontal line describes a relationship where the value of Y does not change at all as the value of X changes. The vertical line describes a relationship where the value of X does not change at all as the value of Y changes. Obviously, there is no relationship between X and Y in either of these cases.

Galton recognized that the strength of the relationship depended on variability. Pearson determined that it would be possible to calculate the value of the correlation coefficient that described the relationship between X and Y using a formula that includes measures of the variability of both X and Y. Box 13.1 shows two formulas for calculating r. Look for the standard deviation in both X and Y in both of these formulas.

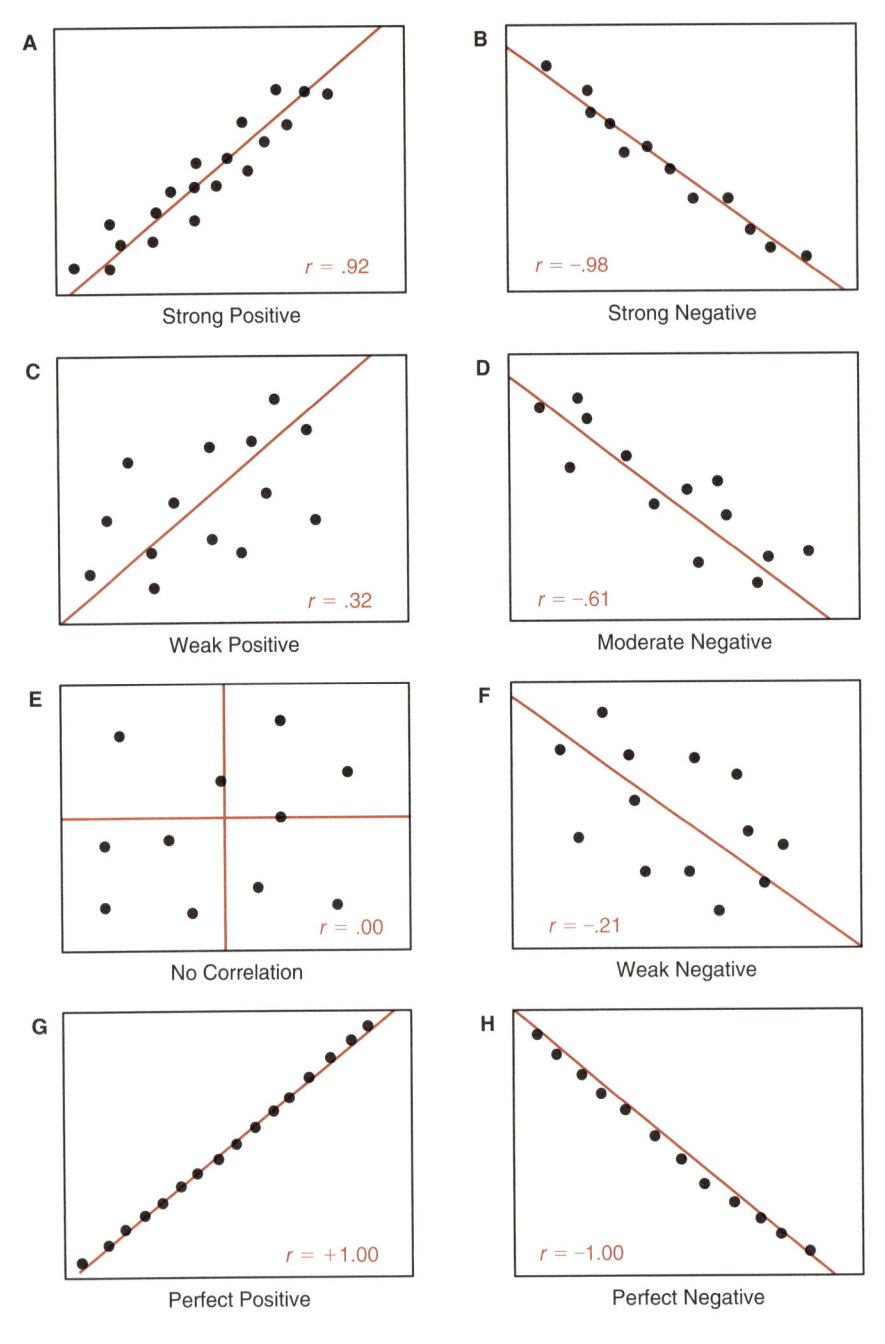

FIGURE 13.5
The strength of the correlation.

BOX 13.1   Two formulas for calculating *r*

The definition formula

$$r = \frac{\frac{\sum XY}{N} - (\bar{X})(\bar{Y})}{(S_X)(S_Y)}$$

XY = each X value times its paired Y value
$\bar{X}$ = mean of the X values
$\bar{Y}$ = mean of the Y values
S_X = standard deviation of X-values
S_Y = standard deviation of Y-values

The "raw data" formula

$$r = \frac{N\sum XY - (\sum X)(\sum Y)}{\sqrt{\left[N\sum X^2 - (\sum X)^2\right]\left[N\sum Y^2 - (\sum Y)^2\right]}}$$

When you don't know the means and standard deviations of X and Y.

Let's take a look at the definition formula first. Notice that we have a variation of the pattern that we saw in *z*-, *t*-, and *F*-tests, where "the difference on the top and the error on the bottom" describes the pattern for each of these tests for "difference." Here, we're looking for how two variables are related to (dependent on) one another rather than how different two or more means are. The top of this formula serves as a measure of how related the *X* and *Y* variables are. We then divide that measure of relatedness by the combined standard deviation of the two variables involved. In the "raw data" formula, we will be doing the same things with the measurements, only with this formula, we do not need to calculate the means and standard deviations first. Both formulas will yield a statistic describing both the direction of the relationship between *X* and *Y* and the strength of that relationship.

CALCULATING *r*

Let's return to our question about the relationship between height and cancer. Box 13.2 contains some imaginary data that is similar to that shown in the published research. The data reflect the heights of 15 randomly selected healthy women and their risk of cancer, where risk of developing cancer is measured as chances out of 100. If you have a risk of 0.14, it means your risk is 14 chances out of 100.

We'll use what we know about the data to calculate our *r*-value. First, take a common-sense look at the data, and decide if it looks as though we have a relationship here. Does it seem as though tall people tend to have higher risk of cancer than

BOX 13.2   Height and risk of cancer in 15 women (hypothetical data)

Height (in Inches)	Risk (Chances Out of 100)	XY Pairs
64	.09	64(.09) = 5.76
64	.04	64(.04) = 2.56
64	.10	6.40
66	.02	1.32
66	.09	5.94
63	.04	2.52
71	.14	9.94
66	.12	7.92
70	.13	9.10
62	.02	1.24
64	.01	0.64
68	.11	7.48
68	.13	8.84
62	.01	0.62
64	.03	1.92
$\Sigma_X = 982$	$\Sigma_Y = 1.08$	$\Sigma_{XY} = 72.20$
$\Sigma_{X^2} = 64{,}394$	$\Sigma_{Y^2} = .1112$	
$S_X = 2.65$	$S_Y = 0.047$	

rho (ρ) The theoretical value of the correlation coefficient in the population.

short people do? The tallest woman in our sample (71 inches tall) has the highest cancer risk (.14), while the two shortest women (62 inches tall) both have fairly low cancer risk (.01 and .02). So, there might be a positive relationship here.

As in all of the other tests we've conducted so far, we can designate a null and an alternative hypothesis about the relationship between height and risk of cancer. The null hypothesis would predict that the r-value should be close to zero (indicating no relationship between these two variables). The alternative hypothesis should be that r is positive (between 0.00 and +1.00) and strong (around .80, if we use the traditional values of r). Box 13.3 shows our hypotheses.

The data in our sample suggest that the relationship, whatever it might be, will likely be positive. If the relationship is strong, we should see that for any one height, all of the women of that height should have roughly the same risk of cancer—in other words, the variability should be small and the data not too spread out. Our data would seem to indicate a strong relationship between height and cancer risk. Box 13.4 shows the calculation of r.

BOX 13.3 Null and alternative hypotheses concerning height and cancer risk

r = correlation coefficient in the sample

ρ (rho) = correlation coefficient in the population

Hypothesis in symbols	Hypothesis in words
$H_0: \rho = 0.00$	There is no relationship between height and cancer risk in the population
$H_1: \rho > 0.00$	There is a positive correlation between height and cancer risk in the population.

We can also calculate the degrees of freedom for our hypotheses. In this case, $f = N - 2$, where N indicates the number of pairs of data points. With 15 women in our sample, $df = 13$.

BOX 13.4 Calculating r using the definition formula and the raw data formula

Using the definition formula

$$r = \frac{\frac{\sum XY}{N} - (\bar{X})(\bar{Y})}{(S_X)(S_Y)}$$

$$= \frac{\frac{72.2}{15} - (65.47)(0.072)}{(2.65)(0.047)}$$

$$= \frac{4.81 - 4.71}{0.125}$$

$$= \frac{0.10}{0.125}$$

$$r = 0.80$$

Using the raw data formula

$$r = \frac{N\sum XY - (\sum X)(\sum Y)}{\sqrt{[N\sum X^2 - (\sum X)^2][N\sum Y^2 - (\sum Y)^2]}}$$

$$= \frac{15(72.2) - (982)(1.08)}{\sqrt{[15(64,394) - 982^2][15(.1112) - 1.08^2]}}$$

$$= \frac{1,083.00 - 1,060.56}{\sqrt{[965,910 - 964,324][1.6680 - 1.1664]}}$$

$$= \frac{22.44}{\sqrt{(1,586)(0.5016)}}$$

$$= \frac{22.44}{\sqrt{795.54}}$$

$$= \frac{22.44}{22.81}$$

$$= 0.795$$

$$r = .80$$

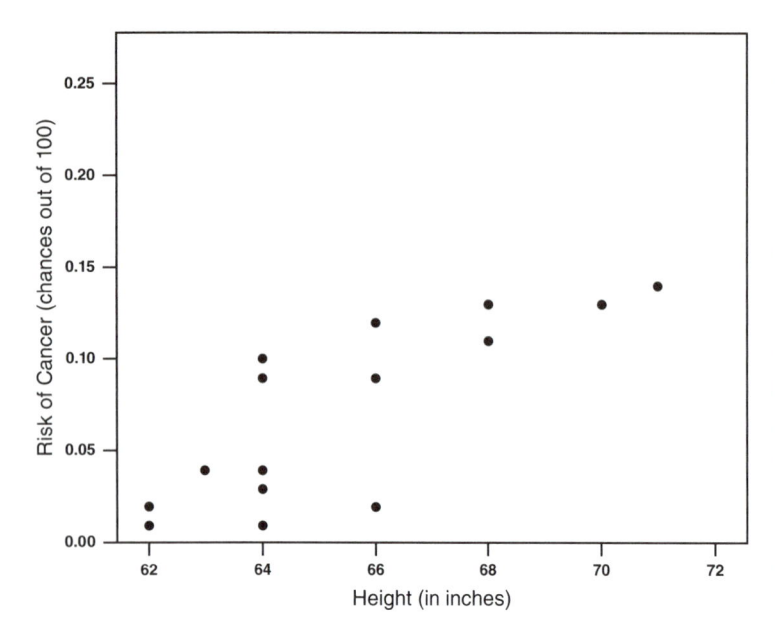

FIGURE 13.6
Scatterplot showing the relationship between height and cancer risk in women.

So, with either formula, our correlation coefficient is .80, indicating a strong positive correlation between height and risk of cancer.* The taller a person is, the higher the risk of cancer for that person. Notice also that the risk of cancer isn't all that high: The highest risk is 14 chances out of 100 for a woman 71 inches tall (5 feet 11 inches) and only 4 chances out of 100 for a woman of average height in the United States (5 feet 3 inches). Nevertheless, the relationship between height and risk is a strong one. If we construct a scatterplot of the two variables, we can get a visual indication of the relationship as well. Figure 13.6 shows the scatterplot of the data.

Remember, each dot represents the height and cancer risk for one person in our study. There are 15 dots, one for each participant in our study, and the cloud of dots tends to rise from the lower left-hand corner of the graph to the upper right—indicative of a positive correlation between the X and Y variables. There is also some variability in the data; it is not a perfect relationship. Take a look at the five women who are 64 inches tall: Their risk of cancer ranges from a low for one woman of .01 (i.e., 1 chance out of 100) to a high of .10 (i.e., 10 chances out of 100). Only two women were 68 inches tall, and their individual risks of cancer were fairly consistent (.11 and .13, respectively).

TESTING HYPOTHESES ABOUT *r*

So, have we supported our null hypothesis, or do we have evidence that the alternative is true? Of course, you know by now that in order to answer a question like this, we need to have some critical *r*-value (that depends on the *df*), and we need to compare our calculated *r*-value with the critical *r*-value that we get from a table of *r*-values. And, if you look at the table in Appendix F, you will see a list of critical Pearson *r*-values. We have 13 *df*, and we proposed a one-tailed test (speculating that the correlation would be positive, or greater than zero). If we choose a level of .05 to start with, we can say that we have a statistically significant correlation, which probably reflects the relationship between height and cancer risk in the population. The critical *r*-value at 13 *df* is .4409, and our observed *r*-value (.80) exceeds the critical value. In fact, we can say that even at a *p*-value of .001, our observed *r*-value still exceeds our critical *r*-value (at $\alpha = .001$, $r_{critical} = .7603$).

---

*I should mention again that while the link between height and cancer risk has been studied, the data we are using in this chapter are hypothetical. The correlation found in the original research was positive and significant, but it was *not* .80. I don't want to scare tall women (I'm fairly tall myself), and it is, as we will discuss later, "just a correlation."

CheckPoint

Stacey notices that those of her friends who smoke cigarettes tend to smoke more while drinking alcohol. The table below displays the average rates of alcohol and cigarette consumption for a 1-hour period for 10 of Stacey's cigarette-smoking friends. Use the data provided in the table to answer the following questions and help Stacey better quantify the relationship between the rates of alcohol and cigarette consumption. Round your answers to two decimal places, where appropriate.

Average number of drinks consumed and cigarettes smoked in 1 hour

Drinks per Hour	Cigarettes per Hour
4	6
2	3
1.5	0
0.5	1
2.5	4
2	1
3	3
0.75	1
1	2
5	7
$\Sigma = 22.5$	$\Sigma = 28$
$n = 10$	$n = 10$

1. Calculate S_X.
2. Calculate S_Y.
3. Calculate r.
4. Calculate df.
5. Identify the critical r-value ($\alpha = .05$), compare it to $r_{observed}$, and interpret your findings.

Answers to these questions are found on page 565.

THE LEAST SQUARES REGRESSION LINE, A.K.A. THE LINE OF BEST FIT

CONNECTING THE DOTS

Galton recognized that he could describe the relationship between the two variables with a straight line—but *which* straight line? Figure 13.7 shows the scatterplot from Figure 13.6 with a number of straight lines drawn on it, each line providing a different description of the data.

Galton maintained there would be one line, and one line alone, that did the best job of describing the data. That line would come as close to each data point as possible without bending. The line would describe the relationship between X and Y not only for the data collected but also for data points that do not show up in the data set. For example, the line will allow us to predict the cancer risk for a woman who is 7 feet (84 inches) tall as well as for a woman who is only 4 feet (48 inches) tall. In other words, the line

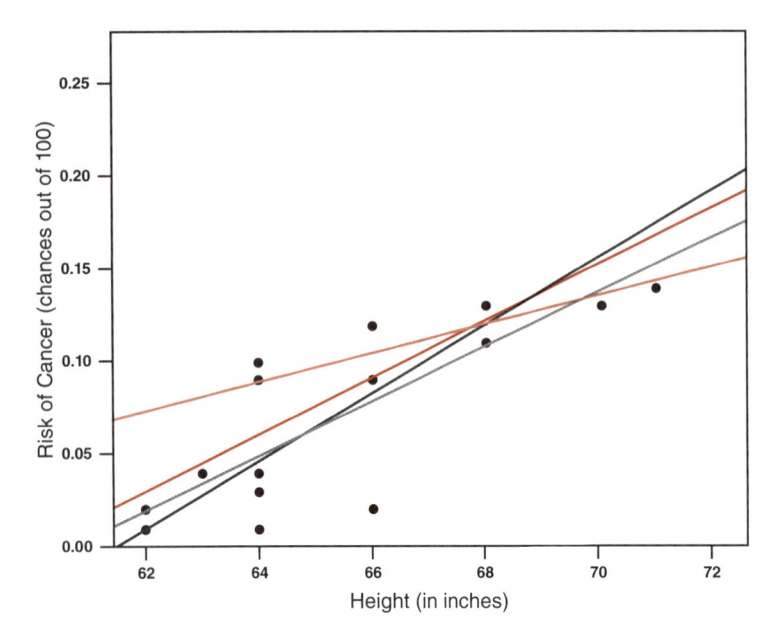

FIGURE 13.7
Possible "lines of best fit" for the data from Figure 13.6.

will show the relationship between X and Y for *theoretical data* as well as for the *empirical data* we've collected. We can predict the future!

Galton and Pearson called this line the **least squares regression line**, or the **line of best fit**. Does "*least squares* regression line" sound familiar? It should: It refers to the one line that minimizes the variability of the data points from the line. "Least squares" refers to standard deviation, or squared deviation of points around the mean. The distance of a data point from the line of best fit is called the *residual*, and the least squares regression line minimizes the residuals. Figure 13.8 shows these residuals for two possible regression lines.

The arrows indicate the deviation of a given data point from a given line. The red and gray lines differ in the description they offer of the relationship between height and cancer risk, and the amount by which a given data point deviates from that lines also varies.

Galton and Pearson showed there is one line, and one line only, that minimizes the deviation of each and every data point from the regression line. But it isn't going to be possible to find that one and only line without a lot of trial and error, just by eye. What we need is a formula for finding a straight line that will guarantee the line we end up with will be the least squares regression line.

FINDING THE REGRESSION LINE

So, we need a formula that will allow us to draw a straight line—and if you remember your high school math classes, you may remember the formula for a straight line looks something like what you see in Box 13.5.

You may also remember this formula as $\hat{y} = mx + b$. If you do, it just means that the symbol for the slope and the *y*-axis intercept are different from what we are using here.

If the formula seems familiar, then you probably remember that the slope of the line can be described as "rise over run." The term *rise* refers to a change on the *y*-axis, and the term *run* refers to a change on the *x*-axis. So, slope is variability in Y divided by variability

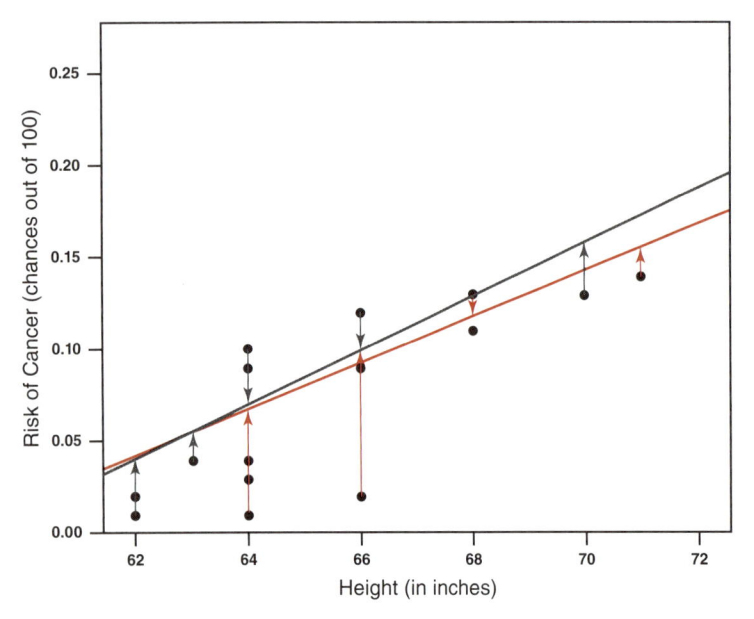

FIGURE 13.8
Residuals, or deviations, from the line of best fit.

BOX 13.5 The formula for a straight line (the regression line)

$$\hat{y} = bx + a$$

where

b = the slope of the line
x = a given X-value

a = the y-axis intercept
$\hat{y}$ = the predicted score on Y for given X

$$b = r\left(\frac{S_Y}{S_X}\right)$$

$$a = \bar{Y} - b(\bar{X})$$

---

in X, times the measure of how related X and Y are. The y-axis intercept is the point where the line would cross, or intersect with, the y-axis. Here, it's being described as the average of the Y variable minus the slope (variability) times the average of the x-axis variable.

Let's calculate the slope and y-intercept for the relationship between height and cancer risk. We will then plug those values into the equation to find the straight line that is the best fit regression line. Box 13.6 shows you these calculations.

To plot our straight line on the scatterplot, we need at least two points that we can connect on the graph. So, let's predict the cancer risk for a woman who is 84 inches tall, and then another Y-value for a woman only 48 inches tall (I picked these two X-values at random). Then, we will put these two predicted data points on our scatterplot and connect them with a straight line. (See Box 13.7 for the calculations.) Once we have two predicted Y-values, we can plot them on the scatterplot and draw our straight line on the graph. Take a look at Figure 13.9 to see how we do this.

The straight line that connects our two predicted Y-values defines the line that minimizes the deviation of any of our empirical points from that line. It is the line that best describes the relationship between height and cancer risk, and any Y-value that we predict with our equation for a straight line will be on this regression line. Notice that the y-intercept that we predicted ($-.85$) does not seem to be on this graph. Do you know why? If you said that it's because the x-axis starts at 48 and not a zero, you would be right. The y-intercept is the value we predict for Y if X is equal to zero. We just can't see it on this graph.

least squares regression line (line of best fit) A straight line that minimizes the deviation of all of the X,Y data points from the line; a straight line that comes as close to all of the X,Y data points possible without bending.

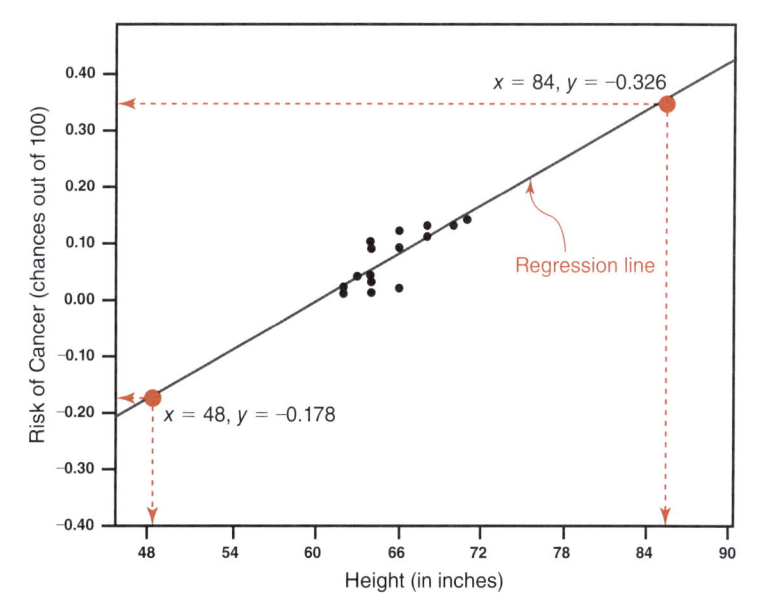

FIGURE 13.9
Drawing the line of best fit on the graph.

BOX 13.6 Calculating the slope and y-intercept

Calculating the slope

$$b = \left(\frac{S_Y}{S_X}\right)$$
$$= .80\left(\frac{.047}{2.65}\right)$$
$$= .80(.0177)$$
$$= .014$$

For every 1 inch of increase in height, there is an increase of 1.4% in cancer risk.

Calculating the y-intercept

$$a = \bar{Y} - b(\bar{X})$$
$$= .072 - .014(65.47)$$
$$= .072 - .92$$
$$= -.85$$

The line of best fit will cross the y-axis at −.85.

The equation for a straight line that describes the relationship between height and cancer risk is

$$\hat{y} = .014(X) + (-.85)$$

Note: If you do not have the standard deviations of the X and Y variables already calculated for you, the slope of the line can be calculated from the raw data using the raw data formula.

The raw data formula for b

$$r = \frac{N\sum XY - (\sum X)(\sum Y)}{N\sum X^2 - (\sum X)^2} = \frac{15(72.2) - (982)(1.08)}{15(64,394) - (982)^2} = \frac{22.44}{1,586} = .014$$

BOX 13.7 Calculating two Y-values

If X = 84 inches, then Y = ...

$$\hat{y} = .014(X) + (-.85)$$
$$= .014(84) + (-.85)$$
$$= 1.176 - .85$$
$$= .326$$

If X = 48 inches, then Y = ...

$$\hat{y} = .014(X) + (-.85)$$
$$= .014(84) + (-.85)$$
$$= .672 - .85$$
$$= -.178$$

A woman who is 7 feet tall would have a cancer risk of 32.6 chances out of 100. (Compare this to the average risk for all of the women in the sample of .047, or 4.7 chances out of 100.)

A woman who is only 4 feet tall would have slightly less than zero chances out of 100 (despite the seeming impossibility of this) of developing cancer. Remember, the line of best fit, and the Y-value it allows us to predict, are both theoretical.

> **CheckPoint Answers**
> *Answers to questions on page 561*
>
> 1. $S_X = 1.37$
> 2. $S_Y = 2.18$
> 3. $r = 0.90$
> 4. $df = 8$
> 5. $r_{obs.} = 90 > r\text{critical} = .549$; this indicates a strong correlation
>
> SCORE: ___/5

Think About It...

WHAT DOES PERFECT MEAN?

Let's consider the following diagrams and think about what r can tell us.

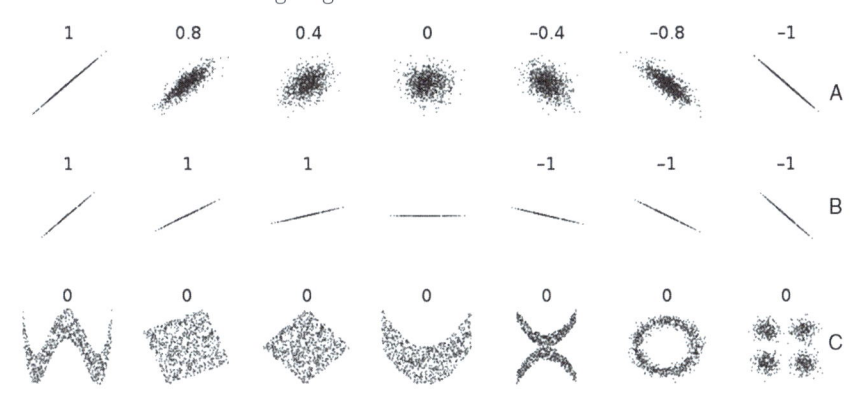

The numbers in row A show the correlation coefficient r calculated for each of seven relationships between X and Y. The correlations vary from a perfect positive correlation ($r = 1.00$) on the far left-hand side to a perfect negative correlation ($r = -1.00$) on the far right-hand side. What can we say about the "cloud" of dots that illustrates each relationship?

1. How does the "cloud" of dots change as the relationship approaches zero?
   _____

2. Row B shows a series of perfect correlations, positive on the left and negative on the right. Why does the r-value of the three positive correlations depicted here not change? Why are they all "perfect" but don't all look alike? _____
   _____

3. What is the difference in the variability of X and Y when the correlation is perfect and when it is not? _____
   _____

   (Hint: Think about variability and consistency in X and Y.)

Think About It... continued

4. What would the value of *r* be for the relationship in the center of the scatterplots in row B? _____

5. Is the relationship in the center of row B also "perfect"? _____

6. Row C shows scatterplots where *r* = 0. What can you say about the relationships in row C? Is there truly no relationship at all between *X* and *Y* in these plots? What does a Pearson *r* of zero mean? _____

_____

Answers to this exercise can be found on page 587.

CheckPoint

Use the information provided in the questions below to calculate the missing values. Round your answers to two decimal places, where appropriate.

1. $a = ?; \bar{Y} = 7; \bar{X} = 6; b = 0.75$
2. $\hat{y} = ?; a = -11.7; b = 3.1; X = 21.6$
3. $\hat{y} = ?; a = 48; b = -6; X = 24$
4. $b = ?; N = 10; \Sigma XY = 89.25; \Sigma X = 22.25; \Sigma X^2 = 68.31; \Sigma Y = 28$
5. $a = ?; \bar{Y} = 43{,}000; \bar{X} = 12; b = 450$
6. $b = ?; r = -.39; S_y = 2.72; S_x = 1.41$
7. $\hat{y} = ?; a = 25; b = .48; X = 368$
8. $a = ?; \bar{Y} = 68; \bar{X} = 7.2; b = 3.6$
9. $b = ?; r = 0.68; S_y = 6.01; S_x = 4.83$
10. $b = ?; N = 400; \Sigma XY = 73{,}160; \Sigma X = 2{,}880; \Sigma X^2 = 18{,}944; \Sigma Y = 9{,}600$

Answers to these questions are found on page 569.

SOME CAUTIONARY NOTES

So, the Pearson product–moment correlation coefficient describes the relationship between *X* and *Y* variables. BUT, there are a few situations where *r* will not provide us with an accurate answer or where *r* is not the appropriate statistic to use. So, let's talk about when *r* won't work for you.

linear relationship A relationship between *X* and *Y* that can best be described with a straight line. For example, as the value of *X* increases, the value of *Y* tends to increase or decrease but does not change direction.

LINEAR VERSUS CURVILINEAR RELATIONSHIPS

First, the Pearson correlation coefficient only describes **linear relationships** between *X* and *Y*. What's a linear relationship? As the name suggests, it is a relationship that

STATISTICS IN CONTEXT

can be described with a straight line, or a relationship that does not change direction. For example, consider the relationship between anxiety and performance on a test. It makes some sense to think of this relationship as linear: The more anxious we are, the poorer our performance on an exam is likely to be. However, the relationship between these two variables (shown in Figure 13.10) is actually not linear at all. It starts off looking like a *positive* linear correlation (i.e., performance actually improves as anxiety increases), but if anxiety continues to increase, performance eventually starts to fall. So, if we are not anxious at all before an exam, our performance is fairly poor. Moderate anxiety tends to be paired with very good performance on exams (apparently, a bit of anxiety if a good thing). At some anxiety level, however, the relationship changes direction. At very high levels of anxiety, performance decreases—too anxious to hold the pencil probably means too anxious to think about the answers on the test.

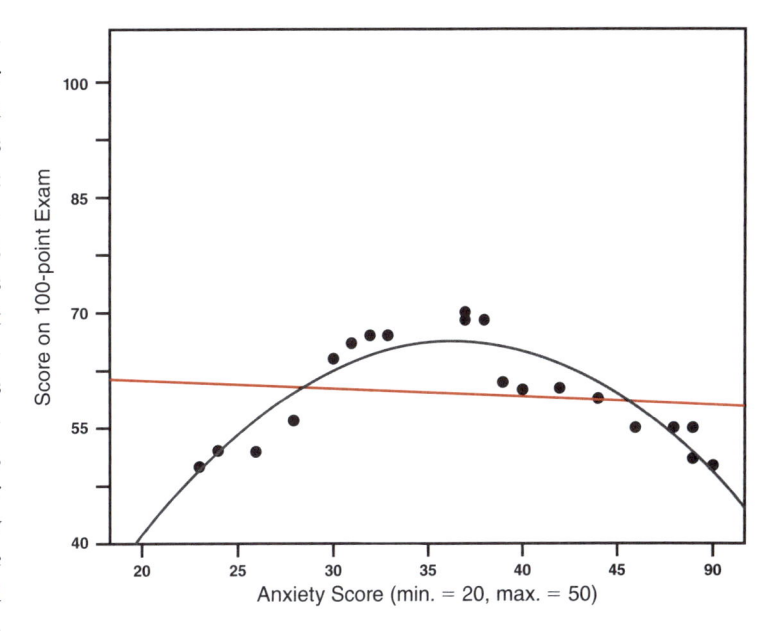

FIGURE 13.10
Curvilinear relationship between anxiety and test performance.

This is an example of a **curvilinear relationship**, and the Pearson coefficient, because it's based on the assumption that the relationship between X and Y can be described with a straight line, will not provide us with an accurate assessment of the real relationship. Notice that while the straight line reduces the relationship to a negative one, the curved line describes a **quadratic relationship**, or a relationship that changes direction once.

SPSS and other statistics software programs will calculate the line of best fit that describes curvilinear relationships. A discussion of the mathematics behind calculating a correlation coefficient for a curvilinear relationship is not suited to an introductory statistics textbook, so we won't be talking about it here. However, because the Pearson coefficient only works with linear relationships, you should graph the relationship first, before you calculate *r*, just to check that the relationship is linear.

curvilinear relationship A relationship between X and Y that can be best described with a curved line. For example, as the value of X increases, the value of Y may increase and then decrease.

quadratic relationship A relationship between X and Y described by a line that changes direction once.

TRUNCATED RANGE

There's another situation in which a Pearson *r* will give you "spurious," or incorrect, results, and that situation arises when you have data with truncated range. **Truncated range** occurs when the variability (remember that the range is a measure of variability) in your sample is much more limited (smaller than) the variability in the population. If the range in your data is truncated, the Pearson *r* that you calculate will tend to underestimate the real value of *r*. The solution to this problem is to go back to the population and select a different sample (one with more variability) and/or select additional participants for your sample to make your *n* larger.

Let's take a look at an extreme example of truncated range. Suppose we were interested in the correlation between high school and college grade point averages

truncated range The situation that occurs when the range of a sample is much smaller than the range of the population the sample was drawn from.

(GPAs). Researchers have suggested that high school GPA does a good job of predicting college GPA: The positive correlation can be as high as .40 (Richardson, Abraham, & Bond, 2012). We go out and collect data from a very tiny sample of first-year college students, as shown in Table 13.1.

TABLE 13.1 High school and college GPAs of four students

X (High School GPA)	Y (First-Year College GPA)
4.00	2.70
4.00	3.50
4.00	4.00
4.00	3.80
$\Sigma x = 16$	$\Sigma y = 14$
$\Sigma x^2 = 64$	$\Sigma y^2 = 49.98$
Range = 0.00	Range = 1.30

Standard deviation = 0.00 (there is no variability in this set of observations).

The range is seriously truncated in our example. In fact, there is no variability at all in the high school GPAs we are using to predict college GPA. What will the correlation look like? I hope you see that the correlation here will be very much smaller than the .40 suggested in the literature. In fact, the correlation will be zero in this case. We can show this using the definition formula for finding r (Box 13.8).

Because the standard deviation of the X values will be zero, the correlation between X and Y will also be zero—we cannot divide by zero. The absence of variability in the X variable (truncation of the range in this variable) has greatly reduced the size of the correlation.

BOX 13.8   **Definition formula for finding *r***

$$r = \frac{\frac{\Sigma XY}{N} - (\bar{X})(\bar{Y})}{(S_X)(S_Y)}$$

STATISTICS IN CONTEXT

> **CheckPoint Answers**
> *Answers to questions on page 566*
>
> 1. $a = 11.5$ ☐
> 2. $\hat{y} = 78.66$ ☐
> 3. $\hat{y} = -96$ ☐
> 4. $b = 1.43$ ☐
> 5. $a = 48{,}400$ ☐
> 6. $b = -.75$ ☐
> 7. $\hat{y} = 201.64$ ☐
> 8. $a = 42.08$ ☐
> 9. $b = 0.85$ ☐
> 10. $b = -.2.25$ ☐
>
> SCORE: /10

COEFFICIENT OF DETERMINATION

After you have found *r*, you have a description of the direction and strength of the relationship between *X* and *Y*. However, a better statistic to use when talking about the strength of the relationship would be something called the *coefficient of determination*. This coefficient is extremely easy to determine because it can be found by squaring the *r*-value you calculate (the symbol for the coefficient of determination is r^2, which is also the "equation" for finding it). The **coefficient of determination** is a statistic that measures the proportion of variability that is shared by both the *X* and *Y* variables. You can also think of the coefficient of determination as a measure of the amount of variability in your dependent measure, or *Y* variable, that can be explained by knowing its relationship to the independent or predictor *X* variable.

So, what does that mean? Let's use the example of the relationship between height and cancer risk. There is measurable variability in both the *X* and the *Y* variables here: The height of the women in the sample varies (from 62 to 71 inches), as does their risk of cancer (from 1 to 14 chances out of 100). But how much variability do these two variables have in common?

If you think about it, you will see that if these two variables are not related to one another at all, then they don't share any variability. If two variables are perfectly related to one another, then *X* and *Y* share *all* of the variability in the data.

Height and cancer risk obviously must share some variability because they are related to one another—they *vary together*. If we square the correlation coefficient, we will get the coefficient of determination, which shows the proportion of variability the two variables share in common. Box 13.9 shows the calculation of r^2.

The stronger the relationship between *X* and *Y*, the more dependent *Y* is on *X*, and the closer the coefficient of determination will be to 1.00. In our case, height and cancer risk share 64% of the variability in the data. So, you could also say that 64% of the variability in cancer risk can be explained by the relationship between cancer risk and height.

coefficient of determination (r^2) A measure of the amount of variability that is shared by both *X* and *Y*; the amount of variability in *Y* that can be explained by knowing its relationship to *X*.

CHAPTER 13 Correlation and Regression: Are We Related?

BOX 13.9

Calculating r^2

Calculation of r^2	Calculation of r^2 if $r = 1.00$	Calculation of r^2 if $r = 0.00$
$r^2 = (r)(r) = .80^2$	$r^2 = (r)(r) = 1.00^2$	$r^2 = (r)(r) = 0.00^2$
$r^2 = .64$	$r^2 = 1.00$	$r^2 = 0.00$
∴ height and cancer risk share 64% of their variance.	∴ height and cancer risk share 100% of their variance.	∴ height and cancer risk share 0% of their variance.

Think About It...

SHARED VARIABILITY AND RESTRICTED RANGE

You can visualize shared variability using Venn diagrams, those overlapping circles so often used to describe how sets intersect. For example, if:

(A)
$r = .05$
$r^2 = .004$ (or 0.4% shared variability)

(B)
$r = .72$
$r^2 = .52$, or 52%

(C)
$r = -.98$
$r^2 = .96$, or 96%

Let's use an example of the correlation between scores on the SAT critical reading section (SAT-CR), mathematics section (SAT-M), and writing section (SAT-W) and the first-year GPA (FYGPA) in college.

1. What kind of correlation would you expect to see?
   Positive   Negative   Zero
   The relationship between SAT scores and FYGPA (from Kobrin, Patterson, Shaw, Matthern, & Barbuti, 2008) is shown below.

2. Calculate the coefficient of determination for the correlation between each SAT section and FYGPA, and then sketch the Venn diagram to illustrate the amount of shared variability between X and Y.

   SAT-CR and FYGPA
   $r = .29$
   $r^2 = $ _____

   SAT-M and FYGPA
   $r = .26$
   $r^2 = $ _____

   SAT-W and FYGPA
   $r = .33$
   $r^2 = $ _____

Venn diagrams:

In their study, these authors noticed that the the SAT scores of students who had been admitted to college was *less variable* than were SAT scores in the larger population. So, they recalculated the correlations using a correction for the restricted range.

3. Recalculate the coefficients of determination, and compare the two sets of measures of the relationship.

   SAT-CR and FYGPA   SAT-M and FYGPA   SAT-W and FYGPA
   $r = .48$ $r = .47$ $r = .51$
   Venn diagrams:

4. Does it make sense to you that the range of SAT scores for admitted students was less variable than the range in the whole population? Why?
   _____

5. What did the restricted range do to the correlations? _____
   _____

Answers to this exercise can be found on page 587.

STATISTICS IN CONTEXT: CORRELATIONS AND CAUSATION

Traditionally, discussions of correlation and regression in introductory statistics classes end with a statement about correlations and causation. That statement is almost always that *correlations cannot tell you about causation*. Users of statistics, both students and professors alike, tend to remember that admonition and interpret their *r*-values as simply descriptive, not causal. Often, researchers will make a related mistake when interpreting the result of a *t*-test, or an ANOVA. They will assume that because they used a *t*- or an *F*-test, the results of that test automatically reveal a cause-and-effect relationship between two variables. In other words, many people assume that it is the *statistic* that tells you about causation, when in fact it is the *methodology used* that determines how you interpret the results of that statistic (see Hatfield, Faunce, & Soames, 2006, for an excellent description of this issue).

Let's use the data that we've been playing with in this chapter—the relationship between height and cancer risk—to illustrate this point. The correlation coefficient that we calculated suggested a strong positive relationship between height and cancer risk: the taller you are, the greater your risk of cancer. The method used to collect the data in this study was observational, meaning we simply observed two characteristics of the women in our sample. We did not control or manipulate any variable in the design; we simply observed them in their natural setting.

Because we didn't control any of the variables, we cannot tell exactly which one is causing the cancer risk to rise. It could be height, or it could be some other variable that we don't know about. The authors of the original study (Green et al., 2011) proposed some possible **hidden variables**, or other variables that might be causal in this relationship. For example, they suggest that early childhood nutrition, frequency of infection during childhood, hormone levels in childhood and adulthood, or the fact that taller people have more cells in their bodies and so a greater chance for these cells to mutate or become malignant could be causing the relationship seen in the data.

Suppose, instead of a correlation, we'd used a different statistical test in our investigation. Suppose we had separated the women into two groups on the basis of height (in statistical parlance, we would say that we had **dichotomized** the height variable—in other words, split it in two). We put all of the tall women (say, 66 inches and taller) in one group and the short women (less than 66 inches) in a second group and then compared mean cancer risk for these two groups using a *t*-test. We still don't have any knowledge of cause and effect, even though we used a statistic that is commonly thought to tell us about what caused a change in the dependent variable, or DV. The data are still observational, and we still have not controlled for possible intervening, or hidden, variables.

The key, then, is how we go about collecting our data. If the data are observational, then the correlation coefficient we calculate will be descriptive, and it will not tell us about causation. If, on the other hand, we can take control of the variables in our study, and if we can manipulate other variables that might be influencing our dependent variable, holding them constant so that we are testing the effect of only the independent variable, or IV, on the DV, then a correlation can tell us as much about causation as any other test.

To be able to determine cause and effect, however, we would need to use a different methodology. We would have to use an experimental method, where one or more independent variables are manipulated in order to see their effects on a dependent variable. Manipulating a variable means that the researchers hold that variable constant, or manipulate how much or how little of that variable is administered to the participants in the study. If we could control how much of a particular hormone circulated in the body, we could measure the relationship between height (hormone level) and cancer risk using a correlation coefficient, and *that r*-value would tell us about cause and effect.

So, to make a long story short, correlations should never be taken as evidence of causation *unless* you have used an experimental design to collect your data. If you have been able to control the variables involved in your study so that you can see the effect of manipulating an IV on the DV, then a correlation will tell you about causation just as well as a *t*-test or an *F*-test will. It's the methodology used to collect the data that matters, not the statistic itself.

---

hidden variable In a correlation, a variable that was not controlled for or used in the research and that might play a causal role in the relationship.

dichotomize To split a continuous variable into two groups.

SUMMARY

Sir Francis Galton developed the idea of a statistic that could measure the relationship between two variables, X and Y, and his colleague Karl Pearson came up with the mathematics necessary to calculate that statistic. This statistic is symbolized by the letter r (in a sample) and is called the Pearson product–moment correlation coefficient.

Correlations can be positive or negative, and they vary in strength. A positive correlation is one where as X increases, so does Y. A negative correlation is one where Y decreases as X increases.

The correlation coefficient that is calculated in a sample can be used to test a hypothesis about the relationship between X and Y in the population. A theoretical line of best fit can be fitted to the data and used to predict Y values from given X values. The equation for a straight line—$y = bx + a$, where b is the slope of the theoretical line and a is the y-intercept—can be calculated from the variability of X and Y and the degree to which X and Y are related to one another.

Finally, a statistic called the coefficient of determination (r^2) can be calculated by squaring the correlation coefficient. The r^2 statistic describes the proportion of variability shared by the two variables X and Y, and it is a measure of the strength of the relationship between the two variables.

Correlations can provide misleading information if they are used with nonlinear relationships or with data sets that suffer from truncated range. And researchers should *never* use correlations as evidence of a causal relationship between X and Y.

TERMS YOU SHOULD KNOW

coefficient of determination (r^2), p. 569
correlation, p. 554
curvilinear relationship, p. 567
dichotomize, p. 572
hidden variable, p. 572
least squares regression line, p. 562
linear relationship, p. 566
line of best fit, p. 562
negative correlation, p. 555

Pearson product–moment correlation coefficient (correlation coefficient) (r), p. 552
positive correlation, p. 554
quadratic relationship, p. 567
regression to the mean, p. 555
rho (ρ), p. 558
truncated range, p. 567

GLOSSARY OF EQUATIONS

Formula	Name	Symbol
$r = \dfrac{\dfrac{\sum XY}{N} - (\bar{X})(\bar{Y})}{(S_X)(S_Y)}$	The Pearson product–moment correlation coefficient (definition formula)	r
$r = \dfrac{N\sum XY - (\sum X)(\sum Y)}{\sqrt{[N\sum X^2 - (\sum X)^2][N\sum Y^2 - (\sum Y)^2]}}$	The Pearson product–moment correlation coefficient (raw score formula)	r

Formula	Name	Symbol
$\hat{y} = bx + a$	Equation for a straight line, where $\hat{y}$ = predicted Y value; b = slope of the line; a = y-intercept	
$b = r\left(\dfrac{S_Y}{S_X}\right)$	Slope of the regression line (definition formula)	b
$b = \dfrac{N\sum XY - (\sum X)(\sum Y)}{N\sum X^2 - (\sum X)^2}$	Slope of the regression line (raw score formula)	b
$a = \bar{Y} - b(\bar{X})$	The y-intercept	a
r^2	The coefficient of determination	r^2

WRITING ASSIGNMENT

There is a well-known relationship between sales of ice cream and the rate of crime (sample data are provided in Table 13.2). Calculate the Pearson r-value, and write a short description of the results, including whether the relationship is significant. In addition, discuss any possible hidden variables that might be responsible for this relationship and how they are related to the two variables being considered here.

Table 13.2 Data for the Chapter 13 writing assignment

Ice Cream Sales (Number Sold per Day)	Crime Rate (per 100,000 People)
100	200
55	197
200	200
237	209
306	400
309	403
300	399
400	405
420	401
410	480
500	524
555	556
569	600
650	638

PRACTICE PROBLEMS

1. Describe the relationship between *X* and *Y* pictured in each of the scatterplots shown in Figure 13.11.

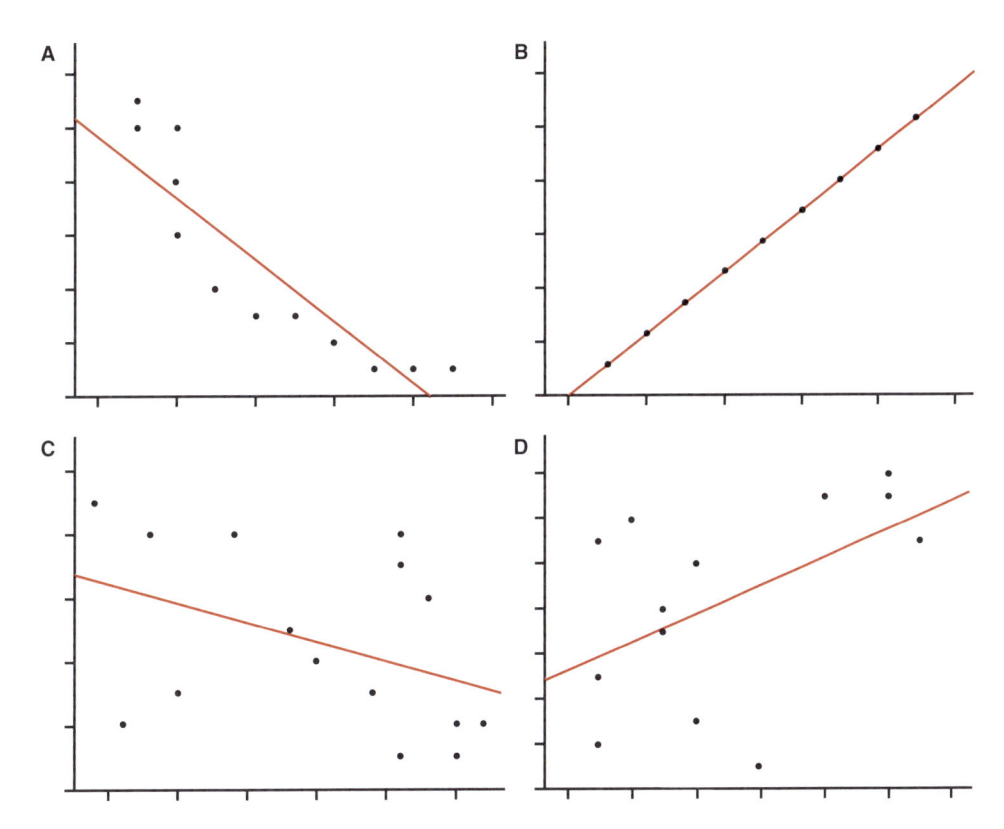

FIGURE 13.11
Scatterplots for question 1.

2. For the six following situations, determine the type of relationship described (positive, negative, linear, or curvilinear).
   a. The depth of the water and the temperature of the water.
   b. The age of a child and the size of the child's feet.
   c. The number of years of employment at a company and annual income.
   d. The number of ambulances in a city and the infant mortality rate.
   e. The value of a piece of furniture and the age of that piece of furniture.
   f. Hours spent playing video games and GPA.

3. The coefficient of determination (r^2) is a measure of effect size. Compare r^2 to eta squared for a one-way ANOVA. How are these two measure of effect size similar?

4. Without calculating *r*, describe the relationship between *X* and *Y* (positive, negative, or no relationship) for the following data sets:

CHAPTER 13 Correlation and Regression: Are We Related?

a.

X	Y
1	16
2	23
3	35
4	28
5	44
6	40
7	22
8	61
9	82

b.

X	Y
1	200
2	187
3	154
4	167
5	153
6	158
7	165
8	127
9	108

c.

X	Y
1	6
2	9
3	17
4	42
5	4
6	6
7	3
8	22
9	3

5. The line of best fit is also known as the "least squares" regression line. What does the term *least squares* mean here?

6. A researcher wants to determine if age and height are correlated. He obtains measures of age and height for 50 individuals from the time of their birth until they are 75 years of age. The researcher uses a Pearson correlation coefficient to determine if these two sets of measurements are correlated. You review his work and suggest that he has made a major mistake. What mistake has he made?

7. Suppose you collect data on the relationship between belief in being lucky and religiosity (both measured on interval scales with a maximum value of 10, where higher numbers indicate higher belief in luck or higher degree of religiosity). Table 13.3 shows the results of your study.

Table 13.3 Belief in luck and religiosity scores

Belief in luck	Religiosity
1	1
2	1
4	1
3	2
5	1
6	2
3	1
8	2
9	3

Before you calculate r, a colleague points out that you may have a problem with your data, and as a result of this problem, your r-value may be biased. Is your colleague correct? If so, what is the problem with your data, and how will your r-value be biased?

8. Use the following data to determine the equation for the line of best fit:

X	Y
−2	−1
1	1
3	2

9. Use the following data to determine the equation for the line of best fit:

X	Y
1	7
2	2
3	3
4	1
5	0

10. Chen and Escarce (2010) wondered if childhood obesity might be related to the number of siblings a child has. To find out, they measured both the Body Mass Index (BMI) for 12 children in a kindergarten class and the number of brothers and sisters each child had. Note that a BMI index of 80% or above is considered obese. The data in Table 13.4 are similar to those described by Chen and Escarce.

Table 13.4 BMI measures and number of siblings for 12 kindergarten-age children

Number of Siblings	BMI
0.00	80.00
1.00	85.00
1.00	71.00
2.00	66.00
1.00	46.00
0.00	37.00
1.00	35.00
2.00	30.00
4.00	12.00

Table 13.4 Continued

1.00	29.00
2.00	17.00
5.00	16.00

   a. Are these two variables correlated with one another? If so, describe that correlation.
   b. Create a scatterplot showing the relationship between these two values.
   c. Draw the line of best fit on the scatterplot. (Show your calculations.)
   d. Predict the BMI for a child who has six siblings and for a child who has nine siblings.

11. We all know that smoking has harmful effects on the smoker, but studies have also shown that exposure to smoking while the infant is still in the womb can have long-term effects on the cognitive development of the child. Rosenthal and Weitzman (2011) reviewed several studies of the effects of early intrauterine exposure to cigarette smoke. We'll look at three variables from among the many measured in these studies: head circumference at birth ($M = 36$ cm, $SD = 2$ cm), the number of months of pregnancy during which the mother smoked, and the child's score on the Bayley Infant Intelligence Test (a standardized test of intelligence in children with a population mean of 100 and a population standard deviation of 16) at the age of 5. The data shown in Table 13.5 are similar to those reported by Rosenthal and Weitzman.

Table 13.5 Head circumference, Bayley Infant IQ Score, and duration of smoking during pregnancy

Head Circumference (in Centimeters)	Bayley Score ($\mu = 100$, $\sigma = 16$)	Mother's Smoking Behavior (Number of Months During Pregnancy)
36.93	90	5.00
37.67	107	7.00
33.83	80	5.00
37.99	129	3.00
37.21	108	3.00
34.51	83	8.00
34.57	75	7.00
31.02	39	4.00
37.81	111	0.00
39.86	104	2.00
34.40	75	2.00
40.27	135	0.00

33.25	99	9.00
37.71	115	1.00
38.84	109	4.00

a. Describe the relationship between number of months that a mother smoked during pregnancy and head circumference. Don't forget to discuss the direction and strength of each of these relationships.
b. Describe the relationship between head circumference and Bayley score.
c. Construct scatterplots for each of these relationships, and determine the lines of best fit for each relationship. Predict head circumference for a woman who smokes for 3.5 months during pregnancy.

12. Table 13.6 shows the results of a study examining the relationship between the age of a mother and the birth weight of her child (in grams). Is there a relationship between the age of the mother and the body weight of her newborn?

Table 13.6 Maternal age at birth and birth weight (in grams) of the newborn

Age of Mother	Birth Weight (in Grams)
19	2,245
33	2,792
20	2,557
21	2,594
18	2,600
21	2,622
22	2,637
17	2,133
29	2,663
26	2,665
19	2,330
19	2,145
22	2,750
30	2,750
18	2,769
$\Sigma x = 334$	$\Sigma y = 38,252$
$\Sigma x^2 = 7,776$	$\Sigma y^2 = 98,256,796$

a. Describe the relationship between body weight and the age of the mother.
b. Is the relationship described in part a significant?
c. Determine the line of best fit, and predict the body weight for a mother who is 40 and for a mother who is 13.

13. You would like to know if there is a relationship between the number of hours spent studying per week and the number of hours spent playing video arcade games. You randomly select five individuals from the Student Center and record the data shown in Table 13.7. Describe the relationship between hours spent studying and hours spent playing video games.

Table 13.7 Hours spent studying and hours spent playing video games for five volunteers

ID	Hours Studying	Hours Playing
1	15	3
2	0	6
3	5	8
4	19	2
5	8	8

14. A researcher wants to know if there is a relationship between the number of hours that nine people exercised per week and the amount of milk (in ounces) these people consumed during the week. The data are shown in Table 13.8.
    a. Is there a significant relationship between these two variables?
    b. If you knew that a person spent 6 hours exercising in a week, how much milk would you predict that person drank?
    c. How much variability in milk consumption is explained by knowing the number of hours a person exercised?

Table 13.8 Exercise and milk consumption

Subject	Hours of Exercise	Milk Consumed (in Ounces)
A	3	48
B	0	8
C	2	32
D	5	64
E	8	10
F	5	32
G	10	56
H	2	72
I	1	48

15. Female college students were given the American College Health Association (ACHA) eating disorders scale together with a questionnaire that measured manic tendencies in personality (both measured on interval-level scales). Researchers wanted to know if there was a relationship between the ACHA eating disorders score and the manic tendencies score. The data are shown in Table 13.9.

Table 13.9 ACHA and mania scores for 20 individuals

ACHA Score	Mania Score
3	4
10	5
9	3
6	4
11	6
2	4
3	5
0	5
4	5
11	5
8	6
13	5
2	5
10	5
8	5
12	4
12	6
8	5
11	4
1	2

a. State the null and alternative hypotheses.
b. Decide whether or not to do a two-tailed test.
c. Calculate r, draw the scatterplot, and plot the line of best fit on the scatterplot.
d. Determine if your r-value is statistically significant.
e. Predict the manic score for a student whose ACHA score was 20.

16. Table 13.10 lists the overall quality scores and the prices for nine types of cordless phones. The overall quality score represents the phone's speech clarity and

convenience (this is an interval scale of measurement, where higher numbers represent higher clarity and convenience). At $\alpha = .05$, can you conclude that there is a correlation between overall score and price? Describe the relationship between these two variables.

Table 13.10 Overall quality score and price of nine cordless phones

Phone ID	Overall Score	Price (in Dollars)
1	81	130
2	78	200
3	72	70
4	68	80
5	66	105
6	65	130
7	67	100
8	63	150
9	57	60

17. The idea of using tests to predict who will do well in college began to emerge around 1900. Many researchers assumed that people with quick reaction times and keen sensory abilities would be quick thinkers with keen intellects—people who would, of course, make good grades. James McKeen Cattell (1896) at Columbia University gathered data on this assumption. The summary statistics below are representative of his findings:

Sensory Ability Score	GPA
$\Sigma x = 3{,}500$	$\Sigma y = 115$
$\Sigma x^2 = 250{,}000$	$\Sigma y^2 = 289$
$\Sigma x(y) = 8{,}084$	
$N = 50$	

Is sensory ability score related to grade point average?

18. Head circumference and body size (height and weight, for example) are anthropometric measures featured in most physical examinations of children and adults. Erica J. Geraedts and her colleagues (2011) were interested in determining if head circumference might be useful as an early marker for potential growth problems in children and in adults. For example, short children born small for gestational age (meaning they

were small at birth) who also have a small head circumference are a greater risk for developing a defect in insulin-like growth factor 1, which causes slow physical growth. "Sotos syndrome" (a.k.a. cerebral gigantism), a genetic disorder characterized by macrocephaly (an abnormally large head), causes very rapid growth in the first 2 to 3 years of life along with some facial and body deformities.

Geraedts et al. examined head circumference, height, and weight for a large group of children, adolescents, and young adults and found that head circumference was positively correlated with both height and body weight. Suppose you want to see if this same relationship holds for the students in your statistics class. You measure head circumference, height, and weight for 27 female students (omitting the male students) currently enrolled in your statistics class and find the data shown in Table 13.11.

a. What is the relationship between height and weight?
b. What is the relationship between height and head circumference?
c. What is the relationship between weight and head circumference?
d. Predict the head circumference for a person whose height is two standard deviations below average.
e. Predict your own head circumference from your height. Is your prediction accurate?

Table 13.11 Height, weight, and head circumference for 20 female college students

ID	Height (in Inches)	Weight (in Pounds)	Head Circumference (in Centimeters)
1	63.00	180.00	61.00
2	63.00	106.00	52.00
3	60.00	102.00	63.00
4	70.75	170.00	60.00
5	64.00	208.00	60.50
6	67.00	180.00	57.00
7	64.50	160.00	55.00
8	69.75	142.00	61.00
9	67.00	150.00	60.00
10	61.00	94.00	55.00
11	65.00	122.50	54.00
12	64.00	120.00	59.00
13	61.50	115.00	57.00
14	62.50	140.00	57.00
15	58.75	106.00	58.00
16	64.00	154.00	56.50
17	64.00	104.00	58.42

Table 13.11   Continued

18	64.00	166.00	67.31
19	63.00	135.00	58.42
20	64.00	120.00	58.42
21	65.50	152.00	60.96
22	66.00	185.00	57.00
23	67.75	142.00	54.30
24	62.00	235.00	60.50
25	66.50	120.00	57.15
26	62.75	112.00	55.25
27	65.00	145.00	59.00

19. Every year, faculty and administrators everywhere debate the pros and cons of student employment outside of the classroom. We are all aware that many students depend on the income these jobs provide. However, we are also aware that working, even only part-time, while also going to school can create tremendous difficulties with study time, writing time, and participation in school activities. Use the data presented in Table 13.12 to determine if there is a relationship between number of hours worked per week at a part-time job and GPA.

Table 13.12   Hours worked per week and GPA for 15 college students

Hours Worked per Week	GPA
21	2.5
10	2.7
20	3.0
17	3.1
13	3.2
13	3.25
20	3.3
15	4.0
17	2.4
19	3.9
20	3.0

13	1.8
13	2.0
14	1.9
14	1.8

20. Sabrina hypothesizes that people who suffer from depression are endowed with enhanced capacities for analytical thinking. Using the Hamilton Rating Scale for Depression (HSRD) and a self-designed measure of analytical capacities, Sabrina documents the severity of depression and intensity of analytical rumination scores for each of her participants. Sabrina's data are displayed in Table 13.13.

Table 13.13 Depression and rumination data

HSRD Score	Intensity of Analytical Rumination
13	4.9
24	9
16	7
21	8.6
8	4.1
17	6.8
13	5.3
19	7.6
22	7.4
26	8.8

a. Calculate the correlation coefficient, and interpret your results.
b. Should Sabrina accept or reject the null hypothesis at the .01 level?

21. Calculate the predicted Y-values using for the line of best fit with $a = -17$ and $b = 3.2$ for the following values of X:
    a. $X = 8$
    b. $X = -10$
    c. $X = 41$
    d. $X = 100$
    e. $X = -16$

22. Complete the crossword shown below.

ACROSS

2. Developer of the idea of *r*
6. Developer of the mathematics for *r*
8. Relationship where *X* increases and *Y* decreases
10. Correlations do not imply this
11. A relationship that is best described by a curved line is this
13. To split a continuous variable into 2 groups
14. A relationship where *X* increases and so does *Y*

DOWN

1. The theoretical value of *r* in the population
3. (2 words) Situation where the range of the sample is much smaller than the range in the population
4. When extreme variables tend to get closer to the mean the second time they are measured
5. (2 words) Straight line that minimizes deviati of all *X*, *Y* pairs is the line called this
7. (2 words) The symbol for the coefficient of determination
9. A relationship between *X* and *Y* that can be described by a straight line
10. When 2 variables (*X* and *Y*) vary together they are said to be this
12. A variable that is not controlled for but might play a role in the relationship is said to be this

586 STATISTICS IN CONTEXT

Think About It...

WHAT DOES PERFECT MEAN?

SOLUTIONS

1. How does the "cloud" of dots change as the relationship approaches zero? **The "cloud" gets less and less linear, and more and more cloud-like ("wider," more variable, and more diffuse), as *r* approaches zero.**
2. Row B shows a series of perfect correlations, positive on the left and negative on the right. Why does the *r*-value of the three positive correlations depicted here not change? Why are they all "perfect," but don't all look alike? **"Perfect" and "predictable" are interchangeable here. A single straight line can describe all of the relationships, but the slope of the line changes. An *r* of zero (center of row B) indicates that *X* and *Y* are not related to one another, so as *X* changes, there is zero change in *Y*. For all the other correlations, a change in *X* predicts a change in *Y*.**
3. What is the difference in the variability of *X* and *Y* when the correlation is perfect and when it is not? **The variability of *Y* given a particular *X* is as small as it can possibly be if $r = +1$ or -1. For the other relationships, other values of *r*, the value of *Y*, given an *X*, varies.**
4. What would the value of *r* be for the relationship in the center of the scatterplots in row B? **$r = 0.00$.**
5. Is the relationship in the center of row B also "perfect"? **In a way. *Y* is perfectly predictable for any given *X*.**
6. Row C shows scatterplots where $r = 0$. What can you say about the relationships in row C? Is there truly no relationship at all between *X* and *Y* in these plots? What does a Pearson *r* of zero mean? **X and Y are related in each of these scatterplots, but the relationship is not strictly linear: It cannot be described with a single straight line.**

SHARED VARIABILITY AND RESTRICTED RANGE

SOLUTIONS

1. What kind of correlation would you expect to see? **Positive.**
2. Calculate the coefficient of determination for the correlation between each SAT section and FYGPA, and then sketch the Venn diagram to illustrate the amount of shared variability between *X* and *Y*.

SAT-CR and FYGPA	SAT-M and FYGPA	SAT-W and FYGPA
$r = .29$	$r = .26$	$r = .33$
$r^2 = .084$ (8.4%)	$r^2 = .068$ (6.8%)	$r^2 = .11$ (11%)

   Venn Diagrams:

3. Recalculate the coefficients of determination and compare the two sets of measures of the relationship.

SAT-CR and FYGPA
$r = .48$
$r^2 = .23$ (23%)

SAT-M and FYGPA
$r = .47$
$r^2 = .22$ (22%)

SAT-W and FYGPA
$r = .51$
$r^2 = .26$ (26%)

Venn Diagrams:

4. Does it make sense to you that the range of SAT scores for admitted students was less variable than the range in the whole population? Why? **Yes. High school seniors who apply to college might well have higher and more consistent SAT scores than do seniors who do not opt to apply.**

5. What did the restricted range do to the correlations? **It made them appear to be lower than they really were.**

REFERENCES

Berberich, G., Berberich, M., Grumpe, A., Wohler, C., & Schreiber, U. (2013). Early results of the three-year monitoring of red word ants' behavioral changes and their possible correlation with earthquake events. *Animals, 3*, 63–84.

Cattell, J. M., and Farrand, L. (1896). Physical and mental measurements of students at Columbia University. *The Psychological Review, 3*(6), 618–648.

Chen, A., & Escarce, J. J. (2010). Family structure and childhood obesity, early childhood longitudinal study—Kindergarten cohort. *Preventing Chronic Disease, Public Health, Research, Practice and Policy, 7*(3), 1–8.

Geraedts, E. J., van Dommelen, P., Caliebe, J., Visser, R., Ranke, M. B., van Buuren, S., Wit, J. M., & Oostdijk, W. (2011). Association between head circumference and body size. *Hormone Research in Paediatrics, 75*, 213–219.

Green, J., Cairns, B. J., Casabonne, D., Wright, F. L., Reeves, G., & Beral, V. (2011). Height and prospective cancer incidence in the Million Women Study: Prospective cohort, and meta-analysis of prospective studies of height and total cancer risk. *Lancet Oncology, 12*, 785–794.

Hatfield, J., Faunce, G. J., & Soames, J. (2006). Avoiding confusion surrounding the phrase "correlation does not imply causation." *Teaching of Psychology, 33*(1), 49–51.

Kobrin, J. L., Patterson, B. F., Shaw, E. J., Mattern, K. D., & Barbuti, S. M. (2008). *Validity of the SAT for predicting first-year college grade point average* (College Board Research Report No. 2008-5). New York, NY: College Board.

Richardson, M., Abraham, C., & Bond, R. (2012). Psychological correlates of university students' academic performance: A systematic review and meta-analysis. *Psychological Bulletin, 138*(2), 353–387.

Rosenthal, D. G., & Weitzman, M. (2011). Examining the effects of intrauterine and postnatal exposure to tobacco smoke on childhood cognitive and behavioral development. *International Journal of Mental Health, 40*(1), 39–64.

Stanton, J. M. (2001). Galton, Pearson, and the peas: A brief history of linear regression for statistics instructors. Retrieved from www.amstat.org/publications/jse/v9n3/stanton.html

Using SPSS and R for Correlation and Regression

The data set we will be using for this tutorial is drawn from the Behavioral Risk Factor Surveillance System (BRFSS), a nationwide health survey carried out by the Centers for Disease Control and Prevention. The set we will be using is based on results of the 2008 questionnaire, completed by 414,509 individuals in the United States and featuring a number of questions about general health, sleep, and anxiety, among other things. The version we will be using is a trimmed-down version of the full BRFSS dataset, which you can find on the companion website for this textbook www.oup.com/us/blatchley.

Pearson Correlation Coefficient in SPSS

If we want to better understand how two variables may be correlated, we can extract the Pearson correlation coefficient (r) from the data. In this case, we would like to examine whether there is a relationship between an individual's age and the number of nights of poor sleep the individual had during the 30 days prior to completing the questionnaire.

Our null and alternative hypotheses are:

H_0: $r = 0$; no relationship exists between the variables.

H_1: $r \neq 0$; a relationship exists between the variables.

STEP 1: Load the data into SPSS.

STEP 2: Set up the test. Click **Analyze**, then **Correlate**, and then **Bivariate...**

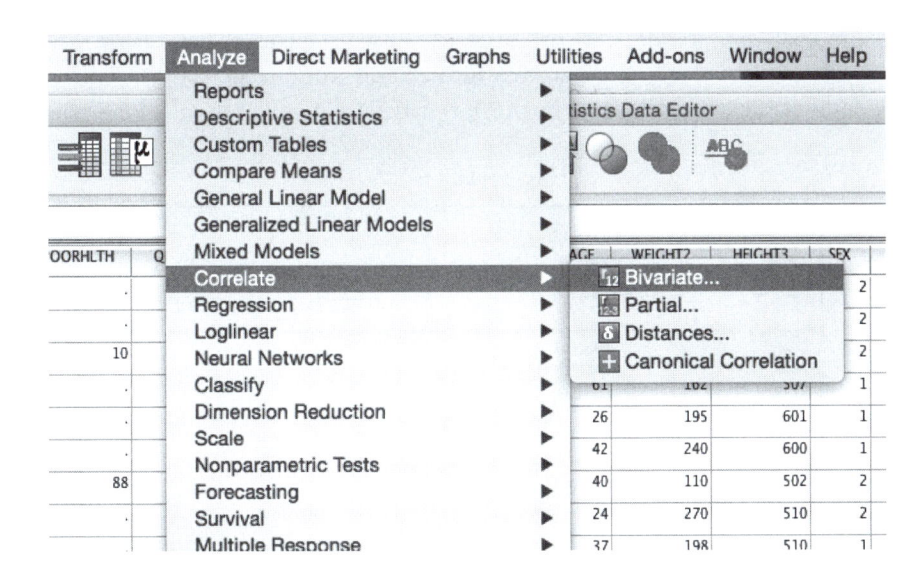

CHAPTER 13 Correlation and Regression: Are We Related?

STEP 3: Select the variables. In the **Bivariate Correlations** dialogue box, select the variables you are interested in testing. In this case, select **How Many Days Did You Get Enough Sleep (QLREST2)** and **Reported Age in Years (Age)**. Pearson is automatically selected in the dialogue box.

Click **OK**.

STEP 4: Analyze the results. The correlation matrix will populate in the **Output** window in SPSS. Here, we are able to see the size of the coefficient, the direction of the relationship between the variables (+ or −), and whether it is statistically significant.

Correlations

		HOW MANY DAYS DID YOU GET ENOUGH SLEEP I	REPORTED AGE IN YEARS
HOW MANY DAYS DID YOU GET ENOUGH SLEEP I	Pearson Correlation	1	−.059**
	Sig. (2-tailed)		.000
	N	260855	258902
REPORTED AGE IN YEARS	Pearson Correlation	−.059**	1
	Sig. (2-tailed)	.000	
	N	258902	410856

**. Correlation is significant at the 0.01 level (2-tailed).

In the example, our two variables—age and number of nights of poor sleep—have a negative relationship: In the sample we studied, higher age values are associated with fewer nights of poor sleep over the previous 30 days. However, the size of the coefficient is small, suggesting only a small relationship between the variables. If we look at the Sig. value (*p*-value), we can see that this result is statistically significant at the .05 level.

This may be a first step in determining what IVs impact our DV. If we wanted to continue to explore this relationship to see the size of the impact of our IV on our DV, we would conduct a regression.

Conducting a Regression in SPSS

We will be building off of our last example to explore whether there is a relationship between the number of nights of poor sleep individuals have and their age. Specifically, we want to see the direction of the relationship (the sign) and the magnitude of the relationship (the coefficient) for the IV in order to understand how the variables move together. We also want to know how closely correlated our variables are; for this, we will use the coefficient of determination or the r^2 statistic.

STEP 1: Load the data into SPSS.

STEP 2: Look at the data. Create a scatterplot to see if you can see any correlation. Click **Graphs**, and then click **Chart Builder**.

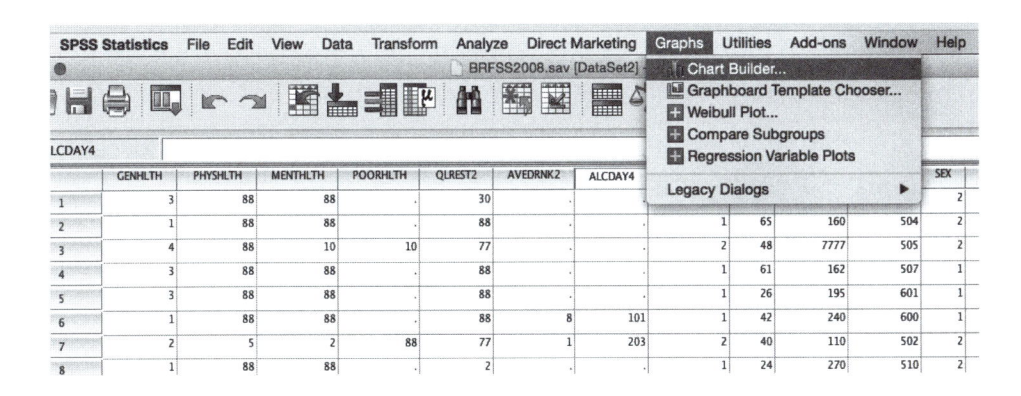

STEP 3: Build your scatterplot. In the **Chart Builder,** menu select **Scatter/Dot**. A set of eight options will populate to the right. Select the **Simple Scatter** option (the top-left choice). Drag and drop the two variables of interest into the chart preview window. The *Y*-value (the DV) is **How Many Days Did You Get Enough Sleep**. The *X*-value (the IV) is **Age**.

Click **OK**.

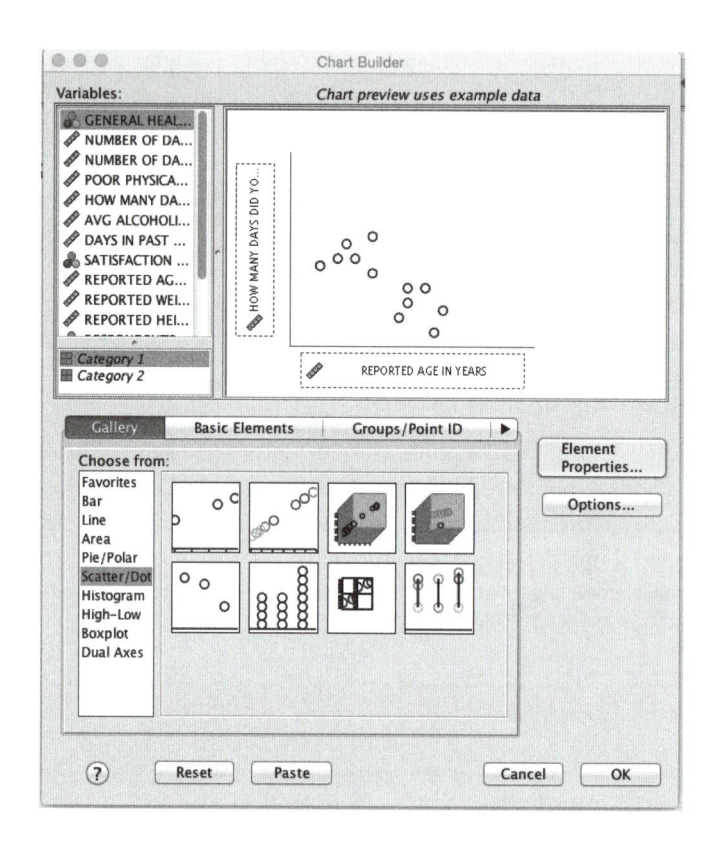

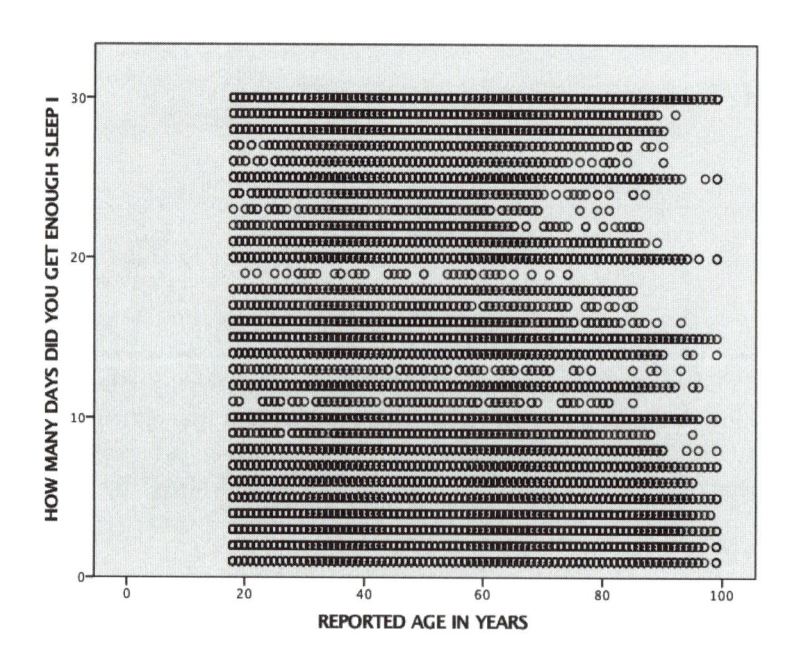

The scatterplot will appear in the **Output** window of SPSS. What does it look like? In this case, we have a large number of observations ($n = 402,000$), so it is hard to decipher a clear picture. With smaller surveys, it is easier to see if there is, or is not, a relationship between the data. From the picture here, it seems as though people of all ages experience a range of sleep patterns, although it does appear that people either experience good sleep (25 days of poor sleep in the last 30 days).

STEP 4: Run a linear regression. Click **Analyze**, then **Regression**, and the **Linear.**

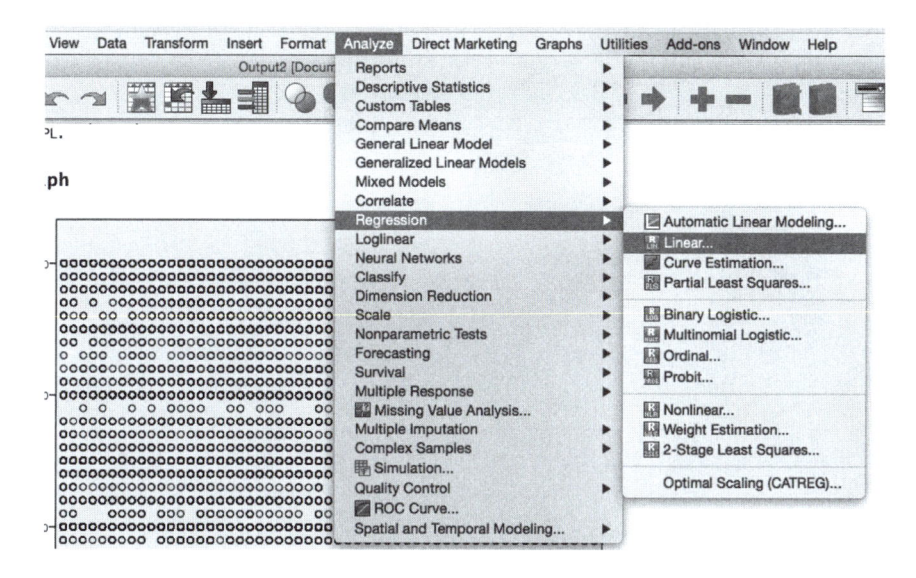

STEP 5: Set up your variables. Select **How Many Days Did You Get Enough Sleep (QLREST2)** as your DV, and move it to the **Dependent** box using the first arrow.

Select **Reported Age in Years** (Age) as your IV, and move it to the **Independent(s)** box using the second arrow.

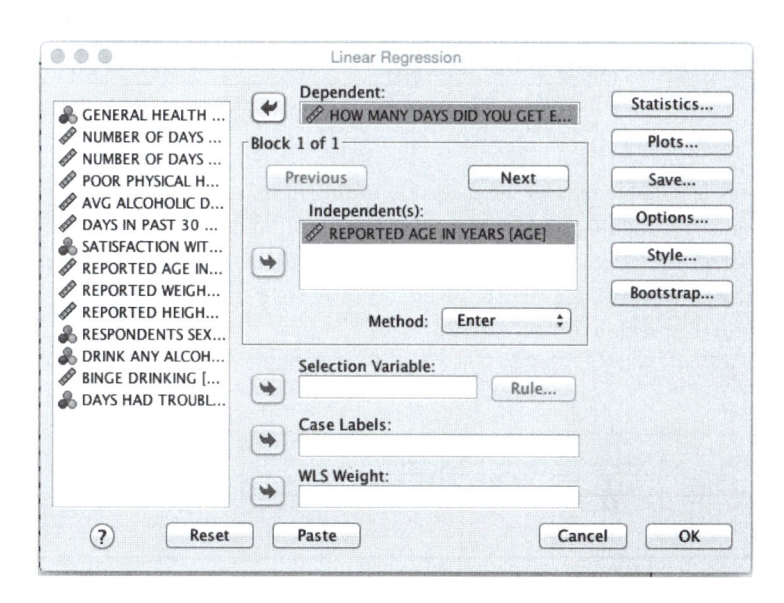

CHAPTER 13 Correlation and Regression: Are We Related? 593

STEP 6: Set up your descriptive statistics. Click **Statistics. . .**, and then select **Estimates** and **Confidence intervals**, as well as **Model fit**.

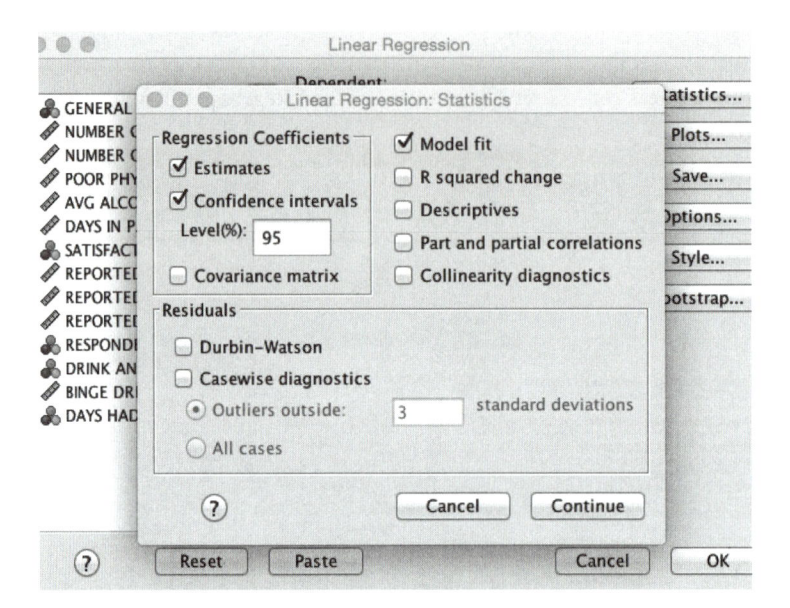

Click **Continue**, and then click **OK**.

STEP 7: Interpret your results. The model summary gives us the r and r^2 values.

With respect to goodness of fit, a linear regression does not appear to fit the data well. We have an r-value of .059 and an r^2 of .003. Remember that the r-values can range between −1.00 and 1.00. These are very low values.

Model Summary

Model	R	R Square	Adjusted R Square	Std. Error of the Estimate
1	.059[a]	.003	.003	10.178

a. Predictors: (Constant), REPORTED AGE IN YEARS

The coefficients table tells us the direction of the linear relationship between our variables (the sign) and the magnitude of the relationship (the coefficient, or the slope of the line). The coefficient −.038 suggests that the number of nights of poor sleep decreases (very slightly) with age. So as the age of respondent increases we would expect the number of poor night's sleep in the previous 30 days to decrease.

Coefficients[a]

Model		Unstandardized Coefficients B	Std. Error	Standardized Coefficients Beta	t	Sig.	95.0% Confidence Interval for B Lower Bound	Upper Bound
1	(Constant)	14.299	.068		209.855	.000	14.165	14.432
	REPORTED AGE IN YEARS	−.038	.001	−.059	−30.045	.000	−.041	−.036

a. Dependent Variable: HOW MANY DAYS DID YOU GET ENOUGH SLEEP I

STATISTICS IN CONTEXT

More than one independent variable is often needed to "explain" the variation in the dependent variable. Adding variables to your model will never decrease the r or r^2 value—that's why the authors of some research papers will adopt a "kitchen sink" model, where they include every variable they think is relevant. The example above is very simple and does not necessarily capture the variation in sleep patterns between individuals in the survey. Other factors would also be important to consider, including (and not limited to) sex, health status, employment status, and number of children. Consider what other IVs may be relevant when constructing your regression.

Pearson Correlation Coefficient in R

Obtaining Pearson's correlation coefficient in R is fairly simple and can be done with a few lines of code. In this example, we will be testing whether there is a statistically significant correlation between an individual's age and the number of poor night's sleep that individual had in the previous 30 days.

First, to run the specific code, we will need to load the Hmisc and foreign libraries.

For this you will need to make sure you have the foreign package loaded and installed. To complete these steps, go to **Packages and Data,** and click **Package Installer.**

In the **Package installer** window click **Get List** search for **Hmisc** and **foreign** separately. Once you have found each individually, click **Install Selected**; make sure that the **Install Dependencies** check box is selected. Exit the window.

Go to **Packages and Data,** and click **Package Manager.**

In the **Package Manager** window search for **Hmisc** and **foreign,** and select the check box beside it, make sure that the status column changes to "loaded."

STEP 1: Load your data into R.

```
>sleep< - read.csv("example.csv")
```

STEP 2: Drop any missing variables. Any missing variables will cause the calculated correlation to be incorrect.

We must recode any missing or not-applicable data as follows:

```
#Remove "Do not know", no response and 0 from the sample of QLREST2.
>sleep$QLREST2[sleep$QLREST2==77]<-NA
>sleep$QLREST2[sleep$QLREST2==88]<-NA
>sleep$QLREST2[sleep$QLREST2==99]<-NA

#Remove "Do not know" and no response
>sleep$AGE[sleep$AGE==7]<-NA
>sleep$AGE[sleep$AGE==9]<-NA
```

STEP 3: Run the test. We want to explore whether there is a correlation between age of our respondents and the number of nights of poor sleep they had in the previous 30 days, if they had at least one poor night's sleep.

```
> rcorr(sleep$QLREST,sleep$AGE, type="pearson")
```

The results will populate automatically. Note that R rounds to two decimals.

```
      x       y
x    1.00   -0.06
y   -0.06    1.00

n
      x        y
x   260855   258902
y   258902   410856

P
      x       y
x             0
y     0
```

The rcorr( ) function provides us with the value of correlation between the *X* and *Y* variable as well as the number of observations and the *p*-value. You can see them listed above.

In this example, we have a small negative correlation between age and number of nights poor sleep. This value is significant at the .05 level.

Conducting a Regression in R

Now that we can confirm there is a statistically significant correlation between age and the number of poor night's sleep in our sample, we want to see what the magnitude of that relationship is. To determine the size of the impact, we must conduct a regression. This will allow us to measure the size of the impact that age has on the number of poor night's sleep, and also the sign of the relationship.

STEP 1: Load your data into R.

```
>sleep<- read.csv("example.csv")
```

STEP 2: Drop any missing variables. Any missing variables will cause the calculated correlation to be incorrect.

We must recode any missing or not-applicable data as follows:

```
#Remove "Do not know", no response and 0 from the sample of
QLREST2.
>sleep$QLREST2[sleep$QLREST2==77]<-NA
>sleep$QLREST2[sleep$QLREST2==88]<-NA
>sleep$QLREST2[sleep$QLREST2==99]<-NA

#Remove "Do not know" and No response
>sleep$AGE[sleep$AGE==7]<-NA
>sleep$AGE[sleep$AGE==9]<-NA
```

STEP 3: Run the regression, and show your results.

```
> # Linear Regression Example
> regression <- lm(QLREST2 ~ AGE, data=sleep)
```

Note that you are able to add any number of independent variables to this code, you simply add "+ 'variable name'" after the ~.

```
# show results
> summary(regression)

Call:
lm(formula = QLREST2 ~ AGE, data = sleep)

Residuals:
    Min      1Q   Median      3Q     Max
-12.609  -8.611   -3.110    7.580  19.498

Coefficients:
              Estimate   Std.Error   t value   Pr(>|t|)
(Intercept)  14.298885    0.068137    209.85    <2e-16 ***
AGE          -0.038351    0.001276    -30.05    <2e-16 ***
---
Signif. codes:  0 '***' 0.001 '**' 0.01 '*' 0.05 '.' 0.1 ' ' 1

Residual standard error: 10.18 on 258900 degrees of freedom
  (155607 observations deleted due to missingness)
Multiple R-squared:  0.003475,   Adjusted R-squared:  0.003471
F-statistic: 902.7 on 1 and 258900 DF,  p-value: < 2.2e-16
```

R will automatically generate an intercept value. The estimated coefficient for AGE is −.0381 (remember that it was −.038 in SPSS; this is just a significant digit). We can see that the *p*-value for a *t*-sided *t*-test shows that the variable is statistically different from zero at the .005 level. Therefore, we can now assign a magnitude and a direction to the correlation between age and the number of poor night's sleep for our respondents.

CHAPTER FOURTEEN

THE CHI-SQUARE TEST

*Thou shalt not answer questionnaires
Or quizzes upon world affairs,
Nor with compliance
Take any test. Thou shalt not sit
with statisticians nor commit
A social science.*

—W. H. AUDEN

Everyday Statistics

DO YOU LIKE HORROR MOVIES?

Some people are die-hard (I apologize) fans of horror movies; others just avoid them.

Whether you're in the former or the latter category, you probably have an idea of how the typical horror movie scene unfolds. It usually features a pretty young woman—the damsel in distress—being chased by a demented killer, with lots of blood-curdling screams ensuing.

Eston Martz (2012), a statistics blogger, used what's known as a chi-square test to find out if there was a "gender gap" in ways that characters in horror movies die. Are the deaths of female characters more gruesome than those of male characters? He examined movies in the *Friday the 13th* and *Halloween* franchises and categorized first the sex of the victim and then the cause of death for that victim: stabbing, blunt-force trauma, removal of essential body parts (like heads), shooting, and "exotic" (a catch-all for more unusual causes of death that didn't fit the other categories).

It turned out that women were *not* more likely to be the victim in these movies; in fact, only about 32% of the victims were female. However, the type of death did tend to differ across gender, with male characters being significantly more likely to suffer death by blunt-force trauma and the removal of body parts and women more likely to be dispatched using some unusual, "exotic" method. I don't know if this reassures me or not. Chapter 14 will investigate the chi-square test, and you will learn why it is so well suited to this kind of analysis.

OVERVIEW

- THE CHI-SQUARE TEST AND WHY WE NEED IT
- THE ONE-WAY CHI-SQUARE TEST FOR GOODNESS OF FIT
- THE TWO-WAY CHI-SQUARE TEST OF INDEPENDENCE
- NONPARAMETRIC TESTS IN CONTEXT: TYPES OF DATA REVISITED

LEARNING OBJECTIVES

Reading this chapter will help you to . . .

- Become familiar with the origins of the questionnaire and the type of data usually collected on a questionnaire. (Concept)
- Identify the differences between parametric and nonparametric inferential testing. (Concept)
- Explain why frequency of response (the mode) is compared in a chi-square test, rather than to the mean. (Concept)
- Calculate a chi-square statistic. (Application)
- Understand the difference between a goodness-of-fit test and a test for independence. (Concept)
- Use the goodness-of-fit test and the test for independence appropriately. (Application)

THE CHI-SQUARE TEST AND WHY WE NEED IT

The chi-square test was devised by Sir Francis Galton, whom we encountered last chapter in the context of pea plants and the Pearson product–moment correlation coefficient. Galton's creation of the chi-square test had nothing to do with pea plants. It stemmed from his work with a survey he had devised in an effort to discover what makes someone a scientist (see *The Historical Context* on p. 601). Galton realized he could not subject the qualitative data yielded by the survey questionnaires to the same rigorous analysis he practiced when using quantitative data, which were ideally suited to calculating means and standard deviations. His solution was the chi-square test and the chi-square statistic this test depends on.

PARAMETRIC VERSUS NONPARAMETRIC TESTING

So far, we have been discussing a set of statistical tests (the z-, t-, F-, and r-tests) that are known as *parametric tests*. If you recognize in this term a word we have used before, you are halfway to understanding what we mean by parametric and nonparametric tests. Recall that a *parameter* is a statistic that describes a characteristic of the population. The mean of a population ("mu," or μ) is a *parameter* that describes the arithmetic average of the population, while $\bar{X}$ is a *statistic* that describes the arithmetic average of a sample. Sigma (σ, or s) is a number that describes standard deviation: If it represents the standard deviation of a population (σ), then the number is a parameter; if it describes the standard deviation of a sample (s), then it is a statistic.

parametric test A statistical analysis that is based on assumptions about the parameters (mean and standard deviation) of the population.

So, a **parametric test** is one that tests assumptions about a population using one of two key parameters: the population mean or the standard deviation. These tests are possible because the quantitative data allow meaningful measures of center and spread (mean and standard deviation) to be calculated.

nonparametric test A statistical analysis that is not based on assumptions about the parameters of the populations stated in the central limit theorem.

You have likely guessed that nonparametric tests do not involve population means or standard deviation, and this is correct: The **nonparametric test** pioneered by Galton is used in cases where means and standard deviation cannot be generated from the available data, and for this reason, nonparametric tests seldom generate findings that are used to make assumptions about a population. A simple way to distinguish between the two kinds of tests is to remember that parametric tests are used with quantitative data and nonparametric tests with qualitative data. Why the nonparametric test is necessary should become clear as you read through the following sections.

The Historical Context

THE QUESTIONNAIRE

In Chapter 13, we discussed Sir Francis Galton, creator of the idea of a correlation, if not the mathematics behind it, and a scientist interested in just about everything (a *polymath*). Galton was particularly interested in the "nature versus nurture" question, which has a long history in psychology. Specifically, he wanted to investigate which component of a person's personal history matters more in determining the person turns out as an adult: the genetic legacy inherited from parents or the environment where a person grew up.

In *English Men of Science: Their Nature and Nurture* (1874/2011), Galton detailed his passionate support for the nature side of the argument. He recounted his attempts to determine what made a man a scientist. In his introduction to the problem, he asks:

> What then are the conditions of nature and the various circumstances and conditions of life—which I have included under the general name of nurture—which have selected that one [to become a scientist] and left the remainder? The object of this book is to answer this question. (Galton, 1874/2011, p. 10)

If this were your question, how would you go about collecting your data? Galton first determined who the most famous living scientists of his day were: He came up with a list of 180 men acknowledged by their peers to be important "men of science" (Galton excluded women as intellectually inferior). Then, he determined what the likely antecedent conditions to becoming a scientist of note might be. He decided that the variables that mattered included a person's race, place of birth, birth order, number of siblings, parents' occupation and social status, and even what he referred to as parents' physical peculiarities, such as hair color, personality, and physical build (corpulent versus thin, for example). Table 14.1 shows some of the categories that Galton was interested in.

Galton then sent each of the men on his list one of the first questionnaires ever to be used in science. It included questions about the men's political and religious affiliations, their hat size (a measure of head size), their birth order, and so on. He also asked his respondents to state why they became interested in science and whether they believed this interest to be innate.

TABLE 14.1 Sample questions from Galton's questionnaire

Question Category	Question
Place of birth	Mention if it was in a large or small town, a suburb, a village, or a house in the country.
Temperament of parents and self	What is the temperament of self, father, and mother, if distinctly nervous, sanguine, bilious, or lymphatic?
Strongly marked mental peculiarities bearing on scientific success	Strong mental peculiarities of self, father, and mother—the following list may serve to suggest: impulsiveness, steadiness, strong feelings and partnerships, social affections, religious bias of thought, love of the new and marvelous, curiosity about facts, love of pursuit, constructiveness of imagination, foresight, public spirit, disinterestedness.
Birth order	State the number of males and those of the females in each of the following degrees of relationship who have attained 30 years of age, or thereabouts: grandparents, both sides; parents, uncles, and aunts, both sides; brothers and sisters; first-cousins of all four descriptions; nephews and nieces. How many brothers and sisters had you older than yourself, and how many younger?

Using the data he collected, Galton came to a number of conclusions about the influence of nature and nurture on the lives of these men of science. For example, he concluded that the majority of these famous men were born in the center of the country, not the coastal areas. Also, they tended to be of "pure" English or "pure" Scottish blood rather than of mixed blood, and they were born interested in science. On this last point, Galton writes: "[A] special taste for science seems frequently to be so ingrained in the constitution of scientific men, that it asserts itself throughout their whole existence" (Galton, 1874/2011, p. 193).

The use of questionnaires to collect scientific data was very new when Galton sent out his lengthy letters, and Galton was (again) an original in thinking that the information he collected this way could be used to answer a question as ambitious and broad as whether nature or nurture caused a person to be interested in science. After all, the answers to his questions did not lend themselves to calculating means and standard deviations. Imagine performing calculations on data gleaned from a question like "Your father and your mother, are they respectively English, Welsh, Scotch, Irish, Jewish, or foreign?" Yet this was exactly what Galton wanted to do.

The *t*-test that Galton's friend and colleague William Gossett was working on would not work with the data Galton had collected on his questionnaire. So, to analyze his data, and this should also be no surprise, Galton came up with an entirely new statistic that would allow him to use the data he had collected to answer his question. The statistic he came up with was the chi square, used in a test called the chi-square test, which is the subject of this chapter.

MAKING ASSUMPTIONS

The tests we have discussed so far start with hypotheses about parameters and then test the probability that these hypotheses are true, using samples drawn from the population as guesses about what the values of the parameters might be. Let's take a null hypothesis as an example. And since we have been talking about sleep for quite a while, let's take a statement that I have heard quite frequently: Americans are, as a nation, quite sleep-deprived. To test that statement, we need to define a population (this one is easy to obtain: it's the current adult population of America) with at least a theoretical mean (8 hours of sleep per night). Next, we draw a representative sample from that population and measure the sleep duration of the members of our sample. We can then use the appropriate test to compare the average sleep duration in the sample to the average in the population in order to see how much they differ.

The null and alternative hypotheses that we have been testing are hypotheses about populations. For example, the null hypothesis would be stated as shown in Box 14.1, and it would refer to the parameters of the population.

BOX 14.1 **Research question: Are Americans sleep-deprived?**

Null hypothesis
$H_0: \mu_{\bar{X}} = 8$
$H_1: \mu - \mu_{\bar{X}} = 0$

Alternative hypothesis
$H_1: \mu_{\bar{X}} \neq 8$
$H_1: \mu - \mu_{\bar{X}} \neq 0$

Notice that we are making our hypotheses about what's happening in the population by using samples as guesses about the population parameters. The data must allow us to calculate a meaningful mean, which is just another way of saying that it must be quantitative data (interval level or ratio level).

With Galton's qualitative data, we are no longer using means and standard deviations. Galton and his partner Pearson determined that they would have to change both the assumptions they were making about the data *and* the statistic they used to model the population. They couldn't use the sample mean to guess about the population mean because they could not calculate a meaningful mean from their data. They could, however, use a different descriptive statistic—for example, the mode. The mode would describe frequency. In the case of Galton's questionnaire, the mode would describe the number of respondents who answered one of Galton's questions in a particular way. Let's take a look at some of the data that Galton collected on the question "What is your father's occupation?" The results are shown in Box 14.2.

BOX 14.2   **Responses to Galton's question, "What is your father's occupation?"**

Responses	Frequency	"Sum"
1 = Noblemen and private gentlemen	9	9
2 = Army or Navy, civil service	18	36
3 = Law or other professionals (teacher, architect, clergy or minister)	34	102
4 = Banker, merchant, or manufacturer	43	172
5 = Farmers	2	10
6 = Other	1	6
TOTAL	107	335

$$\bar{X} = \frac{\sum x}{n}$$
$$= \frac{335}{107}$$
$$= 3.13$$

Does it make sense to calculate the mean occupation of the fathers in this data set? We could assign the numeral 1 to the answer "Noblemen and private gentlemen" and 2 to the answer "Army or Navy, civil service" and so on. We could then add up the number of 1's (there were 9 of them) and 2's (there were 18), and so on, to come up with a sum that we could then divide by 107 and produce a mean. The problem is that the mean is meaningless. We find that the average occupation of the fathers of these prominent scientists was 3.13, which is 0.13 unit (whatever our units are here) between a lawyer/other professional and a banker. Lawyer jokes aside, what would that occupation be?

Galton and Pearson quickly realized that there was more useful information in the mode than in the mean, at least for their purposes. The modal occupation for fathers of the 107 scientists who filled out his questionnaire was a banker, merchant, or manufacturer. The data suggests that 43 out of 107 (or 40.2%) of the men in this sample came from homes where Dad was a banker or a businessman (in today's vernacular). Now that is a meaningful statistic.

Galton and Pearson set out to create a statistic that would allow them to develop hypotheses they could test using the **frequency** of responses in a given category as their data points. They would test their hypotheses in the usual way—by creating null and alternative hypotheses and then determining the probability of the null being false by random chance alone. Since they would not be hypothesizing about the mean response or the variability in responses anymore—since they would not be using parametric tests, in other words—the hypotheses would have to change.

frequency The number of times an event occurs; in our case, the number of times an observation occurs in a data set.

So, let's use Galton's data again to set up our hypotheses. We know the largest number of scientists reported that their fathers were employed as bankers or businessmen. However, would we see this difference in the frequency of an occupation just by random chance, or is this difference the result of our independent variable, the fact that these men are all scientists? Our null hypothesis would be that having a banker or a businessman for a father was no more likely or unlikely than having a father with any other occupation. You could also say that all of the occupations listed were equally likely. If the null hypothesis is true, then all of the occupations listed should be equally likely.

expected frequency The frequency of an event that is predicted by a model or by the rules of probability.

The alternative can be stated, as you might expect, as follows: All of the occupations are *not* equally likely, or some are more or less likely than others. What kind of evidence for or against the null hypothesis should we look for? Galton and Pearson decided they would compare the **expected frequency** of an event (in this case, an event is a paternal occupation) with the frequency that actually shows up in the data, or the **observed frequency**.

observed frequency The frequency of an event that is observed in a sample.

THE ONE-WAY CHI-SQUARE TEST FOR GOODNESS OF FIT

We want to compare the way we expected paternal occupation to be distributed (all occupations should be equally likely if only random chance is operating) with the actual distribution we see in the data (the observed frequencies of each occupation). In other words, we are checking to see if the model we expect actually fits the data we collected. This explains why the full name of the test we'll run to test this hypothesis is the **One-Way Chi-Square Test for Goodness of Fit**.

chi-square test for goodness of fit A nonparametric test that can be used to determine if there is a good fit between expected and observed event frequencies; also known as a one-way chi-square test.

The observed frequency is easy to find: It's the data in Box 14.2. The expected frequency takes a bit of thought. If the null hypothesis is true, then we would expect to see the same number of fathers in each occupational category. We have 107 answers in our data set and six occupational categories. If we divide the number of data points we have by the number of categories, we should wind up with a theoretical, expected frequency for each category. Box 14.3 shows how we can determine the expected frequency for each category.

BOX 14.3 **Expected versus observed frequency**

Responses	Observed (f_o)	Expected (f_e)
Noblemen and private gentlemen	9	17.83
Army or Navy, civil service	18	17.83
Law or other professionals	34	17.83
Banker, merchant, or manufacturer	43	17.83
Farmers	2	17.83
Other	1	17.83

Finding the expected frequency (f_e)

$$f_e = \frac{\text{Total}}{\text{\# Categories}}$$

$$= \frac{107}{6} = 17.83$$

The expected frequencies reflect what the null hypothesis would predict (theoretical data). The observed frequencies are empirical.

It certainly looks as though there is a difference between observed and expected frequencies, although in some categories, observed and expected frequencies are quite similar. Take a look at observed and expected frequencies for the military/civil service category of paternal employment: If the null hypothesis is true, we would expect the number of fathers in this category to be 17.83 (because these are theoretical data, we can have 83 hundredths of a person). We actually saw 18 people in this category. On the other hand, in other categories (farmers, for instance), the number of people we expected to see is very different from what we actually saw.

Galton and Pearson developed a statistic that would reflect the difference between the observed and the expected data. Think about how you would go about finding out if what you expected to see was different from what you actually saw. Then, take a look at the formula for this statistic in Box 14.4.

BOX 14.4 The formula for chi square (χ^2)

$$\chi^2 = \sum \frac{(f_o - f_e)^2}{f_e}$$

where

f_o = observed frequency

f_e = expected frequency

NOTE: χ^2 is chi square, pronounced with a hard *k* as in *kite*, not the soft *ch* as in *Chia pet* or *pumpkin chai latte*.

To see how what is expected differs from what is observed, we simply subtract the two [i.e., $(f_o - f_e)$]. We then square that difference to get rid of any negative signs [i.e., $(f_o - f_e)^2$]. Next, we divide this squared difference by the expected frequency for that category, and then put this value aside. This division standardizes the difference between observed and expected because the expected frequency is what we have proposed to exist in the null hypothesis. We repeat this process for each category in our model, then add up all of these difference values to come up with our statistic. Box 14.5 shows the calculation of χ^2 for the paternal occupation data Galton collected from his questionnaire.

BOX 14.5

Calculating $\chi^2 = \sum \dfrac{(f_o - f_e)^2}{f_e}$

	Nobleman	Military/Civil service	Law/Other Professional	Banker/other Businessman	Farmer	Other
Observed	9.00	18.00	34.00	43.00	2.00	1.00
Expected	**17.83**	**17.83**	**17.83**	**17.83**	**17.83**	**17.83**
$f_o - f_e$	−8.83	0.17	16.17	25.17	−15.83	−16.83
$(f_o - f_e)^2$	77.97	0.029	261.47	633.53	250.59	283.25
$(f_o - f_e)^2/f_e$	4.37	0.0016	14.66	35.53	14.05	15.89

$\chi^2 = 4.37 + 0.0016 + 14.66 + 35.53 + 14.05 + 15.89 = 84.5016$

The gray shaded row in Box 14.5 is the difference between the observed and the expected. The red shaded row at the bottom of the table shows the chi-square values, or difference values, for each category in our design. In the final row, we sum them all to get the overall chi-square statistic.

Suppose the value of your chi-square statistic were zero: What would that tell us about the difference between the observed and expected frequencies? The only way we could get a chi square of zero would be if the observed and expected frequencies were exactly the same, meaning that what we saw in the data was what we expected to see if the null were true.

A small chi-square statistic (close to zero) would indicate very little difference between the observed and expected frequencies. A very large chi-square statistic would indicate a very large difference between the two. So, how far away from zero is far enough away for us to reject the null hypothesis? (Remember, the null hypothesis states there is no difference between the observed and the expected frequencies.) If you said that we need to compare our observed chi-square value with a critical value from a table, as we've been doing right along with parametric tests, you would be right. However, before we consult the almost inevitable table, let's do a quick review of why this works.

Back when we were comparing means (sample and population, or observed mean versus theoretical mean) we were basing our decision about how far out in a distribution of all possible sample means our sampling distribution of the means had to be in order to reject H_0. Our decision depended on the shape of the SDM. The people who studied SDMs in general, and who contributed to the writing of the central limit theorem, said that SDMs would be normal in shape and would have a variability that was equal to the population variability divided by the square root of the sample size used to create the SDM. Determining what the critical value of the test statistic would also depend on the size of the sample (df). For Gossett, $df = n - 1$. For Pearson and Galton, the df for the chi-square test was defined as the number of categories $-$ 1.

Pearson and Galton examined their chi-square statistic and determined how it would be distributed. The assumption at the time was that this statistic, like the t and the z, would be normally distributed—in order words, that most chi-square values should be centered around zero, with very large chi-square values being less frequent or less probable. However, Pearson found that the shape of the distribution of χ^2 was not normal, at least when the data were divided into only a few categories. The distribution of χ^2 (which reflects the difference between observed and expected frequencies) was seriously positively skewed when the number of categories was small. The distribution became more and more normal in shape only as the number of categories was increased.

Figure 14.1 shows the distribution of chi-square statistics. The value of chi square is shown on the *x*-axis, and the frequency of those values is shown on the *y*-axis. Notice that as *df* (the number of categories) increases, the shape of the distribution changes to become more normal.

As you look at the graph of theoretical chi-square values, think about this: Is it possible to have a chi-square value that is less than zero? If so, what would that mean? (See the answers to the practice problems 1 and 2 for more on this.)

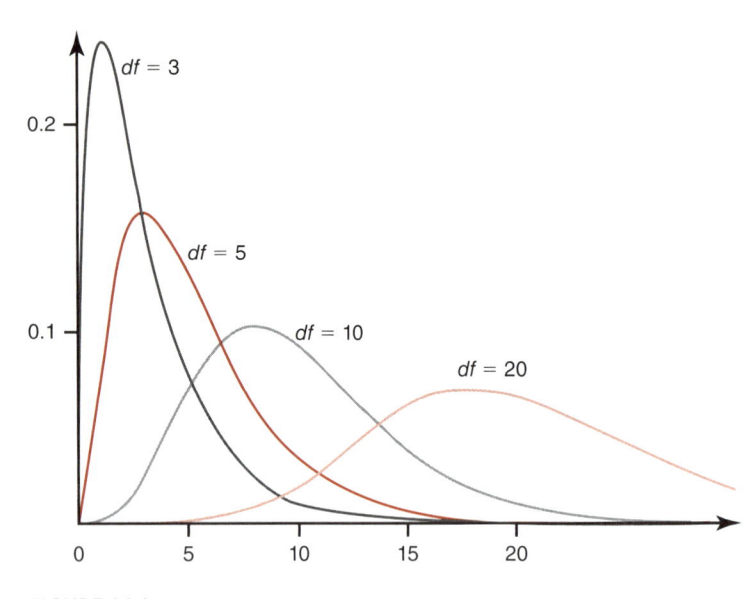

FIGURE 14.1
Chi-square distributions.

So, in our study of fathers' occupations, we have a chi square of 84.50, a nice healthy value pretty far away from zero. It's fairly likely that this chi-square value is unusual if, in fact, the null is true (and all the categories of paternal employment have the same frequency). Is it statistically significant? We already know how to find out: We determine the degrees of freedom for this situation and then compare our observed value with a critical value from the table. The critical value will change as df changes. Box 14.6 shows the critical values and the rejection regions for the two chi-square distributions, at an alpha level of .05, with $df = 5$ (gray line) and $df = 10$ (red line). The table of critical values is shown in Appendix E.

The larger the difference between the observed and expected frequencies, the larger the value of chi square will be. And the larger the value of chi square, the more likely that it will fall in the rejection region of the chi-square distribution. So, we can reject the null hypothesis that predicts all paternal occupations are equally likely. This would suggest that some occupations are, in fact, more likely in this set of important scientists. To find out which ones might be more frequent, we simply go back to the data. We already know that the biggest difference between the observed and expected frequencies was seen for the occupational category of bankers and businessmen: Far more fathers were bankers or businessmen than we expected.

BOX 14.6 **Determining the significance of our observed chi square**

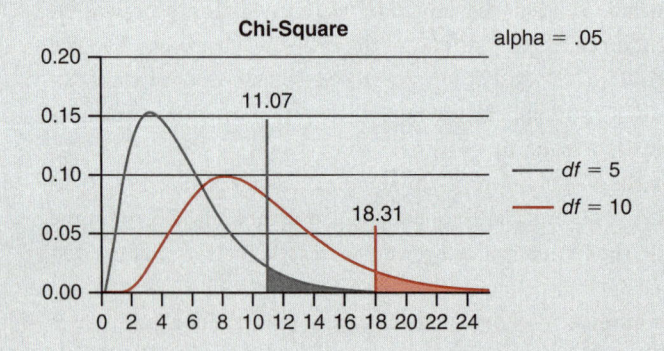

608 STATISTICS IN CONTEXT

CheckPoint

Jennifer works with community groups to organize volunteer opportunities for students at her college. She suspects some faculties are overrepresented with regards to the number of volunteers they yield. To discover whether some faculties yield more volunteers than others, Jennifer creates a contingency table. Using the observed values in the contingency table, answer the following questions and help Jennifer determine whether or not she has statistical evidence that warrants rejecting the null hypothesis that the faculties at her college yield equal numbers of volunteers. Round your answers to two decimal places, where appropriate.

Faculty Affiliation of Volunteer Frequencies

Applied Health Sciences	Education	Humanities	Mathmatics and Science	Social Science
23	15	31	16	40

1. Calculate the expected frequency (f_e).
2. Calculate the chi square (χ^2).
3. Calculate the degrees of freedom (df).
4. Identify the appropriate critical value ($\alpha = .05$), compare it to your calculated χ^2 value, and state whether to reject or accept the null hypothesis that a volunteer is equally likely to come from any of the five college faculties.

Answers to these questions are found on page 611.

THE TWO-WAY CHI-SQUARE TEST OF INDEPENDENCE

Right about now, you're probably thinking that if there is a one-way chi-square test, there is bound to be a two-way chi-square test. And if that is what you're thinking, then you are absolutely right. The two-way chi-square test is used when we have two categorical variables. The question we would answer with a two-way test does not involve how a possible model fits the data. Instead, with a two-way chi-square test, we can answer questions about whether or not the two variables we have are independent of one another. The alternative name for the two-way test is the **test of independence**.

chi-square test of independence
A nonparametric test in which expected and observed frequencies are compared in order to determine if two categorical variables are independent of one another; also known as a two-way chi-square test.

AN EXAMPLE OF THE TWO-WAY TEST: AMERICANS' BELIEF IN GHOSTS BY REGION

Let's take a look at a test of independence. Do you believe in ghosts? Do you think that a belief in ghosts depends on where you live? Let's find out. We will do a survey asking people to answer two questions: *Do you believe in ghosts?* (with the possible answers being "yes," "no," or "maybe"), and *Which part of the country do you live in?* We will use the map shown in Figure 14.2 to put the answers into categories.

FIGURE 14.2
Regions of the United States.

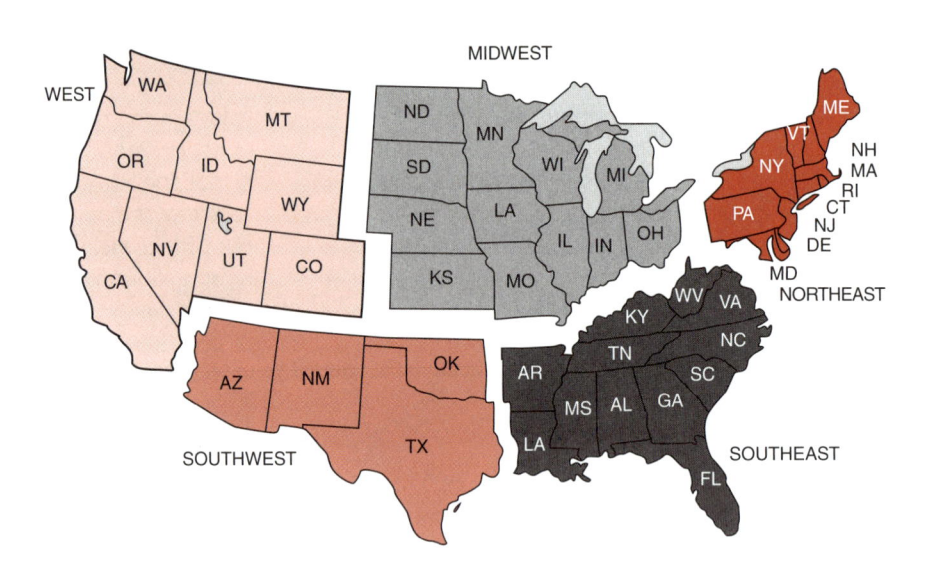

BOX 14.7 **Survey responses: Do you believe in ghosts?**

Do You Believe in Ghosts?	SE	SW	W	MW	NE	Row Totals
Yes	3	5	9	3	2	22
No	9	9	2	7	10	37
Maybe	10	9	11	7	4	41
Column totals	22	23	22	17	16	Overall total = 100

---

contingency table A table showing the distribution of one variable in rows and another in columns; used to study the correlation between the two variables.

dependent Indicating that the occurrence of one event alters the probability of the occurrence the other event.

A table like the one shown in Box 14.7 is called a **contingency table**. It is an example of a frequency distribution that shows two variables at the same time and summarizes the relationships, or dependencies, between these two variables. These are our observed frequencies. We'll use the chi-square test again because we have categorical, nonparametric data that make comparing means pointless. Does it look as though belief in ghosts might be **dependent** on the region of the country one lives in? What would the data look like if these two variables were dependent on one another? What would the data look like if these two variables were completely independent of one another? Let's find out.

First, we need to create our null hypothesis. Remember that when we used the one-way chi-square test, our null hypothesis was that the observed frequencies

610   STATISTICS IN CONTEXT

for each category were all equally likely to occur. Here, our null hypothesis will be that belief in ghosts and region of the country are completely **independent** of one another.

In chi-square terms, we would anticipate that the expected frequencies will be about the same as the observed frequencies if the null hypothesis is true. If the null is false, then we should see the data points bunch up in some of the boxes. If someone who lives in the Northeast were more likely to believe in ghosts than someone living in the Southwest, we should see more of our subjects in that box, and fewer of them distributed across the other boxes.

Next, we need to figure out what our expected frequencies might be. If the null is true, what would we expect to see in each of these boxes? We are going to rely on common sense and probability to find out. Take a look at Box 14.8 in order to see how we'll do this.

Notice that we expect to have something less than five full people who live in the Southeast and also believe in ghosts, but it doesn't matter. The expected frequencies are theoretical, not empirical. It is possible to have 84 hundredths of a person theoretically, even when we obviously cannot have less than a whole person in reality.

independent Indicating that the occurrence of one event does not change the probability of the other event.

A SHORTCUT

You will be relieved to know that there's a shortcut for finding the expected frequencies for each of the cells in a two-way chi-square design shown in Box 14.8. The shortcut allows you to do each step in finding the expected frequencies we took in the previous cell, but now to do those all at once. Box 14.9 shows you this shortcut.

Use the shortcut to find the remaining expected frequencies. You should see what is shown in Box 14.10.

Notice that there are some differences between the observed and expected frequencies. Some of these differences are relatively small—take, for example, the people from the Midwest who are not sure if they believe in ghosts. The expected frequency here differs from the observed frequency by 0.03, or less than one person. Other differences are larger (we expected 8.14 people from the Western states to say they did not believe in ghosts, for example, and we only saw two in this category—a difference of about six people). We will calculate the chi square using the same formula we used in the one-way example. See Box 14.11 for the results.

CheckPoint Answers
Answers to questions on page 609

1. $f_e = 25$ ☐
2. $\chi^2 = 17.84$ ☐
3. $df = 4$ ☐
4. $\chi^2_{obs} = 17.84 > \chi^2_{critical} = 9.488 \therefore$ reject ☐

SCORE: /4

BOX 14.8   **Determining the expected frequencies for our regional belief in ghosts study**

We will find the expected frequencies cell by cell. Let's start with the expected frequency for the highlighted cell (people who believe in ghosts and who live in the Southeast).

Do You Believe in Ghosts?	SE	SW	W	MW	NE	Row Totals
Yes	3 4.84	5	9	3	2	22
No	9	9	2	7	10	37
Maybe	10	9	11	7	4	41
Column totals	22	23	22	17	16	Overall total = 100

1. What is the probability that you will believe in ghosts?

    There are 22 people who believe in ghosts, out of 100 people total, so the probability is 22% (22/100).

2. What is the probability that you will live in the Southeast?

    There are 22 people out of 100 who live in the Southeast, so the probability that you will be from the Southeast is 22% (22/100).

3. What is the probability that you will believe in ghosts and live in the Southeast? (Or, what is the probability that you will be in the highlighted box?)

    To find the probability of one event *and* another, we multiply the individual probabilities. So, this probability should be $0.22 \times 0.22 = 0.0484$. This means that 4.84% of the sample of 100 people should be in this cell if the null is true.

    We have 100 people total in this design, so what is 4.84% of 100?

    We would expect 4.84 people in this cell.

I'm fairly sure that you know what comes next: We need to find the critical value of chi square and compare it to our observed value of 14.35. Again, we need to know our degrees of freedom. In a two-way chi square, the degrees of freedom need to reflect the two variables that we're comparing in our

BOX 14.9 A shortcut for finding expected frequencies in a two-way chi-square test

$$f_e = \frac{(\text{row})(\text{column})}{\text{overall}}$$

So, to find the expected frequency for the people who believe in ghosts and who live in the Southeast, we multiply the total for the row this cell is in (22) by the total for the column this cell is in (22) and then divide this sum by the overall total (100).

$$f_e = \frac{(\text{row})(\text{column})}{(\text{overall})} = \frac{(22)(22)}{100} = \frac{484}{100} = 4.84$$

BOX 14.10 Observed and expected frequencies

Do You Believe in Ghosts?	SE	SW	W	MW	NE	Row Totals
Yes	$f_o = 3$ $f_e = 4.84$	$f_o = 5$ $f_e = 5.06$	$f_o = 9$ $f_e = 4.84$	$f_o = 3$ $f_e = 3.74$	$f_o = 2$ $f_e = 3.52$	22
No	$f_o = 9$ $f_e = 8.14$	$f_o = 9$ $f_e = 8.51$	$f_o = 2$ $f_e = 8.14$	$f_o = 7$ $f_e = 6.29$	$f_o = 10$ $f_e = 5.92$	37
Maybe	$f_o = 10$ $f_e = 9.02$	$f_o = 9$ $f_e = 9.43$	$f_o = 11$ $f_e = 9.02$	$f_o = 7$ $f_e = 6.97$	$f_o = 4$ $f_e = 6.56$	41
Column totals	22	23	22	17	16	Overall total = 100

design. To find the *df*, we multiply the *number of categories* for the row variable (RV) minus one by the number of categories for the column variable (CV) minus one. In our case, the RV is belief in ghosts, and we have three categories possible here (yes, no, and maybe). Our CV is region of the country the participants are from, and there are five categories here. So, $df = (3 - 1)(5 - 1) = 2(4) = 8$.

The table of critical chi-square values (see Appendix E) will show the critical value for this comparison. The critical value equals 15.51, and our observed value

CHAPTER 14 The Chi-Square Test

BOX 14.11 Calculating the chi square

Region

Do You Believe in Ghosts?	SE	SW	W	MW	NE	Row Totals
Yes	$f_o = 3$ $f_e = 4.84$	$f_o = 5$ $f_e = 5.06$	$f_o = 9$ $f_e = 4.84$	$f_o = 3$ $f_e = 3.74$	$f_o = 2$ $f_e = 3.52$	22
No	$f_o = 9$ $f_e = 8.14$	$f_o = 9$ $f_e = 8.51$	$f_o = 2$ $f_e = 8.14$	$f_o = 7$ $f_e = 6.29$	$f_o = 10$ $f_e = 5.92$	37
Maybe	$f_o = 10$ $f_e = 9.02$	$f_o = 9$ $f_e = 9.43$	$f_o = 11$ $f_e = 9.02$	$f_o = 7$ $f_e = 6.97$	$f_o = 4$ $f_e = 6.56$	41
Column totals	22	23	22	17	16	Overall total = 100

Observed (f_o)	Expected (f_e)	$f_o - f_e$	$(f_o - f_e)^2$	$\dfrac{(f_o - f_e)^2}{f_e}$
3	4.84	−1.84	3.39	3.39/4.84 = **0.70**
5	5.06	−0.06	0.004	0.004/5.06 = **0.0008**
9	4.84	4.16	17.31	17.31/4.84 = **3.58**
3	3.74	−0.74	0.55	0.55/3.74 = **0.15**
2	3.52	−1.52	2.31	2.31/3.52 = **0.66**
9	8.14	0.86	0.74	0.74/8.14 = **0.09**
9	8.51	0.49	0.24	0.24/8.51 = **0.03**
2	8.14	−6.14	37.70	37.7/8.14 = **4.63**
7	6.29	0.71	0.50	0.50/6.29 = **0.08**
10	5.92	4.08	16.65	16.65/5.92 = **2.81**
10	9.02	0.98	0.81	0.81/9.02 = **0.09**
9	9.43	−0.43	0.96	0.96/9.43 = **0.10**
11	9.02	1.98	3.92	3.92/9.02 = **0.43**
7	6.97	0.03	0.0009	0.0009/6.97 = **0.0001**
4	6.56	−2.56	6.55	6.55/6.56 = **1.00**

$$\chi^2 = \sum \frac{(f_o - f_e)^2}{f_e} = 14.35$$

STATISTICS IN CONTEXT

CheckPoint

The contingency table below displays the observed frequencies of highest level of education attained by sex for a sample of 1,374 Americans over the age of 18.

Frequencies of educational attainment by sex

	Male	Female	Total
High school	355	354	709
College degree	213	238	451
Advanced degree	101	113	214
TOTAL	669	705	1,374

Use the data in the table to answer the following questions. Round your answers to two decimal places, where appropriate.

1. Calculate f_e.
2. Calculate χ^2.
3. Calculate df.
4. Identify the appropriate critical value ($\alpha = .05$), compare it to your calculated χ^2 value, and state whether you reject or accept the null hypothesis that the independent and dependent variables are independent.

Answers to these questions are found on page 616.

---

is 14.35. Because our observed value is less than the critical value, we cannot reject the null hypothesis. Our data show that belief in ghosts and region of the country a respondent is from are independent of one another.

A SPECIAL CASE FOR CHI SQUARE: THE "2 BY 2" DESIGN

If you are using the chi-square test of independence with two variables, each with only two categories (a "2 by 2" design), there is a shortcut available for calculating the chi-square statistic. If you use this shortcut, you will not have to calculate expected frequencies—a big time-saver. Remember, though, that this only works with a 2 by 2 situation.

Suppose we want to find out if handedness (either left-handedness or right-handedness) is dependent on sex (male or female). We ask 50 men and 50 women if they are right- or left-handed and record the results. Box 14.12 shows the data and the calculation of chi square using the shortcut.

We have a significant dependence here: Our observed chi-square value is 19.87, we have 1 df [$(2 - 1)(2 - 1) = (1)(1) = 1$], so our critical chi-square value is 3.84. We can reject our null hypothesis saying that sex and handedness are independent and instead conclude that handedness is dependent on sex. The data suggest that females are more likely to be left-handed than were males (I made this data up, but it might be fun to check this question out in your statistics class—it's very easy data to collect).

Please note that you will get the same result for chi square using the more traditional formula, and please remember that the shortcut works only with a 2 by 2 design.

BOX 14.12 Data and chi-square calculation for a 2 by 2 design

	Right-Handed	Left-Handed	Row Totals
Male	32 (A)	18 (B)	50 (A + B)
Female	10 (C)	40 (D)	50 (C + D)
Column totals	42 (A + C)	58 (B + D)	Overall total = 100 (N)

$$\chi^2 = \frac{N(AD - BC)^2}{(A+B)(C+D)(A+C)(B+D)}$$

$$= \frac{(100)[(32)(40) - (18)(10)]^2}{(50)(50)(42)(58)}$$

$$= \frac{(100)(1{,}100)^2}{6{,}090{,}000}$$

$$= \frac{(100)(1{,}210{,}000)}{6{,}090{,}000}$$

$$= \frac{121{,}000{,}000}{6{,}090{,}000}$$

$$= 19.87$$

CheckPoint Answers
Answers to questions on page 615

1. Frequencies of educational attainment by gender ☐

	Male	Female	Total
High school	345.21	363.79	709
College degree	219.59	231.41	451
Advanced degree	104.20	109.80	214
TOTAL	669	705	1,374

2. $\chi^2 = 1.12$ ☐
3. $df = 2$ ☐
4. $\chi^2_{obs} = 1.12 < \chi^2_{critical} = 5.991 \therefore$ fail to reject ☐

SCORE: /4

CheckPoint

Have you heard of Belief in a Just World (BJW) theory? It posits that most people see the world as just and will use conscious and unconscious strategies to preserve this belief when it is threatened by evidence of injustice. For example, upon encountering a victim of an unfortunate accident, people might adopt a *positive* interpretation of the victim's suffering (that suffering builds character, for instance), or they might adopt a *negative* interpretation of the character of the victim so as to recast the victim as deserving of suffering.

Loveleen is interested to learn whether coping styles influence the selection of interpretive strategies people use to preserve their just world beliefs. She designs a survey to identify people's coping styles and administers it to a sample of undergraduate students. Among her participants, she has identified a group she calls "repressors" because they tend to unconsciously repress their feelings when facing an emotional situation; the remaining participants are labeled "nonrepressors."

Loveleen shows each participant a video of an unfortunate victim and records their responses in a contingency table. Using the data in the contingency table, answer the following questions and help Loveleen determine whether coping style and selection of BJW preservation strategy are dependent upon one another. Round your answers to two decimal places, where appropriate.

Responses to injustice by coping style

	Repressor	Nonrepressor	Total
Positive interpretation	13	5	18
Negative interpretation	5	7	12
TOTAL	18	12	30

1. Calculate χ^2.
2. Calculate *df*.
3. Identify the appropriate critical value ($\alpha = .05$), compare it to your calculated χ^2 value, and state whether you reject or accept the null hypothesis, which states that the two variables are independent of each other.

Answers to these questions are found on page 619.

Think About It...

THE RISK RATIO

Have you ever read a news story that described the odds of contracting a disease, or of having a disease cured by a particular treatment? If you have, then you've likely seen reference to the "odds ratio" or its cousin, the "risk ratio."

The risk ratio (RR) is a statistical measure of the risk that an event will happen in one group versus another group. You will often see it calculated to describe the risk of getting a disease after exposure to a suspected causal agent versus the risk if you haven't been exposed.

Suppose we conduct a study of the particularly risky behavior of smoking cigarettes and the odds of developing lung cancer. We collect 100 smokers and 100 nonsmokers, and we follow them for a set period of time, counting the number of members of both groups who develop lung cancer.

Think About It... continued

The RR is the ratio of the probability of an event occurring (in this case, developing lung cancer) in an exposed group to the probability of that event occurring in a nonexposed group. As an equation, we would write RR like this:

$$RR = \frac{p(\text{event when exposed})}{p(\text{event when not exposed})}$$

And here is the contingency table for our hypothetical example.

	Developed Lung Cancer	Did Not Develop Lung Cancer	Total (Row)
Smokers	30 (A)	70 (B)	100
Nonsmokers	10 (C)	90 (D)	100
TOTAL (column)	40	160	TOTAL (overall) 200

Remember how to calculate probability? It's the frequency of an event divided by the total number of times that event could have happened. For us, RR would be found like this:

$$RR = \frac{A/(A+B)}{C/(C+D)} = \frac{30/100}{10/100} = \frac{0.30}{0.10} = 3.0$$

So, smokers are three times more likely to develop lung cancer than are nonsmokers.

1. What would an RR of 1.00 indicate? _____

2. What would an RR of less than 1.00 indicate? _____

3. What would an RR of greater than 1.00 indicate? _____

Answers to this exercise are found on page 629.

CheckPoint Answers
Answers to questions on page 617

1. $\chi^2 = 2.80$ ☐
2. $df = 1$ ☐
3. $\chi^2_{obs} = 2.80 < \chi^2_{crit} = 3.841 \therefore$ fail to reject the null ☐

SCORE: /3

NONPARAMETRIC TESTS IN CONTEXT: TYPES OF DATA REVISITED

I am almost certain that each of you reading this has responded to a questionnaire at some point in your life. Questionnaires can be found almost anywhere, from psychology laboratories to the pages of popular magazines, not to mention online (perhaps you've visited RateMyProfessor.com to grade your stats instructor). I'm also certain that at least one of those questionnaires has used a very specific type of scale to assess your response to whatever question it might be asking. That scale, which invites you to indicate how much you agree or disagree with a series of statements, is called a Likert scale.

The Likert scale is named after its inventor, Dr. Rensis Likert, founder of the Institute of Social Research at the University of Michigan. Typically, agreement or disagreement with a particular statement is assessed using an ordered response scale, where responses can range from 1 to 5, or 1 to 7, or even 1 to 9, and are often paired with a pictorial representation of the scale to help the subject in making his/her decision. Figure 14.3 shows the typical Likert scale.

Notice that the scale is "balanced," meaning there are an equal number of "positive" and "negative" positions that surround a "neutral" position. This balance has actually become the subject of disagreement within the community of researchers who make regular use of questionnaires.

The disagreement revolves around the type of measurement scale that the Likert scale represents. Technically, Likert scales are ordinal: Responses are put into categories, and those categories are put into order, from least to most. If you remember our discussion of types of data back in Chapter 2, these are the two characteristics of an ordinal measurement scale. We know that someone who agrees with the test statement and answers the question with a 5 agrees with the statement more than does someone who answers with a 4—but by how much more? With a scale such as this, we can't determine *how much* agreement or disagreement there is in any one category of response.

Technically, then, we should not count this as quantitative data. We should not calculate a mean or standard deviation of responses, and we should not use a parametric test to compare answers. The problem,

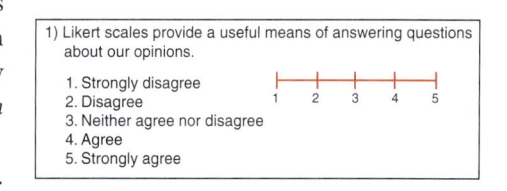

1) Likert scales provide a useful means of answering questions about our opinions.
   1. Strongly disagree
   2. Disagree
   3. Neither agree nor disagree
   4. Agree
   5. Strongly agree

   1 2 3 4 5

FIGURE 14.3
Typical Likert scales.

however, is that quantitative measurement allows us to make quantitative comparisons and come to (what are often seen as) stronger, more number-based conclusions than does qualitative measurement.

This has led researchers to rethink what type of data they have when they use a Likert scale. Some argue that the way the responses are worded on the scale tends to imply that the categories are, in fact, symmetrical, or balanced around a central point, which suggests that when we read our options, we think about our response as if they were on a quantitative, interval scale (with an arbitrary zero point). In addition—the accompanying visual scale where relative levels of agreement are evenly separated one from another as the level of agreement increases—tends to add to the interpretation of the scale as quantitative, interval-level data. This interpretation makes the scale appropriate for use with a parametric test.

If we treat Likert scale data as ordinal, then we can use the median or the mode, but not the mean, as our measure of center. We can measure spread or variability with the interquartile range, but not the standard deviation. And we can compare groups with a nonparametric test like the chi square, but not an ANOVA. Obviously, all of this changes if we consider the measurements in our sample to be on an interval scale. Because if the data are on an interval scale, then we can make use of the parametric analysis that we've been discussing so far, and come to our more traditional conclusion.

More and more often, I see parametric interpretations of Likert scale data in the literature. The arguments in favor of an interval interpretation seem as valid as do those in favor of viewing the data as ordinal. It's up to each individual researcher to determine how the data will be treated and what kind of test to apply.

SUMMARY

We began this chapter with a brief discussion of what are called nonparametric statistics, or statistics that are not based on assumptions about population parameters.

The data used in a nonparametric test are qualitative, making calculation of either population or sample means and standard deviations moot. In nonparametric testing, instead of comparing sample means with population means, we compare frequencies. We discussed two kinds of nonparametric tests that can be used to answer questions with frequency data: the one-way chi-square test and the two-way chi-square test.

The one-way chi-square test is also called the chi-square test for goodness of fit. It compares observed and expected frequencies to allow a conclusion to be made about the probability of the distribution of an event matching or not matching an expected distribution. The expected frequencies in a one-way chi square can come from common sense or from a theoretical model based on previous research.

The two-way chi-square test, also called the chi-square test of independence, compares observed and expected frequencies to allow a conclusion to be made about the probability that two variables are independent or dependent (i.e., contingent) upon one another. The expected frequencies here are generated using common sense and probability theory.

TERMS YOU SHOULD KNOW

chi-square test of goodness of fit, p. 604
chi-square test of independence, p. 609
contingency table, p. 610
dependent, p. 610
expected frequency, p. 604
frequency, p. 604
independent, p. 611
nonparametric tests, p. 600
observed frequency, p. 604
parametric test, p. 600

GLOSSARY OF EQUATIONS

Formula	Name	Symbol
$\sum \dfrac{(f_o - f_e)^2}{f_e}$	Chi square	χ^2
$\chi^2 = \dfrac{N(AD - BC)^2}{(A + B)(C + D)(A + C)(B + D)}$	Shortcut for 2 by 2 design	χ^2

PRACTICE PROBLEMS

1. I mentioned earlier in the chapter that a negative chi square is impossible. Why?

2. What would a negative chi square mean?

3. Suppose you were visiting Las Vegas and wanted to win at dice. Because you want to win, you decide to practice ahead of time in order to familiarize yourself with the game and work on your skills. You find a friendly local craps game, sit down, and watch the action. Soon, though, you become suspicious about the way the game is being played. You notice that one die in particular seems to come up with six pips showing much more often than seems possible if everything were on the up and up. You start to collect data, watching the game for 30 throws of that particular die and writing down which face comes up on each roll. The data you collect are shown in Table 14.2.
   a. Calculate the expected frequencies.
   b. Determine the chi-square value.
   c. Should you be concerned about the dice in this establishment?

Table 14.2 Thirty consecutive roles of a single die

1 pip	2 pips	3 pips	4 pips	5 pips	6 pips
3	3	4	5	5	10

4. Do you think political party affiliation is still distributed across the country the way that it was described during the 2008 presidential election? Let's use the regional map of the United States shown in Figure 14.2 and ask a total of 1,000 people from the five

regions of the country about their political affiliation (Democrat or Republican). The data are shown in Table 14.3.

  a. Which type of chi-square test is appropriate here?
  b. Perform the appropriate chi-square test. State the null and alternative hypotheses, and determine if you can reject the null.
  c. What conclusions can you come to?

Table 14.3 Political party affiliation by region of the country

	SE	SW	W	MW	NE	Row totals
Democrat	98	41	123	70	150	
Republican	102	159	77	130	50	
Column totals						

5. Do you think that attitude toward abortion is in any way dependent on sex? We will interview 50 men and 50 women and ask them if they support the current law making abortion legal, if they do not support the current law, or if they have not made up their minds as of yet. The data are shown in Table 14.4.

Table 14.4 Attitude toward abortion

	Support	Do not Support	Do not yet Know	Row Totals
Women	14	17	19	
Men	16	13	21	
Column Total				

  a. What kind of chi-square test should you do and why?
  b. Perform the appropriate test, and describe the results.
  c. Do the data support the null hypothesis?

6. Suppose we repeated the study we did in question 5, but this time, we don't allow people to answer the question with "I haven't decided yet." All 50 men and 50 women are forced to choose either "Yes—I support the current law that permits abortion" or "No—I do not support the current law." The data are shown in Table 14.5. Perform the appropriate test to answer the question and describe the results of your test.

Table 14.5 Attitude toward abortion: Revised questionnaire

	Support	Do not support	Total
Women	30	20	
Men	10	40	
TOTAL			

7. According to the Red Cross, the four basic blood types (disregarding for the moment the Rh factor) are distributed as shown in Table 14.6. Let's test the hypothesis that the distribution of blood types in the United States is not the same as it is in the worldwide population. We ask 220 Americans what their blood type is—O, A, B, or AB—and get the following results:
   - Type O: 88 people
   - Type A: 84 people
   - Type B: 40 people
   - Type AB: 8 people

   Table 14.6 Worldwide distribution of blood type

Type O	Type A	Type B	Type AB
43.91%	34.80%	16.55%	5.14%

   a. What test should we do in this example and why?
   b. Perform the appropriate test, and discuss the relevant results.

8. A zookeeper hypothesizes that changing the intensity of the light in the primate exhibits will reduce the amount of aggression between the baboons. In Exhibit A, with a lower light intensity, he observes 36 incidences of aggression over a 1-month period. In Exhibit B, with "normal" or moderate light intensity, he observes 42 incidences of aggression. And in Exhibit C, with high-intensity light chosen to mimic the intense sunlight of the open and arid African savanna, he observed 25 instances of aggression. Should he support or reject his hypothesis?

9. An engineering psychologist is interested in the way people use everyday objects. She has been tasked with redesigning the doors in a public building, so she spends some time watching the way customers in this building use the doors that are already in place. She notices that there are three different types of doors in the lobby of the building:

   - a "hands-free" door that opens when someone stands on a pressure pad located in front of the door
   - a door with a "push" handle, which opens when someone pushes on a horizontal bar
   - a door with a "pull" handle that opens the door when the handle is pulled up by the user

   She counts the number of people who choose to use Door A ("hands-free"), Door B ("push bar"), and Door C ("pull handle") and sees the following:
   - Door A was used by 80 people.
   - Door B was used by 66 people.
   - Door C was used by 60 people.

   What kind of door design should she recommend to the owners of the building?

10. Here's a question near and dear to all textbook writers. Suppose you take a random sample of 30 students who are using a brand-new textbook and a second sample of

30 students who are using the older, more traditional text. You compare student achievement on the state test given to all students at the end of the course. The data are shown in Table 14.7. Based on state test performance, would you recommend the new book?

Table 14.7 Student performance on the state achievement test using the new and the old textbook

	Passed	Failed
Using the new textbook	28	2
Using the old textbook	20	10

11. Life Savers candies were introduced to the world in 1912, when Clarence Crane (father of the poet Hart Crane) created a hard candy that could, he hoped, stand up to the summer heat better than chocolate. To make his product stand out from the competition, Crane punched a hole in the center and named the candies after the life preservers they resembled. The candies, with five flavors in a roll, were introduced in 1935. The five flavors were revamped in 2003: They are now cherry, pineapple, raspberry, watermelon, and orange. There are 14 candies to a roll, but a statistician suspects that the fruit flavors are not evenly distributed. He purchases five rolls of the candies and counts the number of the flavors he finds in the rolls. Table 14.8 shows the results of his study. Are the flavors evenly distributed within a roll?

Table 14.8 Distribution of flavors in a five-flavor roll of Life Savers

Cherry	Pineapple	Raspberry	Watermelon	Orange
8	20	12	10	10

12. A researcher in industrial/organizational psychology is interested in shift work and how it affects life satisfaction. He asks nurses if they prefer 8-, 10-, or 12-hour shifts and finds that 100 nurses said they prefer an 8-hour shift, 90 prefer a 10-hour shift, and 35 prefer a 12-hour shift. Do the results deviate significantly from what would be expected due to chance?

13. A statistics professor claims to grade "on the curve," which would mean that grades should be distributed as follows:

- A's: 7%
- B's: 24%
- C's: 38%
- D's: 24%
- F's: 7%

The professor allows the students in his statistics class to use data from his grade book showing the distribution of grades from the very first year he taught the class. These data are shown in Table 14.9. Does the professor grade on a curve?

Table 14.9 Grade distribution for professor in his first year of teaching

Letter grade	A	B	C	D	F	Total
Frequency	11	73	104	52	10	250

14. Many psychology departments require a passing grade in a college algebra course as a prerequisite for the course in introductory statistics. A professor wants to find out if taking algebra helps students do better in the statistics course. She gets the final statistics grades from two sets of students: those who took statistics before the algebra prerequisite was put into place, and those who took statistics after the prerequisite rule was adopted. Table 14.10 shows the number of students who passed and who failed the course and whether they had taken algebra first. Are students more likely to pass the course if they have taken college algebra?

Table 14.10 Student performance in statistics as a function of algebra prerequisite

	Passed	Failed
Algebra	48	10
No algebra	17	25

15. A researcher wants to see if four strategies to combat work fatigue are used equally by office workers of both sexes. She draws a sample of 200 office workers—half of them women, half of them men—and obtains the data shown in Table 14.11. Is there a preference for a particular method of combating drowsiness on the job as a function of gender?

Table 14.11 Methods of combating fatigue, by sex

	Caffeine	Nap	Exercise	Sugar
Female	35	17	27	21
Male	25	20	30	25

16. A sociologist wants to see if the number of years of college a person has completed is related to his or her place of residence (urban, suburban, or rural). A sample of 90 people is selected and classified as shown in Table 14.12. Is there a dependency between years of college completed and the type of community where a person lives?

Table 14.12 Number of years of college completed and residence

Location	No college	Four-year degree	Advanced degree
Urban	8	10	15
Suburban	7	12	10
Rural	15	5	8

17. Following a statistics course, 60 graduate students completed a course-evaluation questionnaire. One question asked the students to evaluate the relative importance of statistics to their proposed specialization in psychology. The results are shown in Table 14.13. Is the level of belief in the importance of statistics dependent on the area of specialization in psychology?

Table 14.13 Importance of statistics to chosen specialization in psychology

"Statistics is Important to my Chosen Specialization"	Clinical/ Counseling	Social Psychology	Developmental Psychology	Neuroscience
Very important	3	6	2	2
Important	6	5	5	5
Somewhat important	5	8	3	2
Not important	1	2	3	2

18. Researchers involved in a longitudinal study observed a sample of children for 9 years, from age 3 to age 12. Periodically during that time, the children's IQs were measured. The researchers found that 56 children showed increases in their IQ scores over this span, while 55 children showed decreases in their IQ scores. Visitors to the homes of all of the children in the study were asked to observe and rate each mother on the extent to which she encouraged and expected intellectual achievement and success. The data are shown in Table 14.14 below. Test the hypothesis that there is no difference in the distribution of maternal encouragement for intellectual success for children who show IQ increases and for those showing IQ declines.

Table 14.14 Maternal encouragement and change in IQ scores

IQ Trend Over Time	Maternal encouragement		
	Low	Medium	High
Increase	12	15	29
Decrease	29	16	10

19. It is a typical weekday morning in the Grimsby household. Five-year-old Tomas is, with great concentration, eating a bowl of cereal, while his older brother, Zane, watches cartoons on a tablet over a piece of toast. They are oblivious to the delicate bit of kitchen choreography their parents are engaged in, opening and closing cupboards, dropping apples and juice boxes into lunch bags, and packing up their own paraphernalia in preparation for another busy workday. They perform the well-rehearsed routine flawlessly until Maria, drawing a yogurt cup from the fridge, catches her husband's elbow as he attempts a reckless pass while sipping from a full cup of coffee. His reaction is swift, sonorous, and not in the best social and child-rearing traditions: "$#*%!" Maria now has two options: She may either censor her husband in front of the children—"Craig, you shouldn't swear like that in front of the kids!"—or she may continue with her preparations without commenting on the profanity. The question is, under which of these maternal response conditions is the child of five more likely to remember the profanity?

Suppose it is possible to perform this study. Some fathers are verbally censored, while others are not. The young child is later asked what his father said in response to spilling his coffee. The responses are scored as either not remembering or remembering the father's profanity. Test the hypothesis that the mother's response to the father's profanity does not influence the ability of the child to remember the profanity (i.e., that the mother's response and the child's ability to remember are independent of one another).

Table 14.15 Memory for father's profanity as a function of censoring the father

Mother's Response	Child's memory of profanity	
	Does not Remember	Remembers
Censors father	16	28
No response	26	17

20. A fundraising lottery advertises that of the 350,000 tickets it sells, 10 will be redeemable for cash prizes, 35 for vehicles, 50 for vacation prizes, and 500 for a variety of smaller prizes, including televisions and bicycles. A consumer protection agency retrieves the production data used to print the lottery tickets and analyzes the distribution of winning tickets in a random sample of 25,000 tickets. The distribution of winning tickets from the sample of 25,000 tickets is displayed in Table 14.16. Does the distribution of winning tickets in the sample warrant further scrutiny of the fundraising lottery?

Table 14.16 Lottery prize frequencies

Cash Prizes	Vehicle Prizes	Vacation Prizes	Smaller Prizes	Nonwinning Tickets
1	1	5	42	24,951

21. Do you think the number of hours the average person spends watching television each week changes from season to season? To find out, suppose the television viewing of a random sample of 40 participants was tracked over a 1-year period. The contingency table shown in Table 14.17 reports seasonal differences is the average number of hours per week participants spent watching television. Do the data in Table 14.17 suggest a relationship of dependence between season and the average number of hours per week spent watching television?

Table 14.17 Average number of hours per week spent watching television, by season

	Spring	Summer	Fall	Winter	Total
0 hours	3	4	0	1	8
1–10 hours	17	17	9	5	48
11–20 hours	11	13	17	12	53
21–30 hours	8	6	11	14	39
>30 hours	1	0	3	8	12
Total	40	40	40	40	160

22. Complete the crossword puzzle shown on page 629.

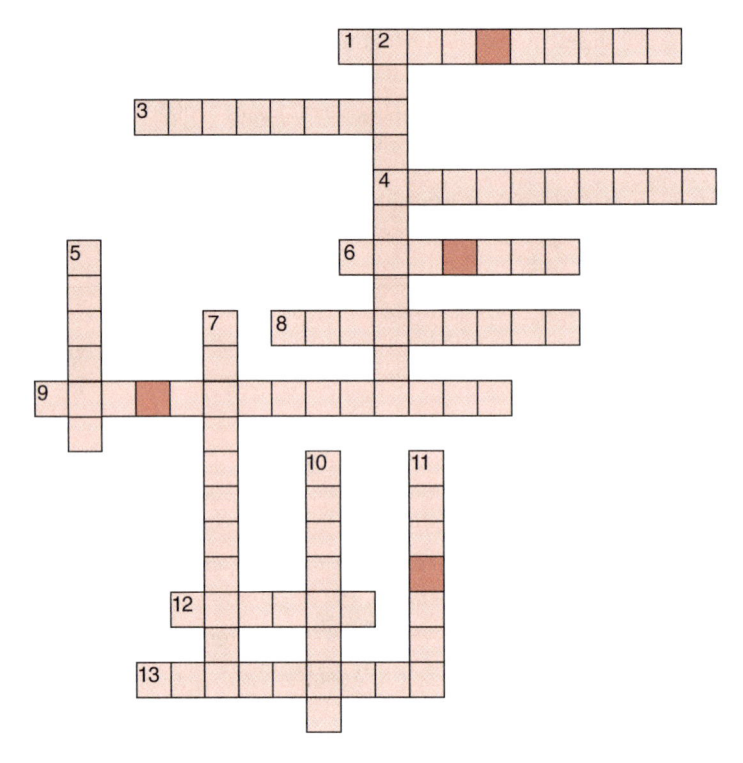

ACROSS

1. (2 words) Statistical measure of the risk of an event happening in one group vs. another
3. The frequency of an event predicted by a model or by the rules of probability
4. Tests based on assumptions about population parameters
6. (2 words) Also called a test of goodness of fit
8. When the occurrence of one event alters the probability of another event
9. (2 words) A statistical analysis not based on assumptions about population parameters
12. A self-report scale often used on questionnaires
13. The number of times an event occurs

DOWN

2. When the occurrence of one event does not alter the probability of another these two events are ___
5. Developer of the idea of a χ^2 test
7. The kind of data a χ^2 test is designed to work with
10. The empirical frequency of an event
11. (2 words) Also called a test of independence

Think About It...

THE RISK RATIO

SOLUTIONS

1. What would an RR of 1.00 indicate? That both groups are equally at risk of developing lung cancer.
2. What would an RR of less than 1.00 indicate? That the second group has a greater risk of developing cancer than the second group.
3. What would an RR of greater than 1.00 indicate? That the first group has a greater risk of developing the disease than the second group.

REFERENCES

Galton, F. (1874/2011). *English men of science: Their nature and nurture*, remastered edn. Lexington, KY: Forgotten Books.

Martz, E. (2012, October 31). Chi square analysis for Halloween: Is there a slasher movie gender gap? Retrieved online at http://blog.minitab.com/blog/understanding-statistics/chi-square-analysis-of-halloween-and-friday-the-13th-is-there-a-slasher-movie-gender-gap

Conducting Chi-Square Tests with SPSS and R

The data set we will be using for this tutorial is drawn from the Behavioral Risk Factor Surveillance System (BRFSS), a nationwide health survey carried out by the Centers for Disease Control and Prevention. The set we will be using is based on results of the 2008 questionnaire, completed by 414,509 individuals in the United States and featuring a number of questions about general health, sleep, and anxiety, among other things. The version we will be using is a trimmed down version of the full BRFSS data set, which you can find on the companion website for this textbook (www.oup.com/us/blatchley).

Chi-Square Test for Goodness-of-Fit in SPSS

In this example, we will use the chi-square test to determine whether there is a difference in the observed and expected frequencies of our data. We are going to see if the distribution of general health status for respondents of the BRFSS differs from the expected distribution. The variable in question is categorical. Respondents were asked to rate their general health on a scale of 1 to 5 (1 = *excellent*, 5 = *poor*).

STEP 1: Load your data into SPSS.

STEP 2: Select the chi-square test. Click **Analyze**, then **Nonparametric Tests**, scroll to **Legacy Dialogs**, and then click **Chi-square.**

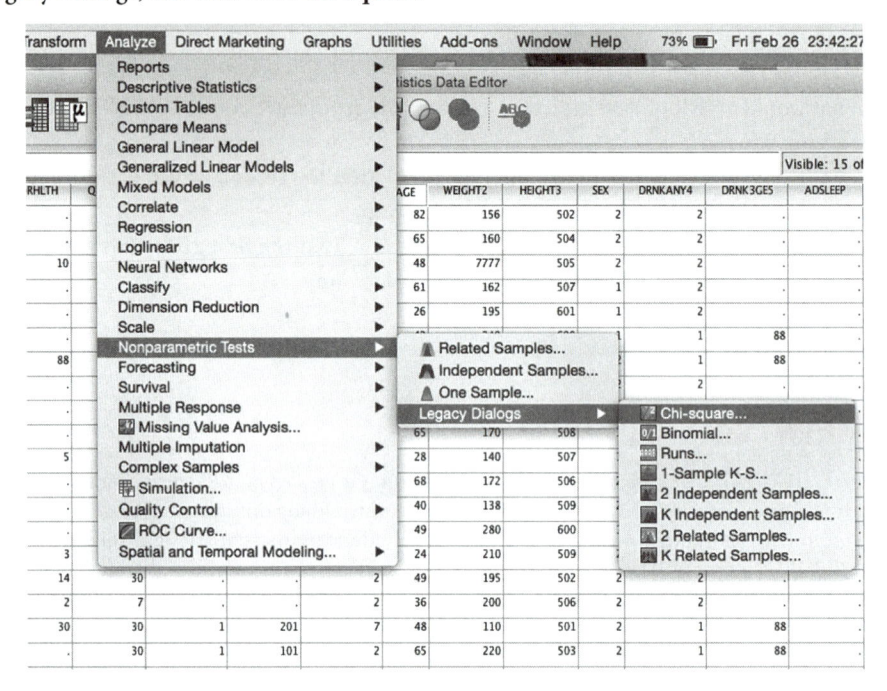

STEP 3: Pick your variable of interest. In the dialogue box that opens, select **General Health (GENHLTH)** from the list of variables and move it to the **Test Variables List** box using the arrow button. Click **OK**.

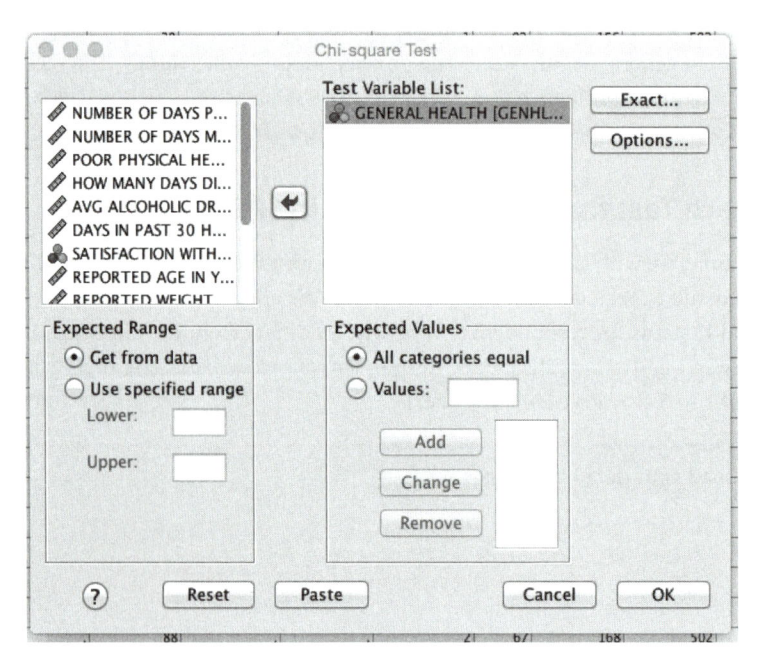

630   STATISTICS IN CONTEXT

STEP 4: Two tables of output will be generated after conducting the chi-square test in SPSS. Interpret the data.

GENERAL HEALTH

	Observed N	Expected N	Residual
1	75152	82527.8	−7375.8
2	131760	82527.8	49232.2
3	126263	82527.8	43735.2
4	55778	82527.8	−26749.8
5	23686	82527.8	−58841.8
Total	412639		

Test Statistics

	GENERAL HEALTH
Chi-Square	103830.33[a]
df	4
Asymp. Sig.	.000

a. 0 cells (0.0%) have expected frequencies less than 5. The minimum expected cell frequency is 82527.8.

The first table compares the actual observed number in each category with the expected number of observations in each category. The "Residual" column shows the difference between the observed and the expected values.

The second table presents the significance statistics. It is how we know whether the observed frequency is equal to the expected frequency or if there is a statistically significant difference between the two. In this case, our chi square is very large, and it is significant at the 0.000 level. So, we can say that our observed and expected frequencies are not the same. More specifically, we would say:

> A chi-square goodness-of-fit test was conducted to determine whether an equal number of participants from each of the general health categories took part in the survey. The chi-square goodness-of-fit test showed that the five general health categories were not equally distributed among our sample.

Two-Way Chi-Square Test for Independence in SPSS

In this example, we will examine whether the sex of respondents and their general health as rated on a scale of 1 to 5 are independent of one another. In other words, we want to see if there is a relationship between the two categorical variables. To do this, we are going to conduct a two-way chi-square test for independence.

STEP 1: Load data into SPSS

STEP 2: Set up your test. Click **Analyze**, scroll to **Descriptive Statistics**, and click **Crosstabs**

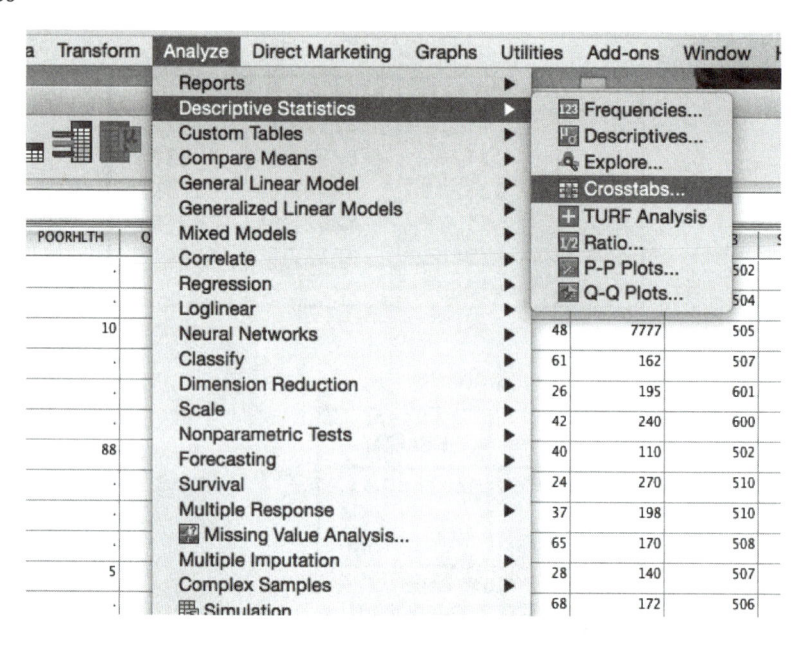

STEP 3: Select your test variables. In the dialogue box that opens, transfer **Respondents Sex (Sex)** into the **Row(s)** box by selecting it and clicking on the first arrow. Transfer **General Health Quality (GENHLT)** into the **Column(s)** box by selecting it and clicking on the second arrow.

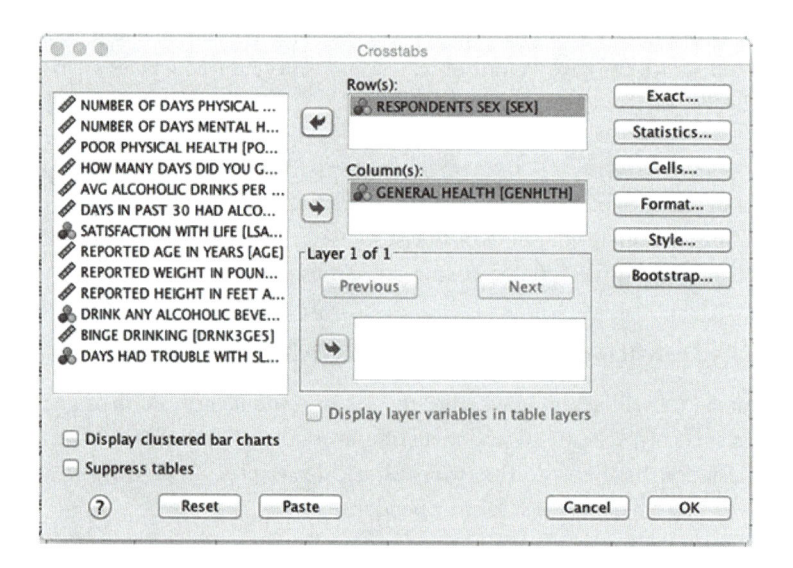

STEP 4: Choose your stats. Click **Statistics . . .** In the dialogue box that opens, select **Chi-square**, and then click **Continue**.

Step 5: Set up your crosstab. Click **Cells . . .** Select **Expected**, **Row**, **Column**, and **Total**. Click **Continue**, and then click **OK**.

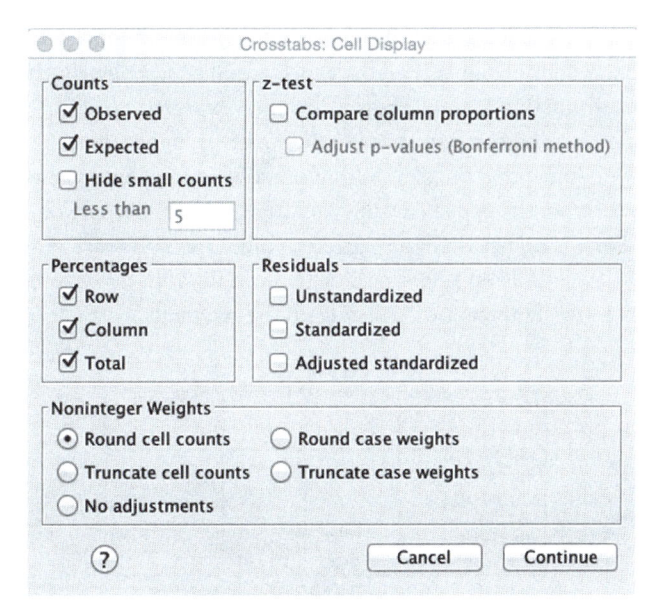

CHAPTER 14 The Chi-Square Test   633

STEP 6: Three tables will be generated in the output window. Interpret the results.

Case Processing Summary

	Cases					
	Valid		Missing		Total	
	N	Percent	N	Percent	N	Percent
RESPONDENTS SEX * GENERAL HEALTH	412639	99.5%	1870	0.5%	414509	100.0%

The first table provides a breakdown of the number of valid cases (the number of observations that exist for both pieces of data).

RESPONDENTS SEX * GENERAL HEALTH Crosstabulation

			GENERAL HEALTH					Total
			1	2	3	4	5	
RESPONDENTS SEX	Male	Count	28533	49592	48337	19985	8545	154992
		Expected Count	28228.0	49490.6	47425.8	20950.9	8896.7	154992.0
		% within RESPONDENTS SEX	18.4%	32.0%	31.2%	12.9%	5.5%	100.0%
		% within GENERAL HEALTH	38.0%	37.6%	38.3%	35.8%	36.1%	37.6%
		% of Total	6.9%	12.0%	11.7%	4.8%	2.1%	37.6%
	Female	Count	46619	82168	77926	35793	15141	257647
		Expected Count	46924.0	82269.4	78837.2	34827.1	14789.3	257647.0
		% within RESPONDENTS SEX	18.1%	31.9%	30.2%	13.9%	5.9%	100.0%
		% within GENERAL HEALTH	62.0%	62.4%	61.7%	64.2%	63.9%	62.4%
		% of Total	11.3%	19.9%	18.9%	8.7%	3.7%	62.4%
Total		Count	75152	131760	126263	55778	23686	412639
		Expected Count	75152.0	131760.0	126263.0	55778.0	23686.0	412639.0
		% within RESPONDENTS SEX	18.2%	31.9%	30.6%	13.5%	5.7%	100.0%
		% within GENERAL HEALTH	100.0%	100.0%	100.0%	100.0%	100.0%	100.0%
		% of Total	18.2%	31.9%	30.6%	13.5%	5.7%	100.0%

The second table shows us the differences between the actual count and the expected count. From this table, we can see that the number of male respondents with General Health equal to 1, 2, or 3 are slightly higher than the expected value. Conversely, the count of males with General Health equal to 4 or 5 is slightly less than expected.

The number of women who responded to the survey is higher than the number of men. From the table, we can see that the observed number of women with General Health equal to 1, 2, or 3 is lower than the expected value, and that the number of women with General Health equal to 4 or 5 is higher than the expected value. The difference in these results suggests there may be a relationship between these two variables. To test whether this suspicion is right and statistically significant we will consult the last table.

Chi-Square Tests

	Value	df	Asymptotic Significance (2-sided)
Pearson Chi-Square	127.233[a]	4	.000
Likelihood Ratio	127.733	4	.000
Linear-by-Linear Association	47.830	1	.000
N of Valid Cases	412639		

a. 0 cells (0.0%) have expected count less than 5. The minimum expected count is 8896.74.

The third table provides the chi-square statistics. The Pearson chi-square test statistic is listed first. From this table, we can see that the respondent's sex and their general health status are not independent. We have a chi-square value of 127.233, which is significant at the 0.000 level.

Chi-Square Test for Goodness-of-Fit in R

To conduct a chi-square test for goodness of fit in R, we need to use a few lines of code. We will be testing to see if the frequency of each category of the GENHLTH variable to see if the differences between observed and expected frequencies are statistically significant. In this case, we will be testing for expected frequencies that are equal.

To run the specific code, we will need to load the pander and foreign libraries. For this, you will need to make sure you have these packages loaded and installed. To complete these steps, go to **Packages and Data,** and click **Package Installer.**

In the **Package installer** window, click **Get List** and search for **pander** and **foreign** separately. Once you have found each individually, click **Install Selected**; make sure that the **Install Dependencies** check box is selected. Exit the window.

Go to **Packages and Data**, and click **Package Manager.**

In the **Package Manager** window, search for **pander** and **foreign,** and select the check box beside each. Make sure that the status column changes to "loaded."

Alternatively, you can type the following code:

```
>library(foreign)
>library(pander)
```

STEP 1: Load your data into R.

```
>health<-read.csv("example.csv)
```

STEP 2: Remove null variables. We will use the subset function to remove the null variables and create a new variable with only the observations of interest.

```
>health2<-subset(health, health$GENHLTH<=5)
```

STEP 3: Tabulate the frequencies, and create your observed variable. We must determine what the observed frequencies in our data are so that we can set up our observed variable correctly.

```
>GenHealth<-factor(health2$GENHLTH)
>t-table(GenHealth)
> pander(t)
-------------------------------------------
   1       2        3       4        5
------  -------  -------  -----  ----------
 75152  131760   126263   55778    23686
------  -------  -------  -----  ----------
```

STEP 4: Create your observed column variable. This will act as a frequency calculator. When the chi-square test is run, it will compare the number of observations we have entered into the column to the percentage we expect.

```
> observed=c(75152, 131760, 126263, 55778, 23686)
```

STEP 5: Set your expected frequencies. Since we do not have any specific population distribution in mind, we will test against the null hypothesis that all of the groups are equally distributed. Since we have five categories and $1/5 = .2$, we will create an expected variable that depicts these frequencies. If you knew the expected frequencies in the population, you would enter those values in the brackets below, in the same order in which the associated variables appear.

```
> expected=c(0.2, 0.2, 0.2, 0.2, 0.2)
```

STEP 6: Run the test, and analyze your results.

```
> result<-(chisq.test(x=observed, p=expected,))
> pander(results)
 Test statistic   df   P value
 ---------------- ---- --------
     103830        4    0 * * *
```

Table: Chi-squared test for given probabilities: `observed`

Based on the results above, we can say that the observed frequencies of life satisfaction in our sample are statistically significant and are distributed differently from our expected population values. In other words, our observed distribution is not equal.

Two-Way Chi-Square Test for Independence in R

Similar to the way we performed the chi-square test for distribution, we can conduct the two-way chi-square test for independence in R using a few lines of code. We are going to test whether the sex and a respondent's general health are independent of one another.

Again, in order to run the specific code, we will need to load the pander and foreign libraries using the instructions provided above.

STEP 1: Load the data into R.

```
>health<-read.csv("example.csv")
```

STEP 2: Create a frequency table. We want to see how our data is distributed over our categories (Sex and General Health). This will give us an idea of what the data look like.

```
> frequencytable<- table(health$GENHLTH,health$SEX)
> pander(frequencytable)
```

```
                  1             2
    **1**     28533         46619
    **2**     49592         82168
    **3**     48337         77926
    **4**     19985         35793
    **5**      8545         15141
    **7**       333           515
    **9**       377           643
```

STEP 3: Remove missing or unneeded variables. We will need to remove the values for nonresponses and for respondents who said "I don't know," which are coded as 9 and 7, respectively. Because the other values in our question of interest are coded between 1 and 5, we can specify that the value is less than or equal to 5.

```
> healthQ<-subset(health, health$GENHLTH<=5)
```

STEP 4: Construct your new frequency table. Make sure that the data are correct and that the excess variables have been dropped.

```
> frequencytable<-table(healthQ$GENHLTH, healthQ$SEX)
> pander(frequencytable)

                  1             2
    **1**     28533         46619
    **2**     49592         82168
    **3**     48337         77926
    **4**     19985         35793
    **5**      8545         15141
```

STEP 5: Conduct the test, and analyze the results. Use the new version of the frequency table to conduct the two-way chi-square test.

```
> Xsq<-chisq.test(frequencytable)
> pander(Xsq)
 Test statistic   df   P value
 ---------------  ---  ---------
      127.2        4   1.52e-26 * * *
```

Table: Pearson's Chi-squared test: `frequencytable`

From the above table, we can see that the respondent's sex and their general health status are not independent. We have a chi-square value of 127.233, which is significant at least at the 0.001 level.

As you can see, the results in R are much more sparse than the results in SPSS. If you want to extend the results you see in R, you will need code to specifically request tables and descriptive and/or test statistics. The coding provided here tells you how to conduct chi-square tests for distribution and frequency and interpret the results.

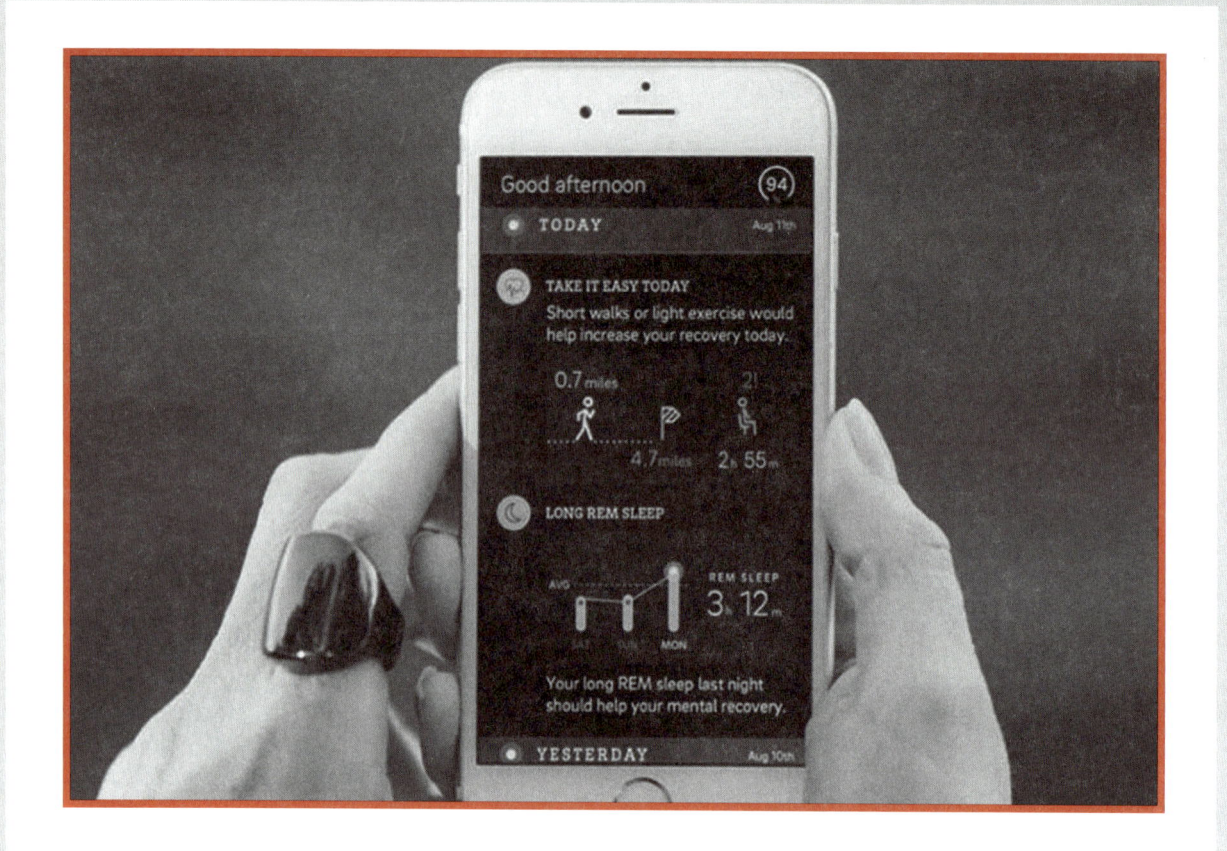

CHAPTER FIFTEEN

NONPARAMETRIC TESTS

Statistics are human beings with the tears wiped off.
—PAUL BRODEUR

Everyday Statistics

HOW TO MAKE STRESS WORK FOR YOU

We all know that stress can be very bad for our physical and emotional health. Too much stress, particularly that long-term stress we feel unable to control, is associated with breakdown of the immune system, damage to the cardiovascular system, and serious decreases in cognitive function, just to name a few of the more serious consequences.

It may come as a surprise, then, to learn that some scientists say we can actually make stress work *for* us, rather than against us. Sin, Chow, and Cheung (2015) examined how stress levels, optimism, and self-efficacy affected the performance of marathon runners.

Measuring psychological characteristics like feeling optimistic or feeling efficacious requires the use of qualitative measurement scales, so the researchers used the nonparametric Spearman correlation in their study. They found that slower runners at the back of the pack tended to have higher stress levels than did their faster counterparts. However, they also found that the best and fastest runners—the ones with low stress levels—were more optimistic and had higher levels of self-efficacy than did the slower, stressed runners. Maybe optimism and the feeling that your effort actually does make a difference can mitigate the effects of stress. Chapter 15 will teach us more about nonparametric testing.

OVERVIEW

NONPARAMETRIC STATISTICS
RANKING THE DATA
THE SPEARMAN CORRELATION (r_s)
THE MANN–WHITNEY U-TEST
THE WILCOXON SIGNED-RANK, MATCHED-PAIRS t-TEST
NON-NORMAL DISTRIBUTIONS
NONPARAMETRIC TESTS IN CONTEXT

LEARNING OBJECTIVES

Reading this chapter will help you to . . .

- Understand eugenics and the importance of looking beyond the statistics when interpreting data. (Concept)

- Grasp the difference between parametric and nonparametric statistics. (Concept)

- Know the purpose of ranking data for nonparametric tests. (Concept)

- Convert qualitative data into ranks for the Spearman correlation (r_s), Mann–Whitney U-test, and Wilcoxon signed-rank t-test. (Application)

- Explain why converting data to ranks is beneficial when working with skewed data. (Concept)

- Calculate statistics for the Spearman correlation tests, Mann–Whitney U-test, and Wilcoxon signed-rank t-test. (Application)

NONPARAMETRIC STATISTICS

REVIEW: HOW PARAMETRIC AND NONPARAMETRIC TESTS DIFFER

nonparametric test Statistical analyses that are not based on assumptions about the parameters of the populations stated in the central limit theorem.

parametric tests Statistical analyses that are based on assumptions about the parameters (mean and standard deviation) of the population.

Spearman's rank-order correlation coefficient, discussed in *The Historical Context* on page 641, is one of the most frequently cited examples of what is known as a *nonparametric test*. A **nonparametric test**, as we saw in the last chapter, is one that does not require any assumptions about parameters and that can be used with *nonquantitative* data. If there are nonparametric tests, it stands to reason that there will be *parametric* ones as well. And logic would also tell us that these parametric tests do require assumptions about parameters. So, let's revisit the assumptions that we've been making as we use parametric and nonparametric tests and see what our options are should we violate those assumptions.

Parametric tests, as the name implies, are tests that involve parameters. And as you recall, parameters describe the characteristics of populations. When we do a parametric statistical test, we are making a set of assumptions about the populations that our data come from. Table 15.1 shows you these assumptions.

Notice how the assumptions that underlie parametric tests include assumptions about the kind of data (the type of measurement scale used to measure our dependent variable) we have collected. If we have quantitative data, then mathematical operations can be used, and means, standard deviations, and so on can be meaningfully calculated. We can use this kind of data to come to a conclusion about the mean of a population (a parameter) with a *t*-statistic, an *F*-statistic, or a Pearson *r*-statistic.

TABLE 15.1 Parametric versus nonparametric tests

Parametric Tests	Nonparametric Tests
• Assume that the characteristic being measured is normally distributed in the population.	• Require no assumptions about the population.
• Assume that the central limit theorem holds and, therefore, that $\mu = \mu_x$ and $\frac{\sigma}{\sqrt{n}} = \sigma_x$.	• Do not depend on the central limit theorem because no assumptions about the population are being made.
• Assume that when two or more groups are being compared, the variance of all groups is approximately equal (called *homoscedasticity*).	• Require no assumptions about variance.
• Parametric tests should be used when . . . • the data are QUANTITATIVE • the data are normally distributed • the groups all have equal variance	• Non-parametric tests should be used when . . . • the data are QUALITATIVE • the data are quantitative and seriously skewed • the groups have unequal variance

STATISTICS IN CONTEXT

If we violate these assumptions, however, then we change the ground rules. Now we need to use a test that does not require normally distributed, quantitative data. Instead, we will compare measurements made with qualitative data. Typically, the data used with nonparametric tests use ordinal-level measurement (although nominal-level measurements can be used), and just as typically, the data are converted to *ranks* for use with the nonparametric inferential tests. When we rank data, we first arrange the data in order, from lowest to highest, and then replace each data point with a number that represents the relative position of that data point in the ordered set—1 represents the lowest value, and a number equal to the size of the sample (*n*) represents the highest value.

factor analysis A statistical procedure used to identify unobserved variables responsible for observed data.

eugenics The science of improving a population by controlled breeding to increase the occurrence of desirable heritable characteristics.

The Historical Context

THE POWER OF STATISTICS IN SOCIETY

Charles Edward Spearman (see Figure 15.1) is one of the more unusual personalities in the history of statistics. Born in 1863 in England, he followed an unusual path to psychology. After obtaining a degree in engineering, he began a career in the military, serving with the Royal Engineers in Burma and India. He later wrote that his years of service with the military were "the greatest mistake of my life" (1930, p. 300), perhaps because his true interests lay elsewhere. At the rather advanced age of 34, he resigned his commission and returned to the halls of academe in Leipzig, Germany, where he eventually received his PhD in the new field of psychology at the age of 42.

Spearman developed several now-famous ideas in statistics. The Spearman rank-order correlation coefficient allows researchers to determine if two *qualitative* variables are correlated with one another. He also pioneered the procedure now known as **factor analysis**, which allows researchers to examine hidden or latent variables that underlie and may explain correlations between other variables. He also described the two-factor theory of intelligence, claiming that intelligence was determined by an underlying factor he called "g," which he saw as common to all mental activity.

Spearman's theory of intelligence and the "g-factor," as well as the statistical techniques he developed to study concepts like it, are well known. His theories on the heritability of intelligence and on the nature/nurture question in psychology in general are considerably less well known, though they are perhaps more influential.

Spearman was an early supporter of the field called **eugenics**—the study of ways to improve the human race through selective breeding. Proposed and developed by Sir Francis Galton, eugenics became extremely popular in both America and Europe during the 1920s and 1930s.

FIGURE 15.1
Charles Edward Spearman.

Galton, Pearson, and Spearman (along with others) believed that intelligence was inherited—the result of genetics alone and therefore not changeable through manipulation or improvement of the environment. If this were the case, then social programs to help the poor and the disadvantaged (the social tier where the "less intelligent," and thus "less productive," members of society wound up) would be a waste of time. Eugenics eventually became identified with the ideas of racial superiority and with social movements, like the National Socialism of the Nazi Party in Germany, that attempted to preserve racial purity in society.

The influence of eugenics was not limited to large, notorious political movements, however. In fact, much of the influence of eugenics was relatively subtle. Compulsory sterilization of the mentally ill in order to prevent the spread of undesirable heritable characteristics was pioneered in the United States (the first law ordering involuntary sterilization was introduced in Michigan in 1897) before spreading worldwide, and the last forced sterilization in the United States occurred as recently in 1981. More modern versions of the ideas proposed by Galton and others have come cloaked in an outer shell of respectable science. They can be seen in the works of Arthur Jensen and Phillipe Rushton, as well as in the controversial book *The Bell Curve* by the late Richard Herrnstein (a psychologist at Harvard University) and Charles Murray (a political scientist at the American Enterprise Institute). Jensen, Rushton, and others studied the heritability of cognitive function, a question worthy of study and posed by any number of serious researchers. So, what's the trouble?

Critics of works like *The Bell Curve* cite a small but important change in the way the question is asked as a problem central to this kind of work. In science, we start with a question and then look for evidence in the data to support a conclusion about that question. A scientific investigation into the heritability of intelligence should begin with asking *Is it inherited?* and then continue with a search for evidence that supports the null or alternative. More recent work seems to have flipped this question around, turning it into a statement about intelligence—*It is inherited*—and then reflexively looking for evidence to support this statement. Without a question being asked, it isn't science.

PERFORMING NONPARAMETRIC TESTS

When we conduct a test using nonparametric data, we still start with a null hypothesis, and we still test that hypothesis. Now, however, we are not coming to a conclusion about how a quantitative measure is distributed in the population; instead, we are coming to a conclusion about how the data are ranked. Values that are very different from one another should have very different rankings.

In this chapter, we will encounter three types of nonparametric inferential tests that should look familiar to you. Remember, we are going to rank the ordinal-level data first and then apply the statistical test to the **ranked data**. First, we will talk about a way to test ranked data in order to see if they are correlated with one another; this procedure involves the Spearman correlation coefficient. Then, we will discuss the Mann–Whitney U-test, a method for determining if two independent groups, measured on an ordinal-level scale, are significantly different from one another. Finally, we will talk about a test that assesses whether two dependent groups, measured qualitatively, are significantly different from one another. This is the Wilcoxon signed-rank, matched-pairs t-test, and mouthful though it may be, the name should give you a hint about where you might have seen these types of tests before.

ranked data Data that have been ordered from least to most, with each data point assigned a number indicating its position in the ordered list.

Each of these tests is the nonparametric version of a parametric test we have already discussed. The Spearman correlation is the nonparametric version of the Pearson correlation coefficient, the Mann–Whitney U-test is the nonparametric version of the independent-samples t-test we encountered in Chapter 10, and the Wilcoxon (the shorthand name for this test) is the nonparametric version of the dependent-samples t-test also discussed in Chapter 10. So, let's begin with Spearman's contribution, the Spearman ranked-order correlation coefficient, and a discussion of the ranking procedure.

STATISTICS IN CONTEXT

RANKING THE DATA

The first step in using a nonparametric test consists of ranking the data. The data are organized from lowest to highest. The smallest, or lowest, value is ranked 1, and the highest value in the set gets the highest ranking. Ranking data is easy enough to do. Suppose you were asked by your local tennis club to rank the age of the three players shown in Box 15.1 from first to last (lowest to highest). Not hard to do at all.

Occasionally, when we're ranking the data, we will come across two or more scores that are identical—in other words, tied for a particular rank. Spearman decided that these scores should receive identical ranks, and he developed a procedure for ranking tied data. First, put all the scores in order, from smallest to largest. Then, assign a rank (first, second, third, and so on) to each *position* in the ordered list. When two or more scores are tied, *compute the average of their ranked positions*, and assign this average rank to each of the tied scores. The data in Box 15.2 illustrate

BOX 15.1 **Ranking three tennis players by age**

Age	Rank
17	1
28	2
43	3

BOX 15.2 **Ranking tennis players with tied scores**

Age	Position	Rank	Calculating Average Rank
16	1	2	Positions $1 + 2 + 3 = 6$
16	2	2	$6 \div 3 = 2$
16	3	2	The average positions 1, 2, and 3 = 2.
17	4	4	Positions 1–3 are now ranked. We start again at position 4.
18	5	5.5	Positions $5 + 6 = 11$
18	6	5.5	$11 \div 2 = 5.5$
22	7	7	
23	8	8	
29	9	9	
27	10	10	

this procedure for dealing with tied ranks. Once again, we are ranking the ages of the players in our local tennis club, but now there are some ties to deal with.

Notice that we have a tie in the data right at the start. The youngest three members of this tennis club are all 16 years old. We need to assign all three of these 16-year-old players the same average rank. The three tied data points take up the first three positions in the data set, so we average these three positions. Mathematically, we add position 1, position 2, and position 3 together, and then we divide by the number of positions occupied by the tied data points (3, in this case). We assign the resulting average rank to each of the tied data points. We then start ranking the rest of the data at position 4 in the ordered set.

THE SPEARMAN CORRELATION (r_S)

We used the Pearson correlation coefficient to determine the extent to which two quantitatively measured variables were related to one another. We will do the same thing with the Spearman correlation coefficient, with one small change. We use the Spearman when *one or both* of the variables we are interested in describing are measured on a qualitative scale. Box 15.3 shows you the formula for calculating the Spearman correlation coefficient.

BOX 15.3   **Calculating the Spearman correlation coefficient**

$$r_s = 1 - \left[\frac{6 \sum D^2}{N(N^2 - 1)} \right]$$

Notice the symbol for the Spearman correlation is again the letter "*r*", but with the addition of a subscript letter "*s*" to indicate that it is a Spearman correlation.

THE SPEARMAN CORRELATION: AN EXAMPLE

Suppose we wanted to see if there was a correlation between the age of a tennis player and that player's level of competitiveness in playing the game. Age is measured quantitatively, but competitiveness is much harder to assign a number to. Typically, an aspect of personality like competitiveness in athletics is measured with a questionnaire. Take the Perception of Competitiveness Scale (PCS), developed by Carter and Weissbrod (2011), to assess differences in competitiveness in men and women. The PCS questionnaire asks participants to evaluate their own competitiveness by answering questions about how much they enjoyed competition and how winning or losing affected their self-perception. The questionnaire measured agreement with a series of 38 statements on a scale from 1 (*disagree*) to 5 (*agree*).

This kind of scale is ordinal: Each response is put into one of five boxes, and these boxes are then put in order from least agreement (1) to most agreement (5). We have measured competitiveness, but on a qualitative scale. Our data are shown in Box 15.4 along with the ranking of both age (our X variable) and PCS score (Y).

BOX 15.4   Data for our age and competitiveness study

Age X	Ranks	PCS Score Y	Ranks
16	2	29	2
16	2	38	7
16	2	29	2
17	4	31	4
18	5.5	34	5
18	5.5	35	6
22	7	29	2
23	8	39	8
25	9	40	9
27	10	45	10

Low rankings on age tend to be paired with low rankings on PCS score.

Notice that any relationship between X and Y in this data set isn't perfect.

High rankings on age tend to be paired with high rankings on PCS.

$H_0: \rho_s = 0.00$ (There is no meaningful correlation between X and Y.)

$H_1: \rho_s \neq 0.00$ (Either a negative or a positive correlation exists between X and Y.)

Suppose there is a positive relationship between age and PCS score. If that's the case, then a score that is ranked low on the age variable would tend to be paired with a score that is ranked low on the PCS variable. The oldest tennis player in our sample (ranked 10 in a set of 10) should tend to be paired with the highest-ranked PCS score (see the arrows in Box 15.4). Conversely, if there is a negative correlation between X and Y, we should see the ages that are ranked low on the X variable paired with the PCS score that are ranked the highest, and vice versa. When we take a look at the data, there seems to be a positive correlation between age and PCS: The older a tennis player in our sample is, the more competitive they perceive themselves to be.

The formula for calculating the Spearman correlation coefficient asks us to compare the ranks of the paired X and Y variables (see Box 15.5). First, we rank the X variable, and then we rank the Y variable. Remember that when we do a correlation, we have two measurements from each individual in our study. The X and Y variables are paired. The equation then tells us to sum the squared D scores. We've encountered D scores before (see Chapter 10 and the calculation the dependent-samples *t*-test using the difference between scores or D scores). Remember that the D stands for "difference"—in this case, the difference in rank of X and Y for each element in the sample (our "elements" are people here). So, we determine the difference for each person in their ranking on the age variable and on the PCS variable.

BOX 15.5 **Calculating the difference in rank between age and competitiveness scores**

$$r_s = 1 - \left[\frac{6\sum D^2}{N(N^2-1)}\right] = 1 - \left[\frac{6(50.05)}{10(100-1)}\right] = 1 - \left[\frac{30.3}{990}\right] = 1 - [0.03] = 0.97$$

	Age		PCS			
X	Rank	Y	Rank	D ($X_{rank} - Y_{rank}$)	D²
16	2	29	2	2 – 2 = 0	0
16	2	38	7	2 – 7 = –5	25
16	2	29	2	2 – 2 = 0	0
17	4	31	4	4 – 4 = 0	0
18	5.5	34	5	5.5 – 5 = 0.5	0.25
18	5.5	35	6	5.5 – 6 = –0.5	0.25
22	7	29	2	7 – 2 = 5	25
23	8	39	8	8 – 8 = 0	0
25	9	40	9	9 – 9 = 0	0
27	10	45	10	10 – 10 = 0	0
$\Sigma = 198$		$\Sigma = 349$		$\Sigma D = 0$	$\Sigma D^2 = 50.5$
$\bar{X} = 19.8$		$\bar{X} = 34.9$			

Once we have calculated the difference in rankings, we square all the rankings and then square each of the D scores (to eliminate the negative D's, much as we did when we were calculating the standard deviation of a set of scores). The sum of the square D scores is multiplied by six (a constant Spearman discovered) and then divided by $n(n^2 - 1)$ Notice how small the D scores are in this example: That tells us that the rankings on the X variable are very similar to the rankings on Y, which we would expect if there were a positive correlation between X and Y.

When we complete the calculation of the portion of the formula in the brackets, we subtract that portion from 1.00, so we will be subtracting a small number from 1.00 and getting a number very close to 1.00 as a result. When the ranks on X and Y are very similar, we likely have a strong positive correlation between the two variables.

THE RESULTS

The outcome of this test tells us there is a strong positive correlation between age and PCS score ($r_s = 0.97$). Older athletes tended to have a stronger sense of themselves as being competitive than did younger athletes. It looks as though our null hypothesis can be rejected—but how can we tell? If you said that we could compare the Spearman correlation coefficient that we obtained from our data with a critical Spearman score found in a table at the back of the book, you would be right. The table of critical r_s-values can be found in Appendix G. First determine if we're asking a one-tailed or two-tailed question (look to our alternative hypothesis for the answer here), and then, we read down the column marked n to find the number of pairs of measurements in our sample. We then read across to find our critical values. With 10 participants (and so 10 pairs of measurements), we need to have a correlation of at least 0.648 in order to call our correlation significant. Our calculations produced a correlation coefficient of 0.97, indicating a very strong positive correlation. The correlation we found between these two variables is significant at the .05 and the .01 level.

CheckPoint

June suspects that people enjoy country music more the older they get. She administers a questionnaire to a sample of 10 randomly selected participants. One question on the questionnaire asks participants to report the degree to which they identify with the statement "I enjoy country music" by selecting one of the following:

1 = *Strongly disagree*

2 = *Disagree*

3 = *Neither agree nor disagree*

4 = *Agree*

5 = *Strongly agree*

The participants' ages and responses to the question are summarized in the table below. Use the data in the table to answer the questions that follow and help June determine whether increases in age correspond to increases in enjoyment of country music. Round your answers to two decimal places, where appropriate.

Age	Enjoyment of Country Music
12	3
62	4
29	1
47	3
54	5
17	3
9	1
17	4
12	2
31	4
$n = 10$	$n = 10$

1. Order and rank the values for each variable.
2. Calculate ΣD^2.
3. Calculate r_s.
4. Identify the appropriate critical value ($\alpha = .05$), compare it to your calculated r_s-value, and state whether you should accept or reject the null hypothesis.

Answers to these questions are found on page 650.

THE MANN–WHITNEY *U*-TEST

The Mann–Whitney *U*-test (or the MWU test) was designed to allow researchers to compare two independent samples when the data collected on both of these samples are qualitative. It can be thought of as the nonparametric version of the independent-samples *t*-test. The MWU test is used quite frequently in psychology and other social sciences because of the kinds of questions researchers in these disciplines ask—questions about opinion, attitude, preference, and so on, which are almost always measured with a qualitative scale.

The test is based on a set of assumptions (as are all of the tests we've been working with). The MWU test assumes that both groups are independent of each other and that the responses are measured on an ordinal scale. The null hypothesis would be that the distributions of ranked data for both groups are equal, and the alternative hypothesis would be that the two distributions are not equal to one another (the two-tailed alternative), or that the distributions differ in a specific direction (one-tailed alternative).

To use this test, we begin by converting the data to ranks. The process we will use to do this is similar to—*but not the same as*—the process we used with the Spearman test. With the Spearman, we first rank the *X*-values, and then we separately we rank the *Y*-values. With the MWU test, we pool the data from both groups together, and then we rank all of the pooled data points, without paying attention to which sample they come from. If the observations in the two groups are very different from one another, then all or most of the observations with low ranks should be in one group and all of the high-ranking observations in the other. Then, when we add up the ranks for each group, they should be very different from one another. On the other hand, if the groups don't differ at all, then the ranks of the observations should be evenly distributed across the two groups, and the sums of the ranks for the two groups should be about the same.

The MWU test differs from the other tests we have used in that once the data are ranked, we put them back into their original samples and then calculate a *U*-statistic *for each sample*. In all of the tests that we've been doing so far, we have calculated a single statistic and then compared that one statistic to a critical value. With this test, we will calculate a *U*-statistic for each sample, and then (and here's the second difference between *U* and *t*), we compare *the smaller of the two U-statistics* with a critical value from (what else?) a table of critical *U*-values. If our observed *U*-statistic is *smaller than the critical value*, we can say we have a significant difference between the two groups. The closer the *U*-statistic is to zero, the *less* likely it is to have been caused by random chance alone. A *U* of zero would indicate a systematic difference between the two groups. We use the smaller of the two *U*'s we calculate—the one closer to zero—as our observed value.

How about we try an example: Do men and women differ in their attitudes toward online dating? We will randomly select 10 men and 10 women and give them an opinion survey designed to measure their attitudes toward online dating. The responses on the survey are measured on an ordinal scale where higher values (the maximum value possible on the scale is 100) indicate a positive attitude and lower values (the minimum value possible on the scale is 20) indicate a negative attitude. Table 15.2 shows the data.

TABLE 15.2 Attitude toward online dating for men and women

Men	Women
69	27
41	93
73	44
70	28
56	85
69	72
76	54
88	22
71	44
35	52

CheckPoint Answers
Answers to questions on page 648

1. ☐

Age	Rank	Enjoyment of Country Music	Rank
9	1	1	1.5
12	2.5	1	1.5
12	2.5	2	3
17	4	3	5
29	5	3	5
31	6	3	5
47	7	4	8
54	8	4	8
62	9	4	8
71	10	5	10

2. $\Sigma D^2 = 37$ ☐
3. $r_s = 0.78$ ☐
4. $r_{sobs} = 0.78 > r_{scritical} = 0.648 \therefore$ reject ☐

SCORE: /4

Our first job is to pool the data from both groups and then rank the data. Take a look at Box 15.6 to see the ranking process. Notice that we deal with ties in the same way we did when we were ranking the Spearman data.

BOX 15.6 **Ranking the data in our survey on online dating**

Men	Women	Pooled Data	Ranked Pooled Data	Ranked Men	Ranked Women
69	27	22	1	4	1
41	93	27	2	5	2
73	44	28	3	10	3
70	28	35	4	11.5	6.5
56	85	41	5	11.5	6.5
69	72	44	6.5	13	8
76	54	44	6.5	14	9
88	22	52	8	16	15
71	44	54	9	17	18
35	52	56	10	19	20
		69	11.5	$\Sigma R_1 = 121$	$\Sigma R_2 = 89$

STATISTICS IN CONTEXT

		69	11.5		
		70	13		
		71	14		
		72	15		
		73	16		
		76	17		
		85	18		
		88	19		
		93	20		

Box 15.7 shows the formula for calculating the *U*-values. We use it with each group. Remember that we will use the *smaller* of the two *U*-values to compare with the tabled critical value.

BOX 15.7 Using the MWU formula to calculate *U*-values

For men

$$U_1 = (n_1)(n_2) + \frac{n_1(n_1+1)}{2} - \sum R_1$$
$$= (10)(10) + \frac{10(11)}{2} - 121$$
$$= 100 + 55 - 121$$
$$= 34$$

For women

$$U_2 = (n_1)(n_2) + \frac{n_2(n_2+1)}{2} - \sum R_2$$
$$= (10)(10) + \frac{10(11)}{2} - 89$$
$$= 100 + 55 - 89$$
$$= 66$$

The smaller of the two *U*-values we calculated is the *U*-statistic for the men ($U = 34$). So, we compare this value to the critical value from the table of critical *U*-values in Appendix H. To be considered significant, our observed *U* needs to be equal to *or smaller than* the critical *U*. In this case, we look down the column on the left side of the table to n_1 and across to n_2. So, we go down 10 rows and across 10 columns to find our critical *U*-values. Our critical values here are either 19 ($\alpha = .02$, for a two-tailed test) or 16 ($\alpha = .01$, for a two-tailed test). Our observed

U-value of 33 is not smaller than either critical U-value, so our MWU test shows that the attitude of men and women toward online dating does *not* differ significantly. If we take a look at the ranked data, it looks as though men had a somewhat more positive attitude toward online dating than did women, but this difference was not significant.

CheckPoint

Two samples—one sample of randomly selected men over the age of 45 and one sample of randomly selected women over the age of 45—completed a questionnaire measuring their levels of interest in politics. Based on their answers, each respondent was assigned a score between 0 (*not at all interested*) and 10 (*very interested*). Using these political interest scores, which are shown in the table below, answer the following questions to determine whether men and women over the age of 45 differ with respect to their levels of interest in politics. Round your answers to two decimal places, where appropriate.

Interest in politics

Men	Women
6.4	10
5.6	3.8
7.8	2.2
6.4	2.4
9.8	5.6
3.4	7.6
5.2	3.6
7.8	4.4
2.4	3.6
9.2	4.8

1. Order and rank the values for each variable.
2. Calculate R_1 and R_2.
3. Calculate U_1 and U_2.
4. Identify the appropriate critical value ($\alpha = .05$), compare it to your calculated U-value, and state whether you should accept or reject the null hypothesis.

Answers to these questions are found on page 654.

THE WILCOXON SIGNED-RANK, MATCHED-PAIRS t-TEST

The Wilcoxon test is the nonparametric version of the dependent-samples t-test. It is used to compare two dependent groups measured with a qualitative measurement scale. Remember that dependent samples are samples where random selection of subjects is not used. Instead, the data have been paired together.

In Chapter 10, we discussed several methods for creating dependent samples (matching, natural-pairs, and repeated-measures designs). Please review this section if you're unsure of the definition of dependent samples. The Wilcoxon test generates a *t*-value that we will again compare to a tabled critical value (see Appendix I). Box 15.8 shows you the "formula" for calculating the Wilcoxon *t*-value. We'll use these steps to compare the two groups in the following example—do you recognize the research?

BOX 15.8 **Steps in the Wilcoxon *t*-test***

1. Find the difference (D) between each pair of scores (subtract score in Group 2 from that subject's score in Group 1).
2. Rank the difference scores. (Pay attention to the sign of the difference scores.)
3. Eliminate ALL ranks of ZERO.
4. Calculate the sum of the negative ranks.
5. Calculate the sum of the positive ranks.
6. Compare the absolute value of the smaller of the two sums to the critical value in the table.

*There is no formula as such for the Wilcoxon *t*-test.

THE WILCOXON TEST: AN EXAMPLE

It is well known in neuroscience that the development of visually guided behavior requires experience with visual stimuli *and* feedback obtained from physical interaction with the stimulus environment. In a classic experiment in neurobiology (Held & Hein, 1963), eight pairs of kittens (littermates) were reared in darkness until they were 10 weeks old. They then received brief visual stimulation each day. Only one kitten in each pair was allowed to interact with the visual environment; the other kitten was just allowed to passively experience the visual world around it (the "passive" kitten rode around the environment in a side-car moved by the "active" kitten). Figure 15.2 shows a sketch of the research environment. Theory predicts that the active kitten (allowed both kinds of visual experience) should show faster development of visually guided behavior than the passive kitten.

CheckPoint Answers
Answers to questions on page 652

1. ☐

Interest Level	Rank	Interest Level	Rank
2.2	1	5.6	11.5
2.4	2.5	5.6	11.5
2.4	2.5	6.4	13.5
3.4	4	6.4	13.5
3.6	5.5	7.6	15
3.6	5.5	7.8	16.5
3.8	7	7.8	16.5
4.4	8	9.2	18
4.8	9	9.8	19
5.2	10	10	20

2. Interest in politics: Ranked data ☐

Men	Women
2.5	1
4	2.5
10	5.5
11.5	5.5
13.5	7
13.5	8
16.5	9
16.5	11.5
18	15
19	20
$\Sigma R_1 = 125$	$\Sigma R_2 = 85$

3. $U_1 = 30$; $U_2 = 70$ ☐

4. $U_{obs} = 30 > U_{critical} = 23 \therefore$ accept ☐

SCORE: /4

FIGURE 15.2
Held and Hein's (1963) experimental setup: The "kitten carousel."

The development of the visual system of the kittens in the study was assessed in a number of ways. We will use an assessment that I've made up to make life a bit easier. Suppose that we scored the visual development using a scale from 1 to 15, where the scores indicate the proficiency of each kitten on a series of tests of visually guided behavior: the higher the score, the more developed the visual system of the kitten. The measurement scale we're using to compare these two, literally yoked-together samples is ordinal, which requires a nonparametric test. Table 15.3 shows the data for the active and passive kittens in this study. Box 15.9 shows the calculation of the Wilcoxon *t*-value.

654   STATISTICS IN CONTEXT

TABLE 15.3 Proficiency of visually guided behavior of kittens

Active	Passive
14	12
13	10
11	10
12	12
15	13
11	12
13	11
15	14

BOX 15.9 **Calculating the Wilcoxon *t*-value**

Active	Passive	D	Ranks
14	12	2	5
13	10	3	6
11	10	1	2
12	12	0	—
15	13	2	5
11	12	−1	−2
13	11	2	5
15	14	1	2

$t_{positive} = \left|\sum_{positive}\right| = 25$

$t_{negative} = \left|\sum_{negative}\right| = 2$

We use the smaller of the two *t*-values as our observed *t*-value.

$t_{observed} = 2$

OUR RESULTS

If our observed *t*-value is 2, what is our critical value? To be considered significant, our observed *t*-value must be *less than* the critical *t*-value. Once again, we determine if we're doing a one- or two-tailed test. Then, we read down the left-hand column of critical *t*-values in Appendix I until we find the number of pairs we have in our study, and then we read across to the critical value. If we use $\alpha = .05$ and a two-tailed test, our critical value is 3. Our observed value is 2, so we have a significant difference in the level of development of visually guided behavior for the active and passive kittens in this study. (This is exactly what Held and Hein concluded about the effect of early visual experience in their actual study.)

CheckPoint

After noticing attitudinal changes in her friends and family around the December holidays, Rose hypothesized that people value their family connections more in December than they do earlier in the year. The following year, Rose tested her hypothesis with a survey she administered twice to 10 participants. The survey, which featured Likert scale questions, measured the degree to which respondents valued family connections.

Respondents were asked to complete the survey in June and again in December. Rose converted the responses she collected into scores, which are reported in the table below. Use these scores to answer the questions that follow and help Rose determine whether people value family more around the December holidays than they do earlier in the year. Round your answers to two decimal places, where appropriate.

Perceived importance of family connections

June	December
4	5
3	1
4	5
1	5
5	5
3	3
2	5
4	4
2	3
3	5

1. Calculate D for each pair of scores.
2. Sign-rank the D-values.
3. Calculate $t_{positive}$.
4. Calculate $t_{negative}$.
5. Identify the appropriate critical value ($\alpha = .05$), compare it to your calculated t-value, and state whether you should accept or reject the null hypothesis.

Answers to these questions are found on page 661.

NON-NORMAL DISTRIBUTIONS

Remember from Table 15.1 that we can use these nonparametric versions of the t-test in two quite different situations. The first situation occurs when we have two variables measured on a qualitative scale and we want to compare average rank on these two variables. The second occurs when we have quantitative data but those data are seriously skewed (i.e., not normally distributed). If the data in our sample are skewed, we have violated one of the key assumptions of parametric testing. Using the MWU test with skewed data will solve this problem.

One of the benefits of ranking our data is that ranking will "unskew" the data. Let's take a look at some seriously skewed *quantitative* data before and after we transform the data into ranks. Table 15.4 shows these data.

TABLE 15.4 Height (in inches) of 18 boys and 18 girls

BOYS		GIRLS	
Heights	Ranks	Heights	Ranks
53	1	47	1
59	2	48	2.5
60	3.5	48	2.5
60	3.5	49	5
61	5.5	49	5
61	5.5	49	5
62	7	50	7.5
63	8	50	7.5
64	9.5	52	9
64	9.5	53	10
65	11.5	55	11
65	11.5	56	12
66	13.5	57	13
66	13.5	58	14
67	15.5	73	15.5
67	15.5	73	15.5
69	17	74	17
70	18	79	18

We're interested in comparing the heights of boys and girls, so we select a random sample of 18 boys and 18 girls, all of them between the ages of 12 and 16. By random chance, our sample of girls has three outliers in it—three girls who are much taller than most of the rest of the girls in the sample. We can calculate a measure of **skew** in both of these samples, where the farther away from zero our measure of skew is, the more skewed the data are. Figure 15.3 shows the descriptive statistics for both samples (including skew) and a frequency histogram for each sample showing the skew visually.

Figure 15.4 shows what happens to the skewed distributions when we convert the raw data to ranks. Notice that the measured skew of both data sets has been reduced by ranking the data.

skew A non-normal, or asymmetrical, distribution of data, in which most of the observations have high values and the outliers have low values (negative skew), or else most of the observations have low values and the outliers have high values (positive skew).

FIGURE 15.3
Histograms of girls' and boys' height data.

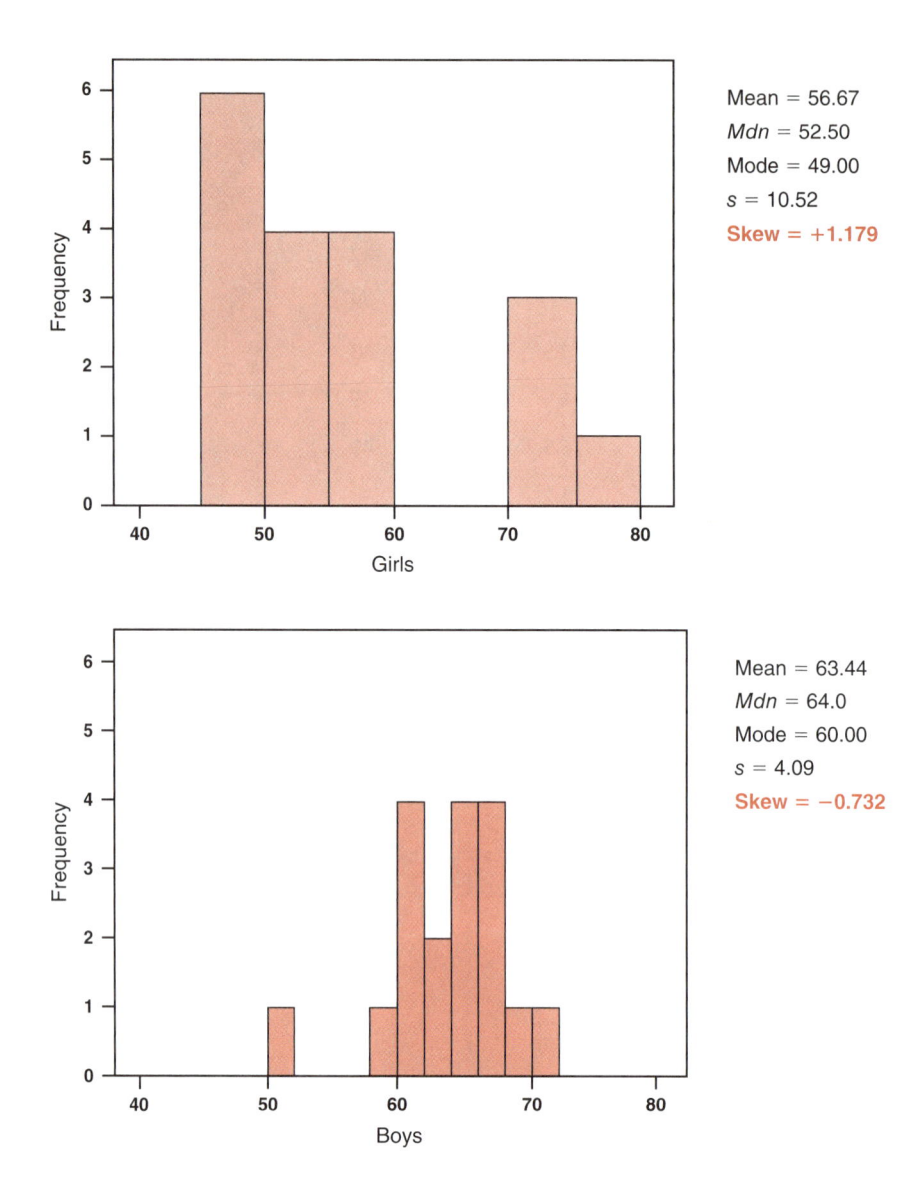

Mean = 56.67
Mdn = 52.50
Mode = 49.00
s = 10.52
Skew = +1.179

Mean = 63.44
Mdn = 64.0
Mode = 60.00
s = 4.09
Skew = −0.732

If you think about it, this reduction in skew should make sense. The tallest girl in the sample had a height of 79 inches (6.58 feet). The mean height for all of the girls in the sample was 56.67 inches (4.72 feet tall), for a difference between the outlier and the mean of 22 inches. When we convert to ranks, the difference between the mean rank (9.5) and the rank of the tallest girl (18) is much smaller (8.5). Ranking the data has pulled the outlier in toward the rest of the data because the difference between any two adjacent ranks is always 1.

To see if these two groups are significantly different, we would pool the data, rank the pooled data, and then compute two *U*-values for the two sets of data. We would take the smaller of the two *U*-values and compare it to the tabled critical value. You can try this example in question 6 of the Practice Problems set at the end of this chapter.

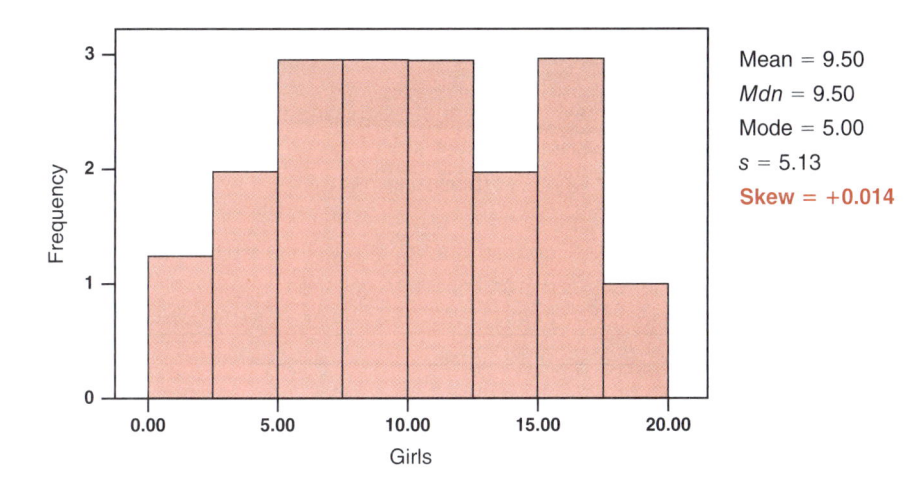

FIGURE 15.4
Histograms of ranked height data for boys and girls

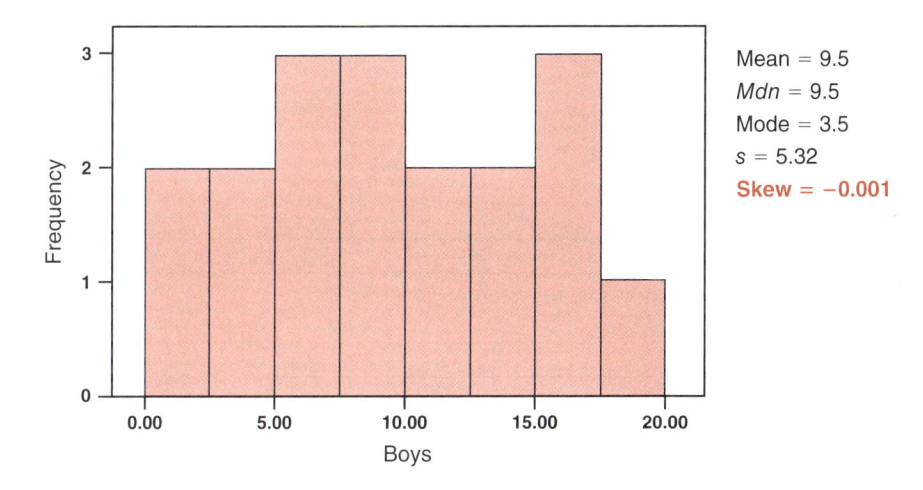

Mean = 9.50
Mdn = 9.50
Mode = 5.00
s = 5.13
Skew = +0.014

Mean = 9.5
Mdn = 9.5
Mode = 3.5
s = 5.32
Skew = −0.001

Think About It...

THINKING AHEAD

This *Think About It* section is going to be a bit different from the previous ones. I am going to ask you to think ahead to what's coming in Chapter 16. Specifically, I want you to think about how you might help yourself make the very important decision about which statistical test you should use, and why that test is the appropriate one.

This decision is a critical one for a number of reasons. The type of test that you run will determine the kind of conclusions you can draw about your data. And as computer data-analysis software gets more and more powerful, running the wrong kind of test gets easier and easier to do.

Think About It... continued

So, let's get started with making the right decisions about tests early. The following questions should be answered before you start collecting your data and will help you make the decision about what test to use:

1. What kind of data (nominal, ordinal, interval, or ratio) are you collecting as your dependent variable? Why is this an important question?
   _____
   _____

2. Are you comparing your sample data to a known population? If so, do you know both the population mean and the population standard deviation? Why are these important questions?
   _____
   _____

3. What are you looking for? List the tests we've discussed for each of the kinds of questions given below.
   a) Is there a difference between means?
   _____
   _____

   b) Is there a relationship between variables?
   _____
   _____

   c) Is there a match between a distribution in a sample and one in a population?
   _____
   _____

   d) Is there an association or dependence between categories?
   _____
   _____

4. Is my sample normally distributed, or is it skewed? Why is this question important?
   _____
   _____

Answers to this exercise are found on page 675.

CheckPoint Answers
Answers to questions on page 656

1. **Perceived importance of family connections** ☐

June	December	D
4	5	−1
3	1	2
4	5	−1
1	5	−4
5	5	0
3	3	0
2	5	−3
4	4	0
2	3	−1
3	5	−2

2. **Perceived importance of family connections** ☐

June	December	D	Rank
4	5	−1	−2
3	1	2	4.5
4	5	−1	−2
1	5	−4	−7
5	5	0	—
3	3	0	—
2	5	−3	−6
4	4	0	—
2	3	−1	−2
3	5	−2	−4.5

3. $t_{Positive} = 4.5$
4. $t_{Negative} = 23.5$
5. $t_{Positive} = 4.5 < t_{critical} = 10 \therefore$ reject ☐

☐
☐
☐

SCORE: /5

NONPARAMETRIC TESTS IN CONTEXT

Remember S. S. Stevens? (If you don't, flip back to Chapter 2 to refresh your memory.) Stevens was the creator of the idea of scales of measurement that we've been using thus far in our exploration of statistics. We have been discussing statistics that are appropriate for use with measurement on an interval or a ratio scale—quantitative statistics (z, t, F, and r). All of these parametric statistical tests rely on several assumptions, which in part rely on Stevens' assumptions about measurement scales. Let's go back and discuss what Stevens meant by quantitative and qualitative data.

In 1946, Stevens wrote a very controversial article. In it, he stated that because of the way these four scales of measurement actually measured data, some kinds of inferential tests were inappropriate for use with some kinds of measurement scales. Qualitative measurement scales (nominal and ordinal) did not measure quantity (amount) and lacked interval strength, meaning there was no consistent interval size between adjacent categories on nominal and ordinal scales. And without intervals of a constant size, it was not possible to apply mathematical operations like addition, multiplication, and subtraction to the data. Therefore, a researcher couldn't calculate a sample mean or standard deviation to compare with any parameter from the population. Sounds tame enough, and you might wonder why this suggestion was controversial.

The reason? Stevens was actually proposing a new theory of measurement, and there was by no means consensus about whether or not his new theory of what it meant to measure something was correct. Classic measurement theory said that only quantitative variables could be measured because measuring meant assigning numbers to questions of "how much," or quantity. Stevens, however, was proposing assigning numbers to quality.

For psychology in particular, Stevens' new theory raised considerable controversy. Psychologists attempt to measure the "immeasurable" all the time. Psychology is full of theories about feelings, attitudes, opinions, and other concepts that have psychological and philosophical weight, but no discernible weight on a balance or a scale. How do you measure the amount or quantity of an opinion? What does it weigh? How big is it?

Stevens' measurement scales gave psychology a means to measure these sorts of concepts, but then he turned around and said that you couldn't use means and standard deviations—and, therefore, all of the by-now-usual inferential statistical tests—with qualitative data. Quantitative assessment tools were used in all of the other sciences, and to suggest that you couldn't apply them to measurement of mental state, or social development, or personality meant that psychology might be "less than" these other disciplines.

The argument over what it means to measure something still rages in science and psychology today. I'm not sure which side is right, although it does make sense that assessing quality is fundamentally different from assessing quantity, and so different statistical tools might be necessary. Way back in Chapter 1, I suggested that statistics can be used to support just about any side of an argument, and that as a consumer of statistics, you will benefit by knowing how statistics can be both correctly used and incorrectly abused.

One of the most common forms of "abuse" of statistics is their application to data where statistics just don't make sense. Let's come full circle and reiterate that you, as the consumer, creator, and interpreter of statistics, need to use your own common sense in the statistical process. Just because a number can be assigned to an idea doesn't mean you have measured something meaningful or meaningfully.

SUMMARY

When we have data measured with a qualitative measurement scale, or we have quantitative data that are seriously skewed, a nonparametric test is appropriate.

We have discussed the nonparametric versions of three very popular parametric tests:

- the Spearman ranked-order correlation coefficient
- the Mann–Whitney U-test
- the Wilcoxon signed-rank, matched-pairs t-test

All three tests require that the raw data be converted to ranks, from smallest to largest; the statistic is then calculated using the ranked data.

The observed statistic calculated with the Mann–Whitney U-test and the Wilcoxon test must be smaller than the tabled critical value in order to be considered statistically significant.

TERMS YOU SHOULD KNOW

eugenics, p. 641
factor analysis, p. 641
nonparametric test, p. 640

parametric tests, p. 640
ranked data, p. 642
skew, p. 657

GLOSSARY OF EQUATIONS

Formula	Name	Symbol
$r_s = 1 - \left[\dfrac{6 \sum D^2}{N(N^2 - 1)} \right]$	Spearman rank-ordered correlation coefficient	r_s
$U_1 = (n_1)(n_2) + \dfrac{n_1(n_1 + 1)}{2} - \sum R_1$ $U_2 = (n_1)(n_2) + \dfrac{n_2(n_2 + 1)}{2} - \sum R_2$	Mann–Whitney U-test	U
$t_{positive} = \left\| \sum_{positive} \right\|$ $t_{negative} = \left\| \sum_{negative} \right\|$	Wilcoxon signed-rank, matched-pairs t-test	t

WRITING ASSIGNMENT

This writing assignment is a bit different from the others. I want you to design a study to examine the relationship between two qualitative variables. Using the lists provided in Table 15.5, choose one X variable and one Y variable. Describe each variable and how you plan to measure it. (If you happen to know of a survey or other instrument that will work, feel free to propose using it.) Explain why your measurement scale requires the use of a Spearman correlation test. Then, use a random number table to generate some imaginary data (keep your N small—say around 10 pairs of X and Y measurements) and actually carry out the test. In APA Style, write a short description of the question you asked and the results you found.

Table 15.5   Data for the Chapter 15 writing assignment

X Variable	Y Variable
Intelligence	Happiness
Self-esteem	Success in life
Reading speed (words per minute)	Success in school
Belief in ghosts (on a scale from 1 to 5 where 1 = strong disbelief, 3 = don't know, 5 = strong belief)	Psychological health

PRACTICE PROBLEMS

1. We discussed the fact that when we are doing a Spearman correlation, there is a pattern between the rankings of the measurements on X and Y that indicates a strong positive correlation between the two variables. What pattern between the rankings of X and Y would you expect to see if there were a strong negative correlation between X and Y?

2. When doing a Spearman correlation, what pattern between the rankings of X and Y would you expect to see if there were no correlation at all between the two variables?

3. Take a look at the table of critical Spearman values in Appendix G. Describe the relationship between the number of pairs of data points (n) and the size of the critical value.

4. The table of critical Spearman values in Appendix G does not list critical values for very small samples (with an n of less than 4), and the critical value for $n = 4$ indicates that we would need a perfect correlation between X and Y for that relationship to be called significant. Does converting the raw data to ranks weaken the relationship between X and Y? Does it change the amount of variability in the data?

5. High school students are asked to estimate the number of hours they spend watching TV per week. TV viewing is presented to one group as a negative behavior (prior to estimations), while TV viewing is presented as a positive behavior to the other group. Does the depiction of TV viewing influence estimation of TV viewing time?

Table 15.6 Estimated weekly TV viewing times

TV Favorable	TV Unfavorable
4	0
5	0
5	1
5	2
10	14
12	42
20	43
49	

  a. What test should you use to answer this question, and why?
  b. Is the test one-tailed or two-tailed?
  c. What is the critical value of the test?
  d. Write an interpretation of the results of the test in APA Style.

6. Let's compare the heights of adolescent boys and girls that we saw in our discussion of non-normal distributions. The data are shown in Table 15.7. Are boys significantly taller than girls?

Table 15.7 Height (in inches) of 18 boys and 18 girls

Boys	Girls
53	47
59	48
60	48
60	49
61	49
61	49
62	50
63	50
64	52
64	53
65	55
65	56
66	57
66	58
67	73
67	73
69	74
70	79

   a. What test should you use to answer this question, and why?
   b. Is the test one-tailed or two-tailed?
   c. What is the critical value of the test?
   d. Write an interpretation of the results of the test in APA Style.

7. A researcher has developed a simple test to measure general intelligence. To evaluate this test, the researcher administers it to 12 seventh-grade students. In addition, the teacher of the seventh-grade class is asked to rank order the students in terms of intelligence, with 1 being the smartest child in the class. The teacher's rankings and the test score for each child are shown in Table 15.8. Is there a relationship between intelligence as measured by the test and intelligence as measured by the teacher?

Table 15.8 Teacher's ranking of student intellectual ability and intelligence test scores

Teacher's Ranking	Test Score
1	41
2	47
3	36
4	42
5	41
6	39
7	40
8	36
9	32
10	39
11	35
12	33

    a. What test should you use to answer this question, and why?
    b. Is the test one-tailed or two-tailed?
    c. What is the critical value of the test?
    d. Write an interpretation of the results of the test in APA Style.

8. Chronic pain is often associated with depression. We want to know if treatment for the pain will alleviate the depression. We select 12 volunteer patients at a local clinic and administer our own test to measures both depression and severity of pain. We call this test the Depression and Pain Scale, or DPS. Scores on the DPS can range from a low of 5 to a high of 20. Higher scores indicate more severe depression and pain. We first assess the patients before treatment for the pain they've been experiencing; then, we assess the patients again after 6 weeks of treatment for the pain. Table 15.9 below presents the results.

Table 15.9 Scores on the DPS before and after treatment for chronic pain

Before Treatment	After Treatment
18	18
19	16
11	5
17	5
14	10

20	7
20	9
13	12
19	10
18	5
15	13
19	20

  a. What test should you use to answer this question, and why?
  b. Is the test one-tailed or two-tailed?
  c. What is the critical value of the test?
  d. Write an interpretation of the results of the test in APA Style.

9. Next time you visit the doctor, make a note of the pain assessment chart that I'll bet is posted in the office somewhere. Pain is a personal event, and measuring the severity of the pain that a particular person experiences is difficult to do. Doctors often ask their patients to use a 10-point ordinal scale to describe the pain they are experiencing, where 0 = *no pain*, 5 = *moderate pain*, and 10 = *the worst pain possible*. In recent years the number of studies on how men and women experience pain has skyrocketed (Fillingim, King, Ribeiro-Dasilva, Rahim-Williams, & Riley, 2009). Suppose 10 men and 10 women, chosen at random from patients who had just had a tooth implant, were asked to report their experience of pain using the 10-point scale. Does the reported pain measurement differ for women and men?

Table 15.10 Rating of pain by men and women

Women	Men
5	4
5	5
6	3
4	4
5	5
2	2
7	8
1	3
8	7
5	7

10. To compensate for the obvious differences in language ability, children are asked to rate their pain experience with a series of cartoons rather than on a 10-point scale (see Figure 15.5). The doctor or nurse then supplies a numeric rating of the pain the child says she or he is experiencing by associating a number with the face.

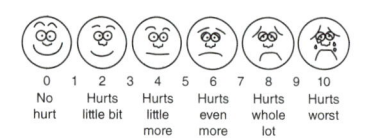

FIGURE 15.5
Cartoon pain rating scale for children.

   Suppose two groups of children are asked to rate the pain of getting a shot using this scale. One group of children is seeing the doctor for the very first time, and the second group of children are old hands at visiting the doctor—they've each been examined by the doctor twice before. Does experience with all that a visit to the doctor entails affect pain rating?

Table 15.11 Rating of pain by first-time and third-time patients of pediatrician

First Time	Third Time
6	7
7	6
6	8
5	6
4	9
6	2
5	2
6	3
9	5
8	2

11. The number of days patients spend in the hospital after coronary artery bypass surgery is measured as a function of the size of the hospital in which the surgery was performed. The length of the postoperative stay for eight patients who had surgery at a very large hospital (400 beds) was compared with the stay for patients at a small, rural hospital (100 beds). The data are shown in Table 15.12. Does the size of the hospital affect the length of stay after surgery?

Table 15.12 Length of postoperative hospital stay (in days)

Large Hospital	Small Hospital
4	5
3	4
4	3

STATISTICS IN CONTEXT

6	15
4	9
3	3
4	4
21	5

12. Olds and Milner (1954) carried out what was arguably one of the most important studies in the history of psychology and neuroscience. Using rats as subjects, they inserted stimulating electrodes into what they thought was a part of the brain called the reticular activating system (RAS) but was actually the nearby septal region. It turned out that the implanted rat would work very hard to get a small electrical stimulation in this area—apparently, it felt good. (You may remember this study from the *Practice Problems* in Chapter 1.)

Suppose you want to find out if rats prefer stimulation of region A in the brain compared to region B. You train eight rats to press a bar when a light comes on. Pressing the bar on the left side of a Skinner box produces stimulation of region A. Pressing the bar on the right side of the cage results in stimulation to region B. After learning the task, the rats are allowed free access to both bars, and the number of bar presses on each bar, left and right, is recorded. The data are shown in Table 15.13. Do the rats in the study show a preference for stimulation of one part of the brain or the other?

Table 15.13 Number of bar presses by eight rats

Rat	Left Bar	Right Bar
1	20	40
2	18	25
3	24	38
4	14	27
5	5	31
6	26	21
7	29	38
8	9	18

13. In a revision of the experiment described in question 12, suppose a researcher trains two groups of rats to press a bar to receive brain stimulation. Group 1 rats receive stimulation to region A, and Group 2 rats receive stimulation to region B. The number of trials needed to learn to press the bar is compared. The date are shown in Table 15.14. Is stimulating one part of the brain or the other a significant factor in the rats' ability to learn the bar press task?

Table 15.14 Number of trials needed to learn criterion by two groups of rats

Rat	Group 1	Group 2
1	3	20
2	12	25
3	11	18
4	15	17
5	12	20
6	16	21
7	10	22

14. Do we get happier as we get older? To find out, suppose we ask 10 randomly selected volunteers to complete a Happiness Scale (HS), which measures happiness on an ordinal scale (from 0 to 6, with higher scores indicating higher happiness levels). We also ask each person how old they are. Is there a relationship between happiness and age?

Table 15.15 Age and score on the HS for 10 individuals

Person	Age	HS Score
1	15	3
2	19	4
3	21	4
4	28	2
5	35	5
6	40	6
7	47	4
8	53	6
9	60	3
10	78	5

15. Americans love their pets—in fact, pets have been described as the only form of love money can actually buy. But do we love them too much? According to the Association for Pet Obesity Prevention (APOP), more than half of the dogs and cats owned by Americans are overweight or obese—a physical condition that leads to early death in both pets and their human owners. Let's suppose you wanted to discover if there was a difference in the average weight of cats owned by families with children and those owned by single individuals. You compare the average weight of 10 cats in each of these two kinds of household and find the data listed in Table 15.16. Is there a difference in the average weights?

Table 15.16 Weights (in pounds) of cats owned by families with children and by single people

Owned by Families	Owned by Singles
9.9	11.9
7	11
10.2	10.2
10.8	12.7
6.3	11.1
9.1	7
9	8.2
6.3	7
9.1	9.1
9.1	8

16. A researcher wants to know if novice observers who are watching children interact during play are seeing and recording the same kinds of behaviors that their more experienced counterparts are. To find out, the researcher asks novice and experienced observers to watch a 20-minute video of children interacting during recess at school. Both sets of observers are told to identify and rank order the 10 most important events they see on the video (1 = *most important event*, 10 = *least important event*). The ranks assigned to the events are then averaged to create an average rank order for the entire group. Table 15.17 shows the data.

Table 15.17 Average rank order for 10 events for novice and experienced observers

	RANK ORDER	
Event	Novice	Experienced
1	2	1
2	4	3
3	1	2
4	6	7
5	5	5
6	3	4
7	10	8
8	8	6
9	7	9
10	9	10

a. What kind of correlation (positive or negative, weak or strong) would you expect if the novice observers were in agreement in their rankings with the more experienced observers?
b. What is the correlation between these two sets of averaged rankings?

17. Do male and females smokers differ in the number of cigarettes they smoke per day? To find out, a researcher asks men and women how many cigarettes they smoked yesterday. The data are shown in Table 15.18. Is there a difference in number of cigarette smoked?

Table 15.18 Number of cigarettes smoked in a day by women and men

By Women	By Men
4	2
5	2
7	5
10	6
20	8
20	18
	20

18. Have you heard of the "Mozart Effect"? Listening to Mozart, especially the piano concertos, is supposed to improve your IQ, at least temporarily (Campbell, 1997). Let's use some hypothetical data about IQ and hours spent listening to Mozart to compare the Spearman and Pearson correlation coefficients. Use the data in Table 15.19 to first calculate Pearson's r statistic and then the Spearman r_s. Do the two statistics tell you the same thing?

Table 15.19 IQ score and number of hours spent listening to Mozart

ID	IQ Score	Hours Spent Listening to Mozart
1	99	0
2	120	2
3	98	25
4	102	14
5	123	45
6	105	20
7	85	15
8	110	19
9	117	22
10	90	4

19. Are strawberries an aphrodisiac? To find out, Sheila generated a control group and an experimental group, each comprising 12 randomly selected participants. Sheila showed the participants images and film clips of possible romantic matches, which varied according to participants' sexual orientations, and recorded the participants' self-reported levels of sexual arousal. Participants in the experimental group were each fed 100 grams of strawberries before the test, while participants in the control group were not. The resultant arousal scores, which range from 1 (*no arousal*) to 10 (*extreme arousal*), are reported below. Does eating strawberries stimulate sexual desire?

Table 15.20   Level of sexual arousal

Control Group	Experimental Group
4	3
6	4
3	2
3	7
5	5
1	4
4	3
1	6
7	5
2	8
3	3
4	6

20. Spoiler alert! Have you ever wondered if learning about the plot of a movie before watching it decreases the enjoyment of watching it? Two groups of randomly selected participants—a control group and an experimental group—were invited to a movie screening. Before entering the theater, the groups were read plot synopses. The version of the synopsis read to the control group was vague and did not reveal any important details about the plot. The synopsis read to the experimental group was very detailed and gave away the ending of the film. After watching the movie, participants self-reported their enjoyment of the movie by completing a postmovie survey composed of Likert scale questions. Table 15.21 below reports enjoyment scores based on the participants' responses to the postmovie survey. Did hearing the spoiler-filled synopsis diminish the experimental groups' enjoyment of the movie?

Table 15.21 Movie enjoyment

Control Group	Experimental Group
4	3
5	3
2	2
3	1
4	2
4	3
3	4
2	5
1	5
3	2

21. Find the answer to the saying.

George Burns said about growing old, "If you can live to be 100, you've got it made. _____ _____ _____ _____ _____ _____ _____ ."

Solution: There seven (7) words in the solution to the riddle. You will find the words in the box below. Each of the words in the solution section is associated with a number. To determine which word you need, solve the problems in the problem section. The word next to the answer to the first problem is the first word in the solution to the riddle. The word next to the answer to the second problem is the second word in the solution, and so on. Round all answers to two decimal places (e.g., 2.645 rounds to 2.65).

Solution to the saying: Words

1	Very	2.5	Interesting	4	That	9	Some
10	Live	6	Die	69	People	280	Surpass
0.98	Forever	7	Few	0.97	Age	286	Past

Problems section: Solve each problem to find a number. Match the number to a word in the word section in order to finish the riddle.

Data set:

X	Y
3	23
5	27

6	26
7	30
7	33
7	34
9	35
10	38
15	40

For variable X:
1. The rank of the number 3 is _____.
2. The rank of the number 7 is _____.
3. The sum of the X variable is _____.

For variable Y:
4. The rank of the number 34 is _____.
5. The sum of the Y variable is _____.

For both variables:
6. The sum of D^2 is _____.
7. The value of r_s is _____.

Think About It . . .

THINKING AHEAD

SOLUTIONS

1. What kind of data (nominal, ordinal, interval, or ratio) are you collecting as your dependent variable? Why is this an important question? **According to Stevens, some kinds of statistical tests are better applied to some kinds of data. If our data are *quantitative* (interval or ratio), then we can calculate meaningful means and standard deviations and use inferential tests based on these statistics (z-, t-, F-, and r-tests). If we have nominal or ordinal data, nonparametric tests are more appropriate (χ^2, r_s, U, and Wilcoxon t).**

2. Are you comparing your sample data to a known population? If so, do you know both the population mean and the population standard deviation? Why are these important questions? **If we know σ and μ and we have quantitative data, we can use the z-test. If we know μ but not σ, we can estimate σ and use a single-sample t-test.**

3. What are you looking for? List the tests we've discussed for each of the kinds of questions listed below.
   a) Is there a difference between means? **If there are only two means, then we can use either an independent-samples or a dependent-samples *t*-test (depending, of course, on whether our groups are independent or dependent). If we have one independent variable and more than two levels, we can use a one-way ANOVA. If we have two independent variables, each with at least two levels, then we can use a two-way ANOVA.**
   b) Is there a relationship between variables? **If we have quantitative measurements of the two variables (*X* and *Y*), we can assess how they are correlated using a Pearson *r*-test. If one or both of our variables are quantitative, we can use the Spearman r_s-test.**
   c) Is there a match between a distribution in a sample and one in a population? **If we know how our qualitative data are distributed in the population and we want to see if the distribution in our sample fits the one in the population, we can use a one-way χ^2.**
   d) Is there an association or dependence between categories? **A two-way χ^2.**
4. Is my sample normally distributed, or is it skewed? Why is this question important? **When we use a parametric statistical test, we make assumptions about the data. One of these assumptions is that the data are at least reasonably normally distributed. If the data are seriously skewed, we have violated that assumption. And if we have violated that assumption, we should use a nonparametric test like the Mann–Whitney *U*-test to "unskew" the data when we compare our groups.**

REFERENCES

Campbell, D. (1997). *The Mozart effect: Tapping the power of music to heal the body, strengthen the mind, and unlock the creative spirit.* New York: Harper Collins.

Carter, M., & Weissbrod, C. S. (2011). Gender differences in the relationship between competitiveness and adjustment among athletically identified college students. *Psychology, 2*(2), 85–90.

Fillingham, R. B., King, C. D., Ribeiro-Dasilva, M. C., Rahim-Williams, B., & Riley, J. L. (2009). Sex, gender, and pain: A review of recent clinical and experimental findings. *Journal of Pain, 10*(5), 447–485.

Held, R., & Hein, A. (1963). Movement-produced stimulation in the development of visually guided behavior. *Journal of Comparative and Physiological Psychology, 6*(5), 872–876.

Olds, J., & Milner, P. (1954). Positive reinforcement produced by electrical stimulation of septal area and other regions of rat brain. *Journal of Comparative and Physiological Psychology, 47*, 419–427.

Sin, E. L. L., Chow, C., & Cheung, R. T. H. (2015). Relationship between personal psychological capitals, stress level, and performance in marathon runners. *Hong Kong Physiotherapy Journal, 33*, 67–77.

Spearman, C. E. (1930). Autobiography. In C. A. Murchison (Ed.), *A history of psychology in autobiography*, Vol. 1. (pp. 299–333). Worchester, MA: Clark University Press.

Stevens, S. S. (1946). On the theory of scales of measurement. *Science, 103*(2684), 677–680.

Conducting Nonparametric Tests with SPSS and R

The data set we will be using for this tutorial is drawn from the Behavioral Risk Factor Surveillance System (BRFSS), a nationwide health survey carried out by the Centers for Disease Control and Prevention. The set we will be using is based on results of the 2008 questionnaire, completed by 414,509 individuals in the United States and featuring a number of questions about general health, sleep, and anxiety, among other things. The version we will be using is a trimmed-down version of the full BRFSS data set, which you can find on the companion site for this textbook (www.oup.com/us/blatchley).

Spearman's Correlation in SPSS

Similar to the examples in Chapter 13, we will use a statistical test to determine if two variables are correlated. This time we will be using Spearman's correlation. For this test, we are going to look at the relationship between respondents' general health based on a scale of 1 to 5 (1 = *excellent*, 5 = *poor*) and their life satisfaction as ranked on a scale of 1 to 4 (1 = *very satisfied*, 4 = *very dissatisfied*). Our null and alternative hypotheses are as follows:

H_0: There is no relationship between the two variables, $r_s = 0$.

H_1: The two variables are related, $r_s \neq 0$.

STEP 1: Load the data into SPSS.

STEP 2: Set up the test. Click **Analyze**, scroll to **Correlate**, and then click **Bivariate**...

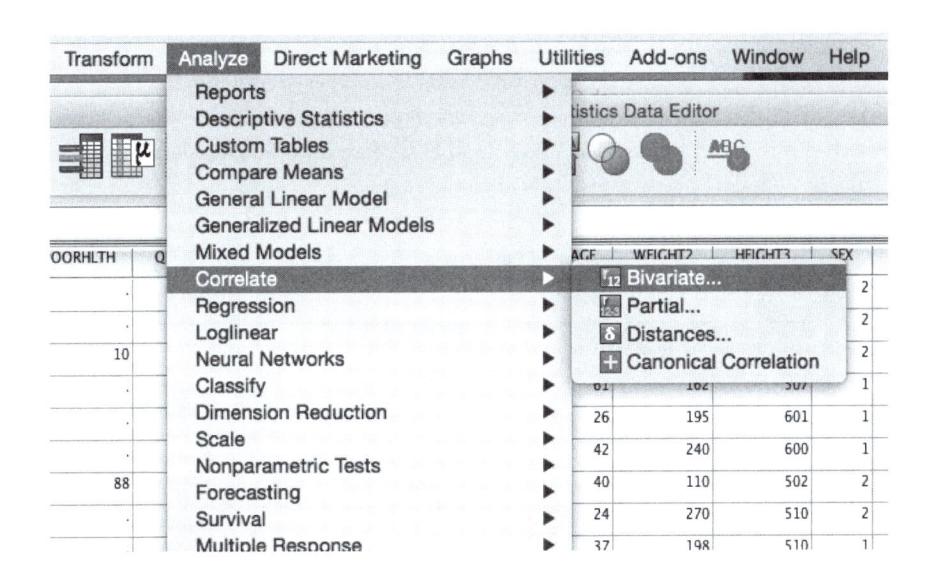

STEP 3: Select the variables. In the **Bivariate Correlations** dialogue box, select the variables you are interested in testing. In this case, select **Satisfaction With Life (LSATISFY)** and **General Health (GENHLT)**.

Pearson is the default setting in the dialogue box; remove the checkmark by clicking the box beside **Pearson** and select **Spearman**. Click **Ok**.

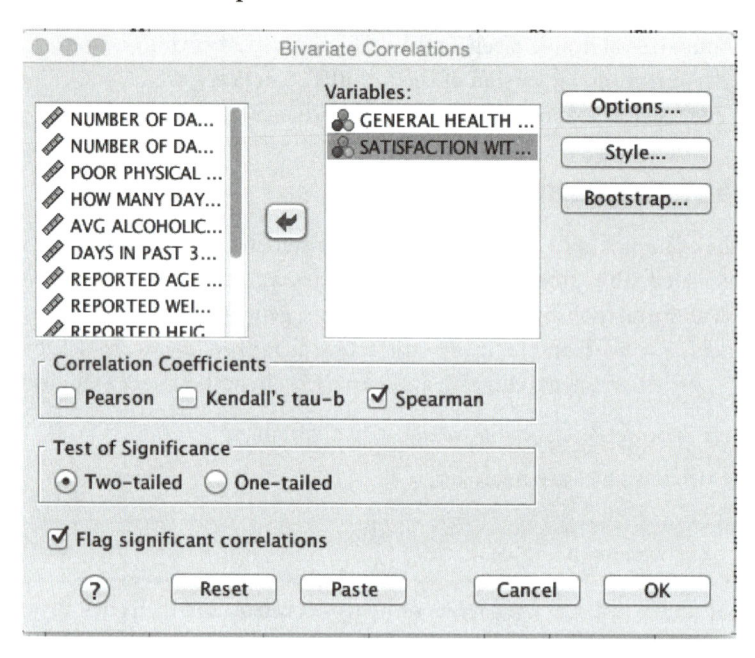

STEP 4: Analyze the results. The Correlations table will be generated in the **Output** window in SPSS. The results will be reported in a matrix of correlations. We are specifically interested with the correlation between General Health and Life Satisfaction (the top right box or the bottom left box—the results will be the same).

Correlations

			GENERAL HEALTH	SATISFACTION WITH LIFE
Spearman's rho	GENERAL HEALTH	Correlation Coefficient	1.000	.317**
		Sig. (2-tailed)	.	.000
		N	412639	402090
	SATISFACTION WITH LIFE	Correlation Coefficient	.317**	1.000
		Sig. (2-tailed)	.000	.
		N	402090	403885

**. Correlation is significant at the 0.01 level (2-tailed).

In this example, the Spearman correlation coefficient (r_s) is .317, which suggests that there is a positive relationship between life satisfaction and general health. The two-tailed Sig. value (*p*-value) shows that the correlation coefficient is statistically significant at the .0005 level (remember that .000 does not mean zero).

678 **STATISTICS IN CONTEXT**

Mann–Whitney *U*-Test in SPSS

The Mann–Whitney *U*-test allows us to compare whether the distribution of two groups is equal. For this test, we are once again going to examine the relationship between men and women and the number of poor night's sleep they had in the previous 30 days. Specifically, we want to examine whether the distributions for the two groups are identical, even if one group has experienced more nights of poor sleep on average.

Our null and alternative hypotheses are as follows:

H_0: The distributions for the two groups are equal.

H_1: The distributions for the two groups are different.

STEP 1: Load your data into SPSS.

STEP 2: Set up the test. Click **Analyze**, scroll to **Nonparametric Tests**, and then click **Independent Samples**.

A dialogue box will open up. In the **Objective** tab, ensure that the option **Automatically compare distributions across groups** is selected (this is the default setting).

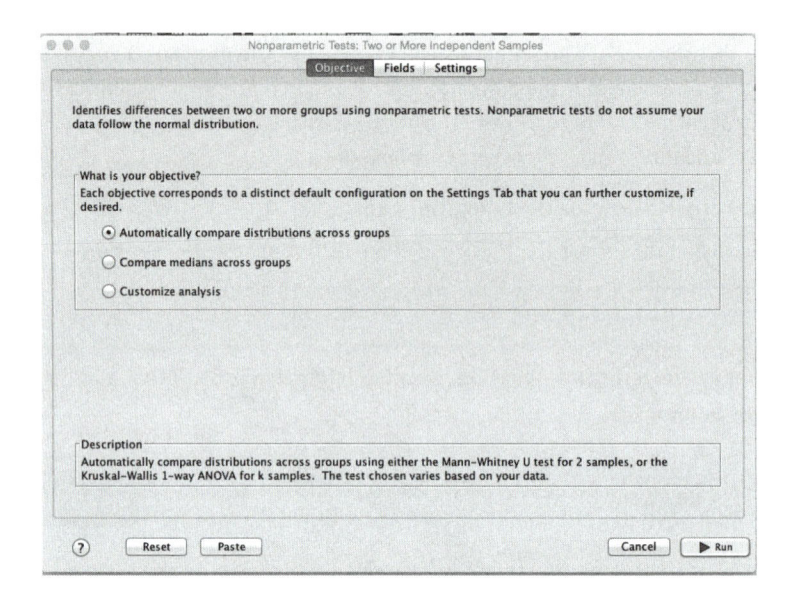

STEP 3: Choose your variables. In the **Fields** tab, we will select our relevant variables. Select **Respondents Sex** to the **Groups** box using the bottom arrow. Select **How many days did you get enough sleep? (QLREST2),** and move it to the **Test Fields** box using the top arrow. Click **Run**.

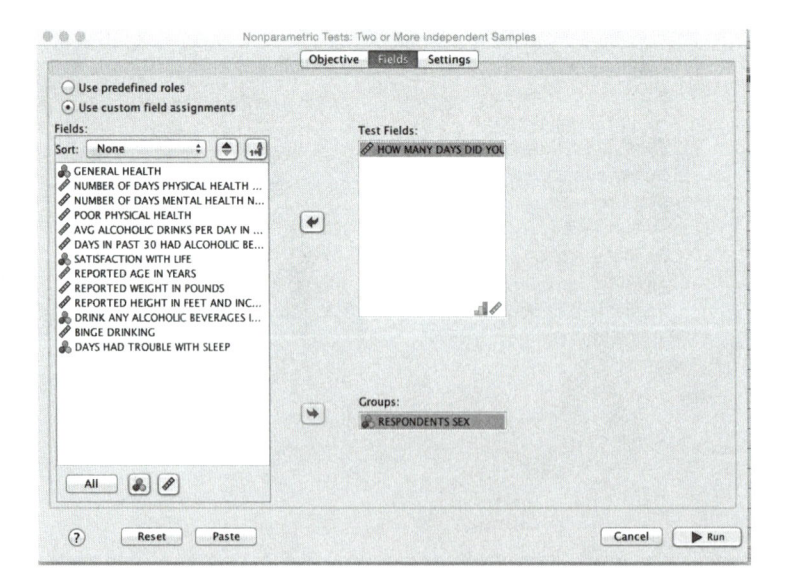

680   STATISTICS IN CONTEXT

STEP 4: Analyze the results. It is very easy to analyze the results of the Mann–Whitney *U*-test in SPSS because there is no guesswork. The results that show up in the **Output** window tell you whether you fail to reject the null hypothesis that the distributions between groups are the same.

Hypothesis Test Summary

	Null Hypothesis	Test	Sig.	Decision
1	The distribution of HOW MANY DAYS DID YOU GET ENOUGH SLEEP I is the same across categories of RESPONDENTS SEX.	Independent-Samples Mann-Whitney U Test	.000	Reject the null hypothesis.

Asymptotic significances are displayed. The significance level is .05.

In this case, we reject the null hypothesis that our distributions across groups are the same.

This test essentially allows us to determine how similar our groups are, which will tell us how to approach a comparison. In a sense, we are trying to determine if we are comparing apples to apples or apples to oranges. If the distributions of the data are not similar, then we cannot infer certain things about other similarities or differences. For instance, we cannot comment on the magnitude of difference between two groups if the distributions are not the same.

In the case that we retain the null hypothesis and the distributions are similar, we could go a step further and examine the medians of our two groups. We would do this in order to determine the magnitude of the difference between each group. Here are the steps required to do this.

STEP 1: Click **Analyze**, scroll to **Compare Means**, and then click **Means...**

CHAPTER 15 Nonparametric Tests

STEP 2: Select the variables we want to compare.

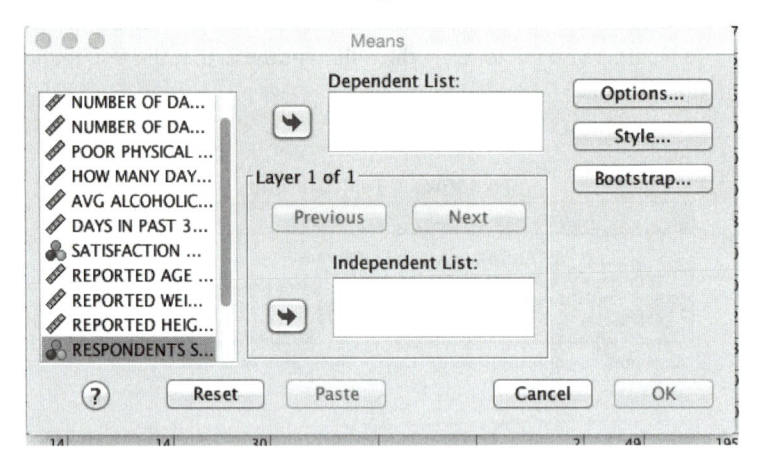

STEP 3: Choose **Median** as your statistic: Click **Options...**, and then, in the **Options** dialogue box, select **Median** from the menu and move it to the **Cell Statistics** box. Remove all other cell statistics, and click **Continue**.

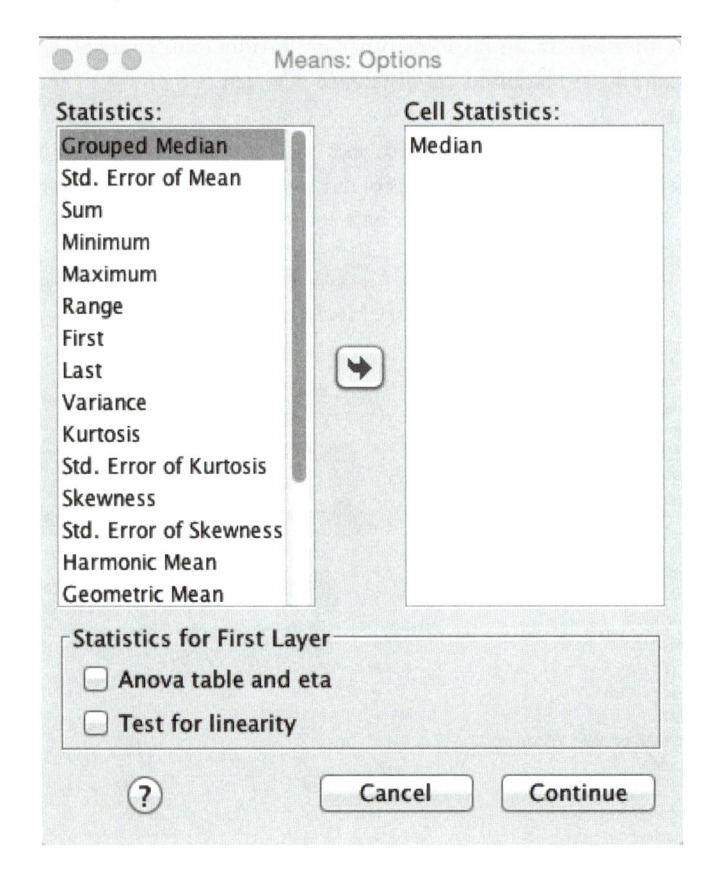

682   STATISTICS IN CONTEXT

STEP 4: Compare your results. The output provided will give you an idea of the magnitude of the differences between groups, and how the medians compare.

Spearman's Correlation in R

Conducting a test for Spearman's correlation with R is very similar to how we test for Pearson's correlation (see Chapter 13). There is just one simple change to the code. In this example, we will look at the relationship between respondents' general health on a scale of 1 to 5 and their overall life satisfaction as rated on a scale of 1 to 4, as we did earlier in this chapter

First, in order to run the specific code, we will need to load the Hmisc and foreign libraries.

For this, you will need to make sure you have the foreign package loaded and installed. To complete these steps, go to **Packages and Data** and click **Package Installer**.

In the **Package Installer** window, click **Get List,** and search for **Hmisc** and **foreign** separately. Once you have found each individually, click **Install Selected**; make sure that the **Install Dependencies** checkbox is selected. Exit the window.

Go to **Packages and Data,** and click **Package Manager**.

In the **Package Manager** window, search for **Hmisc** and **foreign,** and select the checkbox beside each. Make sure that the status column changes to "loaded."

Alternatively, you can type the following code:

```
>library(Hmisc)
>library(foreign)
>library(pander)
```

STEP 1: Load your data into R.

```
#load data
> health<- read.csv("example.csv")
```

STEP 2: Drop any missing variables. Missing variables will cause the calculated correlation to be incorrect.

We must recode any missing or not-applicable data as follows:

```
#Remove "Do not know", no response and 0 from the sample of
LSATISFY.
> health$LSATISFY[health$LSATISFY==7]<-NA
> health$LSATISFY[health$LSATISFY==9]<-NA

#Remove "Do not know" and no response from GENHLTH
> health$GENHLTH[health$GENHLTH==7]<-NA
> health$GENHLTH[health$GENHLTH==9]<-NA
```

STEP 3: Run the test, and analyze the results.

We want to explore whether there is a correlation between an individual's general health and their life satisfaction as rated on a Likert scale of 1 to 5 and 1 to 4, respectively.

```
> rcorr(sleep$QLREST,sleep$AGE, type="spearman")
```

The results will populate automatically. Note that R rounds to two decimals.

```
> spearman<-rcorr(health$LSATISFY, health$GENHLTH,
type="spearman")
>pander(spearman)
```

* **r**:

    ```
    ------------------
        x     y
    -------  ----- -----
    **x**    1     0.315
    **y**    0.315 1
    ------------------
    ```

* **n**:

    ```
    --------------------
        x      y
    -------  ------ ------
    **x**    4e+05  389150
    **y**    389150 403150
    --------------------
    ```

* **P**:

    ```
    ---------------
        x    y
    -------  ---  ---
    **x**    NA   0
    **y**    0    NA
    ---------------
    ```

The results show an r_s-value of 0.32 between a participant's general health rating and their life satisfaction rating. Looking at the *p*-table, we can see that the results are significant at the .005 level.

Mann–Whitney *U*-Test in R

Conducting a Mann–Whitney *U*-test in R requires a few lines of code. We need to set up our variables and then run the test. In this example, we will look at the distribution between male and female respondents to the BRFSS and the number of poor night's sleep they reported in the preceding 30 days. We will explore whether the distribution of the two groups is the same. If this is the case, we are able to make comparisons between the categories that extend beyond the mean.

STEP 1: Load your data into R.

```
#load data
>health<-read.csv("example.csv")
```

STEP 2: Drop missing variables.

```
#Remove "Do not know", no response and 0 from the sample of QLREST2.
> health$QLREST2[health$QLREST2==77]<-NA
> health$QLREST2[health$QLREST2==88]<-A
> health$QLREST2[health$QLREST2==99]<-NA
```

STEP 3: Run the test, and analyze your results.

```
> utest<-wilcox.test(health$QLREST2~health$SEX)
> pander(utest)

---------------------------------------------------------
 Test statistic      P value      Alternative hypothesis
----------------   --------------  ----------------------
    7.565e+09       9.512e-40 * * *        two.sided
---------------------------------------------------------

Table: Wilcoxon rank sum test with continuity correction:
`health$QLREST2` by `health$SEX`
```

By looking at the test statistic and the *p*-value, we are able to reject the null hypothesis that the distribution among the groups is equal. This result means that we are limited in the types of comparison we can do with our data. We know that there is a statistically significant difference in the average number of poor nights of sleep for men and for women, but we cannot compare these differences beyond their average.

The code that we used above was for the comparison on a continuous variable to a binary variable (i.e., male and female) and is written like this:

```
wilcox.test(y~A)
```

If our independent variable was also continuous, we would amend the code as follows:

```
wilcox.test(y,x),
```

where x is the name of the continuous independent variable.

CHAPTER SIXTEEN

WHICH TEST SHOULD I USE, AND WHY?

On two occasions I have been asked [by members of Parliament], "Pray, Mr. Babbage, if you put into the machine wrong figures, will the right answers come out?" I am not able rightly to apprehend the kind of confusion of ideas that could provoke such a question.

—CHARLES BABBAGE

STATISTICS IN CONTEXT

The focus of this chapter is the context in which we use statistics. We are not starting in our usual way, with a section on statistics in the context of history. Instead, we're going to jump right in to statistics in the context of the research that we use them in.

Modern research laboratories have computers and software that will analyze any data you enter into them. The computer, as Mr. Babbage pointed out, operates on what has been called a "GIGO" system—or "Garbage In, Garbage Out." The computer does not care if the data entered are correct. Even if you enter the "wrong" data (garbage in) or ask for a test that is inappropriate to the data you've entered, the computer will give you a result. Right or wrong, you *will* get an answer (potentially, garbage out). Your job is to know what kind of data you have, what kind of test is appropriate to answer your research question, and what the answer you get at the end of the analysis actually means.

Deciding what statistical test to use in a particular situation gives almost everyone who uses statistics some pause. Students are often confused about where and when a particular test is appropriate, but I've been asked the same question by seasoned researchers who have been using statistics for years.

This chapter will provide you with a series of examples of research from laboratories, from the literature, and from my own imagination. Your job will be to figure out what statistical test is appropriate to use in answering the research question being asked. Figure 16.1 shows you a decision tree that you can use to help determine what test is appropriate. This is my version of the decision tree; it works for me because I set it up. I strongly advise you to create your own decision tree and make it work for you. Think carefully about the decision points on the tree—the places where you have to choose between branching one way or another. What should define that decision point? How will you know which way to move through the tree? The more you work with your own knowledge of statistics and your own understanding of how to make this decision, the easier the process of deciding what test to use will become.

On occasion, there might be more than one way to approach the analysis of the data. Take a look at the "what kind of data do I have" decision point, and you will see why this might be the case. If you define the results of a Likert scale as ordinal-level data, then you automatically direct your decision about the test you should use to a particular

> **OVERVIEW**
> STATISTICS IN CONTEXT
> DECISION TREE: WHICH TEST SHOULD I USE
> EXAMPLES
> A FINAL WORD

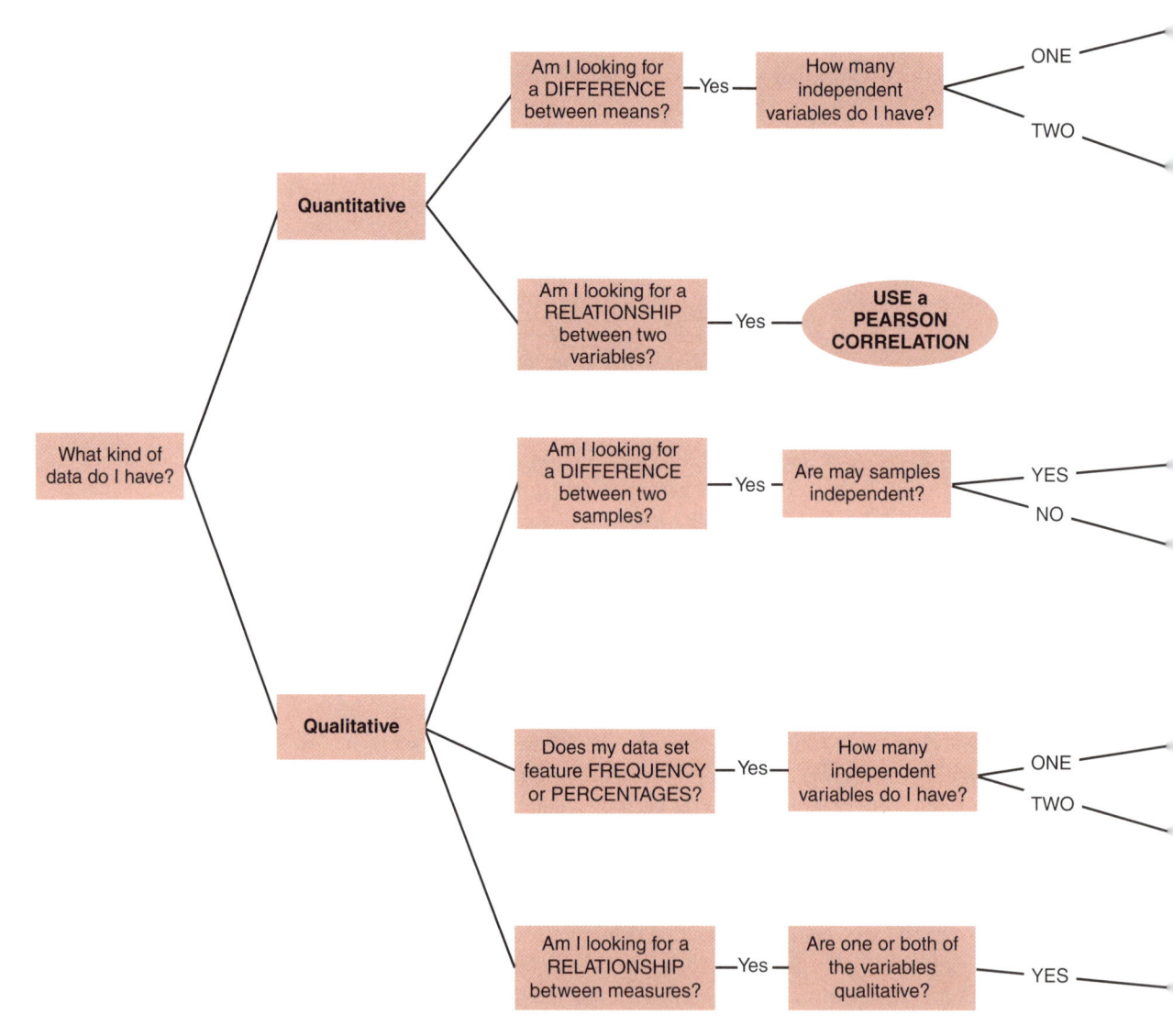

FIGURE 16.1
Decision tree: Which test should I use?

688   **STATISTICS IN CONTEXT**

Flowchart: Which Test Should I Use?

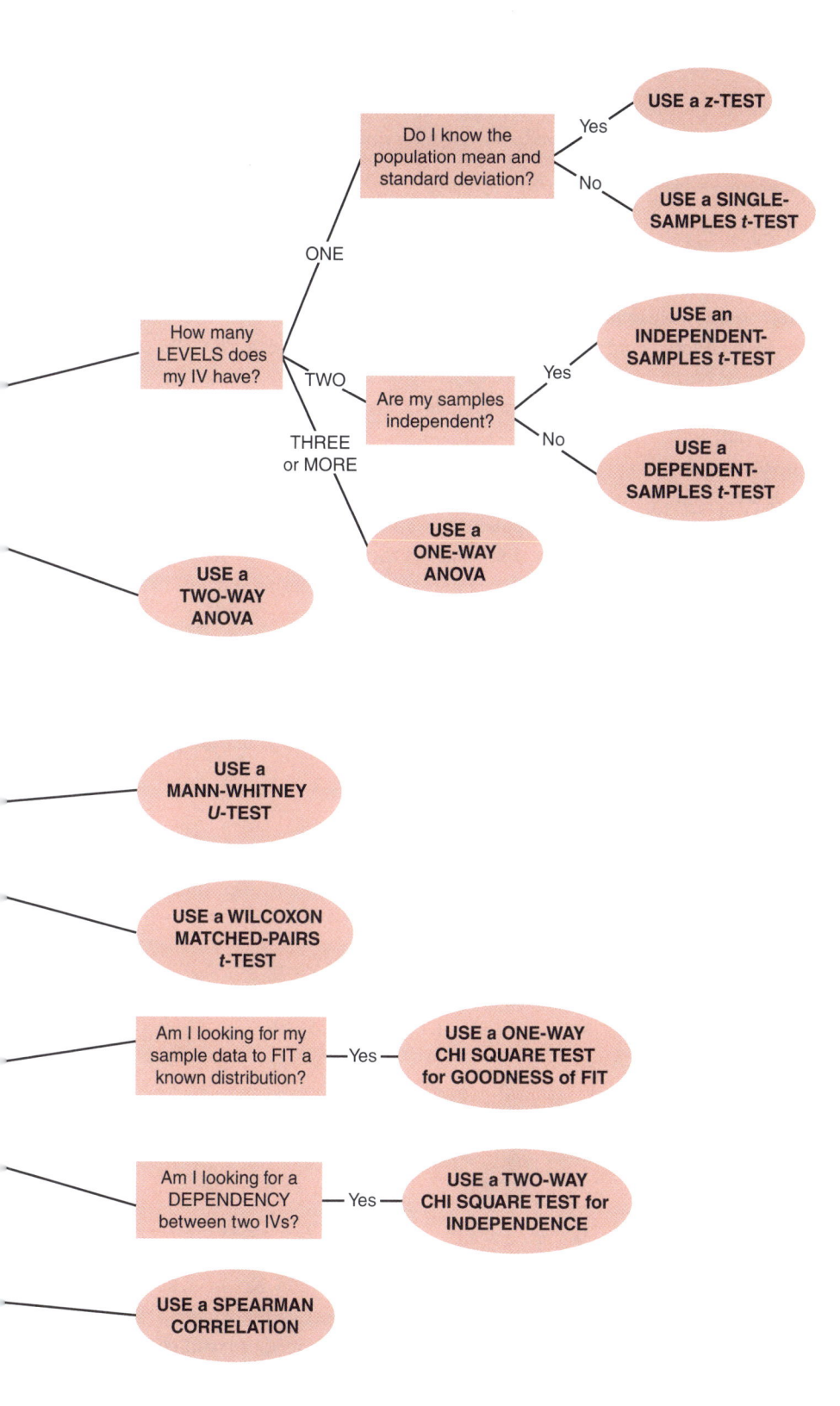

How many LEVELS does my IV have?

- **ONE** → Do I know the population mean and standard deviation?
  - Yes → **USE a z-TEST**
  - No → **USE a SINGLE-SAMPLES t-TEST**

- **TWO** → Are my samples independent?
  - Yes → **USE an INDEPENDENT-SAMPLES t-TEST**
  - No → **USE a DEPENDENT-SAMPLES t-TEST**

- **THREE or MORE** → **USE a ONE-WAY ANOVA**

- **USE a TWO-WAY ANOVA**

- **USE a MANN-WHITNEY U-TEST**

- **USE a WILCOXON MATCHED-PAIRS t-TEST**

- Am I looking for my sample data to FIT a known distribution? — Yes — **USE a ONE-WAY CHI SQUARE TEST for GOODNESS of FIT**

- Am I looking for a DEPENDENCY between two IVs? — Yes — **USE a TWO-WAY CHI SQUARE TEST for INDEPENDENCE**

- **USE a SPEARMAN CORRELATION**

part of the tree. On the other hand, if you define measures on a Likert scale as interval-level data, then you will wind up using a very different kind of test to analyze the data.

So, let's get started. On my decision tree, one of the first questions that you should ask yourself is *What kind of question am I trying to find the answer to?* Asking if two variables are related to one another is quite different from asking if two variables, represented by two means, are different from one another. So, first, make sure you know what question you're asking.

Tests for qualitative data and tests for quantitative data differ quite a bit in how they are carried out, primarily because measuring on a quantitative scale (interval or ratio) tells you something different than does measurement on a qualitative (usually ordinal) scale.

You will also need to know how many groups or levels of your independent variable you are comparing. The single-sample *t*-test will allow you to compare a single-sample mean to a known population mean, while a dependent-samples *t*-test will allow you to compare two sample means that represent two dependent populations. An ANOVA will compare more than two sample means and will also allow you to compare means across two independent variables.

For each example below, determine which of the statistical tests we have discussed is (or are) appropriate for use with the situation described. You should choose from the following tests:

- The *z*-test
- The single-sample *t*-test
- The independent-samples *t*-test
- The dependent-samples *t*-test
- The one-way ANOVA
- The two-way ANOVA
- The Pearson correlation coefficient
- The one-way chi-square test
- The two-way chi-square test
- The Spearman rank-ordered correlation coefficient
- The Mann–Whitney *U*-test
- The Wilcoxon signed-rank, matched-pairs *t*-test

EXAMPLES

1. Learning to read and write is a major milestone of the elementary school curriculum. Suppose you want to test a new teaching method that focuses on having children sound out the words as they read. You ask two different groups of third-graders to read a series of stories, and you measure reading comprehension by asking the children to answer a series of true/false questions about the stories. Reading comprehension is measured in terms of the

number of questions the children answer correctly. One group of children learns to read with your new method, while the control group learns with the standard method used by the school.

   Test(s) _____
   Rationale _____
   _____

2. You have developed a new paper-and-pencil test that (you hope) measures reading comprehension in college graduates. You'd like to be able to report that your new test has what is known as "predictive validity." If it does, then scores on your test of reading comprehension should be able to predict how well a person does on a reading comprehension–based task—perhaps job performance as an editor. To test predictive validity, you measure reading comprehension using your test and then perform a job performance assessment of these same people in their capacity as editors.

   Test(s) _____
   Rationale _____
   _____

3. To test the effects of alcohol consumption on the perception of pain, a researcher administers a dose of alcohol to 25 rats (via their water) and then treats the rats with the tail-flick test of analgesia. In this test, the tails of the animals are exposed to a hot beam of light (intense thermal stimulation), and the latency to flick the tail out of the beam is measured. Longer latencies indicate greater analgesia. Previous research has shown the average latency to flick the tail out of the light beam in sober, healthy rats to be 13.4 seconds.

   Test(s) _____
   Rationale _____
   _____

4. Several studies have shown that exposure to a stressor can induce temporary analgesia in rats and mice. Kurrikoff, Inno, Matsui, and Vasar (2008) used foot shock to produce stress in mice and measured tail-flick latency to assess the effects of stress on analgesia. Five groups of mice were exposed to different intensities of foot shock: Group 1 (control), 0.0 mA; Group 2, 0.2 mA; Group 3, 0.4 mA; Group 4, 0.6 mA; and Group 5, 0.9 mA. Tail-flick latencies (measured in seconds) were measured immediately after presentation of shock.

   Test(s) _____
   Rationale _____
   _____

5. Kurrikoff et al. (2008) added another variable to their study of stress and analgesia. They compared five levels of foot-shock stress in two different strains of mice—wild mice and mice with a genetic mutation that results in

a decreased number of receptors for a hormone called cholecystokinin. This hormone is thought to play an important role in inducing tolerance to opioid drugs like heroin.

Test(s) _____
Rationale _____
_____

6. Have you heard of "pet therapy," also known as "animal-assisted therapy," or AAT? Access to a pet (typically, but certainly not limited to, a dog) has been shown to be associated with beneficial physical responses like lowered blood pressure and reduced anxiety. In a review of the literature on pet therapy, Giaquinto and Valentini (2009) describe a study that compared two groups of heart patients. Group 1 consisted of patients who had high blood pressure and who owned pets. These participants were matched with participants in a control group who did not own pets. Six months later, blood pressure was measured in both groups. They found that both groups showed reduced blood pressure, but that pet owners experienced a larger reduction than did the non–pet owners.

Test(s) _____
Rationale _____
_____

7. Walker, Brakefield, Morgan, Hobson, and Stickgold (2002) examined the relationship between the amount of Stage-2 sleep obtained and performance on a finger-tapping task. They found that performance on this simple motor task (measured in terms of the number of repeated five-keypress sequences completed in a 30-second period) improved as the amount of Stage-2 sleep increased.

Test(s) _____
Rationale _____
_____

8. Harrison and Horne (2000) examined the effects of sleep deprivation on temporal memory. Forty young adults were randomly assigned to four groups. Group 1 was a control group given caffeine. Group 2 was a control group given a placebo. Group 3 participants were sleep-deprived for 36 hours and given caffeine, and Group 4 participants were sleep-deprived for 36 hours and given a placebo. Temporal memory was tested by showing all participants a series of facial images (Series A—12 faces and Series B—a different set of 12 faces). After seeing the photographs, participants were shown a set of 48 faces that included the 24 they had just seen. Participants were asked if they had seen the face before and, if they had, which series of photographs, A or B, it had been in (recency discrimination or temporal memory). The number of correct responses on the temporal memory task were recorded.

Test(s) _____
Rationale _____
_____

9. A study of stress and sleep in 200 college-age students examined year in school (freshman, sophomore, junior, or senior) and quality of sleep (measured on a 5-point scale where 1 = *very poor sleep quality*, 3 = *average quality*, and 5 = *very good sleep quality*).

   Test(s) _____
   Rationale _____
   _____

Table 16.1 Sleep quality of 200 college students, by year in school

Sleep quality

Year in School	1 (Very Poor)	2 (Fairly Poor)	3 (Average)	4 (Fairly Good)	5 (Very Good)	Total
Freshman	3	10	11	15	11	50
Sophomore	5	15	16	7	7	50
Junior	12	4	12	10	12	50
Senior	33	4	5	3	5	50
Total	53	33	44	35	35	200

10. In a study of treatment for depression, a set of 50 patients, all diagnosed with depression, were deprived of dreaming (REM) sleep for one night. Depression was measured in terms of serum levels of serotonin metabolites and compared to a known average level of these metabolites. The results showed that sleep-deprivation therapy reduced depression.

    Test(s) _____
    Rationale _____
    _____

11. A researcher wants to know if there was a relationship between education level (measured on a scale where 1 = *grade school education or less*, 2 = *some high school*, 3 = *high school diploma*, 4 = *some college*, 5 = *college degree*, and 6 = *at least some graduate education*) and average annual salary.

    Test(s) _____
    Rationale _____
    _____

12. The Centers for Disease Control and Prevention measured average Body Mass Index (BMI) in young adults (20–25 years of age). Researchers wanted to know if BMI differed across racial/ethnic groups, so they randomly sampled young adults in non-Hispanic white, non-Hispanic black, and Mexican-American groups.

    Test(s) _____
    Rationale _____
    _____

13. Studies have suggested that very low cholesterol levels might be associated with risk of suicide. Plana et al. (2009) measured levels of cholesterol in 66 randomly selected patients admitted to a psychiatric facility following a suicide attempt. The researchers compared these levels with those obtained from a control group of 54 randomly selected patients with no history of suicide.
    Test(s) _____
    Rationale _____
    _____

14. Is there a relationship between the grade earned in class (where $1 = A$, $2 = B$, $3 = C$, $4 = D$, and $5 = F$) and the overall assessment rating that students give to their professors (on a scale where $1 = $ *poor* and $5 = $ *excellent*)?
    Test(s) _____
    Rationale _____
    _____

15. In a study of childhood obesity, researchers examined the likelihood that children with zero, one, two, or three or more siblings would become obese.
    Test(s) _____
    Rationale _____
    _____

16. A related study examined the BMI of children who came from either single-parent homes or from two-parent homes. Children of single parents were found to have significantly higher BMI measures (Chen & Escarce, 2010).
    Test(s) _____
    Rationale _____
    _____

17. Hope et al. (2010) wanted to know if osteoprotegerin (OPG), a chemical known to be involved in the body's inflammation response, is elevated in patients with severe mental disorders. They measured the levels of OPG in two randomly selected groups of people: patients diagnosed with schizophrenia or bipolar disorder and normal, healthy controls.
    Test(s) _____
    Rationale _____
    _____

18. I want to know if students who take statistics in the early morning (8:00 a.m. class) get better grades on their final exams than do students who take the class in the late afternoon (4:00 p.m. class). I notice that two students in the early morning class get perfect scores (number correct out of 150 total points) on the final. The closest score to these two outliers is a score of 110.

Test(s) _____
Rationale _____
_____

19. If it is true that in spring a young man's fancy turns to thoughts of love, can the same be said of young women? To find out, we assume that if this old saying is true, then attention to academics should decrease in the spring compared to the undistracted fall semester. We compare the average GPA of a random sample of 10 young women at the end of the fall semester and again at the end of the spring semester.
    Test(s) _____
    Rationale _____
    _____

20. We repeat the study in example 19 during the following year with a change in how we measure the young woman's fancy. Instead of using GPA, we use female students enrolled in a two-semester course, and we ask the professor teaching the course to rank the students in the order of their attention to the material in the course.
    Test(s) _____
    Rationale _____
    _____

21. During a recent election, male and female voters were asked whether they agreed or disagreed with a proposal to change the state law governing Sunday sales of alcohol. The frequency of "agree" and "disagree" votes by men and women was counted.
    Test(s) _____
    Rationale _____
    _____

22. Scheduling courses is an always difficult and often contentious task. To make things a bit easier, a psychology department decides to assess enrollments in three different time periods for an introductory psychology class. If the time of day the course is offered makes no difference to the students, then the course can be fitted into the schedule whenever there's a spot for it. Data showing the number of students enrolled in the class in the early morning, mid-afternoon, and late afternoon are collected.
    Test(s) _____
    Rationale _____
    _____

ONE LAST WORD

I hope that these examples have helped you feel more confident about how to go about selecting the appropriate inferential test for your data. Keep your goal in mind, think about the question that you're asking, and you should be fine. The results of any test that you run depend on the question you ask, the method you use to collect and measure your data, and the test you select to analyze the data.

The conclusions you come to also depend on using your own common sense. Remember that you're looking for patterns in the data. If you find that there is no significant difference between your groups, or that there is a statistically significant difference and your results surprise you, don't stop there. Consider the possible confounding variables, think about your sample (its size as well as the selection process you used), and apply your common sense. Do you see where you might have gone wrong, or missed something, or biased your data? The application of common sense to what you are doing has resulted in many a reconsideration of design and produced some very interesting results.

Science involves asking questions and then questioning the answers you get. Your own perception of both the problem and the solution should not be discounted. Trust your own intelligence, and if you have doubts, ask questions of people who might know the answer. Remember, "Asking costs nothing."

REFERENCES

Chen, A. Y., & Escarce, J. J. (2010). Family structure and childhood obesity, early childhood longitudinal study—Kindergarten cohort. *Preventing Chronic Disease, Public Health Research, Practice and Policy, 7*(3), 1–8.

Giaquinto, S., & Valentini, F. (2009). Is there a scientific basis for pet therapy? *Disability and Rehabilitation, 31*, 595–598.

Harrison, Y., & Horne, J. A. (2000). Sleep loss and temporal memory. *Quarterly Journal of Experimental Psychology, 53A*(1), 271–279.

Hope, S., Melle, I., Aukrust, P., Agartz, I., Lorentzen, S., Steen, N. E., Djurovic, S., Ueland, T., & Andreassen, O. A. (2010). Osteoprotegerin levels in patients with severe mental disorders. *Journal of Psychiatry and Neuroscience, 35*(5), 304–310.

Kurrikoff, K., Inno, J., Matsui, T., & Vasar, E. (2008). Stress-induced analgesia in mice: Evidence for interaction between endocannabinoids and cholecystokinin. *European Journal of Neuroscience, 27*, 2147–2155.

Plana, T., Gracia, R., Mendez, I., Pintor, L., Lazaro, L., & Castro-Fornieles, J. (2010). Total serum cholesterol levels and suicide attempts in child and adolescent psychiatric inpatients. *European Child and Adolescent Psychiatry 19*, 615–619.

Walker, M. P., Brakefield, T., Morgan, A., Hobson, J. A., & Stickgold, R. (2002). Practice with sleep makes perfect: Sleep-dependent motor skill learning. *Neuron, 25*, 205–211.

APPENDIX A

THE PROPORTIONS UNDER THE STANDARD NORMAL CURVE

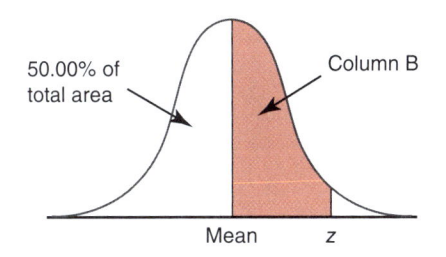

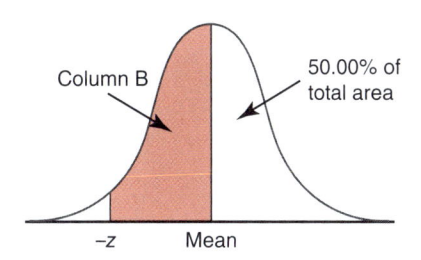

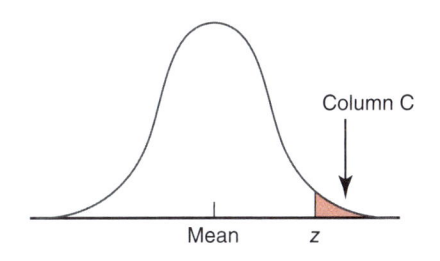

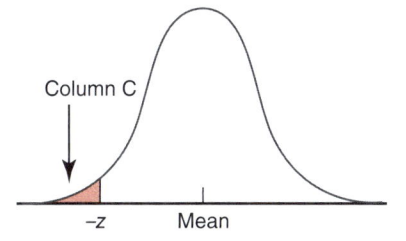

A	B	C	A	B	C
z	Area between mean and z	Area at or above z	z	Area between mean and z	Area at or above z
0.00	.0000	.5000			
0.01	.0040	.4960	0.41	.1591	.3409
0.02	.0080	.4920	0.42	.1628	.3372
0.03	.0120	.4880	0.43	.1664	.3336
0.04	.0160	.4840	0.44	.1700	.3300
0.05	.0199	.4801	0.45	.1736	.3264
0.06	.0239	.4761	0.46	.1772	.3228
0.07	.0279	.4721	0.47	.1808	.3192
0.08	.0319	.4681	0.48	.1844	.3156
0.09	.0359	.4641	0.49	.1879	.3121
0.10	.0398	.4602	0.50	.1915	.3085
0.11	.0438	.4562	0.51	.1950	.3050
0.12	.0478	.4522	0.52	.1985	.3015
0.13	.0517	.4483	0.53	.2019	.2981
0.14	.0557	.4443	0.54	.2054	.2946
0.15	.0596	.4404	0.55	.2088	.2912

A	B	C	A	B	C
z	Area between mean and z	Area at or above z	z	Area between mean and z	Area at or above z
0.16	.0636	.4364	0.56	.2123	.2877
0.17	.0675	.4325	0.57	.2157	.2843
0.18	.0714	.4286	0.58	.2190	.2810
0.19	.0753	.4247	0.59	.2224	.2776
0.20	.0793	.4207	0.60	.2257	.2743
0.21	.0832	.4168	0.61	.2291	.2709
0.22	.0871	.4129	0.62	.2324	.2676
0.23	.0910	.4090	0.63	.2357	.2643
0.24	.0948	.4052	0.64	.2389	.2611
0.25	.0987	.4013	0.65	.2422	.2578
0.26	.1026	.3974	0.66	.2454	.2546
0.27	.1064	.3936	0.67	.2486	.2514
0.28	.1103	.3897	0.68	.2517	.2483
0.29	.1141	.3859	0.69	.2549	.2451
0.30	.1179	.3821	0.70	.2580	.2420
0.31	.1217	.3783	0.71	.2611	.2389
0.32	.1255	.3745	0.72	.2642	.2358
0.33	.1293	.3707	0.73	.2673	.2327
0.34	.1331	.3669	0.74	.2704	.2297
0.35	.1368	.3632	0.75	.2734	.2266
0.36	.1406	.3594	0.76	.2764	.2236
0.37	.1443	.3557	0.77	.2794	.2207
0.38	.1480	.3520	0.78	.2823	.2177
0.39	.1517	.3483	0.79	.2852	.2148
0.40	.1554	.3446	0.80	.2881	.2119
0.81	.2910	.2090	1.21	.3869	.1131
0.82	.2939	.2061	1.22	.3888	.1112
0.83	.2967	.2033	1.23	.3907	.1093
0.84	.2995	.2005	1.24	.3925	.1075
0.85	.3023	.1977	1.25	.3944	.1056
0.86	.3051	.1949	1.26	.3962	.1038
0.87	.3078	.1922	1.27	.3980	.1020
0.88	.3106	.1894	1.28	.3997	.1003
0.89	.3133	.1867	1.29	.4015	.0985
0.90	.3159	.1841	1.30	.4032	.0968
0.91	.3186	.1814	1.31	.4049	.0951
0.92	.3212	.1788	1.32	.4066	.0934
0.93	.3238	.1762	1.33	.4082	.0918
0.94	.3264	.1736	1.34	.4099	.0901
0.95	.3289	.1711	1.35	.4115	.0885
0.96	.3315	.1685	1.36	.4131	.0869
0.97	.3340	.1660	1.37	.4147	.0853
0.98	.3365	.1635	1.38	.4162	.0838
0.99	.3389	.1611	1.39	.4177	.0823
1.00	0.3413	.1587	1.40	.4192	.0808
1.01	.3438	.1562	1.41	.4207	.0793
1.02	.3461	.1539	1.42	.4222	.0778

APPENDIX A The Proportions Under the Standard Normal Curve

1.03	.3485	.1515	1.43	.4236	.0764
1.04	.3508	.1492	1.44	.4251	.0749
1.05	.3531	.1469	1.45	.4265	.0735
1.06	.3554	.1446	1.46	.4279	.0721
1.07	.3577	.1423	1.47	.4292	.0708
1.08	.3599	.1401	1.48	.4306	.0694
1.09	.3621	.1379	1.49	.4319	.0681
1.10	.3643	.1357	1.50	.4332	.0668
1.11	.3665	.1335	1.51	.4345	.0655
1.12	.3686	.1314	1.52	.4357	.0643
1.13	.3708	.1292	1.53	.4370	.0630
1.14	.3729	.1271	1.54	.4382	.0618
1.15	.3749	.1251	1.55	.4394	.0606
1.16	.3770	.1230	1.56	.4406	.0594
1.17	.3790	.1210	1.57	.4418	.0582
1.18	.3810	.1190	1.58	.4429	.0571
1.19	.3830	.1170	1.59	.4441	.0559
1.20	.3849	.1151	1.60	.4452	.0548
1.61	.4463	.0537	2.01	.4778	.0222
1.62	.4474	.0526	2.02	.4783	.0217
1.63	.4484	.0516	2.03	.4788	.0212
1.64	.4495	.0505	2.04	.4793	.0207
1.65	.4505	.0495	2.05	.4798	.0202
1.66	.4515	.0485	2.06	.4803	.0197
1.67	.4525	.0475	2.07	.4808	.0192
1.68	.4535	.0465	2.08	.4812	.0188
1.69	.4545	.0455	2.09	.4817	.0183
1.70	.4554	.0446	2.10	.4821	.0179
1.71	.4564	.0436	2.11	.4826	.0174
1.72	.4573	.0427	2.12	.4830	.0170
1.73	.4582	.0418	2.13	.4834	.0166
1.74	.4591	.0409	2.14	.4838	.0162
1.75	.4599	.0401	2.15	.4842	.0158
1.76	.4608	.0392	2.16	.4846	.0154
1.77	.4616	.0384	2.17	.4850	.0150
1.78	.4625	.0375	2.18	.4854	.0146
1.79	.4633	.0367	2.19	.4857	.0143
1.80	.4641	.0359	2.20	.4861	.0139
1.81	.4649	.0351	2.21	.4864	.0136
1.82	.4656	.0344	2.22	.4868	.0132
1.83	.4664	.0336	2.23	.4871	.0129
1.84	.4671	.0329	2.24	.4875	.0125
1.85	.4678	.0322	2.25	.4878	.0122
1.86	.4686	.0314	2.26	.4881	.0119
1.87	.4693	.0307	2.27	.4884	.0116
1.88	.4699	.0301	2.28	.4887	.0113
1.89	.4706	.0294	2.29	.4890	.0110
1.90	.4713	.0287	2.30	.4893	.0107
1.91	.4719	.0281	2.31	.4896	.0104
1.92	.4726	.0274	2.32	.4898	.0102

A	B	C	A	B	C
z	Area between mean and z	Area at or above z	z	Area between mean and z	Area at or above z
1.93	.4732	.0268	2.33	.4901	.0099
1.94	.4738	.0262	2.34	.4904	.0096
1.95	.4744	.0256	2.35	.4906	.0094
1.96	.4750	.0250	2.36	.4909	.0091
1.97	.4756	.0244	2.37	.4911	.0089
1.98	.4761	.0239	2.38	.4913	.0087
1.99	.4767	.0233	2.39	.4916	.0084
2.00	.4772	.0228	2.40	.4918	.0082
2.41	.4920	.0080	2.81	.4975	.0025
2.42	.4922	.0078	2.82	.4976	.0024
2.43	.4925	.0075	2.83	.4977	.0023
2.44	.4927	.0073	2.84	.4977	.0023
2.45	.4929	.0071	2.85	.4978	.0022
2.46	.4931	.0069	2.86	.4979	.0021
2.47	.4932	.0068	2.87	.4979	.0021
2.48	.4934	.0066	2.88	.4980	.0020
2.49	.4936	.0064	2.89	.4981	.0019
2.50	.4938	.0062	2.90	.4981	.0019
2.51	.4940	.0060	2.91	.4982	.0018
2.52	.4941	.0059	2.92	.4982	.0018
2.53	.4943	.0057	2.93	.4983	.0017
2.54	.4945	.0055	2.94	.4984	.0016
2.55	.4946	.0054	2.95	.4984	.0016
2.56	.4948	.0052	2.96	.4985	.0015
2.57	.4949	.0051	2.97	.4985	.0015
2.58	.4951	.0049	2.98	.4986	.0014
2.59	.4952	.0048	2.99	.4986	.0014
2.60	.4953	.0047	3.00	.4987	.0013
2.61	.4955	.0045	3.01	.4987	.0013
2.62	.4956	.0044	3.02	.4987	.0013
2.63	.4957	.0043	3.03	.4988	.0012
2.64	.4959	.0041	3.04	.4988	.0012
2.65	.4960	.0040	3.05	.4989	.0011
2.66	.4961	.0039	3.06	.4989	.0011
2.67	.4962	.0038	3.07	.4989	.0011
2.68	.4963	.0037	3.08	.4990	.0010
2.69	.4964	.0036	3.09	.4990	.0010
2.70	.4965	.0035	3.10	.4990	.0010
2.71	.4966	.0034	3.11	.4991	.0009
2.72	.4967	.0033	3.12	.4991	.0009
2.73	.4968	.0032	3.13	.4991	.0009
2.74	.4969	.0031	3.14	.4992	.0008
2.75	.4970	.0030	3.15	.4992	.0008
2.76	.4971	.0029	3.16	.4992	.0008
2.77	.4972	.0028	3.17	.4992	.0008
2.78	.4973	.0027	3.18	.4993	.0007
2.79	.4974	.0026	3.19	.4993	.0007
2.80	.4974	.0026	3.20	.4993	.0007

3.21	.4993	.0007	3.61	.4998	.0002
3.22	.4994	.0006	3.62	.4999	.0001
3.23	.4994	.0006	3.63	.4999	.0001
3.24	.4994	.0006	3.64	.4999	.0001
3.25	.4994	.0006	3.65	.4999	.0001
3.26	.4994	.0006	3.66	.4999	.0001
3.27	.4995	.0005	3.67	.4999	.0001
3.28	.4995	.0005	3.68	.4999	.0001
3.29	.4995	.0005	3.69	.4999	.0001
3.30	.4995	.0005	3.70	.4999	.0001
3.31	.4995	.0005	3.71	.4999	.0001
3.32	.4995	.0005	3.72	.4999	.0001
3.33	.4996	.0004	3.73	.4999	.0001
3.34	.4996	.0004	3.74	.4999	.0001
3.35	.4996	.0004	3.75	.4999	.0001
3.36	.4996	.0004	3.76	.4999	.0001
3.37	.4996	.0004	3.77	.4999	.0001
3.38	.4996	.0004	3.78	.4999	.0001
3.39	.4997	.0003	3.79	.4999	.0001
3.40	.4997	.0003	3.80	.4999	.0001
3.41	.4997	.0003	3.81	.4999	.0001
3.42	.4997	.0003	3.82	.4999	.0001
3.43	.4997	.0003	3.83	.4999	.0001
3.44	.4997	.0003	3.84	.4999	.0001
3.45	.4997	.0003	3.85	.4999	.0001
3.46	.4997	.0003	3.86	.4999	.0001
3.47	.4997	.0003	3.87	.4999	.0001
3.48	.4997	.0003	3.88	.4999	.0001
3.49	.4998	.0002	3.89	.4999	.0001
3.50	.4998	.0002	3.90	.5000	.0000
3.51	.4998	.0002	3.91	.5000	.0000
3.52	.4998	.0002	3.92	.5000	.0000
3.53	.4998	.0002	3.93	.5000	.0000
3.54	.4998	.0002	3.94	.5000	.0000
3.55	.4998	.0002	3.95	.5000	.0000
3.56	.4998	.0002	3.96	.5000	.0000
3.57	.4998	.0002	3.97	.5000	.0000
3.58	.4998	.0002	3.98	.5000	.0000
3.59	.4998	.0002	3.99	.5000	.0000
3.60	.4998	.0002	4.00	.5000	.0000

APPENDIX B

THE STUDENT'S TABLE OF CRITICAL *t*-VALUES

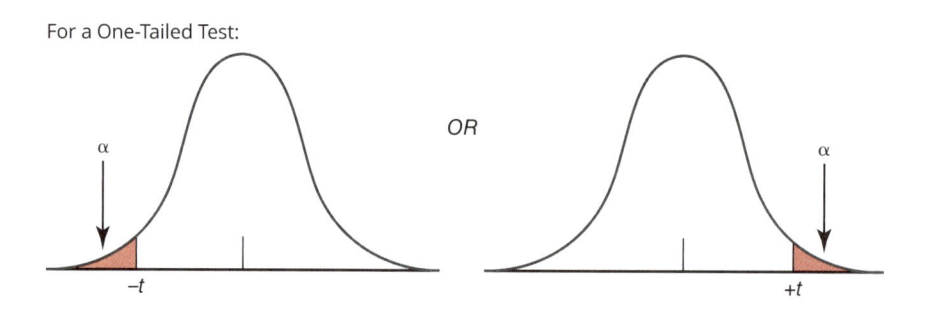

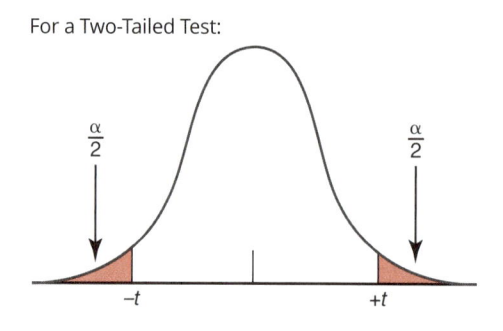

For a One-Tailed Test:

OR

For a Two-Tailed Test:

	Level of Significance (α) For One-Tailed Test					
	.1	.05	.025	.01	.005	.001
	Level of Significance (α) for Two-Tailed Test					
df	.2	.1	.05	.02	.01	.002
1	3.078	6.314	12.706	31.821	63.657	318.313
2	1.886	2.920	4.303	6.965	9.925	22.327
3	1.638	2.353	3.182	4.541	5.841	10.215
4	1.533	2.132	2.776	3.747	4.604	7.173
5	1.476	2.015	2.571	3.365	4.032	5.893
6	1.440	1.943	2.447	3.143	3.707	5.208
7	1.415	1.895	2.365	2.998	3.499	4.782
8	1.397	1.860	2.306	2.896	3.355	4.499
9	1.383	1.833	2.262	2.821	3.250	4.296
10	1.372	1.812	2.228	2.764	3.169	4.143

11	1.363	1.796	2.201	2.718	3.106	4.024
12	1.356	1.782	2.179	2.681	3.055	3.929
13	1.350	1.771	2.160	2.650	3.012	3.852
14	1.345	1.761	2.145	2.624	2.977	3.787
15	1.341	1.753	2.131	2.602	2.947	3.733
16	1.337	1.746	2.120	2.583	2.921	3.686
17	1.333	1.740	2.110	2.567	2.898	3.646
18	1.330	1.734	2.101	2.552	2.878	3.610
19	1.328	1.729	2.093	2.539	2.861	3.579
20	1.325	1.725	2.086	2.528	2.845	3.552
21	1.323	1.721	2.080	2.518	2.831	3.527
22	1.321	1.717	2.074	2.508	2.819	3.505
23	1.319	1.714	2.069	2.500	2.807	3.485
24	1.318	1.711	2.064	2.492	2.797	3.467
25	1.316	1.708	2.060	2.485	2.787	3.450
26	1.315	1.706	2.056	2.479	2.779	3.435
27	1.314	1.703	2.052	2.473	2.771	3.421
28	1.313	1.701	2.048	2.467	2.763	3.408
29	1.311	1.699	2.045	2.462	2.756	3.396
30	1.310	1.697	2.042	2.457	2.750	3.385
31	1.309	1.696	2.040	2.453	2.744	3.375
32	1.309	1.694	2.037	2.449	2.738	3.365
33	1.308	1.692	2.035	2.445	2.733	3.356
34	1.307	1.691	2.032	2.441	2.728	3.348
35	1.306	1.690	2.030	2.438	2.724	3.340
36	1.306	1.688	2.028	2.434	2.719	3.333
37	1.305	1.687	2.026	2.431	2.715	3.326
38	1.304	1.686	2.024	2.429	2.712	3.319
39	1.304	1.685	2.023	2.426	2.708	3.313
40	1.303	1.684	2.021	2.423	2.704	3.307
41	1.303	1.683	2.020	2.421	2.701	3.301
42	1.302	1.682	2.018	2.418	2.698	3.296
43	1.302	1.681	2.017	2.416	2.695	3.291
44	1.301	1.680	2.015	2.414	2.692	3.286
45	1.301	1.679	2.014	2.412	2.690	3.281
46	1.300	1.679	2.013	2.410	2.687	3.277
47	1.300	1.678	2.012	2.408	2.685	3.273
48	1.299	1.677	2.011	2.407	2.682	3.269
49	1.299	1.677	2.010	2.405	2.680	3.265
50	1.299	1.676	2.009	2.403	2.678	3.261
51	1.298	1.675	2.008	2.402	2.676	3.258
52	1.298	1.675	2.007	2.400	2.674	3.255
53	1.298	1.674	2.006	2.399	2.672	3.251
54	1.297	1.674	2.005	2.397	2.670	3.248
55	1.297	1.673	2.004	2.396	2.668	3.245
56	1.297	1.673	2.003	2.395	2.667	3.242
57	1.297	1.672	2.002	2.394	2.665	3.239
58	1.296	1.672	2.002	2.392	2.663	3.237
59	1.296	1.671	2.001	2.391	2.662	3.234
60	1.296	1.671	2.000	2.390	2.660	3.232

	Level of Significance (α) for One-Tailed Test					
	.1	.05	.025	.01	.005	.001
	Level of Significance (α) for Two-Tailed Test					
df	.2	.1	.05	.02	.01	.002
61	1.296	1.670	2.000	2.389	2.659	3.229
62	1.295	1.670	1.999	2.388	2.657	3.227
63	1.295	1.669	1.998	2.387	2.656	3.225
64	1.295	1.669	1.998	2.386	2.655	3.223
65	1.295	1.669	1.997	2.385	2.654	3.220
66	1.295	1.668	1.997	2.384	2.652	3.218
67	1.294	1.668	1.996	2.383	2.651	3.216
68	1.294	1.668	1.995	2.382	2.650	3.214
69	1.294	1.667	1.995	2.382	2.649	3.213
70	1.294	1.667	1.994	2.381	2.648	3.211
71	1.294	1.667	1.994	2.380	2.647	3.209
72	1.293	1.666	1.993	2.379	2.646	3.207
73	1.293	1.666	1.993	2.379	2.645	3.206
74	1.293	1.666	1.993	2.378	2.644	3.204
75	1.293	1.665	1.992	2.377	2.643	3.202
76	1.293	1.665	1.992	2.376	2.642	3.201
77	1.293	1.665	1.991	2.376	2.641	3.199
78	1.292	1.665	1.991	2.375	2.640	3.198
79	1.292	1.664	1.990	2.374	2.640	3.197
80	1.292	1.664	1.990	2.374	2.639	3.195
81	1.292	1.664	1.990	2.373	2.638	3.194
82	1.292	1.664	1.989	2.373	2.637	3.193
83	1.292	1.663	1.989	2.372	2.636	3.191
84	1.292	1.663	1.989	2.372	2.636	3.190
85	1.292	1.663	1.988	2.371	2.635	3.189
86	1.291	1.663	1.988	2.370	2.634	3.188
87	1.291	1.663	1.988	2.370	2.634	3.187
88	1.291	1.662	1.987	2.369	2.633	3.185
89	1.291	1.662	1.987	2.369	2.632	3.184
90	1.291	1.662	1.987	2.368	2.632	3.183
91	1.291	1.662	1.986	2.368	2.631	3.182
92	1.291	1.662	1.986	2.368	2.630	3.181
93	1.291	1.661	1.986	2.367	2.630	3.180
94	1.291	1.661	1.986	2.367	2.629	3.179
95	1.291	1.661	1.985	2.366	2.629	3.178
96	1.290	1.661	1.985	2.366	2.628	3.177
97	1.290	1.661	1.985	2.365	2.627	3.176
98	1.290	1.661	1.984	2.365	2.627	3.175
99	1.290	1.660	1.984	2.365	2.626	3.175
100	1.290	1.660	1.984	2.364	2.626	3.174
120	1.289	1.658	1.980	2.358	2.617	3.373
∞	1.282	1.645	1.960	2.326	2.576	3.090

APPENDIX C

CRITICAL *F*-VALUES

Number of Levels (between)

df for error term	α	1	2	3	4	5	6	7	8
1	.05	161.4	199.50	215.71	224.58	230.16	233.99	236.77	238.88
	.01	4052	4999	5404	5624	5764	5859	5928	5981
2	.05	18.51	19.00	19.16	19.25	19.30	19.33	19.35	19.37
	.01	98.94	99.00	99.17	99.25	99.30	99.33	99.34	99.36
3	.05	10.13	9.55	9.28	9.12	9.01	8.94	8.89	8.85
	.01	34.12	30.82	29.46	28.71	28.24	27.91	27.67	27.49
4	.05	7.71	6.94	6.59	6.39	6.26	6.16	6.09	6.04
	.01	21.20	18.00	16.69	15.98	15.52	15.21	14.98	14.80
5	.05	6.61	5.79	5.41	5.19	5.05	4.95	4.88	4.82
	.01	16.26	13.27	12.06	11.39	10.97	10.67	10.45	10.27
6	.05	5.99	5.14	4.76	4.53	4.39	4.28	4.21	4.15
	.01	13.74	10.92	9.78	9.15	8.75	8.47	8.26	8.10
7	.05	5.59	4.74	4.35	4.12	3.97	3.87	3.79	3.73
	.01	12.25	9.55	8.45	7.85	7.46	7.19	7.00	6.84
8	.05	5.32	4.46	4.07	3.84	3.69	3.58	3.50	3.44
	.01	11.26	8.65	7.59	7.01	6.63	6.37	6.19	6.03
9	.05	5.12	4.26	3.86	3.63	3.48	3.37	3.29	3.23
	.01	10.56	8.02	6.99	6.42	6.06	5.80	5.62	5.47
10	.05	4.96	4.10	3.71	3.48	3.33	3.22	3.14	3.07
	.01	10.04	7.56	6.55	5.99	5.64	5.39	5.21	5.06
11	.05	4.84	3.98	3.59	3.36	3.20	3.09	3.01	2.95
	.01	9.65	7.20	6.22	5.67	5.32	5.07	4.88	4.74
12	.05	4.75	3.89	3.49	3.26	3.11	3.00	2.91	2.85
	.01	9.33	6.93	5.95	5.41	5.06	4.82	4.65	4.50
13	.05	4.67	3.81	3.41	3.18	3.03	2.92	2.83	2.77
	.01	9.07	6.70	5.74	5.20	4.86	4.62	4.44	4.30
14	.05	4.60	3.74	3.34	3.11	2.96	2.85	2.76	2.70
	.01	8.86	6.51	5.56	5.03	4.69	4.46	4.28	4.14
15	.05	4.54	3.68	3.29	3.06	2.90	2.79	2.71	2.64
	.01	8.68	6.36	5.42	4.89	4.56	4.32	4.14	4.00
16	.05	4.49	3.63	3.24	3.01	2.85	2.74	2.66	2.59
	.01	8.53	6.23	5.29	4.77	4.44	4.20	4.03	3.89
17	.05	4.45	3.59	3.20	2.96	2.81	2.70	2.61	2.55
	.01	8.40	6.11	5.18	4.67	4.34	4.10	3.93	3.79
18	.05	4.41	3.55	3.16	2.93	2.77	2.66	2.58	2.51
	.01	8.28	6.01	5.09	4.58	4.25	4.01	3.85	3.71

df for error term	α	\multicolumn{8}{c}{Number of Levels}
		1	2	3	4	5	6	7	8
19	.05	4.38	3.52	3.13	2.90	2.74	2.63	2.54	2.48
	.01	8.18	5.93	5.01	4.50	4.17	3.94	3.77	3.63
20	.05	4.35	3.49	3.10	2.87	2.71	2.60	2.51	2.45
	.01	8.10	5.85	4.94	4.43	4.10	3.87	3.71	3.56
21	.05	4.32	3.47	3.07	2.84	2.69	2.57	2.49	2.42
	.01	8.02	5.78	4.87	4.37	4.04	3.81	3.65	3.51
22	.05	4.30	3.44	3.05	2.82	2.66	2.55	2.46	2.40
	.01	7.94	5.72	4.82	4.31	3.99	3.76	3.59	3.45
23	.05	4.28	3.42	3.03	2.80	2.64	2.53	2.44	2.38
	.01	7.88	5.66	4.76	4.26	3.94	3.71	3.54	3.41
24	.05	4.26	3.40	3.01	2.78	2.62	2.51	2.42	2.36
	.01	7.82	5.61	4.72	4.22	3.90	3.67	3.50	3.36
25	.05	4.24	3.39	2.99	2.76	2.60	2.49	2.41	2.34
	.01	7.77	5.57	4.68	4.18	3.86	3.63	3.46	3.32
26	.05	4.23	3.37	2.98	2.74	2.59	2.47	2.39	2.32
	.01	7.72	5.53	4.64	4.14	3.82	3.59	3.42	3.29
27	.05	4.21	3.35	2.96	2.73	2.57	2.46	2.37	2.31
	.01	7.68	5.49	4.60	4.11	3.79	3.56	3.39	3.26
28	.05	4.20	3.34	2.95	2.71	2.56	2.45	2.36	2.29
	.01	7.64	5.45	4.57	4.07	3.76	3.53	3.36	3.23
29	.05	4.18	3.33	2.93	2.70	2.55	2.43	2.35	2.28
	.01	7.60	5.42	4.54	4.04	3.73	3.50	3.33	3.20
30	.05	4.17	3.32	2.92	2.69	2.53	2.42	2.33	2.27
	.01	7.56	5.39	4.51	4.02	3.70	3.47	3.30	3.17
40	.05	4.08	3.23	2.84	2.61	2.45	2.34	2.25	2.18
	.01	7.31	5.18	4.31	3.83	3.51	3.29	3.12	2.99
50	.05	4.03	3.18	2.79	2.56	2.40	2.29	2.20	2.13
	.01	7.17	5.06	4.20	3.72	3.41	3.18	3.02	2.88
60	.05	4.00	3.15	2.76	2.53	2.37	2.25	2.17	2.10
	.01	7.08	4.98	4.13	3.65	3.34	3.12	2.95	2.82
70	.05	3.99	3.13	2.74	2.50	2.35	2.23	2.14	2.07
	.01	7.01	4.92	4.08	3.60	3.29	3.07	2.91	2.77
80	.05	3.97	3.11	2.72	2.49	2.33	2.21	2.13	2.06
	.01	6.96	4.88	4.04	3.56	3.25	3.04	2.87	2.74
90	.05	3.96	3.10	2.71	2.47	2.32	2.20	2.11	2.04
	.01	6.93	4.85	4.01	3.53	3.23	3.01	2.84	2.72
100	.05	3.95	3.09	2.70	2.46	2.31	2.19	2.10	2.03
	.01	6.90	4.82	3.98	3.51	3.20	2.99	2.82	2.69
120	.05	3.92	3.07	2.68	2.45	2.29	2.18	2.09	2.02
	.01	6.85	4.79	3.95	3.48	3.17	2.96	2.79	2.66
∞	.05	3.84	3.00	2.61	2.37	2.22	2.10	2.01	1.94
	.01	6.64	4.60	3.78	3.32	3.02	2.80	2.64	2.51

APPENDIX D

CRITICAL TUKEY HSD VALUES

df for error term	α	\multicolumn{9}{c}{Number of Levels}
		2	3	4	5	6	7	8	9	10
5	.05	3.64	4.60	5.22	5.67	6.03	6.33	6.58	6.80	6.99
	.01	5.70	6.98	7.80	8.42	8.91	9.32	9.67	9.97	10.24
6	.05	3.46	4.34	4.90	5.30	5.63	5.90	6.12	6.32	6.49
	.01	5.24	6.33	7.03	7.56	7.97	8.32	8.61	8.87	9.10
7	.05	3.34	4.16	4.68	5.06	5.36	5.61	5.82	6.00	6.16
	.01	4.95	5.92	6.54	7.01	7.37	7.68	7.94	8.17	8.37
8	.05	3.26	4.04	4.53	4.89	5.17	5.40	5.60	5.77	5.92
	.01	4.75	5.64	6.20	6.62	6.96	7.24	7.47	7.68	7.86
9	.05	3.20	3.95	4.41	4.76	5.02	5.24	5.43	5.59	5.74
	.01	4.60	5.43	5.96	6.35	6.66	6.91	7.13	7.33	7.49
10	.05	3.15	3.88	4.33	4.65	4.91	5.12	5.30	5.46	5.60
	.01	4.48	5.27	5.77	6.14	6.43	6.67	6.87	7.05	7.21
11	.05	3.11	3.82	4.26	4.57	4.82	5.03	5.20	5.35	5.49
	.01	4.39	5.15	5.62	5.97	6.25	6.48	6.67	6.84	6.99
12	.05	3.08	3.77	4.20	4.51	4.75	4.95	5.12	5.27	5.39
	.01	4.32	5.05	5.50	5.84	6.10	6.32	6.51	6.67	6.81
13	.05	3.06	3.73	4.15	4.45	4.69	4.88	5.05	5.19	5.32
	.01	4.26	4.96	5.40	5.73	5.98	6.19	6.37	6.53	6.67
14	.05	3.03	3.70	4.11	4.41	4.64	4.83	4.99	5.13	5.25
	.01	4.21	4.89	5.32	5.63	5.88	6.08	6.26	6.41	6.54
15	.05	3.01	3.67	4.08	4.37	4.59	4.78	4.94	5.08	5.20
	.01	4.17	4.84	5.25	5.56	5.80	5.99	6.16	6.31	6.44
16	.05	3.00	3.65	4.05	4.33	4.56	4.74	4.90	5.03	5.15
	.01	4.13	4.79	5.19	5.49	5.72	5.92	6.08	6.22	6.35
17	.05	2.98	3.63	4.02	4.30	4.52	4.70	4.86	4.99	5.11
	.01	4.10	4.74	5.14	5.43	5.66	5.85	6.01	6.15	6.27
18	.05	2.97	3.61	4.00	4.28	4.49	4.67	4.82	4.96	5.07
	.01	4.07	4.70	5.09	5.38	5.60	5.79	5.94	6.08	6.20

df for error term	α	\multicolumn{9}{c	}{Number of Levels}
		2	3	4	5	6	7	8	9	10
19	.05	2.96	3.59	3.98	4.25	4.47	4.65	4.79	4.92	5.04
	.01	4.05	4.67	5.05	5.33	5.55	5.73	5.89	6.02	6.14
20	.05	2.95	3.58	3.96	4.23	4.45	4.62	4.77	4.90	5.01
	.01	4.02	4.64	5.02	5.29	5.51	5.69	5.84	5.97	6.09
24	.05	2.92	3.53	3.90	4.17	4.37	4.54	4.68	4.81	4.92
	.01	3.96	4.55	4.91	5.17	5.37	5.54	5.69	5.81	5.92
30	.05	2.89	3.49	3.85	4.10	4.30	4.46	4.60	4.72	4.82
	.01	3.89	4.45	4.80	5.05	5.24	5.40	5.54	5.65	5.76
40	.05	2.86	3.44	3.79	4.04	4.23	4.39	4.52	4.63	4.73
	.01	3.82	4.37	4.70	4.93	5.11	5.26	5.39	5.50	5.60
60	.05	2.83	3.40	3.74	3.98	4.16	4.31	4.44	4.55	4.65
	.01	3.76	4.28	4.59	4.82	4.99	5.13	5.25	5.36	5.45
120	.05	2.80	3.36	3.68	3.92	4.10	4.24	4.36	4.47	4.56
	.01	3.70	4.20	4.50	4.71	4.87	5.01	5.12	5.21	5.30
∞	.05	2.77	3.31	3.63	3.86	4.03	4.17	4.29	4.39	4.47
	.01	3.64	4.12	4.40	4.60	4.76	4.88	4.99	5.08	4.16

APPENDIX D Critical Tukey HSD Values

APPENDIX E

CRITICAL VALUES OF CHI SQUARE

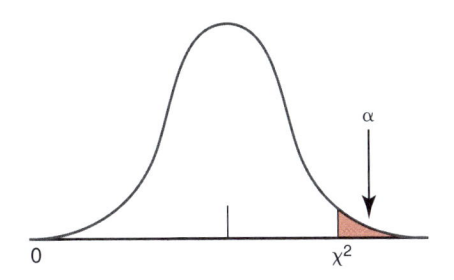

	Level of Significance for Two-Tailed Test				
df	.1	.05	.025	.01	.001
1	2.706	3.841	5.024	6.635	10.828
2	4.605	5.991	7.378	9.210	13.816
3	6.251	7.815	9.348	11.345	16.266
4	7.779	9.488	11.143	13.277	18.467
5	9.236	11.070	12.833	15.086	20.515
6	10.645	12.592	14.449	16.812	22.458
7	12.017	14.067	16.013	18.475	24.322
8	13.362	15.507	17.535	20.090	26.125
9	14.684	16.919	19.023	21.666	27.877
10	15.987	18.307	20.483	23.209	29.588
11	17.275	19.675	21.920	24.725	31.264
12	18.549	21.026	23.337	26.217	32.910
13	19.812	22.362	24.736	27.688	34.528
14	21.064	23.685	26.119	29.141	36.123
15	22.307	24.996	27.488	30.578	37.697
16	23.542	26.296	28.845	32.000	39.252
17	24.769	27.587	30.191	33.409	40.790
18	25.989	28.869	31.526	34.805	42.312
19	27.204	30.144	32.852	36.191	43.820
20	28.412	31.410	34.170	37.566	45.315
21	29.615	32.671	35.479	38.932	46.797
22	30.813	33.924	36.781	40.289	48.268
23	32.007	35.172	38.076	41.638	49.728
24	33.196	36.415	39.364	42.980	51.179
25	34.382	37.652	40.646	44.314	52.620

| | Level of Significance for Two-Tailed Test |||||
df	.1	.05	.025	.01	.001
26	35.563	38.885	41.923	45.642	54.052
27	36.741	40.113	43.195	46.963	55.476
28	37.916	41.337	44.461	48.278	56.892
29	39.087	42.557	45.722	49.588	58.301
30	40.256	43.773	46.979	50.892	59.703
31	41.422	44.985	48.232	52.191	61.098
32	42.585	46.194	49.480	53.486	62.487
33	43.745	47.400	50.725	54.776	63.870
34	44.903	48.602	51.966	56.061	65.247
35	46.059	49.802	53.203	57.342	66.619
36	47.212	50.998	54.437	58.619	67.985
37	48.363	52.192	55.668	59.893	69.347
38	49.513	53.384	56.896	61.162	70.703
39	50.660	54.572	58.120	62.428	72.055
40	51.805	55.758	59.342	63.691	73.402
41	52.949	56.942	60.561	64.950	74.745
42	54.090	58.124	61.777	66.206	76.084
43	55.230	59.304	62.990	67.459	77.419
44	56.369	60.481	64.201	68.710	78.750
45	57.505	61.656	65.410	69.957	80.077
46	58.641	62.830	66.617	71.201	81.400
47	59.774	64.001	67.821	72.443	82.720
48	60.907	65.171	69.023	73.683	84.037
49	62.038	66.339	70.222	74.919	85.351
50	63.167	67.505	71.420	76.154	86.661
51	64.295	68.669	72.616	77.386	87.968
52	65.422	69.832	73.810	78.616	89.272
53	66.548	70.993	75.002	79.843	90.573
54	67.673	72.153	76.192	81.069	91.872
55	68.796	73.311	77.380	82.292	93.168
56	69.919	74.468	78.567	83.513	94.461
57	71.040	75.624	79.752	84.733	95.751
58	72.160	76.778	80.936	85.950	97.039
59	73.279	77.931	82.117	87.166	98.324
60	74.397	79.082	83.298	88.379	99.607
61	75.514	80.232	84.476	89.591	100.888
62	76.630	81.381	85.654	90.802	102.166
63	77.745	82.529	86.830	92.010	103.442
64	78.860	83.675	88.004	93.217	104.716
65	79.973	84.821	89.177	94.422	105.988
66	81.085	85.965	90.349	95.626	107.258
67	82.197	87.108	91.519	96.828	108.526
68	83.308	88.250	92.689	98.028	109.791
69	84.418	89.391	93.856	99.228	111.055
70	85.527	90.531	95.023	100.425	112.317
71	86.635	91.670	96.189	101.621	113.577
72	87.743	92.808	97.353	102.816	114.835

73	88.850	93.945	98.516	104.010	116.092
74	89.956	95.081	99.678	105.202	117.346
75	91.061	96.217	100.839	106.393	118.599
76	92.166	97.351	101.999	107.583	119.850
77	93.270	98.484	103.158	108.771	121.100
78	94.374	99.617	104.316	109.958	122.348
79	95.476	100.749	105.473	111.144	123.594
80	96.578	101.879	106.629	112.329	124.839
81	97.680	103.010	107.783	113.512	126.083
82	98.780	104.139	108.937	114.695	127.324
83	99.880	105.267	110.090	115.876	128.565
84	100.980	106.395	111.242	117.057	129.804
85	102.079	107.522	112.393	118.236	131.041
86	103.177	108.648	113.544	119.414	132.277
87	104.275	109.773	114.693	120.591	133.512
88	105.372	110.898	115.841	121.767	134.746
89	106.469	112.022	116.989	122.942	135.978
90	107.565	113.145	118.136	124.116	137.208
91	108.661	114.268	119.282	125.289	138.438
92	109.756	115.390	120.427	126.462	139.666
93	110.850	116.511	121.571	127.633	140.893
94	111.944	117.632	122.715	128.803	142.119
95	113.038	118.752	123.858	129.973	143.344
96	114.131	119.871	125.000	131.141	144.567
97	115.223	120.990	126.141	132.309	145.789
98	116.315	122.108	127.282	133.476	147.010
99	117.407	123.225	128.422	134.642	148.230
100	118.498	124.342	129.561	135.807	149.449

APPENDIX F

THE PEARSON CORRELATION COEFFICIENT: CRITICAL r-VALUES

df(N − 2)	.05	.01	df(N − 2)	.05	.01
1	.997	1.000	31	.344	.442
2	.950	.990	32	.339	.436
3	.878	.959	33	.334	.430
4	.812	.917	34	.329	.424
5	.755	.875	35	.325	.418
6	.707	.834	36	.320	.413
7	.666	.798	37	.316	.408
8	.632	.765	38	.312	.403
9	.602	.735	39	.308	.398
10	.576	.708	40	.304	.393
11	.553	.684	41	.301	.389
12	.533	.661	42	.297	.384
13	.514	.641	43	.294	.380
14	.497	.623	44	.291	.376
15	.482	.606	45	.288	.372
16	.468	.590	46	.285	.368
17	.456	.575	47	.282	.365
18	.444	.562	48	.279	.361
19	.433	.549	49	.276	.358
20	.423	.537	50	.273	.354
21	.413	.526	60	.250	.325
22	.404	.515	70	.232	.302
23	.396	.505	80	.217	.283
24	.388	.496	90	.205	.267
25	.381	.487	100	.195	.254
26	.374	.479	200	.138	.181
27	.367	.471	300	.113	.148
28	.361	.463	400	.098	.128
29	.355	.456	500	.088	.115
30	.349	.449	1,000	.062	.081

APPENDIX G

CRITICAL r_s VALUES FOR THE SPEARMAN CORRELATION COEFFICIENT

	\multicolumn{9}{c}{Level of Significance (α) for One-Tailed Test}
	.25	.10	.05	.025	.01	.005	.0025	.001	.0005
	\multicolumn{9}{c}{Level of Significance (α) for Two-Tailed Test}
n	.50	.20	.10	.05	.02	.01	.005	.002	.001
4	.600	1.000	1.000						
5	.500	.800	.900	1.000	1.000				
6	.371	.657	.829	.886	.943	1.000	1.000		
7	.321	.571	.714	.786	.893	.929	.964	1.000	1.000
8	.310	.524	.643	.738	.833	.881	.905	.952	.976
9	.267	.483	.600	.700	.783	.833	.867	.917	.933
10	.248	.455	.564	.648	.745	.794	.830	.879	.903
11	.236	.427	.536	.618	.709	.755	.800	.845	.873
12	.217	.406	.503	.587	.678	.727	.769	.818	.846
13	.209	.385	.484	.560	.648	.703	.747	.791	.824
14	.200	.367	.464	.538	.626	.679	.723	.771	.802
15	.189	.354	.446	.521	.604	.654	.700	.750	.779
16	.182	.341	.429	.503	.582	.635	.679	.729	.762
17	.176	.328	.414	.485	.566	.615	.662	.713	.748
18	.170	.317	.401	.472	.550	.600	.643	.695	.728
19	.165	.309	.391	.460	.535	.584	.628	.677	.712
20	.161	.299	.380	.447	.520	.570	.612	.662	.696
21	.156	.292	.370	.435	.508	.556	.599	.648	.681
22	.152	.284	.361	.425	.496	.544	.586	.634	.667
23	.148	.278	.353	.415	.486	.532	.573	.622	.654
24	.144	.271	.344	.406	.476	.521	.562	.610	.642
25	.142	.265	.337	.398	.466	.511	.551	.598	.630
26	.138	.259	.331	.390	.457	.501	.541	.587	.619
27	.136	.255	.324	.382	.448	.491	.531	.577	.608
28	.133	.250	.317	.375	.440	.483	.522	.567	.598
29	.130	.245	.312	.368	.433	.475	.513	.558	.589
30	.128	.240	.306	.362	.425	.467	.504	.549	.580
31	.126	.236	.301	.356	.418	.459	.496	.541	.571
32	.124	.232	.296	.350	.412	.452	.489	.533	.563

33	.121	.229	.291	.345	.405	.446	.482	.525	.554
34	.120	.225	.287	.340	.399	.439	.475	.517	.547
35	.118	.222	.283	.335	.394	.433	.468	.510	.539
36	.116	.219	.279	.330	.388	.427	.462	.504	.533
37	.114	.216	.275	.325	.383	.421	.456	.497	.526
38	.113	.212	.271	.321	.378	.415	.450	.491	.519
39	.111	.210	.267	.317	.373	.410	.444	.485	.513
40	.110	.207	.264	.313	.368	.405	.439	.479	.507
41	.108	.204	.261	.309	.364	.400	.433	.473	.501
42	.107	.202	.257	.305	.359	.395	.428	.468	.495
43	.105	.199	.254	.301	.355	.391	.423	.463	.490
44	.104	.197	.251	.298	.351	.386	.419	.458	.484
45	.103	.194	.248	.294	.347	.382	.414	.453	.479
46	.102	.192	.246	.291	.343	.378	.410	.448	.474
47	.101	.190	.243	.288	.340	.374	.405	.443	.469
48	.100	.188	.240	.285	.336	.370	.401	.439	.465
49	.098	.186	.238	.282	.333	.366	.397	.434	.460
50	.097	.184	.235	.279	.329	.363	.393	.430	.456
51	.096	.182	.233	.276	.326	.359	.390	.426	.451
52	.095	.180	.231	.274	.323	.356	.386	.422	.447
53	.095	.179	.228	.271	.320	.352	.382	.418	.443
54	.094	.177	.226	.268	.317	.349	.379	.414	.439
55	.093	.175	.224	.266	.314	.346	.375	.411	.435
56	.092	.174	.222	.264	.311	.343	.372	.407	.432
57	.091	.172	.220	.261	.308	.340	.369	.404	.428
58	.090	.171	.218	.259	.306	.337	.366	.400	.424
59	.089	.169	.216	.257	.303	.334	.363	.397	.421
60	.089	.168	.214	.255	.300	.331	.360	.394	.418
61	.088	.166	.213	.252	.298	.329	.357	.391	.414
62	.087	.165	.211	.250	.296	.326	.354	.388	.411
63	.086	.163	.209	.248	.293	.323	.351	.385	.408
64	.086	.162	.207	.246	.291	.321	.348	.382	.405
65	.085	.161	.206	.244	.289	.318	.346	.379	.402
66	.084	.160	.204	.243	.287	.316	.343	.376	.399
67	.084	.158	.203	.241	.284	.314	.341	.373	.396
68	.083	.157	.201	.239	.282	.311	.338	.370	.393
69	.082	.156	.200	.237	.280	.309	.336	.368	.390
70	.082	.155	.198	.235	.278	.307	.333	.365	.388
71	.081	.154	.197	.234	.276	.305	.331	.363	.385
72	.081	.153	.195	.232	.274	.303	.329	.360	.382
73	.080	.152	.194	.230	.272	.301	.327	.358	.380
74	.080	.151	.193	.229	.271	.299	.324	.355	.377
75	.079	.150	.191	.227	.269	.297	.322	.353	.375
76	.078	.149	.190	.226	.267	.295	.320	.351	.372
77	.078	.148	.189	.224	.265	.293	.318	.349	.370
78	.077	.147	.188	.223	.264	.291	.316	.346	.368
79	.077	.146	.186	.221	.262	.289	.314	.344	.365
80	.076	.145	.185	.220	.260	.287	.312	.342	.363
81	.076	.144	.184	.219	.259	.285	.310	.340	.361
82	.075	.143	.183	.217	.257	.284	.308	.338	.359

APPENDIX G Critical r_s Values for the Spearman Correlation Coefficient

	Level of Significance (α) for One-Tailed Test								
	.25	.10	.05	.025	.01	.005	.0025	.001	.0005
	Level of Significance (α) for Two-Tailed Test								
n	.50	.20	.10	.05	.02	.01	.005	.002	.001
83	.075	.142	.182	.216	.255	.282	.306	.336	.357
84	.074	.141	.181	.215	.254	.280	.305	.334	.355
85	.074	.140	.180	.213	.252	.279	.303	.332	.353
86	.074	.139	.179	.212	.251	.277	.301	.330	.351
87	.073	.139	.177	.211	.250	.276	.299	.328	.349
88	.073	.138	.176	.210	.248	.274	.298	.327	.347
89	.072	.137	.175	.209	.247	.272	.296	.325	.345
90	.072	.136	.174	.207	.245	.271	.294	.323	.343
91	.072	.135	.173	.206	.244	.269	.293	.321	.341
92	.071	.135	.173	.205	.243	.268	.291	.319	.339
93	.071	.134	.172	.204	.241	.267	.290	.318	.338
94	.070	.133	.171	.203	.240	.265	.288	.316	.336
95	.070	.133	.170	.202	.239	.264	.287	.314	.334
96	.070	.132	.169	.201	.238	.262	.285	.313	.332
97	.069	.131	.168	.200	.236	.261	.284	.311	.331
98	.069	.130	.167	.199	.235	.260	.282	.310	.329
99	.068	.130	.166	.198	.234	.258	.281	.308	.327
100	.068	.129	.165	.197	.233	.257	.279	.307	.326

APPENDIX H

MANN–WHITNEY CRITICAL *U*-VALUES

α = .005 (two-tailed)

$n_1 \backslash n_2$	2	3	4	5	6	7	8	9	10	11	12	13	14	15	16	17	18	19	20
2																			
3											0	0	0	1	1	1	1	1	2
4							0	0	1	1	2	2	3	3	4	4	5	5	5
5				0	0	1	2	3	3	4	5	6	6	7	8	9	9	10
6				0	1	2	3	4	5	6	7	8	9	10	11	12	13	14	15
7				0	2	3	4	5	7	8	9	11	12	13	15	16	18	19	20
8			0	1	3	4	6	7	9	11	12	14	16	17	19	21	22	24	26
9			0	2	4	5	7	9	11	13	15	17	19	21	23	25	27	29	31
10			1	3	5	7	9	11	13	16	18	20	23	25	27	30	32	35	37
11			1	3	6	8	11	13	16	18	21	24	26	29	32	35	37	40	43
12		0	2	4	7	9	12	15	18	21	24	27	30	33	36	39	43	46	49
13		0	2	5	8	11	14	17	20	24	27	30	34	37	41	44	48	51	55
14		0	3	6	9	12	16	19	23	26	30	34	38	41	45	49	53	57	61
15		1	3	6	10	13	17	21	25	29	33	37	41	46	50	54	58	62	67
16		1	4	7	11	15	19	23	27	32	36	41	45	50	54	59	64	68	73
17		1	4	8	12	16	21	25	30	35	39	44	49	54	59	64	69	74	79
18		1	5	9	13	18	22	27	32	37	43	48	53	58	64	69	74	80	85
19		1	5	9	14	19	24	29	35	40	46	51	57	62	68	74	80	85	91
20		2	5	10	15	20	26	31	37	43	49	55	61	67	73	79	85	91	97

α = .01 (two-tailed)

$n_1 \backslash n_2$	2	3	4	5	6	7	8	9	10	11	12	13	14	15	16	17	18	19	20
2																		0	0
3								0	0	0	1	1	1	2	2	2	2	3	3

n1\n2	2	3	4	5	6	7	8	9	10	11	12	13	14	15	16	17	18	19	20
4					0	0	1	1	2	2	3	3	4	5	5	6	6	7	8
5				0	1	1	2	3	4	5	6	7	7	8	9	10	11	12	13
6			0	1	2	3	4	5	6	7	9	10	11	12	13	15	16	17	18
7			0	1	3	4	6	7	8	10	12	13	15	16	18	19	21	22	24
8			1	2	4	6	7	9	11	13	15	17	18	20	22	24	26	28	30
9		0	1	3	5	7	9	11	13	16	18	20	22	24	27	29	31	33	36
10		0	2	4	6	9	11	13	16	18	21	24	26	29	31	34	37	39	42
11		0	2	5	7	10	13	16	18	21	24	27	30	33	36	39	42	45	46
12		1	3	6	9	12	15	18	21	24	27	31	34	37	41	44	47	51	54
13		1	3	7	10	13	17	20	24	27	31	34	38	42	45	49	53	56	60
14		1	4	7	11	15	18	22	26	30	34	38	42	46	50	54	58	63	67
15		2	5	8	12	16	20	24	29	33	37	42	46	51	55	60	64	69	73
16		2	5	9	13	18	22	27	31	36	41	45	50	55	60	65	70	74	79
17		2	6	10	15	19	24	29	34	39	44	49	54	60	65	70	75	81	86
18		2	6	11	16	21	26	31	37	42	47	53	58	64	70	75	81	87	92
19	0	3	7	12	17	22	28	33	39	45	51	56	63	69	74	81	87	93	99
20	0	3	8	13	18	24	30	36	42	46	54	60	67	73	79	86	92	99	105

$\alpha = .02$ (two-tailed)

n^1\n^2	2	3	4	5	6	7	8	9	10	11	12	13	14	15	16	17	18	19	20
2																			
3		—	0	0	0	0	0	1	1	1	2	2	2	3	3	4	4	4	5
4		—	—	0	1	1	2	3	3	4	5	5	6	7	7	8	9	9	10
5		—	0	1	2	3	4	5	6	7	8	9	10	11	12	13	14	15	16
6		—	1	2	3	4	6	7	8	9	11	12	13	15	16	18	19	20	22
7		0	1	3	4	6	7	9	11	12	14	16	17	19	21	23	24	26	28
8		0	2	4	6	7	9	11	13	15	17	20	22	24	26	28	30	32	34
9		1	3	5	7	9	11	14	16	18	21	23	26	28	31	33	36	38	40
10		1	3	6	8	11	13	16	19	22	24	27	30	33	36	38	41	44	47
11		1	4	7	9	12	15	18	22	25	28	31	34	37	41	44	47	50	53
12		2	5	8	11	14	17	21	24	28	31	35	38	42	46	49	53	56	60
13		2	5	9	12	16	20	23	27	31	35	39	43	47	51	55	59	63	67

n^1\n^2	2	3	4	5	6	7	8	9	10	11	12	13	14	15	16	17	18	19	20
14		2	6	10	13	17	22	26	30	34	38	43	47	51	56	60	65	69	73
15		3	7	11	15	19	24	28	33	37	42	47	51	56	61	66	70	75	80
16		3	7	12	16	21	26	31	36	41	46	51	56	61	66	71	76	82	87
17		4	8	13	18	23	28	33	38	44	49	55	60	66	71	77	82	88	93
18		4	9	14	19	24	30	36	41	47	53	59	65	70	76	82	88	94	100
19		4	9	15	20	26	32	38	44	50	56	63	69	75	82	88	94	101	107
20		5	10	16	22	28	34	40	47	53	60	67	73	80	87	93	100	107	114

$\alpha = .05$ (two-tailed)

$n1$\$n2$	2	3	4	5	6	7	8	9	10	11	12	13	14	15	16	17	18	19	20
2							0	0	0	0	1	1	1	1	1	2	2	2	2
3				0	1	1	2	2	3	3	4	4	5	5	6	6	7	7	8
4			0	1	2	3	4	4	5	6	7	8	9	10	11	11	12	13	14
5		0	1	2	3	5	6	7	8	9	11	12	13	14	15	17	18	19	20
6		1	2	3	5	6	7	10	11	13	14	16	17	19	21	22	24	25	27
7		1	3	5	6	8	10	12	14	16	18	20	22	24	26	28	30	32	34
8	0	2	4	6	7	10	13	15	17	19	22	24	26	29	31	34	36	38	41
9	0	2	4	7	10	12	15	17	20	23	26	28	31	34	37	39	42	45	48
10	0	3	5	8	11	14	17	20	23	26	29	33	36	39	42	45	48	52	55
11	0	3	6	9	13	16	19	23	26	30	33	37	40	44	47	51	55	58	62
12	1	4	7	11	14	18	22	26	29	33	37	41	45	49	53	57	61	65	69
13	1	4	8	12	16	20	24	28	33	37	41	45	50	54	59	63	67	72	76
14	1	5	9	13	17	22	26	31	36	40	45	50	55	59	64	67	74	78	83
15	1	5	10	14	19	24	29	34	39	44	49	54	59	64	70	75	80	85	90
16	1	6	11	15	21	26	31	37	42	47	53	59	64	70	75	81	86	92	98
17	2	6	11	17	22	28	34	39	45	51	57	63	67	75	81	87	93	99	105
18	2	7	12	18	24	30	36	42	48	55	61	67	74	80	86	93	99	106	112
19	2	7	13	19	25	32	38	45	52	58	65	72	78	85	92	99	106	113	119
20	2	8	13	20	27	34	41	48	55	62	69	76	83	90	98	105	112	119	127

http://www.real-statistics.com/statistics-tables/mann-whitney-table/

APPENDIX I

CRITICAL VALUES FOR THE WILCOXON SIGNED-RANK, MATCHED-PAIRS *t*-TEST

	Level of Significance (α) for a One-Tailed Test		
	.025	.01	.005
	Level of Significance (α) for a Two-Tailed Test		
n	.05	.02	.01
6	0	—	—
7	2	0	—
8	4	2	0
9	6	3	2
10	8	5	3
11	11	7	5
12	14	10	7
13	17	13	10
14	21	16	13
15	25	20	16
16	30	24	20
17	35	28	23
18	40	33	28
19	46	38	32
20	52	43	38
21	59	49	43
22	66	56	49
23	73	62	55
24	81	69	61
25	89	77	68

ANSWERS TO ODD NUMBERED END-OF-CHAPTER PRACTICE PROBLEMS

CHAPTER 1

1. a. IV = the dose of supplement
b. DV = the number of colds contracted in the subsequent 3 months

3. a. IV = the type of exercise class taken
b. DV = overall fitness level

5. a. IV = the type of movie seen
b. DV = the amount of popcorn consumed during the movie

7. a. IV = the environment in which participants consumed alcohol
b. DV = blood alcohol level. The extraneous variables might be the order in which the students consumed the alcohol: They all drank their alcohol in a social setting first and then alone in their dorm rooms. "Practice" with the testing situation (drinking in a social setting) might affect their behavior in the second environment. (There are solutions to this problem that you'll likely encounter in a research methods class.)

9. a. IV = loud noise vs. no loud noise
b. DV = the appearance of a conditioned response (fear when the white rat is presented)

11. a. IV = the "controllability" of the shocks
b. DV = the number of jumps the dogs made across the low barrier

13. a. IV = electrical stimulation of the RAS (actually the MFB) in corner A
b. DV = the number of times the rat visited the corner where shock was delivered (corner A)

15. a. IV = effect of facial expression on gender identification by male participants
b. DV = accuracy of facial expression identification in males

17. a. IV = the context (range of numbers presented)
b. DV = the number picked (single-digit vs. two-digit numbers)

19. a. IV = what each group was told about the cards
b. DV = the number of correct matches made

21.
1. Levels of IV = 2
2. Levels of IV = 2
3. Levels of IV = 3
4. Levels of IV = 5
5. Levels of IV = 4
6. Levels of IV = 2

7. Levels of IV = 2
8. Levels of IV = 2
9. Levels of IV = 2
10. Levels of IV = 2
11. Levels of IV = 2
12. Levels of IV = 2
13. Levels of IV = 1
14. Levels of IV by experiment:
   a. Levels of IV = 2 (before making the shock contingent on a right turn and after the contingency was established)
   b. Levels of IV = 2 (before making the shock contingent on a left turn and after the contingency was established)
   c. Levels of IV = 2 (before making the shock contingent on stopping in the alleyway and after the contingency was established)
15. Levels of IV = 5
16. Levels of IV = 5
17. Levels of IV = 2
18. Levels of IV = 2
19. Levels of IV = 3
20. Levels of IV = 2

23. a. Problems with the data include the fact there is no mention of the sample size (the number of nurses who were asked to assess "mildness") nor any description of how the nurses were selected to participate in the study. The ad also fails to mention how mildness was measured.
b. The ad does present information beyond the 82% statistic to sway our opinion about the mildness of Cavalier cigarettes. For instance, the before-and-after facial expressions of the nurse are shown (from skeptical to pleased). The cigarette is described as "cooler" and "lighter," but without any mention of how these qualities were measured or compared across brands. The reference to the price of Cavalier cigarettes—that it is "no higher than other leading brands"—identifies Cavalier as a "leading" brand and a good value, points that are irrelevant to the mildness of the cigarette.
c. IV = the type of cigarette smoked (Cavalier compared with one other cigarette brand)
DV = a measure of mildness of the smoke
d. Mildness of the cigarette could be measured using a survey with several short questions assessing "mildness" on a Likert scale ranging from 1 (*very mild*)

to 7 (*extremely harsh*). "Mildness" has been linked to the amount of tar and nicotine in the cigarette, so these measures (provided on the side of the package) could also be used.

CHAPTER 2

1. a. Nominal: Organized data into mutually exclusive, discrete categories.
    b. Ordinal: Organized data into mutually exclusive, discrete, and ordered categories.
    c. Interval: Organized data into continuous, ordered categories with equal intervals between numerals, indicating amount or count relative to an arbitrary zero point.
    d. Ratio: Organized data into continuous, ordered categories with equal intervals between numerals, indicating amount or count relative to a real zero point.

3. a. The frequency distribution should not start at zero because there are no observations below 23.
    b. The first two intervals overlap (5 is included in each category).
    c. Unequal interval sizes: The interval 90–129 has a width of 40; all of the other intervals have a width of 10.

5. a. i) Age group: Ordinal
       ii) Error in time estimation: Ratio
       iii) Participant ID: Nominal
    b. I'm going to use the same intervals for both age groups so that I can compare them. The range of the first age group is 29. If I plan on five groups, I should have 29/5 = 5.8, or six observations per group. I'll use the same organization for the second age group.

Time estimations by two age groups

	f		%		CF	
Interval	Group 1	Group 2	Group 1	Group 2	Group 1	Group 2
−16 to −11	2	0	20	0	2	0
−10 to −5	0	0	0	0	2	0
−4 to 1	3	2	30	20	5	2
2 to 7	1	5	10	50	6	7
8 to 13	4	2	40	20	10	9
14 to 19	0	1	0	10	10	10

   i) Approximately 70% (7 out of 10) of over-50-year-olds made errors of 5 milliseconds or less.
   ii) Approximately 50% (5 out of 10) of the participants in age group 1 made underestimations of the time interval.
   iii) Approximately 20% (2 out of 10) of the over-50-year-olds made estimations of less than zero.
   iv) Approximately 60% of participants in age group 1 and approximately 70% of participants in age group 2 made overestimations.

c. i) Age group: Ordinal
   ii) Errors: Ratio
   iii) Coffee serving size: Ordinal
   iv) Score on the Myers-Briggs Personality Inventory: Interval
   v) Quality of sleep: Ordinal

7. a. IV = musical training (at least 2 years of training or none — 2 levels)
       DV = accuracy (the number of sound correctly identified)
       Scale of measurement = Ratio
    b. IV = the genuineness of the Big Mac (home-made or purchased at McDonald's: 2 levels)
       DV = preference for one burger over the other AND accuracy of identification of the home-made burger
       Scale of measurement for preference = Ordinal
       Scale of measurement for accuracy = Nominal (yes vs. no)
    c. IV = the opportunity to visually identify the objects (some or none — 2 levels)
       DV = accuracy (the number of objects correctly identified)
       Scale of measurement = Ratio
    d. IV = the location on the body tested (palm vs. sole of the foot — 2 levels)
       DV = the two-point threshold (detection of 1 vs. 2 points)
       Scale of measurement = Ratio

9. a. Because I want to compare the two groups, I first determined the range of the biggest value (208 in the control group) and the smallest value (101 in the experimental group). That range is 208 − 101 = 107. If I have 10 groups, I should have about 10 cholesterol points in each group. I started the first group at an even multiple of the width of 10.

Frequency distribution of cholesterol level in adolescents who attempted suicide (experimental group) and depressed adolescents who have not attempted suicide (control group)

Interval	Experimental Group	Control Group
100–109	1	1
110–119	1	1
120–129	2	3
130–139	1	1
140–149	1	2
150–159	3	3
160–169	6	0
170–179	1	1
180–189	1	2
190–199	0	0
200–209	0	3

b. i) Teens who have attempted suicide seem to have higher cholesterol levels than do depressed teens who have not attempted suicide. Three of the teens in the control group had cholesterol levels above 200 mg/dl, which would be considered high. None of the teens in the experimental group had cholesterol levels above 189 mg/dl.
   ii) All of the patients in the experimental group (17 out of 17, or 100%) had cholesterol levels below 200 mg/dl. In the experimental group, 14 out of 17 patients (82.36%) had cholesterol levels below 200 mg/dl.
   iii) The typical cholesterol level for the control group was between 120 and 159 mg/dl (9 out of 17 participants in the control group, or 53%, had a cholesterol level within this range). The experimental group typically had a cholesterol level between 150 and 169 mg/dl (again, 9 out of 17, or 53%, of the participants had cholesterol levels within this range).
c. Measurement scale = Ratio
d. IV = suicide history (attempted vs. did not attempt. There were 2 levels to this IV)
   DV = cholesterol level

11. a. To compare these two different groups (of different sizes), I need to use the same intervals in the frequency distributions for both groups. So, first I calculated the range across both groups.

Elapsed time to first emesis in seconds

	Group 1				Group 2			
Interval	f	%	CF	CRF	f	%	CF	CRF
0–10	0	0%	0	0%	2	7.14%	2	7.14%
11–21	0	0%	0	0%	3	10.71%	5	17.86%
22–32	1	5%	1	5%	1	3.57%	6	21.43%
33–43	0	0%	1	5%	0	0%	6	21.43%
44–54	3	15%	4	20%	0	0%	6	21.43%
55–65	0	0%	4	20%	2	7.14%	8	28.57%
66–76	1	5%	5	25%	2	7.14%	10	35.71%
77–87	1	5%	6	30%	3	10.71%	13	46.43%
88–98	1	5%	7	35%	0	0%	13	46.43%
99–109	0	0%	7	35%	1	3.57%	14	50%
110–120	13	65%	20	100%	14	50%	28	100%
	n = 20				n = 28			

Range: largest value in either group (120) − smallest value in either group (5) = 115. I decided to use 10 groups, which gives me 115 ÷ 10 = 11.5, or 11 scores in each group. I began the lowest group at zero (0).
b. IV = The "choppiness" of the sea (frequency of acceleration)
c. DV = Time to first emesis

d. Scale of measurement = Ratio
e. For both groups, the 90th percentile would be in the largest interval (110–120). The 50th percentile in both groups would also be in this largest interval. This tells me that most of the participants in each group "survived" the entire 2 hours of testing without emesis.
f. In the first condition (the first level), where the "sea" was slowly rolling, the overwhelming majority of participants (65%) survived the entire 2 hours before emesis. In the second condition (the second level), where the "sea" was very choppy and waves were irregular, half the participants vomited before the end of the testing period, and half made it the entire 2 hours before emesis. I would conclude that the acceleration of the motion did have an effect on the participants. The irregular motion in Group 2 resulted in more seasickness than did the smooth and rolling motion in Group 1.

13. a. **Cigarette smoking in schizophrenic and non-schizophrenic adults**

	Schizophrenics		Nonschizophrenics	
Do you smoke?	f	%	f	%
Yes	17	85	5	25
No	3	15	15	75

b. Generally speaking, the men diagnosed with schizophrenia were much more likely to smoke than were the men without the diagnosis of schizophrenia. Eight-five percent of the men who suffered from schizophrenia smoked cigarettes, while only 25% of the nonschizophrenic men were smokers.
c. Scale of measurement = Nominal

15. a. **Grip strength in neutral and blue light**

	f	
Interval	Neutral Light	Blue Light
64–71	0	1
72–79	0	0
80–87	0	0
88–95	1	0
96–103	2	0
104–111	3	6
112–119	0	0
120–127	3	8
128–135	1	2
136–143	4	0
144–151	4	1

b. IV = The color of the light (blue vs. white)
c. DV = Grip strength in pounds
d. Scale of measurement = Ratio
e. In neutral light, most of the grip strength measures are between 96 and 129 pounds. In blue light, most of the grip strength measures are higher, between 120 and 135 pounds. It appears that grip strength is higher when the room is lit with blue light.

17. Attractiveness ratings as a function of four waist-to-hip (WHR) ratios

Attractiveness Rating	WHR 1.00	WHR 0.90	WHR 0.80	WHR 0.70
1	0	0	0	0
2	1	0	0	0
3	2	0	0	0
4	5	2	0	0
5	2	4	2	0
6	0	4	4	1
7	0	0	2	5
8	0	0	2	4

a. Scale of measurement = Ordinal (although many researchers treat Likert scale ratings like this one as Interval)
b. A WHR of 1.00 is perceived as the least attractive.
c. Yes, it appears that figures with the lowest WHRs (an "hourglass" figure) are rated as more attractive than those with higher WHRs (little change in width of waist and hip).

19. Frequency distributions

A	f	B	f	C	f	D	f
2–6	5	63–69	2	15–19	7	99–107	1
7–11	4	70–76	4	20–24	0	108–116	2
12–16	1	77–83	2	25–29	3	117–125	3
17–21	0	84–90	1	30–34	1	126–134	2
22–26	3	91–97	0	35–39	2	135–143	0
27–31	1	98–104	4	40–44	0	144–152	2
		105–111	1	45–49	0	153–161	0
				50–54	1	162–170	1
						171–179	1
						180–188	1
						189–197	1

a. **Data Set A:**
Range = 28   # of Groups = 5   Width = 5.6 or 6
Midpoints = 4, 9, 14, 19, 24, 29
b. **Data Set B:**
Range = 43   # of Groups = 6   Width = 7.2 or 7
Midpoints = 66, 73, 80, 87, 94, 101, 108

c. **Data Set C:**
Range = 35   # of Groups = 7   Width = 5
Midpoints = 17, 22, 27, 32, 37, 42, 47, 52
d. **Data Set D:**
Range = 92   # of Groups = 10   Width = 9.2 or 9
Midpoints = 103, 112, 121, 130, 139, 148, 157, 166, 175, 184, 193

21. a. Plague
b. 35
c. 84
d. Wenn, Poysened, Shingles and Swine Pox, Leprosie, and Calenture
e. .088%
f. .00057
g. 70%

23. a. IV = consumption of Pomegranate Juice (2 levels: 3 ounces/day and 0 ounces/day)
DF = risk of cancer over a 10-year period
Scale of measurement = Ratio
b. IV = walking with the shoes (2 levels: with the shoes and with other walking shoes)
DF = score on the Ashworth Scale of Muscular Resistance
Scale of measurement = Ordinal
c. IV = dose of Airborne (2 levels: daily vs. none)
DF = number of colds experienced in a 1-year period
Scale of measurement = Ratio
d. IV = dose of Activia (2 levels: daily vs. none)
DF = gastric juice pH measurement using Heidelberg Gastric Analysis Test
Scale of measurement = Ratio

25. a.

Type of Questions	f
Target	17
Nontarget	0

b. Superstitious behavior was only elicited in the target (stress-inducing) condition. No superstitious behavior was seen when participants were asked the nonstressful, distracting questions.
c. Given that superstitious behavior only happened when the participants were asked about stressful events, it would appear that magical thinking is related to feelings of stress.

CHAPTER 3

1. a. **Sleep quality ratings by 50 students**

Sleep Quality Rating	f
1	9
2	14
3	11
4	10
5	6

Cortisol level in 50 students

Cortisol Level (ng/g)	f
22–32	3
33–43	4
44–54	2
55–65	1
66–76	2
77–87	6
88–98	2
99–109	5
110–120	2
121–131	2
132–142	5
143–153	5
154–164	0
165–175	4
176–186	2
187–197	3
198–208	2

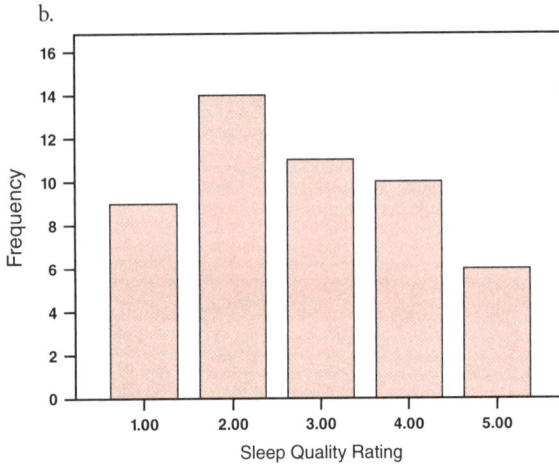

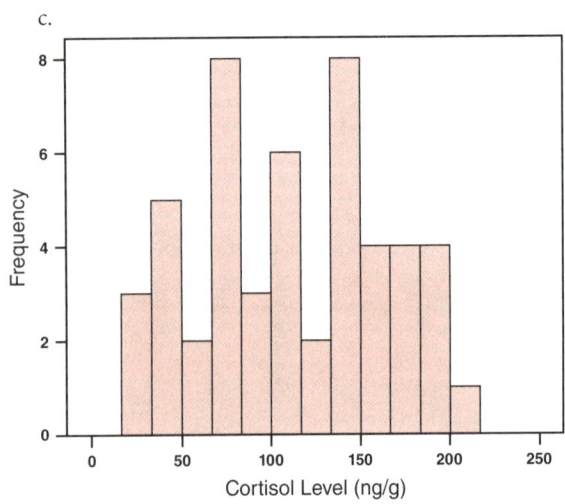

b.

c.

3.

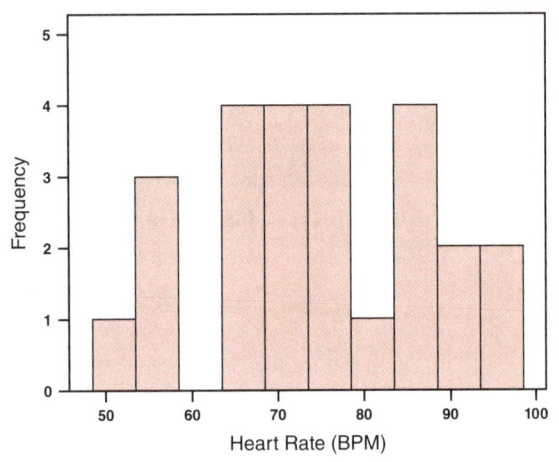

a. There are two main points to the graph. First, there is a great deal of variability in the heart rate data. The slowest heart rate was 51 BPM, and the fastest was 98 BPM. Second, there appears to be at least two groups in this data set: a group with a relatively slow heart rate and another, larger group with a faster heart rate.

Grouped frequency distribution

Interval	f
51–60	4
61–70	6
71–80	7
81–90	4
91–100	4

Graph of grouped frequency distribution

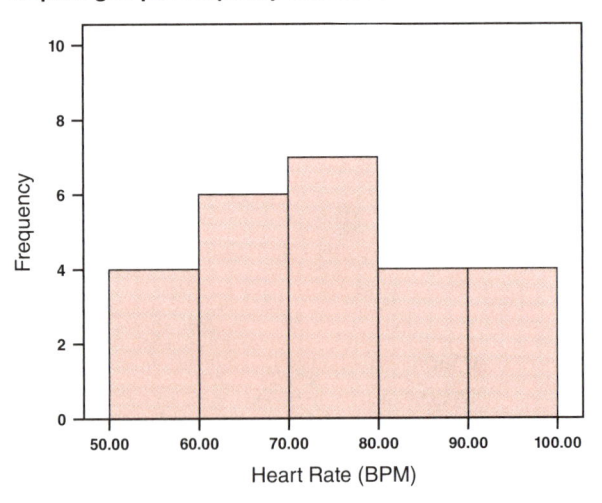

724 ANSWERS TO ODD NUMBERED END-OF-CHAPTER PRACTICE PROBLEMS

Grouping the data obscures the details: The slow group is now subsumed in the larger data set and cannot be seen anymore. The ungrouped frequency distribution is the better, more informative graph.

b. The graphs suggest that two groups are included in the data. One group consists of four people with relatively slow heart rates (between 51 and 57 BPM); the other group features people with faster heart rates (between 65 and 98 BPM). The smaller, slower group might be men.

c. **Heart rate (BPM) stem-and-leaf plot**

Stem	Leaf
5	1,6,6,7
6	5,5,5,7,9
7	0,1,3,4,5,6,7
8	0, 4, 5, 5, 8
9	2, 2, 6, 8

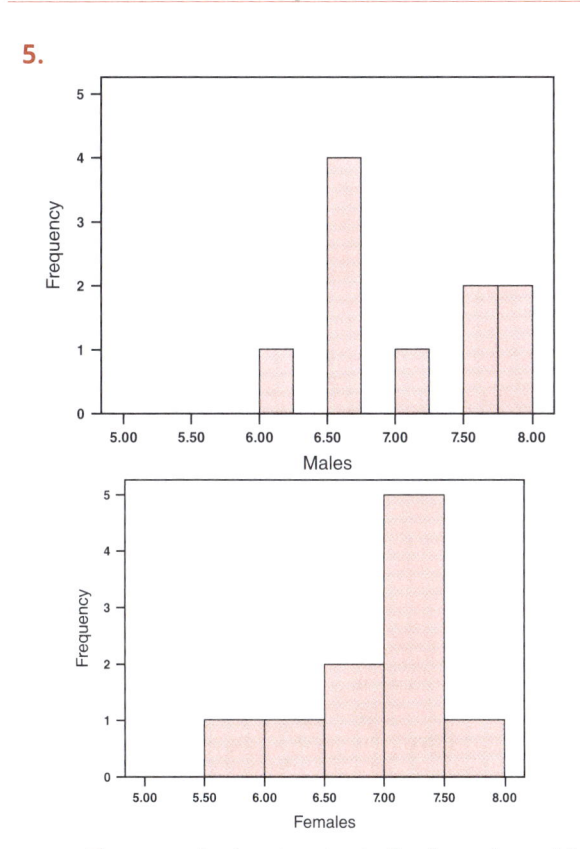

5.

The range of values is quite similar for males and for females, although there appear to be some high-value outliers in the distribution for males. There may be more variability in the measurements for the males compared to those for females. The two means are most likely not meaningfully different from one another.

7.

Recidivism for 500 male convicts who re-offend by cause

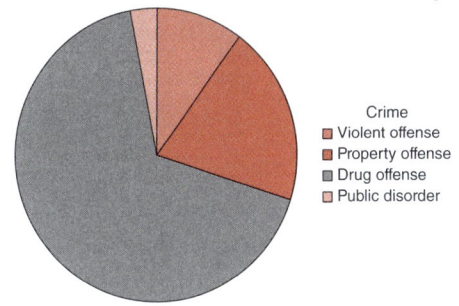

Overwhelmingly, if a male convict is going to reoffend, he will commit a drug offense.

9. a. There is one outlier in the list of golfers: Bubba Watson drives the ball much farther than do any of the other top 25 male golfers on the PGA tour.

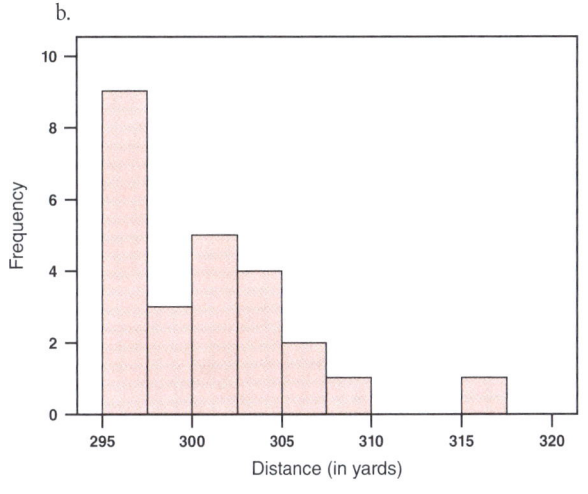

b.

c. For those golfers who are not Mr. Watson, the typical drive would be around 300 yards or so. For comparison, a great drive for a male amateur golfer is, according to the PGA, 214 yards.

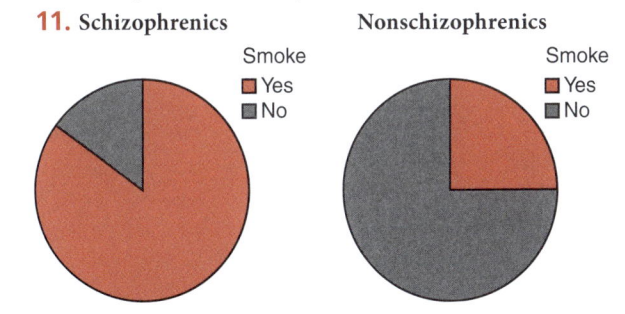

11. Schizophrenics   Nonschizophrenics

Schizophrenics are much more likely to smoke than are nonschizophrenics. The two graphs are almost reverse images of one another.

ANSWERS TO ODD NUMBERED END-OF-CHAPTER PRACTICE PROBLEMS   725

13.

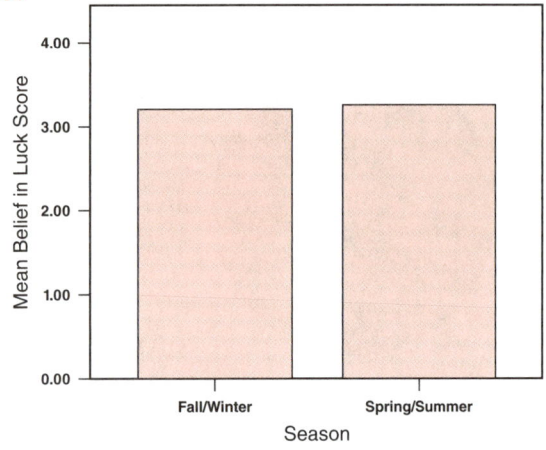

There is a slight tendency for people born in the Spring/Summer to have a higher belief in luck.

15. Yes, it appears that the color of the grid does have an effect on perception of the "illusory dot." Specifically, the color yellow seemed to make it more difficult to see the illusion.

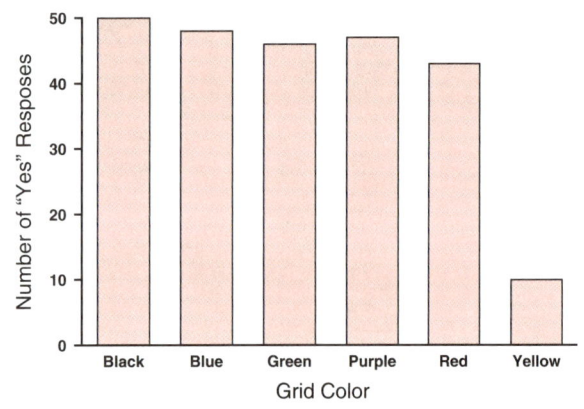

17. As age increases, errors in time estimation decrease. The older the participant is, the better his or her estimation of time.

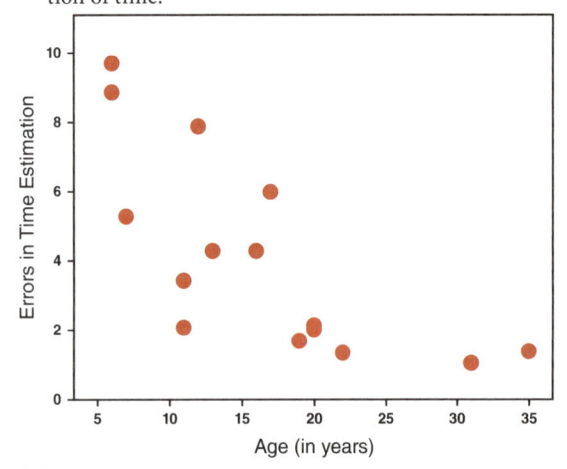

19. a. Researchers have found that of any population of people, those of European descent have the greatest variety in eye color. If you want to study eye color in a group of people, using a group with a large amount of variability makes sense. Comparing eye color in a group where eye color varies a great deal to one where eye color tends not to vary much would bias the results of the study.
b. 11,000 inmates were studied.
c. 6,270 inmates with light-colored eyes were in the sample.
d. 4,730 inmates with dark-colored eyes were in the sample.
e. 4,430 inmates had a history of alcohol abuse.
f. **Light-eyed inmates**   **Dark-eyed inmates**

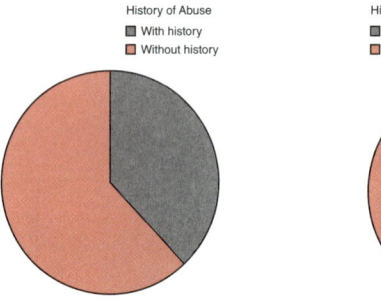

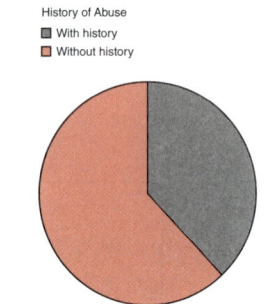

g. The percentage of inmates with a history of alcohol abuse problems was approximately the same in both eye color groups. There does not seem to be a relationship between eye color and alcohol dependency.

History of Alcohol Abuse Problems	Light-Colored Eyes	Dark-Colored Eyes	Row Totals
With	2,633 (41.99%)	1,797 (37.99%)	4,430
Without	3,637 (58.01%)	2,933 (62.01%)	6,570
Column totals	6,270	4,730	11,00

21. a. Using archival data means relying on unknown others to collect, categorize, and measure the data. Since there is no way of telling who collected the data, there's no way to determine if the variables were measured correctly. For example, the way that eye color is determined would have a serious impact on the data, but there is no way of ascertaining how this variable was measured.
b. Self-report data are notoriously suspect, and with good reason. People often suffer from bias when they are measuring aspects of their own behavior. The bias is only amplified when the participant is aware that the behavior being considered is socially unacceptable.

23. a. The highest concentrations of deaths from cholera are clustered around the Broad Street Pump. The farther away from the Broad Street Pump a household is, the lower the number of deaths from cholera in that home.

b. Since most households likely obtained their drinking water from the pump located nearest to their home (water is very heavy), this pattern of deaths would suggest that the Broad Street Pump is contaminated.

CHAPTER 4

1. a. Ordered x values: 9, 8, 7, 5, 3, 3, 3, 2, 1
Mean: $\sum x = 41$; $n = 9$; $\overline{X} = \sum x/n = 41/9 = 4.55$
Position of median $= (n+1)/2 = (9+1)/2 =$ value in the fifth position in the ordered set
Median $= 3.00$
Mode $= 3.00$

b. Ordered x values: 205, 205, 167, 150, 145, 139, 104, 20
Mean: $\sum x = 1{,}135$; $\overline{X} = \sum x/n = 1{,}135/8 = 141.875$
Position of median: $n + 1/2 = 8 + 1/2 =$ value half way between values in the fourth and fifth positions in ordered set
Median $=$ halfway between 150 and 145:
$150 + 145/2 = 295/2 = 147.5$
Mode $= 205$

c. The values are already in order.
Mean: $\sum x = 919$; $\overline{X} = \sum x/n = 919/27 = 34.04$
Position of median: $n + 1/2 = 27 + 1/2 =$ value in the fourteenth position of the ordered set
Median $= 29$
Mode $= 27$

3. Mean: $\sum x = 2{,}234{,}000$;
$$\overline{X} = \frac{\sum x}{n} = \frac{2{,}234{,}000}{37} = \$60{,}378.38$$
Position of the median: $\frac{n+1}{2} = \frac{37+1}{2} =$ the value in the nineteenth position
Median: $\$50{,}000$
Mode: $\$47{,}000$

5. If the distribution were completely normal in shape, then trimming off the top and bottom 5% should not change the value of the mean at all.

7. a. Group 1 Because the data was quite different for the two age groups, I split the data set into the two groups and calculated the measures of center for each group.
Ordered: 13, 13, 11, 9, 5, 1, 1, −1, −11, −16
$\sum x = 25$; $\overline{X} = \frac{25}{10} = 2.5$
Position of median: $\frac{n+1}{2} = \frac{11}{2} = 5.5$
Median: $\frac{1+5}{2} = 3.00$

Mode: Bimodal, 13 and 1
Mean < median, so there is a negative skew

b. Group 2
Ordered set: 14, 11, 10, 7, 5, 5, 4, 3, 1, −3
$\sum x = 57$ $\overline{X} = \frac{57}{10} = 5.7$
Position of Median: $\frac{n+1}{2} = \frac{11}{2} = 5.5$
Median: $\frac{5+5}{2} = 5.00$
Mode: 5
Mean slightly > than median, so a slight positive skew

9. a. To find the measures of center, first put the data in an ordered set:

Ordered reaction time data

Position	TMS	No TMS
1	460	480
2	460	500
3	490	500
4	490	500
5	490	520
6	490	520
7	500	520
8	500	520
9	520	530
10	520	540
11	580	550
	$\sum x = 5500$	$\sum x = 5680$
	$n = 11$	$n = 11$

	TMS Condition	No TMS Condition
Mean	$\overline{X} = \frac{5{,}500}{11} = 500$ ms	$\overline{X} = \frac{5{,}680}{11} = 516.36$ ms
	Position of median $= \frac{n+1}{2} = \frac{12}{2} =$ sixth position in the ordered set for both conditions	
Median	490 ms	520 ms
First quartile	490 ms	500 ms
Third quartile	520 ms	530 ms

ANSWERS TO ODD NUMBERED END-OF-CHAPTER PRACTICE PROBLEMS

b. The TMS distribution is slightly positively skewed. The value of 580 milliseconds is a high outlier in this distribution that pulls the mean up from 490 to 500 milliseconds. The No TMS distribution is slightly negatively skewed. The outlier in this distribution is low (480 milliseconds) compared to the mean and median values.
c. TMS resulted in a slightly faster reaction time. The average reaction time in the TMS condition was 500 milliseconds compared to an average reaction time of 516.36 milliseconds in the No TMS distribution.

11. a. Some students in the group are probably studying slightly less than the average amount of time: **True**. (**The key word here is "probably."**)
b. There are some students who study significantly more than average in the set: **True**.
c. This data set has outliers at the low end of the distribution: **False**.
d. This data set is positively skewed: **True**.
e. This data set is normally distributed: **False**.
f. The mean is being pulled up in value by outliers at the high end of the distribution: **True**.
g. Fifty percent of the students in the sample study 5 hours or more: **True**.
h. The high value outlier in this data set happened only once: **Unable to determine**.

13. a. $\bar{X} = \dfrac{\sum x}{n} = \dfrac{307k}{10} = 30.7k$ ($30,700)
b. The mean is not the best measure of central tendency because the distribution of salaries at the bakery is seriously skewed. There are two extreme outliers that are pulling the mean up.
c. Because the median is less affected by outliers than is the mean, the median is a better measure of center in a skewed distribution.
d. $\dfrac{n+1}{2} = \dfrac{11}{2}$ = the score in the 5.5th position. The median in this distribution is halfway between a salary of $15,000 and $16,000. The median salary is $15,500.
e. The difference between the mean ($30,700) and the median ($15,500) is $15,200.

15. Hypothetical data set for calculating the 10% trimmed mean

x	Frequency	Top/bottom 5 Scores Removed
7	1	
8	1	
9	1	
10	3	1
11	4	4
12	6	6
13	6	6

x	Frequency	Top/bottom 5 Scores Removed
14	13	13
15	15	10

a. $n = 50$
b. 10% of 50 = 5
c. Lowest five scores are 7, 8, 9, 10, and 10
  Highest five scores are 15, 15, 15, 15, and 15
d. $\sum x = 655$, n = 50
  Mean = $\dfrac{655}{50} = 13.10$
e. $\sum x = 536$, $n = 40$ (the lowest and highest five scores have been removed, so n decreases by 10).
  Mean = $\dfrac{536}{40} = 13.4$
f. The trimmed mean is slightly higher than the untrimmed mean. Trimming removed all of the lowest value observations and five of the highest values. There are still high value outliers in the set.

17. a. **Diurnal type score and maternal response: Measures of center**

	DTS		Depression	
	Milk	No Milk	Milk	No Milk
Mean	21.08	19.00	1.67	1.83
Median	21	19	1	1.5
Mode	22	15	1	1

b. Yes, drinking milk did affect diurnal preference. Children who drank milk at breakfast had a somewhat higher DTS score ($M = 21.08$) compared to children who did not drink milk in the morning ($M = 19.00$), indicating a stronger preference for the morning in children who regularly consumed milk in the morning. The distribution of DTS scores for the children who did not drink milk in the morning was negatively skewed. The high value outlier of 28 in this group pulled the mean up away from the median (18.5) and the mode (15).
c. Yes, drinking milk did affect frequency of depression in these children—slightly. Depression scores were somewhat lower for children who drank milk in the morning ($M = 1.67$, median = 1.5, mode = 1) than for the children who did not drink milk in the morning ($M = 1.83$, median = 1.5, mode = 1), indicating a slightly higher frequency of depression in the non-milk-drinking sample.

19. a. Yes: The distribution is negatively skewed.
b. i) A few teachers have significantly fewer contact hours than do the majority: TRUE. The low outlier is pulling the mean down.

ii) A few teachers have significantly more contact hours than do the majority: FALSE—or at least, not necessarily true. Some teachers would have more contact hours than was average in either a perfectly normal distribution or in a skewed one.

21. Mode = 6
Median = 5.5
Data: 6, 2, 8, ?, 2, 6, 5, 1, 7

x	f	Position
1	1	1st
2	2	2nd
5	1	4th
6	1	
6	2	5th
7	1	7th
8	1	8th

The missing data point is greater than the median; in fact, it should be 6. If the missing data point is 6, then the most frequently occurring score is 6 (the frequency of a score of 6 is now 3). The median of the eight known data points is the score in the 4.5th position, which would be halfway between the score in the 4th position (a score of 5) and the score in the 6th position (a score of 6), which would be a score of 5.5. If the missing data point is 6, then the position of the median is the score in the 5th position (there are now nine total scores, which moves the median slightly up to 6).

23. Overall mean = 94.47 (change of 1.17)
Overall median = 92.00 (no change)
The overall mean was most affected by the missing data

Mean and median by team:
Red mean = 92.50 (change of 0.20)
Red median = 90.00 (change of 1.00)
Green mean = 95.78 (change of 1.88)
Green median = 96.00 (change of 3.5)

For the red and the green teams, the median was changed most by the missing data. Most of the missing data came from the red team, so the position of the middle of the data set would be most affected the loss of data points.

Mean and median by position:
Defense mean = 94.83 (change of 1.73)
Defense median = 94.50 (change of 3.00)
Forward mean = 94.22 (change of 0.72)
Forward median = 96.00 (change of 4.00)

For the two positions, the median was changed most by the missing data. Most of the missing data came from the forwards, so the position of the middle of the data set would be most affected the loss of data points.

Mean and median by team and position:
Red Defense mean = 89.00   Red Defense median = 89.00 (change of 1.00)
(change of 3.80 in mean and 1.00 in median)
Red Forward mean = 93.20   Red Forward median = 91.00
(change of 0.60 in mean and 0 in median)
Green Defense mean = 96.00   Green Defense median = 96.00
(change of 2.60 in mean and 3.00 in median)
Green Forward mean = 95.50   Green Forward median = 97.00
(change in 1.10 in mean and 5.00 in median)

There was only one defense team member left on the red team after the missing data were removed, so both the mean and median are based on only one observation.

25. Graphically, you can arrange the data (e.g., using a histogram) and look for a peak in the graph with a frequency of 30%.

27. Summary of results: Reaction time test

Group	Mean	Median	Trimmed Mean	Trimmed Median
Control	5.31	4.78		
Caffeine	5.33	4.32	3.77	4.05

There is an extreme score (19.34) in the caffeine group. If that score is removed, the means change drastically. With the outlier removed, the mean reaction time for the caffeine group is quite different from the control group. Caffeine seems to shorten reaction times.

29.

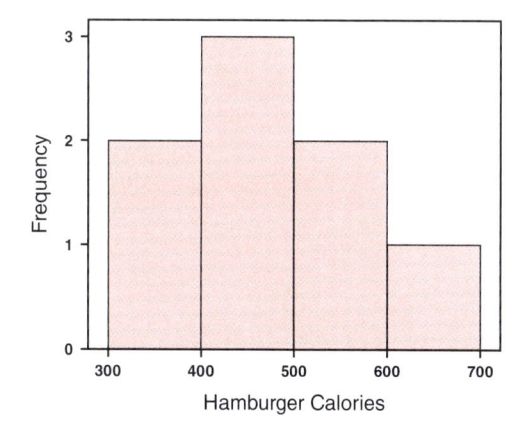

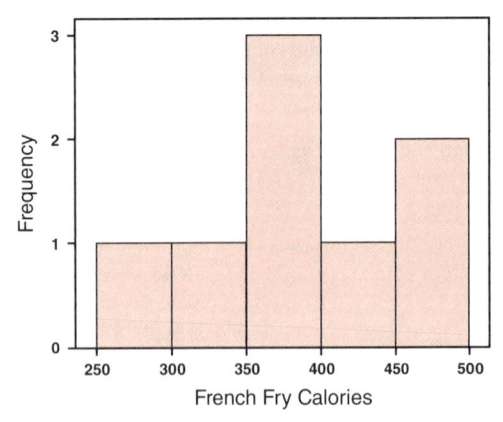

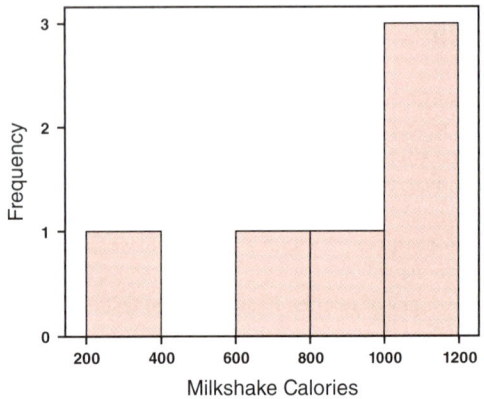

CHAPTER 5

1. Estimates of variability for the data in Table 5.4

	A	B	C
Σx	110.86	118.24	121.32
n	15	15	15
Mean	7.39	7.88	8.09
Σx^2	852.65	933.62	1405.16
Estimated variability	Moderate	Small	Great

Calculations:

A. $\sigma^2 = \dfrac{\Sigma x^2 - \dfrac{(\Sigma x)^2}{N}}{N} = \dfrac{852.65 - \dfrac{(110.86)^2}{15}}{15}$

$= \dfrac{852.65 - \dfrac{12,289.94}{15}}{15} = \dfrac{852.65 - 819.33}{15} = 2.22$

$\sigma = \sqrt{\sigma^2} \sqrt{2.22} = 1.49$

range = maximum − minimum = 9.75 − 5.15 = 4.6

B. $\sigma^2 = \dfrac{\Sigma x^2 - \dfrac{(\Sigma x)^2}{N}}{N} = \dfrac{933.62 - \dfrac{(118.24)^2}{15}}{15}$

$= \dfrac{933.62 - \dfrac{13,980.70}{15}}{15} = \dfrac{933.62 - 932.05}{15} = 0.105$

$\sigma = \sqrt{\sigma^2} \sqrt{.105} = 0.32$

range = maximum − minimum = 8.12 − 7.07 = 1.05

C. $\sigma^2 = \dfrac{\Sigma x^2 - \dfrac{(\Sigma x)^2}{N}}{N} = \dfrac{1405.16 - \dfrac{(121.32)^2}{15}}{15}$

$= \dfrac{1405.16 - \dfrac{14,718.54}{15}}{15} = \dfrac{1,405.16 - 981.24}{15}$

$= 28.26$

$\sigma = \sqrt{\sigma^2} \sqrt{28.26} = 5.32$

range = maximum − minimum = 18.22 − (−0.34) = 18.56
Estimates were accurate. Distribution C had the most variability (s = 5.32), and distribution B and the vleast variability (s = 0.32). The variability in distribution A was in between that of distributions C and B (s = 1.49).

3. The mean of the distribution in Table 5.5 was 9.15 pounds with a standard deviation of 1.81 pounds. Mr. Peebles' weight of 3 pounds is way below average (9.15 − 3.00 = 6.15 pounds below average). If we use the standard deviation as a "measuring stick" in this distribution, we can say that Mr. Peebles' weight is slightly more than three standard deviations below average (6.15 ÷ 1.81 = 3.40 standard deviations below average). A cat weighing exactly three standard deviations less than average would weigh 3.72 pounds (9.15 − 1.81 = 7.34 − 1.81 = 5.53 − 1.81 = 3.72 pounds), so Mr. Peebles weighs slightly more than three standard deviations below average in this distribution.

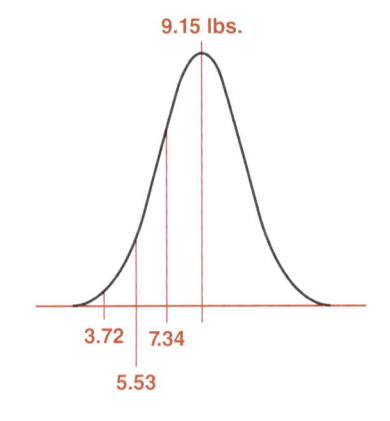

730   **ANSWERS TO ODD NUMBERED END-OF-CHAPTER PRACTICE PROBLEMS**

5. Time needed to complete a visual search task while smelling unpleasant or pleasant odor: Mean, median, mode, and standard deviation

Unpleasant Odor	Pleasant Odor
$\sum x = 1{,}517$	$\sum x = 1{,}983$
$n = 24$	$n = 24$
$\overline{X} = 63.21$	$\overline{X} = 82.63$
$\sum x^2 = 103{,}359$	$\sum x^2 = 174{,}947$
$\sigma^2 = \dfrac{103{,}359 - \dfrac{(1{,}517)^2}{24}}{24} = 311.33$	$\sigma^2 = \dfrac{174{,}947 - \dfrac{(1{,}983)^2}{24}}{24} = 462.57$
$\sigma = \sqrt{\sigma^2} = \sqrt{311.33} = 17.64$	$\sigma = \sqrt{\sigma^2} = \sqrt{462.57} = 21.51$
Median observation is 12.5th position = halfway between 57 and 60 = 58.5	Median observation is 12.5th position = 75
Mode = 48 and 55	Mode = 71

IQR (Unpleasant)	IQR (Pleasant)
25th quartile = 48.00	25th quartile = 70.25
Median (50th quartile) = 58.50	Median (50th quartile) = 75.00
75th quartile = 81.75	75th quartile = 85.75
IQR = 81.75 − 48 = 33.75	IQR = 85.75 − 70.25 = 15.5

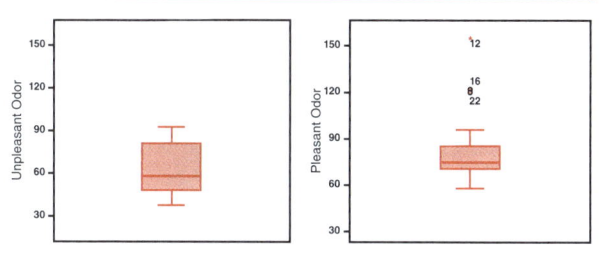

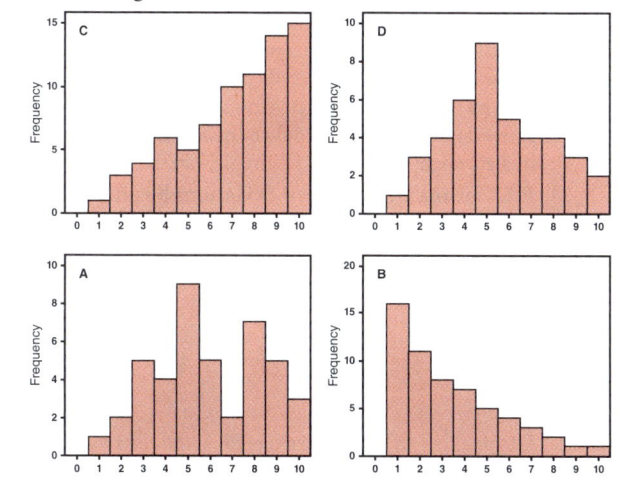

7. Histograms

The skewness measurement for a distribution that is perfectly normal in shape would be 0.00.

9. Volume of the amygdala and size of the social network for the study's 20 participants

Amygdala Size	Social Network Size
$\overline{X} = \dfrac{66.10}{20} = 3.305 \text{ mm}^3$	$\overline{X} = \dfrac{589}{20} = 29.45 \text{ contacts}$
$\sigma = \sqrt{\dfrac{228.69 - \dfrac{(66.10)^2}{20}}{20}}$	$\sigma = \sqrt{\dfrac{20{,}701 - \dfrac{(589)^2}{20}}{20}}$
$= \sqrt{\dfrac{228.69 - 218.46}{20}}$	$= \sqrt{\dfrac{20{,}701 - 17{,}346}{20}} .05$
$= \sqrt{\dfrac{10.23}{20}} = \sqrt{0.512}$	$= \sqrt{\dfrac{13{,}354{,}95}{20}} = \sqrt{167.75}$
$= 0.72 \text{ mm}^3$	$= 12.95 \text{ contacts}$
Median = 3.30 mm³	Median = 30.50 contacts
Mode = 3.30 mm³	Mode = 44.00 contacts
Range = 2.30 mm³	Range = 15.41 contacts

The distribution of amygdala sizes is relatively normal in shape (the mean, median, and mode are similar values). The distribution of social network sizes is negatively skewed. The mean is being pulled downward by the low outlier values of 5 and 9. The mean (29.45 contacts) is lower than the mode (44.00 contacts).

11. The average IQ scores for all of the groups were higher than average. The scores for Group 1 appear to be the highest. The median IQ score for Group 1 was slightly more than one standard deviation above average, despite the fact that there was an outlier more than one standard deviation below average in this group. The Interquartile Range for Groups 4 and 5 encompass the mean IQ score of 100, suggesting that these two groups contained scores that were lower than the scores in Groups 1 through 3. Scores in Group 4 were the most variable (least consistent) of all five groups. Scores in Group 3 were the most consistent of the set of five groups.

13. a–c. Mean, range, IQR, variance, and standard deviation of each set

A	B	C
$\overline{X} = \dfrac{50}{5} = 10$	$\overline{X} = \dfrac{50}{5} = 10$	$\overline{X} = \dfrac{51}{5} = 10.2$
Range = 6	Range = 0	Range = 19
IQR = 12 − 8 = 4	IQR = 10 − 10 = 0	IQR = 19.5 − 1 = 18.5
$s^2 = \dfrac{520 - \dfrac{(50)^2}{5}}{5}$ = 4.00	$s^2 = \dfrac{500 - \dfrac{(50)^2}{5}}{5}$ = 0.00	$s^2 = \dfrac{863 - \dfrac{(51)^2}{5}}{5}$ = 8.56
$s = \sqrt{4.00} = 2.00$	$s = \sqrt{0.00} = 0.00$	$s = \sqrt{8.56} = 2.93$

d. Data set B has no variability at all (all of the scores are exactly the same).

ANSWERS TO ODD NUMBERED END-OF-CHAPTER PRACTICE PROBLEMS

e. *SD* for each set, as determined using the range rule:

A	B	C
6 ÷ 4 = 1.5	0 ÷ 4 = 0.00	19 ÷ 4 = 4.75

f. When the amount of variability in a set is very low, the range rule does a fairly good job of estimating the standard deviation. Data set B has the least variability (in fact, it has none at all), and both the calculated value and range rule value of the standard deviation agree perfectly. Data set A has the next lowest variability, and the standard deviation and range rule estimate differ by only 0.50. Finally, data set C has the most variability, and the two measures of standard deviation differ by quite a bit: The calculated value of standard deviation is 2.93, and the range rule estimate is 4.75, for a difference of 1.82.

15 a–d.

Table 5.10a   A small data set

x	x − x̄	(x − x̄)²	x − Median	(x − Median)²	(x − Mode)	(x − Mode)²
2	−3	9	−1	1	0	0
2	−3	9	−1	1	0	0
3	−2	4	0	0	1	1
4	−1	1	1	1	2	4
14	9	81	11	121	12	144

$\bar{X} = \frac{25}{5} = 5$ median = 3.00   mode = 2.00

$\sum(x - \bar{X})^2 = 104$ $\sum(x - \text{median})^2 = 124$ $\sum(x - \text{mode})^2 = 149$

$\sum(|x - \bar{X}|) = 18$ $\sum(|x - \text{median}|) = 14$ $\sum(|x - \text{mode}|^2) = 149$

$\sigma = \sqrt{\frac{\sum(x-\bar{X})^2}{N}} = 4.56$ $\sigma = \sqrt{\frac{\sum(x-\text{median})^2}{N}} = 4.98$ $\sigma = \sqrt{\frac{\sum(x-\text{mode})^2}{N}} = 5.46$

$\sigma = \sqrt{\frac{\sum(|x-\bar{X}|)^2}{N}} = 3.60$ $\sigma = \sqrt{\frac{\sum(|x-\text{median}|)^2}{N}} = 2.8$ $\sigma = \sqrt{\frac{\sum(|x-\text{mode}|)^2}{N}} = 3.00$

Table 5.10b   Results

Using the mean	
SD = 4.56	MD = 3.6
Using the median	
SD = 4.98	MD = 2.8
Using the mode	
SD = 5.46	MD = 3.00

e. The smallest measures of variability are obtained when the mode is used as the measure of center. However, because the mode is the least robust of the three measures of center, this would also be the least robust (and least descriptive) of the three measures of variability. The standard deviation (*SD*) and average deviation (*MD*) are fairly similar when the mean is used (they differ by only 0.96 units). When then median is used, the difference between the *SD* and *MD* is larger (2.18 units), and when the mode is used, the difference between *SD* and *MD* is larger still (2.46).

17. a. $\bar{X} = \frac{616}{14} = 44.00$

$S = \sqrt{\frac{28{,}214 - \frac{(616)^2}{14}}{14}} = 8.90$

b. **Small data set with constant of 10 added to each observation**

x	f	x + 10
25	1	35
33	1	43
41	1	51
42	2	52
43	4	53
45	2	55
52	1	62
54	1	54
65	1	65

c. ☑ The original mean + 10
d. ☑ The same standard deviation as before a constant was added to each data point
e. The mean (the arithmetic average of the data) will increase by the size of the constant added to each data point. Adding a constant changes the sum of the individual data points but does not affect the number of data points, so the mean increases by the value of the constant added to each point. Measures of variability (average deviation from the mean) will not change when a constant is added to each data point because adding the constant does not change the position of each data point relative to the mean.

19. a. All of the dogs are very big (they weigh a lot). Their high weights are pulling the standard deviation upward: **False**.
b. The weights of the dogs may vary a great deal. Some are small dogs and some are very large, making the standard deviation quite large: **True**.
c. A few dogs in the sample could be quite large. These outliers might be creating positive skew in the data: **True**.
d. A few dogs in the sample could be quite small. These outliers might be creating positive skew in the data: **False**.

21. a. **ERA means and standard deviations by pitcher**

Pitcher 1	Pitcher 2	Pitcher 3
Mean = 4.04	Mean = 4.10	Mean = 2.50
$s = \sqrt{\frac{163.79 - \frac{(40.43)^2}{10}}{10}}$	$s = \sqrt{\frac{205.40 - \frac{(41)^2}{10}}{10}}$	$s = \sqrt{\frac{110.20 - \frac{(24.88)^2}{10}}{10}}$
= 0.18	= 1.93	= 2.20

Pitcher 3 has the lowest average ERA and, therefore, the best ERA in the set of three pitchers.

b. Pitcher 1 is the most consistent: His standard deviation is the smallest (0.18).
c. Pitcher 3 is the least consistent: His standard deviation is the largest (2.20).
d. I would not start Pitcher 3 against Team G, because his ERA is very high against this team. Compared to his performance against all of the other teams, his ERA against team G is 9.09, his own personal worst performance and the highest of the three pitchers in the set.

23. **Data set: Grouped**

	f	$x - \bar{X}$
1.20	1	4.76
1.50	1	4.46
2.30	1	3.66
6.80	1	0.84
6.90	1	0.94
7.30	1	1.34
7.60	1	1.64
7.90	1	1.94
8.40	1	2.44
9.70	1	3.74

a. The mean of the 10 data points is 5.96.
b. The standard deviation:

$$s = \sqrt{\frac{440.94 - \frac{(59.60)^2}{10}}{10}} = 2.93$$

c. The mean absolute deviation:

$$MD = \frac{\sum(|x - \bar{X}|)}{n} = \frac{|25.76|}{10} = 2.576$$

The standard deviation tends to emphasize bigger deviations from the mean (because we're squaring all of the deviations, so big deviations make for bigger squares) compared to the absolute value of the deviations. However, big deviations from the mean are unusual and less likely to happen than are small deviations from the mean. Fisher showed that the standard deviation of a sample is a more consistent and more accurate estimate of the variability in the population than is the mean deviation.

25. There were equal numbers of males and females in the data set (15 males and 15 females). The age of the participants ranged from 65 to 96 years of age (range = 31 years, IQR =14 years). The average of the participants in the study 76.70 years with a variance of 81.18 years and a standard deviation of 9.01 years. The median age was 75.5 years, and the modal age was 67 years.

On average, bone density scores (T-scores) were 2.02 standard deviations below that of an average healthy male. The variance and standard deviation of the T-scores were 0.28 and 0.53, respectively. The median T-score was –2.05, and the modal score was –2.70. T-scores had a range of 1.90 and an IQR of 0.80.

Education level ranged from no high school to a postsecondary degree. Seven of the 30 participants had no high school education at all, while 14 had at least some high school ($n = 7$) or had graduated from high school ($n = 7$). The remaining nine participants had some postsecondary education ($n = 5$) or had completed a postsecondary degree ($n = 4$).

Means, variances, and standard deviations of nominal (categorical) variables (education and gender) are not useful in describing the data and were not calculated.

27. Females deviated less from average than did males, and T-scores from males were more inconsistent than were scores from females (the IQR for males was larger than the IQR for females).

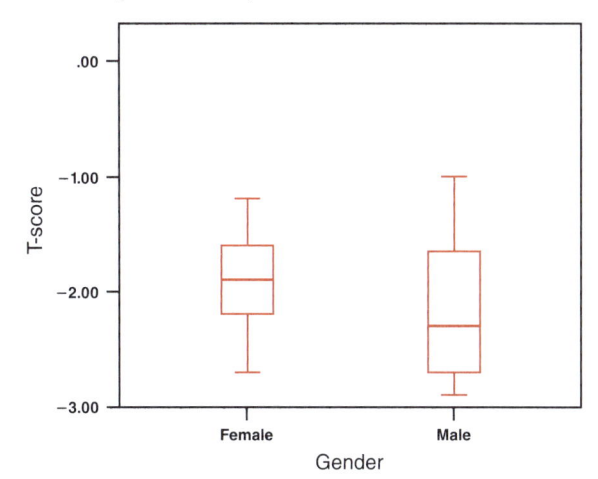

CHAPTER 6

1. a. $\sum x = 271$
   $\sum x^2 = 3,787$
   $\bar{X} = \frac{\sum x}{n} = \frac{271}{20} = 13.55$
   median = score in the 10.5th position = 14
   mode = 15

   $$s = \sqrt{\frac{3,787 - \frac{271^2}{20}}{20}} = 2.40$$

   b. A score of 20 has a deviation score of $20 - 13.55 = 6.45$.
   c. A score of 9 has a deviation score of $9 - 13.55 = -4.55$.

3. Because the mean (13.55), median (14), and mode (15) are all close together in value, we can conclude that the data are normally distributed.

5. $z = \dfrac{x - \overline{X}}{s}$

$\dfrac{140 - 100}{16} = 2.50$

$\dfrac{120 - 100}{16} = 1.25$

$\dfrac{110 - 100}{16} = 0.625$

$\dfrac{90 - 100}{16} = -0.625$

$\dfrac{80 - 100}{16} = -1.25$

$\dfrac{70 - 100}{16} = -1.875$

If you've taken the Stanford-Binet test and know your own score, convert your score to a z-score and see where you sit on the scale of intelligence proposed by Louis Terman.

7. a. $z = \dfrac{650 - 500}{100} = \dfrac{50}{100} = 0.50$

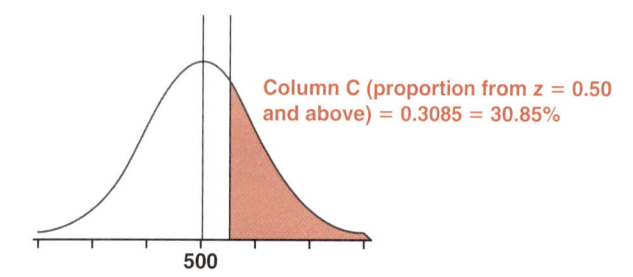

Column C (proportion from z = 0.50 and above) = 0.3085 = 30.85%

b. $z = \dfrac{766 - 500}{100} = \dfrac{266}{100} = 2.66$

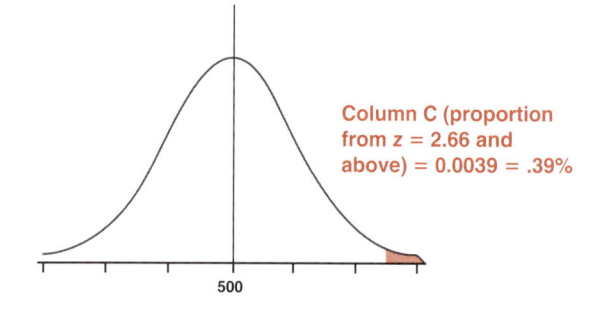

Column C (proportion from z = 2.66 and above) = 0.0039 = .39%

c. The score at the 75th percentile cuts off the top 25% (expressed as a proportion = 0.2550) of the normal distribution. The corresponding z-score that cuts off the top 5% in column C is $z = 0.67$. We need to convert this z-score back into a GRE score.

$x = s(z) + \overline{X}$ $x = 100(0.67) + 500$

$x = 67 + 500 = 567$

So, a score of 567 would be at the 75th percentile. The score at the 80th percentile cuts off the top 20% (expressed as a proportion = 0.2000) of the normal distribution. The corresponding z-score that cuts off the top 20% in column C is $z = 0.84$. We need to convert this z-score back into a GRE score.

$x = s(z) + \overline{X}$ $x = 100(0.84) + 500$

$x = 84 + 500 = 584$

So, a score of 584 would be at the 80th percentile.

d. $z = \dfrac{400 - 500}{100} = \dfrac{-100}{100} = -1.00$

So, your score of 400 on the GRE quantitative section is exactly one standard deviation below average. We need to find out what score on the hot dog test would be in the same relative position. There are two solutions to this problem. First, we can convert a z of -1.00 into an x-score on the on the hot dog test. (Note: We need to use the mean and standard deviation from the hot dog test to do this.)

$x = s(z) + \overline{X}$ $x = 25(-1) + 75$

$x = -25 + 75 = 50$

The second way to solve this problem is to use common sense. Your score on the GRE was one standard deviation below average. What score would be one standard deviation below average on the hot dog test? One standard deviation = 25, so 75 − 25 = 50.

9. a. The general pattern is for BMD to rise until the age of about 40 and to decline after that. The pattern is the same for all three ethnic groups.

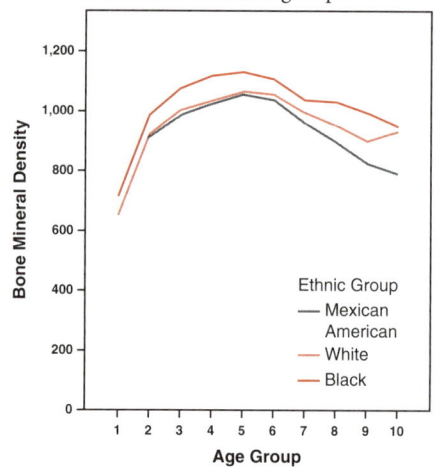

b.

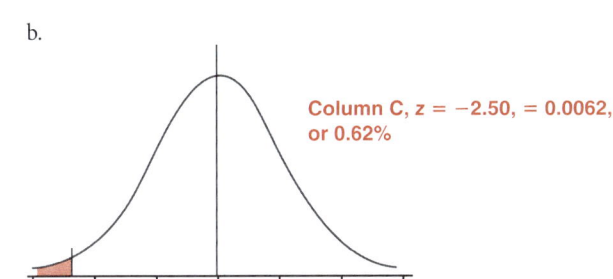

Column C, $z = -2.50$, = 0.0062, or 0.62%

c. **BMD measure of 720 mg/cm² converted to a z-score**

Ethnic Group	Age	z-Score Equivalent (BMD = 720)	Diagnosis of Osteoporosis? (Yes or No)
Mexican-American	35	$z = \dfrac{720 - 1056}{110} = -3.05$	Yes
	60	$z = \dfrac{720 - 895}{138} = -1.27$	No
	83	$z = \dfrac{720 - 791}{218} = -.33$	No
White (non-Hispanic)	35	$z = \dfrac{720 - 1065}{110} = -3.14$	Yes
	60	$z = \dfrac{720 - 952}{142} = -1.63$	No
	83	$z = \dfrac{720 - 932}{141} = -1.50$	No
Black (non-Hispanic)	35	$z = \dfrac{720 - 1130}{119} = -3.45$	Yes
	60	$z = \dfrac{720 - 1029}{163} = -1.90$	No
	83	$z = \dfrac{720 - 948}{222} = -1.03$	No

d. In each ethnic group listed, a BMD of 720 mg/cm² would indicate osteoporosis in the youngest age group listed (35 years). That same BMD would NOT indicate osteoporosis in the two older age groups regardless of ethnicity.

11. a. Convert the 700 to a z-score in the 2013 population, then look up the proportion on the table.

$$z = \dfrac{700 - 496}{115} = 1.77$$

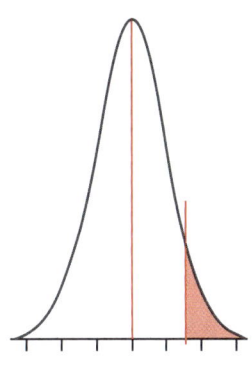

From column C, we find the proportion of scores that would probably be above a z of 1.77 and the mean = .0392, or 3.92%, of the students who took the test in 2013 would be expected to score a 700 or higher on the Reading test.

b. Convert 710 to a z-score in the 2013 population, then look up the proportion on the table.

$$z = \dfrac{710 - 514}{118} = 1.66$$

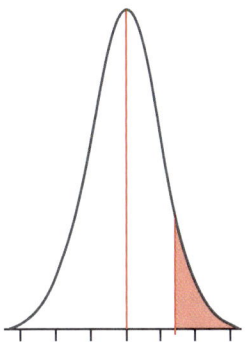

From column C, we find the proportion of scores that would probably above a z of 1.66 and the mean = .0485, or 4.85%, of the students who took the test in 2013 would be expected to score a 710 or higher on the Math test.

c. Convert 680 to a z-score in the 2013 population, then look up the proportion on the table.

$$z = \dfrac{680 - 488}{114} = 1.68$$

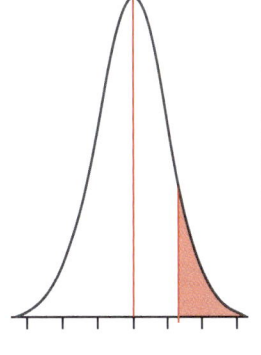

From column C, we find the proportion of scores that would probably above a z of 1.68 and the mean = .0465, or 4.65%, of the students who took the test in 2013 would be expected to score a 680 or higher on the Writing test.

d. i) Reading test scores: 3.92% of the students who took the test would be eligible for entry into BSU (Writing test scores of 700 or above). So, 3.92% of 1,660,045 students who took the test = 65,074 students.

ANSWERS TO ODD NUMBERED END-OF-CHAPTER PRACTICE PROBLEMS

ii) Mathematics test scores: Slightly more (4.85%) of the students who took the test would be eligible for entry into BSU (math test score of 710 or above). So, 4.85% of 1,660,045 students who took the test = 80,512 students.

iii) Writing test scores: 4.65% of the students who took the test would be eligible for entry into BSU (reading test scores of 680 or above). So, 4.65% of 1,660,045 students who took the test = 77,192 students.

e. i) Reading test:

$$z = \frac{600 - 496}{115} = 0.90 \qquad z = \frac{500 - 496}{115} = 0.03$$

$z = 0.90$; column B = 0.3159
$z = 0.03$; column B = 0.0120
Subtracting = 0.3039, or 30.39%

ii) Mathematics test:

$$z = \frac{600 - 514}{118} = 0.73 \qquad z = \frac{500 - 514}{118} = -0.12$$

$z = 0.73$; column B = 0.2673
$z = -0.12$; column B = 0.0478
Sum = 0.3151, or 31.51%

iii) Writing test:

$$z = \frac{600 - 488}{114} = 0.98 \qquad z = \frac{500 - 488}{114} = .11$$

$z = 0.98$; column B = 0.3365
$z = 0.11$; column B = 0.0438
Sum = 0.3803, or 38.03%

f. The easiest way to solve for these percentages is to take the answers you got for question (d) above (the proportion who would be admitted) and subtract those proportions from 1.00 (the whole set). The remainder, after those admitted are subtracted, would be those not admitted.
  i) Reading test: $1 - .0392 = 0.9608$, or 96.08%
  ii) Mathematics test $1 - .0485 = 0.9515$, or 95.15%
  iii) Writing test $1 - .0465 = 0.9535$, or 95.35%

g. The 75th percentile would be a score that cuts off the upper 25% of the scores on the test (75% of the scores at least this high). So, first, find the z-score that cuts off the upper 25% of the distribution (as close to 0.2500 as you can get in column C), then convert that z-score to a raw score in each distribution. The z-score that cuts off the top 25% of the scores in a normal distribution is 0.67 (cuts of .2514, or the upper 25.14%, of the scores).
  i) Reading test:

  $$x = z(s) + \bar{X} = 0.67(115) + 496 = 573.05$$

  ii) Mathematics test:

  $$x = z(s) + \bar{X} = 0.67(118) + 514 = 593.06$$

  iii) Writing test:

  $$x = z(s) + \bar{X} = 0.67(114) + 488 = 564.38$$

13. a. First, convert 750 to a z-score to see where it is in the distribution of New test scores:

$$z = \frac{x - \bar{x}}{s} = \frac{750 - 500}{100} = 2.50$$

Next, convert the z-score of 2.50 into a raw score on the Stanford-Binet test to find the IQ score that would be in the same position in the distribution of Stanford-Binet test scores:

$$x = z(s) + \bar{X} = 2.50(16) + 100 = 140$$

Both scores, 750 on the New test and 140 on the Stanford-Binet test, are 2.50 standard deviations above the mean.

b. First, convert 525 to a z-score to see where it is in the distribution of New test scores:

$$z = \frac{x - \bar{X}}{s} = \frac{525 - 500}{100} = 0.25$$

This score is ¼ of a standard deviation above average. Find the raw IQ score that sits in the same position (¼ of a standard deviation above average):

$$x = z(s) + \bar{X} = 0.25(16) + 100 = 104$$

c. $z = \dfrac{x - \bar{X}}{s} = \dfrac{78 - 100}{16} = -1.38$

A score of 78 on the Stanford-Binet test is 1.38 standard deviations below average. Converting to the New test gives us

$$x = z(s) + \bar{X} = -1.38(100) + 500 = 362$$

d. $z = \dfrac{x - \bar{X}}{s} = \dfrac{490 - 500}{100} = -0.10$

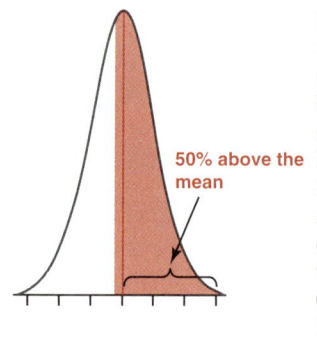

Using column B from the table, we find that 0.0398, or 3.98%, of the scores are between a z of −0.10 and the mean. All of these scores are higher than 490. Half the scores are above the mean as well as above a score of 490 so we need to add them in as well.
$0.5000 + 0.0398 = 0.5398$ or 53.98% of the scores on the new test will be above 490. Slightly more than half of the New test-takers.
53098% of 4,000 = 2,159 people

15. a. Column B = 0.2123
    b. Column B = 0.3925

c.

Column C, z = 1.00 = 0.1587
Column C, z = 1.34 = 0.0901
Subtracting = 0.0686

1.00 1.34

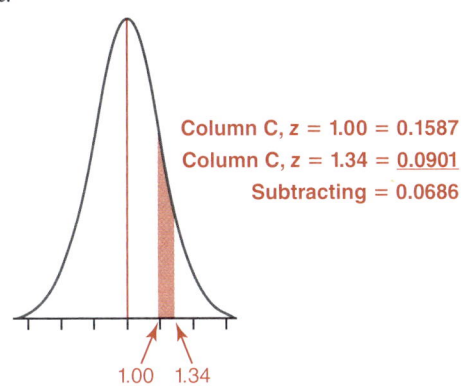

d.

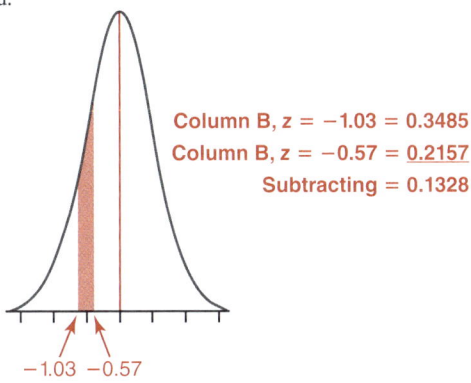

Column B, z = −1.03 = 0.3485
Column B, z = −0.57 = 0.2157
Subtracting = 0.1328

−1.03 −0.57

17. a. Convert both values (45 and 50) to z-scores:

$$z = \frac{x - \bar{X}}{s} = \frac{45 - 38}{5} = 1.40$$

$$z = \frac{x - \bar{X}}{s} = \frac{50 - 38}{5} = 2.40$$

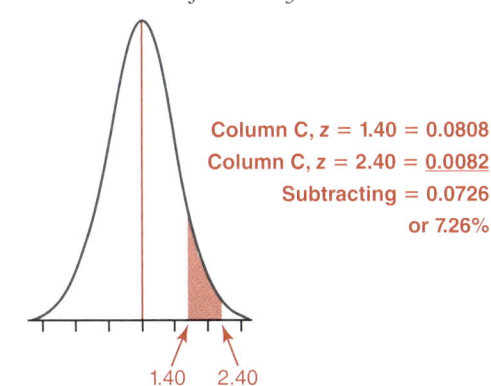

Column C, z = 1.40 = 0.0808
Column C, z = 2.40 = 0.0082
Subtracting = 0.0726
or 7.26%

1.40 2.40

b. There are approximately 160 class days possible (since we don't know if the class meets daily, three times/week, twice per week, etc., we'll assume we're dealing with the total number of days possible that we *do* know about). Half of 160 = 80 days. So, we want to know the percentage of students who will miss 80 days or more.

$$z = \frac{x - \bar{X}}{s} = \frac{80 - 38}{5} = -8.40$$

The z-table in Appendix A does not go past a z = 4.00. So, we'll have to assume that the percentage of students who miss half the possible class days or more is less than 0.003% (column C, z = 4.00).

c. First, we have to find the z that cuts off the outer 15% of the distribution: This is 1.03 (column C). We want the lower 15%, so the z we need is actually −1.03. Then, we need to convert this z back into a raw number of days missed.

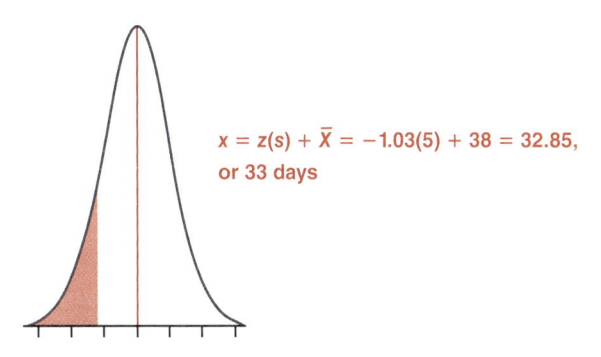

x = z(s) + $\bar{X}$ = −1.03(5) + 38 = 32.85, or 33 days

19. a. Convert both 14 and 18 to z-scores:

$$z = \frac{x - \bar{X}}{s} = \frac{14 - 15}{4} = -0.25$$

$$z = \frac{x - \bar{X}}{s} = \frac{18 - 15}{4} = 0.75$$

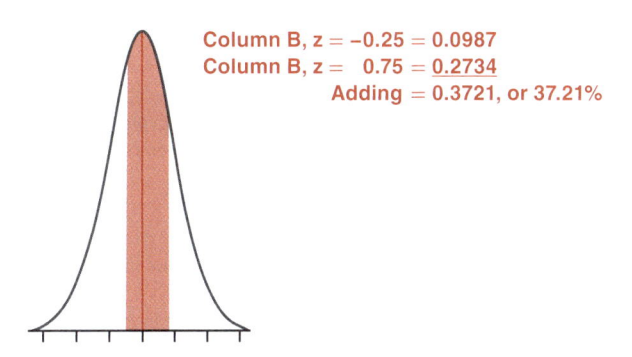

Column B, z = −0.25 = 0.0987
Column B, z =   0.75 = 0.2734
Adding = 0.3721, or 37.21%

b. 20 miles per hour converts to a z-score of

$$z = \frac{x - \bar{X}}{s} = \frac{20 - 15}{4} = 1.25$$

ANSWERS TO ODD NUMBERED END-OF-CHAPTER PRACTICE PROBLEMS

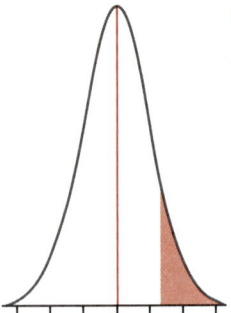

Column C, $z = 1.25 = 0.1056$, or 10.56%
10.56% of 200 = 21.12, or 21 people out of 200.

c. First, we need to find the z-score that cuts off the outer 5% of the distribution: Column C, $z = 1.65$. Then, we convert the z-score of 1.65 back into the number of miles per hour over the speed limit the car was going.

$$x = z(s) + \bar{X} = 1.65(4) + 15 = 21.56$$

So, if the posted speed limit was 45 miles per hour, the speed exceeded by only 5% of the speeders would be $45 + 21.56$, or 66.56 miles per hour.

21. Converted z-scores for five students

Student ID	Normalized Score	Raw Scores
201454746	−0.59	43.82
201419939	2.14	49.28
201462630	1.05	47.10
201491160	−2.81	39.38
201492113	0.75	46.50

Because the variance is 4.00, we know that the standard deviation of the junior professor's student evaluations is 2.00 (the square root of 4). So, now we can convert the normalized scores back into raw scores given a mean of 45 and a standard deviation of 2. Remember that $x = z(s) + \bar{X}$.

23. a. Find the z-score that cuts off the upper 8% of the distribution of shoe sizes, and convert this z into a shoe size. That shoe size will be the largest shoe Sarah's machine should make.

The z that cuts off the top 8% (looking in column C) is 1.40.

$$x = z(s) + \bar{X}$$
$$x = 1.40(3) + 9 = 13.2$$

The largest shoe that Sarah's machine should make is a size 13.

b. $x = 1.40(3) + 9 = 4.80$

The smallest shoe Sarah's machine should make is a size 5.

c. We need to find the z-scores (both positive and negative) that cut off the upper and lower 4% of the distribution (for a total of the outer 8% of the distribution) to leave us with the middle 92%. Then, convert these z's to raw scores.

The z-score that cuts off the upper 4% of the distribution is ±1.75.

$$x = -1.75(3) + 9 = 3.75$$
$$x = 1.75(3) + 9 = 14.25$$

Sarah's machine should make shoes between a size 4 and a size 14 to ensure that the middle 92% of the shoe-buying public has shoes.

25. First, we convert each x-value into a z-score. Then, we use the z-table in Appendix A to calculate the appropriate percentages.

Mean	s	x_1	x_2	$x_1(z)$	$x_2(z)$	%
6	2	5.5	6.3	−0.25	0.15	15.83
243	17	253	265	0.59	1.29	17.91
75.4	21.2	62.8	76.1	−0.59	0.03	23.44
500	100	250	750	−2.50	2.50	98.76
0.453	0.014	0.449	0.453	−0.29	0	11.41

27. Convert McKenzie's arrival time into a z-score in this distribution, and then determine what percentage of trains will arrive at or before this time. First, we need to convert time into a more workable number. Convert 8 hours and 1 minute past midnight into minutes past midnight ($60 \times 8 = 480 + 1 = 481$). Then, do the same for 8 hours and 3 minutes to get 483. Remember that the question gives us the train departure variance, not the standard deviation.

$$z = \frac{481 - 483}{\sqrt{4}} = \frac{2}{2} = -1.00$$

McKenzie's arrival time is one standard deviation below the average departure time. Column C tells us that 15.87% of the trains will arrive before McKenzie arrives at the station. In 1,000 trips to the city, McKenzie will likely miss the train 158.7, or 159 times.

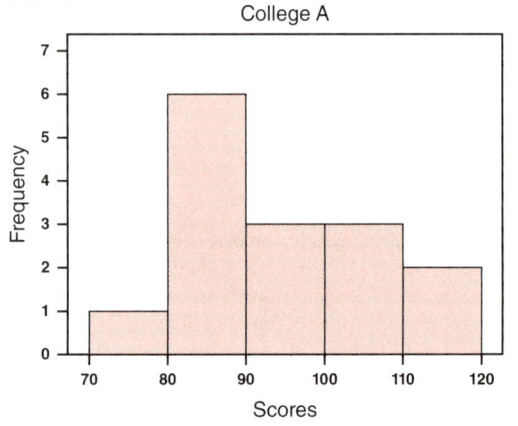

29. a.

College A

738 ANSWERS TO ODD NUMBERED END-OF-CHAPTER PRACTICE PROBLEMS

[College B histogram]

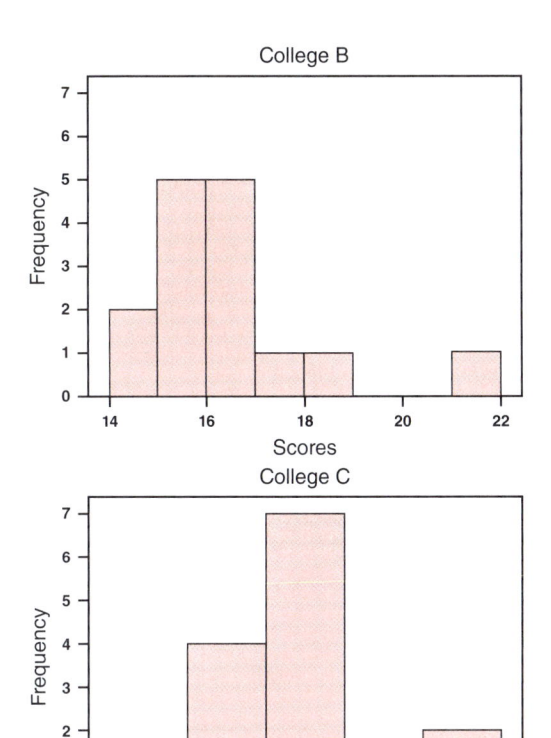

[College C histogram]

The distributions for Colleges A and C seem to be reasonably normal, while College B's distribution looks skewed. In fact, if you ask SPSS to calculate a measure of the skew of all three of the distributions (where a skew measure of zero indicates perfectly normal, and measurements above 1.00 indicate serious skew), both College A (skew = 0.58) and College C (skew = 0.577) are relatively normal, but the distribution for College B has a lot of skew (skew = 2.00). There is an outlier (21.3) in the distribution for College B.

b. **Mean and standard deviation for each college**

	College A	College B	College C
Mean	94.733	16.31	235.37
SD	11.62	1.62	9.89

c. *z*-score for each science student in each college

Student ID	College A	College B	College C
1	−0.18	0.30	−0.65
2	−1.19	−0.19	0.45
3	−1.18	−0.69	2.22
4	−0.54	0.18	0.57
5	1.93	1.10	−1.23
6	1.13	−0.01	−0.80
7	−0.82	0.06	0.30
8	−0.85	−1.12	1.68
9	−0.42	−0.62	0.29
10	0.55	0.49	−0.79
11	−1.37	−0.25	−0.54
12	1.53	−0.62	−0.17
13	0.32	−0.75	−1.71
14	0.44	−0.93	0.04
15	0.65	3.08	0.35

d. The mean of each set of *z*-scores is zero, and the standard deviation of each set of *z*-scores is 1.00

e. A *z*-score of 1.28 cuts off the top 10% of the distribution for each college. We can count the number of students with a *z*-score equivalent at or above +1.28 in each distribution.

College A	College B	College C
2	1	2

31. a. Convert the poverty threshold income into a *z*-score relative to the mean in each county. Then, look up the expected proportion in column C for that cutoff *z*-score.

County	Mean	s	Threshold z	Percentage
A	$34,495.00	$12,632.00	−1.38	8.38
B	$45,667.00	$9,812.00	−2.92	0.18
C	$58,634.00	$11,723.00	−3.55	0.02
D	$29,187.00	$4,521.00	−2.70	0.34
E	$69,455.00	$23,931.00	−2.19	1.43

b. Find the *z*-score that cuts off the upper 1% of the distribution and convert that back to a family income in each county. A *z* of +2.32 cuts off the top 1% of the distribution.

County	Mean	s	99th Percentile Income
A	$34,495.00	$12,632.00	$x = 2.32(12,632) + 34,495$ = $63,801
B	$45,667.00	$9,812.00	$x = 2.32(9,812) + 45,667$ = $68,431
C	$58,634.00	$11,723.00	$x = 2.32(11,723) + 58,634$ = $85,831
D	$29,187.00	$4,521.00	$x = 2.32(4,521) + 29,187$ = $39,676
E	$69,455.00	$23,931.00	$x = 2.32(23,931) + 69,435$ = $138,895

CHAPTER 7

1. Primarily because the mean is the "best" measure of center. Unlike the median and the mode, the mean is the measure of center that responds to, or reflects, all of the observations in a set.

3. a. $96/8 = 12$ cards in each suit.
b. $12/96 + 12/95 = .125 \times .126. = .016$ (note that the total number of cards in the deck changes after the first card is drawn)
c. $12/196 + 12/96 = .125 + .125. = .25$

5. a. 1: $p = 3{,}407/20{,}000 = .17$ 4: $p = 2{,}916/20{,}000 = .15$
2: $p = 3{,}631/20{,}000 = .18$ 5: $p = 3{,}448/20{,}000 = .17$
3: $p = 3{,}716/20{,}000 = .16$ 6: $p = 3{,}422/20{,}000 = .17$
b. It's most likely the die was fair because the probabilities of each roll are approximately equal. If the die were biased, I would expect unequal probabilities.

7. Table 7.4 Age, smoking and heart disease in a sample of male physicians

	Coronary Deaths		
Age	Nonsmokers	Smokers	Totals
35–44	2	32	34
45–54	12	104	126
55–64	28	206	234
65–74	28	186	214
75–84	31	102	133
Total	101	630	731

a. $101 + 630 = 731$ doctors were surveyed
b. 630 doctors reported that they smoked; 101 doctors reported that they did not smoke
c. probability that a person 35–44 will die of coronary disease $= 2/34 = .06$
probability that a smoker 35–44 will die of coronary disease $= 32/34 = .94$
d. probability that a smoker 75–84 will die of coronary disease $= 102/133 = .77$
e. probability of dying of coronary disease is highest for the youngest group of smokers (35–44 years of age)

9. a. $z = \dfrac{6.39 - 2.67}{1.70} = \dfrac{3.72}{1.70} = 2.19$
b. probability of a PSQI score equal to or higher than the sample mean $= .0143$, or 1.43%
c. $z = \dfrac{6.92 - 8.00}{1.50} = \dfrac{-1.08}{1.50} = -0.72$
probability of a sleep duration equal to or higher than the sample mean $= .7642$, or 76.42%

d.

Sample mean PSQI

$\mu_{(PSQI)} = 2.67$
$\sigma = 1.70$

Sample mean Sleep Duration

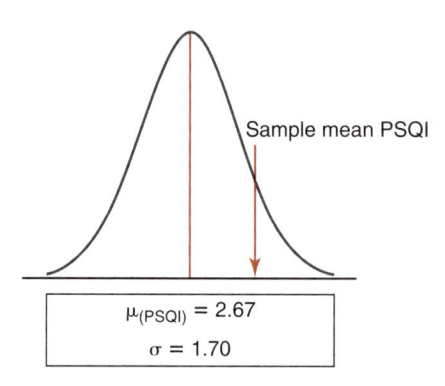

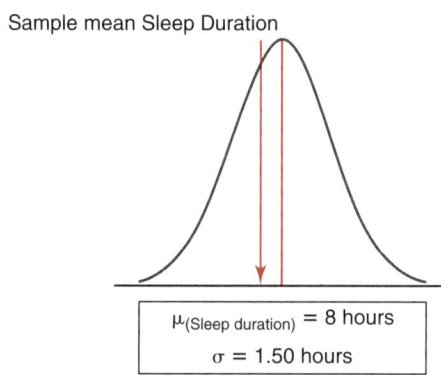

$\mu_{(Sleep\ duration)} = 8$ hours
$\sigma = 1.50$ hours

e. High PSQI scores (closer to the maximum 21 possible) indicate poor sleep quality. It looks as though the quality of sleep the medical students were getting is poorer than what is typically seen in the population, despite the fact that the number of hours of sleep the students were getting was probably not significantly different than the 8 hours seen in the population.

11. a. probability that a student selected at random will be a freshman $= 500/1{,}400 = .36$
b. probability that a student selected at random will be a senior $= 200/1400 = .14$
c. probability that a student selected at random will be a sophomore or a junior
$= 400/1{,}400 + 300/1{,}400 = .29 + .21 = .50$
alternative solution: $p = 400/1{,}400 + 300/1{,}400 = 700/1{,}400 = .50$
d. probability that a student selected at random will not be a senior $= 1 - 200/1{,}400 = 1 - .14 = .86$

13. a. probability of drawing the number $17 = 1/100 = .01$
b. probability of drawing the number $92 = 1/100 = .01$
c. probability of drawing either a 2 or a $4 = 1/100 + 1/100 = 2/100 = .02$
d. probability of drawing an even number and then a number from 3 to 19: Half the numbers are even, so the probability of drawing an even number is $50/100 = .50$. There are 17 numbers between 3 and 19

740   ANSWERS TO ODD NUMBERED END-OF-CHAPTER PRACTICE PROBLEMS

(including the numbers 3 and 19), so the probability of drawing a number between 3 and 19 = $17/100$ = .17. The AND rule says we should multiply these two probabilities together to find the joint probability: $.50 \times .17 = .085$.

e. probability of drawing a number from 96 to 100 or a number from 70 to 97: There are 5 numbers between 96 and 100 (including 96 and 100) and 28 numbers between 70 and 97 (including the "edges" of the interval). The probability of drawing a number from 96 to 100 is $5/100$ = .05. The probability of drawing a number from 70 to 97 is $28/100$ = .28. The OR rule says we should add the individual probabilities together: $.05 + .28 = .33$.

15. a. probability of rolling a 1 on each die = $1/6 \times 1/6 = .17 \times .17 = .03$.

b. probability that the sum of the two dice will be a number larger than 8 if the number on the first die is 6.: First, figure out what the possible sums of two dice would be if the first number is 6:

Combination	Sum
6 and 1	7
6 and 2	8
6 and 3	9
6 and 4	10
6 and 5	11
6 and 6	12

This is an example of the OR rule. We're asking for the probability of rolling a 3 or a 4 or a 5 or a 6 on the second die. There are four possible combinations of rolls where the sum will be greater than 8 given that the first number is 6. So, the probability of getting this roll is $1/6 + 1/6 + 1/6 + 1/6 = 4/6 = 0.67$.

c. The probability of rolling a 6 is $1/6$. The probability of rolling a 2 is also $1/6$, as is the probability of rolling a 3 ($1/6$). So, the probability of rolling a 6 on one die and either a 2 or a 3 on the other is

$$\frac{1}{6} \times \left(\frac{1}{6} + \frac{1}{6}\right) = \frac{1}{6} \times \frac{2}{6} = .17 \times .33 = .06$$

17. a. probability that a student's grade is greater than 80: $z = x - \bar{X}/s = 80 - 75/5 = 1.00$

Column C, z = 1.00 = 0.1587

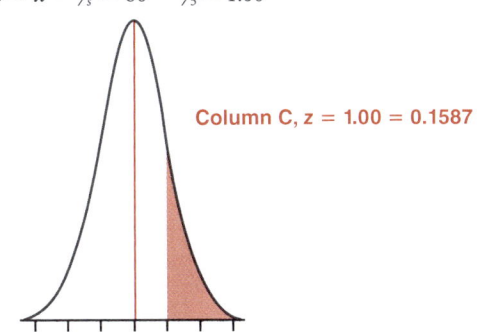

b. probability that a student's grade is less than 50: $z = x - \bar{X}/s = 50 - 75/5 \frac{50-75}{5} = -5.00$. The table does not go past a z of -4.00, so the probability will be less than 0.00003.

c. probability that a student's grade is between 50 and 80: 50% of the scores on the test will be between 50 and 75. We need to add the probability of the score being between 75 and 80. We already know that 80 is exactly one standard deviation above the mean ($z = 1.00$) so add the proportion in column B to .5000 to get the total.

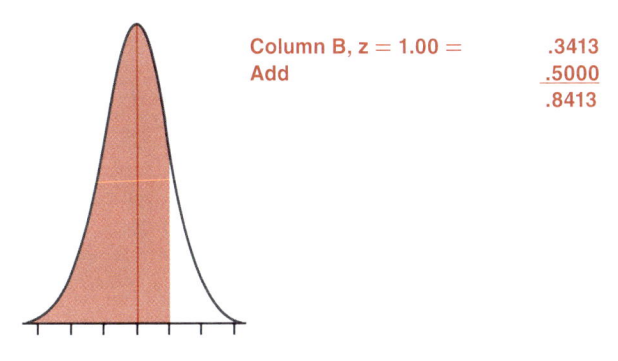

Column B, z = 1.00 =	.3413
Add	.5000
	.8413

d. 15.87% of the students will probably have a grade greater than 80: $200 \times 0.1587 = 31.74$, or 32, students will likely have a grade greater than 80.

19. a. Probability of randomly selecting a puppy that has brown fur, but no black fur: $p = .20$.

b. Probability of randomly selecting a puppy that is has brown or black fur, but not brown and black fur: $p = .30$.

c. Probability of randomly selecting a puppy that has both brown and black fur: p. = .70.

d. Probability of randomly selecting two puppies that are white and brown only on the first two selections: $p = .02$.

e. Proportion of Border Collie litters yielding 10 puppies or more: .0228.

CHAPTER 8

1. Step 1: Collect all possible samples of a given size.
Step 2: Calculate the mean of each sample.
Step 3: Construct a frequency distribution of the sample means. This distribution is the sampling distribution of the means.

3. A distribution with no variability would look like a straight vertical line. All of the values in the distribution would be identical.

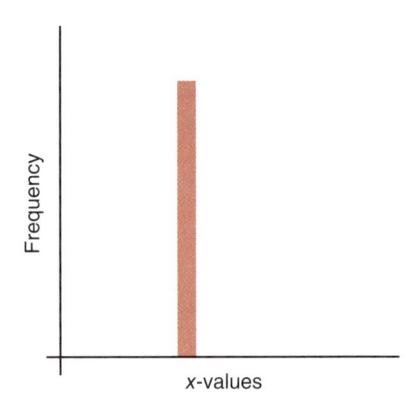

5. a. The null hypothesis: Heart rates of participants stressed in the lab will be the same as the heart rates of participants stressed in natural setting.
b. The alternative hypothesis: Heart rates of participants stressed in the lab will be higher than heart rates of participants stressed in a natural setting.
c. The null: $H_0: \mu_{(lab)} = \mu_{(natural\ setting)}$
The alternative: $H_1: \mu_{(lab)} > \mu_{(natural\ setting)}$

7. a. $\bar{X} = \dfrac{\sum x}{n} = \dfrac{33}{10} = 3.30$

$s = \sqrt{\dfrac{\sum x^2 - \dfrac{(\sum x)^2}{n}}{n}} = \sqrt{\dfrac{117 - \dfrac{(33)^2}{10}}{10}} = \sqrt{\dfrac{117 - 108.9}{10}}$

$= \sqrt{0.81} = 0.90$

$\mu_{\bar{X}} = \mu = 20$

$\sigma_{\bar{X}} = \dfrac{\sigma}{\sqrt{n}} = \dfrac{15}{\sqrt{10}} = \dfrac{15}{3.16} = 4.75$

b. $\bar{X} = \dfrac{\sum x}{n} = \dfrac{226}{16} = 14.13$

$s = \sqrt{\dfrac{\sum x^2 - \dfrac{(\sum x)^2}{n}}{n}} = \sqrt{\dfrac{6{,}638 - \dfrac{(226)^2}{16}}{16}}$

$= \sqrt{\dfrac{6{,}638 - 3{,}192.25}{16}} = \sqrt{215.36} = 14.67$

$\mu_{\bar{X}} = \mu = 20$

$\sigma_{\bar{X}} = \dfrac{\sigma}{\sqrt{n}} = \dfrac{15}{\sqrt{16}} = \dfrac{15}{4} = 3.75$

c. $\bar{X} = \dfrac{\sum x}{n} = \dfrac{1270.533}{50} = 25.41$

$s = \sqrt{\dfrac{\sum x^2 - \dfrac{(\sum x)^2}{n}}{n}} = \sqrt{\dfrac{33{,}706 - \dfrac{1{,}270.533^2}{50}}{50}}$

$= \sqrt{\dfrac{33{,}706 - 32{,}285.08}{50}} = \sqrt{28.42} = 5.33$

$\mu_{\bar{X}} = \mu = 20$

$\sigma_{\bar{X}} = \dfrac{\sigma}{\sqrt{n}} = \dfrac{15}{\sqrt{50}} = \dfrac{15}{7.07} = 2.12$

9. a. $\mu = \dfrac{1+2+3+4+5+6}{6} = 3.50$

b.
$\mu_{\bar{x}} = \dfrac{3{,}407 + (3{,}631 \times 2) + (3{,}176 \times 3) + (2{,}916 \times 4) + (3{,}448 \times 5) + (3{,}422 \times 6)}{20{,}000}$

$= \dfrac{69{,}636}{20{,}000} = 3.48$

c. They are very close in value but not identical. The sampling distribution of the means contains only 20,000 rolls of the die—less than there are in the population—so it is incomplete and the two means are not exactly the same.

11. a. According to the CLT, as n increases, $\sigma_{\bar{X}}$ also increases: **False**.
b. According to the CLT, the standard deviation (s) of a single sample always equals the standard deviation of the population (σ): **False**.
c. According to the CLT, the mean of a single sample ($\bar{X}$) always equals the mean of the population (μ): **False**.
d. Statisticians typically use μ to estimate $\bar{X}$: **False**.
e. As σ increases, $\sigma_{\bar{X}}$ also increases: **True**.
f. According to the CLT, the sampling distribution of the means will be normal in shape: **True**.
g. The CLT says that $\mu_{\bar{X}} = \mu$: **True**.
h. The CLT says that there will be less variability in the SDM than in the population from which all the samples were drawn: **True**.

13. c. $\mu_{\bar{X}} > \mu$
d. $\mu_{\bar{X}} < \mu$

15. a. mean sample mean in the SDM = 300
b. standard error of the SDM = 2.24
$(\sigma_{\bar{x}} = \dfrac{\sigma}{\sqrt{n}} = \dfrac{10}{\sqrt{20}} = 2.24)$
c. percentage of sample means in the SDM within one standard error above and below $\mu_{\bar{X}} = 68\%$
d. percentage of sample means in the SDM between $\mu_{\bar{X}}$ and one standard error above the mean = 34%
e. percentage of sample means in the SDM at least two standard errors below $\mu_{\bar{X}} = 14\%$

17. a. $\sigma_{\bar{x}} \times \sqrt{n} = 10(5) = 50$
b. $\sigma_{\bar{x}} \times \sqrt{n} = 0.5(5) = 2.50$
c. $\sigma_{\bar{x}} \times \sqrt{n} = 2(5) = 10$

19. a. smallest sum of two dice possible = 2
b. largest sum of two dice possible = 12
c. theoretical probability of rolling two dice and getting a 7: There are six combinations that will result in a sum of 7, and 36 possible combinations of the two dice. So, theoretically, the probability of rolling two dice and getting a 7 is 6/36, or .17.

	Die 1					
	1	2	3	4	5	6
Die 2 1	2	3	4	5	6	7
2	3	4	5	6	7	8
3	4	5	6	7	8	9
4	5	6	7	8	9	10
5	6	7	8	9	10	11
6	7	8	9	10	11	12

CHAPTER 9

1. a. The SDM is the distribution used in this formula. We know this because the resulting z will tell us the position of a given sample mean from the mean of all the sample means in standard error units.
b. The top of the equation tells us the deviation of the sample mean from the mean of all of the sample means.
c. The bottom of the equation tells us the standard error of the sampling distribution of the means.
d. Since $\sigma_{\bar{x}} = \sigma/\sqrt{n}$, increasing the variability in the population (σ) would make $\sigma_{\bar{x}}$ larger.
e. Since $\sigma_{\bar{x}} = \sigma/\sqrt{n}$, increasing n would make $\sigma_{\bar{x}}$ smaller.

3. a. i) A Type I error would be rejecting the null when in fact it should be accepted. The null says that the drug has no effect on tumor size. A Type I error would be saying that the drug reduces tumor size when in fact it does not.
ii) A Type II error would be failing to reject the null when in fact it should be rejected. A Type II error would be saying that the drug does not work when in fact it does reduce tumor size.
iii) A Type I error would likely result in more lawsuits as the drug would not significantly affect tumor size but would create some serious detrimental side effects.
iv) Both types of errors carry serious risk for the patients. A Type I error would mean that the patient taking the drug would suffer the negative side effects without experiencing the reduction in tumor size promised by the drug manufacturers. A Type II error would mean that the patient would miss out on a drug that might potentially be effective in reducing tumor size.
v) To reduce the risk of a Type I error, I would reduce the chance of rejecting the null hypothesis by setting the alpha level at less than the outer 5% ($\alpha = .01$ or .001). To reduce the risk of a Type II error, I would increase sample size as much as possible.

b. i) A Type I error would be rejecting the null when in fact it should not be rejected. The consequence is concluding that the drug reduces depression when in fact it does not.
ii) A Type II error would be Accepting the null when in fact it should be rejected. The consequence is concluding that the drug does not reduce depression when in fact it does.

c. A Type I error would be concluding that the drug reduces high blood pressure when in fact it does not. A Type II error would be concluding that the drug does not reduce high blood pressure when in fact it does.

d. i) A Type I error would be concluding that the null is false (gender does affect time needed to react to an emergency) when in fact it should not be rejected.
ii) A Type II error would be concluding that the null cannot be rejected (gender does not affect time needed to react to an emergency) when in fact it should be.

5. a $\mu_{\bar{x}} = \mu = 100$

$$\sigma_{\bar{x}} = \frac{\sigma}{\sqrt{n}} = \frac{16}{\sqrt{15}} = 4.13$$

$$\bar{X} = \frac{\sum x}{n} = \frac{1{,}457}{15} = 97.13$$

$$\sigma = \sqrt{\frac{\sum x^2 - \frac{(\sum x)^2}{N}}{N}} = \sqrt{\frac{143{,}553 - \frac{(1{,}457)^2}{15}}{15}}$$

$$= \sqrt{\frac{143{,}553 - 141{,}523.27}{15}} = \sqrt{\frac{2{,}029.73}{15}}$$

$$= \sqrt{135.32} = 11.63$$

b. There is a slight decrease in average IQ score. The decrease is less than one standard error below average.
c. The research question is whether or not hypocholesterolemia affects performance on the Stanford–Binet Intelligence test. The null hypothesis is that hypocholesterolemia has no effect on IQ scores ($\mu_{\bar{x}} = 100$).

The alternative hypothesis is that hypocholesterolemia affects IQ scores $(\mu_{\bar{x}} \neq 100)$. A two-tailed test is appropriate.

d. $z = \dfrac{\bar{X} - \mu_{\bar{x}}}{\sigma_{\bar{x}}} = \dfrac{97.13 - 100}{\frac{16}{\sqrt{15}}} = \dfrac{-2.87}{4.13} = -0.69$

e.

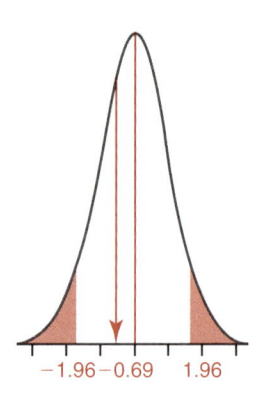

f. IQ scores were lower than average in patients with hypocholesterolemia, but the difference was not statistically significant. We must retain the null hypothesis.

7. a. $z = \dfrac{105 - 100}{\frac{30}{\sqrt{50}}} = \dfrac{5}{4.24} = 1.18$; $z_{(critical)} = \pm 1.96$

   The difference between sample mean and population mean is not statistically significant.

   b. $z = \dfrac{45 - 50}{\frac{13}{\sqrt{20}}} = \dfrac{5}{2.91} = 1.72$; $z_{(critical)} = 2.33$

   The difference between sample mean and population mean is not statistically significant.

   c. $z = \dfrac{200 - 207}{\frac{25}{\sqrt{9}}} = \dfrac{7}{8.33} = 0.84$; $z_{(critical)} = 3.10$.

   The difference between sample mean and population mean is not statistically significant.

9. a. mean GPA for the RTC students: $\bar{X} = \dfrac{31.10}{10} = 3.11$

   b. The null hypothesis says that the average GPA of RTC students is not different from the average GPA of non-RTC students $(H_0 : \mu_{\bar{X}} = 2.90)$. The alternative hypothesis is that the average GPA of RTC students is higher than the average GPA of non-RTC students $(H_1 : \mu_{\bar{X}} > 2.90)$. This is a one-tailed test.

   c. $z = \dfrac{\bar{X} - \mu_{\bar{X}}}{\sigma_{\bar{X}}} = \dfrac{3.10 - 2.90}{\frac{0.89}{\sqrt{10}}} = \dfrac{0.20}{0.28} = 0.71$

   $z_{(critical)} = 1.65$

Conclusion: Although the average GPA of RTC students is higher than the average GPA of non-RTC students, the difference is not statistically significant. We cannot reject the null hypothesis.

11. a. The researcher is claiming that students in her district are brighter than average, so their IQ scores should be higher than average. We should do a one-tailed test.

    b. A Type I error would mean that the researcher concludes that the IQ scores of students in her district are higher than average when in fact they are not.

    c. $z = \dfrac{\bar{X} - \mu_{\bar{X}}}{\sigma_{\bar{X}}} = \dfrac{111.7 - 100}{\frac{16}{\sqrt{10}}} = \dfrac{11.7}{5.06} = 2.31$

    If $\alpha = .05$, then $z_{(critical)} = 1.65$. Our observed z is in the rejection region. We should reject the null hypothesis and accept the alternative hypothesis: Average IQ scores for students in this district are higher than the average score in the population.

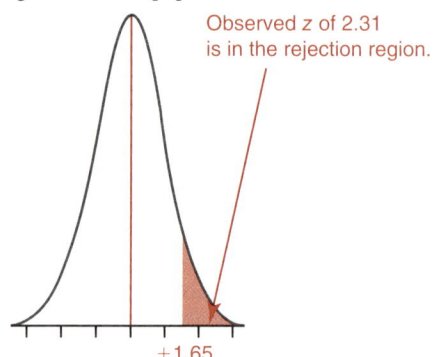

13. a. The independent variable is the color of the emergency indicator light (there are two levels, red and yellow). The dependent variable (the variable that is being measured) is reaction time.

    b. This is a two-tailed question because we're looking for any difference in reaction time (increase or decrease) when the yellow light is used. The null hypothesis is that reaction time is the same for yellow and red light ($H_0 : \mu_{\bar{x}} = 205$ milliseconds). The alternative hypothesis is that the reaction time to yellow light is not the same as the reaction time to red light ($H_1 : \mu_{\bar{x}} \neq 205$ milliseconds). Our critical z-value will be ± 1.96 if we set our α level to .05.

    c. $z = \dfrac{\bar{X} - \mu_{\bar{x}}}{\sigma_{\bar{X}}} = \dfrac{195 - 205}{\frac{20}{\sqrt{25}}} = \dfrac{-10}{4.00} = -2.50$

    Our observed z is in the rejection region (farther out in the tail of the SDM than the critical z). We can reject the null hypothesis and conclude that reaction time is significantly faster when a yellow light is used compared to reaction time when a red light is used.

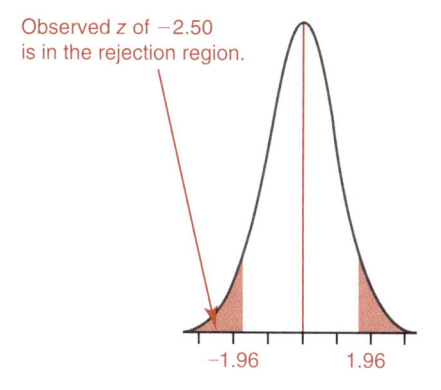

Observed z of −2.50 is in the rejection region.

−1.96   1.96

15. a. The independent variable is smoking (two levels: smoking and nonsmoking). The dependent variable is birth weight.
b. This is a one-tailed question. The researcher is looking for an average birth weight of babies born to mothers who smoke that is lower than the average birth weight in the population. The critical z-value is −1.65.
c. $z = \dfrac{\overline{X} - \mu_{\overline{X}}}{\sigma_{\overline{X}}} = \dfrac{5.24 - 6.39}{1.43/\sqrt{14}} = \dfrac{-1.15}{0.38} = -3.03$

Our observed z (−3.03) is in the rejection region (farther out in the tail of the SDM than the critical z (−1.65). We can reject the null hypothesis and conclude that birth weight in babies born to mothers who smoke is significantly lower than in nonsmoking mothers.

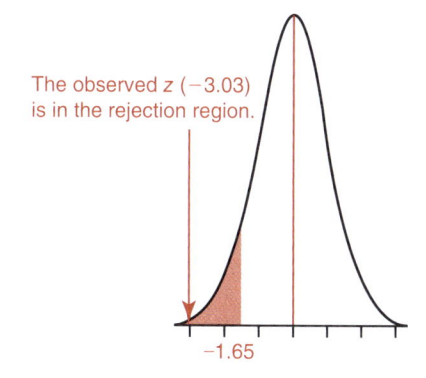

The observed z (−3.03) is in the rejection region.

−1.65

17. a. The null hypothesis is that the new preservative will keep the hot dogs for an average of 35 days (H_0: $\mu_{\overline{X}} = 35$ days). The alternative is that the new preservative will keep the hot dogs for either more or less than 35 days (H_1: $\mu_{\overline{X}} \neq 35$ days).

b. The independent variable is the preservative (two levels: new preservative and old preservative). The dependent variable is the number of days the hot dogs will stay fresh with refrigeration.

c. $z = \dfrac{\overline{X} - \mu_{\overline{X}}}{\sigma_{\overline{X}}} = \dfrac{32 - 35}{\dfrac{4.5}{\sqrt{50}}} = \dfrac{-3.00}{0.64} = -4.05$

The observed z-value is within the rejection region, meaning we can reject the null hypothesis. The new preservative does not support the manufacturer's 35-day freshness guarantee. The number of days the hot dogs are preserved is significantly less than average.

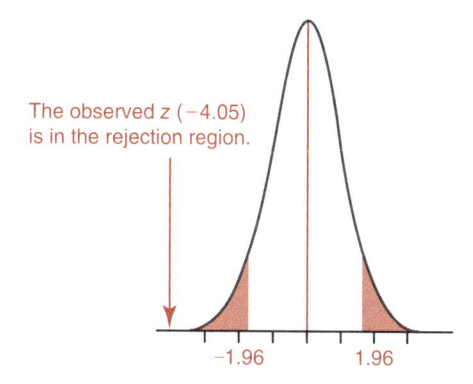

The observed z (−4.05) is in the rejection region.

−1.96   1.96

19. a. The null hypothesis is that the amount of sugar consumed by Americans is 156 pounds per year (H_0: $\mu_{\overline{X}} = 156$ pounds). The alternative hypothesis is that the amount of sugar consumed by Americans has changed, and that it is not 156 pounds per year (H_1: $\mu_{\overline{X}} \neq 156$ pounds).
b. This is a two-tailed question so the critical z-value is ± 1.96, $\alpha = .05$.
c. $z = \dfrac{\overline{X} - \mu_{\overline{X}}}{\sigma_{\overline{X}}} = \dfrac{148 - 156}{9/\sqrt{25}} = \dfrac{-8.00}{1.80} = -5.13$
d. The observed z-value exceeds the critical z-value so we can reject the null hypothesis: The amount of sugar consumed has decreased significantly.

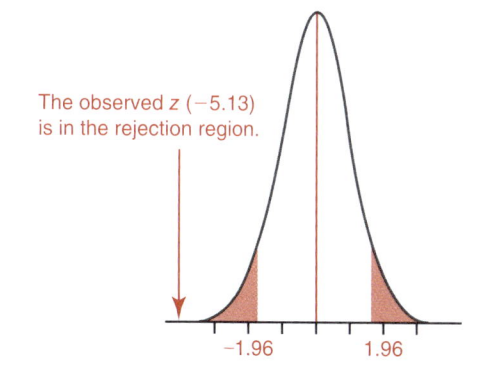

The observed z (−5.13) is in the rejection region.

−1.96   1.96

ANSWERS TO ODD NUMBERED END-OF-CHAPTER PRACTICE PROBLEMS

21. a. $H_0: \mu_{\bar{X}} = 6.4$
$H_0: \mu_{\bar{X}} \neq 6.4$
b. $1.33 < 1.96$
We conclude that the difference between the mean subjective well-being scores of Americans and Americans who own dogs is not statistically significant at the .05 level. We cannot reject the null hypothesis.

CHAPTER 10

1. $z = \dfrac{\bar{X} - \mu_{\bar{X}}}{\sigma_{\bar{X}}}$ $t = \dfrac{\bar{X} - \mu_{\bar{X}}}{s_{\bar{X}}}$ $t = \dfrac{\bar{X}_C - \bar{X}_E}{s_{\bar{X}_C} - s_{\bar{X}_E}}$

$t = \dfrac{\bar{X}_1 - \bar{X}_2}{s_{\bar{D}}}$

The formula for each test features the difference between two means (either a sample mean and a population mean or two sample means). Each test divides the difference between the two means by a measure of variability (standard error, estimated standard error, or standard error of the difference). There's a pattern here—namely, the difference between two means or two measures on the top and a measure of error on the bottom of the ratio. The error term in the denominator of each ratio is different for each test.

3. As n increases, the critical value of t decreases (moves closer to the mean of the sampling distribution of the means, a.k.a. the population mean). The critical values are decreasing because the sampling distribution of the means is getting "tighter" and narrower as n increases. Our estimate of the standard error of the means is dependent on the size of the sample (n). As n increases, our estimate of $\sigma_{\bar{X}}$ gets better and better. It also gets smaller and smaller. Remember that $s_{\bar{X}} = \hat{s}/\sqrt{n}$.

5. a. We should use a t-test for independent samples. We have two group means to compare, and we do not know either the population mean or the population standard deviation. So, we will have to estimate both of them. As far as it is possible to determine, the two groups were selected at random, so they are independent groups.
b. A Type I error in this case would result from rejecting the null hypothesis and concluding that education level significantly affects the number of words recalled from the middle of the list when in fact it does not. Since this is not a particularly serious Type I error, we can safely use an alpha of .05.
c. Both groups remembered the same number of words from the list, so we can conclude that our independent variable (number of years of education) did not have a significant effect on our dependent variable (number of words recalled from the middle of the list).
d. $H_0: \mu_{\bar{X}_1} = \mu_{\bar{X}_2}$ (mean number of words recalled is the same for both groups)

$H_1: \mu_{\bar{X}_1} \neq \mu_{\bar{X}_2}$ (mean number of words recalled significantly different for the two groups)

$\bar{X}_1 = \dfrac{138}{10} = 13.8$

$\bar{X}_2 = \dfrac{122}{10} = 12.2$

$s_{(\bar{X}_1 - \bar{X}_2)} = \sqrt{\dfrac{\sum x_1^2 - \dfrac{(\sum x_1)^2}{n_1} + \sum x_2^2 - \dfrac{(\sum x_2)^2}{n_2}}{n_1(n_2 - 1)}}$

$= \sqrt{\dfrac{1{,}928 - \dfrac{(138)^2}{10} + 1{,}826 - \dfrac{(122)^2}{10}}{10(10-1)}}$

$= \sqrt{\dfrac{1{,}928.0 - 1{,}904.4 + 1{,}826.0 - 1{,}488.4}{90}}$

$= \sqrt{\dfrac{23.6 + 337.6}{90}} = \sqrt{\dfrac{361.2}{90}}$

$= \sqrt{4.013} = 2.003$

$t = \dfrac{\bar{X}_1 - \bar{X}_2}{s_{(\bar{X}_1 - \bar{X}_2)}} = \dfrac{13.8 - 12.2}{2.003} = \dfrac{1.60}{2.003} = 0.798$, or 0.80

Observed $t = 0.798$, $df = 18$, critical $t = \pm 2.101$
Conclusion: Cannot reject the null hypothesis. Number of years of education (the independent variable) does not affect the number of words recalled from the middle of the list (the dependent variable). Both groups remembered the same number of words from the list.

7. a. Because the participants have been matched for the severity of their insomnia, the two samples are dependent. There is no specific direction specified for the test, so use a two-tailed, dependent-samples t-test.
b. The consequences of a Type I error (concluding that the sleeping pill has an effect on dreaming sleep when in fact it does not) has some serious consequences, so use an alpha level of .01.
c. $H_0: \mu_{\bar{X}_1} - \mu_{\bar{X}_2} = 0$ (mean time spent in REM is the same for both groups)
$H_1: \mu_{\bar{X}_1} - \mu_{\bar{X}_2} \neq 0$ (mean time spent in REM is significantly different for the two groups)

$t = \dfrac{\bar{D}}{s_{\bar{D}}} = \dfrac{-5.88}{4.47} = -1.32$

$s_{\bar{D}} = \dfrac{\sqrt{\dfrac{\sum D^2 - \dfrac{(D)^2}{n}}{n-1}}}{\sqrt{n}} = \dfrac{\sqrt{\dfrac{1{,}395 - \dfrac{(-47)^2}{8}}{7}}}{\sqrt{8}}$

$= \dfrac{\sqrt{\dfrac{1{,}395 - 276.13}{7}}}{2.83} = \dfrac{12.64}{2.83} = 4.47$

The *t*-critical value with *df* = 7, α = .01, and a two-tailed test is ± 3.499. The observed *t*-value of −1.32 does not exceed the critical *t*-value, so we must retain the null hypothesis. There is no significant difference in the time spent sleeping between these two matched groups.

  d. A dependent-samples *t*-test was used to compare the mean time spent in REM sleep for two groups matched for the severity of their insomnia. There was no statistically significant difference in time spent in REM sleep between the two groups: $t_{(7)} = -1.32, p > .01$.

Number of minutes of REM sleep in patients matched for insomnia.

Placebo	Sleeping Pill	D	D²
110	99	11	121
89	84	5	25
86	94	−8	64
87	87	0	0
85	87	−2	4
87	106	−19	361
74	102	−28	784
85	91	−6	36
		ΣD = −47.00	ΣD² = 1,395

$$\bar{D} = \frac{-47}{8} = -5.88$$

On average, REM sleep duration is 5.88 minutes higher in Group 2 (participants who took the new sleeping pill) than in Group 1 (participants who took the placebo).

9. a. Use a single-sample *t*-test to compare your sample mean (52) with the population mean (50). The null and alternative hypotheses have not changed.

$$\hat{s} = \sqrt{\frac{\sum x^2 - \frac{(\sum x)^2}{n}}{n-1}} = \sqrt{\frac{56,208 - \frac{(1,040)^2}{20}}{19}}$$

$$= \sqrt{\frac{56,208 - 54,080}{19}} = \sqrt{112} = 10.58$$

$$t = \frac{\bar{X} - \mu_{\bar{x}}}{\frac{\hat{s}}{\sqrt{n}}} = \frac{52 - 50}{\frac{10.58}{\sqrt{20}}} = \frac{2}{2.37} = 0.84$$

  b. Once again, we cannot reject the null hypothesis.
  c. There was no significant change in the anxiety scores of the participants in the study, $t_{(19)} = 0.84, p > .01$.

11. We're comparing a control group with an experimental group with an independent-samples *t*-test. We have different *n*'s for both samples, so we should use the formula for different *n*'s to calculate our error term for the *t*-test. Since we have no information about a specified direction for the test, we'll use a two-tailed test and an alpha level of .05. Our critical *t*-value, with 11 *df*, will be ±2.201.

$H_0: \mu_{\bar{X}_1} = \mu_{\bar{X}_2}$ (means are the same for both groups)
$H_0: \mu_{\bar{X}_1} \neq \mu_{\bar{X}_2}$ (means are not the same for both groups)

$$t = \frac{\bar{X}_1 - \bar{X}_2}{\sqrt{\frac{\sum x_1^2 - \frac{(\sum x_1)^2}{n_1} + \sum x_2^2 - \frac{(\sum x_2)^2}{n_2}}{(n_1 + n_2) - 2}} \left(\frac{1}{n_1} + \frac{1}{n_2} \right)}$$

$$= \frac{39.71 - 57.33}{\sqrt{\frac{11,478 - \frac{278^2}{7} + 20,520 - \frac{344^2}{6}}{7 + 6 - 2}} (0.31)}$$

$$= \frac{-17.62}{\sqrt{\frac{11,478 - 11,040.57 + 20,520 - 19,722.67}{11}} (0.31)}$$

$$= \frac{-17.62}{\sqrt{112.25(0.31)}}$$

$$= \frac{-17.62}{\sqrt{34.80}} = \frac{-17.62}{5.90} = -2.99$$

Our critical *t*-value is ±2.201, and our observed *t*-value is −2.99, which is in the rejection region (it exceeds the critical value). We can reject the null hypothesis. The mean for the experimental group is significantly smaller than the mean for the control group: $t_{(11)} = -2.99, p < .05$.

13.

$S_{\bar{X}_C - \bar{X}_E}$	estimated standard error of the difference
df	degrees of freedom (generally = *n* − 1)
α	alpha level
σ	standard deviation of a population
s	standard deviation of a sample
μ	population mean
$S_{\bar{X}}$	estimated standard error of the mean
$\mu_{\bar{X}}$	standard error of the mean
n_c	number of observations in the control group
D	difference score

ANSWERS TO ODD NUMBERED END-OF-CHAPTER PRACTICE PROBLEMS

15. Critical *t*-values and results

Observed *t*-Value	Group 1 (*n* =)	Group 2 (*n* =)	Type of Test	Alpha Level	Critical *t*-Value	Interpretation (Reject or Retain H_0)
1.533	9	9	Two-tailed	.05	±2.120	Retain H_0
−2.069	10	6	One-tailed	.01	−2.624	Retain H_0
3.820	5	5	Two-tailed	.01	±3.355	Reject H_0
3.000	9	9	One-tailed	.05	2.120	Reject H_0
−5.990	29	29	Two-tailed	.05	±2.021	Reject H_0
1.960	61	61	One-tailed	.05	1.658	Reject H_0

17. Results in APA Style

Observed *t*-Value	Group 1 (*n* =)	Group 2 (*n* =)	Type of Test	Alpha Level	APA Style Result
1.533	9	9	Two-tailed	.05	$t_{(16)} = 1.533$, $p > .05$
−2.069	10	6	One-tailed	.01	$t_{(14)} = -2.069$, $p > .01$
3.820	5	5	Two-tailed	.01	$t_{(8)} = 3.82$, $p < .01$
3.000	9	9	One-tailed	.05	$t_{(16)} = 3.00$, $p < .05$
−5.990	29	29	Two-tailed	.05	$t_{(56)} = -5.99$, $p < .05$
1.960	61	61	One-tailed	.05	$t_{(120)} = 1.96$, $p < .05$

19. Use a dependent-samples *t*-test (each pair of cats are littermates and natural pairs) and a one-tailed *t*-test (looking for an increase in alcohol consumption) with $df = 6$. Set $\alpha = .05$. Critical *t*-value = −1.943 (we are expecting a negative observed *t* if the null is false).

$H_0: \mu_{\bar{X}_1} - \mu_{\bar{X}_2} = 0$ (mean alcohol consumption for neurotic cats the same as for controls)

$H_1: \mu_{\bar{X}_1} - \mu_{\bar{X}_2} \neq 0$ (mean alcohol consumption is higher for the neurotic cats than for controls)

Amount of alcohol-spiked milk consumed (in ml)

Littermates	No Experimental Neurosis	Experimental Neurosis	D	D²
1	63	88	−25.00	625
2	59	90	−31.00	961
3	52	74	−22.00	484
4	51	78	−27.00	729
5	46	78	−32.00	1024
6	44	61	−17.00	289
7	38	54	−16.00	256
	$\Sigma x = 353$	$\Sigma x = 523$	$\Sigma D = -170$	$\Sigma D^2 = 4{,}368$

$$s_{\bar{D}} = \frac{\sqrt{\dfrac{\Sigma D^2 - \dfrac{(\Sigma D)^2}{n}}{n-1}}}{\sqrt{n}} = \frac{\sqrt{\dfrac{4{,}368 - \dfrac{(-170)^2}{7}}{6}}}{\sqrt{7}}$$

$$= \frac{\sqrt{\dfrac{4{,}368 - 4{,}128.57}{6}}}{2.65} = \frac{6.32}{2.65} = 2.38$$

$$t = \frac{\bar{D}}{s_{\bar{D}}} = \frac{\dfrac{-170}{7}}{2.38} = \frac{-24.29}{2.38} = -10.20$$

Conclusion: Reject the null hypothesis (the observed *t*-value of −10.20 exceeds the critical *t*-value of −1.943). Experimentally induced neurosis significantly increases the consumption of alcohol in cats: $t_{(6)} = -10.20$, $p < .05$.

21. Use a dependent-samples *t*-test (the participants have been matched for body weight) to compare the mean number of errors made on the driving test for the two groups. The test is one-tailed because we're looking for a deterioration of driving skills. We expect that the number of errors will be higher for the group that consumed alcohol than for the control group so the critical *t*-value, with 9 *df* and an alpha level of .05, will be negative: −1.833.

$H_0: \mu_{\bar{X}_C} - \mu_{\bar{X}_E} = 0$ (mean number of errors made is the same for both groups)

$H_1: \mu_{\bar{X}_C} - \mu_{\bar{X}_E} < 0$ (mean number of errors made is higher for the experimental group than for the control group)

$$s_{\bar{D}} = \frac{\sqrt{\dfrac{\Sigma D^2 - \dfrac{(\Sigma D)^2}{n}}{n-1}}}{\sqrt{n}} = \frac{\sqrt{\dfrac{761 - \dfrac{(-81)^2}{10}}{9}}}{\sqrt{10}}$$

$$= \frac{\sqrt{\dfrac{761 - 656.1}{9}}}{3.16} = \frac{3.41}{3.16} = 1.08$$

$$t = \frac{\bar{D}}{s_{\bar{D}}} = \frac{\dfrac{-81}{10}}{1.08} = \frac{-8.1}{1.08} = -7.50$$

Conclusion: The observed *t*-value (−7.50) exceeds the critical *t*-value (−1.833). The observed *t*-value is in the rejection region, so we can reject the null hypothesis. Drinking alcohol significantly increases the number of errors made on the driving simulator, indicating that driving performance has deteriorated: $t_{(9)} = -7.50, p < .05$.

Number of errors made on a driving simulator

Control Group	Experimental Group	D	D²
19	27	−8.0	64
16	28	−12.0	144
16	28	−12.0	144
19	25	−6.0	36
23	28	−5.0	25
21	29	−8.0	64
19	28	−9.0	81
18	23	−5.0	25
14	27	−13.0	169
17	20	−3.0	9
		$\Sigma D = -81$	$\Sigma D^2 = 761$

23. Use a dependent-samples *t*-test (this is a repeated-measures, before-and-after design) to compare the mean blood pressure (BP) before and after the drug. The test is one-tailed because we are looking for a decrease in BP. We expect that the average BP will be lower in the "after" group than it is in the "before" group, so the critical *t*-value, with 6 *df* and an alpha level of .01 (used because Type I errors might be dangerous when testing a new drug), will be positive (3.143).

$H_0: \mu_{\bar{X}_B} - \mu_{\bar{X}_A} = 0$ (mean BP is the same for both groups)

$H_1: \mu_{\bar{X}_B} - \mu_{\bar{X}_A} < 0$ (mean BP is higher for the before group than for the after group).

$$s_{\bar{D}} = \frac{\sqrt{\frac{\Sigma D^2 - \frac{(\Sigma D)^2}{n}}{n-1}}}{\sqrt{n}} = \frac{\sqrt{\frac{9,009 - \frac{(217)^2}{7}}{6}}}{\sqrt{7}}$$

$$= \frac{\sqrt{\frac{9,009 - 6,727}{6}}}{2.65} = \frac{19.50}{2.65} = 7.36$$

$$t = \frac{\bar{D}}{s_{\bar{D}}} = \frac{\frac{217}{7}}{7.36} = \frac{31}{7.36} = 4.21$$

Conclusion: The observed *t*-value (4.21) exceeds the critical *t*-value (3.143). The observed *t*-value is in the rejection region, so we can reject the null hypothesis. The new drug significantly reduces diastolic pressure: $t_{(6)} = 4.21, p < .01$.

Supine diastolic pressure (in mm Hg)

Patient	Before Drug	After Drug	D	D²
1	98	82	16.00	256.00
2	96	72	24.00	576.00
3	140	90	50.00	2,500.00
4	120	108	12.00	144.00
5	130	72	58.00	3,364.00
6	125	80	45.00	2,025.00
7	110	98	12.00	144.00
			$\Sigma D = 217$	$\Sigma D^2 = 9,009$

25. $t = -1.81; df = 18$
$-1.81 < -1.734$, or $|1.81| > |1.734|$

Conclusion: We reject the null hypothesis and conclude that employees' perceived fairness of workplace policies is higher when employees are allowed to exercise some control in the process of creating workplace policies.

CHAPTER 11

1. The general formula for *F* is the ratio of variability between the groups (caused by the effect of the IV and random chance) to variability within groups (caused by random chance):

$$F = \frac{\text{effect of the IV} + \text{effect of random chance}}{\text{effect of random chance}}$$

F can be 1.00 (which would indicate that the IV had no effect and that random variability between groups was the same as random variability within groups), or it can be less than 1.00 (indicating that there is less random variability between the groups than there is within the groups). However, it cannot be negative (less than zero). There will always be some random variability in any model, so *F* will always be positive. Because *F* is a measure of variability and it is impossible to have less than zero variability, a negative *F* value is impossible as well. Another way to think of this: *F* is a ratio of variance, which by definition must be positive as it is the squared result of the standard deviation.

3. You know that there was the same amount of variability being caused by random chance between the groups (MS_b) as there was within the groups (MS_w).

5. In an independent-samples *t*-test, there is one factor (one IV) with two levels. Typically, these levels are the control and experimental groups. The average of the DV is compared across levels.

7. **Data for question 7**

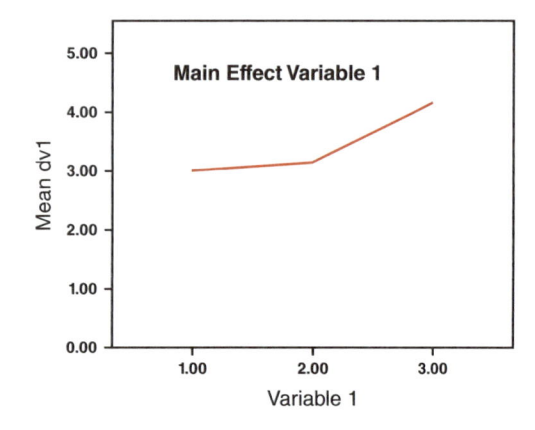

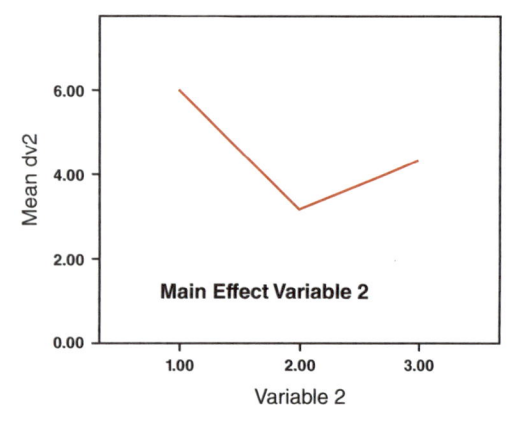

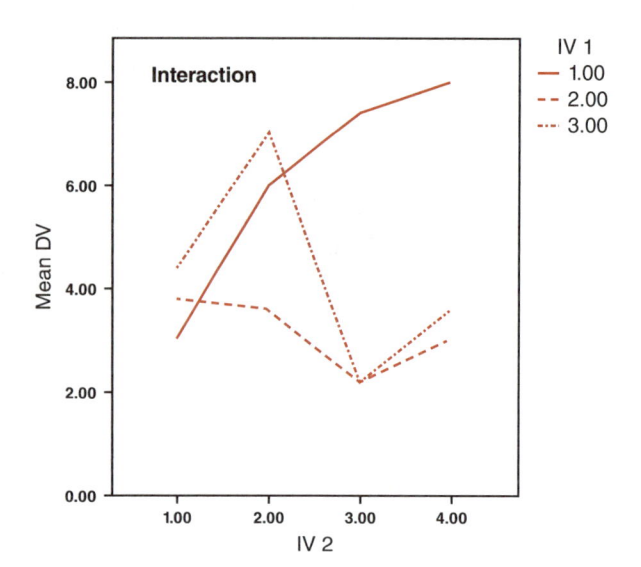

	1	2	3	4	Row Totals
1	$\Sigma = 15$ $n = 5$ $\bar{X} = 3.00$	$\Sigma = 30$ $n = 5$ $\bar{X} = 6.00$	$\Sigma = 37$ $n = 5$ $\bar{X} = 7.40$	$\Sigma = 40$ $n = 5$ $\bar{X} = 8.00$	$\Sigma = 122$ $N_{(R1)} = 20$ $\bar{X}_{row} = 3.00$
2	$\Sigma = 19$ $n = 5$ $\bar{X} = 3.80$	$\Sigma = 18$ $n = 5$ $\bar{X} = 3.60$	$\Sigma = 11$ $n = 5$ $\bar{X} = 2.20$	$\Sigma = 15$ $n = 5$ $\bar{X} = 3.00$	$\Sigma = 63$ $N_{(R2)} = 20$ $\bar{X}_{row} = 3.15$
3	$\Sigma = 22$ $n = 5$ $\bar{X} = 4.40$	$\Sigma = 35$ $n = 5$ $\bar{X} = 7.00$	$\Sigma = 11$ $n = 5$ $\bar{X} = 2.20$	$\Sigma = 15$ $n = 5$ $\bar{X} = 3.60$	$\Sigma = 83$ $N_{(R3)} = 20$ $\bar{X}_{row} = 4.15$
Column totals	$\Sigma = 56$ $N_{(C1)} = 15$ $\bar{X}_{col} = 3.73$	$\Sigma = 83$ $N_{(C2)} = 15$ $\bar{X}_{col} = 5.53$	$\Sigma = 59$ $N_{(C3)} = 15$ $\bar{X}_{col} = 3.93$	$\Sigma = 70$ $N_{(C4)} = 15$ $\bar{X}_{col} = 4.67$	$\Sigma = 268$ $N_{(tot)} = 60$ $\bar{X}_{tot} = 4.47$

9. a. Looks like there is a significant main effect for the variable represented in red but no significant main effect for the variable represented in gray. There might be a significant interaction between the two variables.
   b. There is no significant main effect for either the red or the gray variable. There is a significant main effect for the pink variable. There is likely a significant interaction.
   c. There are significant main effects for both the red and gray variables, but there is likely no significant interaction between them.
   d. There are significant main effects for all three variables (red, pink, and gray) but likely no significant interactions.

11.
$$SS_b = \left[\frac{(253)^2}{6} + \frac{(394)^2}{9} + \frac{(392)^2}{6}\right] - \frac{(253 + 394 + 392)^2}{21}$$
$$= [(10,668.17) + (17,248.44) + (25,610.67)] - 51,405.76$$
$$= [53,527.28] - 51,405.76$$
$$= 2,121.52$$

$$SS_w = \left[\left(11,139 - \frac{(253)^2}{6}\right) + \left(17,628 - \frac{(394)^2}{9}\right) + \left(25,752 - \frac{(392)^2}{6}\right)\right]$$
$$= [(11,139 - 10,668.17) + (17,628 - 17,248.44) + (25,752 - 25,610.67)]$$
$$= [470.83 + 379.56 + 141.33]$$
$$= 991.72$$

	SS	df	MS	f
Between	2,121.52	2	1,060.76	19.25
Within	991.72	18	55.10	
Total	3113.24	20		

750 **ANSWERS TO ODD NUMBERED END-OF-CHAPTER PRACTICE PROBLEMS**

F-critical with 2 and 18 df, α = .05 = 3.56
F-critical with 2 and 18 df, α = .01 = 6.01

The observed F-value of 19.25 is significant at the .01 level. We can reject the null hypothesis that the mean scores for all three groups are the same. Alcohol has an effect on manual dexterity.

Tukey post-hoc test

	1 n = 6 $\bar{X} = 42.17$	2 n = 9 $\bar{X} = 43.78$	3 n = 6 $\bar{X} = 65.33$
1		1.61 $s_{\bar{x}} = 2.78$	23.16 $s_{\bar{x}} = 3.06$
		HSD = $\frac{1.61}{2.78}$ = 0.58	HSD = $\frac{23.16}{3.06}$ = 7.57*
2			21.55 $s_{\bar{x}} = 2.78$
			HSD = $\frac{21.55}{2.78}$ = 7.75*
3			

$$\text{Tukey} = \frac{\bar{X}_1 - \bar{X}_2}{s_{\bar{x}}}$$

$$s_{\bar{x}} = \sqrt{\frac{MS_w}{2}\left(\frac{1}{n_1} + \frac{1}{n_2}\right)}$$

$$1 \text{ vs. } 2 = \sqrt{\frac{55.10}{2}\left(\frac{1}{6} + \frac{1}{9}\right)} = \sqrt{27.55(0.28)} = \sqrt{7.71} = 2.78$$

$$s_{\bar{x}} = \sqrt{\frac{MS_w}{2}\left(\frac{1}{n_1} + \frac{1}{n_2}\right)}$$

$$1 \text{ vs. } 3 = \sqrt{\frac{55.10}{2}\left(\frac{1}{6} + \frac{1}{6}\right)} = \sqrt{27.55(0.34)} = \sqrt{9.37} = 3.06$$

The critical Tukey statistic, with three treatment groups and 18 df within, is 3.609 at α = .05. The difference between the mean of Group 1 (3 ounces of alcohol consumed, M = 42.17) and Group 3 (0 ounces of alcohol consumed, M = 65.33) is statistically significant (HSD = 7.57) at the .05 level. The difference between the mean of Group 2 (1 ounce of alcohol consumed, M = 43.78) and the control group was also statistically significant at the .05 level (HSD = 7.75). There was no significant difference in manual dexterity between the group consuming 3 ounces of alcohol and the group consuming 1 ounce. Alcohol impaired manual dexterity.

13. $SS_b = \left[\frac{(435)^2}{8} + \frac{(330)^2}{8} + \frac{(203)^2}{8} + \frac{(139)^2}{8}\right]$

$- \frac{(435 + 330 + 203 + 139)^2}{32}$

$= [23,653.13 + 13,612.5 + 5,151.13 + 2,415.13]$

$- \frac{(1,107)^2}{32}$

$= 44,831.89 - 38,295.28$

$= 6,536.61$

$SS_W = \left[\left(24,739 - \frac{(435)^2}{8}\right) + \left(14,012 - \frac{(330)^2}{8}\right)\right.$

$\left. + \left(5,427 - \frac{(203)^2}{8}\right) + \left(2,455 - \frac{(139)^2}{8}\right)\right]$

$= [(24,739 - 23,653.13) + (14,012 - 13,612.5)$

$+ (5,427 - 5,151.13) + (2,455 - 2,415.13)]$

$= (1,085.87 + 399.5 + 275.87 + 39.87)$

$= 1,801.11$

	SS	df	MS	f
Between	6,536.61	3	2,178.87	33.87
Within	1,801.11	28	64.33	
Total	8,337.72	31		

The association value of the CVC trios significantly affects the time needed to learn the list: $F_{(3,28)} = 33.87$, $p < .01$. A Tukey post-hoc test will tell us about which pairs differed significantly. Since all groups are of equal size, we only need to calculate one error term for each HSD test.

$$\text{Tukey} = \frac{\bar{X}_1 - \bar{X}_2}{s_{\bar{x}}}, \text{ where } s_{\bar{x}} = \sqrt{\frac{MS_W}{N_{total}}}$$

$$s_{\bar{x}} = \sqrt{\frac{MS_W}{N_{total}}} = \sqrt{\frac{64.33}{8}} = \sqrt{8.04} = 2.84$$

Tukey post-hoc test

	1 $n=8$ $\bar{x}=54.38$	2 $n=8$ $\bar{x}=41.25$	3 $n=8$ $\bar{x}=25.38$	4 $n=8$ $\bar{x}=17.38$
1		13.13 $s_{\bar{x}}=2.84$ $HSD=\dfrac{13.13}{2.84}$ $=4.62^*$	29 $s_{\bar{x}}=2.84$ $HSD=\dfrac{29}{2.84}$ $=10.21^*$	37 $s_{\bar{x}}=2.84$ $HSD=\dfrac{37}{2.84}$ $=13.03^*$
2			15.87 $s_{\bar{x}}=2.84$ $HSD=\dfrac{15.87}{2.84}$ $=5.59^*$	23.87 $s_{\bar{x}}=2.84$ $HSD=\dfrac{23.87}{2.84}$ $=8.40^*$
3				8 $s_{\bar{x}}=2.84$ $HSD=\dfrac{8}{2.84}$ $=2.82$
4				

The critical Tukey statistic, with four treatment groups and 28 df within, is 3.861 at $\alpha=.05$. There were statistically significant differences the time needed to learn the lists for all comparisons except Group 3 versus Group 4.

15. a. $\text{Tukey}=\dfrac{\bar{X}_1-\bar{X}_2}{s_{\bar{x}}}$,

where $s_{\bar{x}}=\sqrt{\dfrac{MS_W}{N_{\text{total}}}}=\sqrt{\dfrac{65.9}{5}}=\sqrt{13.18}=3.63$

Tukey post-hoc test

	1 $n=5$ $\bar{x}=38.2$	2 $n=5$ $\bar{x}=20.6$	3 $n=5$ $\bar{x}=14.80$
1		17.60 $s_{\bar{x}}=3.63$ $HSD=\dfrac{17}{3.63}=4.85^*$	23.40 $s_{\bar{x}}=3.63$ $HSD=\dfrac{23.4}{3.63}=7.09^*$
2			5.80 $s_{\bar{x}}=3.63$ $HSD=\dfrac{5.80}{3.63}=1.60$
3			

The critical Tukey statistic, with three treatment groups and 12 df within, is 3.773 at $\alpha=.05$. Beck depression scores were significantly higher in the low-dose group compared to the medium- and high-dose groups. The difference between the medium- and high-dose groups depression scores was not statistically significant.

b. $s_{\bar{x}}=\sqrt{\dfrac{MS_w}{2}\left(\dfrac{1}{n_1}+\dfrac{1}{n_2}\right)}=\sqrt{\dfrac{8.926}{2}\left(\dfrac{1}{6}+\dfrac{1}{6}\right)}$

$=\sqrt{4.463(0.33)}=\sqrt{1.47}=1.21$

$s_{\bar{x}}=\sqrt{\dfrac{MS_w}{2}\left(\dfrac{1}{n_2}+\dfrac{1}{n_3}\right)}=\sqrt{\dfrac{8.926}{2}\left(\dfrac{1}{6}+\dfrac{1}{5}\right)}$

$=\sqrt{4.463(0.37)}=\sqrt{1.65}=1.28$

Tukey post-hoc test

	1 $n=6$ $\bar{x}=19.17$	2 $n=6$ $\bar{x}=16.33$	3 $n=5$ $\bar{x}=9.20$
1		$HSD=\dfrac{19.17-16.33}{1.21}$ $HSD=2.35$	$HSD=\dfrac{19.17-9.20}{1.28}$ $HSD==7.79^*$
2			$HSD=\dfrac{16.33-9.20}{1.28}$ $HSD=5.57^*$
3			

The critical Tukey statistic, with three treatment groups and 14 df within, is 3.701 at $\alpha=.05$. Number of trials needed to learn the maze differed significantly for animals reared in enriched environments (Group 3) and both those reared in standard environments (Group 2) and in impoverished environments (Group 1). The difference in the number of errors between rats reared in impoverished and standard environments was not statistically significantly significant.

c. $\text{Tukey}=\dfrac{\bar{X}_1-\bar{X}_2}{s_{\bar{X}}}$,

where $s_{\bar{x}}=\sqrt{\dfrac{MS_W}{Nt}}=\sqrt{\dfrac{4.18}{8}}=\sqrt{0.52}=0.72$

Tukey post-hoc test

	1 $n = 8$ $\bar{x} = 6$	2 $n = 8$ $\bar{x} = 4$	3 $n = 8$ $\bar{x} = 3$
1		$HSD = \frac{6-4}{0.72}$ $HSD = 2.78$	$HSD = \frac{6-4}{0.72}$ $HSD = 417*$
2			$HSD = \frac{4-3}{0.72}$ $HSD = 1.39$
3			

The critical Tukey statistic, with three treatment groups and 21 *df* within, is 3.565 at $\alpha = .05$. Only the difference in the number of nonsense syllables recalled between participants studying in silence (Group 1) and those studying with unpredictable changes in sound volume (Group 3) was statistically significantly different.

d. $s_{\bar{X}} = \sqrt{\frac{MS_w}{2}\left(\frac{1}{n_1} + \frac{1}{n_2}\right)} = \sqrt{\frac{26.986}{2}\left(\frac{1}{10} + \frac{1}{10}\right)}$
$= \sqrt{13.49(0.20)} = \sqrt{2.70} = 1.65$

$s_{\bar{X}} = \sqrt{\frac{MS_w}{2}\left(\frac{1}{n_2} + \frac{1}{n_3}\right)} = \sqrt{\frac{26.986}{2}\left(\frac{1}{10} + \frac{1}{11}\right)}$
$= \sqrt{13.49(0.19)} = \sqrt{2.56} = 1.60$

$s_{\bar{X}} = \sqrt{\frac{MS_w}{2}\left(\frac{1}{n_2} + \frac{1}{n_3}\right)} = \sqrt{\frac{26.986}{2}\left(\frac{1}{10} + \frac{1}{9}\right)}$
$= \sqrt{13.49(0.21)} = \sqrt{2.83} = 1.68$

$s_{\bar{X}} = \sqrt{\frac{MS_w}{2}\left(\frac{1}{n_2} + \frac{1}{n_3}\right)} = \sqrt{\frac{26.986}{2}\left(\frac{1}{11} + \frac{1}{9}\right)}$
$= \sqrt{13.49(0.20)} = \sqrt{2.70} = 1.65$

Tukey post-hoc test

	1 $n = 10$ $\bar{x} = 80$	2 $n = 10$ $\bar{x} = 85$	3 $n = 11$ $\bar{x} = 46$	4 $n = 9$ $\bar{x} = 71$
1		$HSD = \frac{80-85}{1.65}$ $HSD = 3.05$	$HSD = \frac{80-46}{1.60}$ $HSD = 21.25*$	$HSD = \frac{80-71}{1.68}$ $HSD = 5.36*$
2			$HSD = \frac{85-46}{1.60}$ $HSD = 24.38*$	$HSD = \frac{85-46}{1.68}$ $HSD = 23.31*$
3				$HSD = \frac{46-71}{1.65}$ $HSD = 15.24*$
4				

The critical Tukey statistic, with four treatment groups and 36 *df* within, is 3.809 at $\alpha = .05$. Heart rate, measured in beats per minute, differed significantly for all comparisons except for the comparison of controls (Group 1) and participants performing the task with a good friend present (Group 2).

17. a. largest critical *F*-value = 6,055.85
  i. $\alpha = .01$
  ii. $df_b = 1, df_w = 10$
  iii. Two groups are being compared [df_b (number of groups $- 1) = 1$, ∴ number of groups being compared = 2]
  iv. 12 participants were used [$df_w = N - k$; $10 = N - 2$; ∴ $N = 12$]
 b. smallest critical *F*-value = 1.93
  i. $\alpha = .05$
  ii. $df_b = 100, df_w = 10$
  iii. 101 groups are being compared [df_b (number of groups $- 1) = 100$ \ number of groups being compared = 101]
  iv. 111 participants were used [$df_w = N - k$; $10 = N - 101$; \ ∴ $N = 111$]
 c. Find *F*-critical = 1.96
  i. df_b (number of groups $- 1) = 73$ or 74 or 75 or 76 or 77; $df_w = 10$
  ii. $\alpha = .05$
  iii. 72, 73, 74, 75, or 76 groups are being compared at this critical *F*-value
  iv. between 83 and 87 participants would have been used [$df_w = N - k$; $10 = N - 73$; ∴ $N = 83, 84, 85, 86$, or 87]

19. a. $SS_b = 193.4$; $SS_w = 3,991.8$
 b. $df_b = 2$; $df_w = 27$
 c. $MS_b = 96.7$; $MS_w = 147.84$
 d. $F = 0.65$
 e. $0.65 < 3.35$ ∴ we cannot reject the null hypothesis and conclude that we do not have statistically significant evidence that academic standing varies in accordance with where students sit in relation to the front of the classroom.

21. Variance in the numerator of the *F*-statistic (MS_b) is caused by the effect of the independent variable and by random chance. Variance in the denominator of the *F*-statistic (MS_w) is caused by random chance.

ANSWERS TO ODD NUMBERED END-OF-CHAPTER PRACTICE PROBLEMS

CHAPTER 12

1. Because the interval surrounds the mean, extending both below and above it, a two-tailed critical value that defines an interval above and below a given point must be used.

3. a. 95% CI $= \bar{X} \pm (z_{critical})(\sigma_{\bar{X}}) = 50 \pm 1.96(12/\sqrt{5})$
   $= 45.30–54.70$
   b. 99% CI $= \bar{X} \pm (z_{critical})(\sigma_{\bar{X}}) = 50 \pm 2.36(2.4)$
   $= 44.35–55.66$
   c. 68% CI $= \bar{X} \pm (z_{critical})(\sigma_{\bar{X}}) = 50 \pm 1.00(2.4)$
   $46.70–52.40$
   d. 50% CI $= \bar{X} \pm (z_{critical})(\sigma_{\bar{X}}) = 50 \pm 0.67(2.4)$
   $= 48.39–51.61$
   e. 34% CI $= \bar{X} \pm (z_{critical})(\sigma_{\bar{X}}) = 50 \pm 0.95(2.4)$
   $47.72–52.28$
   f. 60% CI $= \bar{X} \pm (z_{critical})(\sigma_{\bar{X}}) = 50 \pm 0.51(2.4)$
   $48.78–51.22$
   g. The width of the CI increases as the critical z-value increases.

5. a. We should use a single-sample t-test. We are comparing two groups that are selected at random, and we know the population mean (μ) but not the population standard deviation (σ).

   $\hat{s} = \sqrt{\dfrac{\sum x^2 - \dfrac{(\sum x)^2}{n}}{n-1}} = \sqrt{\dfrac{128.51 - \dfrac{(34.63)^2}{10}}{9}}$

   $= \sqrt{\dfrac{128.51 - 119.92}{9}} = \sqrt{0.95} = 0.97$

   $s_{\bar{X}} = \dfrac{\hat{s}}{\sqrt{n}} = \dfrac{0.97}{\sqrt{10}} = \dfrac{0.97}{3.16} = 0.31$

   $t = \dfrac{\bar{X} - \mu_{\bar{X}}}{s_{\bar{X}}} = \dfrac{3.46 - 2.52}{0.31} = \dfrac{0.94}{0.31} = 3.03$

   $t_{critical}$ with $df = 9$; two-tailed, $\alpha = .05 = \pm 2.262$

   b. We can reject the null hypothesis that says that the average OPG level in schizophrenics is the same as the average in healthy individuals. The data suggest that the average OPG level in schizophrenics is significantly higher than 2.52 ng/ml.

   c. 95% CI $= (\bar{X}) \pm (t_{critical})(s_{\bar{X}})$
   $= 3.46 \pm 2.262(0.31)$

   Lower limit of CI $= 2.76$
   Upper limit of CI $= 4.16$
   We are 95% confident that the average OPG level in schizophrenics is between 2.76 and 4.16 ng/ml. The effect of schizophrenia on OPG level is large ($d = 0.97$).

   $d = \dfrac{\bar{X} - \mu_{\bar{X}}}{\hat{s}} = \dfrac{0.94}{0.97} = 0.97$

7. a. Theoretically, 95% of the samples should include the mean of the population. If you count the number of lines that cross the dashed line representing the mean of the population, you will find that all but three of them include the population mean. This means that 47 out of 50 include the population mean, which means that 94% include it.
   b. Again, theoretically, just 5% of the means should indicate that the mean of the sample is significantly different from the mean of the population. Empirically, three of these CIs do not include the population mean, and 3 out of 50 = 6%.
   c.

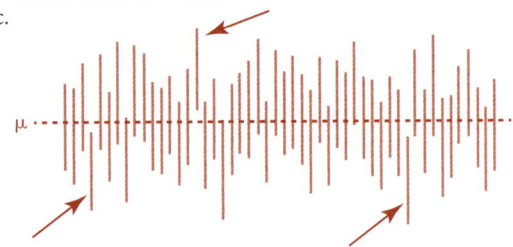

9. $\sum x = 28$
   $\bar{X} = 4.00$
   $\hat{s} = 2.16$
   $n = 7; df = 6$
   CI $= \bar{X} \pm (t_{critical})(s_{\bar{X}})$

   $s_{\bar{X}} = \dfrac{\hat{s}}{\sqrt{n}} = \dfrac{2.16}{\sqrt{7}} = 0.82$

   CI $= 4.00 \pm (2.447)(0.82)$
   Lower limit of CI $= 1.99$
   Upper limit of CI $= 6.01$
   We can be 95% sure that the mean of the true population is between 1.99 and 6.01.

11. CI $= (\bar{X}) \pm (t_{critical})(s_{\bar{X}})$

    $s_{\bar{X}} = \dfrac{\hat{s}}{\sqrt{n}} = \dfrac{\sqrt{\dfrac{1,980 - \dfrac{(156)^2}{13}}{12}}}{\sqrt{13}} = \dfrac{\sqrt{\dfrac{1,980 - 1,872}{12}}}{3.61}$

    $= \dfrac{\sqrt{9}}{3.61} = \dfrac{3}{3.61} = 0.83$

    $df = n - 1 = 12$

    CI $= \dfrac{156}{13} \pm (2.179)(0.83) = 12 \pm 1.81$

    Lower limit of CI $= 12 - 1.81 = 10.19$
    Upper limit of CI $= 12 + 1.81 = 13.81$
    We can be 95% confident that the mean of the population this sample comes from is between 10.19 and 13.81.

13. $\bar{X} = \dfrac{8,820}{12} = 735$

    $H_0: \mu_{\bar{X}} = 750$ hours
    $H_1: \mu_{\bar{X}} \neq 750$ hours (two-tailed)

    $s_{\bar{X}} = \dfrac{\hat{s}}{\sqrt{n}} = \dfrac{\sqrt{\dfrac{6,487,100 - \dfrac{(8,820)^2}{12}}{11}}}{\sqrt{12}}$

$$= \sqrt{\dfrac{\dfrac{6{,}487{,}100 - 6{,}482{,}700}{11}}{3.46}} = \dfrac{\sqrt{400}}{3.46} = \dfrac{20}{3.46} = 5.78$$

$t = \dfrac{735 - 750}{5.78} = -2.59;\ df = 12 - 1 = 11;\ t_{\text{critical}} = \pm 2.201$

We can reject the null hypothesis that the average lifespan of the lightbulbs is 750 hours. The average life span of the bulbs in the sample ($M = 735$, $SD = 20$) is significantly less than 750 hours: $t_{(11)} = -2.59$, $p < .05$.

$\text{CI} = (\bar{X}) \pm (t_{\text{critical}})(s_{\bar{x}}) = 735 \pm (2.201)5.78$

Lower limit of CI $= 735 - 12.72 = 722.28$ hours
Upper limit of CI $= 735 + 12.72 = 747.72$ hours

We are 95% confident that the actual average lifespan of the bulbs in the population is between 722.28 and 747.72 hours, which is significantly less than the advertised 750 hours.

15.

	Shown	Not Shown
	$\Sigma x = 80$	$\Sigma x = 118$
	$\Sigma x^2 = 1{,}120$	$\Sigma x^2 = 2{,}392$
	$\bar{X}_1 = 13.33$	$\bar{X}_2 = 19.67$

$H_0: \mu_{\bar{X}_1} - \mu_{\bar{X}_2} = 0$

$H_1: \mu_{\bar{X}_1} - \mu_{\bar{X}_2} \neq 0$ (two-tailed)

$\hat{s}_1 = \sqrt{\dfrac{1{,}120 - \dfrac{80^2}{6}}{5}} = 3.27$

$\hat{s}_2 = \sqrt{\dfrac{2{,}392 - \dfrac{118^2}{6}}{5}} = 3.78$

$s_{\bar{X}_1 - \bar{X}_2} = \sqrt{\dfrac{\left(1{,}120 - \dfrac{80^2}{6}\right) + \left(2{,}392 - \dfrac{118^2}{6}\right)}{6(5)}}$

$= \sqrt{\dfrac{53.33 + 71.33}{30}} = \sqrt{\dfrac{124.66}{30}} = \sqrt{4.16} = 2.04$

$t = \dfrac{\bar{X}_1 - \bar{X}_2}{s_{\bar{X}_1 - \bar{X}_2}} = \dfrac{13.33 - 19.67}{2.04} = \dfrac{-6.34}{2.04} = -3.11$;

$df = 10$; $t_{\text{critical}} = \pm 2.228$

We can reject the null hypothesis that says the mean time to retrieve the food for the cats that were shown the location of the food is the same as the time needed to retrieve the food by the cats that were not shown the location of the food. In fact, the animals shown the location of the food took significantly less time to locate it ($M = 13.33$ seconds, $SD = 3.27$ seconds) than did the cats in the control group ($M = 19.67$ seconds, $SD = 3.78$ seconds): $t_{(10)} = -3.11$, $p < .05$.

$\text{CI} = (\bar{X}_1 - \bar{X}_2) \pm (t_{\text{critical}})(s_{\bar{X}_1 - \bar{X}_2})$

$= (-6.34) \pm (2.228)(2.04)$

$d = \dfrac{\bar{X}_1 - \bar{X}_2}{\hat{s}}$, where $\hat{s} = \sqrt{n}(s_{\bar{X}_1 - \bar{X}_2})$

$d = \dfrac{6.34}{\sqrt{6}(2.04)} = \dfrac{6.34}{2.45(2.04)} = \dfrac{6.34}{4.99} = 1.27$

We are 95% confident that the difference in time needed to retrieve the food for animals shown the location and animals not shown the location of the food is between -10.89 seconds and -1.79 seconds. Showing the animals the location of the food had a large effect on time needed to retrieve the food ($d = 1.27$).

17.

Sample a	Sample b
$z = \dfrac{\bar{X} - \mu}{\sigma_{\bar{x}}} = \dfrac{62 - 70}{8/\sqrt{16}}$	$z = \dfrac{\bar{X} - \mu}{\sigma_{\bar{x}}} = \dfrac{62 - 70}{8/\sqrt{400}}$
$= \dfrac{-8.0}{2.0} = -4.00$	$= \dfrac{-8.00}{0.40} = 20$

Assuming a two-tailed test with $\alpha = .05$, the z-critical is ± 1.96 for both tests. Both observed z-values exceed the critical value, so both are statistically significant.

Effect size (a and b): $d = \dfrac{\bar{X} - \mu}{8} = \dfrac{-8.00}{8.00} = 1.00$

The independent variable has a large effect in both situations. The value of the observed z-statistic is much larger in sample b because the sample size is so large, but since we don't need to estimate the population standard deviation (σ), the size of the effect does not change as the size of the sample changes.

19.

Source	f
Music	7.03*
Water	0.67 n.s.
Music × water	0.14 n.s.

Effect of music:

$\eta p^2 = \dfrac{SS_{\text{effect}}}{SS_{\text{effect}} + SS_{\text{within}}} = \dfrac{536.34}{536.34 + 1{,}220.75} = \dfrac{536.34}{1{,}757.10}$

$= 0.31$

Effect of water:

$\eta p^2 = \dfrac{SS_{\text{effect}}}{SS_{\text{effect}} + SS_{\text{within}}} = \dfrac{34.32}{34.32 + 1{,}220.75} = \dfrac{34.32}{1{,}255.07}$

$= 0.027$

Effect of interaction:

$\eta p^2 = \dfrac{SS_{\text{effect}}}{SS_{\text{effect}} + SS_{\text{within}}} = \dfrac{22.07}{22.07 + 1{,}220.75} = \dfrac{22.07}{1{,}242.82}$

$= 0.02$

ANSWERS TO ODD NUMBERED END-OF-CHAPTER PRACTICE PROBLEMS

Music has a large and significant effect on plant growth. Water and the interaction between water and music have very small and nonsignificant effects on plant growth.

21. a. $t = -1.75$
$-1.75 < -1.734$, or $|1.75| > |1.734| \therefore$ we reject the null hypothesis and conclude that the mean walking speed of people who receive age priming is slower than the walking speed of those who are not primed.
b. $-1.75 < -2.552$, or $|1.75| > |2.552| \therefore$ we fail to reject the null hypothesis and conclude that the difference between the mean walking speeds of people who receive age priming and of those do not is not statistically significant.
c. 95% CI $= -5.8 \pm 6.97$
99% CI $= -5.8 \pm 9.55$

CHAPTER 13

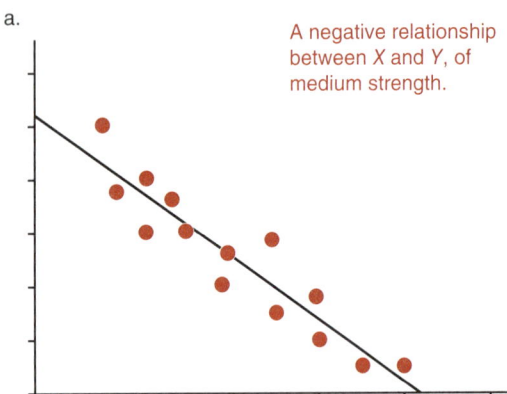

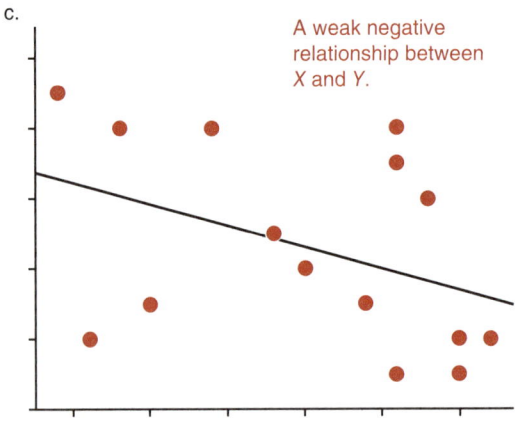

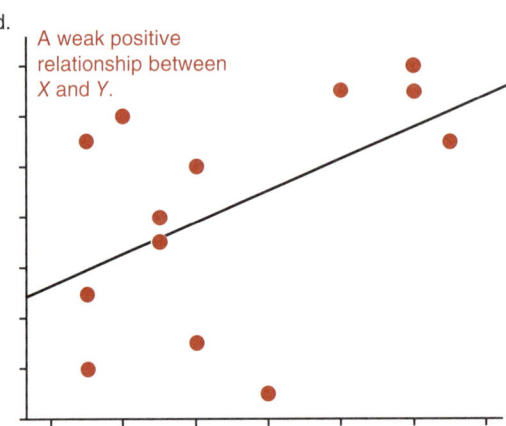

1.
a. A negative relationship between X and Y, of medium strength.

b. A perfect positive relationship between X and Y.

c. A weak negative relationship between X and Y.

d. A weak positive relationship between X and Y.

FIGURE 13.11
Scatter plots for question 1.

3. Both statistics (r^2 and eta squared) are standardized, meaning they consist of a measure of difference (the difference between means in an ANOVA, which is the effect we are interested in) or a measure of relatedness (in a correlation, because the effect we're interested in is the degree of relatedness of the two variables).

Eta squared $\left(\eta^2 = \dfrac{ss_b}{ss_{total}} \right)$ measures variability between groups (how different the groups are) divided by overall total variability in the model. The statistic r is a measure of the degree of relatedness between X and Y divided by the combined variability in both X and Y.

5. "Least squares" refers to variability in correlations as well as in other situations where the phrase is used. Here, least squares—smallest deviation—refers to minimizing the variability of each X, Y pair from the regression line.

7. $\Sigma X = 41$, $\Sigma Y = 14$
$\Sigma X^2 = 245$, $\Sigma Y^2 = 26$, $\Sigma XY = 76$

$$r = \frac{N\sum XY - (\sum X)(\sum Y)}{\sqrt{N\sum X^2 - (\sum X)^2}}$$

$$= \frac{9(76) - (41)(14)}{\sqrt{[9(245) - (41)^2][9(26) - (14)^2]}}$$

$$= \frac{684 - 574}{\sqrt{[2,205 - 1,681][234 - 196]}} = \frac{110}{\sqrt{[524][38]}}$$

$$= \frac{110}{\sqrt{19,912}} = \frac{110}{141.11} = .78$$

There is a strong positive correlation between belief in luck and religiosity. The stronger one's belief in luck, the more religious one is. Sixty-one percent ($r^2 = .61$) of the variability in religiosity can be accounted for by knowing how lucky a person believes he or she is. The problem with the data, correctly identified by our colleague, is that the Y variable suffers from restricted range. There is very little variability in the measure of religiosity. This could result in a spuriously low correlation between x and y.

9. $\Sigma X = 15$, $\Sigma Y = 13$
$\Sigma X^2 = 55$, $\Sigma Y^2 = 63$, $\Sigma XY = 24$

$$b = \frac{N\sum XY - (\sum X)(\sum Y)}{N\sum X^2 - (\sum X)^2} = \frac{5(24) - (15)(13)}{5(55) - (15)^2}$$

$$= \frac{120 - 195}{275 - 225} = \frac{-75}{50} = -1.50$$

$$a = \bar{Y} - b(\bar{X}) = \frac{13}{5} - (-1.50)\left(\frac{15}{5}\right) = 2.60 - (-1.50)(3)$$

$$= 2.60 - (-4.50) = 7.10$$

Answer: $\hat{y} = bx + a = (-1.50)x + 7.10$

11. a. $r = \frac{N\sum XY - (\sum X)(\sum Y)}{\sqrt{[N\sum X^2 - (\sum X)^2][N\sum Y^2 - (\sum Y)^2]}}$

$$= \frac{15(2,126.08) - (545.87)(60)}{\sqrt{[15(19,963.01) - 545.87^2][15(352) - 60^2]}}$$

$$= \frac{31,891.20 - 32,752.20}{\sqrt{[299,445.15 - 297,974.06][5,280 - 3,600]}}$$

$$= \frac{-861}{\sqrt{[1,471.09][1,680]}} = \frac{-861}{\sqrt{2,471,431.2}}$$

$$= \frac{-861}{1,572.08} = -.5477 \text{ or } -.55$$

There is a strong negative correlation between the number of months a woman smokes and the head circumference of her infant: The longer Mom smokes during pregnancy, the smaller the infant's head circumference. This correlation is statistically significant. Smoking behavior accounts for 30% of the variability in head circumference ($r^2 = .3025$, or 30%).

b.
$$r = \frac{N\sum XY - (\sum X)(\sum Y)}{\sqrt{[N\sum X^2 - (\sum X)^2][N\sum Y^2 - (\sum Y)^2]}}$$

$$= \frac{15(53,867.80) - (545.87)(1459)}{\sqrt{[15(19,963.01) - 297,974.06][15(150,183) - 2,128,681]}}$$

$$= \frac{11,592.67}{\sqrt{[1,471.09][124,064]}} = \frac{11,592.67}{\sqrt{182,509,309.8}}$$

$$= \frac{11,592.67}{13,509.60} = 0.858, \text{ or } 0.86$$

There is a strong positive correlation between head circumference and score on the Bayley test. The larger the infant's head, the higher the infant's score on the Bayley. This correlation is statistically significant. Head circumference accounts for 74% of the variability in the Bayley score ($r^2 = .7396$, or 74%).

c. $b = \frac{N\sum XY - (\sum X)(\sum Y)}{N\sum X^2 - (X)^2} = \frac{-861}{1,471.09}$

$$= -.585, \text{ or } -.59$$

$$a = \bar{Y} - b(\bar{X}) = 4.00 - (-.59)(36.39)$$

$$= 4.00 - (-21.47) = 25.47$$

$$\hat{Y} = bX + a = (-.59)(3.5) + 25.47$$

$$= -2.065 + 25.47 = 23.41$$

A woman who smoked for 3.5 months during her pregnancy would be predicted to have an infant with a head circumference of 25.47 centimeters.

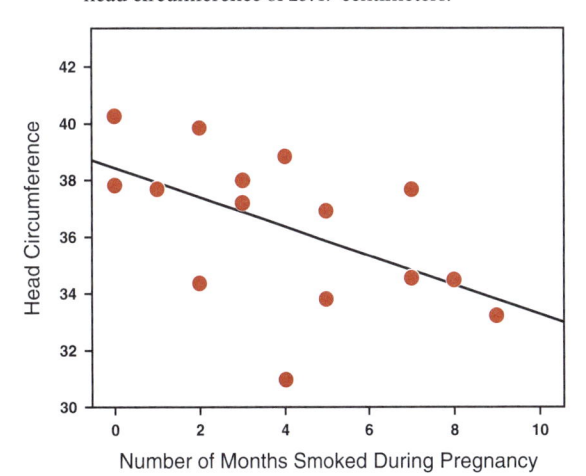

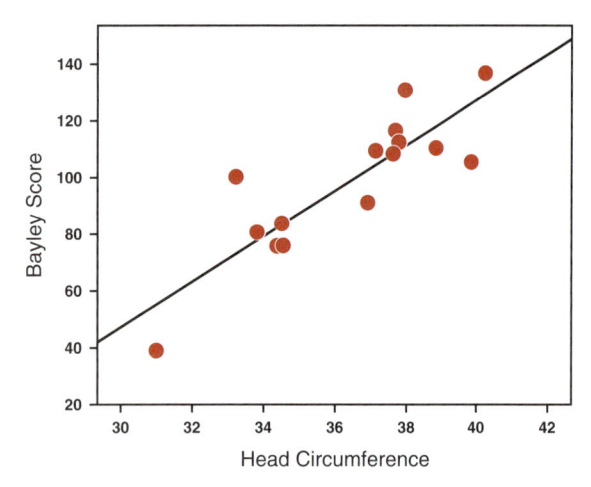

13. $\Sigma X = 57, \Sigma Y = 27$
$\Sigma X^2 = 775, \Sigma Y^2 = 177, XY = 247$

$$r = \frac{N\sum XY - (\sum X)(\sum Y)}{\sqrt{\left[N\sum X^2 - (\sum X)^2\right]\left[N\sum Y^2 - (\sum Y)^2\right]}}$$

$$= \frac{1,235 - 1,539}{\sqrt{[3,875 - 3,249][885 - 729]}} = \frac{-304}{\sqrt{626(156)}}$$

$$= \frac{-304}{\sqrt{97,656}}$$

$$= \frac{-304}{312.50} = -.97$$

There is a very strong negative correlation between time spent studying and time spent playing video games: The more time is spent studying, the less time is spent playing video games. Almost all of the variability in time spent playing video games (94%) can be explained by knowing the amount of time spent studying. However, the relationship is not statistically significant, likely because of the very small sample size (critical r-value with 3 df is .87, $\alpha = .05$).

15. a. $H_0: \rho = 0.00$
$H_1: \rho \neq 0.00$
b. Because the researchers have not predicted a direction for the correlation, we should do a two-tailed test to look for either a positive or a negative correlation.
c. $\Sigma X = 144, \Sigma Y = 93$
$\Sigma X^2 = 1,372, \Sigma Y^2 = 451, XY = 696$

$$r = \frac{N\sum XY - (\sum X)(\sum Y)}{\sqrt{\left[N\sum X^2 - (\sum X)^2\right]\left[N\sum Y^2 - (\sum Y)^2\right]}}$$

$$= \frac{20(696) - (144)(93)}{\sqrt{[20(1,372) - 144^2][20(451) - 93^2]}}$$

$$= \frac{13,920 - 13,392}{\sqrt{[27,440 - 20,736][9,020 - 8,649]}}$$

$$= \frac{528}{\sqrt{[6,704][371]}}$$

$$= \frac{528}{\sqrt{2,487,184}} = \frac{528}{1,577.08} = .33$$

There is a moderate positive correlation between ACHA score and manic tendencies score: The stronger the indication of an eating disorder, the stronger the indication of manic personality tendencies. The score on the ACHA test accounted for 10.89% (or 11%) of the variability in the manic tendencies score.
d. The correlation between ACHA and mania scores is not statistically significant (critical r-value with 18 df is .444, $\alpha = .05$). We cannot reject the null hypothesis.

e. $b = \dfrac{N\sum XY - (\sum X)(\sum Y)}{\left[N\sum X^2 - (\sum X)^2\right]} = \dfrac{528}{6,704} = 0.08$

$a = \bar{Y} - b\bar{X} = 4.65 - 0.08(7.2) = 4.07$

$\bar{Y} = bX + a = 0.08(20) + 4.07 = 1.6 + 4.07 = 5.67$
We would predict a manic tendencies score of 5.67 for a student who scored a 20 on the ACHA test.

17.
$$r = \frac{N\sum XY - (\sum X)(\sum Y)}{\sqrt{\left[N\sum X^2 - (\sum X)^2\right]\left[N\sum Y^2 - (\sum Y)^2\right]}}$$

$$= \frac{50(8,084) - (3,500)(115)}{\sqrt{[50(250,000) - 3,500^2][50(289) - 115^2]}}$$

$$= \frac{404,200 - 402,500}{\sqrt{[250,000][1,225]}} = \frac{1,700}{\sqrt{306,250,000}}$$

$$= \frac{1,700}{17,500} = .10$$

There is a weak positive correlation between sensory ability score and GPA. This relationship is not statistically significant (critical r-value with 48 df is .279, $\alpha = .05$).

19. $\Sigma X = 239, \Sigma Y = 41.85$
$\Sigma X^2 = 3,973, \Sigma Y^2 = 123.9, XY = 678.15$

$$r = \frac{N\sum XY - (X)(\sum Y)}{\sqrt{\left[N\sum X^2 - (\sum X)^2\right]\left[N\sum Y^2 - (\sum Y)^2\right]}}$$

$$= \frac{15(678.15) - (239)(41.85)}{\sqrt{[15(3,973) - 239^2][15(123.9) - 41.85^2]}}$$

$$= \frac{10,172.25 - 10,002.15}{\sqrt{[2,474][107.08]}} = \frac{170.10}{\sqrt{264,915.92}}$$

$$= \frac{170.10}{514.70} = 0.33$$

There is a moderate positive correlation between number of hours worked and GPA: The more hours a student works, the higher that student's GPA is predicted to be. However, the correlation is not statistically significant (critical *r*-value with 13 *df* is .514, α = .05). The number of hours worked accounts for 10.89% of the variability in GPA.

21. a. $X = 8, Y = 8.6$
    b. $X = -10, Y = -49$
    c. $X = 41, Y = 114.2$
    d. $X = 100, Y = 303$
    e. $X = -16, Y = -68.2$

CHAPTER 14

1. The formula for chi square is $\chi^2 = \sum \frac{(f_o - f_e)^2}{f_e}$, the ratio of observed to expected frequencies. Negative frequencies are impossible (something can't happen fewer than zero times). Because observed frequencies cannot be zero, the ratio of observed to expected also cannot be less than zero.

3. Thirty consecutive roles of a single die

1 pip	2 pips	3 pips	4 pips	5 pips	6 pips
3	3	4	5	5	10
$f_e = 5$	$f_e = 5$	$f_e = 5$	$f_e = 5$	$f_e = 5$	$f_e = 5$

a. Expected frequency would be the number of faces on the die that could show up (there are six faces on a single die). With a fair die, we would expect for each of the faces to show up equally often, so the expected frequency of each face would be 30/6 = 5. In other words, we would expect each face to show up five times on a fair die.

b. $\chi^2 = \sum \frac{(f_o - f_e)^2}{f_e} = \left[\frac{(3-5)^2}{5}\right] + \left[\frac{(3-5)^2}{5}\right]$
$+ \left[\frac{(4-5)^2}{5}\right] + \left[\frac{(5-5)^2}{5}\right] + \left[\frac{(5-5)^2}{5}\right] + \left[\frac{(10-5)^2}{5}\right]$
$= \left[\frac{[-2]^2}{5}\right] + \left[\frac{[-2]^2}{5}\right] + \left[\frac{[-1]^2}{5}\right] + \left[\frac{[0]^2}{5}\right] + \left[\frac{[0]^2}{5}\right] + \left[\frac{[5]^2}{5}\right]$
$= 0.8 + 0.8 + 0.2 + 0 + 0 + 5 = 6.80$

We have 6 − 1 or five degrees of freedom, so if our alpha level is 0.05, then our critical chi-square value is 11.07. Our observed chi-square value of 6.80 does not exceed the critical value, so we have to retain the null hypothesis and conclude that the die is a fair one. Random chance accounts for the tendency for sixes to come up more often than we expected.

c. No, we should not be concerned about the dice in this establishment: We have no evidence that the dice are unfair.

5. **Attitude toward abortion**

	Support	Do not Support	Do not yet Know	Row Totals
Women	14 $f_e = 15$	17 $f_e = 15$	19 $f_e = 15$	50
Men	16 $f_e = 15$	13 $f_e = 15$	21 $f_e = 15$	50
Column totals	30	30	40	100

a. We should use a two-way chi-square test of independence to see if gender and opinion about abortion are independent of one another.

b. Expected frequencies: $EF = \frac{(\text{row})(\text{column})}{\text{overall}}$

Row/Column		
1/1	1/2	1/3
$\frac{(50)(30)}{100}$ $= 15$	$\frac{(50)(30)}{100}$ $= 15$	$\frac{(50)(40)}{100}$ $= 20$
2/1	2/2	2/3
$\frac{(50)(30)}{100}$ $= 15$	$\frac{(50)(30)}{100}$ $= 15$	$\frac{(50)(40)}{100}$ $= 20$

$\chi^2 = \sum \frac{(f_o - f_e)^2}{f_e} = \left[\frac{(14-15)^2}{15}\right] + \left[\frac{(17-15)^2}{15}\right]$
$+ \left[\frac{(19-20)^2}{20}\right]$
$\left[\frac{(16-15)^2}{15}\right] + \left[\frac{(13-15)^2}{15}\right] + \left[\frac{(21-20)^2}{20}\right]$
$= 0.067 + 0.27 + 0.05 + 0.067$
$+ 0.27 + 0.05 = 0.774$

c. The null hypothesis, that gender and opinion about abortion legalization are independent of one another, cannot be rejected. The observed chi-square value of 0.774 does not exceed the critical value of 9.448 [$df = (3-1)(3-1)$, or 4; α = .05]. Gender and opinion about the legalization of abortion are independent of one another.

7. a. We should do a one-way chi-square test of goodness of fit. We know the percentages we would expect to see for each type and can compare them with the observed frequencies. The null hypothesis is that the distribution of blood type is the same in the United States as it is elsewhere in the world. The alternative is that the distribution of blood types seen worldwide does not fit the distribution seen in the United States.

b. **Worldwide distribution of blood type**

Type O	Type A	Type B	Type AB
43.91%	34.80%	16.55%	5.14%
$f_e = 96.60$	$f_e = 76.56$	$f_e = 36.41$	$f_e = 11.31$
$f_o = 88.00$	$f_o = 84.00$	$f_o = 40.00$	$f_o = 8.00$

$$\chi^2 = \sum \frac{(f_o - f_e)^2}{f_e} = \left[\frac{(88 - 96.60)^2}{96.60}\right]$$
$$+ \left[\frac{(84 - 76.56)^2}{76.56}\right] + \left[\frac{(40 - 36.41)^2}{36.41}\right] + \left[\frac{(8 - 11.31)^2}{11.31}\right]$$
$$= 0.77 + 0.72 + 0.35 + 0.97 = 2.81$$

The critical chi-square value with $df = (4 - 1)$, or 3, is 7.815 ($\alpha = .05$). Our observed chi-square value does not exceed the critical value, so we cannot reject the null hypothesis. There is no evidence that the distribution of blood types in the United States differs from the distribution seen worldwide.

9. Door A (used by 80 people): **Expected frequency = 68.67**
Door B (used by 66 people): **Expected frequency = 68.67**
Door C (used by 60 people): **Expected frequency = 68.67**
The null hypothesis for the one-way, chi-square test for goodness of fit is that people are equally likely to use each of the three different door types being compared. The alternative hypothesis is that a rectangular distribution of door use does not fit the data: People are more likely to use at least one door than they are the others. The total number of participants in the study was $80 + 66 + 60 = 206$. The expected frequency is 206/3, or 68.67, for each of the three door types tested.

$$\chi^2 = \sum \frac{(f_o - f_e)^2}{f_e} = \left[\frac{(80 - 68.67)^2}{68.67}\right] + \left[\frac{(66 - 68.67)^2}{68.67}\right]$$
$$+ \left[\frac{(60 - 68.67)^2}{68.67}\right] = 1.87 + 0.10 + 1.09 = 3.06$$

The critical chi-square value with $(3 - 1)$, or 2, df is 5.991 ($\alpha = .05$). Our observed chi-square value is 3.06, less than the critical value, so we cannot reject the null hypothesis. It appears that each type of door is equally likely to be used.

11. **Distribution of flavors in a five-flavor roll of Life Savers**

Cherry	Pineapple	Raspberry	Watermelon	Orange
8	20	12	10	10
$f_e = 14$	$f_e = 14$	$f_e = 14$	$f_e = 14$	$f_e = 14$

The null hypothesis is that each of the five flavors is as likely as the others to be found in a roll of Life Savers. The alternative hypothesis is that the flavors are not equally distributed. Five rolls were tested, each with 14 candies in a roll, for a total of 70 candies. The expected frequency of each flavor would then be $70/5 = 14$ per flavor.

$$\chi^2 = \sum \frac{(f_o - f_e)^2}{f_e} = \left[\frac{(8-14)^2}{14}\right] + \left[\frac{(20-14)^2}{14}\right]$$
$$+ \left[\frac{(12-14)^2}{14}\right] + \left[\frac{(10-14)^2}{14}\right] + \left[\frac{(10-14)^2}{14}\right]$$
$$= 2.57 + 2.57 + 0.29 + 1.14 + 1.14 = 7.71$$

The critical chi-square value with $(5 - 1)$, or 4, df is 9.488 ($\alpha = .05$). Our observed chi-square value does not exceed the critical value, so we have failed to reject the null hypothesis: We do not have evidence that the five flavors are not evenly distributed.

13. **Grade distribution for the professor in his first year of teaching**

Letter Grade	A	B	C	D	F	Total
Frequency	11	73	104	52	10	250
	$f_e = 17.5$	$f_e = 60$	$f_e = 95$	$f_e = 60$	$f_e = 17.5$	

Our null hypothesis is that the professor grades on a curve (that a normal distribution fits the observed data). This would mean that 7% of the 250 students would earn an A, and 7% would earn an F (7% of $250 = 17.5$); 24% of the 250 students should earn either a B or a D (24% of $250 = 60$), and 38% (95 out of 250) should earn a C.

$$\chi^2 = \sum \frac{(f_o - f_e)^2}{f_e} = \left[\frac{(11 - 17.5)^2}{17.5}\right] + \left[\frac{(73 - 60)^2}{60}\right]$$
$$+ \left[\frac{(104 - 95)^2}{95}\right] + \left[\frac{(52 - 60)^2}{60}\right] + \left[\frac{(10 - 17.5)^2}{17.5}\right]$$
$$= 2.41 + 2.82 + 0.85 + 1.07 + 3.21 = 10.36$$

The critical chi-square value with $(5 - 1)$, or 4, df is 9.488 ($\alpha = .05$). Our observed chi-square value exceeds the critical value, so we can reject the null hypothesis. The normal curve does not fit the data. The professor is assigning fewer A's and F's, and more C's than the normal curve would predict.

15. **Methods of combating fatigue, by sex**

	Caffeine	Nap	Exercise	Sugar	Row Totals
Female	35	17	27	21	100
	$f_e = 30$	$f_e = 18.5$	$f_e = 28.5$	$f_e = 23$	
Male	25	20	30	25	100
	$f_e = 30$	$f_e = 18.5$	$f_e = 28.5$	$f_e = 23$	
Column totals	60	37	57	46	200

The null hypothesis is that the method used to combat fatigue is independent of gender. The alternative is that the method used to combat fatigue depends on gender.

Expected frequencies: $EF = \dfrac{(\text{row})(\text{column})}{\text{overall}}$

Row/Column			
1/1	1/2	1/3	1/4
$\dfrac{(100)(60)}{200}$ $= 30.0$	$\dfrac{(100)(37)}{200}$ $= 18.5$	$\dfrac{(100)(57)}{200}$ $= 28.5$	$\dfrac{(100)(46)}{200}$ $= 23.0$
2/1	2/2	2/3	2/4
$\dfrac{(100)(60)}{200}$ $= 30.0$	$\dfrac{(100)(37)}{200}$ $= 18.5$	$\dfrac{(100)(57)}{200}$ $= 28.5$	$\dfrac{(100)(46)}{200}$ $= 23.0$

$$\chi^2 = \sum \dfrac{(f_o - f_e)^2}{f_e} = \left[\dfrac{(35-30)^2}{30}\right] + \left[\dfrac{(17-18.5)^2}{18.5}\right]$$
$$+ \left[\dfrac{(27-28.5)^2}{28.5}\right] + \left[\dfrac{(21-23)^2}{23}\right] + \left[\dfrac{(25-30)^2}{30}\right]$$
$$+ \left[\dfrac{(20-18.5)^2}{18.5}\right] + \left[\dfrac{(30-28.5)^2}{28.5}\right] + \left[\dfrac{(25-23)^2}{23}\right]$$
$$= 0.83 + 0.12 + 0.08 + 0.17 + 0.83 + 0.12 + 0.08 + 0.17$$
$$= 2.40$$

The critical chi-square value with $(4-1)(4-1)$, or 9, df is 16.919 ($\alpha = .05$). Our observed chi-square value does not exceed the critical value, so we must retain the null hypothesis: The method used to combat fatigue is independent of gender.

17. Importance of statistics to chosen specialization in psychology

"Statistics is Important to My Chosen Specialization"	Clinical/ Counseling	Social Psychology	Developmental Psychology	Neuroscience	Row Totals
Very important	3 $f_e = 3.25$	6 $f_e = 4.55$	2 $f_e = 2.82$	2 $f_e = 2.38$	13
Important	6 $f_e = 5.25$	5 $f_e = 7.35$	5 $f_e = 4.55$	5 $f_e = 3.85$	21
Somewhat important	5 $f_e = 4.50$	8 $f_e = 6.30$	3 $f_e = 3.90$	2 $f_e = 3.30$	18
Not important	1 $f_e = 2.00$	2 $f_e = 2.80$	3 $f_e = 1.73$	2 $f_e = 1.47$	8
Column totals	15	21	13	11	60

The null hypothesis is that the importance of statistics is independent of specialization in psychology.

The alternative hypothesis is that the importance of statistics is dependent on the field of specialization in psychology.

$$\chi^2 = \sum \dfrac{(f_o - f_e)^2}{f_e} = \left[\dfrac{(3-3.25)^2}{3.25}\right] + \left[\dfrac{(6-4.55)^2}{4.55}\right]$$
$$+ \left[\dfrac{(2-2.82)^2}{2.82}\right] + \left[\dfrac{(2-2.38)^2}{2.38}\right] + \left[\dfrac{(6-5.25)^2}{5.25}\right]$$
$$+ \left[\dfrac{(5-7.35)^2}{7.35}\right] + \left[\dfrac{(5-4.55)^2}{4.55}\right] + \left[\dfrac{(5-3.85)^2}{3.85}\right]$$
$$+ \left[\dfrac{(5-4.50)^2}{4.50}\right] + \left[\dfrac{(8-6.30)^2}{6.30}\right] + \left[\dfrac{(3-3.90)^2}{3.90}\right]$$
$$+ \left[\dfrac{(2-3.30)^2}{3.30}\right] + \left[\dfrac{(1-2.00)^2}{2.00}\right] + \left[\dfrac{(2-2.80)^2}{2.80}\right]$$
$$+ \left[\dfrac{(3-1.73)^2}{1.73}\right] + \left[\dfrac{(2-1.47)^2}{1.47}\right] = 0.02 + 0.46 + 0.24$$
$$+ 0.06 + 0.11 + 0.75 + 0.04 + 0.34 + 0.06 + 0.46$$
$$+ 0.21 + 0.05 + 0.50 + 0.23 + 0.93 + 0.19 = 4.65$$

The critical value of chi square with $(4-1)(4-1)$, or 9, df ($\alpha = .05$) is 16.919. Our observed chi-square value does not exceed the critical value, so we must retain the null hypothesis. The importance of statistics is independent of field of specialization in psychology.

19. Memory for father's profanity as a function of censorship of the father

Mother's Response	Child's Memory of Profanity	
	Child Does not Remember	Child Remembers
Censors father	16 **(A)**	28 **(B)**
No response	26 **(C)**	17 **(D)**

The null hypothesis is that memory for the profanity is independent of maternal response. The alternative is that memory for profanity depends on the way that the mother responds. A total of 87 children were tested.

$$\chi^2 = \dfrac{N(AD - BC)^2}{(A+B)(C+D)(A+C)(B+D)}$$
$$= \dfrac{87(272 - 728)^2}{(44)(43)(42)(45)} = \dfrac{18,090,432}{3,575,880} = 5.06$$

The critical value of chi square with $(2-1)(2-1)$, or 1, df ($\alpha = .05$) is 3.841. Our observed chi-square value exceeds the critical value, so we may reject the null hypothesis. Memory for the profanity is dependent on maternal response: Children were more likely to remember the profanity if the father was censored by the mother.

21. $\chi^2 = 32$

$32 > 21.026$ ∴ we reject the null hypothesis and conclude that the average number of hours spent watching television in a week is dependent upon season.

CHAPTER 15

1. If there were a strong negative correlation between X and Y, we would expect to have high-ranking X values paired with the lowest-ranking Y values.

3. The larger the number of pairs (N), the smaller the critical correlation needed to reject the null hypothesis. When there are only five pairs and we're doing a two-tailed test ($\alpha = .05$), the correlation between X and Y has to be perfect (1.00 or -1.00) in order to reject H_0. When there are 100 pairs, a correlation of at least .197 or larger will allow the researcher to reject H_0.

5. Estimated weekly TV viewing times (hours)

TV Favorable	Rank	TV Unfavorable	Rank
4	5	0	1.5
5	7	0	1.5
5	7	1	3
5	7	2	4
10	9	14	11
12	10	42	13
20	12	43	14
49	15		
$n_1 = 8$	$\Sigma R_1 = 72$	$n_2 = 7$	$\Sigma R_2 = 48$

a. The null hypothesis is that presenting TV viewing as favorable/unfavorable will not affect TV viewing time. The alternative is that the way TV viewing is presented will have an effect on TV viewing time. We should use a Mann–Whitney U-test because the data, although they are quantitative (measuring time), are extremely skewed (the skew for Group 1 is 2.214, and the skew for group 2 is 1.00), which violates the normality assumption needed for an independent-samples t-test.

b. This is a two-tailed test: We're looking for any difference between the TV watching times of the two groups.

c. With $n_1 = 8$ and $n_2 = 7$, the critical value will be 10 ($\alpha = .05$) or 6 ($\alpha = .01$).

$U_1 = (n_1)(n_2) + \dfrac{n_1(n_1+1)}{2} - \Sigma R_1 \qquad U_2 = (n_1)(n_2) + \dfrac{n_1(n_1+1)}{2} - \Sigma R_1$

$= (8)(7) + \dfrac{8(9)}{2} - 72 \qquad\qquad = (8)(7) + \dfrac{8(9)}{2} - 48$

$= 56 + 36 - 72 \qquad\qquad\qquad = 56 + 36 - 48$

$= 20 \qquad\qquad\qquad\qquad\qquad = 44$

The lowest value of U is 20, so we compare this value with the critical values. To reject the null hypothesis, our observed U-value (20) must be smaller than the critical value (10). $U_{observed}$ is not smaller than $U_{critical}$, so we cannot reject the null.

d. Presenting TV viewing as favorable or unfavorable did not significantly affect the time spent watching TV, $U = 20$, $p > .05$.

7. Teacher's ranking of student intellectual ability and intelligence test scores

Teacher's Ranking*	Test Score	Rank of Test Score	D	D²
1	41	3.5	−2.50	6.25
2	47	1	1.00	1.00
3	36	8.5	−5.50	30.25
4	42	2	2.00	4.00
5	41	3.5	1.50	2.25
6	39	6.7	−0.70	0.49
7	40	5	2.00	4.00
8	36	8.5	−0.50	0.25
9	32	12	−3.00	9.00
10	39	6.7	3.30	10.89
11	35	10	1.00	1.00
12	33	11	1.00	1.00
*Notice that the teacher's scores are already ranks				$\Sigma D^2 = 70.38$

a. We should conduct a Spearman correlation because we're looking for a relationship between a qualitative variable (the teacher's ranking of the students) and a quantitative variable.

b. This is a two-tailed test because we are looking for any relationship between the two variables.

c. The critical value of r_s with 12 pairs is .587 ($\alpha = .05$)

d. $r_s = 1 - \left[\dfrac{6 \Sigma D^2}{N(N^2-1)}\right] = 1 - \left[\dfrac{6(70.38)}{12(143)}\right]$

$= 1 - \left[\dfrac{422.28}{1716}\right] = 1 - .25$

$= .75$

There was a strong positive correlation between the teacher's ranking of intelligence and the scores made on the intelligence test. The correlation was statistically significant: $r_s(12) = 0.75$, $p < .05$.

9. We should do a Mann–Whitney U-test to compare these two random and independent groups. The data are ordinal. The null hypothesis is that women and men experience pain in the same way. The alternative is that there is a difference in the ratings of pain given by women compared to those provided by men.

Rating of pain by men and women

Women	Rank	Men	Rank
5	11.5	4	7
5	11.5	5	11.5
6	15	73	4.5
4	7	4	7
5	11.5	5	11.5
2	2.5	2	2.5
7	17	8	19.5
1	1	3	4.5
8	19.5	7	17
5	11.5	7	17
$n_1 = 10$	$\Sigma R_1 = 108$	$n_2 = 10$	$\Sigma R_2 = 102$

$U_1 = (n_1)(n_2) + \dfrac{n_1(n_1+1)}{2} - \Sigma R_1$ $\quad U_2 = (n_1)(n_2) + \dfrac{n_1(n_1+1)}{2} - \Sigma R_1$

$= (10)(10) + \dfrac{10(11)}{2} - 108$ $\quad = (10)(10) + \dfrac{10(11)}{2} - 102$

$= 100 + 55 - 108$ $\quad = 100 + 55 - 102$

$= 47$ $\quad = 53$

The smaller of the two U-values is 47, so we compare that to the tabled critical value. With 10 participants in each group (both n_1 and $n_2 = 10$) and a one-tailed test, the critical value for U is 27 ($\alpha = .05$) or 19 ($\alpha = .01$). Our observed value (47) is larger than the critical value, so we fail to reject the null: There is no significant difference in the pain ratings provided by men and women.

11. The data are quantitative but seriously skewed: SPSS measures the skew of the large hospital data at 2.704 and that of the small hospital data at 1.882. Skew measures greater than ±1.00 indicate serious skew in the data. We need to do a Mann–Whitney U-test to compare the two samples. The critical U-value is 15.

Length of postoperative hospital stay (in days)

Large Hospital	Rank$_1$	Small Hospital	Rank$_2$
4	7.5	5	11.5
3	2.5	4	7.5
4	7.5	3	2.5
6	13	15	15
4	7.5	9	14
3	2.5	3	2.5
4	7.5	4	7.5
21	16	5	11.5
$n_1 = 8$	$\Sigma R_1 = 64$	$n_2 = 8$	$\Sigma R_2 = 72$

$U_1 = (n_1)(n_2) + \dfrac{n_1(n_1+1)}{2} - \Sigma R_1$ $\quad U_2 = (n_1)(n_2) + \dfrac{n_1(n_1+1)}{2} - \Sigma R_1$

$= (8)(8) + \dfrac{8(9)}{2} - 64$ $\quad = (8)(8) + \dfrac{8(9)}{2} - 72$

$= 64 + 36 - 64$ $\quad = 64 + 36 - 72$

$= 36$ $\quad = 28$

The smaller of the two U-values is 28. The observed U-value is larger than the critical value, so we cannot reject the null hypothesis: The size of the hospital does not affect time spent in the hospital after surgery.

13. We need to compare these two groups with a Mann–Whitney U-test (the two samples are independent and the data are once again skewed). The critical U-value is 8.

Number of trials needed to learn criterion by two groups of rats

Rat	Group 1	Rank$_1$	Group 2	Rank$_2$
1	3	1	20	10.5
2	12	4.5	25	14
3	11	3	18	9
4	15	6	17	8
5	12	4.5	20	10.5
6	16	7	21	12
7	10	2	22	13
	$n_1 = 7$	$\Sigma R_1 = 28$	$n_2 = 7$	$\Sigma R_2 = 77$

$U_1 = (n_1)(n_2) + \dfrac{n_1(n_1+1)}{2} - \Sigma R_1$ $\quad U_2 = (n_1)(n_2) + \dfrac{n_1(n_1+1)}{2} - \Sigma R_1$

$= (7)(7) + \dfrac{7(8)}{2} - 28$ $\quad = (7)(8) + \dfrac{7(8)}{2} - 77$

$= 49 + 28 - 28$ $\quad = 49 + 28 - 77$

$= 49$ $\quad = 0$

The smallest of the two U-values is zero, so we compare this value with the tabled critical value. Our observed U-value is smaller than the critical value (8), so we can reject the null hypothesis: Rats in Group 2 required significantly more trials to learn criterion, showing that the region of the brain stimulated was a significant factor in learning.

15. The data are quantitative but skewed, so we should use a Mann–Whitney U-test to compare the two means. The critical value of U is 23.

ANSWERS TO ODD NUMBERED END-OF-CHAPTER PRACTICE PROBLEMS

Weights (in pounds) of cats owned by families with children or by single people

Owned by Families	Rank₁	Owned by Singles	Rank₂
9.9	13	11.9	19
7	4	11	17
10.2	14.5	10.2	14.5
10.8	16	12.7	20
6.3	1.5	11.1	18
9.1	10.5	7	4
9	8	8.2	7
6.3	1.5	7	4
9.1	10.5	9.1	10.5
9.1	10.5	8	6
$n_1 = 10$	$\Sigma R_1 = 90$	$n_2 = 10$	$\Sigma R_2 = 120$

$$U_1 = (n_1)(n_2) + \frac{n_1(n_1+1)}{2} - \Sigma R_1 \qquad U_2 = (n_1)(n_2) + \frac{n_1(n_1+1)}{2} - \Sigma R_1$$

$$= (10)(10) + \frac{10(11)}{2} - 90 \qquad = (10)(10) + \frac{10(11)}{2} - 120$$

$$= 100 + 55 - 90 \qquad = 100 + 55 - 120$$

$$= 65 \qquad = 35$$

The smaller of the two U-values is 35, so we compare this to the critical U-value of 23. The observed U-value is not smaller than the critical value, so we cannot reject the null hypothesis: The average weight of cats owned by single people and those owned by families with children does not differ significantly.

17. Because of the skew of the quantitative data, we should use the Mann–Whitney U-test to compare the two independent groups. The critical U-value is 6.

Number of cigarettes smoked in one day by women and by men

By women	Rank₁	By men	Rank₂
4	3	2	1.5
5	4.5	2	1.5
7	7	5	4.5
10	9	6	6
20	12	8	8
20	12	18	10
		20	12
$n = 6$	$\Sigma R_1 = 47.5$	$n = 7$	$\Sigma R_2 = 43.5$

The smaller of the two U-values is 15.5. We compare this value to the critical value of U from the table. We cannot reject the null hypothesis because the observed value is not smaller than the critical value. Male and female smokers do not differ in the number of cigarettes they smoke.

$$U_1 = (n_1)(n_2) + \frac{n_1(n_1+1)}{2} - \Sigma R_1 \qquad U_2 = (n_1)(n_2) + \frac{n_1(n_1+1)}{2} - \Sigma R_1$$

$$= (6)(7) + \frac{6(7)}{2} - 47.5 \qquad = (6)(7) + \frac{6(7)}{2} - 43.5$$

$$= 42 + 21 - 47.5 \qquad = 42 + 21 - 43.5$$

$$= 15.5 \qquad = 19.5$$

19. $U_2 = 49.5$

49.5 > 42 ∴ we accept the null hypothesis and conclude there is no statistically significant difference in levels of arousal between the experimental group, who ate strawberries, and the control group, who did not.

21. "If you can live to be 100, you've got it made. Very few people die past that age."

Data set

X	Rank_X	Y	Rank_Y	D	D²
3	1	23	1	0	0
5	2	27	3	−1	1
6	3	26	2	1	1
7	5	30	4	1	1
7	5	33	5	0	0
7	5	34	6	−1	1
9	7	35	7	0	0
10	8	38	8	0	0
15	9	40	9	0	0

For variable X:
1. rank of the number 3 = 1 ∴ the first word in the riddle is **very**
2. rank of the number 7 = 5 ∴ the second word in the riddle is **few**
3. sum of the X variable = 69 ∴ the third word in the riddle is **people**

For variable Y:
4. rank of the number 34 = 6 ∴ the fourth word in the riddle is **die**
5. sum of the Y variable = 286 ∴ the fifth word in the riddle is **past**

For both variables:
6. sum of D² = 4 ∴ the sixth word in the riddle is **that**
7. value of r_s = .97 ∴ the seventh word in the riddle is **age**

$$r_s = 1 - \left[\frac{6\Sigma D^2}{N(N^2-1)}\right]$$

$$= 1 - \left[\frac{6(4)}{9(80)}\right]$$

$$= 1 - \left[\frac{24}{720}\right]$$
$$= 1 - .03$$
$$= .97$$

CHAPTER 16

1. Test(s): Independent-samples *t*-test.
   Rationale: One independent variable with two levels. Quantitative dependent variable. Looking for a difference between two independent sample means.

3. Test(s): Single sample *t*-test.
   Rationale: One independent variable with one level. Quantitative dependent variable. Looking for a difference between a sample mean and a known population mean. Population standard deviation is unknown.

5. Test(s): Two-way ANOVA.
   Rationale: Two independent variables, one with five levels and one with two levels. Quantitative dependent variable. Looking for a difference between five independent sample means (main effect of IV 1), a difference between two sample means (main effect of IV 2), and the interaction between the two IVs.

7. Test(s): Pearson correlation.
   Rationale: Two quantitative dependent variables. Looking for a relationship between the two sets of measurements.

9. Test(s): Two-way chi-square test of independence.
   Rationale: Two independent variables, one with four levels, one with five levels. Qualitative dependent variable (frequency). Looking for an association (dependency) between the IVs.

11. Test(s): Spearman correlation.
    Rationale: Two dependent variables, one qualitative and one quantitative. Looking for a relationship between the two sets of measurements.

13. Test(s): Independent-samples *t*-test.
    Rationale: One independent variable with two levels. Quantitative dependent variable. Looking for a difference between two independent sample means.

15. Test(s): One-way chi-square test for goodness of fit.
    Rationale: One independent variable with four levels. Looking for fit with hypothetical (in this case, rectangular) distribution.

17. Test(s): Independent-samples *t*-test.
    Rationale: One independent variable with two levels. Quantitative dependent variable. Looking for a difference between two independent sample means.

19. Test(s): Dependent-samples *t*-test.
    Rationale: One independent variable with two levels. Quantitative dependent variable. Looking for a difference between two dependent (repeated-measures) sample means.

21. Test(s): Two-way chi-square test of independence.
    Rationale: Two independent variables, each with two levels. Qualitative dependent variable (frequency). Looking for an association (dependency) between the IVs.

CREDITS

Chapter 1

p. 1 (photo): Zoonar GmbH / Alamy Stock Photo

p. 2 (Fig. 1.1): photo courtesy of www.ancienttouch.com; (Fig. 1.2): Pictorial Press Ltd / Alamy Stock Photo

p. 7 (Fig. 1.3): Photo Josse/Scala, Florence

p. 8 (Fig. 1.4): Photo Franca Principe. Museo Galileo, Florence

p. 24 (Fig. 1.5): R.J. Reynolds Tobacco Company

Chapter 2

p. 32 (photo): Zoonar GmbH / Alamy Stock Photo

p. 33 (figure) Reprinted from World Health Statistics, Chapter 3, MONITORING THE HEALTH GOAL – INDICATORS OF OVERALL PROGRESS, Page 6, Copyright (2016).

p. 92 (software image): Reprint Courtesy of International Business Machines Corporation, © International Business Machines Corporation. IMP SPSS Statistics software (SPSS). SPSS Inc. was acquired by IBM in October, 2009. (IBM® / SPSS®).

p. 93 (software images): Reprint Courtesy of International Business Machines Corporation, © International Business Machines Corporation.

p. 94 (software images): Reprint Courtesy of International Business Machines Corporation, © International Business Machines Corporation.

p. 95 (software images): Reprint Courtesy of International Business Machines Corporation, © International Business Machines Corporation.

p. 96 (software images): Reprint Courtesy of International Business Machines Corporation, © International Business Machines Corporation.

p. 97 (software images): RStudio and Shiny are trademarks of RStudio, Inc.

p. 98 (software image): RStudio and Shiny are trademarks of RStudio, Inc.

p. 99 (software image): RStudio and Shiny are trademarks of RStudio, Inc.

Chapter 3

p. 100 (photo): robertharding / Alamy Stock Photo

p. 101 (photo): Nathan Yau, FlowingData; (screen image): Reprint Courtesy of International Business Machines Corporation, © International Business Machines Corporation. International underground IBM image

p. 103 (Fig. 3.1): The Commercial and Political Atlas, 1786 (3th ed. Edition 1801). William Playfair.

p. 110 (figure): From the New York Times, 2015, The New York Times. All rights reserved. Used by permission and protected by the Copyright Laws of the Univted States. The printing, copying, redistribution, or retransmission of this Content without expression written permission is prohibited.

p. 127 (Fig. 3.12): Republished with permission of University of Chicago Press – Books, from How to Lie With Maps, 1991, permission conveyed through Copyright Clearance Center, Inc.

p. 128 (Fig. 3.13): Edward Tufte, The Visual Display of Quantitative Information (Cheshire, CT: Graphics Press).

p. 130 (software image): Reprint Courtesy of International Business Machines Corporation, © International Business Machines Corporation.

p. 131 (software images): Reprint Courtesy of International Business Machines Corporation, © International Business Machines Corporation.

p. 132 (software images): Reprint Courtesy of International Business Machines Corporation, © International Business Machines Corporation.

p. 133 (software images): Reprint Courtesy of International Business Machines Corporation, © International Business Machines Corporation.

p. 134 (software images): Reprint Courtesy of International Business Machines Corporation, © International Business Machines Corporation.

p. 135 (software images): Reprint Courtesy of International Business Machines Corporation, © International Business Machines Corporation.

p. 136 (software images): Reprint Courtesy of International Business Machines Corporation, © International Business Machines Corporation.

p. 137 (software images): Reprint Courtesy of International Business Machines Corporation, © International Business Machines Corporation.

p. 138 (software images): Reprint Courtesy of International Business Machines Corporation, © International Business Machines Corporation.

p. 139 (software images): Reprint Courtesy of International Business Machines Corporation, © International Business Machines Corporation.

p. 140 (software images): Reprint Courtesy of International Business Machines Corporation, © International Business Machines Corporation.

p. 141 (software images): Reprint Courtesy of International Business Machines Corporation, © International Business Machines Corporation.

p. 142 (software images): Reprint Courtesy of International Business Machines Corporation, © International Business Machines Corporation.

p. 143 (software image): Reprint Courtesy of International Business Machines Corporation, © International Business Machines Corporation.

p. 145 (software image): RStudio and Shiny are trademarks of RStudio, Inc.

p. 146 (software images): RStudio and Shiny are trademarks of RStudio, Inc.

p. 147 (software images): RStudio and Shiny are trademarks of RStudio, Inc.

p. 148 (software image): RStudio and Shiny are trademarks of RStudio, Inc.

p. 149 (software image): RStudio and Shiny are trademarks of RStudio, Inc.

Chapter 4

p. 150 (photo): Frances Roberts / Alamy Stock Photo

p. 151 (figure): NOAA National Centers for Environmental information, Climate at a Glance: U.S. Time Serious, Average Temperature, published March 2017, retrieved on March 24. http://www.ncdc.noaa.gov/cag/

p. 153 (Fig. 4.1): Peter Horree / Alamy Stock Photo

Chapter 5

p. 190 (photo): ASSOCIATED PRESS

p. 232 (software image): Reprint Courtesy of International Business Machines Corporation, © International Business Machines Corporation.

p. 233 (software images): Reprint Courtesy of International Business Machines Corporation, © International Business Machines Corporation.

p. 234 (software images): Reprint Courtesy of International Business Machines Corporation, © International Business Machines Corporation.

p. 235 (software images): Reprint Courtesy of International Business Machines Corporation, © International Business Machines Corporation.

p. 236 (software images): Reprint Courtesy of International Business Machines Corporation, © International Business Machines Corporation.

p. 237 (software image): Reprint Courtesy of International Business Machines Corporation, © International Business Machines Corporation.

p. 239 (software images): Reprint Courtesy of International Business Machines Corporation, © International Business Machines Corporation.

p. 240 (software images): Reprint Courtesy of International Business Machines Corporation, © International Business Machines Corporation.

p. 241 (software images): Reprint Courtesy of International Business Machines Corporation, © International Business Machines Corporation.

p. 247 (software images): RStudio and Shiny are trademarks of RStudio, Inc.

p. 248 (software image): RStudio and Shiny are trademarks of RStudio, Inc.

p. 250 (software image): RStudio and Shiny are trademarks of RStudio, Inc.

p. 251 (software image): RStudio and Shiny are trademarks of RStudio, Inc.

Chapter 6

p. 262 (photo): RubberBall / Alamy Stock Photo

Chapter 7

p. 294 (photo): Portland Press Herald / Contributor / Getty Images

p. 297 (Fig. 7.1): MagMos / iStock

Chapter 8

p. 322 (photo): JeffG / Alamy Stock Photo

Chapter 9

p. 354 (photo): Blue Jean Images / Alamy Stock Photo

p. 358 (Fig. 9.1): Bloomberg / Contributor / Getty Images

Chapter 10

p. 388 (photo): dima_sidelnikov / iStock

p. 393 (Fig. 10.1): PD-1923

p. 438 (software images): Reprint Courtesy of International Business Machines Corporation, © International Business Machines Corporation.

p. 439 (software images): Reprint Courtesy of International Business Machines Corporation, © International Business Machines Corporation.

p. 440 (software image): Reprint Courtesy of International Business Machines Corporation, © International Business Machines Corporation.

Chapter 11

p. 442 (photo): Getty 491524968

p. 444 (Fig. 11.1): SCIENCE PHOTO LIBRARY

p. 472 (Fig. 11.3): Bandura, A.; Ross, D.; Ross, S. A. "Transmission of aggression through the imitation of

CREDITS 767

aggressive models". *Journal of Abnormal and Social Psychology*, (3): 575–582, 1961, American Psychological Association, reprinted with permission.

p. 497 (software image): Reprint Courtesy of International Business Machines Corporation, © International Business Machines Corporation.

p. 498 (software images): Reprint Courtesy of International Business Machines Corporation, © International Business Machines Corporation.

p. 499 (software images): Reprint Courtesy of International Business Machines Corporation, © International Business Machines Corporation.

p. 500 (software images): Reprint Courtesy of International Business Machines Corporation, © International Business Machines Corporation.

p. 501 (software images): Reprint Courtesy of International Business Machines Corporation, © International Business Machines Corporation.

p. 502 (software images): Reprint Courtesy of International Business Machines Corporation, © International Business Machines Corporation.

p. 503 (software images): Reprint Courtesy of International Business Machines Corporation, © International Business Machines Corporation.

p. 504 (software images): Reprint Courtesy of International Business Machines Corporation, © International Business Machines Corporation.

p. 505 (software image): RStudio and Shiny are trademarks of RStudio, Inc.

Chapter 12

p. 512 (photo) Misty Mountains/Bloom/Silver/Kobal/REX/Shutterstock

p. 515 (photo): Dmitriy Yashin / Shutterstock

p. 516 (Fig. 12.1): G. Paul Bishop - Portraits

Chapter 13

p. 550 (figure): Copyright Guardian News & Media Ltd 2017

p. 553 (Fig. 13.1): Chronicle / Alamy Stock Photo; (Fig. 13.2): Republished with permission of Genetics Society of America, from *Evolution by Jumps: Francis Galton and William Bateson and the Mechanism of Evolutionary Change*, Nicholas W. Gillham, 159, 4, 2001; permission conveyed through Copyright Clearance Center, Inc.

p. 565 (figure): DenisBoigelot;

p. 589 (software image): Reprint Courtesy of International Business Machines Corporation, © International Business Machines Corporation.

p. 590 (software images): Reprint Courtesy of International Business Machines Corporation, © International Business Machines Corporation.

p. 591 (software image): Reprint Courtesy of International Business Machines Corporation, © International Business Machines Corporation.

p. 592 (software images): Reprint Courtesy of International Business Machines Corporation, © International Business Machines Corporation.

p. 593 (software images): Reprint Courtesy of International Business Machines Corporation, © International Business Machines Corporation.

p. 594 (software images): Reprint Courtesy of International Business Machines Corporation, © International Business Machines Corporation.

Chapter 14

p. 598 (photo): Image Source Plus / Alamy Stock Photo

p. 630 (software image): Reprint Courtesy of International Business Machines Corporation, © International Business Machines Corporation.

p. 631 (software images): Reprint Courtesy of International Business Machines Corporation, © International Business Machines Corporation.

p. 632 (software image): Reprint Courtesy of International Business Machines Corporation, © International Business Machines Corporation.

p. 633 (software images): Reprint Courtesy of International Business Machines Corporation, © International Business Machines Corporation.

p. 634 (software images): Reprint Courtesy of International Business Machines Corporation, © International Business Machines Corporation.

p. 635 (software image): Reprint Courtesy of International Business Machines Corporation, © International Business Machines Corporation.

Chapter 15

p. 638 (photo): ASSOCIATED PRESS

p. 641 (Fig. 15.1): PD-scan

p. 654 (Fig. 15.2): Held and Hein, "Movement-Produced Stimulation in the Development of Visually Guided Behavior," *Journal of Comparative and Physiological Psychology*, 56, 5, p. 872-6, 1963, APA, reprinted with permission.

p. 677 (software image): Reprint Courtesy of International Business Machines Corporation, © International Business Machines Corporation.

p. 678 (software images): Reprint Courtesy of International Business Machines Corporation, © International Business Machines Corporation.

p. 679 (software image): Reprint Courtesy of International Business Machines Corporation, © International Business Machines Corporation.

p. 680 (software images): Reprint Courtesy of International Business Machines Corporation, © International Business Machines Corporation.

p. 681 (software images): Reprint Courtesy of International Business Machines Corporation, © International Business Machines Corporation.

p. 682 (software images): Reprint Courtesy of International Business Machines Corporation, © International Business Machines Corporation.

Chapter 16

p. 690 (photo): Bob Daemmrich / Alamy Stock Photo

INDEX

A

Abscissa, 103
Alpha (a) levels, 361–63
Alternative hypothesis, 338–39
Alternative population, 517
Analysis of variance (ANOVA), 442–82
    assessing between-group and within-group variability, 447–48
    in context, 480–82
    and effect size, 533–35
    factors in, 448
    levels in, 448
    one-way, 448, 449
        assumptions in, 471
        balanced, 534
        comparing F-distribution and t-distribution, 456–60
        models of F, 470–71
        post-hoc testing, 460–70
        sums of squares, 449–56
    in R
        one-way, 504–8
        two-way, 508–11
    in SPSS, 496–504
        one-way, 496–501
        two-way, 501–4
    two-way (factorial designs), 471–80
        graphing main effects, 474–76
        interpreting results of, 479–80
        logic of, 476–77
        using source table, 477–79
AND rule, 302–5
    combining OR rule and, 303–5
    defined, 302
APA Style, for descriptive statistics, 210–12
Arithmetic average, 156. *See also* Mean(s)
Assumptions, 411
    in hypothesis testing, 359–60
    in one-way analysis of variance, 471
    in testing, 602–4
    with t-tests, 411–12
Astragalus, 5–6
Average, 151, 152, 156. *See also* Mean(s)
Average deviation from the mean, 198–200
Average man concept, 153

B

Bar charts, 103–6
    with R, 144–45
    with SPSS, 130–32
Before-and-after technique, 401
Between-group variability, 447–48, 453
Bimodal distributions, 155
Binet, Alfred, 255
Box-and-whisker plot, 197

C

Causation, correlation and, 571–72
Central limit theorem, 326–37
    law of large numbers and, 327–28
    measuring distribution of large sets of events, 325–27
    samples from populations
        random sampling, 333–37
        sampling distribution of the means, 329–30
        statements that make up CLT, 330–32
    using, 337
Central limit theorem (CLT), 327
Central tendency, measures of. *See* Measures of central tendency
Chance error, 11
Chi-square tests, 598–620
    critical values, 709–12
    for goodness of fit (one-way), 604–9
        in R, 635–36
        with SPSS, 630–32
    for independence (two-way), 609–17
        in R, 636–37
        with SPSS, 632–35
    need for, 600–604
    2 x 2 design, 615–16
Coefficient of determination ($r2$), 569
Cohen's d, 529–30, 532
Confidence, 514, 515
Confidence intervals (CIs)
    estimates and, 515–17
    and independent- and dependent-samples t-tests, 524–28
    inferential tests vs., 535–36
    and single-sample t-test, 522–23
    and z-test, 518–20
Confounding variable, 8
Consistency (constancy) in data, 192–94
Context, 15
Contingency table, 610
Continuous data, 105–6
Control, 8
Control group, 400
Correlation, 552–60
    causation and, 571–72
    coefficient of determination, 569
    correlation coefficient, 552–61
        calculating, 557–60
        critical values, 712
        positive and negative correlations, 552–55
        Spearman, 641, 644–48, 713–15
        testing hypotheses about, 560
        truncated range, 567–68
        weak and strong correlations, 555–57
    linear vs. curvilinear relationships, 566–67
    positive and negative, 552–55
    in R, 595–96
    in SPSS, 589–91
    weak and strong, 555–57
Correlation coefficient (r), 552–61
    calculating, 557–60
    critical values, 712
    positive and negative correlations, 552–55
    Spearman, 644–48
        critical values, 713–15
        development of, 641
    testing hypotheses about, 560
    weak and strong correlations, 555–57
Critical values
    chi square, 709–12
    correlation coefficient, 712
    F-values, 705–6
    Mann–Whitney U-test, 716–18
    Spearman correlation, 713–15
    Tukey HSD test, 464–65, 707–9
    t-values, 398–99, 702–4
    Wilcoxon signed-rank, matched-pairs t-test, 719
    of z (critical z-value), 359–64
        assumptions in hypothesis testing, 359–60
        finding z-value in nondirectional hypothesis, 363–64
        rejection region, 360–62
CSV files, 92
Cumulative frequency (CF), 47–48
Cumulative relative frequency (CRF), 48
Curvilinear relationships, 566–67

D

Data, 4, 32–62
    conclusions about, 13
    consistency and inconsistency in, 192–94
    continuous, 105–6
    CSV files, 92
    discrete, 104–5
    entering, 92
        with R, 97–99
        with SPSS, 92–96
    grouped
        finding centers with, 163–64
        shapes of distributions, 163–65
    interval, 167
    measures of center, 156, 159, 166–67
    nominal, 167
    ordinal, 167
    organizing, 42–59

qualitative, 34
  measures of center with, 166–67
  nonparametric versions of *t*-test for, 656
  scales of measurement for, 35–38
quantitative, 34
  nonparametric versions of *t*-test for, 656–59
  scales of measurement for, 39–40
ranked, 642
  in nonparametric tests, 643–44
  quantitative, 656–59
ratio, 167
scales of measurement, 34–42
  for qualitative data, 35–38
  for quantitative data, 39–40
visualizing patterns in, 102 (*See also* Graphics)
Data point (datum), 4
Data set, 4
Datum, 4. *See also* Data
Degrees of freedom
defined, 396
t-tests, 396–99
Demographics, 310
Dependent events, 301, 610
defined, 301
more than two, 305
Dependent samples, 401
Dependent samples *t*-tests, 400, 411, 413–17
calculating *t* in, 414–16
and confidence intervals, 524–28
Dependent variables (DVs), 8–11
Descriptive statistics, 5–6, 42
APA Style example of, 210–12
in context, 210–12
with R, 243–51
with SPSS, 232–42
standard scores, 254–77
  benefits of, 270–75
  comparing scores from different distributions, 270–71
  in context, 277
  converting percentile ranks into raw scores, 273–75
  converting *z*-scores into raw scores, 272–73
  proportions in standard normal curve, 262–69
  range rule, 276
  3-sigma rule, 259–62
  z score, 256–59
Deviation (deviation score), 198
average deviation from the mean, 198–200
equation for, 213
mean, 208–9
standard deviation, 200–209
  in context, 207–9
  finding, 204–6
  in a population, 202–4
  in a population vs. in a sample, 204
  and variance, 200–202
z- score, 256–59

Dichotomize, 572
Directional hypotheses, 339–42
Direct (positive) correlation, 552–55
Discrete data, 104–5
Disjoint sets, 301
Distributions. *See* Frequency distributions

E
Earth, shape of, 193
Effect size, 529–35
  and ANOVA, 533–35
  Cohen's *d*, 529–30, 532
Empirical distributions, 324
Empirical set of observations, 306–8
Error(s), 8, 324–25
  chance, 11
  estimated standard, 392
  inferential statistics, 324
  scientific method and, 324
  standard, 331, 415
    of the difference, 403–4
    estimated, 392
  Type I and Type II, 371–73
  z-test and, 356–57
Estimated standard error, 392
Estimates, 514–29
  confidence intervals
    and independent- and dependent-samples *t*-tests, 524–28
    inferential tests vs., 535–36
    and single-sample *t*-test, 522–23
    and *z*-test, 518–20
  interval, 514–15
  of measures of center, 160
  of parameters, 338
  point, 514–15
  statistics as, 370
Eta-squared statistic, 533–35
Eugenics, 641–42
Events, 297–99
  dependent, 301, 610
    defined, 301
    more than two, 305
  distribution of large sets of, 325–27
  frequency of, 604
  independent, 301, 611
  probabilities of, 5, 294–312
    combining, 301–6
    conditional, 300–301
    in context, 310–12
    and frequency, 296–306
    set theory, 296–300
    using probabilities, 306–10
Expected frequency, 604
Experimental group, 400
Experimentation, 7
Extraneous variable, 8

F
Factor analysis, 641
Factorial model (design), 471. *See also* Two-way ANOVA

Factors
  in ANOVA, 448
  independent variables and, 470
F-distribution, comparing *t*-distribution and, 456–60
Fisher, Ronald Aylmer, 445
Fixed-factor/fixed-effect ANOVA, 471
Frequency, 6, 604
  cumulative, 47–48
  cumulative relative, 48
  expected, 604
  observed, 604
  probability and, 296–306
    combining probabilities, 301–6
    conditional probability, 300–301
    set theory, 296–300
  relative, 46–47
  typical events, 44
Frequency distributions, 42–59
  bimodal, 155
  comparing scores from, 270–71
  in context, 59–61
  empirical, 324
  grouped, 51–58
  interval width, 53–54
  multimodal, 155
  negatively skewed, 162
  non-normal, 656–59
  normal, 152, 162
  positively skewed, 162
  probability, 312
  as probability distributions, 312
  shapes of, 161–66
    with grouped data, 163–65
    normal, 162
    skew, 161–62, 657
  theoretical, 324, 327, 330
  trimodal, 155
  ungrouped, 45–51, 58–59
    cumulative frequency, 47–48
    percentile rank, 48–49
    relative frequency, 46–47
  unimodal, 155
Frequency histograms, 105. *See also* Histograms
Frequency polygons, 107–8
  with R, 145–46
  with SPSS, 133–34
F-statistics, 450, 460, 470–71
  calculating, 450
  critical values, 705–6
  interaction effect, 471, 474
  in one-way ANOVA, 470–71
  tests of main effects, 471, 474
  in two-way ANOVA, 471
F-values
  critical, 705–6
  significant, 458

G
Galton, Francis, 552–54, 601–2
Gambler's fallacy, 297–98
Ggplot2 (in R), 144

INDEX 771

Gossett, William, 390–93
Graphics, 100–149
    bar charts, 103–6, 130–32, 144–45
    continuous data, 105–6
    discrete data, 104–5
    frequency polygons, 107–8, 133–34, 145–46
    graphing with R, 143–49
        bar charts, 144–45
        frequency polygons, 145–46
        ggplot2, 144
        histograms, 146–47
        pie charts, 148
        scatterplots, 147–48
        time series, 148–49
    graphing with SPSS, 130–43
        bar charts, 130–32
        frequency polygons, 133–34
        histograms, 134–35
        pie charts, 139–41
        scatterplots, 136–39
        time series, 141–43
    histograms, 103–6, 134–35, 146–47
    interpreting, 110
    of interquartile range, 197
    of means, 111–12
    pie charts, 108–9, 139–41, 148
    in publishing, 102–3
    of relationships, 112–13
    rules for creating good graphs, 114–15
    scatterplots, 112–13, 136–39, 147–48
    stem-and-leaf graphs, 106–7
    of time, 113
    time series, 141–43, 148–49
    visualizing patterns in data, 102
Graphs
    of means, 111–12
    of relationships, 112–13
    rules for creating, 114–15
    of time, 113
Grouped data
    finding centers with, 163–64
    shapes of distributions, 163–65
Grouped frequency distributions, 51–58

H

Hidden variables, 572
Histograms, 103–6
    frequency, 105
    with R, 146–47
    with SPSS, 134–35
Hypotheses, 8, 338
    alternative, 338–39
    directional, 339–42
    finding z-value in, 363–64
    mutually exclusive, 339
    nondirectional, 340, 363–64
Hypothesis testing, 338–44
    about correlation coefficient, 560
    assumptions in, 359–60
    in context, 342–43
    directional hypotheses, 339–42
    null and alternative hypotheses, 338–39

I

Independent events, 301, 611
Independent samples, 400–401
Independent samples t-tests, 400, 402–10
    and confidence intervals, 524–28
    finding difference
        between means with unequal samples, 406–10
        between two means, 404–6
    standard error of the difference, 403–4
Independent variables (IVs), 8–11
    defined, 8
    factors and, 470
Indirect (negative) correlation, 552–55
Inferential statistics, 6–7, 324
    in context, 373–75
    error in, 356–57
    t-tests, 388–422
        assumptions with, 411–12
        compared to z-test and z-scores, 394
        in context, 419–21
        critical t-values, 398–99, 702–4
        degrees of freedom, 396–99
        dependent-samples, 400, 413–17, 524–28
        development of, 390–92
        independent-samples, 400, 402–10, 524–28
        with R, 440–41
        and sample size, 411–12
        single-sample, 394–96, 522–23
        with SPSS, 437–40
        Student's, 391, 392–400
        and variability, 417–19
        when both σ and μ are unknown, 400–402
        Wilcoxon signed-rank, matched-pairs, 652–56, 719
    z-test, 354–76
        and confidence intervals, 518–20
        critical values and p, 359–64
        error, 356–57
        p-values, 361–62, 373
        and statistics as estimates, 370
        Type I and Type II errors, 371–73
Inferential tests
    confidence intervals vs., 535–36
    selecting, 481
Intelligence testing, 255
Interaction effect, 471, 474
Interquartile range (IQR), 195–97, 213
Interval data, measures of center with, 167
Interval estimate, 514–15
Interval scales, 39–40

K

Kinnebrook, David, 325

L

Law of large numbers (LLV), 327–28
Least squares regression line (line of best fit), 561–65

Lemma, 516
Levels, 8
    in ANOVA, 448
    of factors, 470
Likert scale, 619–20
Linear relationships, 566–67
Line of best fit, 561–65
Lower limit (LL), 43

M

Mann–Whitney U-test, 648–52
    critical values, 716–18
    in R, 684–85
    in SPSS, 679–83
Matched-samples method, 401
Mean deviation (MD), 208–9
Mean(s), 38, 154, 156–59
    average deviation from, 198–200
    comparing more than two (See Analysis of variance [ANOVA])
    graphing, 111–12
    with grouped data, 163–64
    independent-samples t-tests
        finding difference between, 404–6
        finding difference between means with unequal samples, 406–10
    of samples in large sets, 327
    sampling distribution of the difference between the means, 403
    of sampling distribution of the means, 331
    trimmed, 167
Mean sample mean, 331
Mean squares between groups (MSb), 449
Mean squares within groups (MSw), 449, 450
Measurement
    properties of, 41
    scales of (See Scales of measurement)
Measures of center
    estimating, 160
    mean, 154, 156–59
    median, 155–56, 158–59
    mode, 154–55, 158–59
    variance, 200–202
Measures of central tendency, 150–68
    in context, 166–67
    mean, 156–59
    median, 155–56, 158–59
    mode, 154–55, 158–59
    shapes of distributions, 161–66
        with grouped data, 163–65
        normal, 162
    typical, 152–54
Measures of variability, 194–200
    average deviation from the mean, 198–200
    interquartile range, 195–97
    range, 194–95
    standard deviation, 200–209
        in context, 207–9
        finding, 204–6
        in a population, 202–4

in a population vs. in a sample, 204
and variance, 200–202
variance, 198
finding, 204–6
in a population, 200–202
Median, 154, 155, 158–59
finding position of, 155–56
with grouped data, 163–64
Midpoint (MP), in grouped distributions, 56, 163–64
Mode, 154, 158–59
finding position of, 154–55
with grouped data, 163–64
Models, 342
Multimodal distributions, 155
Multiple-comparison tests, 462. *See also* Post-hoc tests
Multivariate ANOVA (MANOVA), 471
Mutually exclusive hypotheses, 339
Mutually exclusive (or disjoint) sets, 301

N

Natural-pairs technique, 402
Negative (indirect) correlation, 552–55
Negatively skewed distributions, 162
Neyman, Jerzy, 516
Neyman–Pearson theory/lemma, 516
Nominal data, measures of center with, 167
Nominal scales, 35–36
Nondirectional hypothesis, 340
defined, 340
finding z-value in, 363–64
Non-normal distributions, 656–59
Nonparametric tests, 600, 638–62
chi-square tests for goodness of fit (one-way), 604–9
in R, 635–36
with SPSS, 630–32
chi-square tests for independence (two-way), 609–17
in R, 636–37
with SPSS, 632–35
in context, 619–20, 661–62
Mann–Whitney U-test, 648–52
critical values, 716–18
in R, 684–85
in SPSS, 679–83
non-normal distributions, 656–59
parametric tests vs., 600, 640–41
performing, 642
in R
Mann–Whitney U-test, 684–85
Spearman correlation, 683–84
ranking data, 643–44
Spearman correlation, 644–48
critical values, 713–175
in R, 683–84
in SPSS, 677–78
in SPSS
Mann–Whitney U-test, 679–83
Spearman correlation, 677–78

Wilcoxon signed-rank, matched-pairs t-test, 652–56, 719
Normal distributions, 152
probability in, 359
proportions in, 262–69, 697–701
shape of, 162
standard scores, 254–77
benefits of, 270–75
comparing scores from different distributions, 270–71
in context, 277
converting percentile ranks into raw scores, 273–75
converting z-scores into raw scores, 272–73
proportions in standard normal curve, 262–69
range rule, 276
3-sigma rule, 259–62
z score, 256–59
Null hypothesis, 338–39, 372
Null population, 339, 515
Numbers, 35
Numerals, 35

O

Observation(s)
empirical set of, 306–8
observed frequency, 604
theoretical set of, 306–8
typical, 6
Observed frequency, 604
One-way ANOVA, 448, 449
assumptions in, 471
balanced, 534
comparing F-distribution and t-distribution, 456–60
models of F, 470–71
post-hoc testing, 460–70
Tukey HSD critical values, 464–65, 707–9
Tukey HSD test with equal n's, 462–66
Tukey HSD test with unequal n's, 466–70
in R, 504–8
source table, 454–55
in SPSS, 496–501
sums of squares, 449–56
One-Way Chi-Square Test for Goodness of Fit, 604–9
in R, 635–36
in SPSS, 630–32
Ordinal data, measures of center with, 167
Ordinal scales, 36–38
Ordinate, 103
Organizing data, 42–59
descriptive statistics, 42
frequency distributions, 42–59
in context, 59–61
grouped, 51–58
ungrouped, 45–51, 58–59
OR rule, 303

combining AND rule and, 303–5
defined, 303
Outliers, 161

P

Paired-samples t-test, 411. *See also* Dependent-samples t-tests
Pairwise comparisons, 460–61
Parameters, 7, 600
defined, 7
estimating, 338
Parametric tests
defined, 600, 640
nonparametric tests vs., 600, 640–41
Patterns in data, visualizing, 102
Pearson product-moment correlation coefficient (r), 552, 712. *See also* Correlation coefficient
Percentage, 46, 268–69. *See also* relative frequency
Percentile rank, 48–49
converting into raw score, 273–75
as cumulative relative frequency, 48
Pie charts, 108–9
in R, 148
in SPSS, 139–41
Playfair, William, 102–3
Point estimate, 514–15
Population(s), 6
alternative, 517
null, 339, 515
random sampling of, 333–37
standard deviation in, 202–4
as universal set, 299
variance in, 200–202
Positive (direct) correlation, 552–55
Positively skewed distributions, 162
Post-hoc tests, 460–70
Tukey HSD critical values, 464–65, 707–9
Tukey HSD test with equal n's, 462–66
Tukey HSD test with unequal n's, 466–70
Precision, 514, 515
Probability distributions, 312
Probability(-ies), 5, 294–312
and card games, 309–10
combining, 301–6
OR rule, 303–5
AND rule, 302–5
conditional, 300–301
in context, 310–12
and frequency, 296–306
combining probabilities, 301–6
conditional probability, 300–301
set theory, 296–300
in normal distribution, 359
set theory, 296–300
using, 306–10
Proportion(s)
relative frequency as, 46
under standard normal curve, 262–69, 697–701
p-values, 361–63, 373

Q

Quadratic relationship, 567
Qualitative data, 34
    measures of center with, 166–67
    nonparametric versions of *t*-test for, 656
    scales of measurement for, 35–38
        nominal scales, 35–36
        ordinal scales, 36–38
Quantitative data, 34
    nonparametric versions of *t*-test for, 656–59
    scales of measurement for, 39–40
        interval scales, 39–40
        ratio scales, 39–40
Quartiles, 195–96. *See also* Interquartile range (IQR)
Questionnaires, 601–2
Quetelet, Adolphe, 153
Quincunx, 373–75

R

R (software)
    analysis of variance
        one-way, 504–8
        two-way, 508–11
    chi-square tests
        for goodness of fit (one-way), 635–36
        for independence (two-way), 636–37
    descriptive statistics with, 243–51
    entering data, 97–99
    graphing with, 116, 143–49
        bar charts, 144–45
        frequency polygons, 145–46
        ggplot2, 144
        histograms, 146–47
        pie charts, 148
        scatterplots, 147–48
        time series, 148–49
    Mann–Whitney *U*-test, 684–85
    regression in, 596–97
    Spearman correlation, 683–84
    t-tests with, 440–41
Random error, 11
Random sampling, 329, 333–37
Range, 53, 194–95
    equation for, 213
    interquartile, 195–97
    restricted, 570–71
    truncated, 567–68
Range rule, 206–7, 276
Ranked data, 642
    in nonparametric tests, 643–44
    quantitative, 656–59
Ratio, 6
Ratio data, measures of center with, 167
Ratio scales, 39–40
Raw scores
    converting into *z*-scores, 270–71
    converting percentile ranks into, 273–75
    converting *z*-scores into, 272–73
Regression
    least squares regression line, 561–65
    in R, 596–97
    in SPSS, 591–95
Regression line, 553, 561–65
Regression to the mean, 555
Rejection region, 360–62
Relationships. *See also* Correlation
    graphing, 112–13
    linear vs. curvilinear, 566–67
Relative frequency
    cumulative, 48
    ungrouped frequency distributions, 46–47
Repeated-measures design, 9, 401
Representative sample, 7
Restricted range, 570–71
Rho (ρ), 558
Risk ratio (RR), 617–18
Robust tests, 411

S

Sample(s), 6
    dependent, 401
    independent, 400–401
    random sampling, 333–37
    representative, 7
    sampling distribution of the means, 329–30
    standard deviation in, 204
    statements that make up CLT, 330–32
    as a subset, 299
Sample size, 11
    and *t*-distribution, 397–98
    t-tests and, 411–12
Sampling, 329
    random, 329, 333–37
    representative, 7
Sampling distribution of the difference between the means (SDDM), 403
Sampling distribution of the means (SDM), 329–31, 356
    assumptions in, 359–60
    variability in, 329
SAT scores, 277
Scales of measurement, 34–42
    in context, 59–61
    for qualitative data, 35–38
        nominal scales, 35–36
        ordinal scales, 36–38
    for quantitative data, 39–40
        interval scales, 39–40
        ratio scales, 39–40
Scatterplots, 112–13
    with R, 147–48
    with SPSS, 136–39
Scientific method, 324
Sets, 296–99
    empirical, 306–8
    mutually exclusive (or disjoint), 301
    theoretical, 306–8
    universal, 297
Set theory, 296–300
Shapes of distributions, 161–66
    with grouped data, 163–65
    negatively skewed, 162
    normal, 162
    positively skewed, 162
Shared variability, 569–70
Significant differences, 361, 481
    effect size and, 529–35
    in inferential analysis, 481
    Tukey HSD tests, 462
        critical values, 707–9
        with equal *n*'s, 462–66
        with unequal *n*'s, 466–70
Single-sample *t*-test, 394–96, 522–23
Skew, 657
    negative, 161–62
    positive, 161–62
Source table, 454
    one-way ANOVA, 454–55
    two-way ANOVA, 477–79
Spearman, Charles Edward, 641–42
Spearman rank-order correlation coefficient (r_s), 644–48
    critical values, 713–15
    development of, 641
SPSS (Statistical Package for the Social Sciences), 28–31
    analysis of variance, 496–504
        one-way, 496–501
        two-way, 501–4
    chi-square tests
        for goodness of fit (one-way), 630–32
        for independence (two-way), 632–35
    correlation in, 589–91
    descriptive statistics with, 232–42
    entering data, 92–96
    graphing with, 116, 130–43
        bar charts, 130–32
        frequency polygons, 133–34
        pie charts, 139–41
        scatterplots, 136–39
        time series, 141–43
    Mann–Whitney *U*-test, 679–83
    regression in, 591–95
    Spearman correlation, 677–78
    t-tests with, 437–40
Standard deviation (SD), 200–209
    calculation and definition formulas, 203, 213
    in context, 207–9
    finding, 204–6
    in a population, 202–4
    in a population vs. in a sample, 204
    range rule, 206–7
    of sampling distribution of the means, 332
    standard error and, 415
    and variance, 200–202
Standard error, 331, 415
    of the difference, 403–4
    estimated, 392
Standard normal distribution
    proportions under, 262–69, 697–701
    z-scores in, 270
Standard scores, 254–77
    benefits of, 270–75
    comparing scores from different distributions, 270–71
    in context, 277

converting percentile ranks into raw scores, 273–75
converting z-scores into raw scores, 272–73
proportions in standard normal curve, 262–69
range rule, 276
SAT scores, 277
3-sigma rule, 259–62
z score, 256–59
Statistics, 2–16
   cautionary notes on, 13–15
   in context, 15–16, 687–96
   deciding which test to use, 686–96
   descriptive, 5–6, 42
      APA Style example of, 210–12
      in context, 210–12
      with R, 243–51
      with SPSS, 232–42
      standard scores, 254–77
   development of, 4–5
   as estimates, 370
   inferential, 6–7, 324
      in context, 373–75
      error in, 356–57
      t-tests, 388–422
      z-test, 354–76
   in society, 641–42
   types of, 4–5
   using, 12–13
   variables, 7–12
Stem-and-leaf graphs, 106–7
Stevens, S. S., 35–36
Steven's power law, 34–35
Strong correlation, 555–57
Student's table of critical t-values, 702–4
Student's t-test, 391–400
Stylometrics, 13
Subject mortality, 406
Subsets, 297, 299
Sum of squares, 200, 449–56. *See also* Variance
Sum of squares between groups (SSb), 449
Sum of squares within groups (SSw), 449
Sums of squares, 449–56

T

Target variable, 8
t-distribution, 456
   comparing F-distribution and, 456–60
   sample size and, 397–98
Tests of main effects, 471, 474
Theoretical distributions, 324, 327, 330
Theoretical set of observations, 306–8
3-sigma rule, 259–62
Time series graphs, 113
   with R, 148–49
   with SPSS, 141–43
Treatment variable, 400
Trial of the Pyx, 358
Trimmed mean, 167
Trimodal distributions, 155
Truncated range, 567–68
t-tests, 388–422

assumptions with, 411–12
compared to z-test and z-scores, 394
in context, 419–21
critical t-values, 398–99, 702–4
degrees of freedom, 396–99
dependent-samples, 400, 413–17
   calculating t in, 414–16
   and confidence intervals, 524–28
development of, 390–92
independent-samples, 400, 402–10
   and confidence intervals, 524–28
   finding difference between means with unequal samples, 406–10
   finding difference between two means, 404–6
   standard error of the difference, 403–4
with R, 440–41
and sample size, 411–12
single-sample, 394–96, 522–23
with SPSS, 437–40
Student's, 391, 392–400
and variability, 417–19
when both σ and μ are unknown, 400–402
Wilcoxon signed-rank, matched-pairs, 652–56, 719
Tukey honestly significant difference (HSD) test, 462
   critical values, 464–65, 707–9
   with equal n's, 462–66
   with unequal n's, 466–70
Two-way ANOVA, 449, 471–80
   graphing main effects, 474–76
   interpreting results of, 479–80
   logic of, 476–77
   in R, 508–11
   in SPSS, 501–4
   using source table, 477–79
Two-Way Chi-Square Test of Independence, 609–17
   in R, 636–37
   in SPSS, 632–35
   2 x 2 design, 615–16
Type I errors, 371–73
Type II errors, 371–73
Typical observation, 6

U

Ungrouped frequency distributions, 45–51, 58–59
   cumulative frequency, 47–48
   percentile rank, 48–49
   relative frequency, 46–47
Unimodal distributions, 155
Universal set, 297
Upper limit (UL), 43

V

Variability, 190–212
   between-group, 447–48
   data consistency and inconsistency, 192–94
   measures of, 194–200

   average deviation from the mean, 198–200
   interquartile range, 195–97
   range, 194–95
   standard deviation, 200–209
   variance, 198
   measuring, 192
   in sampling distribution of the means, 332
   shared, 569–71
   t-tests and, 417–19
   within-group, 447–48
Variables, 7–12
   chance error in, 11
   confounding, 8
   dependent, 8–11
   dichotomized, 572
   extraneous, 8
   hidden, 572
   independent, 8–11, 470
   target, 8
   treatment, 400
Variance, 200. *See also* Analysis of variance (ANOVA)
   calculation and definition formulas, 201–2, 213
   definition formula for, 449
   finding, 204–6
   in a population, 200–202
   and standard deviation, 200–202
Venn diagrams, 570–71

W

Weak correlation, 555–57
Wilcoxon signed-rank, matched-pairs t-test, 652–56, 719
Within-group variability, 447–48, 453

Z

z-formula, 365
z-scores, 256–59, 356–57
   compared to z-test and t-test, 394
   converting into raw scores, 272–73
   converting percentile rank into raw score, 273–75
   converting raw scores into, 270–71
   proportions in standard normal curve, 262–69
z-statistic, 368
z-test, 354–76
   compared to z-score and t-test, 394
   and confidence intervals, 518–20
   critical values and p, 359–64
      assumptions in hypothesis testing, 359–60
      finding z-value in nondirectional hypothesis, 363–64
      rejection region, 360–62
   error, 356–57
   p-values, 361–62, 373
   and statistics as estimates, 370
   Type I and Type II errors, 371–73